Principles of Colloid
and Surface Chemistry

UNDERGRADUATE CHEMISTRY

A Series of Textbooks

edited by
J. J. Lagowski
Department of Chemistry
The University of Texas at Austin

Principles of Colloid and Surface Chemistry

Second Edition
Revised and Expanded

Paul C. Hiemenz

Chemistry Department
California State Polytechnic University

MARCEL DEKKER, INC.   New York and Basel

Library of Congress Cataloging in Publication Data

Hiemenz, Paul C
   Principles of colloid and surface chemistry, 2nd ed.

   (Undergraduate chemistry; v. 9)
   Includes bibliographical references and index.
   1. Colloids.   2. Surface chemistry.   I. Title.
QD549.H53   541'.345   76-55600
ISBN 0-8247-7476-0

MARCEL DEKKER, INC.
270 Madison Avenue, New York, New York, 10016

Current printing (last digit):
10 9 8 7 6 5 4 3 2 1

PRINTED IN THE UNITED STATES OF AMERICA

PREFACE TO THE SECOND EDITION

In the preface to the first edition, I stated that this is "primarily a textbook, written with student backgrounds, needs, and objectives in mind." This continues to be true of the second edition, and many of the revisions I have made are attempts to make the book even more useful than its predecessor to student readers. In addition, colloid and surface science continue to flourish. In preparing the second edition I have also attempted to "open up" the coverage to include some of the newer topics from an ever-broadening field.

A number of differences between the first and second editions can be cited which are readily traceable to either one or both of the foregoing considerations.

Two new chapters have been added which explore—via micelles and related structures and metal surfaces under ultra-high vacuum—both "wet" and "dry" facets of colloids and surfaces. Although neither of these areas is new, both are experiencing an upsurge of activity as new instrumentation is developed and new applications are found.

A number of chapters have been overhauled so thoroughly that they bear only minor resemblance to their counterparts in the first edition. The thermodynamics of polymer solutions is introduced in connection with osmometry and the drainage and spatial extension of polymer coils is discussed in connection with viscosity. The treatment of contact angle is expanded so that it is presented on a more equal footing with surface tension in the presentation of liquid surfaces. Steric stabilization as a protective mechanism against flocculation is discussed along with the classical DLVO theory.

Solved problems have been scattered throughout the text. I am convinced that students must work problems to gain mastery of the topics we discuss. Including examples helps bridge the gap between the textbook presentation of theory and student labors over the analysis and mechanics of problem solving.

SI units have been used fairly consistently throughout the book. Since the problems at the ends of the chapters are based on data from the literature and since cgs units were commonly used in the older literature, the problems contain a wider assortment of units. A table of conversions between cgs and SI units is contained in Appendix B.

I am very much aware of the many important topics that the book fails to cover or, worse yet, even mention. However, lines must be drawn somewhere both to keep the book manageable in size and cost and to have it useful as the basis for a course. As it is, I have added a good deal of new material without deleting anywhere near as much of the old. I have tried to select for inclusion topics of fundamenal importance which could be developed with some internal coherence and with some continuity from a (prerequisite) physical chemistry course.

A number of users of the first edition communicated with me, pointing out errors and offering suggestions for improvement. I appreciate the feedback of these correspondents, and hope that the revisions I have made at least partially reflect their input.

I want to express my thanks to Carol Truett for expertly drawing the new illustrations and giving a "new look" to the old ones. I also appreciate the assistance of Lisa Scott in the preparation of the manuscript. Thanks, also, to Reuben Martinez for helping me with the proofreading and indexing. Lastly, let me again invite users to call errors and/or obscurities to my attention, and to thank them in advance for doing so.

Paul C. Hiemenz

PREFACE TO THE FIRST EDITION

Colloid and surface chemistry occupy a paradoxical position among the topics of physical chemistry. These are areas which have traditionally been considered part of physical chemistry and are currently enjoying more widespread application than ever due to their relevance to environmental and biological problems. At the same time, however, colloid and surface chemistry have virtually disappeared from physical chemistry courses. These topics are largely absent from the contemporary general chemistry course as well. It is possible, therefore, that a student could complete a degree in chemistry without even being able to identify what colloid and surface chemistry are about.

The primary objective of this book is to bridge the gap between today's typical physical chemistry course and the literature of colloid and surface chemistry. The reader is assumed to have completed a course in physical chemistry, but no prior knowledge of the topics under consideration is assumed. The book is, therefore, introductory as far as the topic subjects are concerned, although familiarity with numerous other aspects of physical chemistry is required background.

Since physical chemistry is the point of departure for this presentation, the undergraduate chemistry major is the model reader toward whom the book is addressed. This in no way implies that these are the only students who will study the material contained herein. Students majoring in engineering, biology, physics, materials science, and so on, at both the undergraduate and graduate levels will find aspects of this subject highly useful. The interdisciplinary nature of colloid and surface chemistry is another aspect of these subjects that contributes to their relevance in today's curricula.

This is primarily a textbook, written with student backgrounds, needs, and objectives in mind. There are several ways in which this fact manifests itself in the organization of this book. First, no attempt has been made to review the literature or to describe research frontiers in colloid and surface chemistry. A large literature exists which does these things admirably. Our purpose is to provide the beginner with enough background to make intelligible the journals and monographs which present these topics. References have been limited to monographs, textbooks, and reviews which are especially comprehensive and/or accessible. Second, where derivations are presented, this is done in sufficient detail so that the reader should find them self-explanatory. In areas in which undergraduate chemistry majors have minimal backgrounds or have chronic difficulties—for example, fluid mechanics, classical electromagnetic theory, and electrostatics—the presentations begin at the level of general physics, which may be the student's only prior contact with these topics. Third, an effort has been made to facilitate calculations by paying special attention to dimensional considerations. The cgs-esu system of units has been used throughout, even though this is gradually being phased out of most books. The reason for keeping these units is the stated objective of relating the student's experiences to the existing literature of colloid and surface chemistry. At present, the cgs-esu system is still the common denominator between the two. A fairly detailed list of conversions between cgs and SI units is included in Appendix C. Finally, a few problems are included in each chapter. These provide an opportunity to apply the concepts of the chapter and indicate the kinds of applications these ideas find.

Not all who use the book will have the time or interest to cover it entirely. In the author's course, about two-thirds of the material is discussed in a one-quarter course. With the same level of coverage, the entire book could be completed in a semester. To cover the amount of material involved, very little time is devoted to derivations except to answer questions. Lecture time is devoted instead to outlining highlights of the material and presenting supplementary examples.

The underlying unity which connects the various topics discussed here is seen most clearly when the book is studied in its entirety and in the order presented. Time limitations and special interests often interfere with this ideal. Those who choose to rearrange the sequence of topics should note the subthemes that unify certain blocks of chapters. Chapters 1 through 5 are primarily concerned with particle characterization, especially with respect to molecular weight; Chapters 6 through 8, with surface tension/free energy and adsorption; and Chapters 9 through 11, with flocculation and the electrical double layer. Subjects of special interest to students of the biological sciences are given in Chapters 2 to 5, 7, and 11.

Colloid chemistry and surface chemistry each span virtually the entire field of chemistry. The former may be visualized as a chemistry whose "atoms" are considerably larger than actual atoms; the latter, as a two-dimensional chemistry. The point is that each encompasses all the usual subdivisions of chemistry: reaction chemistry, analytical chemistry, physical chemistry, and so on. The various subdivisions of physical chemistry are also represented: thermodynamics, structure elucidation, rate processes, and so on. As a consequence, these traditional categories could be used as the basis for organization in a book of this sort. For example, "The Thermodynamics of Surfaces" would be a logical chapter heading according to such a plan of organization. In this book, however, no such chapter exists (although not only chapters but entire volumes on this topic exist elsewhere). The reason goes back to the premise stated earlier: These days most undergraduates know more about thermodynamics than about surfaces, and this is probably true regardless of their thermodynamic literacy/illiteracy! Accordingly, this book discusses surfaces: flat and curved, rigid and mobile, pure substances and solutions, condensed phases and gases. Thermodynamic arguments are presented—along with arguments derived from other sources—in developing an overview of surface chemistry (with the emphasis on "surface"). A more systematic, formal presentation of surface thermodynamics (with the emphasis on "thermodynamics") would be a likely sequel to the study of this book for those who desire still more insight into that aspect of two-dimensional chemistry. Similarly, other topics could be organized differently as well. Only time will tell whether the plan followed in this book succeeds in convincing students that chemistry they have learned in other courses is also applicable to the "in between" dimensions of colloids and the two dimensions of surface chemistry.

The notion that molecules at a surface are in a two-dimensional state of matter is reminiscent of E. A. Abbott's science fiction classic, *Flatland*.* Perusal of this little book for quotations suitable for Chapters 6, 7, and 8 revealed other parallels also: the color revolt and light scattering, "Attend to Your Configuration" and the shape of polymer molecules, and so on. Eventually, the objective of beginning each chapter with a quote from *Flatland* replaced the requirement that the passage cited have some actual connection with the contents of the chapter. As it ends up, the quotes are merely for fun: Perhaps those who are not captivated by colloids and surfaces will at least enjoy this glimpse of *Flatland*.

---

*E. A. Abbott, *Flatland* (6th ed.), Dover, New York, 1952. Used with permission.

Finally, it is a pleasure to acknowledge those whose contributions helped bring this book into existence. I am grateful to Maurits and Marcel Dekker for the confidence they showed and the encouragement they gave throughout the entire project. I wish to thank Phyllis Bartosh, Felecia Granderson, Jennifer Woodruff, and, especially, Mickie McConnell and Lynda Parzick for making my sloppy manuscript presentable. My appreciation also goes to Bob Marvos, George Phillips, and, especially, Dottie Holmquist for their work on the figures, which are such an important part of any textbook. I also wish to thank Michael Goett for helping with proofreading and indexing. Finally, due to the diligence of the class on whom this material was tested in manuscript form, the book has 395 fewer errors than when it started. For the errors that remain, and I hope they are few in number and minor in magnitude, I am responsible. Reports from readers of errors and/or obscurities will be very much appreciated.

Paul C. Hiemenz

CONTENTS

51

3. Solution Thermodynamics: Osmotic and Donnan Equilibria   **115**

49

4. The Viscosity of Dilute Dispersions   **169**

Contents

Contents

Contents

Principles of Colloid
and Surface Chemistry

1

COLLOID AND SURFACE CHEMISTRY:
Their Scope and Variables

Next... come the Nobility, of whom there are several degrees, beginning at Six-Sided Figures,... and from thence rising in the number of their sides till they receive the honorable title of Polygonal.... Finally when the number of sides becomes so numerous, and the sides themselves so small, that the figure cannot be distinguished from a circle, he is included in... the highest class of all.

[From Abbott's *Flatland*]

1.1 Introduction

"Yesterday, I couldn't define colloid chemistry; today, I'm doing it." This variation of an old quip could apply to many recent chemistry graduates upon entering employment in the "real world." Two facts underlie this situation. First, colloid and surface chemistry, although traditional parts of physical chemistry, have largely disappeared from introductory physical chemistry courses. Second, in research, technology, and manufacture, countless problems are encountered which fall squarely within the purview of colloid and surface chemistry. Later in this section we shall enumerate some examples which illustrate this statement.

The paradoxical situation just described means that it is entirely possible for a chemistry student to have completed a course in physical chemistry and still not have any clear idea of what colloid and surface chemistry are about. A book like this one is therefore in the curious position of being simultaneously "advanced" and "introductory." Our discussions are often advanced in the sense of building on topics from physical chemistry. At the same time, we shall have to describe the phenomena under consideration pretty much from scratch, since they are largely unfamiliar. In keeping with this, this chapter is concerned primarily with a broad description of the scope of colloid and surface chemistry and the kinds of variables with which they deal. In subsequent chapters different specific phenomena are developed in detail.

Our first tasks are to define what we mean by colloid chemistry and how this is related to surface chemistry. For our purposes, any particle which has some linear dimension between 10^{-8} m (10 Å) and 10^{-6} m (1 µm or 1 µ)* is considered a colloid. For us, linear dimensions rather than particle weights or the number of atoms in a particle will define the colloidal size range; however, other definitions may be encountered elsewhere. It should be emphasized that these limits are rather arbitrary. Smaller particles are considered within other branches of chemistry and larger ones are considered within sciences other than chemistry. The preceding statement may be expanded still further. Colloid chemistry is interdisciplinary in many respects; its field of interest overlaps physics, biology, materials science, and several other disciplines. It is the particle dimension—not the chemical composition (organic or inorganic), sources of the sample (biological or mineralogical), or physical state (one phase or two)—that consigns it to our attention. With this in mind, it is evident that colloid chemistry is the science of both large molecules and finely subdivided multiphase systems.

It is in systems of more than one phase that colloid and surface chemistry meet. The word *surface* is thus used in the chemical sense of a phase boundary, rather than in a strictly geometrical sense. Geometrically, a surface has area but not thickness. Chemically, however, it is a region in which the properties vary from those of one phase to those of the adjoining phase. This transition occurs over distances of molecular dimensions at least. For us, therefore, a surface has a thickness which we may imagine as shrinking to zero when we desire a purely geometric description.

It is self-evident that the more finely subdivided a given weight of material is, the higher the surface area will be for that weight of sample. In the following section we discuss this in considerable detail, since it is the basis for combining a discussion of surface and colloid chemistry in a single book.

In subsequent sections of this chapter, we shall discuss further the distinction between macromolecular colloids and multiphase dispersions

---

*In this book SI units are used fairly consistently, in keeping with current practice. Some quantities are traditionally expressed in hybrid units—for example, the specific area is usually measured in $m^2 \ g^{-1}$—and we continue this practice. The older literature uses cgs units almost exclusively, so the reader must be cautious in consulting other sources. Appendix B contains a list of conversion factors between SI and cgs units.

(Sect. 1.3), the use of the term *stability* in colloid chemistry (Sect. 1.4), the size and shape of colloidal particles (Sect. 1.6), states of aggregation among particles (Sect. 1.7), and the distribution of particle sizes that is typical of virtually all colloidal preparations (Sect. 1.8). The fact that particles in the colloidal size range are not all identical in size also requires a preliminary discussion of statistics, which is the subject matter of several sections at the end of this chapter.

One of the basic premises underlying the selection of topics included in this book is that areas of similarity between diverse fields should be stressed. This is not to say, of course, that differences are unimportant. Rather, it seems more valuable to point out to the beginner that useful methods and insights are frequently part of the well-established procedures of other disciplines which deal with related phenomena. Too provincial a viewpoint, especially at the beginning, is apt to isolate the worker from many potentially valuable sources of information. In the long run, this seems like a greater loss than the loss of time that occurs when a worker concludes too hastily that a technique which works well in one system should work equally well in another system where the particles are larger or smaller by several orders of magnitude. Errors of this last sort are generally discovered quickly enough!

Any attempt to enumerate the areas in which surface and colloid chemical concepts find applicability is bound to be incomplete and quite variable with time because of changing technology. Nonetheless, we shall conclude this section with a partial listing of such applications. If there is any difficulty in doing this, it is because of the abundance rather than scarcity of such examples.

Some areas of science and technology in which particles in the colloidal size range are regularly encountered are the following:

1. Analytical chemistry: adsorption indicators, ion exchange, nephelometry, precipitate filterability, chromatography, and decolorization
2. Physical chemistry: nucleation; superheating, supercooling, and supersaturation; and liquid crystals
3. Biochemistry and molecular biology: electrophoresis; osmotic and Donnan equilibria and other membrane phenomena; viruses, nucleic acids, and proteins; and hematology
4. Chemical manufacturing: catalysis, soaps and detergents, paints, adhesives, and ink; paper and paper coating; pigments; thickening agents; and lubricants
5. Environmental science: aerosols, fog and smog, foams, water purification and sewage treatment; cloud seeding; and clean rooms
6. Materials science: powder metallurgy, alloys, ceramics, cement, fibers, and plastics of all sorts

7. Petroleum science, geology, and soil science: oil retrieval, emulsification, soil porosity, flotation, and ore enrichment
8. Household and consumer products: milk and dairy products, beer, waterproofing, cosmetics, and encapsulated products

It is evident from this partial list how many materials or phenomena of current scientific or everyday interest touch on colloid and surface chemistry to some extent. Many of these areas, of course, have enormous technological and/or theoretical facets which are totally outside our perspective. Nevertheless, all share a common interest in small particles and/or large molecules.

1.2 The Importance of the Surface for Small Particles

The contemporary science student is probably aware that the concept of the atom is traceable to early Greek philosophers, notably Democritus. More than likely, however, few have bothered to follow through the hypothetical subdivision process that led to the original concept of an atom. The time has come to remedy this situation, since the colloidal size range lies between microscopic chunks of material and individual atoms.

Consider a spherical particle of some unspecified material in which the sphere has a convenient radius of, say, 1.0 cm. What we propose to do is to "reapportion" this fixed quantity of material by subdividing it, first, into an array of sphere, each with a radius half that of the original sphere. In a second subdivision, the radius of each of these spheres will be cut in half again. In the third "cut" the radii will be halved again, and so on. The results of such an exercise are summarized in Table 1.1.

The first line in Table 1.1 shows the volume and area of the original sphere whose radius is 1.0 cm. The second line in the table defines the symbols we use to signify these and other quantities in the exercise. The next entries in the table show the number of particles, the volume per particle, the area per particle, and the total area of all the particles after five successive reductions by a factor of one-half in the radius. The arithmetical relationships between these quantities should be noted. The volume of a particle varies with the cube of its radius R. Therefore halving the radius decreases the volume by a factor of 8. Since the total amount of material is unchanged, the number of spheres must be increased by a factor of 8 by this cut. The area varies with R^2; therefore halving the radius decreases the area per sphere by a factor of 4. Since the number of particles is increased eightfold by the same cut, however, the total area of the array is increased by a factor of 2. These generalizations are

Table 1.1 The Radius, Area, and Volume per Particle, Number of Particles, and Total Area for Any Array of Spheres After n "Cuts" Where a Cut is Defined to be the Reapportionment of Material into Particles Whose Radius is Half the Starting Value

Cut number	Radius (cm)	Number of spheres	Volume per sphere (cm^3)	Area per sphere (cm^2)	Total area (cm^2)
Original	1	1	4.19	1.26×10^1	1.26×10^1
Original, symbol	R_0	N_0	V_0	A_0	$A_{T,0}$
1	5×10^{-1}	8	5.24×10^{-1}	3.14	2.51×10^1
2	2.5×10^{-1}	6.4×10^1	6.55×10^{-2}	7.86×10^{-1}	5.03×10^1
3	1.25×10^{-1}	5.12×10^2	8.18×10^{-3}	1.96×10^{-1}	1.01×10^2
4	6.25×10^{-2}	4.10×10^3	1.02×10^{-3}	4.91×10^{-2}	2.01×10^2
5	3.13×10^{-2}	3.28×10^4	1.28×10^{-4}	1.23×10^{-2}	4.02×10^2
$\cdots$	$\cdots$	$\cdots$	$\cdots$	$\cdots$	$\cdots$
n	$(\tfrac{1}{2})^n R_0$	$8^n N_0$	$(\tfrac{1}{8})^n V_0$	$(\tfrac{1}{4})^n A_0$	$2^n A_{T,0}$
$\cdots$	$\cdots$	$\cdots$	$\cdots$	$\cdots$	$\cdots$
13.29	10^{-4}	10^{12}	4.2×10^{-12}	1.26×10^{-7}	1.26×10^5
16.61	10^{-5}	10^{15}	4.2×10^{-15}	1.26×10^{-9}	1.26×10^6
19.93	10^{-6}	10^{18}	4.2×10^{-18}	1.26×10^{-11}	1.26×10^7
23.25	10^{-7}	10^{21}	4.2×10^{-21}	1.26×10^{-13}	1.26×10^8
26.58	10^{-8}	10^{24}	4.2×10^{-24}	1.26×10^{-15}	1.26×10^9

summarized in the next line of entries in Table 1.1, which gives general formulas relating each quantity after n halvings to its initial value. In the final entries of the table the various quantities have been evaluated for successive R values which differ by an order of magnitude.

These results are purely geometrical and therefore are independent of any characteristic of the material—almost. It is implied in the calculations reported in Table 1.1 that the density of the material, whatever it may be, remains the same throughout the subdivision process. By assuming that mass and volume bear a constant proportion to one another, the number of particles increases by the same factor that describes the decrease in particle volume. We shall return to this assumption of constant density presently.

We know from other chemical studies that atomic dimensions are of the order of 10^{-8} cm; therefore, from a chemical point of view, the calculations of Table 1.1 have definitely run into trouble by the cut shown as the final entry in the table. To pursue this point a little further, assume that the material from which these spheres are made is water. Taking the density of water to be exactly 1.0 g cm^{-3}, it is easy to convert the particle volumes in Table 1.1 into the number of water molecules per sphere after each cut. This information is shown in Table 1.2 for the same cuts that were used in Table 1.1. It is evident that by the time we reach spheres of radius 10^{-8} cm, we have begun chopping up water molecules. As a matter of fact, calculating by this procedure, we reach one water molecule per sphere after 25.62 halvings of all spheres, starting from one sphere of $R = 1.0$ cm. If we use the formulas from Table 1.1 to evaluate the radius of the spheres, we find the radius of the water molecules themselves to be 0.193 nm. The radius of a water molecule from the van der Waals b value is about 0.145 nm, so a substantial discrepancy arises when a bulk property, such as density, is applied all the way down to molecular dimensions. Our purpose here is not to arrive at a precise estimate of molecular dimensions, so we shall not worry about this difference. However, the discrepancy does point out the fact that the characterization of a material may be sensitive to the size of the sample under consideration. A property such as density depends not only on the mass and volume of the molecules, but also on their packing in a bulk sample.

Next, suppose we calculate the average number of water molecules which reside at the surface of the spheres in Tables 1.1 and 1.2. To do this, we estimate the area occupied by each molecule at the surface to be about 0.10 nm^2. If we divide this figure into the total area at various stages of subdivision from Table 1.1, we obtain an estimate of the number of molecules in the surface at each stage of the process. Our interest is in the order of magnitude of these quantities, so we need not worry about the

Table 1.2 Total Number of Water Molecules per Sphere and Number at Surface for Spheres of Water After n Cuts (also Total Surface Energy of the Array of Spheres of Water)

Cut number	Radius (cm)	Number of water molecules per sphere	Number of water molecules at surface	Fraction of total water molecules at surface	Total surface energy (J)
0	1.0	1.38×10^{23}	1.26×10^{16}	9.13×10^{-8}	9.07×10^{-5}
1	5.0×10^{-1}	1.75×10^{22}	3.14×10^{15}	1.79×10^{-7}	1.81×10^{-4}
2	2.5×10^{-1}	2.18×10^{21}	5.03×10^{16}	3.64×10^{-7}	3.62×10^{-4}
3	1.25×10^{-1}	2.73×10^{20}	1.01×10^{17}	7.32×10^{-7}	7.27×10^{-4}
4	6.25×10^{-2}	3.41×10^{19}	2.01×10^{17}	1.46×10^{-6}	1.45×10^{-3}
5	3.13×10^{-2}	4.27×10^{18}	4.02×10^{17}	2.91×10^{-6}	2.89×10^{-3}
$\cdots$	$\cdots$	$\cdots$	$\cdots$	$\cdots$	$\cdots$
13.29	10^{-4}	1.40×10^{11}	1.26×10^{20}	9.13×10^{-4}	9.07×10^{-1}
16.61	10^{-5}	1.40×10^{8}	1.26×10^{21}	9.13×10^{-3}	9.07
19.93	10^{-6}	1.40×10^{5}	1.26×10^{22}	9.13×10^{-2}	9.07×10^{1}
23.25	10^{-7}	1.40×10^{2}	1.26×10^{23}	9.13×10^{-1}	9.07×10^{2}
26.58	10^{-8}	1.40×10^{-1}	1.26×10^{24}	9.13	9.07×10^{3}

cross-sectional shape or the surface packing efficiency of the water molecules in these calculations. Table 1.2 shows the number of water molecules at the surface calculated in this manner; this quantity is also reported as a fraction of the total number of molecules present. Note that this fraction approaches unity as molecular dimensions are approached.

Finally, let us consider the last column in Table 1.2. In this column the total surface energy of the array of spherical water droplets is reported at each stage of the process. We shall see in Chap. 6 that the surface tension of a substance measures the energy required to make a unit area of new surface. For water, this quantity is about 72 mJ m^{-2} at room temperature. If we multiply the surface tension of water by the total areas from Table 1.1, we obtain the values listed under "total surface energy" in Table 1.2. It should be recalled that a fixed amount of material (4.19 cm^3 water = 4.19 g water = 0.23 mole water) is involved throughout this entire process. It should also be noted that surface tension, like density, is a macroscopic property; its applicability is highly dubious for very small particles. Nevertheless, as the dimensions of the subdivided units decrease, the total energy associated with the formation of surface takes on values comparable to other chemical energies. At $R = 10^{-4}$ cm, the surface energy is about 0.9 J/0.23 mole or about 4 J mole^{-1}. By $R = 10^{-7}$ cm, this quantity equals 4 kJ mole^{-1}.

The increasing importance of the surface area as the linear dimensions of particles decrease is stated concisely in a quantity known as the specific area of a substance, A_{sp}. This quantity is determined as the ratio of the area divided by the mass of an array of particles. If the particles are uniform spheres, as we have assumed throughout this section, this ratio equals

$$A_{sp} = \frac{A_{tot}}{m_{tot}} = \frac{n4\pi R^2}{n\frac{4}{3}\pi R^3 \rho} \tag{1}*$$

where n is the number of spheres having a radius R and made of a material of density ρ. Simplifying Eq. (1)* leads to the result

$$A_{sp} = \frac{3}{\rho R} \tag{2}$$

This formula generalizes the conclusion reached in Tables 1.1 and 1.2. It shows clearly that for a fixed amount of material, the surface area is inversely proportional to the radius for uniform, spherical particles. At

---

*This manner of referencing is used for equations occurring in the same chapter.

the same time the formula reminds us that some lower limit for R must be imposed, since the relationship is undefined for $R = 0$. If SI units were used consistently, A_{sp} would be expressed in $m^2 \, kg^{-1}$; however, $m^2 \, g^{-1}$ are the most commonly used units for this quantity. In the event of nonuniform or nonspherical particles, alternate expressions to Eq. (2) have to be used. The following example considers the case of cylindrical particles.

Example 1.1　A material of density ρ exists as uniform cylindrical particles of radius R and length L. Derive an expression for A_{sp} for this material and examine the limiting forms when either R or L is very small.

Solution　The area of each cylindrical particle equals the sum of the areas of both ends and the edges: $A = 2(\pi R^2) + 2\pi RL$.

The volume of each cylindrical particle equals $\pi R^2 L$ and its mass is given by $\rho \pi R^2 L$.

For an array of n cylindrical particles, the total area per total mass equals A_{sp} and is given by

$$A_{sp} = n(2\pi R^2 + 2\pi RL)/n\rho\pi R^2 L = [2(R^2 + RL)]/\rho R^2 L = (2/\rho)(1/R + 1/L)$$

For a thin rod, $L \gg R$:

$$A_{sp} \simeq \frac{2}{\rho} \frac{1}{R}$$

For a flat disc, $R \gg L$:

$$A_{sp} \simeq \frac{2}{\rho} \frac{1}{L}$$

Note that in both of these limits, it is the *smaller* dimension that affects A_{sp}, with the latter increasing as the smaller dimension decreases.

•

Both Example 1.1 and Eq. (2) show that the surface plays an increasingly important role as the dimensions of the particles decrease. The concept of specific area defined by Eq. (1) is important because this is a quantity which can be measured experimentally for finely divided solids without any assumptions as to the shape or uniformity of the particles. We shall discuss the use of gas adsorption to measure A_{sp} in Chap. 9. If the particles are known to be uniform spheres, this measured quantity may be interpreted in terms of Eq. (2) to yield a value of R. If the actual system consists of nonuniform spheres, an average value of the radius may be evaluated by Eq. (2). Finally, even if the particles are nonspherical, a quantity known as the radius of an equivalent sphere may be extracted from experimental A_{sp} values. This often proves to be a valuable way of characterizing an array of irregularly shaped particles. We shall have a good deal more to say about average dimensions later in this chapter, and about equivalent spheres in Chap. 2.

In the foregoing discussion we emphasized two-phase colloidal systems, in which the concept of "surface" plays an important role. Our definition of the colloidal range is based on the linear dimensions of particles, however, and there are numerous natural and synthetic polymer molecules whose dimensions, considered individually, fall within this range. It is clear when we deal with single molecules that the concept of "surface" is greatly different than when we consider a particle made of many molecules. If, in a given situation, we tend to concentrate on the surface characteristics of a material, then we are working in surface chemistry. If, however, we look at the subdivided sample as an array of particles, then we are working in colloid chemistry. As far as we are concerned, these two fields differ primarily in point of view. Their mutual concern with finely subdivided material is the common denominator that connects the two disciplines.

1.3 Lyophilic and Lyophobic Colloids

In the preceding section, we saw that either large molecules or finely subdivided bulk matter could be considered colloids inasmuch as both may consist of particles in the range 10^{-9}–10^{-6} m in dimension. The difference between these two situations lies in the relationship that exists between the colloidal particle and the medium in which it is embedded. Macromolecular colloids give true solutions in the thermodynamic sense with the medium which surrounds them. Subdivided bulk matter, on the other hand, forms a two-phase (at least) system with the medium. We have already noted that the word *surface* connotes the existence of a phase boundary and therefore has a specific chemical meaning in the multiphase case which is inapplicable to macromolecular colloids.

The terms *lyophilic* and *lyophobic* are used to distinguish between one-phase and two-phase colloidal systems, respectively. These terms mean, literally, "solvent loving" and "solvent fearing." When water is the medium or solvent, the terms *hydrophilic* or *hydrophobic* are often used.

Although the distinction between lyophilic and lyophobic colloids seems quite clear-cut in terms of these definitions, in the world of small

Table 1.3 Summary of Some of the Descriptive Names Used to Designate Two-Phase Colloidal Systems

Continuous phase	Dispersed phase	Descriptive names
Gas	Liquid	Fog, mist, aerosol
Gas	Solid	Smoke, aerosol
Liquid	Gas	Foam
Liquid	Liquid	Emulsion
Liquid	Solid	Sol, colloidal solution, gel, suspension
Solid	Gas	Solid foam
Solid	Liquid	Gel, solid emulsion
Solid	Solid	Alloy

dimensions—where colloid chemistry is practiced—the classification is not always so easy. An example is the case of micelles, which are clusters of small molecules that form spontaneously in aqueous solutions (mostly) of certain compounds. The onset of micellization occurs at a well-defined concentration—the critical micelle concentration—which makes micelle formation very much like a phase separation. However, the individual small molecules retain their identity in the micelle—the latter are not covalent entities like polymers—so their surface is problematic. We shall see in Chap. 8 that both chemical equilibrium and phase equilibrium can be used to discuss micelle formation, so the classification of micelles remains fuzzy on that basis also. By the time we reach Chapter 8, however, we will be more experienced with colloids and less dependent on the "black and white" categories of this section.

The intent of this book is to discuss as wide a variety of colloidal phenomena as possible from a unified point of view and in a single set of terms. We use the words *continuous* and *dispersed* to refer to the medium and to the particles in the colloidal size range, respectively. It should be understood that these are distinctly different phases in lyophobic systems and are solvent and solute in lyophilic systems. In micellar systems, the micelles are dispersed in an aqueous continuous phase. Furthermore, the system as a whole is generally called a dispersion when we wish to emphasize the colloidal nature of the dispersed particles. This terminology is by no means universal. Lyophilic dispersions are true solutions and may be called such, although this term ignores the colloidal size of the solute molecules. Lyophobic colloids are known by a variety of terms, depending on the nature of the phases involved. Some of these are listed in Table 1.3. Some of the terms (e.g., aerosol, gel) are somewhat ambiguous, so the reader is warned to make certain that the system is fully understood, particularly when the original literature is consulted. Remember that a common feature of all systems we consider is that some characteristic linear dimension of the dispersed particles falls in the range defined in Sect. 1.1. When we deal with two-phase colloids in this book, we are primarily concerned with systems in which the dispersed phase is solid and the continuous phase is liquid.

Next, we consider another difference between lyophobic and lyophilic colloids in addition to the presence or absence of surfaces between the continuous and dispersed species. This difference deals with the "stability" of the dispersion. We shall examine the meaning(s) of this term in more detail in the next section. For now, the following distinction is sufficient.

Lyophilic colloids form true solutions, and true solutions are produced spontaneously when solute and solvent are brought together. In the absence of chemical changes or changes of temperature, a solution is stable indefinitely. Finely subdivided dispersions of two phases do not form spontaneously when the two phases are brought together. As a matter of fact, if such a dispersion is allowed to stand long enough, the reverse process would spontaneously occur. For example, oil and water can be vigorously mixed to form a nontransparent, heterogeneous mass; however, on standing, the mixture will separate into two clear, homogeneous layers. We know from thermodynamics that spontaneous processes occur in the direction of decreasing Gibbs free energy. Therefore we may conclude that the separation of a two-phase dispersed system to form two distinct layers is a change in the direction of decreasing Gibbs free energy. In connection with the discussion of Table 1.2, we can say that there is more surface energy in a two-phase system when the dispersed phase is in a highly subdivided state than when it is in a coarser state of subdivision. This suggests a correlation between the inherent instability of a highly dispersed lyophobic system and the thermodynamics of the surface. This is discussed in more detail in Chapter 6. For the present it is sufficient to note that lyophobic systems "dislike" their surroundings enough to want to separate out. Lyophilic systems, on the other hand, are perfectly "happy" in a solution. The two categories differ radically in their thermodynamic stability.

It should also be remembered that thermodynamics has nothing to say about the rate at which processes occur. It is a fact that many two-phase dispersions appear unchanged over very long periods of time. The situation is analogous to the thermodynamic instability of diamond with respect to graphite. The kinetics of the diamond-graphite reaction are slow enough that the thermodynamic instability is of very little practical consequence. Likewise, many colloidal dispersions have kinetic stability, even though they are unstable thermodynamically. This type of colloidal "solution" (Table 1.3) resembles a true solution. This is one way in which two-phase dispersions and solutions of macromolecules are very similar and explains, in part, how the two can be grouped together in our study of colloidal phenomena.

1.4   Stable and Unstable Systems

In the preceding section, we saw that two-phase dispersions will always spontaneously change into a smaller number of large particles, given sufficient time. However, many solutions of macromolecules do not undergo spontaneous separation into two phases. Common usage tempts us to describe the first as "unstable" and the second as "stable." Although these terms are very frequently used in colloid chemistry, the reader should realize that the words are meaningless unless the process to which they are applied has been clearly defined. The situation is somewhat analogous to the insistence in thermochemistry that one always keep in mind the balanced chemical equation to which the thermodynamic quantities (ΔH, ΔC_p, etc.) apply. The coarsening described here is only one of a variety of possible processes that a dispersion might undergo.

The following examples will illustrate this caveat and at the same time help define some of the common processes in colloid chemistry. To begin with, we shall rarely be concerned with chemical reactions in the conventional sense, although there are some important areas such as heterogeneous catalysis in which chemical reactions at surfaces are of primary interest. Our approach to colloid chemistry in this book is directed mainly to the physical chemistry rather than the reaction chemistry of these systems. As a matter of fact, chemical reactions are often unintended side effects in colloidal systems. For example, a protein molecule may undergo hydrolysis to alter the size, shape, and/or the solubility of the molecule. In such a case, we may see a molecule that forms a solution which is thermodynamically stable with respect to phase separation but which is not stable with respect to a chemical change. For the most part, however, we shall regard the systems we discuss as stable with respect to chemical change.

In the preceding section, we noted that two-phase colloids are thermodynamically unstable with respect to the coarsening process. It is important that we define this coarsening in more detail. Specifically, we need to differentiate between coalescence and flocculation. By coalescence we mean the process whereby two (or more) small particles fuse together to form a single larger particle. The central feature of coalescence is the fact that total surface area is reduced. Flocculation is the process whereby small particles clump together like a bunch of grapes (a floc) but do not fuse into a new particle. In flocculation there is no reduction of surface, although certain surface sites may be blocked at the points at which the smaller particles touch.

When small particles coalesce, all evidence of the smaller particles is erased: Only the new, larger particle remains. With flocculation, however, the small particles retain their identity; only their kinetic independence is lost: The floc moves as a single unit. The terms *aggregation* and *coagulation* are also used to describe the process we have called flocculation. Likewise, the clusters which form as the products of the process may be called flocs or aggregates. The individual particles from which the flocs are assembled are called primary particles. A system may be relatively stable in the kinetic sense with respect to one of these processes, say, coalescence, and be unstable with respect to the other, flocculation.

Finally, the word *stability* is also used to describe the extent to which small particles remain uniformly distributed throughout a sample. In Chap. 2, for example, we shall examine the tendency of heavier particles to settle to the bottom of the container, a process known as sedimentation. It is quite possible for a system to be unstable with respect to sedimentation, but relatively stable (kinetically) with respect to flocculation or coalescence. On the other hand, an array of primary particles may be small enough to be relatively stable with respect to sedimentation but unstable with respect to flocculation.

Systems that are "stable" and "unstable" in any of these processes are not equally esteemed in all areas of application. A dispersion which is too fine to settle out may be a source of great frustration in some parts of a manufacturing process. A dispersion that settles too rapidly may be equally troublesome to those who have to pump the stuff around. The chemist doing gravimetric analyses wants coarse precipitates; the one who uses adsorption indicators wants finely subdivided particles to form.

In summary, then, we must keep several things in mind when the word *stability* is used in colloid chemistry. First, whenever we describe a two-phase dispersion in these terms, the words are being used relatively and in a kinetic sense. Second, there is little unanimity among workers about the nomenclature of various processes. Finally, whether a stable or unstable system is desirable depends entirely on the context.

1.5 Microscopy

Because of the particle sizes involved, the microscope is an instrument which is ideally suited for the study of some lyophobic colloids. For a

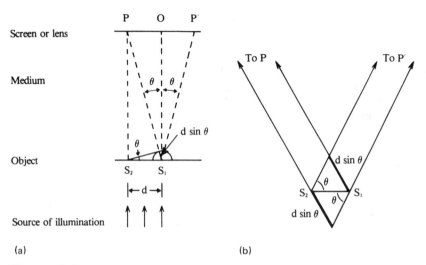

Figure 1.1 (a) The geometry upon which the resolving power d of a microscope is based. (b) Detail showing how light from both sources must be intercepted by the lens to become part of the image.

particle to be visible, there must be an acceptable difference between its refractive index and that of its surroundings. This requirement has nothing to do with particle size: A glass rod can be made to "disappear" by immersing it in a liquid of matching refractive index. The fact that a technique is not applicable to all possible systems does not invalidate it; this only means that the worker must have access to more than one technique in order to deal with a variety of problems.

The first thing we tend to think of in connection with microscopes is the magnification they achieve. More important, however, is a quantity known as the resolving power, or limit of resolution, of the microscope. Magnification determines the size of an image, but the resolving power determines the amount of distinguishable detail. Enlargement without detail is of little value. For example, a row of small spherical particles will appear simply as a line if it is enlarged with an instrument of poor resolving power. Further magnification would increase the thickness of the "line," but would not reveal its particulate nature. If the resolving power is increased, however, the individual spheres would be discernible. One may then choose a magnification which is convenient. Both the depth and area of the in-focus field decrease as the magnification is increased, so one pays a price for enlargement, even though the amount of perceptible detail is not affected much by the magnification.

A beam of light is always diffracted at the edges of an object to produce a set of images of the edge known as a diffraction pattern. The diffracted light is what our eye receives from an object and from which our image of the object is constructed. The diffracted light contains sufficient information to assemble an image; any light that is not incorporated into the image will result in a loss of detail in the image. To estimate the efficiency with which an image reproduces the object, let us consider the situation in which all first-order diffracted light is intercepted by the lens, but light from all higher orders of diffraction is assumed to be lost.

Suppose we have a set of pinholes in an opaque shield and that this shield is illuminated by a source which is far enough away that the incident light may be regarded as a set of parallel rays, as sketched in Fig. 1.1a. To an observer on the opposite side of the shield, each pinhole will function as a light source from which a hemispherical wave front seems to emerge. If adjoining pinholes are separated by a distance d, the wave from one source, say, S_1, must travel a longer distance than light from S_2 to reach a distant screen at point P. If the distance between the shield and the screen is large compared to the wavelength, the extra distance may be equated with $d \sin \theta$. Positive reinforcement through diffraction occurs whenever such an extra distance equals some integral number of wavelengths of light. For first-order diffraction, then, we require

$$d \sin \theta = \lambda' \tag{3}$$

where λ' is the wavelength in the medium between the screen and the shield. If the medium has a refractive index n, then $\lambda' = \lambda/n$, where λ is the wavelength under vacuum. Therefore Eq. (3) becomes

$$d \sin \theta = \frac{\lambda}{n} \tag{4}$$

Equation (3) is called the Bragg equation after the father-and-son team of W. H. and W. L. Bragg (Nobel Prize, 1915); it is the underlying relationship for all diffraction phenomena. We shall encounter the Bragg equation again in Chapter 10 when we discuss the diffraction of low-energy electrons by surface atoms.

Recall that our objective is to consider an image constructed from only first-order diffracted light. To do this, we identify the perforated shield $S_1 S_2$ as the object, the screen as the objective lens of the microscope, and the distance PP' as the diameter of the objective. It is clear from Fig. 1.1b that light originating at S_1 must travel a distance $d \sin \theta$ longer than the light from S_2 to form an image at P; likewise, light originating at S_2 must travel a distance $d \sin \theta$ longer than light from S_1 to form an image at P'. Thus, to be intercepted by the lens and thereby become part of the image, rays from different parts of the source must travel paths which differ by $2d \sin \theta$. In order for this to happen and still consist of only first-order diffracted light, the difference in path lengths must equal λ'. The resolving power is defined to be the magnitude of the separation between objects that is required to produce discernibly different images when the angle subtended by the microscope is 2θ. Therefore the resolving power is identical to d in Fig. 1.1 and is estimated to be

$$d = \frac{\lambda}{2n \sin \theta} \tag{5}$$

which is sometimes written

$$d = \frac{\lambda}{2(\text{NA})} \tag{6}$$

where NA (i.e., $n \sin \theta$) is called the numerical aperture of the lens.

The significance of Eq. (5) is that any points which are closer together than the distance d will produce a first-order diffraction image with light having a wavelength equal to λ/n at some angle greater than θ. This light would not be intercepted by the lens PP', so a significant amount of detail about distances of this magnitude is lost. It is the wave nature of light that imposes a limit to the amount of detail an image may possess. Equation (5) shows that the resolving power is decreased by increasing θ or by decreasing λ/n. The subtended angle 2θ is increased by increasing the diameter of the lens and by decreasing the distance between the object and the lens; the design of the lens limits the range of these parameters. The ratio λ/n may be decreased by decreasing λ or by increasing n. Although shorter wavelengths improve resolving power, visible light is almost always used in microscopy, primarily because of the absorption of shorter wavelengths by glass. Since the refractive index of some oils is 50% higher than that of air, a significant improvement in the resolving power is achieved by filling the gap between object and lens with so-called immersion oil.

We may use Eq. (5) to estimate the particle size which will be clearly resolved microscopically. As a numerical example, we may take λ to be 500 nm, $n = 1.5$, and $2\theta = 140°$, all of which are attainable but optimal values for these quantities. This leads to a resolving power of 142 nm, which represents the lower limit of resolved particle size under completely ideal circumstances. Under closer to average circumstances, a figure twice this value may be more typical. These figures reveal that direct microscopic observation is feasible only for particles at the upper end of the colloidal size range, and then only if the particle contrasts sufficiently with its surroundings in refractive index.

A variation of direct microscopic examination extends the range of microscopy considerably but at the expense of much detail. By this technique, known as dark-field microscopy, particles as small as 5 nm may be detected under optimum conditions, with about 20 nm as the lower limit under average conditions. In dark-field microscopy, the sample is illuminated from the side, rather than from below as in an ordinary microscope. If no particles were present, no light would be deviated from the horizontal into the microscope and the field would appear totally dark. The presence of small dispersed particles, however, leads to the scattering of some light from the horizontal into the microscope objective, a phenomenon sometimes called the Tyndall effect. The presence of colloidal particles is indicated by minute specks of light in an otherwise dark field.

In dark-field microscopy, the particles are only a blur; no details are distinguishable at all. Some rough indication of the symmetry of the particles is afforded by the twinkling that accompanies the rotation of asymmetrical particles, but this is a highly subjective observation. However, the technique does permit the rate of particle diffusion to be followed. We shall see in Chap. 2 how to relate this information to particle size and shape. The number of particles per unit volume may also be determined by direct count once the area and depth of the illuminated field have been calibrated. This is an important technique for the study of kinetics of flocculation, a topic we discuss in Chap. 11. In dark field, as in direct observation microscopy, the refractive index difference between the particles and the medium is one of the crucial factors that influences the feasibility of the method. Very good resolving power is obtained with metallic particles which offer maximum contrast in refractive index. The technique of dark-field microscopy played an important historic role in colloid chemistry, particularly in the study of metallic colloids. R. Zsigmondy (Nobel Prize, 1925), for example, made extensive use of this technique in his study of colloidal gold. Note that it is the deviation of a beam of light or scattered light that is utilized in dark-field microscopy. We shall have a good deal more to say about light scattering in Chapter 5.

It is clear from Eq. (5) that the best prospect for extending the range of microscopy lies in the extension by orders of magnitude of the wavelength used to produce the image. The wave–particle duality principle of modern physics shows us how to make this extension. According to the de Broglie equation, the wavelength of a particle is inversely proportional to its momentum:

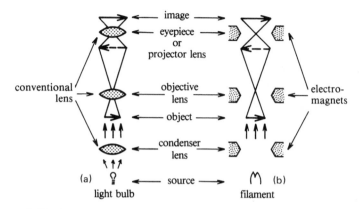

Figure 1.2 Schematic comparison of (a) light and (b) electron microscopes showing components which perform parallel functions in each.

$$\lambda = \frac{h}{mv} \tag{7}$$

where h is Planck's constant and m and v are the particle mass and velocity, respectively. In an electron microscope a beam of electrons replaces light in producing an image. An electron beam is produced by a hot filament, accelerated by an electron gun, and focused by electric or magnetic fields which function as lenses. In this section, we consider the image produced by the transmitted electron beam. Accordingly, the method is called transmission electron microscopy. In Sect. 10.3, we shall take up electron microscopy again, this time as scanning electron microscopy. The latter is useful when it is the surface of a solid and not its state of dispersion that is of interest. A schematic comparison of light and electron microscopes is shown in Fig. 1.2. Although wavelengths on the order of 10 pm are easily achieved in electron microscopes, the numerical apertures of such microscopes are low. Accordingly, the resolving power of a conventional electron microscope is generally about 1 nm. Because of this remarkable resolution, however, small features may be enormously magnified without loss of detail. An easy conversion factor to remember when examining electron micrographs is that an object 1 μm in length appears 1 in. long when it is magnified by 25,400 [10^{-4} cm × (1 in./2.54 cm) × 25,400 = 1 in./μm].

The intensity of the electron beam that is transmitted through the specimen under observation in an electron microscope depends on the thickness of the sample and the concentration of atoms in the sample. Thus, unless there are large differences in these characteristics between the particles of the sample and the support on which they rest, a very low

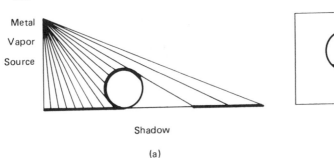
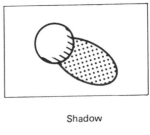

(a)   (b)

Figure 1.3 (a) Side and (b) top views of a spherical particle shadowed by metal vapor.

contrast image will be produced. A poor-quality image results from these low-contrast situations; the picture is analogous to a landscape viewed from an airplane with the sun directly overhead. The most common way of overcoming this difficulty is by means of a technique known as shadow casting. The sample to be examined in the electron microscope is placed in a chamber which is then evacuated. Next some gold or other metal is vaporized in the same chamber. The vapor condenses on all cooler surfaces, including the surface of the sample. If the vapor source is positioned off to the side of the sample, the condensed vapor will deposit unevenly, effectively casting a shadow over the sample. The situation is shown schematically in Fig. 1.3; Fig. 1.9 shows actual photomicrographs enhanced by this technique. With shadow casting, the field shows much more detail, just as the view from an airplane does when the sun is lower in the sky.

When the angular position of the vapor source relative to the sample is known, the thickness of the particle may be evaluated from the length of its shadow by a simple trigonometric calculation. Alternatively, the length of shadow cast by a spherical reference particle, introduced for calibration, may be used as a standard to calculate the thickness of particles.

At the beginning of this section the microscope was proposed as an ideal tool for the study of colloidal particles. The light microscope is limited in range, but the electron microscope clearly has access to the entire colloidal size range. However, a very real disadvantage still persists even with the electron microscope. The ordinary electron microscope requires that the sample be placed in an evacuated enclosure for examination by the electron beam. Therefore a dispersion must be evaporated to dryness before examination. The result, although still informative, bears about as much resemblance to the original two-phase

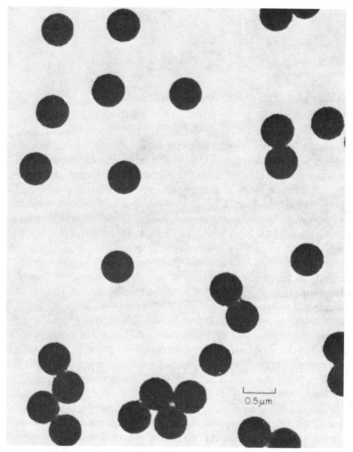

Figure 1.4 Electron micrograph of cross-linked monodisperse polystyrene latex particles. The latex is a commercial product ($\bar{d}$ = 0.500 μm) sold as a calibration standard. (Photograph courtesy of R. S. Daniel and L. X. Oakford, California State Polytechnic University. Pomona, Ca.)

dispersion as a pressed flower bears to a blossom on a living plant. In subsequent chapters we shall discuss some in situ techniques for the characterization of colloidal particles, especially with respect to particle weight.

Despite the limitations just noted, there is no other branch of chemistry that can reasonably expect to see directly the particles of interest in that field, however appealing that prospect may seem to chemists. However, with some colloids this is a possibility. We shall not

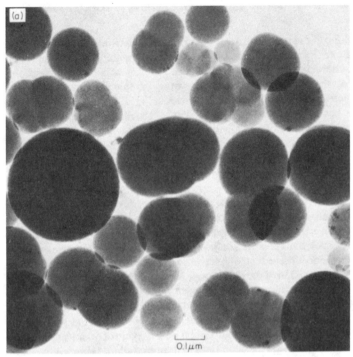

Figure 1.5 Electron micrograph (150,000×) of carbon black particles (a) before and (b) after heating to 2700°C in the absence of oxygen. [From F. A. Heckman, *Rubber Chem. Technol.*, *37*: 1243 (1964), used with permission.]

pass up the opportunity to take a look! Figures 1.4, 1.5, 1.9, and 1.10 are electron micrographs of lyophobic colloids. We shall comment on these individually in the following sections.

1.6 Particle Size and Shape

Figure 1.4 shows an electron micrograph of latex particles made from polystyrene cross-linked with divinylbenzene. Although these are polystyrene particles, this is clearly a two-phase system. The latex particles are not the same as simple polystyrene molecules in a true solution. These particles display a remarkable degree of homogeneity with respect to particle size. Such a sample is said to be monodisperse, in contrast to polydisperse systems, which contain a variety of particle sizes. We shall have a good deal more to say about polydisperse systems in later sections of this chapter. The particles of Fig. 1.4 are also perfectly spherical. Except for the nature of the material involved, this figure could be a photograph of the hypothetical array of spheres discussed in Sect. 1.2.

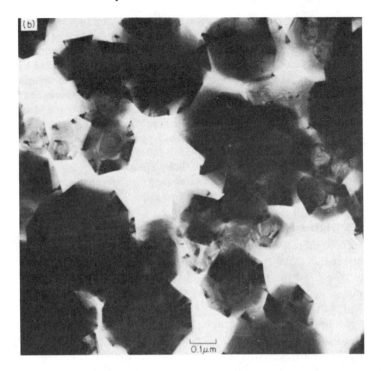

It is difficult to imagine an array of particles which would be easier to describe than the latex particles of Fig. 1.4. A single parameter, such as the radius of the spheres, characterizes the dispersed phase in terms of a linear dimension. If the density of the material is known, the mass of the individual particles is fully determined also. The radius of the particles in Fig. 1.4 is 0.25 μm; therefore a sphere has a volume of 6.54×10^{-3} μm^3. The density of polystyrene is about 1.05 g cm^{-3}; therefore the mass of an individual particle in Fig. 1.4 is 6.87×10^{-17} kg.

When we deal with lyophobic colloids, we shall generally indicate the mass of the individual particle in kilograms per particle. When we deal with lyophilic colloids, we shall use the molecular weight in grams per mole just as we would for a low molecular weight substance. Aside from the incidental difference between grams and kilograms, the two units of measure differ by Avogadro's number. Since the particle weights are so different from those we are accustomed to in ordinary chemistry, we must be especially careful to consider the units (i.e., per particle or per mole) in which a result is reported, since we may not immediately recognize the number itself.

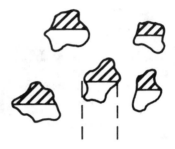

Martin Diameter

Figure 1.6 A schematic illustration of Martin's diameters for irregular particles.

Monodisperse spheres are not only uniquely easy to characterize but also very rarely encountered. Polymerization under carefully controlled conditions allows the preparation of the polystyrene latex shown in Fig. 1.4. Latexes of this sort are used as standards for the size calibration of optical and electron micrographs.

In the majority of two-phase colloidal systems, the particles are neither spherical nor monodisperse. Figure 1.5 shows micrographs of carbon black particles. Broadly speaking, this material is soot, but a great deal of control over its properties may be accomplished by varying the conditions of its preparation. Figure 1.5a shows what is known as a thermal black, in which both discrete and partially fused particles may be seen. Figure 1.5b shows the same carbon black preparation (not the same field of particles) after heat treatment at 2700°C in the absence of oxygen. The particles take on a distinctly polyhedral shape with this treatment, which is known as graphitization.

Many solid particles are not actually spherical but are characterized by a high degree of symmetry like a sphere, and are often approximated as spheres. For example, a polyhedron approximates a sphere more and more closely as the number of its faces increases. The primary particles of the graphitized thermal black shown in Fig. 1.5b are sufficiently symmetrical to be approximated as spheres. Likewise, many substances which display irregular but symmetrical particles are often described by a characteristic dimension which is called a "diameter." This terminology does not necessarily mean that the particles are spherical.

The forms sketched in Fig. 1.6 represent some irregularly shaped particles as they might be observed in a light or electron micrograph. The length of a line which bisects the projected area of a particle is a parameter known as Martin's diameter. The direction along which the

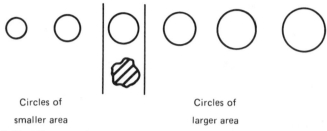

Circles of
smaller area

Circles of
larger area

Figure 1.7 The use of a graticule to estimate a characteristic dimension of an irregular particle.

Martin diameter is measured is arbitrary, but it should be used consistently to avoid subjective bias. The lines sketched in the figure are intended to represent this quantity. The Martin diameter is most easily measured on photographs, although movable crosshairs in the eyepiece of a microscope also permit such distances to be measured by direct observation.

Another method of characterizing irregular particles consists of reporting the diameter of a circle which projects the same cross section as the particle in question. A technique for doing this is to insert an object known as a graticule into the eyepiece of a microscope. This is merely an assortment of circles of different sizes etched on a transparent slide. The observer decides which circle most closely approximates the projected area of the particle. Figure 1.7 illustrates how a graticule might be used to size an irregular particle. Graticules are available with circles whose diameters increase in a $\sqrt{2}$ progression. Both the graticule dimension and the Martin diameter are extremely tedious to evaluate, since a large number of particles must be examined for the values to have any statistical significance.

Fortunately, instruments are available which take the place of the observer and classify images by size automatically. These methods also work on the size-matching principle; therefore the fact that the labor can be done electronically does not decrease the importance of recognizing what is involved in particle sizing. As a matter of fact, these devices also carry out data reduction by calculating means, standard deviations, and so on. This capability does not exempt us from some understanding of statistics either. We shall return to the statistical aspects of particle sizing presently.

The sphere is favored above all other geometrics as a model for actual particles because it is characterized by a single parameter. Sometimes, however, the particles of a dispersion are so asymmetrical that no *single*

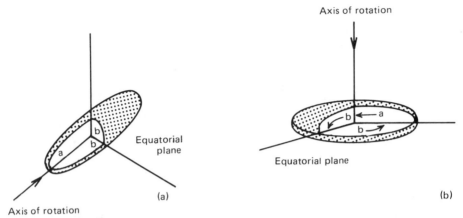

Figure 1.8 (a) Prolate ($a > b$) and (b) oblate ($a < b$) ellipsoids of revolution, showing the relationship between the semiaxes and the axis of revolution.

parameter, however defined, can begin to describe the particle. In this case the next best thing is to describe the particle as an ellipsoid of revolution. An ellipsoid of revolution is that three-dimensional body which results from the complete rotation of an ellipse around one of its axes. We shall define a as the "radius" of the ellipsoid measured along the axis of rotation and b as the "radius" measured in the equatorial plane. Obviously, if these two measurements of radius are equal for a particle, that particle is spherical. If $a > b$, the particle is called a prolate ellipsoid; if $a < b$, it is an oblate ellipsoid. These two geometries are illustrated in Fig. 1.8. The ratio a/b, called the axial ratio of the ellipsoid, is frequently used as a measure of the deviation from sphericity of a particle. It plays an important role, for example, in our discussions of sedimentation and viscosity in Chaps. 2 and 4, respectively. In the event that $a \gg b$, the prolate ellipsoid approximates a cylinder and, as such, is often used to describe rod-shaped particles such as the tobacco mosaic virus particles shown in Fig. 1.9a. Likewise, if $a \ll b$, the oblate ellipsoid approaches the shape of a disk. Thus even the irregular clay platelets of Fig. 1.9b may be approximated as oblate ellipsoids.

It should be fairly obvious that the two-parameter ellipsoidal geometry is far more accurate than the single-parameter spherical geometry to describe asymmetrical particles. We shall see below that just as two parameters are generally used to describe asymmetrical particles, two parameters are also preferable to characterize polydisperse systems. One might argue that in both cases more than two parameters would be better yet. Sometimes this might be true, but often experimental difficulties or uncertainties make this infeasible.

One other particle geometry deserves mention. Suppose we were to take a length of string or other flexible material and allow it to tumble freely for a while in a large container. The string would certainly be expected to emerge from this treatment as a tangled jumble. Many long-chain molecules have sufficient flexibility to take on a random configuration like this under the influence of thermal jostling. This "random coil" is likely to be symmetrical rather than stretched out. We shall, accordingly, refer to the radius of such a coil. The random coil is discussed in detail in Sect. 2.12.

1.7 Aggregation

We have already introduced the idea that the primary particles of a dispersed system tend to associate into larger structures known as flocs or aggregates. The nature of the interparticle forces responsible for this aggregation is one of the most interesting areas of colloid chemistry. In the absence of interactions colloidal dispersions would be analogous in many ways to ideal gases in which the individual particles are also independent of each other. However indebted physical chemistry may be to ideal gases, it is clear that nonideal gases are more interesting! Likewise, colloidal systems which are unstable with respect to flocculation provide a lot more information as to the nature of interparticle forces than stable systems. We shall defer our discussion of the flocculation process until Chapter 11, but a few remarks about flocs—the kinetic units that result from that process—are in order at this time.

In many situations the dispersed phase is present as flocs, not as primary particles. It is of little importance in such systems that the flocs could be disrupted further without subdividing the primary particles or that they might aggregate more to form still larger flocs. The particles that are actually dispersed in the continuous phase are flocs. It is the size, shape, and concentration of the units actually present which determine the properties of the dispersion itself. As a matter of fact, some substances—for example, those carbon blacks known as channel and furnace blacks—possess rigidly fused, floclike structures as their primary particles. Figure 1.10 shows an example of such a particle.

When an electron micrograph shows evidence of aggregation, we must remember that this may be an artifact arising from the preparation of the sample for microscopy. In other words, the amount of flocculation that a colloid displays in its dispersed state and the amount that appears

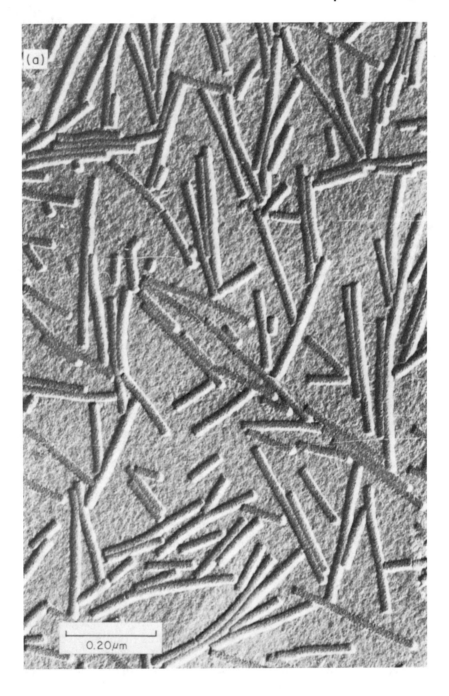

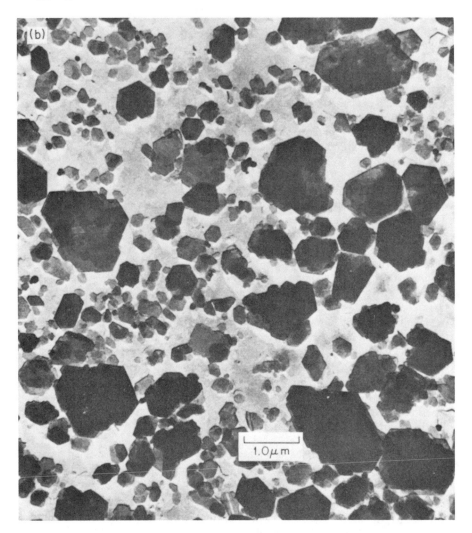

Figure 1.9 Electron micrograph of two different types of particles which represent extreme variations from spherical particles. (a) Tobacco mosaic virus particles. (Photograph courtesy of Carl Zeiss, Inc., New York.) (b) Clay particles (sodium kaolinite) of mean diameter 0.2 μm (by matching circular fields). [From M. D. Luh and R. A. Bader, *J. Colloid Interface Sci., 33*: 539 (1970) used with permission.] In both (a) and (b), contrast has been enhanced by shadow casting.

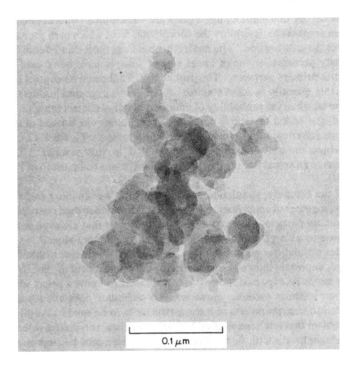

Figure 1.10 Transmission electron micrograph (500,000×) of an individual carbon black (furnace black) particle (Vulcan 6, ISAF, N-220) showing the fused-floc structure of the primary particle of this material. (Photograph courtesy of F. A. Heckman, Cabot Corporation, Billerica, Mass. 01821.)

in an electron micrograph made from the same preparation may be quite different. Optical microscopy is safer in this regard, since the actual dispersion may be examined without first evaporating the continuous phase to dryness. Also, in both optical and transmission electron microscopy, it is the projected image of the particle that is observed, and microscopic observation alone is often inadequate to distinguish a particle such as that shown in Fig. 1.10 from a true floc in which the structure is fairly readily disrupted.

Next, suppose we wish to measure the characteristic dimensions of flocs or flocklike particles using micrographs. As a first approximation, we might expect the flocculation process to result in irregular but, on the average, symmetrical particles just as the random coil does. Thus we might characterize a floc in terms of the dimension of an inscribing boundary, such as the Martin diameter or the diameter of an equivalent

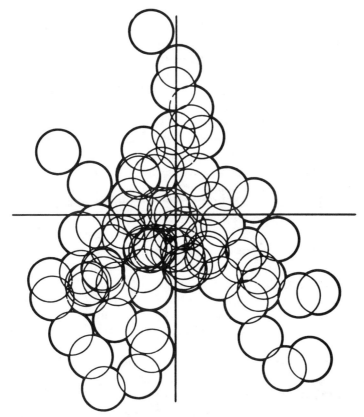

Figure 1.11 Computer simulation of a floc consisting of 76 spherical primary particles. The solid point shows the center of gravity. [From M. J. Vold, *J. Colloid Sci., 18*: 684 (1963), used with permission.]

circle determined from a graticule. Then suppose we wish to evaluate the mass of the floc enclosed within this boundary. To convert a linear dimension into a particle mass, the particle shape and density must be known. We have already commented on the approximations involved in treating irregular particles as spheres. Here, however, we encounter an additional problem besides the geometrical approximation already discussed. The question is what do we use for the density of a floc to convert the particle volume into its mass. If the "floc" dimensions have been measured, it is clearly the "floc" density which must be used. The latter is intermediate between the density of the dispersed and continuous phases, the exact value depending on the structure of the floc. Of course, one might attempt to estimate the number of primary particles in an aggregate and then use their size and density as an alternate means of evaluating the mass of a floc. The main point of this, however, is the following. Whenever the dispersed particles are flocculated, the properties of the dispersed units are intermediate between those of the two different phases involved. We shall encounter this difficulty again in Chapter 2, where the density of the settling unit, whatever it may be, is involved in sedimentation.

A very interesting approach to research concerning flocs is the technique of computer simulation. By this method, aggregates are "assembled" by a computer which uses random numbers to determine the coordinates from which each primary particle approaches the growing floc. The model that has been most studied consists of spherical primary particles, although linear sets of spheres have been used to simulate asymmetrical primary particles. The probability of adhesion on contact has been made a variable quantity in some studies. As might be expected intuitively, more open flocs result when the probability of adhesion at initial contact is high. If, on the other hand, the added particle is permitted to roll along the surface of the growing floc before adhering, a more compact structure results. Figure 1.11 is a sketch of the projection of a random floc assembled in this manner. The resemblance it bears to an actual floclike particle is evident from a comparison of Fig. 1.11 with Fig. 1.10.

A variation of the computer simulation procedure which is even more realistic permits the joining together of small flocs to form larger ones, rather than restricting the addition to primary particles only. This leads to structures that are even more expanded than those resulting from the addition of primary particles alone.

These computer simulations permit the density of primary particles within the floc to be evaluated, important information for relating the properties of the floc to its composition. As might be expected, however, it is difficult to know a priori what model to use for a particular system. However, this technique does allow some interesting a posteriori interpretations of known structures to be made. Another closely related problem that has been studied by computer simulation is the volume occupied by a sediment. As with flocs, is it found that sediments become more voluminous as the probability of adhesion on contact increases.

This discussion of flocs leads us to another important characteristic of dispersions which we have not yet considered in sufficient detail: polydispersity. Monodisperse systems are the exception rather than the rule. Even in those rare cases in which a monodisperse system exists, any flocculation that occurs will result in a distribution of particle sizes because of the random nature of the flocculation process.

1.8 Polydispersity

The only realistic attitude to take toward the dispersed systems we are interested in is to assume that they are polydisperse. Even the particles in Fig. 1.4, which appear remarkably uniform, have a narrow distribution of particle sizes. There are rare cases in which the distribution of dimensions is of negligible width, but, generally speaking, a statistical approach is required to describe a colloidal dispersion.

Students of the physical sciences generally encounter statistics in two different places. One of these deals with the treatment of experimental data. From this viewpoint, all measured quantities contain some error which raises questions concerning the best way to report the results of multiple measurements. This is obviously related to our problem of describing, for example, the "characteristic dimension" of the particles in Fig. 1.5 or 1.9. Another place in which the science student encounters statistics is in the theoretical description of large populations or populations that change through a large number of states, as, for example, in the kinetic molecular theory of gases. We shall be concerned with this aspect of statistics also. For example, the randomly coiled piece of string we considered changes size and shape continually while it is being shaken.

These two applications of statistics, from our point of view, will differ primarily in the kind of information we have available about our system. Sometimes, as when measuring micrographs, we have individual information on a large number of particles. Our question under these circumstances is how to condense these data into a few key parameters. In other circumstances, the experimental quantity itself will be an "average" quantity. Our question, then, is what kind of distribution is consistent with this average. In both cases, the underlying fact is the existence of a distribution of values for the quantity in question. We shall consider some aspects of these statistical topics here; references in statistics should be consulted if additional information is needed.

Suppose we have just measured the diameters of a field of polydisperse spheres in an electron micrograph. Our objective is to devise reasonable ways of presenting a description of the system in terms of the measured data. A fairly large number of observations is required for any statistical approach to be valid; therefore to merely tabulate the measurements is inadequate. Some condensation of the data is clearly required. Generally, the first step along these lines is a device known as classification of the data. Classification consists of sorting the observed quantities into 10–20 categories called classes. Having fewer than 10 categories results in a loss of detail in the description of the distribution;

having more than 20 categories does not improve the representation in proportion to the extra effort it requires. If, then, we observed in an electron micrograph that all the particles were less than 1.2 μm in diameter, it would be reasonable to sort them into classes 0.1 μm wide; that is, all particles less than 0.1 μm in diameter would fall into one class, those for which 0.1 μm ⩽ d < 0.2 μm in the second, and so on until class 12, which would include particles for which 1.1 μm ⩽ d < 1.2 μm. The frequency distribution of such a sample is a tabulation of the number of particles in each class. Table 1.4 represents the frequency distribution for a hypothetical array of spheres; all the numerical examples of this section are based on this sample of 400 particles. Each class is represented by the midpoint of the interval, a quantity called the class mark, symbolized by d_i for class i. Similarly, we define the number of particles in each class as n_i.

A common graphical representation of a frequency distribution is the histogram, a bar graph in which the class marks are plotted as the abscissa and the height of the bar is proportional to the number of particles in the class. Sometimes the ordinate is defined as the fraction of particles in the class, f_i. Figure 1.12a is a plot of the histogram of the data in Table 1.4. Obviously, as the number of classes approaches infinity, the width of each interval approaches zero and the histogram approaches a smooth curve. Analytical distribution functions give the equation for such smooth curves. However, in practice, a bar graph is a convenient approximation to the smooth function.

Another way in which these kinds of data are sometimes represented is as a cumulative curve in which the total number (or fraction) of particles $n_{T,i}$ having diameters less (sometimes more) than and including a particular d_i are plotted versus d_i. Figure 1.12b shows the cumulative plot for the same data shown in Fig. 1.12a as a histogram. The cumulative curve is equivalent to the integral of the frequency distribution up to the specified class mark. Cumulative distribution curves are discussed in Chapter 2 in connection with sedimentation.

1.9 The Average Diameters

Although the histogram is a convenient pictorial way to present data, a more concise representation is often required. The mean and standard deviation are the most familiar numerical parameters used for this purpose. The following equations define the mean diameter $\bar{d}$,

$$\bar{d} = \frac{\Sigma_i n_i d_i}{\Sigma_i n_i} \tag{8}$$

Table 1.4 A Hypothetical Distribution of 400 Spherical Particles[a]

Class boundaries $\leq d < (\mu m)$	Class mark, d_i (μm)	Number of particles, n_i	Fraction of total number in class, f_i	Total number with $d \leq d_i$, $n_{T,i}$
0–0.1	0.05	7	0.018	7
0.1–0.2	0.15	15	0.038	22
0.2–0.3	0.25	18	0.045	40
0.3–0.4	0.35	28	0.070	68
0.4–0.5	0.45	32	0.080	100
0.5–0.6	0.55	70	0.175	170
0.6–0.7	0.65	65	0.163	235
0.7–0.8	0.75	59	0.148	294
0.8–0.9	0.85	45	0.113	339
0.9–1.0	0.95	38	0.095	377
1.0–1.1	1.05	19	0.048	396
1.1–1.2	1.15	4	0.010	400

[a]The data are classified into 12 classes: The class marks, numbers, and fractions of particles per class and the total number of particles up to and including each class are listed.

and the standard deviation σ,

$$\sigma = \left(\frac{\Sigma_i n_i (d_i - \bar{d})^2}{\Sigma_i n_i - 1} \right)^{1/2} \tag{9}$$

of the distribution. Since Σn_i represents the total population size and is constant for a given problem, Eq. (8) may be written

$$\bar{d} = \sum_i f_i d_i \tag{10}$$

since $f_i = n_i / \Sigma_i n_i$. A similar result may be written for the standard deviation, provided the total number of particles is sufficiently large to justify the approximation $n_T - 1 = n_T$:

$$\sigma = \left(\sum_i f_i (d_i - \bar{d})^2 \right)^{1/2} \tag{11}$$

The significance of the standard deviation should be noted. The quantity $d_i - \bar{d}$ is the deviation of a particular value from the mean. Since such deviations can be either positive or negative, this quantity is squared prior to averaging. The square root of the average, then, is a measure of the spread of the data in a particular sample. From a computational point of view, the following formula provides an easier means for evaluating σ:

$$\sigma = [\overline{d^2} - (\overline{d})^2]^{1/2} \tag{12}$$

At this point it is convenient to introduce the concept of a moment of a distribution function. The general definition of the kth moment of the distribution about a point d_0 is given by the equation

$$k\text{th moment} = \sum_i f_i(d_i - d_0)^k \tag{13}$$

where the numerical value of the exponent determines precisely which moment we have. This definition is analogous to the definition of "moments" in physics, except that here the "weighting factor" f_i rather than mass appears in the formula. The mean is the first moment about the origin: Eq. (10) results when $d_0 = 0$ and $k = 1$ in Eq. (13). Likewise, the standard deviation is the square root of the second moment ($k = 2$) about the mean ($d_0 = \overline{d}$). Therefore the first moment of a distribution measures the location of a distribution and the second moment measures the spread of the distribution. Higher moments also convey certain information about the shape of a distribution. For example, the third moment is a measure of the "skewness" or lopsidedness of a distribution. It equals zero for symmetrical distributions and is positive or negative, depending on whether a distribution contains a higher proportion of particles larger or smaller than the mean. The fourth moment (called "kurtosis") purportedly measures peakedness, but this quantity is of questionable value.

The mean and the standard deviation for the data of Table 1.4 are 0.64 and 0.24 μm, respectively; $\overline{d}$ has been marked in Fig. 1.12 also.

The general formulas for calculating averages are given by Eqs. (8) and (10). These two differ from one another inasmuch as the weighting factor is normalized by definition in the latter and not in the former. A weighting factor is said to be normalized if the sum of weighting factors for the whole population equals unity, a consideration that must be introduced at some point in the calculation. When fractions of the whole are used as weighting factors, normalization is introduced through the definition of the weighting factor. If the content of a class is used as a weighting factor, the summation of products must be divided by the sum of weighting factors so that the condition of normalization is eventually included.

In colloid chemistry, a single property is often measured for a large number of particles—such as the surface area of a fine powder—rather than on individual particles as in microscopy. In this case the experiment

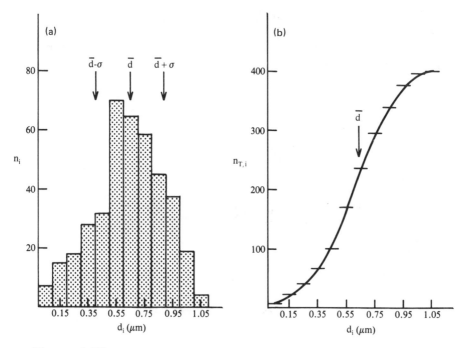

Figure 1.12 Graphical representation of the data in Table 1.4. Data are presented as (a) a histogram and (b) a cumulative distribution curve.

may yield an average size parameter directly, rather than a distribution of sizes which must be reduced mathematically to a single parameter. This is the situation, for example, when Eq. (2) is used to evaluate the radius of an equivalent sphere from experimental A_{sp} values. In this case we do not have to worry about data reduction; instead, we have the opposite problem: We have lost detailed information about the particle size distribution.

It turns out that different experimental procedures "see" polydisperse systems differently and hence measure different averages for a particular distribution of particle sizes. This is most easily seen by considering a numerical example.

Example 1.2　Using the mean diameter of the distribution of spheres in Table 1.4, calculate the average surface area of a sphere in this population. Next calculate the area of each size sphere in the distribution and average these areas, using the number of spheres in each class as weighting factors. How do the two areas compare?

Solution   For spheres $A = \pi d^2$; therefore, using $\bar{d} = 0.64$ μm we can readily calculate the "average" area to be $\pi(0.64)^2 = 1.29$ μm².

To calculate the area of each size sphere, we apply the same formula to each class in Table 1.4:

d_i (μm)	A_i (μm²)	n_iA_i (μm²)
0.05	7.85×10^{-3}	5.50×10^{-2}
0.15	7.07×10^{-2}	1.06
0.25	1.96×10^{-1}	3.53
0.35	3.85×10^{-1}	1.08×10^1
0.45	6.36×10^{-1}	2.04×10^1
0.55	9.50×10^{-1}	6.65×10^1
0.65	1.33	8.63×10^1
0.75	1.77	1.04×10^2
0.85	2.27	1.02×10^2
0.95	2.84	1.08×10^2
1.05	3.46	6.58×10^1
1.15	4.15	1.66×10^1

$$\Sigma = 5.85 \times 10^2$$

Therefore the "average" area is $5.85 \times 10^2/400 = 1.46$ μm².

These calculations show that the "average" area calculated using the mean diameter is less than that calculated by averaging the areas contributed by the various classes. If the latter area is converted to an equivalent diameter, a characteristic linear dimension is obtained which is larger than the mean diameter:

$$d = (A/\pi)^{1/2} = (1.46/\pi)^{1/2} = 0.68 \text{ μm}$$

•

Example 1.2 illustrates a very general feature of polydisperse systems: Different experimental approaches (e.g., measuring d's versus measuring areas) give different averages for what might nominally be called the same quantity. The fallacy here is that the two averages are not the same. In one case the raw data are averaged directly; in the other a datum is squared, then averaged, and then the square root is taken. In this context the latter is called the surface average diameter $\bar{d}_s$. Both the mean and the surface average diameter are reasonable estimates for a characteristic linear dimension of a polydisperse system, but they are not the same.

This divergence between average values determined by different methods can be a cause of consternation for the uninitiated, who might expect corroboration of a previous result determined by a different method. For those who are familiar with these differences, however, the various averages evaluated by different experimental procedures offer a clue as to the breadth of the distribution. This is particularly useful where molecular weights are concerned, as we shall see in Sect. 1.11. For now we continue to focus attention on particle diameters.

The results of the previous example can be generalized as follows. We calculate the area contribution of each size class of spheres, add together the contribution of each class weighted by the number of particles in the class, and then finally divide this sum by the total number of particles. This generates the following equation for the average area per particle, $\bar{A}$:

$$\bar{A} = \frac{\pi \Sigma_i n_i d_i^2}{\Sigma_i n_i} = \pi \sum_i f_i d_i^2 \tag{14}$$

Except for π, this is the same as the second moment about the origin [$k = 2$, $d_0 = 0$ in Eq. (13)] for the distribution. The diameter of a sphere having this "average" area is the surface average diameter $\bar{d}_s$ and equals

$$\bar{d}_s = \left(\sum_i f_i d_i^2 \right)^{1/2} \tag{15}$$

The surface average diameter is always larger than the mean for a polydisperse system, since the larger diameters contribute relatively more to the sum of the squares than they would if totalled directly. Following the same logic as used in Example 1.2, the volumes of various classes of particles could be determined and then averaged. When this is done for the distribution in Table 1.4, the mean volume is found to be 0.19 μm^3; the "average" diameter which corresponds to this volume is given by $(6V/\pi)^{1/3}$ and equals 0.72 μm in this case. Again, we note that this is larger than the mean diameter (0.64 μm) for the same distribution. The diameter of the sphere having this average volume is called the volume average diameter $\bar{d}_v$. It is apparent from the method of calculation used that

$$\bar{V} = \frac{\frac{1}{6}\pi \Sigma_i n_i d_i^3}{\Sigma_i n_i} = \frac{1}{6}\pi \Sigma_i f_i d_i^3 \tag{16}$$

The diameter of a sphere having this average volume is the volume average diameter $\bar{d}_v$:

$$\bar{d}_v = \left(\sum_i f_i d_i^3 \right)^{1/3} \tag{17}$$

We see that $\bar{d}_v$ is larger than the surface average diameter because the cubing leads to a greater contribution from the larger classes.

The various average diameters we have discussed can be described in terms of the moments of the particle size distribution as follows:

1. The mean $\bar{d}$ is the first moment of the distribution about the origin. Since the number fraction of particles in a class is used as the weighting factor, this is also called a number average, $\bar{d}_n$.
2. The surface average is the square root of the second moment about the origin.
3. The volume average is the cube root of the third moment about the origin.

These parameters are summarized in Table 1.5. Note that the relative magnitude of the number, surface, and volume averages are given by the sequence

$$\bar{d}_n < \bar{d}_s < \bar{d}_v \tag{18}$$

for a polydisperse system. Only for a monodisperse system would all three parameters have identical values. Therefore, the divergence from unity of the ratio of any two of these, measured independently, is often taken as an indication of the polydispersity of the dispersion. We shall see in the next section that this comparison of averages evaluated by different techniques finds particular application in the characterization of molecular weight distributions.

Table 1.5 also lists the radius of gyration. This is an average dimension which is often used in colloid chemistry to characterize the spatial extension of a particle. We shall see that this quantity can be measured for polydisperse systems by viscosity (Chap. 4) and light scattering (Chap. 5). It is therefore an experimental quantity which quantifies the dimensions of a disperse system and deserves to be included in Table 1.5. Since the typical student of chemistry has probably not heard much about the radius of gyration since general physics, a short review seems in order.

We assume that the particle whose radius of gyration is under discussion may be subdivided into a number of volume elements of mass m_i. Then the moment of inertia I about the axis of rotation of the body is given by

$$I = \sum_i m_i r_i^2 \tag{19}$$

where r_i is the distance of the ith volume element from the axis of rotation.

Table 1.5 Some of the More Widely Encountered Size "Averages" in Surface and Colloid Chemistry, Including Their Defining Equations

Name	Symbol	Definition	Quantity averaged	Weighting factor
Number average, or mean	$\bar{d}_n$ or $\bar{d}$	$\dfrac{\Sigma_i n_i d_i}{\Sigma_i n_i}$	Diameter	Number in class
Second moment about origin	$\bar{d^2}$	$\dfrac{\Sigma_i n_i d_i^2}{\Sigma_i n_i}$	Square of diameter	Number in class
Surface average	$\bar{d}_s$	$(\bar{d^2})^{1/2}$	Square of diameter	Number in class
Third moment about origin	$\bar{d^3}$	$\dfrac{\Sigma_i n_i d_i^3}{\Sigma_i n_i}$	Cube of diameter	Number in class
Volume average	$\bar{d}_v$	$(\bar{d^3})^{1/3}$	Cube of diameter	Number in class
Radius of gyration	$(\bar{R_g^2})^{1/2}$	$\left(\dfrac{\Sigma_i m_i r_i^2}{\Sigma_i m_i}\right)^{1/2}$	Square of radius	Mass in class

Regardless of the shape of the particle, there is a radial distance at which the entire mass of the particle could be located such that the moment of inertia would be the same as that of the actual distribution of the mass. This distance is the radius of gyration R_g. According to this definition, it is clear that

$$R_g^2 \sum_i m_i = I = \sum_i m_i r_i^2 \tag{20}$$

Therefore we see that the value of R_g^2 for an array of volume elements is the second moment about the origin of r, with mass fraction (rather than number fraction) as the weighting factor:

$$R_g^2 = \frac{\Sigma_i m_i r_i^2}{\Sigma_i m_i} \tag{21}$$

We shall see in Sect. 1.11 that this type of weighting factor gives one of the common molecular weight averages for polydisperse systems.

The radius of gyration is a parameter which characterizes particle size without the need to specify particle shape. The relationship between R_g

and the actual dimensions of a particle depends on the shape of the particle. Such relationships are derived in most elementay physics texts for rigid bodies of various geometries. To translate a radius of gyration into an actual geometrical dimension, some shape must be assumed. For example, for a sphere of radius R, $R_g^2 = \frac{3}{5}R^2$. It should be emphasized, however, that the radius of gyration in itself is a perfectly legitimate way of describing the dimensions of a particle. The specification of particle geometry is really optional.

1.10   Size Exclusion Chromatography

Lyophilic colloids—especially those of synthetic origin—also possess a distribution of particle sizes, but these are generally characterized by molecular weight rather than by linear dimension (except where R_g is measured for such systems). We shall discuss the experimental methods used for molecular weight determination of particles in the colloidal size range in Chaps. 2–5. We shall discuss the kinds of molecular weight averages these methods yield in Sect. 1.11. First, however, let us consider a colloidal solution, say, polystyrene in toluene, and discuss a chromatographic method for its fractionation.

To begin with, a polymer like this is formed by the addition reaction in which styrene monomers $CH_2 = CHC_6H_5$ combine into long chains that may be thousands of repeat units long. This reaction is one of random addition, so not all chains are identical in length. Furthermore, chains containing slightly different numbers of repeat units will not differ enough in their properties to allow complete separation. Nevertheless, procedures exist in which an initially polydisperse system is divided into fractions of narrower molecular weight distribution. One such method is called size exclusion chromatography (SEC), because it segregates molecules in the distribution on the basis of their spatial extensions.

In SEC a chromatographic column is packed with porous particles and then filled with solvent. Next a portion of the colloidal solution is layered on the solvent and the liquid is allowed to pass through the column. The eluted liquid is monitored for the colloid by an appropriate detection method; spectrophotometry and refractive index are probably the most widely used methods of detection. What results is a chromatogram showing the detector output as a function of the volume of liquid eluted through the column. For a synthetic colloid, like polystyrene, a broad peak results. Samples of biological origin often contain several components of quite different molecular weight. Under optimum conditions these may emerge as distinct peaks in the chromatogram.

1.11 Molecular Weight Averages

In Sect. 1.10 we saw that a properly calibrated size exclusion chromatogram could be interpreted to give number fractions and weight fractions of particles in various molecular weight classes. From this sort of information the two most common molecular weight averages are readily available.

Suppose a dispersion is classified into a set of categories in which there are n_i particles of molecular weight M_i in the ith class. Then the number average molecular weight $\overline{M}_n$ equals

$$\overline{M}_n = \frac{\Sigma_i n_i M_i}{\Sigma_i n_i} = \sum_i f_i M_i \tag{24}$$

where f_i is the number fraction of particles in class i. Alternatively, we could define the average in such a way that the weight of particles in each class w_i rather than their number is used as the weighting factor. This results in an average known as the weight average molecular weight $\overline{M}_w$:

$$\overline{M}_w = \frac{\Sigma_i w_i M_i}{\Sigma_i w_i} \tag{25}$$

Note that $w_i/\Sigma_i w_i$ is the weight fraction of particles in class i. The weight of material in a particular size class is given by the product of the number of particles in the class and their molecular weight, however; so Eq. (25) may be written

$$\overline{M}_w = \frac{\Sigma_i (n_i M_i) M_i}{\Sigma_i n_i M_i} = \frac{\Sigma_i n_i M_i^2}{\Sigma_i n_i M_i} = \frac{\Sigma_i f_i M_i^2}{\Sigma_i f_i M_i} \tag{26}$$

The weight average molecular weight is thus seen to equal the ratio of the second moment of the distribution about the origin to the first moment of the distribution about the origin. We shall see that measurement of the osmotic pressure of a polydisperse system permits the experimental evaluation of $\overline{M}_n$ (Chap. 3), and light scattering experiments enable us to measure $\overline{M}_w$ (Chap. 5). It follows from the definition of these various averages that

$$\frac{\overline{M}_w}{\overline{M}_n} \geqslant 1 \tag{27}$$

by analogy with the inequalities in Eq. (18). Only when the system is monodisperse does the equality apply in Eq. (27). Here too the deviation of this ratio from unity may be taken as a measure of polydispersity.

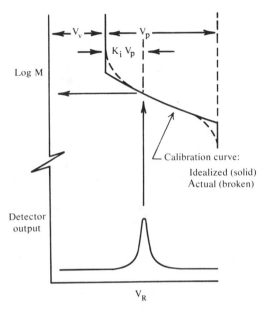

Figure 1.13 Plot of log M and detector output versus retention volume for SEC. Also shown is the relation between V_R, V_v, V_p, and $K_i V_p$ as discussed a little further on in the text. (Reprinted with permission from P. C. Hiemenz, *Polymer Chemistry: The Basic Concepts*, Marcel Dekker, New York, 1984.)

The volume at which a particular colloidal fraction emerges from the column is called the retention volume V_R of that fraction. Normally, a particular experimental system is calibrated with a colloidal solute of known molecular weight M. A plot of log M versus V_R is generally linear over several orders of magnitude in molecular weight. Not only the solute but also the solvent, the column packing, and the operating conditions affect the calibration. Therefore the calibration is set up on a system-by-system basis, using molecular weights that have been determined by one of the absolute methods we shall discuss later in this book. Thus the calibration process can be time-consuming, but, once set up, SEC provides a rapid method for fractionating polydisperse systems. Incidentally, this method is known by various names among workers in different fields; In biochemistry the technique is called gel filtration chromatography (GFC), and in polymer chemistry, it is gel permeation chromatography (GPC).

Figure 1.13 shows how log M and the detector output vary with V_R. By reading from the chromatogram to the calibration curve, the molecular weight of the fraction emerging at a particular value of V_R can be determined. Furthermore, for a molecular weight fraction M_i, the height h_i of the detector output signal is directly proportional to the mass m_i of the solute eluted in that particular fraction. Since $m_i \propto h_i$, $\Sigma_i h_i \propto \Sigma_i m_i$ and $m_i/\Sigma_i m_i = h_i/\Sigma_i h_i$. Thus the weight fractions of different molecular weight classes in the distribution are readily obtained by this method. Likewise, if n_i is the number of moles of component i, $h_i/M_i \propto n_i$ and $(h_i/M_i)/\Sigma_i h_i/M_i = n_i/\Sigma_i n_i$. If we assume that suitable calibration exists, the chromatogram produced by SEC can be interpreted in terms of either number fractions or weight fractions of the various molecular weight classes. In Sect. 1.11 we shall take up the calculation of molecular weight averages using this kind of information.

Now, however, let us briefly consider the mechanism whereby molecules are fractionated by SEC. The basic idea is very simple: Large molecules cannot penetrate into the pores of the packing medium (the gel) and hence are eluted first. The smaller particles are distributed in the pores as well as the voids between the gel particles and emerge later.

The total volume of solvent in the column can be divided into two categories: V_v is the volume of solvent in the voids between the gel particles, and V_p is the volume of solvent in the pores. Not all of the pore volume is accessible to particles in a given molecular weight class. We define $K_i V_p$ as the pore volume into which molecules from the ith class can permeate. Thus K_i describes what fraction of a pore is accessible to molecules in class i. The fraction K_i is zero for very large particles and unity for very small particles and varies over this range for particles of intermediate size.

The retention volume for molecules in class i is given by

$$V_R = V_v + K_i V_p \tag{22}$$

Figure 1.13 shows the relationship between V_R, V_v, V_p, and K_i.

It is fairly easy to arrive at a theoretical expression for K_i based on a simple model of the permeation process. We picture the colloidal particle as a sphere of radius R and the pore as a cylinder of radius a and length l, as shown in Fig. 1.14. The center of the solute particle cannot approach any closer than a distance R from the wall of the pore. Therefore the radius of the pore that is accessible to the colloidal particle is a-R. The layer of solution adjacent to the walls of the pore is "off limits" to the solute, so the concentration of the colloid in the pore is only a fraction of its value in the bulk solution. This fraction is K_i for the sphere of radius R

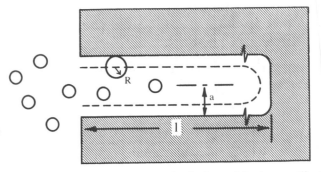

Figure 1.14 Size exclusion of a spherical particle in a cylindr (Reprinted with permission from P. C. Hiemenz, *Polymer Chemistry: Concepts*, Marcel Dekker, New York, 1984.)

and equals the ratio of the accessible volume to the total volum pore:

$$K_i = \frac{\pi(a - R)^2 (l - R)}{\pi a^2 l} \simeq \left(\frac{a - R}{a} \right)^2 = \left(1 - \frac{R}{a} \right)^2$$

In the approximate version of Eq. (23) the pore is assumed to b enough so that $1 - R \simeq 1$. Note that this fraction ranges between ze unity as required for R equalling a and 0, respectively, and thus gi parameter K_i as a function of particle size.

The next stage of a developing theory would be to relate the rac the particle to the molecular weight. Thus expanded, Eqs. (22) an would provide a relationship between M and V_R which—if the were successful—would eliminate the need for empirical calibrati SEC. For some systems, such as rigid globular proteins, establishin connection between R and M is not difficult. For synthetic poly however, the flexible chains exist as random coils whose exte depends on the interaction between polymer and solvent. In this relating R to M is a complicated matter, so we will be content empirical calibration to interpret V_R in terms of M. This, of course, i the only modification that needs to be examined in a fully devel theory for K_i. Models based on more complicated geometries for both colloidal particles and the pores have been considered, but we shall pursue these. Reference 10 discusses additional theoretical models experimental procedures in detail for those desiring more information this topic.

Table 1.6 Number of Moles and Molecular Weights for Eight Classes of a Hypothetical Fractionated Polymer (Remaining Quantities Calculated in Example 1.3)

(1) n_i (mole)	(2) M_i (g mole^{-1})	(3) w_i (g)	(4) $w_i M_i \times 10^6$	(5) $(M_i - \overline{M})^2 \times 10^{-8}$	(6) $n_i (M_i - \overline{M})^2 \times 10^{-6}$
0.003	30,000	90	2.70	3.72	1.12
0.007	35,000	245	8.58	2.04	1.43
0.015	40,000	600	24.0	0.86	1.29
0.024	45,000	1080	48.6	0.18	0.43
0.040	50,000	2000	100	0.00	0.00
0.032	55,000	1760	96.8	0.32	1.02
0.010	60,000	600	36.0	1.14	1.14
0.005	65,000	325	21.1	2.50	1.25
$\Sigma = 0.136$		$\Sigma = 6700$	$\Sigma = 337.8$		$\Sigma = 7.68$

The relationship between the ratio $\overline{M}_w/\overline{M}_n$ and the standard deviation of the molecular weight distribution is easily seen as follows. From Eq. (26) it is clear that

$$\sum_i f_i M_i^2 = \overline{M}_w \sum_i f_i M_i = \overline{M}_w \overline{M}_n \tag{28}$$

From the general procedure for defining the mean, the left-hand side of Eq. (28) may also be written as $\overline{M^2}$. Substituting this result into Eq. (12) permits us to rewrite the latter as

$$\sigma = (\overline{M}_n \overline{M}_w - \overline{M}_n^2)^{1/2} = \overline{M}_n \left(\frac{\overline{M}_w}{\overline{M}_n} - 1 \right)^{1/2} \tag{29}$$

Therefore the square root of the amount by which the molecular weight ratio exceeds unity measures the standard deviation of the distribution relative to the number average molecular weight.

The following example illustrates these relationships for a hypothetical polymer.

Example 1.3 Columns (1) and (2) of Table 1.6 list the number of moles and the molecular weight, respectively, for eight fractions of a synthetic polymer. Calculate $\overline{M}_n$ and $\overline{M}_w$ from these data and evaluate σ using both Eqs. (9) and (29).

Solution For each class of molecules calculate the quantities listed in columns (3)–(6) in Table 1.6.

The sum of the values in column (1) equals $\Sigma_i n_i = 0.136$ mole.

The product of columns (1) and (2) is w_i and values for this quantity are listed in column (3). The sum of the entries in column (3) equals $\Sigma_i n_i M_i = 6700$ g.

Dividing 6700 g by 0.136 mole gives $\overline{M}_n = 49,300$ g mole^{-1}.

The product of columns (2) and (3) equals $w_i M_i$, and values for this quantity are listed in column (4); $\Sigma_i w_i M_i = 337.8 \times 10^6$ g^2 mole^{-1}.

Dividing 3.378×10^8 by 6700 gives $\overline{M}_w = 50,400$ g mole^{-1}.

The square of the difference between M_i and $\overline{M}_n$ is given in column (5), and column (6) lists $n_i(M_i - \overline{M}_n)^2$; $\Sigma_i n_i(M_i - \overline{M}_n)^2 = 7.68 \times 10^6$.

Dividing 7.68×10^6 by 0.136 gives σ^2 by Eq. (9); therefore $\sigma^2 = 5.65 \times 10^7$ and $\sigma = 7500$.

The ratio $\overline{M}_w/\overline{M}_n = 1.022$ and, by Eq. (29), $\sigma/\overline{M}_n = (0.022)^{1/2}$, or $\sigma = 0.148(49,300) = 7300$.

The discrepancy between the two values of σ is a matter of significant figures and is not meaningful.

•

Table 1.7 The Most Common Molecular Weight Averages, Their Definitions, and Their Methods of Determination

Average	Definition	Methods
$\overline{M}_n$	$\dfrac{\Sigma_i n_i M_i}{\Sigma_i n_i}$	Osmotic pressure and other colligative properties
$\overline{M}_w$	$\dfrac{\Sigma_i n_i M_i^2}{\Sigma_i n_i M_i}$	Light scattering Sedimentation velocity
$\overline{M}_v$	$\left(\dfrac{\Sigma_i n_i M_i^{1+a}}{\Sigma_i n_i M_i} \right)^{1/a}$	Intrinsic viscosity

Finally, we define still another average molecular weight, the viscosity average $\overline{M}_v$:

$$\overline{M}_v = \left(\frac{\Sigma_i n_i M_i^{1+a}}{\Sigma_i n_i M_i} \right)^{1/a} \tag{30}$$

The exponent a in this definition is called the Mark–Houwink coefficient; generally $0.5 < a < 1.0$. As its name implies, $\overline{M}_v$ is a molecular weight average determined from viscosity measurements on polydisperse polymer samples. The exponent is characteristic of the polymer–solvent–temperature conditions of the experiment. We shall see how these experiments are conducted and the significance of the Mark–Houwink coefficient in Chap. 4. For now we may take Eq. (30) as merely defining another kind of molecular weight average. Note that $\overline{M}_v = \overline{M}_w$ when $a = 1$.

The various molecular weight averages we have discussed, their definitions, and the experimental methods that measure them are listed in Table 1.7. All these different "averages" are admittedly confusing. Without this information, however, it would be far more confusing to try to rationalize the discrepancy between two different molecular weight determinations on the same sample by methods which yield different averages. The divergence between such values is a direct consequence of polydispersity in the sample. As we have seen, it is not only unavoidable but also informative as to the extent of polydispersity.

1.12 Theoretical Distribution Functions

It was noted earlier that histograms approach smooth distribution curves as the number of classes is increased to a very large number. Sometimes it is desirable to represent a distribution function by an analytical expression which applies continuously to the measured variable. The most familiar of such functions is the normal, or Gaussian, distribution function:

$$f(x)dx = \frac{1}{\sigma\sqrt{2\pi}}\exp\left[-\frac{1}{2}\left(\frac{x-\bar{x}}{\sigma}\right)^2\right]dx \qquad (31)$$

In this equation $f(x)dx$ expresses the fraction of particles having x values between x and $x + dx$; it replaces f_i, which plays a corresponding role in discrete distributions. Some characteristics of the normal distribution are the following:

1. The function $f(x)$ has its maximum value at $x = \bar{x}$ and drops off exponentially with the square of the deviation of a value from the mean, where such deviations are measured as fractions or multiples of the standard deviation.
2. The pre-exponential factor accomplishes the normalization of the function; that is, the integral under the curve over all possible values of x ($-\infty$ to ∞) equals unity. In a broad distribution σ is large and the exponential does not drop off as rapidly as in a narrow distribution (recall that all deviations are measured relative to the standard deviation).
3. Since the area under the curve is always unity, a narrow distribution will show larger values of $f(x)$ at the maximum, whereas a broader distribution will have a smaller value for the function at the maximum. This is why the standard deviation appears in the denominator of the pre-exponential normalization factor.

The normal distribution is the "curve" over which students and teachers alike agonize in connection with course grades. We discuss this distribution function in greater detail in Chapter 2. For the present we are concerned only with its descriptive capabilities. For this purpose it is sufficient to note that tables are available (e.g., in the *Handbook of Tables for Mathematics*, Chemical Rubber Company, Cleveland, OH) which supply the value of this function in terms of the standard unit, $t = (x - \bar{x})/\sigma$:

$$f(t)dt = \frac{1}{\sqrt{2\pi}}\exp\left(-\frac{t^2}{2}\right)dt \qquad (32)$$

The normal distribution is commonly encountered in the cumulative form, that is, as the fraction of particles larger (oversized) or smaller (undersized) than a particular t_i value. Since the total area under the normal curve equals unity, the area under one "tail" of the curve from t_i to ∞ gives the fraction of the population having t values greater than the integration limit t_i:

$$f_{t > t_i} = \frac{1}{\sqrt{2\pi}} \int_{t_i}^{\infty} \exp\left(-\frac{t^2}{2}\right) dt \tag{33}$$

Likewise, the cumulative fraction of particles smaller than t_i equals

$$f_{t < t_i} = 1 - \frac{1}{\sqrt{2\pi}} \int_{t_i}^{\infty} \exp\left(-\frac{t^2}{2}\right) dt \tag{34}$$

Tables are also available (e.g., CRC *Handbook of Tables for Mathematics*) for the area under the normal curve between $\bar{x}(t = 0)$ and one value of t_i (i.e., they apply to one tail only). This area is known as the error function and is often symbolized as $\mathrm{Erf}(t_i)$:

$$\mathrm{Erf}(t_i) = \frac{1}{\sqrt{2\pi}} \int_{0}^{t_i} \exp\left(-\frac{t^2}{2}\right) dt \tag{35}$$

In terms of the error function, Eq. (33) becomes

$$f_{t > t_i} = \tfrac{1}{2} \pm \mathrm{Erf}(t_i) \tag{36}$$

where the positive value is used if t is negative and the negative value is used if t is positive. These signs are reversed if the cumulative fraction of undersized particles is to be determined. Consulting the tables, we see that in a normally distributed sample 15.87% of the particles will have t values greater than $+1.0$. This is the percentage of particles for which the deviation from the mean is greater than one standard deviation unit.

The following example considers a numerical application of these ideas.

Example 1.4 A sample of 75 particles follows the normal distribution with respect to particle diameters with $\bar{d} = 1.140$ μm and $\sigma = 0.311$ μm. Consult a table of the normal distribution to evaluate the fraction of particles in the population having diameters less than 1.0 μm. What is the probability that a particle picked at random from this population lies in the class for which $0.95 < d < 1.05$ in the histogram of this distribution?

Solution   Evaluate t for $d = 1.00$: $t = (d - \bar{d})/\sigma = (1.00 - 1.14)/0.311 =$ -0.450. Tables of the normal distribution* indicate that the area under the curve is 0.1736 at $t = 0.45$. To interpret this, picture the familiar bell-shaped curve centered at $t = 0$ ($d = \bar{d} = 1.140$); we are interested in an abscissa value of $t = -0.45$ ($d = 1.00$). The figure 0.1736 gives the fraction of the area under the whole curve that lies between $t = -0.45$ and $t = 0.00$. This means that $0.5000 - 0.1736 = 0.3264$ is the fraction of the area which lies beyond $t = -0.45$. This means that about 33% of the particles have diameters less than 1.00 μm.

For $d = 0.95$, $(d - \bar{d})/\sigma = -0.61$; the area between $t = -0.61$ and $t = 0.00$ is 0.229 according to the tables. For $d = 1.05$, $(d - \bar{d})/\sigma = -0.29$; the area between $t = -0.29$ and $t = 0.00$ is 0.114. Therefore the probability of a particle falling between 0.95 and 1.15 is $0.229 - 0.114 = 11.5\%$.

•

Suppose a polydisperse system is investigated experimentally by measuring the number of particles in a set of different classes of diameter or molecular weight. Suppose further that these data are believed to follow a normal distribution function. To test this hypothesis rigorously, the chi-squared test from statistics should be applied. A simple graphical examination of the hypothesis can be conducted by plotting the cumulative distribution data on probability paper as a rapid, preliminary way to evaluate whether the data conform to the requirements of the normal distribution.

Probability paper is a commercially available graph paper which has one coordinate subdivided in ordinary arithmetic units and the other coordinate subdivided into cumulative probability units. The latter are spaced in such a way that normally distributed data will produce a straight line graph when the cumulative percentage of undersized (or oversized) particles is plotted on the probability coordinate and the size variable (diameters, weights, etc.) is plotted on the arithmetic scale. Figure 1.15 shows schematically how (a) normally distributed data are transformed when replotted as (b) a cumulative distribution and, finally, when (c) graphed on probability paper. The x value corresponding to $y = 50\%$ on probability paper gives the mean value. The x value at $y = 15.87\%$ gives $\bar{x} - \sigma$, and the x value at $y = 84.13\%$ ($100 - 15.87$) gives $\bar{x} + \sigma$ when the percentage of undersized particles is plotted (the signs are reversed when the percent of oversized particles is plotted). From these values σ may be determined. Thus a linear plot on probability paper

---

Handbook of Tables for Mathematics, Chemical Rubber Company, Cleveland.

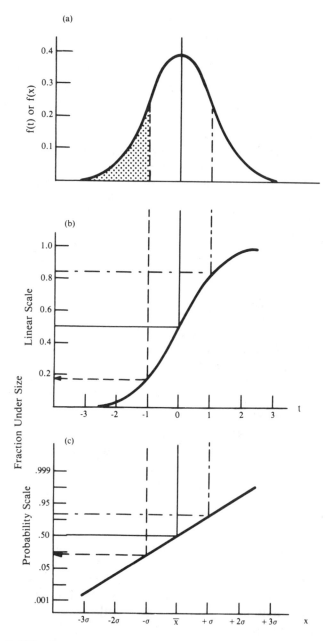

Figure 1.15 A normal, or Gaussian, distribution represented as (a) a frequency function, (b) a cumulative function, and (c) a cumulative function linearized by plotting on probability paper.

suggests conformity to the normal distribution and also permits the graphical evaluation of $\bar{x}$ and σ.

As we shall see in Chapter 2, the normal distribution comes about when a large number of purely random factors are responsible for the distribution. It is mainly applicable to particles which are produced by condensation, precipitation, or polymerization processes, which are purely random.

Dispersions which are produced by comminution—the mechanical subdivision of larger chunks—are more likely to produce a linear graph on probability paper if the logarithm of the variable rather than the variable itself is plotted against the probability. Graph paper graduated this way is called log-probability paper. The logarithmic scale implies a much wider range of values for the variable, and also an asymmetrical distribution function may be written by analogy with Eq. (31):

$$f(\ln x) = \frac{1}{\ln \sigma_g \sqrt{2\pi}} \exp\left[-\frac{1}{2}\left(\frac{\ln x_i - \ln \bar{x}_g}{\ln \sigma_g} \right)^2 \right] \tag{37}$$

However, an important difference also emerges from this analogy. The quantities which are normally distributed are logarithms of variables, not the variables themselves. This means that the mean and standard deviation which are obtained from log-probability plots are geometric averages rather than arithmetic averages. This is the significance of the subscript g in Eq. (37). This is most easily understood by writing the expression for the number average value for the quantity $\ln x_i$:

$$\overline{\ln x} = \sum_i f_i \ln x_i = \sum_i \ln x_i^{f_i} \tag{38}$$

Taking the antilog of this quantity converts the summation into a product over all terms, indicated by Π_i:

$$\text{antiln}(\overline{\ln x}) = \prod_i x_i^{f_i} = \bar{x}_g \tag{39}$$

When the averaging is carried out in this way, the result is known as the geometric mean $\bar{x}_g$. The coordinate corresponding to the 15.87% undersized y value equals $\ln \bar{x}_g - \ln \sigma_g$. Because of the properties of logarithms, this is the same as the logarithm of the ratio $\bar{x}_g/\sigma_g$. From this, σ_g may be evaluated. We shall not concern ourselves further with these geometrical averages except to note that

$$\bar{x}_g < \bar{x}_n \tag{40}$$

for any polydisperse system. The validity of this relationship is easily demonstrated by calculating the two averages for a hypothetical distribution.

References

1. R. D. Cadle, *Particle Size*, Reinhold, New York, 1965.
2. G. Herdan, *Small Particle Statistics*, 2nd ed., Academic, New York, 1960.
3. B. Jirgensons and M. E. Straumanis, *Colloid Chemistry*, Macmillan, New York, 1962.
4. H. R. Kruyt (ed.), *Colloid Science*, Vols. 1 and 2, Elsevier, Amsterdam, 1949, 1952.
5. A. I. Medalia, in *Surface and Colloid Science*, Vol. 4 (E. Matijević, ed.), Wiley, New York, 1971.
6. K. J. Mysels, *Introduction to Colloid Chemistry*, Wiley, New York, 1959.
7. H. F. Schaeffer, *Microscopy for Chemists*, Dover, New York, 1953.
8. D. J. Shaw, *Introduction to Colloid and Surface Chemistry*, Butterworths, London, 1966.
9. R. D. Vold and M. J. Vold, *Colloid and Interface Chemistry*, Addison-Wesley, Reading, Mass., 1983.
10. W. W. Yau, J. J. Kirkland, and D. D. Bly, *Modern Size Exclusion Liquid Chromatography*, Wiley, New York, 1979.

Problems*

1. The specific area of dust particles from the air over Pittsburgh, Pennsylvania, has been determined by gas adsorption†:

Treatment	$A_{sp}(m^2g^{-1})$
4 h under vacuum at 200°C	5.61
8 h under vacuum at 25°C	2.81

(a) Calculate the radius of these particles if they are assumed to be uniform spheres of density 2.2 g cm^{-3}.

(b) Propose an explanation for the effect of degassing on particle size.

(c) What kind of average is obtained for the radius by this procedure?

---

*The data for many of the problems in this book are taken from graphs appearing in the original literature. As a result, the values given do not necessarily reflect the accuracy of the original experiments. Likewise, the number of significant figures cited may not be justified in terms of the approximations involved in graph reading.

†M. Corn, T. L. Montgomery, and R. J. Reitz, *Science, 159*: 1350 (1967).

2. Colloidal palladium particles in an alumina matrix catalyze the hydrogenation of ethene to ethane. The following data describe various catalyst
   preparations*:

Diameter of Pd particles (Å)	55	75	75	115	145
ppm Pd in catalyst	170	250	200	250	250
% conversion per 25 mg catalyst	50	45	40.5	38.5	29

   (a) Calculate the activity of these catalysts on the basis of the weight of Pd
   and the area of Pd in the preparations.
   (b) Does the catalytic role of Pd seem to be a bulk or surface phenomenon?
3. The accompanying table shows how the trace metal content of coal-ash
   aerosols depends on particle size†:

Range of diameters (μm)	Trace element (μg/g ash)					
	Pb	Tl	Sb	Cd	Se	As
30–40	300	5	9	<10	<15	160
5–10	820	20	25	<10	<50	800
1.1–2.1	1600	76	53	35	59	1700

   Discuss the implications of these results on human health in view of the
   following considerations: (a) the intrinsic toxicity of these trace elements; (b)
   small particles travel to the lung, larger particles are stopped in nose,
   pharynx, and so on; (c) trace elements are absorbed from the alveoli 7–10
   times more efficiently than from the upper respiratory spaces; (d) a possible
   mechanism for the effect of particle size on trace element content.
4. Select a field containing about 30 particles from Fig. 1.9b and measure the
   Martin diameters of the population parallel to the bottom of the photograph
   (a photocopy can be made and the particles checked off to avoid duplication
   and omissions). Classify the data and calculate the mean and standard
   deviation. Repeat, measuring the Martin diameter parallel to the side of the
   photograph. Discuss the agreement or discrepancy between the two means
   in terms of (a) bias in the choice of the field, (b) systematic orientation
   effects, and (c) the size of the population.
5. Suppose that the particles in Fig. 1.4 were actually oblate ellipsoids (all lying
   in their preferred orientation) rather than spheres. Would their volume be
   over- or underestimated if the particles were assumed to be spheres? In terms
   of their axial ratio, calculate the factor by which the mass is under- or
   overestimated when the particles are assumed to be spheres. (Consult a
   handbook for the volume of an ellipsoid.)

---

*J. Turkevich and G. Kim, *Science, 169*: 873 (1970).
†D. F. S. Natusch, J. R. Wallace, and C. A. Evans, Jr., *Science, 183*: 202 (1974).

6. Mixtures of 50 g of ZnO–TiO$_2$ are each shaken with 250 ml of water and allowed to settle. After 14 days of equilibration, it is found that the sediment volume is 1.65 times larger when the weight ratio of ZnO to TiO$_2$ is 1.0 than when the ratio is 100.* Some particle characteristics are the following:

	Diameter (μm)	Density (g ml^{-1})	Charge in water
ZnO	1.0	5.6	Positive
TiO$_2$	2.2	4.2	Negative

(a) Assuming the particles are uniform spheres, calculate the ratio of the number of particles of ZnO to TiO$_2$ for each of the weight ratios given.
(b) Propose an explanation for the more voluminous sediment that results when the weight ratio is 1.0 than when it is 100.

7. The following data describe the particle size distribution in a dispersion of copper hydrous oxide particles in water†:

d_i (μm)	0.426	0.401	0.376	0.351	0.326	0.301	0.276	0.251	0.226	0.201
n_i	1	0	6	6	17	14	11	12	6	6

Calculate $\bar{d}_n$, σ, $\bar{d}_s$, and $\bar{d}_v$ for this dispersion.

8. A graticule was used to size sand particles and glass spheres.‡ The percentage by weight of particles less than the stated size was found to be as follows:

d (μm)	0.4	0.8	1.6	2.4	3.0	4.0	8.0	12.0
Sand	0.01	0.07	0.23	0.56	1.23	2.35	11.77	18.06
Glass	0.01	0.11	0.26	0.43	0.72	1.43	17.84	28.07

d (μm)	16.0	20.0	30.0	40.0	60.0	75.0	90.0	120.0
Sand	24.62	32.11	52.33	64.14	83.60	—	98.17	100.0
Glass	36.89	47.68	59.47	61.78	91.49	100.0		

Plot the results on probability and log-probability coordinates. Use the best representation to evaluate (the appropriate) average and standard deviation for the samples.

9. Particle size distributions were measured on aerosols collected in a New York highway tunnel and the following results were obtained§:

---

*L. H. Princen and M. J. DeVena, *J. Am. Oil Chem. Soc.,* *39*: 269 (1962).
†P. McFadyen and E. Matijević, *J. Colloid Interface Sci.,* *44*: 97 (1973).
‡G. L. Fairs, *Chem. Ind.,* *62*: 374 (1943).
§R. E. Lee, Jr., *Science,* *178*: 567 (1972).

Cumulative % mass $< d_i$ (μm)	30	40	50	60	70	80
d_i, weekend	0.5	1.0	2.5	5.0	—	—
d_i, weekday	—	—	0.07	0.2	0.9	4.0

Plot these results on log-probability coordinates and estimate the mean and standard deviation for each distribution. What kind of "averages" are these quantities? How do the weekend and weekday particle size distributions compare with respect to the location and width of the maximum? The weekend results are attributed to automobile exhaust, whereas the weekday results are assumed to be "diluted" by aerosols from outside the tunnel.

10. The mass of bull sperm heads in a sample was determined by interference microscopy and the following results were obtained*:

n_i	4	2	27	37	32	26	20	8	3	3	1
$w_i \times 10^{12}$ (g)	5	6	7	8	9	10	11	12	13	14	15

Calculate the number average and weight average weights of these particles.

11. Yau and co-workers† used polystyrene samples of known molecular weight in tetrahydrofuran to calibrate the retention volume of a size exclusion chromatograph. They obtained the following results:

$M \times 10^{-3}$(g mole^{-1})	4.0	50.0	110	179	390	1800
V_R (ml)	16.3	14.5	13.9	13.6	13.1	12.1

Do these results display the expected correlation between M and V_R? The flow rate in the instrument used in this research was 1.0 ml min^{-1}; how long would it take for samples of molecular weight 10^4, 10^5, and 10^6 to emerge from the column? Is this order of elution times qualitatively consistent with Eqs. (22) and (23)?

---

*G. F. Bahr and E. Zeitler, *J. Cell Biol.,* *21*: 175 (1964).

†W. W. Yau, M. E. Jones, C. R. Ginnard, and D. D. Bly, in *Size Exclusion Chromatography* (T. Provder, ed.), American Chemical Society, Washington, D.C., 1980.

2

SEDIMENTATION AND DIFFUSION
AND THEIR EQUILIBRIUM

Even if you had completed your third year ... in the University, and were perfect in the theory of the subject, you would still find that there was need of many years of experience, before you could move in a fashionable crowd without jostling against your betters.

[From Abbott's *Flatland*]

2.1 Introduction

In this chapter we shall examine the effects on particles in the colloidal size range of sedimentation and diffusion, first taken separately and then combined. Sedimentation occurs under the influence of gravity and, considerably faster, in a centrifuge. Both gravitational and centrifugal sedimentation are discussed herein. A key relationship in understanding the rate of sedimentation is Stokes' law, a hydrodynamic equation that will appear again when we discuss the kinetics of flocculation in Chap. 11 and electrokinetic phenomena in Chap. 13.

Our analysis of diffusion entails statistical arguments, some of which may be familiar from kinetic molecular theory. In discussing diffusion, we focus our attention on the diffusion coefficient. The latter is defined experimentally by Fick's laws and theoretically by two equations derived by Einstein.

As our educated intuition might lead us to expect, larger particles sediment more rapidly and diffuse more slowly than smaller particles. The effects of sedimentation and diffusion, therefore, are not of comparable magnitude for all sizes of particles. There is a range of particle sizes, however, for which the two are comparable and equilibrium between them is established. This equilibrium between sedimentation and diffusion has been extensively studied by means of the ultracentrifuge. Since many of the particles thus investigated are of biological importance, we shall frequently use protein molecules as our examples in this chapter.

Much of our discussion of sedimentation alone and sedimentation combined with diffusion is focused on the determination of the mass or molecular weight of the dispersed particles.

2.2　Sedimentation and the Mass-to-Friction Factor Ratio

To see how sedimentation comes about, consider the gravitational forces which operate on a particle of volume V and density ρ_2 which is submerged in a fluid of density ρ_1. The situation is shown in Fig. 2.1a for a spherical particle, but the discussion is independent of the actual shape of the particle. The particle experiences a force F_g due to gravity, taken to be positive in the downward direction. At the same time a buoyant force F_b acts in the opposite direction. A net force equal to the difference between these forces results in the acceleration of the particle:

$$F_{net} = F_g - F_b = V(\rho_2 - \rho_1)g \tag{1}$$

This force will pull the particle downward—that is, F_{net} will have the same sign as g—if $\rho_2 > \rho_1$, and the particle is said to sediment. If, on the other hand, $\rho_1 > \rho_2$, then the particles will move upward, which is called "creaming." As the net velocity of the particle is increased, the viscous force F_v opposing its motion increases also. Soon this force, shown in Fig. 2.1b, equals the net driving force responsible for the motion. Once the forces acting on the particle balance, the particle experiences no further acceleration and a stationary state velocity is reached. It may be shown that under stationary state conditions and for small velocities the force of resistance is proportional to the stationary state velocity v:

$$F_v = fv \tag{2}$$

where the proportionality constant f is called the friction factor. The friction factor has the dimensions mass time^{-1}, kg s^{-1} in SI units. We shall consider some aspects of the proof of Eq. (2) for spherical particles in the following section. However, Eq. (2) is independent of any particular geometry. The stationary state velocity is positive for sedimentation and negative for creaming.

Since the net force of gravity and the viscous force are equal under stationary state conditions, Eqs. (1) and (2) may be equated to give

$$V(\rho_2 - \rho_1)g = fv \tag{3}$$

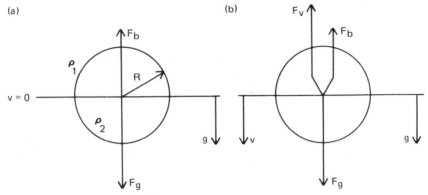

Figure 2.1 The forces acting on a spherical particle due to (a) gravity alone and (b) gravity and the viscosity of the medium ($\rho_2 > \rho_1$).

The stationary state is quite rapidly achieved, so Eq. (3) describes the velocity of a settling particle over much of its fall. Equation (3) may also be written

$$m\left(1 - \frac{\rho_1}{\rho_2}\right) g = f\upsilon \tag{4}$$

where m is the mass of the particle ($\rho_2 V$). This equation has the following features:

1. It is independent of particle shape.
2. It assumes that the bulk density of the pure components applies to the settling units (i.e., no solvation).
3. It permits the evaluation of υ for a situation in which m/f is known.
4. It permits the evaluation of m/f in a situation where υ is known.

The stationary state sedimentation velocity of a particle is an experimentally accessible quantity for some systems, so item (4) summarizes much of our interest in sedimentation.

Unfortunately, it is the ratio m/f rather than m alone that is obtained from sedimentation velocity in the general case of particles of unspecified geometry. The situation is comparable to the result of the classical experiment of J. J. Thomson in which the charge-to-mass ratio of the electron was determined. What is needed is a method for arriving at one of the quantities in the ratio independently. This information, in addition to the ratio, permits the evaluation of both the mass and the friction factor of the particle. In general, it takes two experiments to numerically evaluate these two parameters which characterize the dispersed particles. There are two ways of proceeding to overcome this impasse:

1. The particle may be assumed to be a sphere, in which case its friction factor may be calculated theoretically and thus eliminated from Eq. (4). We shall discuss the friction factor of spherical particles in Sect. 2.3.
2. An experiment may be conducted which permits the evaluation of f from measured quantities. Diffusion studies are ideally suited for this purpose, as we shall see later in this chapter.

Before turning to these two procedures to eliminate f from the sedimentation equation, one other consideration inherent in the use of Eq. (4) should be discussed. As previously noted, Eq. (4) uses the bulk density of the pure components. If the continuous and bulk phases are totally noninteracting, this may be justified. However, for flocs or solvated lyophilic particles, the density of the settling unit is intermediate between the densities of the two pure components. In these cases choosing an appropriate density for the settling particle can be a real problem.

Equation (3) shows that the sedimentation velocity increases with the density difference between the particle and the medium. Any situation which brings the density of the settling unit closer to that of the solvent will decrease the sedimentation velocity. To an observer who is unaware of its derivation, however, the smaller velocity would be interpreted by Eq. (4) as indicating a smaller value of m/f. Since the actual mass of colloidal material is unaffected by the solvation, it is more correct to attribute the reduced sedimentation velocity to an increase in the value of the fraction factor.

In the next section, we shall see how to deal quantitatively with solvation. Until now, the friction factor has been merely a proportionality factor of rather ill-defined origin. We shall not undertake a derivation of Eq. (2) in any general sense. In the next section, however, we shall outline the derivation of an important result due to G. G. Stokes—the friction factor for an unsolvated sphere.

2.3 The Stokes Equation

We begin our discussion of the Stokes equation by considering a single spherical particle in a state of relative motion with respect to the surrounding fluid. For the purposes of this derivation, it does not matter

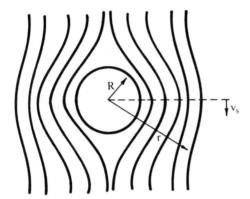

Figure 2.2 Distortion of flow streamlines around a spherical particle. (Reprinted with permission from P. C. Hiemenz, *Polymer Chemistry: The Basic Concepts*, Marcel Dekker, New York, 1984.)

whether the fluid is stationary and the particle is moving through it or whether the particle is stationary with the fluid moving around it. Although the former describes sedimentation, it is somewhat more convenient to discuss the phenomenon in terms of the fluid flowing in the $+z$ direction with a velocity v_z; this is the same as the particle moving in the $-z$ direction with a velocity of the same magitude. This situation is represented schematically in Fig. 2.2.

If the spherical particle were not present in Fig. 2.2, the volume elements of the flowing fluid would move upward in straight lines. In the presence of the particle, however, the flow profile is distorted around the sphere in the manner suggested by Fig. 2.2. It is apparent that the velocity of any volume element passing the sphere is a function of both time and location and must be described as such in any quantitative treatment. The trajectory of such a volume element is called the flow streamline function. For spherical particles, this was analyzed by G. G. Stokes in 1850.

The streamline function is the solution of a differential equation, the details of which we shall not pursue. We will have more to say about flow streamlines in our discussion of viscosity in Chap. 4. For the present, however, it is sufficient to note that an important part of solving any differential equation is the adequate incorporation of the boundary conditions that describe the problem. Since the disturbance of the flow streamlines in Fig. 2.2 is centered at the particle, it is convenient to discuss the boundary conditions in terms of the spatial variable r originating at the center of the sphere. Several aspects of the problem can be described in terms of this variable:

1. The thickness of each successive layer in the fluid is the infinitesimal mathematical increment dr. Since the streamlines follow the outline of the particle and are thin compared to the radius of the sphere R, we can think of each flow layer as moving tangentially to the surface of the sphere.

2. The velocity varies from layer to layer in the fluid near the surface of the sphere; this variation can be described in terms of a velocity gradient dv/dr. Any integration over this gradient must be able to assign two different values of velocity to two different r values.

3. When $r = \infty$, the disturbing influence of the particle has been damped away and the streamlines behave as if the particle were not present.

4. When $r = R$, the velocity is zero. This is described as a non-slip condition between the stationary surface of the particle and the layer of fluid adjacent to it. There is nothing particularly self-evident about the non-slip condition between the solid and the fluid, but it is an experimental fact. Some commonplace evidence that suggests this is the layer of dust that accumulates on the blades of a fan. However stiff the breeze may be some distance in front of the blades, the air is still and travels with the surface of the blades.

The fact that the velocity of a fluid changes from layer to layer is evidence of a kind of friction between these layers. The layers are mathematical constructs, but the velocity gradient is real and a characteristic of the fluid. That property of a fluid which describes the internal friction or resistance to flow is the viscosity of the material. Chapter 4 is devoted to a discussion of the measurement and interpretation of viscosity. For now, it is enough for us to recall that this property is quantified by the coefficient of viscosity η of a material. The coefficient of viscosity has dimensions mass length^{-1} time^{-1}, kg m^{-1} s^{-1} in SI units. In actual practice, the cgs unit of viscosity, the poise (P), is widely used. Note that pure water at 20°C has a viscosity of about 0.01 P $= 10^{-3}$ kg m^{-1} s^{-1}.

Because of internal friction, some of the translational kinetic energy of flow dissipates over a period of time. For the physical situation pictured in Fig.2.2, Stokes was able to show that the rate of energy dissipation is equal to $6\pi v_z^2 R$. The product of the viscous force F_v exerted by the particle on the fluid and the velocity v_z of the fluid also equals the rate at which work is done on the fluid by the particle. Remember, a force times a distance equals energy, so a force times a velocity equals the rate of energy dissipation. Thus we have two different ways of expressing the rate of energy dissipation associated with the presence of a spherical particle in the flowing liquid. Equating these two expressions gives

$$F_v v_z = 6\pi\eta v_z^2 R \tag{5}$$

It is clear that the force acting on the fluid acts in opposition to the flow. Therefore, if the fluid is flowing in the positive direction, the sphere resists the flow by a force in the negative direction. Conversely, if the particle is moving in one direction, the fluid resists the movement by a force in the opposite direction. This force is given by Stokes' law,

$$F = 6\pi\eta R v \tag{6}$$

where the subscripts are no longer needed, provided we remember that F and v are in opposing directions.

Equation (6) expresses, for a spherical particle, the general result presented as Eq. (2). Two aspects of the result are important:

1. For small stationary state velocities the viscous force on a particle is directly proportional to the velocity.
2. For a spherical particle the friction factor is given by

$$f = 6\pi\eta R \tag{7}$$

It was just noted that the case of spherical particles is one in which the friction factor can be eliminated from Eq. (4), yielding a result which permits the mass of a spherical particle to be evaluated from sedimentation data alone. To see how this works, we return to Eq. (3), using Eq. (7) to evaluate f and $\frac{4}{3}\pi R^3$ as a substitution for V. Thus for a spherical particle, Eq. (3) becomes

$$\tfrac{4}{3}\pi R^3(\rho_2 - \rho_1)g = 6\pi\eta R v \tag{8}$$

Equation (8) may be solved for the stationary state sedimentation velocity

$$v = \frac{2}{9}\frac{R^2(\rho_2 - \rho_1)g}{\eta} \tag{9}$$

and for the radius of the spherical particle

$$R = \left(\frac{9\eta v}{2(\rho_2 - \rho_1)g}\right)^{1/2} \tag{10}$$

Since the particle to which it applies is a sphere, Eq. (10) may also be used to calculate the mass and friction factor of the particle:

$$m = \rho_2 \frac{4}{3}\pi R^3 = \frac{4}{3}\pi\rho_2\left(\frac{9\eta v}{2(\rho_2 - \rho_1)g}\right)^{3/2} \tag{11}$$

and

$$f = 6\pi\eta \left(\frac{9\eta v}{2(\rho_2 - \rho_1)g} \right)^{1/2} \tag{12}$$

Equation (8) is an important result, since it describes the relationship between R, v, η, and $\Delta\rho$, the density difference. Any one of these quantities may be evaluated by Eq. (8) when the other three are known. Thus, Eq. (8) can be used to determine the density difference between two phases or to determine the viscosity of a liquid. In this chapter, however, our interest is in the characterization of colloidal particles by means of observations of their sedimentation behavior. Therefore we shall be primarily concerned with Eqs. (10) and (11), which are specifically directed toward this objective.

Stokes' law and the equations developed from it apply to spherical particles only, but the dispersed units in systems of actual interest often fail to meet this shape requirement. Equation (11) is sometimes used in these cases anyway. The lack of compliance of the system to the model is acknowledged by labeling the mass, calculated by Eq. (11), as the mass of an "equivalent sphere." As the name implies, this is a fictitious particle with the same density as the unsolvated particle which settles with the same velocity as the experimental system. If the actual settling particle is an unsolvated polyhedron, the equivalent sphere may be a fairly good model for it, and the mass of the equivalent sphere may be a reasonable approximation to the actual mass of the particle. The approximation clearly becomes poorer if the particle is asymmetrical or solvated, or both. Characterization of dispersed particles by their mass as equivalent spheres at least has the advantage of requiring only one experimental observation, the sedimentation rate, on the system. We shall see in later sections that the equivalent sphere calculations still play a useful role, even in systems for which supplementary diffusion studies have also been conducted.

Actual particles can deviate from the Stokes model by being either solvated, asymmetrical, or both, in which case f increases so that a particle of mass m will display a smaller sedimentation velocity than it would if it were an unsolvated sphere. The justification for this statement is based on Fig. 2.3. Figure 2.3 shows an unsolvated sphere and represents its radius as R_0. If this particle were solvated, it would swell to a larger radius R associated with a larger friction factor. Small asymmetrical particles rotate through every possible orientation during settling because of Brownian motion. Figure 2.3 shows that an encompassing sphere drawn around the tumbling particle will also have a radius larger than R_0.

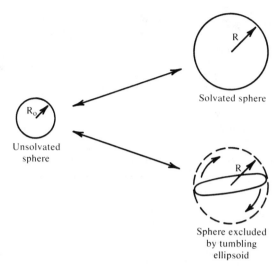

Figure 2.3 Illustration of how solvation and asymmetry have qualitatively equivalent effects in increasing the effective radius of an unsolvated sphere of radius R_0. (Reprinted with permission from P. C. Hiemenz, *Polymer Chemistry: The Basic Concepts*, Marcel Dekker, New York, 1984.)

An increase in f is associated with this situation as well. To deal with these ideas quantitatively, it is convenient to consider the following ratios:

$$f/f_0 = (f/f^*)(f^*/f_0) \qquad (13)$$

1. In Eq. (13) f is the friction factor of the actual particle.
2. The f_0 is the friction factor for an unsolvated sphere given by Stokes' law to equal $6\pi\eta R_0$ in the present notation. This is the lowest friction factor possible for a particle with the required mass.
3. The ratio f/f_0 measures the amount by which the actual friction factor exceeds the minimum value.
4. The f^* is the friction factor for a spherical particle having the same volume as the *solvated* particle of mass m.
5. The ratio f^*/f_0 measures the increase in f due to solvation.
6. The ratio f/f^* measures the increase in f due to asymmetry.

We shall see presently that both of the ratios on the right-hand side of Eq. (13) can be expanded in terms of appropriate models, so the experimental f value—compared to f_0 for a particle of the same mass—can be analyzed in terms of solvation and asymmetry.

Before, however, let us consider how the mass of an equivalent sphere can be extracted from experimental data, since this quantity is an important concept in its own right.

2.4   Sedimentation Under Gravity: Experimental

We begin our discussion of the experimental aspects of sedimentation by considering settling caused by gravity; in Sect. 2.5 we shall examine centrifugation. Substituting numerical values into Eq. (9) reveals that the range of particle sizes that can be studied by sedimentation under gravity is quite limited. For example, taking $\Delta\rho/\eta = 100$ g cm^{-3} P^{-1} yields values of v between 2.2×10^{-12} and 2.2×10^{-6} m s^{-1} for spherical particles of R between 1.0 μm and 1.0 nm. This value of $\Delta\rho/\eta$ corresponds physically to sulfur particles in water at 20°C—where $\Delta\rho$ and v are both positive—and to gas bubbles in water—where $\Delta\rho$ and v are negative. Although sedimentation under gravity is feasible only at the upper size limit of the usual colloidal range, there are still a number of important applications of the technique. We will consider only two of the procedures that have been developed for such studies.

If a particle is large enough to be visible to the unaided eye or in a traveling microscope, its velocity can be measured directly; however, smaller particles present more of a problem. If a dispersion consists of particles of uniform size, all will settle at the same rate, so their positions relative to each other do not change until they reach the bottom of the container. This means that a sharp boundary will exist between the domain occupied by the settling particles and that part of the system which has already been swept free of particles. Even though the individual particles may not be visible, this boundary may be visible if the particles differ sufficiently in color or refractive index from the continuous phase. The velocity of the boundary and the velocity of the particles are the same. If the particles differ in size, no such boundary will develop. Each size fraction in a polydisperse system will settle at its own characteristic velocity, so that at any given time one end of the column may be free of some size fractions but not of others. The "boundary zone" between the domain that clearly contains settling particles and the domain that is totally free of particles will appear to be a diffuse region across which the concentration of the disperse phase gradually diminishes. Since no sharp boundary develops, some other method must be devised to follow sedimentation velocity. Clearly, we would like as much information as possible about the particle size distribution, since the system is polydisperse. The following procedure shows how this information may be obtained.

Suppose the pan of a balance is positioned at some convenient location below the surface of a dispersion. As sedimentation occurs, the settled material collects on the balance pan. The total weight W of the material on the pan is measured at various times t, either by adding counterweights to a second pan or by noting the displacement of a calibrated fiber which supports the pan.

Figure 2.4a shows the type of data obtained when this method is applied to a clay sample. Clearly, at any given time the weight of material that has accumulated equals the weight of all particles large enough to have settled onto the pan in the time of the experiment. Such particles come from two categories; they include all those particles large enough to have fallen through the full length of the container *and* a fraction of smaller particles which have only fallen from a fraction of the column but still land on the pan. Both of these contribute to the total weight observed at any time. Since the cumulative weight equals the sum of two contributions, we may write

$$W = w + t\,\frac{dW}{dt} \tag{14}$$

where w is the weight of the particles which have fallen through the full height of the column at that particular time, and $t(dW/dt)$ represents the cumulative contribution of smaller size particles at that time. The quantity dW/dt is the slope of the cumulative curve at a particular point. As such, it measures the rate of incidence of particles smaller than the cutoff size associated with that time: The larger particles, after all, have already settled out. Furthermore, these smaller particles have been collecting on the pan throughout the entire time of the run t. Therefore their total weight equals $t(dW/dt)$. The graphical significance of Eq. (14) is shown in Fig. 2.4a. This shows that the intercepts defined by tangent lines drawn at different times give w directly, without actually requiring the calculation suggested by Eq. (14).

Next we plot w versus time to obtain a graph which shows the weight contribution at any time due to particles larger than the cutoff size associated with that time. Such a plot for the data of Fig. 2.4a is shown in Fig. 2.4b. This is the integrated size distribution curve for all particles larger than the cutoff size, that is, oversized particles. Since Fig. 2.4b is an integrated distribution curve, it follows that its derivation gives a distribution function for particle sizes. This must be a graphical differentiation, since we do not know the analytical expression for the cumulative curve.

Figure 2.4c shows the results of a graphical differentiation of Fig. 2.4b; that is, the ordinate values in Fig. 2.4c are slopes from Fig. 2.4b, dw/dt. In

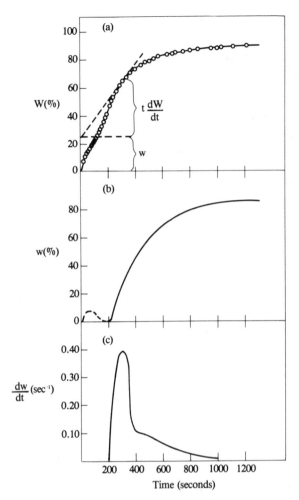

Figure 2.4 Sedimentation of clay particles: (a) cumulative weight (as percent) versus time, (b) cumulative weight of oversized material (as percent) versus time, and (c) frequency by weight versus time. [Data from S. Oden, *Proc. R. Soc. Edinburgh, 36*:219 (1915).]

effect, then, this is the second time the original data have been differentiated: once to prepare Fig. 2.4b and again for Fig. 2.4c. That dw/dt is the second derivative of the original data may also be shown by differentiating Eq. (14) and rearranging to give $dw/dt = -(1/t)(d^2W/dt^2)$. Since the second derivative is negative for a curve which opens downward, such as Fig. 2.4a, dw/dt is a positive quantity.

There is a loss of accuracy with each graphical differentiation. Therefore the original data must be very good for a particle size distribution such as Fig. 2.4c to be prepared from cumulative sedimentation results. At both long and short times, the total weight increases almost linearly, which means that the second derivative cannot be determined with any accuracy at all at either end of the distribution. In spite of these limitations, Fig. 2.4c shows clearly that the distribution is quite sharply peaked at a particle size which settles in about 300 s.

Until now, no assumptions whatsoever about the shape of the settling units have been made in the analysis of Fig. 2.4. If we wish to go further, we may translate settling times into the radius of equivalent spheres by means of Eq. (10). However, at this point an assumed model enters the picture. In the experiment from which the data of Fig. 2.4 were obtained, the height of the column was 0.20 m and $\Delta\rho$ was 2.17 g cm^{-3}. With this information, the 300-s settling time of particles at the maximum in Fig. 2.4c corresponds to a radius of 1.19×10^{-5} m. The equivalent sphere is a very poor model for the system in Fig. 2.4, since clay particles are generally dispersed as platelets, as illustrated in the electron micrograph of Fig. 1.9b. This type of analysis gives not only the radius of the most probable equivalent sphere but also an indication of the distribution of particle sizes.

A slight variation of this method for analyzing particle size distributions is particularly convenient if the system contains particles of several discretely different particle sizes rather than a continuous distribution of sizes. A thin band of such a dispersion is layered on top of a column of pure solvent. Particles in the different size categories will settle through the solvent at different rates so that, if the column is long enough, the first fraction collected at the bottom of the column will contain only the largest particles, with smaller sizes showing up in progressively later fractions. The clear solvent is literally the column within which the various particle sizes are resolved. The "resolving power" of such an arrangement increases with the length of the column, at least up to some optimum length. It might be noted that there is a certain formal similarity between this situation and column chromatography.

Provided the column is long enough to separate the different sizes of particles, this method gives w versus time directly. Any additional interpretations of the results are made in a manner entirely analogous to the one just described. Since only a narrow band of the dispersion is used in this method, the weight of the dispersed phase in each fraction will be relatively low. Fairly sensitive analytical techniques are required for the fractions collected.

Sedimentation runs should be conducted at constant temperature, not only so that $\Delta\rho$ and η are known but also to minimize disturbances due to convection. Any sort of disturbance will obviously disrupt the segregation of the particles by size which has occurred as a result of sedimentation. An intrinsic difficulty with the balance method lies in the fact that the liquid below the balance pan is less dense than the liquid with dispersed particles above the pan. Thus there is a tendency for a counterflow of pure solvent to arise, which would introduce an error in the particle size analysis.

The methods just discussed are only two of a wide variety of techniques which provide essentially the same kinds of information. In general, any measurement which gives (1) the amount of suspended material a fixed distance below the surface at various times or (2) the amount of material at various depths at any one time can be interpreted in terms of particle size distribution. Pressure, density, and absorbance are additional measurements that have been analyzed this way.

Instruments are commercially available which automatically carry out the analysis presented above. Although much of the tedium of the experiment is thereby relieved, automation does not alter the basic assumptions and/or approximations of the method. In one commercial instrument a collimated beam of light or low-energy x-rays scans the settling compartment, measuring the absorbance of the suspended particles at various depths. Because the detector does not have to wait for settling particles to arrive at a fixed position, this method is more rapid than direct observation of sedimentation. A built-in computer converts the absorbance at a particular distance below the top of the sample to concentration after some time has elapsed from the start of settling. The distance–time combination is reduced to a velocity and the latter is converted into the radius of an equivalent sphere by Eq. (10). All of this is calculated automatically, and a recorder plots the cumulative weight percent of settling material versus the equivalent radius. The sample compartment is thermostated and suspending media with a range of known densities and viscosities are used; the latter are part of the input data of the operation. By utilizing the full range of operating parameters, this instrument allows equivalent diameters from 0.1 to 100 μm to be analyzed. The following example considers the application of Stokes' law to the output of such a particle analyzer.

Example 2.1   A titanium dioxide pigment of density 4.12 g cm^{-3} is suspended in water at 33°C. At this temperature the density and viscosity of water are 0.9947 g cm^{-3} and 7.523 × 10^{-3} P, respectively. A particle size analyzer* plots the following data for cumulative weight percent versus equivalent spherical radius:

Cumulative wt %	95	90	80	70	60	50	40	30	20	10	5
$R_{eq.\ sph.}$ (μm)	0.60	0.45	0.33	0.37	0.28	0.27	0.24	0.22	0.21	0.18	0.15

To what sedimentation velocity does the most abundant component in this distribution correspond?

Solution   These cumulative percentages are of the same form as in Fig. 2.4b; therefore the particle size distribution peaks where the cumulative curve increases most steeply. This occurs at about 0.29 μm for this distribution. Equation (9) permits the sedimentation velocity for particles of this size to be calculated:

$$v = \frac{2R^2(\rho_2 - \rho_1)g}{9\eta} = \frac{2(2.9 \times 10^{-5})^2(4.12 - 0.99)(980)}{9(7.52 \times 10^{-3})}$$

$$= 7.62 \times 10^{-5} \text{ cm s}^{-1} = 7.62 \times 10^{-7} \text{ m s}^{-1}.$$

The distance–time combination corresponding to this weight percent is determined by the particle size analyzer and converted to R.

•

A basic limitation of all these methods is the narrow range of particle sizes that can be investigated by sedimentation under gravity. Therefore we turn next to a consideration of centrifugation, particularly the ultracentrifuge, as a means of extending the applicability of sedimentation measurements.

2.5   Sedimentation in a Centrifugal Field

A well-known fact from elementary physics is that a particle traveling in a circular path of radius R at an angular velocity ω (in radians per second)

---

*SediGraph, Micromeritics Instrument Corp., Norcross, Ga. 30093.

is subject to an acceleration in the radial direction equal to $\omega^2 R$. Since there are 2π rad per revolution, ω equals the number of revolutions per second (rps) times 2π, or the number of revolutions per minute (rpm) times the ratio $2\pi/60$. It is not particularly difficult to produce accelerations by centrifugation which are many times larger than the acceleration due to gravity. It is conventional, in fact, to describe the radial acceleration achieved by a centrifuge as some multiple of the standard gravitational acceleration or as being so many g's, where g is about 9.8 m s^{-2}.

The ultracentrifuge is an instrument in which a cell is rotated at very high speeds in a horizontal position. As we shall see presently, the gravitational acceleration is easily increased by a factor of 10^5 in such an apparatus. Accordingly, the particle size which may be studied by sedimentation is decreased by the same factor. The ultracentrifuge has been used extensively for the characterization of colloidal materials, particularly those of biological origin, such as proteins, nucleic acids, and viruses.

Because of the extreme importance of the ultracentrifuge, it seems appropriate to describe it in some detail. Although the particulars differ from instrument to instrument, the essential features are present in all ultracentrifuges. The actual sedimentation takes place in a cell mounted within an aluminum or titanium rotor. The cell is sector shaped; its side walls converge toward the center along radial lines. Since the radial acceleration is proportional to the distance from the axis of rotation, we see that this quantity varies from top to bottom in the cell. Although this variation is considered explicitly in a later section, it is sufficient for the present to consider the average acceleration at the midpoint of the cell, which is typically located 6.5×10^{-2} m from the center of the rotor. For speeds of 10,000, 20,000, and 40,000 rpm, accelerations of 7.13×10^4, 2.85×10^5, and 1.14×10^6 m s^{-2} are produced at this radius. These correspond to 7.27×10^3, 2.91×10^4, and 1.16×10^5 times the acceleration of gravity, respectively.

An important part of the ultracentrifuge is the optical system which makes observation during operation a possibility; it is shown schematically in Fig. 2.5. The sample compartment fits into a hole in the rotor and is positioned so as to intersect the light path of the optics. Those faces of the cell which are perpendicular to the light path are transparent to visible and ultraviolet light, so that various optical methods of chemical analysis may be employed to measure the distribution of material along the sedimentation path. Schlieren refractometry, interferometry, and spectrophotometry are all employed for this purpose, the last being particularly useful when very low concentrations are involved. The

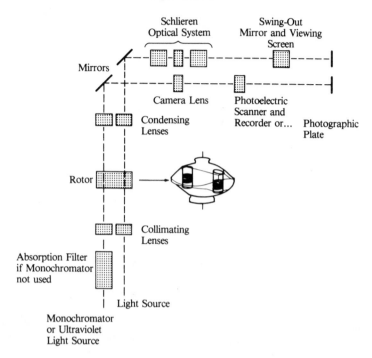

Figure 2.5 Schematic of optical systems in the Spinco Model E Ultracentrifuge. (Beckman Instruments, Inc., Spinco Division, Palo Alto, Ca.)

location of the settling particles may be followed by one of these methods, and the results recorded either photographically or on a chart recorder. Figure 2.6a shows how the concentration profile varies with radial location in the ultracentrifuge as the particles settle. Although the boundary between the solution and the solvent that has been swept free of solute seems clear in this representation, the use of schlieren refractometry makes the sedimentation even more easy to visualize.

The schlieren system of optics is an analytical method that is particularly well suited to following the location of a chemical boundary with time. It is routinely employed in ultracentrifuges and also in electrophoresis experiments, as we shall see in Chapter 13. Schlieren optics product an effect which depends on the way the refractive index varies with position, the refractive index gradient, rather than on the refractive index itself. Therefore the schlieren effect is the same at all locations along the axis of sedimentation, except at any place where the refractive index is changing. In such a region it will produce an optical

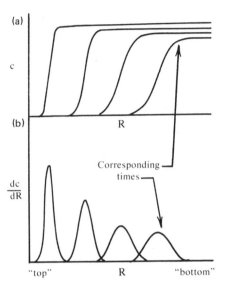

Figure 2.6 Location of the boundary for particles settling in an ultracentrifuge: (a) concentration profile and (b) derivative profile as revealed by schlieren optics. (Reprinted with permission from P. C. Hiemenz, *Polymer Chemistry: The Basic Concepts*, Marcel Dekker, New York, 1984.)

effect which is proportional to the refractive index gradient. The boundary between two layers is thus perceived as a sharp peak on a flat baseline by schlieren optics. The displacement of such a peak with time measures the velocity of the boundary. Likewise, a band of settling particles is seen as a broad schlieren peak whose width and velocity measure the width and velocity of the band. The schlieren optical system works by using a diaphragm to cut off from a photographic plate that light which is deviated from the optical path by the refractive index gradient. The same effect may be produced on a ground-glass screen for instantaneous viewing by an ingenious system which combines a diagonal slit and a cylindrical lens to produce an image of the refractive index gradient. Figure 2.6b shows how the same data as presented in Fig. 2.6a would appear with schlieren optics. It is thus an easy matter to monitor the peak location as the sedimentation proceeds.

The usual precautions regarding temperature and vibration control also apply to the ultracentrifuge. To overcome air resistance and the attendant frictional heating, the compartment in which the rotor spins is evacuated and may be thermostated over a wide range of temperatures.

The rotor is mounted on a flexible drive shaft which minimizes the need for precise balancing as a requirement for vibration-free operation. Finally, the rotor assembly is enclosed in an armored steel chamber for safety. At these speeds, a runaway rotor is deadly!

2.6 The Sedimentation Coefficient

The results of a sedimentation experiment in a centrifugal field are conventionally reported as the sedimentation coefficients. This quantity is defined as

$$s = \frac{dR/dt}{\omega^2 R} \qquad (15)$$

that is, it equals the sedimentation velocity per unit of centrifugal acceleration. In the SI system, this ratio has the units m s^{-1}/m s^{-2}, or seconds. In practice, the quantity 10^{-13} s is defined to be 1.0 svedberg (1.0S) after T. Svedberg, the originator of the ultracentrifuge and pioneer in its use (Nobel Prize, 1926). Sedimentation coefficients are usually reported in this unit. If the location of a particle along its settling path is measured as a function of time, the sedimentation coefficient is readily evaluated by integrating Eq. (15), using the following limits of integration: The component is at radial position R_1 at time t_1 and at R_2 at t_2. Therefore

$$\omega^2 s \int dt = \int \frac{dR}{R} \qquad (16)$$

or

$$s = \frac{\ln(R_2/R_1)}{\omega^2(t_2 - t_1)} \qquad (17)$$

The sedimentation coefficient is dependent on concentration; consequently it is usually measured at several different concentrations and the results are extrapolated to zero concentration. It is customary to designate this limiting value by a superscript zero. Experimental values are also generally labeled with respect to temperature, so a value listed as s_{20}^0 corresponds to a sedimentation coefficient measured at (or corrected to) 20°C and extrapolated to zero concentration.

Under stationary state conditions, the force due to the centrifugal field and the viscous force of resistance will be equal. Therefore $\omega^2 R$ replaces g in Eq. (4) to give

$$m\left(1 - \frac{\rho_1}{\rho_2}\right)\omega^2 R = f\,\frac{dR}{dt} \tag{18}$$

where f is the friction factor for the settling particle. Equations (18) and (15) may be combined to yield

$$\frac{m}{f}\left(1 - \frac{\rho_1}{\rho_2}\right) = s \tag{19}$$

which shows that the sedimentation coefficient is directly proportional to the ratio of the mass-to-friction factor. As with sedimentation under the acceleration of gravity, any further interpretation of m/f depends either on independent determination of the friction factor from diffusion or on the assumption of spherical particles, with Eq. (7) used to evaluate f. Thus experimental sedimentation coefficients may be analyzed to yield the mass, radius, and friction factor of an equivalent sphere. In the absence of supplementary data, this is as far as sedimentation alone can be interpreted. The following example illustrates the use of these relationships.

Example 2.2   What should be the speed of an ultracentrifuge so that the boundary associated with the sedimentation of a particle of molecular weight 60,000 g mole^{-1} moves from $R = 6.314$ cm to 6.367 cm in 10 min? The densities of the particle and the medium are 0.728 and 0.998 g cm^{-3}, respectively, and the friction factor of the molecule is 5.3×10^{-11} kg s^{-1}.

Solution   First we calculate the particle mass by dividing the molecular weight by Avogadro's number: $m = 60{,}000/6.02 \times 10^{23} = 9.97 \times 10^{-20}$ g = 9.97×10^{-23} kg molecule^{-1}.
    Equation (19) allows us to calculate the sedimentation coefficient:

$$s = (m/f)(1 - \rho_1/\rho_2) = (9.97 \times 10^{-23}/5.3 \times 10^{-11})(1 - 0.998/0.728)$$

$$= 6.98 \times 10^{-13}\ \text{s} = 6.98\text{S}$$

Equation (17) can now be solved for ω^2:

$$\omega^2 = \frac{\ln(R_2/R_1)}{s(t_2 - t_1)} = 1.997 \times 10^7\ \text{s}^{-2}$$

or
$\omega = 4.47 \times 10^3$ rad s^{-1}. Dividing by 2π converts this to revolutions per second: $\omega = 711$ rps, or 42,700 rpm.

•

In the preceding sections of this chapter we have considered sedimentation as if it were the only process which influenced the spatial distribution of particles. If this were the case, all systems of dispersed particles, even gases, would eventually settle out. In practice, convection currents arising from temperature differences keep many systems well stirred. Even in carefully thermostated laboratory samples, however, there is another factor operating which prevents the complete sedimentation of small particles, namely, diffusion. Diffusion and sedimentation are opposing processes inasmuch as the former tends to keep things dispersed, whereas the latter tends to collect them in one place. Diffusion is more important for smaller particles: Remember that sedimentation is negligible for gases. For larger particles, diffusion is negligible. Of course, there is a range of particle sizes in which both effects are comparable; we will examine the combined effects of sedimentation and diffusion in later sections. For the present, however, let us examine the phenomenon of diffusion by itself.

2.7 Diffusion and Fick's Laws

If external forces such as gravity can be neglected, the composition of a single equilibrium phase will be macroscopically uniform throughout. This means that the concentration or density is constant throughout the phase. It should be noted that we are talking about the macroscopic description of the phase. At the molecular level there will be local fluctuations from the mean value, a fact important in, for example, light scattering, as we shall see in Chap. 5. However, for the present, we restrict ourselves to the macroscopic description. Fundamentally, it is the second law of thermodynamics that is responsible for the uniform distribution of matter at equilibrium, since entropy is maximum when the molecules are distributed randomly throughout the space available to them. If for some reason a nonuniform distribution of matter should exist, the particles of the system will experience a force which tends to distribute them uniformly. Consider, for example, the situations shown in Fig. 2.7a in which two solutions of different concentrations are separated by a hypothetical porous plug of thickness dx. We can set $dx = 0$; in this case there will be a migration of solute from the high-concentration side to the low-concentration side until a uniform distribution is obtained, that is, equilibrium is reached.

Now let us define the following quantities. Suppose Q represents the amount of material that flows through a cross section of area A in Fig. 2.7. The rate of change in the quantity Q/A is called the flux of solute across the boundary; that is, the flux J is defined as

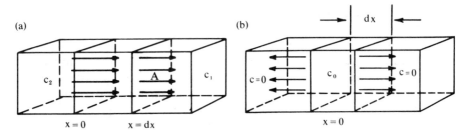

Figure 2.7 Schematic of a diffusion process through a volume element of thickness dx: (a) The volume element separates two solutions of concentrations c_1 and c_2 and (b) the element contains a solution at concentration c_0 and separates two regions of pure solvent.

$$J = \frac{d(Q/A)}{dt} \tag{20}$$

Therefore the amount of material which crosses A in a time interval Δt is

$$Q = A J \, \Delta t \tag{21}$$

The phenomenological equation which relates the flux of material across the boundary to the concentration gradient at that location is given by Fick's first law:

$$J = -D \frac{\partial c}{\partial x} \tag{22}$$

where D is defined to be the diffusion coefficient of the solute. Assuming that the amount of material is measured in the same units both in J and in c, Eqs. (20) and (22) show that the SI units of D are square meters per second.

Now, instead of a boundary of zero thickness, let us consider the concentration changes which occur within a zone of cross section A which has a thickness Δx as shown in Fig. 2.7b. Any change in the amount of material in this zone will equal the difference between the amount of material which enters the zone and the amount of material which leaves it:

$$\Delta Q = Q_{in} - Q_{out} = (J_{in} - J_{out}) A \, \Delta t \tag{23}$$

The quantity ΔQ also equals the product of the volume of the zone and the concentration change that occurs in it:

$$\Delta Q = A \, \Delta x \, \Delta c \tag{24}$$

Equating Eqs. (23) and (24) and substituting Fick's first law [Eq. (22)], we obtain

$$- D \frac{[(\partial c/\partial x)_{x=0} - (\partial c/\partial x)_{x=\Delta x}]}{\Delta x} = D \frac{\Delta(\partial c/\partial x)}{\Delta x} = \frac{\Delta c}{\Delta t} \tag{25}$$

assuming that the cross section is uniform throughout the compartment and that D is independent of small changes of concentration. Finally, if we take the limit of Eq. (25) as Δx and Δt approach zero, we obtain

$$\frac{\partial c}{\partial t} = D \frac{\partial^2 c}{\partial x^2} \tag{26}$$

a result known as Fick's second law.

Equation (26) is a differential equation whose solution describes the concentration of a system as a function of time and position. The solution depends on the boundary conditions of the problem as well as on the parameter D. This is the basis for the experimental determination of the diffusion coefficient. Equation (26) is solved for the boundary conditions which apply to a particular experimental arrangement. Then the concentration of the diffusing substance is measured as a function of time and location in the apparatus. Fitting the experimental data to the theoretical concentration function permits the evaluation of the diffusion coefficient for the system under consideration. Rather than getting deeply involved in the mathematics of differential equations, we shall use a statistical model to find a solution to Eq. (26) for a system with simple boundary conditions. This will be sufficient to illustrate the experimental technique whereby diffusion coefficients are determined and will also lead to a better understanding of the random processes underlying diffusion. This statistical discussion and the experimental procedure it suggests are taken up in Sect. 2.10.

Next it will be helpful to anticipate a description of experimental procedures and consider the magnitude of measured diffusion coefficients. The self-diffusion coefficients for ordinary liquids with small molecules are of the order of magnitude 10^{-9} m^2 s^{-1}; for colloidal substances they are typically of the order of 10^{-11} m^2 s^{-1}. In the next section we shall see that for spherical particles the diffusion coefficient is inversely proportional to the radius of the sphere. Therefore every increase by a factor of 10 in size decreases the diffusion coefficient by the same factor. Qualitatively, this same inverse relationship applies to nonspherical particles as well. Once again, we see that diffusion decreases in importance with increasing particle size, precisely those conditions for which sedimentation increases in importance. For larger particles, where D is very small, the diffusion coefficient also becomes harder to measure. For spherical particles the time required for the particle to diffuse a unit distance is directly proportional to its radius. For small molecules this time is experimentally accessible; it becomes experimentally inconvenient for particles at the upper end of the colloidal range.

2.8   The Diffusion Coefficient and the Friction Factor

As already noted, the driving force underlying diffusion is primarily thermodynamic in origin. A very general way to describe a force is to write it as the negative gradient of a potential. In the context of diffusion, the potential to be used is the chemical potential μ_i, the partial molal Gibbs free energy of the component of interest. Thus the driving force per particle behind diffusion may be written

$$F_{\text{diff}} = \frac{-1}{N_A} \frac{\partial \mu_i}{\partial x} \tag{27}$$

It is necessary to divide by Avogadro's number N_A since μ_i is a molar quantity. Thermodynamics shows that

$$\mu_i = \mu_i^0 + RT \ln a_i = \mu_i^0 + RT \ln(\gamma_i c_i) \tag{28}$$

where a_i, c_i, and γ_i are the activity, concentration, and activity coefficient, respectively, of the ith component. Since we are interested in infinitely dilute systems, the activity coefficient may be assumed to equal unity. Substituting Eq. (28) into Eq. (27) gives

$$F_{\text{diff}} = -\,kT\,\frac{\partial \ln c_i}{\partial x} = \frac{-kT}{c_i}\,\frac{\partial c_i}{\partial x} \tag{29}$$

where k is Boltzmann's constant R/N_A. Under stationary state conditions this force will be equal to the force of viscous resistance, given by $f\upsilon$ according to Eq. (2). Therefore the velocity of diffusion equals

$$\upsilon = \frac{-kT}{fc}\,\frac{\partial c}{\partial x} \tag{30}$$

where the subscript has been omitted from the concentration of the solute c because this is now the only quantity involved in the relationship. Finally, we make the following observation. The flux of material through a cross section equals the product of its concentration and its diffusion velocity:

$$J = c v_{\text{diff}} \tag{31}$$

Combining Eqs. (30) and (31) and comparing with Eq. (22) leads to the important result

$$D = \frac{kT}{f} \tag{32}$$

It should be noted that this derivation contains no assumptions about the shape of the particles.

Many of the relationships of this chapter have involved the friction factor f, which, until now, has been an unknown quantity except for spherical particles. Equation (32) breaks this impasse and points out the complementarity between sedimentation and diffusion measurements. For example, substitution of Eq. (32) into (4) gives

$$m = \frac{kTv}{D(1 - \rho_1/\rho_2)g} \tag{33}$$

and substituting Eq. (32) into Eq. (19) yields

$$m = \frac{kTs}{D(1 - \rho_1/\rho_2)} \tag{34}$$

Equations (33) and (34) show that diffusion studies combined with sedimentation studies, either under the force of gravity or in a centrifuge, yield information about particle masses with no assumptions about the shape of the particle.

Human hemoglobin, for example, has a sedimentation coefficient of 4.48S and a diffusion coefficient of 6.9×10^{-11} m^2 s^{-1} in aqueous solution at 20°C. The density of this material is 1.34 g cm^{-3}. Substituting these values into Eq. (34) shows the particle mass to be

$$m = \frac{(1.38 \times 10^{-23})(293)(4.48 \times 10^{-13})}{(6.9 \times 10^{-11})(1 - 1.0/1.34)} = 1.03 \times 10^{-22} \text{ kg particle}^{-1} \tag{35}$$

or, in terms of molecular weight,

$$M = (1.03 \times 10^{-22})(6.02 \times 10^{23}) = 62,300 \text{ g mol}^{-1} \tag{36}$$

Since the friction factor was measured experimentally, this value is correct regardless of the state of solvation or ellipticity of the hemoglobin molecules in solution. We shall see presently how the combination of these two experimental approaches may be interpreted further to yield some information about solvation and ellipticity.

We have already seen that the ratio f/f_0 describes the effect on the friction factor of either solvation, ellipticity, or both. This ratio equals unity for an unhydrated sphere and increases with both the amount of bound solvent and the axial ratio of the particles. We are now in a position to see how this ratio may be evaluated experimentally. The steps of the procedure are the following:

1. The diffusion coefficient and sedimentation velocity (or sediment-ation coefficient) are used to evaluate m by Eq. (33) or (34).
2. The friction factor is evaluated from the diffusion coefficient by Eq. (32).
3. If we assume the particle to be unsolvated, we can determine its volume by dividing its mass by the density of the dry material.
4. The radius of the particle is calculated from its volume, if we assume the particle to be a sphere.
5. The friction factor f_0 of the equivalent unsolvated sphere is evaluated by Eq. (7).
6. The ratio of the experimental friction factor to f_0 is determined.

As noted, the larger this ratio is than unity, the more the particle deviates from the unsolvated spherical shape assumed to calculate f_0. Although this statement is qualitatively accurate, we realize that the f/f_0 value reflects some actual quantitative condition of the particles, and we search for additional ways to interpret this ratio.

The factoring of f/f_0 into two contributions in Eq. (13) was introduced with this interpretation in mind. It turns out that the effect of solvation is easily handled by a physically reasonable model through $f*/f_0$. Dealing with particle asymmetry through $f/f*$ is more complicated, but this has also been analyzed according to a model which is plausible for many situations. Let us now consider each of these developments in turn.

When particles are solvated, a certain volume of the solvent must be counted as part of the dispersed phase rather than the continuous phase. In dilute solutions the effect of this reclassification of some solvent is negligible for the remaining solvent (component 1), but the effect on the solute (component 2) may be considerable. The effect of the attached solvent on the volume of the solute particles may be calculated if some model is assumed for the mode of attachment. To assume the solvation occurs uniformly throughout the particle is a plausible model for the solvation of lyophilic colloids. For example, protein molecules have ionic and polar substituent groups distributed at random along the polymer chain. Such groups are expected to be extensively hydrated in aqueous solution.

If we assume that solvation occurs at numerous positions along the molecule, the volume of the solvated particle may be written as the sum of the volume of the unsolvated particle and the volume of the bound (subscript b) solvent:

$$\text{volume of solvated particle} = V_2 + V_{1,b} \tag{37}$$

This can be rearranged as

$$V_2 + V_{1,b} = V_2\left(1 + \frac{V_{1,b}}{V_2}\right) = V_2\left[1 + \frac{m_{1,b}}{m_2}\left(\frac{\rho_2}{\rho_1}\right)\right] \tag{38}$$

where $m_{1,b}$ is the mass of bound solvent and the last equality requires the density of the bound solvent to be identical to that of the free solvent. In the event that the solvated specie differs in density from its bulk counterpart, the ratio ρ_2/ρ_1 may be replaced by $\bar{v}_1/\bar{v}_2$, where the $\bar{v}_i$'s are the partial specific volumes of the components. The latter measure the volume per unit mass of the indicated components in the solution at the indicated concentration and hence reflect any volume changes which accompany the interaction. We shall continue to discuss solvation in terms of Eq. (38), using the more easily visualized densities.

We are explicitly looking for the effect of solvation on the friction factor of a sphere; particle asymmetry is handled by the second ratio f/f_0. By Stokes' law, the ratio of the friction factors of solvated to unsolvated spheres, f^*/f_0, is equal to the ratio of their radii, R_{solv}/R_0. This ratio equals the cube root of the volume ratio of the solvated to unsolvated spheres, or

$$\frac{f^*}{f_0} = \left[1 + \frac{m_{1,b}}{m_2}\left(\frac{\rho_2}{\rho_1}\right)\right]^{1/3} \tag{39}$$

The ratio f^*/f_0 equals unity when $m_{1,b} = 0$ and increases with increasing solvation, as required. We shall examine an application of this relationship presently.

Asymmetry as well as solvation can cause a friction factor to have a value other than f_0. Next let us consider the ratio f/f^*, which, according to Eq. (13), accounts for the effect of particle asymmetry on the friction factor. We saw in Sect. 1.6 that ellipsoids of revolution are reasonable models for many asymmetric particles.

F. Perrin derived expressions for the ratio f/f^* for ellipsoids of revolution in terms of the ratio of the equatorial semiaxis to the semiaxis of revolution, b/a. The following expressions were obtained:

1. For prolate ellipsoids ($b/a < 1$),

$$\frac{f}{f^*} = \frac{\left[1 - \left(\frac{b}{a}\right)^2\right]^{1/2}}{\left(\frac{b}{a}\right)^{2/3} \ln\left(\frac{1 + [1 - (b/a)^2]^{1/2}}{b/a}\right)} \tag{40}$$

2. For oblate ellipsoids ($b/a > 1$),

$$\frac{f}{f^*} = \frac{\left[\left(\frac{b}{a}\right)^2 - 1\right]^{1/2}}{\left(\frac{b}{a}\right)^{2/3} \tan^{-1}\left[\left(\frac{b}{a}\right)^2 - 1\right]^{1/2}} \tag{41}$$

Equations (40) and (41) allow the ratio f/f^* to be evaluated for any axial ratio. The reverse problem, finding an axial ratio that is consistent with an experimental f/f^* ratio, is best accomplished by interpolating from plots of Eqs. (40) and (41). These two equations, along with Eqs. (13) and (39), allow the states of solvation and ellipticity compatible with an experimental friction factor to be established. The following example considers this for the human hemoglobin molecule.

Example 2.3   The diffusion coefficient of the human hemoglobin molecule at 20°C is 6.9×10^{-11} m^2 s^{-1}. Use this value to determine f for this molecule. Evaluate f_0 for hemoglobin using the particle mass calculated in Eq. (35). Indicate the possible states of solvation and ellipticity that are compatible with the experimental f/f_0 ratio.

Solution   According to Eq. (32), $f = kT/D = (1.38 \times 10^{-23})(293)/6.9 \times 10^{-11} = 5.86 \times 10^{-11}$ kg s^{-1}.

Equation (35) shows that the mass of one of these molecules is 1.03×10^{-22} kg; dividing this by the density (1.34 g cm^{-3}) gives 7.69×10^{-26} m^3 as the particle volume. The radius of an unsolvated sphere of this volume is given by $R = (3V/4\pi)^{1/3} = 2.64 \times 10^{-9}$ m.

Using the viscosity of water, 0.01 P, the friction factor f_0 is calculated by Eq. (7): $f_0 = (6\pi)(10^{-3})(2.64 \times 10^{-9}) = 4.98 \times 10^{-11}$ kg s^{-1}.

Table 2.1 A Summary of the Characterization of Human Hemoglobin at 20°C Based on Sedimentation and Diffusion Measurements[a]

Quantity	Value	Determination
Sedimentation coefficient, s	4.48×10^{-13}s	Experimental
Diffusion coefficient, D	6.9×10^{-11} m^{2}s^{-1}	Experimental
Density, ρ	1.34 g cm^{-3}	Experimental
Mass of particle, m	1.03×10^{-22} kg molecule^{-1}	Eq. (35)
Molecular weight	6.23×10^4 g mol^{-1}	Eq. (36)
f	5.86×10^{-11} kg s^{-1}	Eq. (32)
Volume of particle	7.69×10^{-26} m^3	$V_{unsolv} = m/\rho$
Radius of equivalent sphere	2.64×10^{-9} m	$R_{sph} = (3V/4\pi)^{1/3}$
f_0	4.98×10^{-11} kg s^{-1}	Eq. (7)
f/f_0	1.18	—
$m_{1,b}/m_2$, if spherical	0.48 g H$_2$O (g protein)$^{-1}$	Eq. (39)
a/b if unsolvated and prolate	4.0	Eq. (40)
a/b if unsolvated and oblate	0.24	Eq. (41)

[a]Data from Ref. 2.

Since these calculations involve only empirical data, we can say the experimental value of $f/f_0 = 5.86 \times 10^{-11}/4.98 \times 10^{-11} = 1.18$.

Next we recall Eq. (13), which interprets f/f_0 as the product $(f/f^*)(f^*/f_0)$.

If $f/f^* = 1.00$, the particle is a sphere and $f^*/f_0 = 1.18$. Using the bulk density of water and the protein and Eq. (39), we can show that $m_{1,b}/m_2 = 0.48$ g water per g hemoglobin.

If $f^*/f_0 = 1.00$, the particle is unsolvated and $f/f^* = 1.18$. This corresponds to an axial ratio $b/a = 0.24$, according to Eq. (40), if the particle is a prolate ellipsoid, or to $b/a = 4.0$, according to Eq. (41), if the particle is oblate. These values are summarized in Table 2.1.

•

It is apparent that there are many combinations of f/f^* and f^*/f_0 that form the product 1.18, which we attempted to explain in Example 2.3. Setting either one of these contributions equal to unity merely establishes an upper limit for the other. Intermediate cases between solvated spheres and unsolvated ellipsoids can also be calculated which are consistent with a given axial ratio. Figure 2.8 is a plot of possible combinations of hydration and asymmetry for protein particles in water. Similar curves could be drawn for other materials as well. For the human hemoglobin molecule discussed in Table 2.1 the combination of sedimentation and

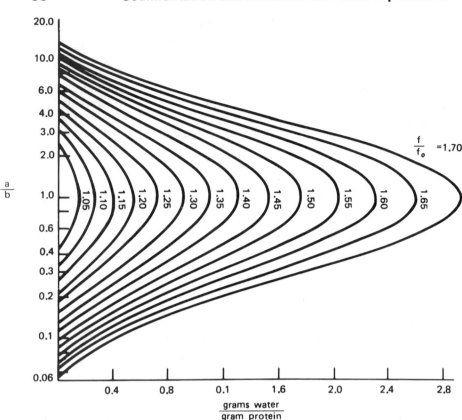

Figure 2.8 Variations of the ratio f/f_0 with asymmetry and hydration for aqueous protein dispersions. [Redrawn from L. Oncley, *Ann. N.Y. Acad. Sci.,* *41*:121 (1941), used with permission.]

diffusion measurements gives an f/f_0 value which lies within the domain defined by the 1.15 and 1.20 contours of Fig. 2.8. The current picture of the structure of human hemoglobin, deduced from x-ray diffraction studies, suggests that the molecule may be regarded as an ellipsoid with height, width, and depth equal to 6.4, 5.5, and 5.0 nm, respectively. Applying these dimensions to the dispersed unit leads us to describe the particle as being hydrated to the extent of about 0.4–0.5 g water (g protein)$^{-1}$.

Sedimentation and diffusion data allow for the unambiguous determination of particle mass and also allow the suspended particles to be placed on a contour in a plot such as that of Fig. 2.8. This is as far as these experiments can take us toward the characterization of the particles. Of course, additional data from other sources, such as the x-ray diffraction results just cited, may lead to still further specification of the system. One such source of information is intrinsic viscosity data for the same dispersion. In Chap. 4 we shall discuss the complementarity between viscosity data and sedimentation-diffusion results.

2.9 The Random Walk

A liquid that is totally homogeneous on a macroscopic scale undergoes continuous fluctuations at the molecular level. As a result of these fluctuations, the density of molecules at any location in the liquid varies with time and at any time varies with location in such a way that the mean density of the sample as a whole has its bulk value. This pattern of "flickering" molecular densities will produce continually varying pressures on the surface of any particle submerged in the liquid. Since the fluctuations are confined to domains of the order of molecular dimensions, this randomly variable pressure is quite small. A small particle will be displaced, however, by the resulting force unbalance at its surface. The pattern of its displacements will also be a totally random thing, a reflection of the fluctuations which cause the motion. As the size of the submerged particles increases, the effect of fluctuations on them decreases. In this section we shall consider the trajectory of a particle in the colloidal size range which is engaged in this pattern of motion. Such movements have been studied microscopically and are called Brownian motion after Robert Brown, who described them in 1828.

In general, a dispersed particle is free to move in all three dimensions. For the present, however, we are going to restrict our consideration to the motion of a particle undergoing random displacements in one dimension only. The model used to describe this motion is called a one-dimensional random walk. Its generalization to three dimensions is taken up in a later section.

To begin with, the statistical nature of this phenomenon should be apparent. We might watch the pattern of displacements of thousands of otherwise identical particles and find no uniformity in the zigzag steps they follow. Only statistical quantities such as the "average" displacement after a certain number of steps or after a certain elapsed time make any sense. Let us consider how to calculate such a quantity.

Suppose we consider a game in which a marker is moved back and forth along a line, say, the x axis, in a direction determined by the toss of a coin. The rules of the game provide that we move the marker one distance l in the plus direction every time "heads" is tossed, and one distance l in the negative direction every time "tails" is tossed. If n_+ and n_- represent the number of heads and tails, respectively, in a game consisting of a total of n tosses, then we may write

$$n_+ + n_- = n \qquad (42)$$

We may also write

$$x = (n_+ - n_-)l \qquad (43)$$

where x is the net displacement of the marker after n tosses. These two equations may be solved simultaneously for n_+ and n_- to give

$$n_+ = \frac{1}{2}\left(n + \frac{x}{l}\right) \qquad (44)$$

and

$$n_- = \frac{1}{2}\left(n - \frac{x}{l}\right) \qquad (45)$$

The problem of interest may now be expressed as follows: What is the probability $P(n, x)$ that the marker will be at position x after n moves? The answer to this problem is supplied by the well-known binomial distribution formula:

$$P(n, x) = \frac{n!}{n_+! n_-!} p_+^{n_+} p_-^{n_-} \qquad (46)$$

in which p_+ and p_- represent the probability of throwing a head or a tail, respectively, in a single toss. Although this formula is often used, its validity is not always fully appreciated. Accordingly, let us briefly examine the origin of Eq. (46).

Each time we toss an unbiased coin, the outcome is independent of all other tosses. The probability of tossing n_+ heads is therefore $p_+^{n_+}$, since the probabilities compound by multiplication when we require a specified set of outcomes. Therefore n_+ events, each of which occur with a probability p_+, will occur with a probability $p_+^{n_+}$. If we specify the outcome further by requiring n_- tails in addition to the heads already specified, then the probability is given by $p_+^{n_+} p_-^{n_-}$. What we have calculated is the probability of a particular, fully specified sequence of outcomes such as the following for $n = 6$: HHHTTT. However, the same net number of heads and tails could come about in other ways. For example, HTTTHH, HHTHTT, and HTHTHT are also consistent with $n = 6$ and $n_+ = n_- = 3$, along with the distribution previously given and other possibilities. Therefore we must multiply the probability of one specified sequence by the number of other sequences which have the same net composition of heads and tails.

At first glance this factor may appear to be $n!$, the number of different ways the n items can be rearranged, giving a probability of $n!p_+^n p_-^n$. This quantity overcounts, however, since not all rearrangements are recognizably different. For example, we can interchange the first two members of the sequence TTHHHT to produce the sequence TTHHHT; the interchange obviously leaves the sequence unaltered. Therefore we must divide $n!p_+^n p_-^n$ by the number of ways that identical members of the series can be interchanged without altering the sequence. Any of the n_+ heads (or the n_- tails) may be interchanged among themselves with this result. There are $n_+!$ permutations of the heads and $n_-!$ permutations of the tails. Therefore these are the factors by which the previous result overcounted. Dividing by these factors leaves us with the binomial equation. The student should be so familiar with this argument as to be able to justify each factor in Eq. (46).

Since the probability of tossing either a head or a tail is equal to $\frac{1}{2}$ for a fair coin, Eq. (46) may be rewritten

$$P(n, x) = \frac{n!}{[\frac{1}{2}(n + x/l)]![\frac{1}{2}(n - x/l)]!} \left(\frac{1}{2}\right)^n \tag{47}$$

by incorporating Eqs. (44) and (45). The length l by which the marker is moved in our hypothetical game is equivalent to the displacement of a particle in a single fluctuation. Because of the nature of the fluctuations underlying the whole process, these individual steps are very small. Observable diffusion is always the result of a very large number of such steps. For the case in which n is a large number, the factorials of Eq. (47) may be expanded according to Sterling's approximation:

$$\ln y! = y \ln y - y \tag{48}$$

which is valid for large values of y. Taking the logarithm of Eq. (47) and applying approximation (48) leads to the result

$$-\ln P = \frac{nl + x}{2l} \ln \left(1 + \frac{x}{nl}\right) + \frac{nl - x}{2l} \ln \left(1 - \frac{x}{nl}\right) \tag{49}$$

The *net* displacement x is always small compared to the total distance traveled nl, since a good deal of the back-and-forth motion cancels out. Therefore the logarithmic terms in Eq. (49) may be expanded as a power series (see Appendix A) in x/nl, with all terms higher than second order in x/nl regarded as negligible. This leads to the result

$$\ln P \simeq -\frac{x^2}{2nl^2} + \cdots \tag{50}$$

or

$$P(n, x) = k \exp\left(- \frac{x^2}{2nl^2} \right) \tag{51}$$

where the factor k represents a normalization constant.

When we discussed normalization in Sect. 1.12, we saw that a well-behaved probability function adds up to unity when the probabilities for all possible outcomes are totaled. Therefore, in order to evaluate the constant k in Eq. (51), we should integrate (51) over all possible values of x—which is equivalent to summing the probabilities—and set the result equal to unity. This leads to the expression

$$k = \left[\int_{-\infty}^{\infty} \exp\left(- \frac{x^2}{2nl^2} \right) dx \right]^{-1} \tag{52}$$

The integral is a gamma function, the value of which can be determined from the list of related integrals in Table 2.2. Evaluating Eq. (52) gives

$$k = (2\pi n l^2)^{-1/2} \tag{53}$$

Substitution of this result into Eq. (51) gives a continuous analytical expression for $P(n, x)$, which equals the binomial result for large values of n:

$$P(n, x) = (2\pi n l^2)^{-1/2} \exp\left(- \frac{x^2}{2nl^2} \right) \tag{54}$$

Strictly speaking, we should multiply both sides of Eq. (54) by dx. The equation now expresses the probability of a displacement between x and $x + dx$ after n random steps of length l.

2.10   Random Walk Statistics and the Diffusion Coefficient

The application of Eq. (54) to at least one set of boundary conditions for a diffusion problem is easy. We shall let a modification of Fig. 2.7a describe the system. This time, instead of having a solution on one side of the barrier and pure solvent on the other side, suppose we imagine that both sides of the barrier are filled with solvent. Furthermore, suppose that the solute under investigation is introduced into the system in the pores of

Table 2.2 Integral Table of Gamma Functions Encountered in This Chapter

$$\int_0^{\infty} x^n e^{-ax^2} \, dx =$$

$\frac{1}{2}\sqrt{\pi/a}$ if $n = 0$	$\frac{1}{2a}$ if $n = 1$
$\frac{1}{4a}\sqrt{\pi/a}$ if $n = 2$	$\frac{1}{2a^2}$ if $n = 3$
$\frac{3}{8a^2}\sqrt{\pi/a}$ if $n = 4$	$\frac{1}{a^3}$ if $n = 5$

the plug, assuming the latter to be infinitesimally thin. Thus at the beginning of the experiment all the material is present at $x = 0$, in the notation of the derivation above, at a concentration c_0. With the passage of time the material will gradually diffuse outward; the number of diffusion steps taken will be directly proportional to the elapsed time:

$$n = Kt \tag{55}$$

Equation (54) may be transformed into an expression that gives the probability as a function of x and t by substituting Eq. (55) into Eq. (54):

$$P(x, t)dx = (2\pi Ktl^2)^{-1/2} \exp\left(-\frac{x^2}{2Ktl^2}\right) dx \tag{56}$$

This suggests that the concentration as a function of x and t in the diffusion cell just described is given by multiplying c_0 by $P(x, t)$:

$$c(x, t)dx = c_0(2\pi Ktl^2)^{-1/2} \exp\left(-\frac{x^2}{2Ktl^2}\right) dx \tag{57}$$

Since $P(x, t)$ is normalized, its integral over all values of x equals unity. Likewise, integrating Eq. (57) over all values of x gives c_0: The same quantity of solute is present at all times whether it is concentrated at the origin or is spread out from $-x$ to $+x$. The function given by Eq. (57) is a solution to Fick's second law [Eq. (26)] for a one-dimensional problem in which all the material is present initially at $x = 0$ and at a concentration c_0.

The only ambiguity remaining at this point is the value of K in Eqs. (55)–(57). This constant is easily evaluated as follows. If Eq. (57) is indeed a solution to (26), then the right- and left-hand sides of the latter must be equal when the indicated differentiations are carried out. Differentiating Eq. (57), substituting into Eq. (26), and simplifying leads to the result

$$K = \frac{2D}{l^2} \tag{58}$$

Substituting this expression into Eq. (57) yields

$$\frac{c}{c_0} dx = (4\pi Dt)^{-1/2} \exp\left(-\frac{x^2}{4Dt}\right) dx \tag{59}$$

This relationship describes the diffusion of solute in the x direction when it is concentrated initially in an infinitesimally thin layer at the origin.

Equation (59) may be devloped somewhat further as follows. We define a parameter z to be

$$z = \frac{x}{(2Dt)^{1/2}} \tag{60}$$

then rewrite Eq. (59) in terms of this quantity:

$$\frac{c}{c_0} dx = \frac{1}{\sqrt{2\pi}} \exp\left(\frac{-z^2}{2}\right) dz = P(z)dz \tag{61}$$

Thus the concentration ratio c/c_0 is seen to be described at all times as a function of the single parameter z. The function $P(z)$ defined by Eq. (61) is the normal distribution function, Eq. (1.32).* The following example considers how the concentration profile of the diffusing species changes with time according to the normal distribution function.

Example 2.4 By consulting tables of the normal distribution function, draw curves which show the broadening of a band of material with time if the substance is initally at concentration c_0 and in a plug of infinitesimal thickness at $x = 0$. Assume the diffusion coefficient has the value $5 \times 10^{-11} \ m^2 \ s^{-1}$ for this material. Use $t = 10^6$ and $t = 3 \times 10^6$ s to see how the concentration profile changes with time.

---

*This manner of referencing is used to describe equations occurring in other chapters. The number to the left of the point is the chapter number.

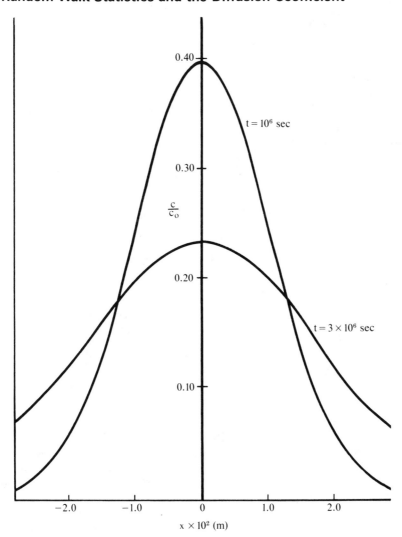

Figure 2.9 Variations of c/c_0 with distance from the origin at $t = 10^6$ and $t = 3 \times 10^6$ s if $D = 5 \times 10^{-11}$ m^2 s^{-1} (plot of data developed in Example 2.4).

Solution   Tables of the normal distribution function* list values of $P(z)$ for different values of z. As examples, we cite the following entries from the tables:

z	0.3	0.7	1.0	1.3	1.7	2.0	2.3	2.7
$P(z)$	0.3814	0.3123	0.2420	0.1714	0.0941	0.0540	0.0283	0.0104

Actual displacements x are obtained by multiplying z by $(2Dt)^{1/2}$; this means that the ordinate must be divided by this factor so that the area under the curve remains normalized:

For $t = 10^6$ s, $(2Dt)^{1/2} = [2(5 \times 10^{-11})(10^6)]^{1/2} = 10^{-2}$ m. Therefore $x = 10^{-2} z$.

x (m)	0.003	0.007	0.010	0.013	0.017	0.020	0.023	0.027
$P(x) \times 10^{-2}$	0.381	0.312	0.242	0.171	0.094	0.054	0.028	0.010

For $t = 3 \times 10^6$ s, $(2Dt)^{1/2} = 1.73 \times 10^{-2}$ m. Again, z is multiplied by this quantity to give x, and $P(z)$ is divided by the same factor to give the relative value of the ordinate, $P(x)$:

x (m)	0.0052	0.0121	0.0173	0.0225	0.0294	0.0346	...	...
$P(x) \times 10^{-2}$	0.221	0.181	0.140	0.099	0.054	0.031	...	...

These values are plotted in Fig. 2.8, where the factor 10^{-2} for $P(x)$ has been used as a scale factor for the ordinate to show the relative shape of the curves.

•

Figure 2.9 indicates a gradual approach toward equilibrium by the fact that the concentration tends to be more nearly uniform as time increases. In Chap. 1 the function z is defined to be δ/σ, where δ is the deviation of a particular value from the mean of a distribution and σ is the standard deviation of the distribution. Since the net displacement of a diffusing particle is analogous to δ, we may infer that the quantity $\sqrt{2Dt}$ is also analogous to the standard deviation. The point of this is the following. It is well known that the width of the normal error curve at its inflection point (where the second derivative changes sign, or the curve changes from concave to convex) is equal to σ. Therefore we can conclude that the width of the curves shown in Fig. 2.9, measured at their inflection points, increases in proportion to $\sqrt{t}$. This is one method whereby D could be measured in an experiment which corresponds to these boundary conditions.

Handbook of Tables for Mathematics, Chemical Rubber Company, Cleveland.

With the mathematics of the one-dimensional random walk as background, we may visualize the following experimental arrangement whereby D could be measured. Suppose that a narrow band of dispersion is layered between two portions of solvent in a long tube. We shall not worry about the practical difficulties in doing this, since, in practice, other initial conditions which are easily obtained are actually used. The narrow band we have pictured, however, approximates the initial state of the system described by Eq. (59). We might further imagine observing this band by means of a schlieren optical system (Sect. 2.5). The two edges of the band, where the refractive index gradient is large, would define the positively and negatively sloped branches of the schlieren peak. With the passage of time, the material in the band diffuses outward; the schlieren peak would also broaden. In other words, the schlieren pattern observed at successive times would generate a family of curves—for example, Fig. 2.5b—which very much resemble the theoretical curves of Fig. 2.9.

The curves in Fig. 2.8 are drawn for an arbitrary value of the diffusion coefficient. The experimental profiles produced by the schlieren optics are characterized by the diffusion coefficient of the experimental system. The remaining question is how to extract the appropriate D value from the experimental observations.

Remember that the normal distribution function has an inflection point (where the second derivative changes sign) at $z = 1.0$. Therefore the x value at which the inflection point occurs at any time equals $\sqrt{2Dt}$, according to Eq. (60). By locating the inflection point at different times during a diffusion experiment, the appropriate D value may be evaluated for the diffusing species. For example, on the $t = 10^6$ s curve in Fig. 2.8, the inflection point lies at $x = 0.010$ m. Substituting $x = 0.010$ m and $t = 10^6$ s when $z = 1$ into Eq. (60) enables us to calculate the value of the diffusion coefficient that was used ($D = 5 \times 10^{-11}$ m^2 s^{-1}) to draw the curves in Fig. 2.8. A similar analysis may be conducted on experimental curves obtained by the schlieren method.

As an experimental procedure, this method is less precise than others which have been developed for the evaluation of D. It does point out, however, the intimate connection between diffusion and the random events at the molecular level which cause it.

A more practical experimental method for the determination of D is based on Fig. 2.7b. The theoretical arrangement represented by Fig. 2.7b is implemented experimentally in an apparatus like that sketched in Fig. 2.10. One side of the sintered glass barrier contains solution; the other side contains pure solvent. The entire apparatus is thermostated and both compartments are magnetically stirred. Samples are withdrawn at various times and the quantity of material that has diffused into the

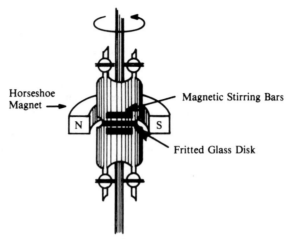

Figure 2.10 Laboratory apparatus equivalent to that in Fig. 2.7b. The entire apparatus is rotated between the poles of a magnet for stirring.

solvent compartment is measured. The primary difficulty with this procedure is the tendency of the pores to plug owing to the entrapment of air bubbles, clogging by solid particles, or adsorption of the diffusing molecules themselves. Nevertheless, what is obtained experimentally is a record of the approach toward a uniform distribution of material on both sides of the barrier.

The situation represented by the apparatus in Fig. 2.10 has also been analyzed theoretically. The function $c(x, t)$ which satisfies Fick's second law when there is a solution of concentration c_0 on one side of a boundary at $t = 0$ and pure solvent on the other side is given by

$$c = \frac{c_0}{2} \left[1 - 2 \int_0^z P(z)dz \right] = \frac{c_0}{2} [1 - 2 \, \text{Erf} \, (z)] \qquad (62)$$

We can verify the plausibility of this expressions as follows. Recall that z and t are inversely related and that the integral gives the area under one-half of the error curve, between its midpoint ($z = 0$) and some specified value of z. Therefore, at $t = 0$, z is infinite, the integral equals $\frac{1}{2}$, and $c = 0$. At $t = \infty$, z is zero, the integral equals zero, and $c = \frac{1}{2}c_0$. Thus Eq. (62) makes sense at either extreme. The detailed profile of c/c_0 versus z is obtained by reading values of the integral in Eq. (62) from tables which give the area under the normal error curve. The results of this procedure are plotted in Fig. 2.11, where c/c_0 is shown versus x at several different

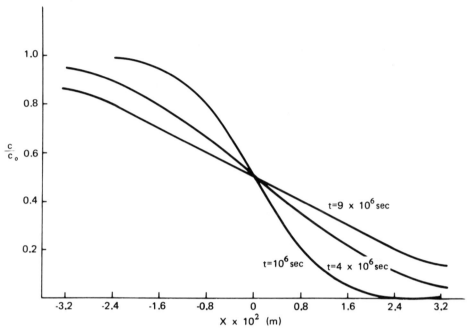

Figure 2.11   Diffusion of solute from solution into pure solvent. Here c/c_0 plotted versus distance at $t = 10^6$, 4×10^6, and 9×10^6 s for $D = 5 \times 10^{11}$ m^2 s^{-1}. The plot was drawn according to Eq. (62).

times. The approach toward a uniform distribution of material is evident.

As in the case of diffusion from an initially thin layer, experimental concentration data and theoretical concentration profiles may be compared. From this comparison, the value of D which is consistent with the observed diffusion behavior may be evaluated. The diffusion coefficient is a function of concentration; therefore it is measured at a series of different concentrations and extrapolated to zero concentration; the limiting value is given the symbol D^0. Diffusion coefficients are temperature dependent and it is sometimes necessary to adjust a value measured at one temperature to some other temperature. The following example considers how this is done.

Example 2.5   Suppose the diffusion coefficient of a material is measured in an experiment (subscript ex) at some temperature T_{ex} where the viscosity of the solvent is η_{ex}. Show how to correct the value of D to some standard (subscript s) conditions where the viscosity is η_s. Evaluate D_{20}^0 for a solute which displays a D^0 value of 4.76×10^{-11} m² s⁻¹ in water at 40°C. The viscosity of water at 20 and 40°C is 1.0050×10^{-2} and 0.6560×10^{-2} P, respectively.

Solution   By Eq. (32), D is proportional to T/f and, by Eq. (7), f is proportional to η. Therefore $D \propto T/\eta$ or

$$\frac{D_s}{D_{ex}} = \frac{T_s/\eta_s}{T_{ex}/\eta_{ex}} = \frac{T_s \eta_{ex}}{T_{ex} \eta_s}$$

Applying this relationship to the data provided gives

$$D_{20}^0 = \frac{293}{313}\frac{0.6560}{1.0050}(4.76 \times 10^{-11}) = 2.91 \times 10^{-11} \text{ m}^2 \text{ s}^{-1}$$

The temperature dependence of the viscosity is seen to have the largest effect on the temperature dependence of D.

•

2.11 "Average" Displacements from Random Walk Statistics

Having examined the connection between the phenomenological equations of diffusion and the statistics of the random walk, let us now return to the random walk as a model for Brownian motion. The problem we wish to consider is the "average" displacement of a marker after an n-step, one-dimensional random walk. The foregoing discussion already supplies the answer to this problem: We have seen that the probability function for the one-dimensional walk is symmetrical about the origin, implying a mean displacement of zero. This simply reflects the fact that on the average the number of heads and tails will be equal. While we accept this result, we feel that there is something unsatisfactory about it. The average displacement is zero because positive and negative displacements are equally probable and effectively cancel one another. It is not because the marker scarcely moves from its initial location, a conclusion suggested by the value of the mean displacement.

This shows that the mean displacement is simply not a useful parameter to describe the trajectory of the particle, and suggests we should seek an alternate quantity. Instead of averaging the displacements directly, suppose we first square them to eliminate the differences in sign, then average them and take the square root. This quantity, called the root-mean-square (rms) displacement, will give a better measure of the meanderings of the marker, since the sign differences have been eliminated.

The calculation of the rms displacement is quite simple. Any average is calculated by multiplying the quantity to be averaged by an appropriate probability function and then integrating the result over all possible values of the variable. This is a direct extension of Eq. (1.10) using a continuous function for the weighting factor. Applying this procedure to the problem at hand gives the result

$$\overline{x^2} = \int_{-\infty}^{\infty} x^2 P(n, x)\, dx \tag{63}$$

in which $P(n, x)$ is given by Eq. (54). Making this substitution gives

$$\overline{x^2} = (4\pi Dt)^{-1/2} \int_{-\infty}^{\infty} x^2 \exp\left(\frac{-x^2}{4Dt}\right) dx \tag{64}$$

the value of which is found in Table 2.2. Evaluating the integral leads to the conclusion that

$$\overline{x^2} = 2Dt \tag{65}$$

This important equation, also derived by Einstein, provides us with a means for measuring the diffusion coefficient for particles which are visible in a microscope. This is a particle size range for which measurement of D by following concentration changes with time is very difficult. Instead, the actual displacement of a particle in a time t is measured microscopically. The rms displacement, evaluated from a statistically meaningful number of observations, permits D to be calculated. J. Perrin (Nobel Prize, 1926) interpreted observations of Brownian motion in terms of Eqs. (7), (32), and (65) as a means of determining the first precise value of Avogadro's number.

Equation (65) also permits us to assign a physical interpretation to the diffusion coefficient in addition to the macroscopic meaning it has from Fick's laws. Rearranging and factoring in a way that admittedly ignores the averaging procedure, we write Eq. (65) as

$$D = \frac{1}{2}\left(\frac{x}{t}\right) x \tag{66}$$

Since D is a constant for a particular substance, the ratio x/t must vary inversely with x. The distance traveled by a diffusing particle divided by the time required for the displacement, x/t, gives the apparent diffusion velocity of the particle. Equation (66) shows that the diffusion velocity is inversely proportional to the length of path over which it is measured. This apparently paradoxical conclusion becomes less mysterious when we concentrate on the distinction between the diffusion velocity and the actual velocity of the particle as it travels along its zigzag path. The latter quantity reflects the average translational kinetic energy of the particle; its average value depends on the absolute temperature but is independent of the distance traveled. The diffusion velocity, on the other hand, decreases as the distance traveled increases. This simply means that large displacements are so much less probable than small displacements that they require disproportionately longer times to occur. Note that if $x = 1$ m, D has the significance of being equal to half the diffusion velocity measured over that distance. This is a result that was anticipated at the end of Sect. 2.7.

2.12   The Random Coil and Random Walk Statistics: A Digression

The dimensions of a randomly coiled polymer molecule are a topic which appears to bear no relationship to diffusion; however, both the coil dimensions and diffusion can be analyzed in terms of random walk statistics. Therefore we may take advantage of the statistical argument we have developed to consider this problem.

Suppose we visualize a polymer molecule as consisting of n segments of length l connected by a completely flexible linkage at each joint. Imagine, furthermore, that the placement of each successive segment is determined by some sort of purely random criterion. We could anchor one end of the chain to the origin of a hypothetical coordinate system, for example, and position successive segments as follows. Suppose we toss a single die and agree to orient the next segment in the $+x$ direction if the die shows a 1, and in the $-x$ direction if a 6 shows. Likewise, 2, 5, 3, and 4 correspond to $+y$, $-y$, $+z$, and $-z$, respectively. Our problem is this: *On the average*, what will be the end-to-end distance $\bar{R}$ for a chain of n segments? To calculate this, we must assume that n is very large and we must ignore the sites excluded by previously positioned segments.

We recognize first that the end-to-end distance can be resolved into x, y, and z components which obey the relationship

$$R^2 = x^2 + y^2 + z^2 \qquad (67)$$

The probability that the loose end of the chain is in a volume element located at x, y, and z is given by

$$P(x, y, z)dx\,dy\,dz = P(x)P(y)P(z)dx\,dy\,dz \qquad (68)$$

(a)   (b)

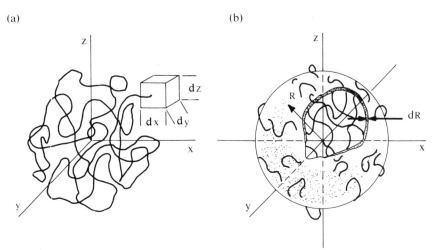

Figure 2.12 Coordinate systems in which one end of a flexible random coil lies at the origin and the other end is (a) in a volume element $dx\,dy\,dz$ and (b) in a spherical volume element $4\pi R^2\,dR$.

where $P(x)$ is the probability that the x coordinate has a value between x and $x + dx$, with similar definitions for $P(y)$ and $P(z)$. Equation (54) gives the expression for $P(x)$, except we must remember that now only $n/3$ segments will be aligned with the x axis, since each of the directions is independent and equally probable. Since the x, y, and z directions are equivalent, the same expression holds for $P(y)$ and $P(z)$ with the appropriate change of variables. Therefore, incorporating Eq. (54), Eq. (68) becomes

$$P(x,\,y,\,z)dx\,dy\,dz = \left(2\pi\,\frac{n}{3}\,l^2\right)^{-3/2} \exp\left(\frac{-3R^2}{2nl^2}\right)dx\,dy\,dz \qquad (69)$$

This expression gives the probability that the loose end of the molecule is in a volume element located at some particular values of x, y, and z, as shown in Fig. 2.12a. Our interest is not in any *specific* x, y, z coordinates, but in all combinations of x, y, z coordinates which result in the end of the chain being a distance R from the origin. This can be evaluated by changing the volume element in Eq. (69) to spherical coordinates and then integrating over all angles. This amounts to replacing the volume element in Eq. (69) with the volume of a spherical shell of radius R and thickness dR:

$$dx\,dy\,dz \rightarrow 4\pi R^2\,dR \qquad (70)$$

A geometrical representation of this situation is shown in Fig. 2.12b. Incorporating this expression into Eq. (69) gives the probability that the loose end of the molecule lies between R and $R + dR$, irrespective of the direction:

$$P(R)dR = 4\pi \left(2\pi \frac{n}{3} l^2 \right)^{-3/2} R^2 \exp \left(\frac{-3R^2}{2nl^2} \right) dR \tag{71}$$

With this distribution function, it is an easy matter to calculate average quantities. Again, since positive and negative displacements are equally probable, we shall evaluate the average of R^2. Following the same procedure as used in Eq. (63), we write

$$\overline{R^2} = \left(\frac{2\pi nl^2}{3} \right)^{-3/2} 4\pi \int_0^\infty R^4 \exp \left(\frac{-3R^2}{2nl^2} \right) dR \tag{72}$$

Evaluation of this integral (see Appendix B) leads to the result

$$\overline{R^2} = nl^2 \tag{73}$$

which is the desired quantity.

Equation (73) shows that the rms end-to-end distance in a polymer coil equals the square root of the number of steps times the length of each step. We might ask, therefore, what is the physical significance of n and l for a polymer molecule? A general formula for a typical vinyl polymer is

$$\left(\begin{array}{cc} H & H \\ | & | \\ -C-C- \\ | & | \\ H & X \end{array} \right)_n$$

where n is called the degree of polymerization. Since this quantity measures the number of repeated segments in the chain, it seems reasonable to equate this with the number of steps in the three-dimensional random walk. If M represents the molecular weight of the polymer and M_0 is the molecular weight of the vinyl monomer, then M/M_0 equals n. Since M_0 and l are constants, we see that the radius of the coil is predicted to be proportional to $M^{1/2}$.

The physical significance of l in Eq. (73) is somewhat harder to define. At first glance it appears to be the length of the repeating unit, about 0.25

nm for a vinyl polymer. We must remember, however, that the derivation of Eq. (73) assumed that the coil was connected by completely flexible joints. Molecular segments are attached at definite bond angles, however, so an actual molecule has less flexibility than the model assumes. Any restriction on the flexibility of a joint will lead to an increase in the dimensions of the coil. The effect of fixed bond angles on the dimensions of the chain may be incorporated into the model as follows.

A 360° rotation around any carbon–carbon bond in a vinyl chain will cause the *next* bond in the chain to trace out a cone with one carbon atom at the apex and the other carbon atom along the rim of the cone. Ignoring hindered rotation for the moment, we see that each position on the rim of such a cone is an equally probable site for the apex of the cone generated by the next bond in the chain. This situation is illustrated schematically in Fig. 2.13. This effect has been shown to increase the actual length of the repeating unit, l_0, by the factor

$$l^2 = l_0^2 \left(\frac{1 - \cos \theta}{1 + \cos \theta} \right) \tag{74}$$

where θ is the bond angle. For a tetrahedral bond angle, $\cos \theta = -\frac{1}{3}$, so the additional factor in Eq. (74) equals 2.

In Fig. 2.13 it is implied that the terminal carbon atom can occupy any position on the rim of the cone; that is, there is assumed to be perfectly free rotation around the penultimate carbon–carbon bond. This is equivalent to saying that all values of ϕ, the angle that describes the rotation (see Fig. 2.13), are equally probable. Any hindrance to free rotation will block certain configurations, expanding the coil dimensions still further.

Finally, interactions between the polymer segments and the solvent molecules may introduce a bias which tends to position segments as close or as far as possible from other segments, depending on the nature of the interaction. As a matter of fact, those properties of polymer solutions which are sensitive to the dimensions of the molecules, such as viscosity, vary widely with solvent "goodness." A good solvent may be defined as one in which polymer–solvent contacts are favored; a poor solvent is one in which polymer–polymer contacts are favored. Therefore the value of l in Eq. (74) will be increased in a good solvent and decreased in a poor solvent. In discussing size exclusion chromatography in Sect. 1.10 we anticipated this conclusion by noting that polymer–solvent interactions complicate the relationship between M and R for a random coil.

The statistical relationships of this section can also be applied to Eq.

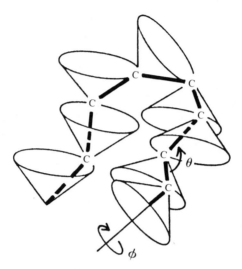

Figure 2.13 The effect of fixed bond angle in restricting the flexibility of a polymer coil. For tetrahedral bonds $\theta = 109°$.

(1.21), which defines the radius of gyration. If the masses in that equation are identified with the mass of the repeat units of the polymer m_0—of which there are n in the molecule—then Eq. (1.21) can be written

$$R_g^2 = \frac{m_0 \Sigma_i r_i^2}{n m_0} = \frac{1}{n} \Sigma_i r_i^2 \tag{75}$$

Next, the summation over discrete values of r_i^2 is replaced by the integral of $r^2 P(r) dr$. This sounds exactly like the procedure that results in Eq. (72), but there are some differences. To evaluate the radius of gyration, we consider the masses relative to the center of mass of the swarm. If an arbitrary segment, say, segment j, is considered to be the center of mass, then r in Eq. (75) is measured outward in both directions from j. Using k to index the segments outward from j gives the following expression for the radius of gyration relative to segment j:

$$(R_g^2)_j = \frac{1}{n} \left(\sum_{k=1}^{j} P(k, r) + \sum_{k=1}^{n-j} P(k, r) \right) r^2 \tag{76}$$

Finally, this latter quantity must be added up (i.e., integrated) for all values of j between 1 and n, since any one of the segments might lie at the center of mass of the coil. The latter summation must also be divided by n, since we are effectively determining the average value for j. When all of these steps are carried out—using Eq. (69) as the expression for $P(r)$—the radius of gyration for the random coil may be shown to equal

$$R_g^2 = \tfrac{1}{6} nl^2 \tag{77}$$

which is simply one-sixth the value of R^2 as given by Eq. (73). Since the radius of gyration can be measured by viscosity and light scattering for random coils, it is apparent that Eq. (77) may be used to measure l. In this way the effects of hindered rotation around bonds and of imbibed solvent may be evaluated quantitatively.

In concluding this section, we note that all the statistical equations of this chapter could have been borrowed directly from kinetic molecular theory by simply changing the variables. We illustrate this now by going in the opposite direction. For example, if we replace the quantity $3/nl^2$ by m/kT and replace R by v in Eq. (69), we obtain the Boltzmann distribution of molecular velocities in three dimensions. If we make the same substitutions in Eq. (73), we obtain an important result from kinetic molecular theory:

$$\overline{v^2} = \frac{3kT}{m} \tag{78}$$

This shows that Eq. (73) occupies a position for the random coil which is analogous to the position of average kinetic energy in the kinetic molecular theory. This is not just a fortuitous similarity, but a reflection of the statistical basis of both. A little self-test: Did you recognize the similarity between the formalisms of the last few sections and kinetic theory as we went along?

2.13 Equilibrium Between Sedimentation and Diffusion

We have already noted that sedimentation and diffusion are opposing processes, the one tending to collect and the other to scatter. Let us now consider the circumstances under which these two tendencies equal each other. Once this condition is reached, of course, there will be no further macroscopic changes; the system is at equilibrium. In order to formulate this problem, consider the unit cross section shown in Fig. 2.14, in which the x direction is in the direction of either a gravitational or a centrifugal field. Suppose this field tends to pull the particles in the $-x$ direction. Gradually the concentration of the particles will increase in the region below the cross section of interest.

Back-diffusion occurs at a rate which increases with the buildup of a concentration gradient. When equilibrium is finally reached, we may write

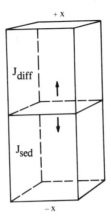

Figure 2.14 The relationship between the flux due to sedimentation and that due to diffusion. At equilibrium the two are equal.

$$J_{sed} = -J_{diff} \tag{79}$$

where J_{sed} is the flux across the area due to sedimentation and J_{diff} is the flux due to diffusion. The latter is given by Eq. (22) and the former by

$$J_{sed} = vc \tag{80}$$

in which v is the sedimentation velocity and c is the concentration at the plane. Substituting Eqs. (22) and (80) into (79) gives

$$vc = D \frac{dc}{dx} \tag{81}$$

If we substitute the value for the rate of sedimentation under gravity, Eq. (4), we obtain

$$\frac{m}{f} \left(1 - \frac{\rho_1}{\rho_2} \right) gc = D \frac{dc}{dx} \tag{82}$$

If the sedimentation occurs in a centrifugal field, on the other hand, g must be replaced by $\omega^2 x$ in Eq. (4):

$$\frac{m}{f} \left(1 - \frac{\rho_1}{\rho_2} \right) \omega^2 xc = D \frac{dc}{dx} \tag{83}$$

Equations (82) and (83) are easily integrated to produce expressions which give c as a function of x at equilibrium. Defining c_1 and c_2 to be the equilibrium concentrations at x_1 and x_2 and then integrating, we obtain for Eq. (82)

$$\frac{m}{fD}\left(1 - \frac{\rho_1}{\rho_2}\right) g(x_2 - x_1) = \ln\left(\frac{c_2}{c_1}\right) \tag{84}$$

and for (83)

$$\frac{m}{2fD}\left(1 - \frac{\rho_1}{\rho_2}\right) \omega^2(x_2^2 - x_1^2) = \ln\left(\frac{c_2}{c_1}\right) \tag{85}$$

Finally, we recall Eq. (32), which permits us to substitute kT for Df, giving

$$\ln\left(\frac{c_2}{c_1}\right) = \frac{m}{kT}\left(1 - \frac{\rho_1}{\rho_2}\right) g(x_2 - x_1) \tag{86}$$

for sedimentation equilibrium under gravity and

$$\ln\left(\frac{c_2}{c_1}\right) = \frac{m}{2kT}\left(1 - \frac{\rho_1}{\rho_2}\right) \omega^2(x_2^2 - x_1^2) \tag{87}$$

for sedimentation equilibrium in a centrifuge. Note that sedimentation equilibrium studies permit the evaluation of particle mass with no assumptions about particle shape.

It should now be apparent why the ultracentrifuge is such an important tool in molecular biology. Although the method is in no way restricted to particles of biological significance, these particles are of a size and density which are ideally suited to the ultracentrifuge. An ultracentrifuge permits the evaluation of particle mass through equilibrium studies [Eq. (87)] and the evaluation of the ratio m/f through studies of the rate of sedimentation [Eq. (18)]. Combining these data permits the separate evaluation of f. From the mass and density of the material, the volume and radius of an equivalent sphere and hence f_0 can be calculated. Then Fig. 2.6 may be consulted to determine particle characterizations which are consistent with the ratio f/f_0. Although both utilize the same instrument, sedimentation rate and sedimentation equilibrium are two different experiments which complement one another very nicely.

It might be noted that sedimentation equilibrium is approached very slowly; however, techniques which permit equilibrium conditions to be estimated from pre-equilibrium measurements have been developed by W. J. Archibald. Equations (86) and (87) predict a linear semilogarithmic plot of c versus x or x^2, for gravitational and centrifugal studies, respectively. The slope of such a plot is proportional to the mass of the particles involved. Remember that monodispersity was assumed in the derivation of these equations. If this condition is not met for an experimental system, the plot just described will not be linear. If each particle size present is at equilibrium, however, each component will follow the equations and the experimental plot will be the summation of several straight lines. Under certain conditions these may be resolved to give information about the polydispersity of the system. In any event, nonlinearity implies polydispersity once true equilibrium is reached.

We conclude this chapter with a final observation about Eq. (86). If the particles in question are gas molecules instead of suspended particles, then the concentration ratio equals the ratio of pressures measured at two locations and the particle mass requires no correction for buoyancy. Under these conditions, Eq. (86) becomes

$$\ln\left(\frac{p_2}{p_1}\right) = \frac{mg(x_2 - x_1)}{kT} = \frac{Mg(x_2 - x_1)}{RT} \tag{88}$$

This familiar equation gives the variation of barometric pressure with elevation. Once again we are reminded of the connection between the material of this chapter and kinetic molecular theory.

References

1. R. B. Bird, W. E. Stewart, and E. N. Lightfoot, *Transport Phenomena*, Wiley, New York, 1960.
2. E. J. Cohn and J. T. Edsall, *Proteins, Amino Acids and Peptides*, American Chemical Society Monograph, reprinted by Hafner, New York, 1965.
3. J. M. Dallavalle, *Micromeritics*, Pitman, New York, 1948.
4. A. Einstein, *Investigations in the Theory of the Brownian Movement*, Dover, New York, 1956.
5. K. J. Mysels, *Introduction to Colloid Chemistry*, Wiley, New York, 1959.
6. T. Svedberg and K. O. Pederson, *The Ultracentrifuge*, Oxford University Press, London, 1940.

Problems

1. The following results describe the rate of accumulation of rutile (TiO_2; $\rho = 4.2$ g cm^{-3}) particles on a submerged balance pan.*

---

*W. F. Sullivan and A. E. Jacobson, *Symposium on Particle Size Measurement*, ASTM Publication No. 234, 1959.

Time t (min)	W (g)	Time t (min)	W (g)
0	0	35	0.310
1.5	0.045	60	0.403
2.0	0.060	84	0.480
2.5	0.075	127	0.610
3.0	0.090	159	0.702
4.0	0.110	226	0.880
5.5	0.125	274	0.998
8.0	0.160	327	1.110
14	0.210	384	1.240
24	0.260	420	1.320

(a) Plot W versus t and use this to evaluate w as a function of time. Prepare a plot of w versus t.

(b) At what time (t_{max}) does the greatest increment in w appear to occur?

(c) Calculate the radius of an equivalent sphere corresponding to t_{max} ($\rho_{soln} = 0.997$ g cm^{-3}, $\eta = 0.00894$ P, $h = 12$ cm).

2. A preparation of reduced and carboxymethylated Mouse-Elberfeld virus protein particles in water reached sedimentation equilibrium at 25°C after 40 h at 12,590 rpm.* The following data show the recorder displacement (proportional to concentration) versus r for this protein:

c (arbitrary units)	2.29	2.51	2.79	3.09	3.51	3.89	4.47	5.01	5.89	6.61	7.41	8.51
r (cm)	6.55	6.58	6.60	6.65	6.67	6.69	6.71	6.74	6.76	6.79	6.81	6.84

(a) Use these results to evaluate the mass of the particles present ($\rho_{protein} = 1.370$ g cm^{-3}) and estimate R_0 and f_0 for the molecules. Does the sample appear to be monodisperse?

(b) The sedimentation coefficient is known to be 2.7S for this preparation. Evaluate f and f/f_0.

(c) What can be said about the possible axial ratio–hydration combinations of this protein in terms of Fig. 2.8.

3. Southern bean mosaic virus (SBMV) particles are centrifuged at 12,590 rpm and the absorbance at 260 nm is measured along the settling direction as a function of time. The center of the absorption band varies with distance from the center of the rotor as follows†:

---

*R. R. Rueckert, *Virology,* 26: 345 (1965).

†J. Vinograd, R. Bruner, R. Kent, and J. Weigle, *Proc. Nat. Acad. Sci., U.S.A.,* 49: 902 (1963).

$t(\text{min})$	$R(\text{cm})$
16	6.22
32	6.32
48	6.42
64	6.52
80	6.62
96	6.72
112	6.82
128	6.92
144	7.02

Calculate the sedimentation coefficient of SBMV particles from these results.

4. Phosphatidylcholine micelles are spherical particles having a molecular weight of 97,000 g mole^{-1}. Assuming that the density of the dry lipid ($\rho = 1.018$ g cm^{-3}) applies to the micelles, calculate R and D for these particles in water at 20°C. The experimental value of the diffusion coefficient is 6.547×10^{-7} cm^2 s^{-1} under these conditions*. Evaluate f/f^* and estimate the hydration of the lipid.

5. Calculate the diameter of a spherical particle ($\rho = 4$) for which the rms displacement due to diffusion at 25°C is 1% the distance of sedimentation in a 24-h period through a medium for which $\rho = 1$ g cm^{-3} and $\eta = 9 \times 10^{-3}$ P. For what diameter is the diffusion distance 10% of the settling distance?

6. The diffusion of alkyl ammonium ions into clay pellets has been studied by bringing the pellet into superficial contact with an isotopically labeled salt, and then, after a suitable time, using a microtome to slice the pellet. The radioactivity is then measured in successive thin slices of the pellet. Assuming that Eq. (62) describes the diffusion process, estimate how long it would take for 1% of the initial activity of each of the ions to appear in the 15th slice inward from the exposed surface of the dry clay if each slice is 40 μm thick. The diffusion coefficients for the methyl and trimethyl ammonium cations under these conditions are 7.03×10^{-12} and 2.65×10^{-11} cm^2 s^{-1}, respectively.†

7. Suppose two reservoirs 4 cm apart are cut into an agar gel in a Petri dish. Solutions of Pb(NO$_3$)$_2$ and Na$_2$CrO$_4$ are introduced simultaneously into the two reservoirs. At what distance into the gel does PbCrO$_4$ precipitate if the diffusion coefficients of Pb^{2+} and CrO$_4^{2-}$ in agar are 0.657×10^{-5} and 0.752×10^{-5} cm^2 s^{-1}, respectively?‡ Where would Prussian blue precipitate if the reservoirs contained Fe^{3+} ($D = 0.434 \times 10^{-5}$ cm^2 s^{-1}) and Fe(CN)$_6^{4-}$ ($D = 0.557 \times 10^{-5}$ cm^2 s^{-1})?

---

*H. Hasser, in *Water, a Comprehensive Treatise*, Vol. 4 (F. Franks, ed.), Plenum, New York, 1975, Chapter 4.
†R. G. Gast and M. M. Morfland, *J. Colloid Interface Sci.*, 37:80 (1971).
‡R. E. Lee and F. R. Meeks, *J. Colloid Interface Sci.*, 35:584 (1971).

8. Use a compass to inscribe a circle around the floc in Fig. 1.11 such as a graticule might be used to define an average diameter for the floc. Compare the radius of the inscribed circle to that of the primary particles in the floc. How does the ratio of these radii compare with the value that would be obtained for the case $n = 76$ if the floc were built up according to a random walk model? Compare the features of the two models with the objective of accounting for the relative magnitude of the floc radius relative to the primary particle radius in the two cases.

9. The molecular weights and sedimentation coefficients of human plasminogen and plasmin ($\rho = 1.40$ g cm^{-3}) are as follows[*]:

	Plasminogen	Plasmin
M (g mole^{-1})	81,000	75,400
s at 20°C (S)	4.2	3.9

(a) Calculate the diffusion coefficient for each.
(b) Prepare a graph such as that of Fig. 2.9 showing quantitatively how an initially thin band of these proteins widens with time. Show at least three different times.

10. Colloids (casein micelles) of two different particle sizes are isolated from skim milk by centrifugation under different conditions. The sedimentation and diffusion coefficients of the two preparations are as follows[†]:

Preparation conditions	$D \times 10^8$ (cm^2 s^{-1})	$s \times 10^{13}$ (s)
5 min at 5,000 rpm	0.97	2200
40 min at 20,000 rpm	2.82	800

Calculate the mass per particle and the gram molecular weight of the two micelle fractions, assuming $\rho = 1.43$ g cm^{-3} for the dispersed phase.

11. The following data give the number of gold particles (as $\log_{10}$) versus depth beneath the surface for an aqueous dispersion allowed to reach sedimentation equilibrium under the influence of gravity[‡]:

Depth (mm)	4.44	5.06	5.67	6.30	6.90	7.53	8.15	8.65
$\log n$	10.36	10.51	10.63	10.75	10.89	11.05	11.22	11.39

Calculate the radius of the gold particles ($\rho_{Au} = 19.3$ g cm^{-3}), treating the dispersed units as equivalent spheres.

12. At equilibrium at 20°C the concentration of tobacco mosaic virus (TMV) shows a linear semilogarithmic graph when plotted against the square of the distance to the axis of rotation in a centrifuge. Evaluate the molecular weight

[*]G. H. Barlow, L. Summaria, and K. C. Robbins, *J. Biol. Chem.,* *244*:1138 (1969).
[†]C. V. Morr, S. H. C. Lin, R. K. Dewan, and V. A. Bloomfield, *J. Dairy Sci.,* *56*:415 (1973).
[‡]C. M. McDowell and F. L. Usher, *Proc. R. Soc. London,* *138A*:133 (1932).

of the TMV particles if $\ln c = -1.7$ at $r^2 = 44.0$ cm^2 and $c = -2.8$ at $r^2 = 41.2$ cm^2 when the dispersion is centrifuged at 6.185 rps. The density of the TMV is 1.36 g cm^{-3}.*

13. Verify that the expansion (see Appendix A) of Eq. (49) leads to Eq. (50). Retain no terms higher than second order in expansions. Verify that the integration (see Table 2.2) of Eqs. (64) and (72) leads to Eqs. (65) and (73), respectively.

---

*F. N. Weber, Jr., R. M. Elton, H. G. Kim, R. D. Rose, R. L. Steere, and D. W. Kupke, *Science,* *140*:1090 (1963).

3

SOLUTION THERMODYNAMICS
Osmotic and Donnan Equilibria

All faults or defects, . . . Pantocyclus attributed to some deviation from perfect Regularity in the bodily figure, caused perhaps . . . by some collision in a crowd; by neglect to take exercise, or by taking too much of it; or even by a sudden change of temperature.

[From Abbott's *Flatland*]

3.1 Introduction

Our concern in this chapter is the equilibrium thermodynamics of solutions of colloidal solutes. This statement makes it clear that our subject matter concerns lyophilic colloids, since finely divided two-phase systems are thermodynamically unstable. Osmotic pressure is the experimental quantity upon which our discussion centers, since this can be measured accurately for colloidal solutes. One molecular parameter of interest that is readily determined by osmometry is the number average molecular weight of the solute. Molecular weights determined by osmometry are absolute values: No calibration with known standards or any assumed theoretical models are required. Even the assumption of solution ideality—which is involved—is not a problem, since results are extrapolated to zero solute concentration before calculations are made.

The nonideality of colloidal solutions can be appreciable, since the solvent and solute particles are so different in size. Pure thermodynamics allows this nonideality to be quantified in terms of the so-called virial coefficients. If we turn from phenomenological thermodynamics to statistical thermodynamics, then we can interpret the second virial coefficient in terms of molecular parameters via a model. We shall pursue this approach for two different models: the excluded volume model for solute molecules with rigid structures and the Flory–Huggins model for polymer chains.

We conclude the chapter with a discussion of the thermodynamic behavior of charged colloids, particularly with respect to osmotic pressure and molecular weight determination. Although this is the first place in this book that we have devoted any attention to charged particles, it is not the last. Chapters 12 and 13, in particular, devote a good deal of attention to such systems.

3.2 The Thermodynamics of Osmotic Pressure

An extremely useful quantity in the thermodynamic treatment of multicomponent phase equilibria is the chemical potential. The chemical potential for component i, μ_i, is the partial molal Gibbs free energy with respect to component i at constant pressure and temperature:

$$\mu_i = \left(\frac{\partial G}{\partial n_i} \right)_{p, T, n_j \pm i} \tag{1}$$

Although the notation of this mathematical definition of chemical potential is somewhat cumbersome, the physical significance is fairly clear. The chemical potential is the coefficient which describes the way the Gibbs free energy of a system changes per mole of component i if the temperature, pressure, and number of moles of all components other than the ith are held constant. Although it is expressed on a molar basis, it is important to note that μ_i is a differential quantity; that is, it represents the local slope of the line which shows the variation of G with n_i. The line itself arises from slicing across the complex surface which describes G at the specified values of p, T, and n_j. For a pure substance, μ_i is identical to the Gibbs free energy per mole of that substance. For the present we shall regard the pure substance as the standard state for μ_i and shall represent it by the symbol μ_i°.

The great utility of the chemical potential in phase equilibrium problems arises in the following way. In open systems where the number of moles of any component may increase or decrease, any change in the Gibbs free energy of the system as a whole may be expressed as the sum of the following contributions:

$$dG = \left(\frac{\partial G}{\partial T} \right)_{p,n} dT + \left(\frac{\partial G}{\partial p} \right)_{T,n} dp + \sum_i \left(\frac{\partial G}{\partial n_i} \right)_{p,T,n_j} dn_i \tag{2}$$

This equation may be written

$$dG = -S \, dT + V \, dp + \sum_i \mu_i dn_i \tag{3}$$

by substituting into Eq. (2) the definition of chemical potential and the familiar relationships

$$\left(\frac{\partial G}{\partial T}\right)_{p,n} = -S \tag{4}$$

and

$$\left(\frac{\partial G}{\partial p}\right)_{T,n} = V \tag{5}$$

For any equilibrium which occurs at constant temperature and pressure the first two terms from the right-hand side of Eq. (3) equal zero, reducing the equation to the form

$$dG = \sum_i \mu_i dn_i \qquad \textit{Gibbs Duhem.} \tag{6}$$

If the total system consists of several different phases, which we shall designate by Greek subscripts $\alpha, \beta, \ldots$, then we may also write

$$dG = dG_\alpha + dG_\beta + \cdots \tag{7}$$

Finally, the equilibrium condition requires that

$$dG = 0 \tag{8}$$

Therefore the equilibrium between two phases means that

$$dG = 0 = dG_\alpha + dG_\beta = \sum_i \mu_{i\alpha} dn_{i\alpha} + \sum_i \mu_{i\beta} dn_{i\beta} \tag{9}$$

That is, for each of the components the following holds:

$$\mu_{i\alpha} dn_{i\alpha} + \mu_{i\beta} dn_{i\beta} = 0 \tag{10}$$

If the entire system consists of only two phases, the hypothesis here, the conservation of matter requires that any substance lost from one phase must appear in the other:

$$dn_{i\alpha} = -dn_{i\beta} \tag{11}$$

Combining these last two results leads to the conclusion that the condition for phase equilibrium at constant temperature and pressure is

$$\mu_{i\alpha} = \mu_{i\beta} \tag{12}$$

It is important to realize that the chemical potential of each component must be the same in all equilibrium phases, although the value for this quantity will, in general, be different for each component.

Perhaps the most basic equation involving the chemical potential is the one which relates this quantity to the activity of component i in the solution, a_i:

$$\mu_i = \mu_i^0 + RT \ln a_i \tag{13}$$

For the present a_i is expressed in mole fraction units. We see, therefore, that μ_i approaches μ_i^0 as a_i approaches unity. Furthermore, since μ_i is the partial molal Gibbs free energy, Eq. (5) also applies to μ_i, provided we replace V with $\bar{V}_i$, the partial molal volume of component i:

$$\left(\frac{\partial \mu_i}{\partial p} \right)_{T,n} = \bar{V}_i \tag{14}$$

Suppose we apply these relationships to the equilibrium of a liquid mixture and its vapor. At equilibrium μ_i must have the same value for each component in both the liquid and vapor phase. Therefore

$$\mu_{i,L} = \mu_{i,V} \tag{15}$$

or

$$\left(\frac{\partial \mu_{i,L}}{\partial p} \right)_{T,n} = \left(\frac{\partial \mu_{i,V}}{\partial p} \right)_{T,n} = \bar{V}_i = \frac{RT}{p_i} \tag{16}$$

where the last relationship assumes the vapor to behave ideally. If Eq. (13) is used to evaluate the left-hand side of this equation, we obtain

$$RT \, \partial \ln a_i = RT \frac{\partial p_i}{p_i} \tag{17}$$

Recalling that $p_i = p_i^0$, the normal vapor pressure of the pure liquid, when $a_i = 1$, we may integrate Eq. (17) to give

$$a_i = \frac{p_i}{p_i^0} \tag{18}$$

This relationship constitutes the basic definition of the activity. If the solution behaves ideally, $a_i = x_i$ and Eq. (18) defines Raoult's law. Those four solution properties that we know as the colligative properties are all based on Eq. (12); in each, solvent in solution is in equilibrium with pure solvent in another phase and has the same chemical potential in both phases. This can be solvent vapor in equilibrium with solvent in solution (as in vapor pressure lowering and boiling point elevation) or solvent in solution in equilibrium with pure, solid solvent (as in freezing point depression). Equation (12) also applies to osmotic equilibrium as shown in Fig. 3.1.

Figure 3.1 shows schematically two liquid phases—one solution, the other pure solvent—separated by a partition known as a semipermeable membrane. This membrane is in many ways the central feature of osmometry; we shall have more to say about it, but for now it is sufficient to define a semipermeable membrane as one which is permeable to the solvent and impermeable to the solute. Thus the semipermeable membrane in Fig. 3.1 prevents the two liquids from mixing and at the same time allows both sides of the membrane to come to equilibrium.

In order for solvent and solution to be in equilibrium in an apparatus such as that shown in Fig. 3.1, the solution side must be at a higher pressure than the solvent side. This excess pressure is what is known as the osmotic pressure of the solution. If no external pressure difference is imposed, solvent will diffuse across the membrane until an equilibrium hydrostatic pressure head has developed on the solution side. In order to prevent too much dilution of the solution as a result of the solvent flow into it, the column in which the pressure head develops is generally of very narrow diameter. We shall return to the details of osmotic pressure experiments in the next section. First, however, the theoretical connection between this pressure and the concentration of the solution must be established.

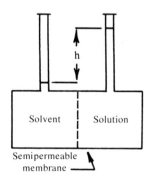

Figure 3.1   Schematic of an osmotic pressure experiment.

Since the two sides of the membrane are in true isothermal equilibrium in an osmotic pressure experiment, the chemical potential of the solvent must be the same on both sides of the membrane. On the side containing pure solvent, μ_1 equals μ_1^0. On the solution side of the membrane, the chemical potential of the solvent must also equal the same value, according to the equilibrium criterion of Eq. (12).

Equation (13) reminds us that the chemical potential has its greatest value for a pure substance, μ_i^0. Any value of a_i less than unity will cause μ_i to be altered from μ_i^0 by an amount $RT \ln a_i$, which will be negative for $a_i < 1$. Second, any pressure on a liquid which exceeds p_i^0 increases μ_i above μ_i^0. This is seen from the combination of Eqs. (13) and (18). Thus consideration of the chemical potential of the solvent makes it clear how osmotic equilibrium comes about. The presence of a solute lowers the chemical potential of the solvent. This is offset by a positive pressure on the solution, the osmotic pressure π, so that the net chemical potential on the solution side of the membrane equals that of the pure solvent on the other side of the membrane. This is summarized by the expression

$$\mu_1^0 = \mu_1^0 + RT \ln a_i + \int_{p_1^0}^{p_1^0 + \pi} \bar{V}_1 dp \tag{19}$$

in which subscript 1 indicates the solvent. We shall use the subscripts 1 and 2 to indicate solvent and solute, respectively, throughout this chapter. In order to relate π to the concentration of the solution, then, we must find a way to integrate Eq. (19). The easiest way of doing this is to assume that $\bar{V}_1$ is constant. This approximation is justified because the solution is a condensed phase and shows negligible compressibility. Making this assumption and integrating Eq. (19) gives

$$\ln a_i = -\frac{\pi \bar{V}_1}{RT} \tag{20}$$

Combining Eqs. (13) and (20) permits us to express the chemical potential in terms of osmotic pressure instead of activity:

$$\mu_1 = \mu_1^0 - \pi \bar{V}_1 \tag{21}$$

This relationship will be useful in Chap. 5.

Equation (20) provides the relationship we have sought between osmotic pressure and concentration. If the solution is ideal, we may replace activity by mole fraction. Then Eq. (20) becomes

$$\ln x_1 = -\frac{\pi \bar{V}_1}{RT} = \ln(1 - x_2) \simeq -x_2 - \frac{x_2^2}{2} - \cdots \tag{22}$$

where the approximation arises from the expansion of the logarithm (see Appendix A). Therefore for ideal, two-component solutions, we write

$$x_2 = \frac{\pi \bar{V}_1}{RT} \tag{23}$$

Since real solutions tend toward ideality as the solute concentration decreases,

$$x_2 = \frac{n_2}{n_1 + n_2} \simeq \frac{n_2}{n_1} \tag{24}$$

in which the n terms equal the number of moles of the indicated component. The approximate form of Eq. (24) applies in the case of dilute solutions for which $n_2 \ll n_1$. Introducing the dilute solution approximation of Eq. (24) into Eq. (23) yields

$$n_2 = \frac{n_1 \pi \bar{V}_1}{RT} = \frac{\pi V_1}{RT} \tag{25}$$

where V_1 is simply the volume of the solvent in the solution, $n_1 \bar{V}_1$. This relationship, known as the van't Hoff equation (after J. H. van't Hoff, first Nobel Prize in Chemistry, 1901), is analogous in form to the ideal gas law and, like the latter, is a limiting law which applies perfectly only in the limit of $\pi \to 0$ or $n_2/V_1 \to 0$. Equation (25) shows that osmotic pressure experiments provide a means of measuring the number of solute particles in a solution. If the weight of solute in the solution is also known, this information may be used to evaluate molecular weights. We shall consider the application of osmometry to problems of molecular weight determination in Example 3.1. For now it is sufficient to note that even high molecular weight solutes—for which the number of molecules per weight of sample is orders of magnitude less than for low molecular weight compounds—produce appreciable osmotic pressures. For example, if a 1% aqueous solution is assumed to be ideal at 25°C, then Eq. (25) shows that solutes of molecular weight 10^3, 10^4, 10^5, and 10^6 have osmotic pressures of 2530, 253, 25.3, and 2.53 mm of solution, respectively. Thus even very high molecular weight solutes generate pressures which result in easily measured liquid columns. We shall see presently how to get around the assumption of ideality in Eq. (25) so this result can be applied with confidence to real solutions.

3.3 Experimental Osmometry

To carry out an osmotic pressure experiment, we need to prepare a solution, find a suitable semipermeable membrane, achieve isothermal equilibrium, and measure the equilibrium pressure. Aside from noting that the pressures produced by colloidal solutes are large enough to be measurable, we have not yet considered any of the experimental aspects of osmometry. This is our present task.

First, a suitable solvent and membrane must be found. The solvent must dissolve enough solute to produce an adequate pressure. The results of measurements made at relatively high concentrations may be extrapolated to zero concentration, so we need not worry about the effects of nonideality. We shall discuss the extrapolation procedure in Sect. 3.4. As low a solvent viscosity as possible is desirable to minimize the time required for equilibration.

It is important that the materials be free from contaminants in osmotic pressure experiments. Suppose, for example, that the solvent contains a small amount of impurity which, like the colloidal solute, is retained by the membrane. Then, as far as the osmotic pressure is concerned, that impurity will contribute to the osmotic pressure in the same way that the colloid does. Since the osmotic pressure responds to the number of solute particles present, a low molecular weight impurity in extremely small amounts may contain as many or more *particles* as a dilute solution of a colloidal solute of very high molecular weight. Quite large errors in molecular weight may arise in this way. The confusion may be compounded if the same system is investigated using a different membrane material. It is conceivable that another membrane would be permeable to the impurity. This would result in a different apparent molecular weight for the colloid. We have approached the issue of impurities from the viewpoint of the solvent. Actually, the colloid is more likely to be the source of the impurity, since these materials are often difficult to purify. We shall discuss some additional aspects of this problem in the sections on average molecular weight (Sect. 3.6), charged colloids (Sect. 3.10), and dialysis (Sect. 3.11).

The membrane is the source of most of the difficulties in osmometry. There is no general way to select a membrane material that will be permeable to one component and impermeable to another for any conceivable combination of chemicals. Very high molecular weight components are generally more easily retained, however, so we have a slight advantage in this regard. The membrane must be sufficiently thin to permit equilibration at a reasonable rate. At the same time, it must be strong enough to withstand the considerable pressure differences which may exist across it. This problem may generally be overcome, at least in part, by suitable mechanical support of the membrane. A more serious problem is the preparation of thin membranes that are free from minute imperfections which would constitute a "leak" between the two compartments. Such a leak would totally invalidate the experimental results. One peculiarity of membranes is their tendency to display what is known as an asymmetry pressure; that is, an equilibrium pressure difference may exist across a membrane even when there is solvent on both sides. This must be measured and subtracted as a "blank correction."

Equilibrium osmometry is a thermodynamic phenomenon. As such, it makes no difference what mechanism the membrane uses to retain the solute, nor do we learn anything about the mechanism from equilibrium studies. It is easy to visualize that some solutes, especially those in the colloidal size range, are retained by a sieve effect; that is, the molecules are simply too large to pass through the pores in the membrane material. Another possibility is a mechanism whereby the membrane displays selective solubility. This means that the membrane dissolves the solvent but not the solute. In this way the solvent can pass through the membrane, while the solute is retained. An analogous mechanism for charged particles may arise by the membrane repelling (and thus retaining) particles of one particular charge.

A great many different materials have been used in osmotic pressure experiments. Various forms of cellophane and animal membranes are probably the most common membrane materials. Various other polymers, including polyvinyl alcohol, polyurethane, and polytrifluorochloroethylene, have also been used along with such inorganic substances as $CuFe(CN)_6$ precipitated in a porous support.

Once a suitable membrane and solvent are selected, an experimental arrangement must be devised which measures the equilibrium pressure under isothermal conditions. Many variations in apparatus design have been studied. Two particularly instructive pieces of apparatus are shown in Fig. 3.2. The assembly shown in Fig. 3.2a consists of an inner solution compartment with a relatively large opening at the membrane end and a capillary at the small end. The entire solution chamber is then immersed in a tube containing the solvent. Once assembled, the entire apparatus is placed in a constant temperature bath for equilibration.

Another osmometer design is shown assembled in Fig. 3.2b and in detail in Fig. 3.2c. The membrane is pressed between two grooved faces which contain solvent on one side and solution on the other. The grooves are attached to filling tubes and to capillaries where the pressure head develops. This apparatus permits a large contact area between liquids with a minimum volume of liquids involved.

Several times in this discussion we have noted the importance of experimental conditions which permit as rapid an equilibration as possible. The implication of these remarks is that osmotic equilibrium is reached slowly. In some cases as much as 1 week may be required for

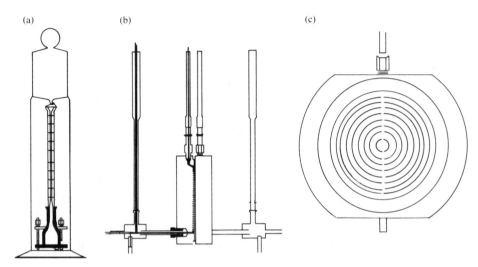

Figure 3.2 Two osmometers: (a) The solution compartment is submerged in solvent [reprinted with permission from D. M. French and R. H. Ewert, *Anal. Chem.*, *19*:165 (1947), copyright by the American Chemical Society] and (b) the solution and solvent occupy grooves on opposite faces of the central unit, as shown in detail in (c) [reprinted with permission from R. M. Fuoss and D. J. Mead, *J. Phys. Chem.*, *47*:59 (1943), copyright by the American Chemical Society].

equilibrium to be achieved. To shorten this time, procedures based on measuring the rate of approach to equilibrium have been developed. The osmometer of Fig. 3.2b is especially suited for this procedure. In successive runs the capillary on the solution side of the membrane is filled with solution to some initial setting which will be above or below the equilibrium location of the meniscus. At various times after the initial settings, the height of the liquid column is measured. It is found that the ascending and descending branches of the curve converge to the same point—a value which equals the equilibrium osmotic pressure. This is shown in Fig. 3.3. The rate at which equilibrium is achieved decreases as equilibrium is approached. Therefore it is desirable to bracket the true value as narrowly as possible to take full advantage of this approach-to-equilibrium extrapolation procedure. With some judicious planning, the time for an osmotic experiment may be shortened considerably by this dynamic method.

   Once equilibrium has been reached, the height difference between the two liquid surfaces is all that remains to be measured. The primary factor

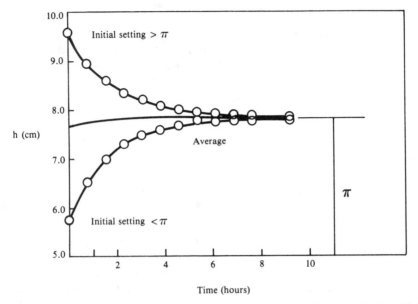

Figure 3.3 Data showing the approach to osmotic equilibrium from initial settings above and below the equilibrium column height. (From Ref. 2, redrawn with permission.)

to note here is that capillaries are used to minimize the dilution effects. This means that corrections for capillary rise must be taken into account unless the apparatus allows the difference between two carefully matched capillaries to be measured. We shall discuss capillary rise in Sect. 6.2. Finally, there is an extremely important practical reason for good thermostating in osmometry experiments in addition to the theoretical requirement of isothermal conditions. The apparatus consists of a large liquid volume attached to a capillary and therefore has the characteristics of a liquid thermometer: The location of the meniscus is quite sensitive to temperature fluctuations.

3.4 Osmotic Pressure of Nonideal Solutions

In Sect. 3.2 we abandoned a completely general discussion of osmotic pressure after reaching Eq. (20) in favor of the simpler assumption of ideality. The ideal result applies to real solutions in the limit of infinite dilution. The objective of this section is to examine the extension of Eq. (20) to nonideal solutions or, more practically, to solutions whose concentrations are greater than infinitely dilute.

Since both the osmotic pressure of a solution and the pressure-volume–temperature behavior of a gas are described by the same formal relationship [Eq. (25)], it seems plausible to approach nonideal solutions along the same lines that are used in dealing with nonideal gases. The behavior of real gases may be written as a power series in one of the following forms for n moles of gas:

$$\frac{pV}{nRT} = 1 + Bp + Cp^2 + \cdots \tag{26}$$

or

$$\frac{pV}{nRT} = 1 + B\left(\frac{n}{V}\right) + C\left(\frac{n}{V}\right)^2 + \cdots \tag{27}$$

Equations of this type are known as virial equations, and the constants they contain are called the virial coefficients. It is the second virial coefficient B which describes the earliest deviations from ideality. It should be noted that B would have different but related values in Eqs. (26) and (27), even though the same symbol is used in both cases. One must be especially attentive to the form of the equation involved, particularly with respect to units, when using literature values of quantities such as B. The virial coefficients are temperature dependent and vary from gas to gas. Clearly, both Eqs. (26) and (27) reduce to the ideal gas law as $p \rightarrow 0$ or as $n/V \rightarrow 0$. Finally, it might be recalled that the second virial coefficient in Eq. (27) is related to the van der Waals a and b constants as follows:

$$B = b - \frac{a}{RT} \tag{28}$$

This last relationship points out that for gases the second virial coefficient arises both from the finite size and from the interactions of the molecules of the gas, since these are the origins of b and a, respectively.

As we extend these ideas to nonideal solutions, a similar set of statements will apply to the resulting power series:

1. The second virial coefficient is our primary concern, since we shall focus attention on the first deviations from ideality.
2. The value of B will depend in part on the units chosen for concentration, as well as on the temperature and the nature of the system.

3. The virial equation for osmotic pressure must reduce to the van't Hoff equation in the limit of infinite dilution.
4. The second virial coefficient may be expected to reflect both the finite size and the interactions of the molecules.

Each of these points is taken up in the following discussion.

The easiest way to extend these considerations to the osmotic pressure of nonideal solutions is to return to Eq. (22), which relates π to a power series in mole fraction. This equation applies to ideal solutions, however, since ideality is assumed in replacing activity by mole fraction in the first place. To retain the form yet extend its applicability to nonideal solutions, we formally include in each of the concentration terms a correction factor defined to permit the series to be applied to nonideal solutions as well:

$$\frac{\pi \bar{V}_1}{RT} = A'x_2 + \frac{1}{2}B'x_2^2 + \cdots \tag{29}$$

The coefficients $A', B', \ldots$, must all equal unity in ideal solutions in order to recover Eq. (22). Since the van't Hoff equation is a limiting law, the coefficient A' must equal unity in *all* solutions. Therefore Eq. (29) becomes

$$\frac{\pi \bar{V}_1}{RT} = x_2 + \frac{1}{2}B'x_2^2 + \cdots \tag{30}$$

Next, let us consider the transformation of mole fraction concentration units into other units more appropriate for use with solutes whose molecular weight might be unknown. In working with unknowns, mass volume^{-1} units offer the greatest flexibility, since only a balance and a volumetric flask are needed to quantitatively characterize a solution in these units. We shall represent this unit of concentration by c in this chapter. For the reasons presented here, this same concentration unit is used in viscosity and light scattering work with solutes of unknown molecular weight. If V is the volume of a solution of components 1 and 2, it may be written $n_1 \bar{V}_1 + n_2 \bar{V}_2$, where the $\bar{V}_i$'s are partial molar volumes. Remember, the latter give the volume occupied by a mole of the indicated component in a mixture; their precise value depends on the concentration of the solution. Therefore, if m_2 is the mass of solute in the solution,

$$c = \frac{m_2}{n_1 \bar{V}_1 + m_2 \bar{V}_2} \simeq \frac{m_2}{n_1 \bar{V}_1} \tag{31}$$

where the approximation applies to dilute solutions in which $n_2 \ll n_1$. Since m_2 is equal to the product of n_2 and M_2,

$$c = \frac{n_2 M_2}{n_1 \bar{V}_1} = x_2 \frac{M_2}{\bar{V}_1} \qquad \longrightarrow \quad x_2 = \frac{c \, \bar{V}_1}{M_2} \qquad (32)$$

in which the dilute solution form of Eq. (24) has been used to introduce x_2. Substituting this result into Eq. (30) gives

$$\frac{\pi \bar{V}_1}{RT} = \frac{\bar{V}_1}{M_2} c + \frac{1}{2} B' \left(\frac{\bar{V}_1}{M_2} \right)^2 c^2 + \cdots \qquad (33)$$

Equation (33) may be rearranged to yield

$$\frac{\pi}{RTc} = \frac{1}{M_2} + \frac{1}{2} \frac{B' \bar{V}_1}{M_2^2} c = \frac{1}{M_2} + Bc \qquad (34)$$

The quantity π/c is called the reduced osmotic pressure, and $(\pi/c)_0$ the limiting reduced osmotic pressure. Osmotic pressure results are most commonly encountered as plots of reduced osmotic pressure—with or without the RT—versus c. This form suggests that a plot of π/RTc versus c should be a straight line, the intercept and slope of which have the following significance:

$$\text{intercept} = \left(\frac{\pi}{RTc} \right)_0 = \frac{1}{M_2} \qquad (35)$$

$$\text{slope} = B = \frac{1}{2} \frac{B' \bar{V}_1}{M_2^2} \qquad (36)$$

A plot of π/RTc versus c is the usual way in which osmotic pressure results are presented.

Note that the value of the intercept, the value of π/RTc at infinite dilution, obeys the van't Hoff equation [Eq. (25)]. At infinite dilution even nonideal solutions reduce to this limit. The value of the slope is called the second virial coefficient by analogy with Eq. (27). Note that the second virial coefficient is the composite of two factors, B' and $\frac{1}{2} \bar{V}_1/M_2^2$. The factor B' describes the first deviation from ideality in a solution; it equals unity in an ideal solution. The second cluster of constants in B arises from the conversion of practical concentration units to mole fractions. Although it is the nonideality correction in which we are primarily interested, we shall discuss it in terms of B rather than B', since the former is the quantity which is measured directly. We shall return to an interpretation of the second virial coefficient in Sect. 3.6.

Figure 3.4 shows examples of two plots of π/RTc versus concentration. In Fig. 3.4a the data all describe different molecular weight fractions of the same solute, cellulose acetate, in acetone solutions. Since the lines in this plot all have essentially the same slope, B must be the same for each.

Figure 3.4b shows data for a sample of nitrocellulose in three different solvents. All show the same intercept corresponding to a single molecular weight as required. Note, however, that the slopes are different, including even a negative slope, indicating that wide variations in B are possible. We shall examine the factors that determine B in Sect. 3.6.

The following example considers the molecular weights of the polymers in Fig. 3.4 in terms of Eq. (25).

Example 3.1 To what molecular weights do the limiting reduced osmotic pressures obtained from Fig. 3.4 correspond? The data in Fig. 3.4 are presented in such a way that unit problems do not enter the picture. We are not usually so lucky! Consider some possible combinations of units for π, V, and R that are compatible with the units of the ordinate in Fig. 3.4.

Solution According to Eq. (25), the molecular weights of the various polymers are simply the reciprocals of the limiting reduced osmotic

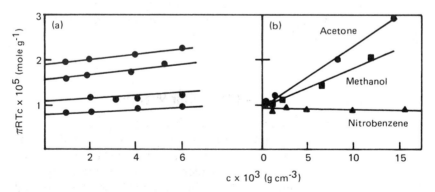

Figure 3.4 Plots of π/RTc versus concentration for (a) various cellulose acetate fractions in acetone [data from A. Bartovics and H. Monk, *J. Am. Chem. Soc.*, 65:1901 (1943)] and (b) nitrocellulose in three different solvents [data from A. Dobry, *J. Chem. Phys.*, 32:50 (1935)].

pressures in the units that have been used for the ordinate in Fig. 3.4, namely, mole g^{-1}. Therefore

Part in Fig. 3.4	$(\pi/RTc)_0 \times 10^{-5}$ (mole g^{-1})	M (g mole^{-1})
(a)	1.92	52,000
(a)	1.59	63,000
(a)	1.09	92,000
(b)	0.79	126,000
	0.90	111,000

Osmotic pressures may be expressed in any of the usual pressure units (and then some!) and volumes are often—but not always—expressed in cubic centimeters (milliliters); compatible units of R must be chosen:

If π is in atm and V in cm^3, use $R = 82.05$ atm cm^3 K^{-1} mole^{-1}.

If π is in Torr and V in cm^3, use $R = 62,360$ Torr cm^3 K^{-1} mole^{-1}.

If π is in mm of solution and V in cm^3, use $R = 62,360(\rho_{Hg}/\rho_{soln})$ mm cm^3 K^{-1} mole^{-1}, where the ρ's are densities.

If π is in dyne cm^{-2} and V in cm^3, use $R = 8.314 \times 10^7$ erg K^{-1} mole^{-1}.

If π is in N m^{-2} and V in m^3, use $R = 8.314$ J K^{-1} mole^{-1}

If the masses (in c) are expressed in g and kg, respectively, in the last two situations, then the units of the ordinate reduce to cm^2 s^{-2} and m^2 s^{-2}, respectively, which do not bear much resemblance to molecular weight units.

Note, too, that the acceleration of gravity in appropriate units can be factored out of these latter quantities to leave units of cm and m, respectively, for reduced osmotic pressure.

•

3.5　The Number Average Molecular Weight

As we saw in Chap. 1, the condition of polydispersity is quite normal with colloidal solutes. Our discussion of osmometry has shown that it is possible to evaluate the molecular weight of a colloidal solute by osmotic pressure measurements. Next we consider the fact that the sample on which such a measurement is made is more than likely polydisperse and that the molecular weight obtained from such an experiment is some average quantity. The objective of this section is to show that it is the number average molecular weight which is determined in an osmotic pressure experiment on a polydisperse system. For the purposes of this demonstration, the solution is assumed to be ideal.

Experimental results from a polydisperse system may be related as follows:

$$\pi_{exp} = \frac{c_{exp}RT}{\overline{M}} \tag{37}$$

where π_{exp} and c_{exp} represent the experimental osmotic pressure and concentration, respectively, and $\overline{M}$ is the average molecular weight. It is the method of averaging in Eq. (37) that we seek to determine. This relationship applies to the observable quantities. Precisely the same equation may be written, however, for each of the molecular weight fractions of the solute, designated here by the subscript i:

$$\pi_i = \frac{c_i RT}{M_i} \tag{38}$$

Two additional relationships are fairly evident. The experimental osmotic pressure is the sum of the pressure contributions of the individual components:

$$\pi_{exp} = \sum_i \pi_i \tag{39}$$

and the experimental concentration is the sum of the concentrations of the components:

$$c_{exp} = \sum_i c_i \tag{40}$$

Next, Eqs. (37)–(40) can be combined as follows:

$$\pi_{exp} = \sum_i \pi_i = RT \sum_i \frac{c_i}{M_i} = RT \frac{c_{exp}}{\overline{M}} = RT \frac{\Sigma_i c_i}{\overline{M}} \tag{41}$$

Equation (41) may be simplified to yield

$$\overline{M} = \frac{\Sigma_i c_i}{\Sigma_i (c_i/M_i)} \tag{42}$$

This result still fails to resemble any of the standard averages listed in Table 1.7. However, if we introduce the expression

$$c_i = \frac{n_i M_i}{V} \tag{43}$$

we obtain

$$\overline{M} = \frac{(1/V)\Sigma_i n_i M_i}{(1/V)\Sigma_i (n_i M_i / M_i)} = \frac{\Sigma_i n_i M_i}{\Sigma_i n_i} \tag{44}$$

Equation (44) shows the average molecular weight determined from osmometry to be the number average molecular weight as defined by Eq. (1.24).

This same conclusion may also be reached by the following argument. The product $n_i M_i$ in Eq. (43) equals the weight of component i in the solution; the total weight of solute in the solution equals $\Sigma_i n_i M_i$. The experimental osmotic pressure depends on and therefore measures the total number of moles of solute $\Sigma_i n_i$. The ratio of the total weight to the total number of moles of solute defines the number average molecular weight.

Any experiment that "counts" the number of molecules in a given weight of polydisperse material can be interpreted in terms of a number average molecular weight. All of the colligative properties have this as a feature of their shared thermodynamic origin. Another technique, known as end group analysis, may be used to determine the molecular weight of certain polymers. The following example makes it clear that this too yields a number average molecular weight for polydisperse samples.

Example 3.2 Polycaprolactam—otherwise known as nylon 6—with a degree of polymerization n has the following molecular structure: $H_2N(CH_2)_5CO[NH(CH_2)_5CO]_{n-2}NH(CH_2)_5COOH$. A 1.06-g sample of this material is dissolved in an appropriate solvent and titrated with an alcoholic KOH solution of normality 0.0250 N. Exactly 5.00 ml of base are required to neutralize the carboxyl groups of the sample. What is the molecular weight of the sample? What is n? Criticize or defend the following proposition: This method can only be used for molecules which contain just one of the analyzed functional groups per molecule.

Solution This problem is really no different than the molecular weight determinations of unknown acids that are often conducted in general chemistry lab courses. What is important to recognize is that there is one carboxyl group per molecule or one equivalent per mole. Therefore the molecular weight of the polymer is given by

$$\overline{M} = \frac{1.06\ g}{(5.00)(0.0250)mEq} \times \frac{10^3\ mEq}{Eq} \times \frac{1\ Eq}{1\ mole} = 8480\ g\ mole^{-1}$$

This is the same average molecular weight that would be determined by an osmotic pressure experiment on the same sample. Since the molecular weight of the repeat unit is 113 g mole^{-1}, the degree of polymerization $n = 8480/113 = 75$.

The proposition recognizes the importance of knowing the number of analyzed groups per molecule but incorrectly requires that there be only one such group. Some polymers might contain the same functional group at both ends of a linear chain. In this case there are 2 Eq/mole. A Y-shaped molecule could have three such groups, an X-shaped molecule four, and so forth. These last cases would have 3 and 4 Eq/mole, respectively. The number of groups does not matter as long as that number is known.

•

That end group analysis and osmometry give number averages for polydisperse samples is simply because the number of solute molecules is counted in each case. We shall see in Chap. 5 that light scattering effectively weighs the molecules rather than counts them. Hence light scattering gives a weight average value for M. The fact that osmotic pressure "counts" solute particles has some interesting consequences for charged colloids, as we shall see in Sect. 3.10.

3.6 The Second Virial Coefficient: Excluded Volume

By analogy with Eq. (28), we might expect the second virial coefficient to depend on the size and/or the interactions of the molecules in solution. Although this expectation is basically correct, we must not take the form of Eq. (28), a gas equation, too literally in discussing solutions. Furthermore, it must be recalled that in gases only interactions between the molecules of the gas are possible. In solution, we may consider solvent–solvent, solute–solute, and solvent–solute interactions. This is a good reminder that the analogy with gases cannot be pushed too far.

By itself, thermodynamics provides us with no information on the molecular origin of the second virial coefficient. The latter is merely a phenomenological coefficient from an exclusively thermodynamic viewpoint. Statistical thermodynamics undertakes the task of providing molecular interpretations to such quantities. The added complication is that the statistical thermodynamic approach requires the use of models, and models inevitably oversimplify things. Nevertheless, we have arrived at that point where we will get additional insight only to the extent that we do some modeling. Thus, for example, we can learn something about solute–solute interactions and solute–solvent interactions from experimental B values, but only if we are willing to accept the models upon which these interpretations are based. Toward this end, it is important to know what goes into these models. A model that is very plausible in one system may not make any sense for another.

In this section we shall examine a model for the second virial coefficient which is based on the concept of the excluded volume of the solute particles. A solute–solute interaction arising from the spatial extension of particles is the premise of this model. Therefore the potential exists for learning something about this extension for systems where the model is applicable. In the next section we consider a model which considers the second virial coefficient in terms of solute–solvent inter- action. This approach offers a quantitative measure of such interactions through B. In both instances we shall only outline the pertinent statistical thermodynamics; a somewhat fuller development of these ideas is given in Ref. 3. Finally, we should note that some of the ideas of this section are going to reappear in our discussion of flocculation in Chapter 11.

Before considering how the excluded volume affects the second virial coefficient, let us first review what we mean by excluded volume. We encountered this concept in our model for size exclusion chromato- graphy in Sect. 1.10. The development of Eq. (1.23) is based on the idea that the center of a spherical particle cannot approach the walls of a pore any closer than a distance equal to its radius. A zone of this thickness adjacent to the pore walls is a volume from which the particles— described in terms of their centers—are denied entry because of their own spatial extension. The volume of this zone is what we call the excluded volume for such a model. The van der Waals b constant in Eq. (28) measures the excluded volume of gas molecules; for spherical molecules it equals four times the actual volume of the sphere, as shown in most physical chemistry texts.

A point of entry for statistical considerations into thermodynamics is the Boltzmann entropy relationship

$$S = k \ln \Omega \tag{45}$$

in which Ω is called the thermodynamic probability of a state and k is the Boltzmann constant. The latter is replaced by R when we are working on a "per mole" rather than a "per molecule" basis. The thermodynamic probability is the tricky part of Eq. (45); it counts the number of ways a particular state can come about. On a qualitative basis, chemists learn to use this relationship almost intuitively. We associate higher entropy with states that are more disordered, and disordered states can come about in a larger number of ways than states of higher order. The familiar analogy that is often used here is to note that there are more ways for a deck of cards to exist as "shuffled" than as "arranged by suits." We can readily see that Eq. (45) encompasses the third law of thermodynamics: There is only one way to organize a perfect crystal at absolute zero, hence $\Omega = 1$ and $S = 0$.

Entropy changes can also be developed in terms of Eq. (45):

$$S_2 - S_1 = k \ln \left(\frac{\Omega_2}{\Omega_1} \right) \tag{46}$$

Our interest is in the entropy of mixing—that is, ΔS for $1 + 2 \rightarrow$ mixture—which may be written

$$\Delta S_m = k \ln \left(\frac{\Omega_{mix}}{\Omega_1 \Omega_2} \right) \tag{47}$$

by virtue of Eq. (46). Note that we use the subscripts m for the mixing process and mix for the mixture itself. What remains to be done is to count the ways in which N_1 molecules of component 1 and N_2 molecules of component 2 exist when mixed together, or more specifically, the factor by which this number exceeds the number of ways the separate molecules can exist.

The derivation we shall follow proceeds through four stages:

1. We must devise an expression for Ω_{mix}.
2. From Ω_{mix} a straightforward application of Eq. (45) is required to obtain an equation for S_{mix}. It takes a bit of additional argument to convert this to an expression for ΔS_m.
3. A single approximation will enable us to go from ΔS_m to ΔG_m, but the implications of this assumption deserve some comment.
4. Some mathematical manipulations convert ΔG_m into an expression for $\mu_1 - \mu_1^0$, which is directly related to π through Eq. (21).

We begin by considering the number of ways a solute molecule of excluded volume u can be placed in an otherwise empty volume V. Suppose we imagine V to be sectioned off into a number of sites each of which can accommodate the volume u. Since V is intended to represent the volume of the solution—a macroscopic quantity—and u is a molecular parameter, the number of such sites is large. The first solute molecule to enter V can occupy any one of these sites and we represent by ω_1 the number of possible placements for the first solute molecule. Because the total volume is partitioned into these sites, it follows that

$$\omega_1 = KV \tag{48}$$

where K is an appropriate proportionality constant.

The second molecule to be placed can go into any one of the *remaining* sites; hence the number of ways to place the second molecule is given by

$$\omega_2 = K(V - u) \tag{49}$$

For the third molecule this quantity is given by $\omega_3 = K(V - 2u)$, and for the jth it is

$$\omega_j = K[V - (j - 1)u] \tag{50}$$

The number of ways of placing the first *and* the second *and* the third and so on up to the ith is

$$\Omega \propto \omega_1 \omega_2 \omega_3 \cdots \omega_i \tag{51}$$

We have written Eq. (51) as a proportion rather than as an equality, since the given expression actually overcounts the number of possible placements. If there are a total of N_2 solute molecules, then there are $N_2!$ different permutations of these molecules and Eq. (51) should be divided by $N_2!$ because of the interchangeability of the molecules. Incorporating this idea and Eq. (50) into Eq. (51) gives

$$\Omega = \frac{1}{N_2!} \prod_{j=1}^{N_2} K[V - (j - 1)u] \tag{52}$$

where Π represents the product of terms. Letting $i = j - 1$ allows Eq. (52) to be written more concisely as

$$\Omega = \frac{1}{N_2!} \prod_{i=0}^{N_2-1} K(V - iu) = \frac{1}{N_2!} \prod_{i=0}^{N_2-1} KV\left(1 - i\,\frac{u}{V}\right) \tag{53}$$

After the N_2 solute molecules are placed, all remaining sites are filled with solvent molecules. Since our model is specifically interested in the solute-solute excluded volume effect, we may say that there is only one way for the solvent molecules to be placed. Although this overlooks details about the solvent, we shall see presently that such details would eventually be subtracted away, so we lose nothing by this simplification. Equation (53) therefore gives the expression for Ω_{mix} which we sought as the first step of our derivation.

Substituting Eq. (53) into Eq. (45) gives a statistical expression for the solute contribution to the configurational entropy of the mixture:

$$\frac{S_{\text{mix}}}{k} = -\ln N_2! + \sum_{i=0}^{N_2-1} \ln\left[KV\left(1 - i\,\frac{u}{V}\right)\right] \tag{54}$$

The summation replaces the product in going from Eq. (53) to Eq. (54), since we are dealing with logarithms in the latter. Note that the configurational entropy refers explicitly to the entropy associated with the mixture itself; the internal entropy of the molecules themselves is clearly not included. Next a series of mathematical manipulations will transform Eq. (54) into a more useful form:

1. Divide the summation in Eq. (54) into two terms:

$$\sum \ln \left[KV \left(1 - i \frac{u}{V} \right) \right] = \sum \ln(KV) + \sum \ln \left(1 - i \frac{u}{V} \right)$$

2. Note that $\sum \ln(KV) = N_2 \ln(KV)$, since there are N_2 identical terms in the summation.
3. Since $u/V \ll 1$ as already noted, $\ln(1 - iu/V)$ can be expanded as a series (see Appendix A) with only the leading term retained to give

$$\sum \ln \left(1 - i \frac{u}{V} \right) \simeq \sum \left(- i \frac{u}{V} \right) \simeq - \frac{u}{V} \sum i$$

4. The sum of integers from 0 to y is $y(y + 1)/2$; hence $\sum_{i=0}^{N_2-1} i = (N_2 - 1)N_2/2 \mp N_2^2/2$, since N_2 is large.
5. We can change the equation from dealing with numbers of molecules (N) to numbers of moles (n) by dividing both sides by Avogadro's number N_A and writing $kN_A = R$.

Applying these manipulations to Eq. (54) gives

$$\frac{S_{mix}}{R} = - \frac{1}{N_A} \ln N_2! + n_2 \ln(KV) - \frac{1}{2} N_A n_2^2 \frac{u}{V} \tag{55}$$

Replacing V by the mole-weighted sum of the partial molar volumes, as was done in developing Eq. (31), we write

$$\frac{S_{mix}}{R} = \frac{-1}{N_A} \ln N_2! + n_2 \ln[K(n_1 \bar{V}_1 + n_2 \bar{V}_2)] - \frac{1}{2} \frac{N_A n_2^2 u}{n_1 \bar{V}_1 + n_2 \bar{V}_2} \tag{56}$$

If we set n_2 equal to zero in Eq. (56), the physical system described corresponds to pure solvent and the equation reduces to $S_1 = 0$. If we set n_1 equal to zero, we are describing pure solute and Eq. (56) becomes

$$S_2 = n_2 \ln K + \ln(n_2 \bar{V}_2) - \frac{1}{N_A} \ln N_2! - \frac{N_A n_2 u}{2 \bar{V}_2} \tag{57}$$

Subtracting these expressions for S_1 and S_2 from S_{mix} gives ΔS_m, which was the goal of the second stage of our derivation:

$$\frac{\Delta S_m}{R} = n_2 \ln\left(\frac{n_1\bar{V}_1 + n_2\bar{V}_2}{n_2\bar{V}_2}\right) - \frac{1}{2}N_A u \frac{n_2^2}{n_1\bar{V}_1 + n_2\bar{V}_2} + \frac{1}{2}N_A \frac{n_2 u}{\bar{V}_2}$$

(58)

Note that whatever configurational entropy we had attributed to the solvent would have disappeared at this point.

As outlined above, our next objective is to write an expression for ΔG_m. Since $\Delta G = \Delta H - T\,\Delta S$ for a constant temperature process such as this, we can immediately write

$$\Delta G_m = -RT\left[n_2 \ln\left(\frac{n_1\bar{V}_1 + n_2\bar{V}_2}{n_2\bar{V}_2}\right) - \frac{1}{2}N_A u\frac{n_2^2}{n_1\bar{V}_1 + n_2\bar{V}_2} \right.$$

$$\left. + \frac{1}{2}N_A \frac{n_2 u}{\bar{V}_2} \right]$$

(59)

by assuming $\Delta H_m = 0$. In making this assumption we imply that the energetic interactions between solvent–solvent pairs, solute–solute pairs, and solvent–solute pairs are all the same. Thus, removing a solvent molecule from a bulk sample of the liquid and replacing it with a molecule of the solute does not involve an enthalpy change. It is the introduction of this assumption that leaves the excluded volume as the sole contributor to the second virial coefficient. In the next section we shall assume a nonzero value for ΔH_m and examine the effect this has on B. For now it is enough to note that the present model attributes solution nonideality to molecular size, and not to the energetics of molecular interactions. The model is clearly inappropriate for any system in which the latter are important.

The final stage of our derivation converts Eq. (59) to an expression for $\mu_1 - \mu_1^0$ by differentiating Eq. (59) with respect to n_1 according to Eq. (1):

$$\mu_1 - \mu_1^0 = -RT\left[\frac{n_2\bar{V}_1}{V} + \frac{1}{2}N_A\bar{V}_1\left(\frac{n_2}{V}\right)^2 u \right]$$

(60)

Equation (21) shows how to convert Eq. (60) into an expression for osmotic pressure:

$$\pi = \frac{RT}{\bar{V}_1}\left[\frac{n_2\bar{V}_1}{V} + \frac{1}{2}N_A u\bar{V}_1\left(\frac{n_2}{V}\right)^2\right] \tag{61}$$

Since $n_2/V = c/M_2$, Eq. (61) may be written in practical concentration units:

$$\pi = RT\left(\frac{c}{M} + \frac{1}{2}\frac{N_A u}{M^2}c^2\right) \tag{62}$$

Comparing this result with Eq. (34) shows that

$$B = \frac{1}{2}\frac{N_A u}{M^2} \tag{63}$$

according to this model.

The essence of this model for the second virial coefficient is that an excluded volume is defined by surface contact between solute molecules. As such, the model is more appropriate for molecules with a rigid structure than for those with more diffuse structures. For example, protein molecules are held in compact forms by disulfide bridges and intramolecular hydrogen bonds; by contrast, a randomly coiled molecule has a constantly changing outline and imbibes solvent into the domain of the coil to give it a very "soft" surface. The present model, therefore, is much more appropriate for the former configuration than for the latter. The following example applies the excluded volume interpretation of B to an aqueous protein solution.

Example 3.3 A plot of π/c versus c for an aqueous solution of the bovine serum albumin molecule at 25°C and pH = 5.37 is shown in Fig. 3.5. The molecule is known to be nearly spherical and uncharged at this pH. Evaluate the molecular weight and the excluded volume of this protein from the intercept and slope of this line, 0.268 Torr (g kg^{-1})$^{-1}$ and 1.37×10^{-3} Torr kg^2 g^{-2}, respectively. From the particle mass and volume, estimate the partial specific volume of the solute in solution. The specific volume of the unsolvated protein is about 0.75 cm^3 g^{-1}; does the solute appear to be solvated?

Solution First, convert to SI units, assuming the density of the solution to be 1.0 g cm^{-3}.

For the intercept

$$0.268 \text{ Torr (g kg}^{-1}\text{)}^{-1} \times \frac{1.01 \times 10^5 \text{N m}^{-2}}{760 \text{ Torr}} \times \frac{10^{-3} \text{ m}^3}{\text{kg}} \times \frac{10^3 \text{ g}}{\text{kg}}$$

$$= 35.6 \text{ N m}^{-2} \text{ (kg m}^{-3}\text{)}^{-1}$$

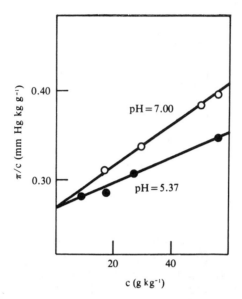

Figure 3.5 Plot of π/c versus concentration for bovine serum albumin in 0.15 M NaCl at pH 7.00 and 5.37. [From G. Scatchard, A. C. Batchelder, and A. Brown, *J. Am. Chem. Soc.*, *68*:2320 (1946), copyright by the American Chemical Society.]

For the slope

$$1.37 \times 10^{-3} \text{ Torr kg}^2 \text{ g}^{-2} \times \frac{1.01 \times 10^5 \text{ N m}^{-2}}{760 \text{ Torr}} \times \left(\frac{10^{-3} \text{ m}^3}{\text{kg}}\right)^2$$

$$\times \left(\frac{10^3 \text{ g}}{\text{kg}}\right)^2 = 0.182 \text{ N m}^4 \text{ kg}^{-2}$$

Calculate the molecular weight by Eq. (35):

$M = RT(\pi/c)_0^{-1} = (8.314)(298)/35.6 = 69.6 \text{ kg mole}^{-1} = 69{,}600 \text{ g mole}^{-1}$

The mass per particle is obtained by dividing by N_A:

$69.6/6.02 \times 10^{23} = 1.16 \times 10^{-22} \text{ kg molecule}^{-1}$

Divide the slope by RT to get B:

$0.182/(8.314)(298) = 7.35 \times 10^{-5} \text{ m}^3 \text{ kg}^{-2} \text{ mole}$

Use Eq. (63) to calculate the excluded volume:

$$u = \frac{2BM^2}{N_A} = \frac{2(7.35 \times 10^{-5})(69.6)^2}{6.02 \times 10^{23}} = 1.18 \times 10^{-24} \text{ m}^3 \text{ molecule}^{-1}$$

The actual volume of a spherical particle is one-quarter the excluded volume; therefore the volume of the molecule is 2.95×10^{-25} m^3 molecule^{-1}.

The partial specific volume is given by the ratio of the molecular volume to the molecular mass:

$$2.95 \times 10^{-25}/1.16 \times 10^{-22} = 2.54 \times 10^{-3} \text{ m}^3 \text{ kg}^{-1} = 2.54 \text{ cm}^3 \text{ g}^{-1}.$$

Comparing this with the nonsolvated value of 0.75 cm^3 g^{-1} definitely suggests that the particle is hydrated.

Note that the volume of the spherical molecule may also be converted to a molecular radius; the latter equals 4.11 nm for this molecule.

•

The second set of data points in Fig. 3.5—measured for the same solute but at a pH = 7.00—shows a steeper slope and therefore a larger B value. It would be incorrect to analyze this slope by the procedure used in the example above, however. The reason for this is that the protein acquires a charge in going from pH 5.37 to 7.00. The charge must be explicitly taken into account to interpret B in this case. We shall take this up in Sect. 3.10. It is pertinent to note, however, that charged particles require counter-ions to give them electrical neutrality. These counter-ions occupy a region in the solution that surrounds the charged colloid; we describe them as setting up an ion atmosphere around the central particle. It makes sense that the particle plus its ion atmosphere should have a larger excluded volume than the uncharged particle which entrains no ion atmosphere.

3.7 The Flory–Huggins Theory

In many colloidal solutions the solute is best described as a random coil in which the domain of the molecule contains both polymer segments and solvent molecules. Vinyl-type synthetic polymers in organic solvents are especially well suited for this representation. Depending on the energetics of the interaction between solvent molecules and polymer chain segments, the solvent may be imbibed into the coil domain to a greater or lesser extent, and the spatial extension of the chain depends on these interactions, as well as on the size of the molecule itself. A solvent which swells the coil dimensions is called a good solvent, and one which causes the coil to shrink is called a poor solvent. The former situation arises when solvent–solute contacts are favored, and the latter when solute–solute contacts are favored. The "goodness" of a solvent for a particular polymer depends not only on the nature of the species involved, but also on the temperature. Generally speaking, lowering the temperature causes a decrease in solvent goodness.

In this section we shall look at a statistical model for a solution which allows for both the random coil geometry of the solute molecule and the variability of interactions between the solute and different solvents. We shall merely sketch the general outline of the theory; in the next section we examine the implications of this model for interpreting the second virial coefficient as a measure of solvent goodness. The approach we adopt is generally known as the Flory–Huggins theory, after P. J. Flory and M. L. Huggins, whose independent efforts merged in the final result. Flory was awarded the Nobel Prize in 1974 for his numerous contributions to polymer chemistry.

The Flory–Huggins theory begins with a model for the polymer solution which visualizes the solution as a three-dimensional lattice of N sites of equal volume. Each lattice site is able to accommodate either one solvent molecule or one polymer segment, since both of these are assumed to be of equal volume. The polymer chains are assumed to be mondisperse and to consist of n segments. Thus, if the solution contains N_1 solvent molecules and N_2 solute (polymer) molecules, the total number of lattice sites is given by

$$N = N_1 + nN_2 \qquad\qquad (64)$$

Without spelling out the geometry of the lattice explicitly, we further assume that each site in the lattice is surrounded by z neighboring sites, making z the coordination number of the lattice.

On the basis of this model, an expression for ΔS_m can be derived. We will not go through the details of the derivation, but merely note the following similarities and differences between this derivation and the one which leads to Eq. (58) for the excluded volume model:

1. Instead of counting the ways to place solute molecules, this derivation counts the ways to place solute *segments*
2. The placement of successive segments is subject to a restriction that did not arise in the previous derivation; namely, segments from the same chain must occupy adjacent lattice sites because they are covalently bonded together.
3. The derivation considers the number of ways ω_i of placing each of the

n segments in, say, the ith solute molecule, then goes on to count the number of ways of scaling this up for N_2 solute molecules. The thermodynamic probability of the mixture Ω_{mix} results from these steps.

4. The entropy of the mixture is calculated from this by the Boltzmann entropy equation [Eq. (45)]. By separately letting N_2 and N_1 equal zero, the configurational entropies of the solvent and the solute, respectively, are obtained from S_{mix}. Finally, by subtracting S_1 and S_2 from S_{mix} an expression is obtained for ΔS_m.

Except for the complication of positioning connected segments in adjacent sites, the general outline of this derivation clearly parallels the previous derivation.

For a mole of solution—i.e., $N_1 + N_2 = N_A$—the theory outlined above results in the following expression for ΔS_m:

$$\frac{\Delta S_m}{R} = -\left[\frac{N_1}{N_A} \ln\left(\frac{N_1}{N}\right) + \frac{N_2}{N_A} \ln\left(\frac{nN_2}{N}\right) \right] \qquad (65)$$

The ratios N_1/N_A and N_2/N_A are mole fractions of the two components, since N_A is the total number of molecules in a mole of solution. On the other hand, N is the total number of lattice sites and N_1 and nN_2 are the numbers of sites occupied by solvent and solute, respectively. Since all of the lattice sites have equal volumes, the ratios in the logarithms are the volume fractions occupied by the two components. Letting x_i be the mole fraction of component i and ϕ_i the volume fraction of that component, Eq. (65) can be written

$$\Delta S_m = -R[x_1 \ln \phi_1 + x_2 \ln \phi_2] \qquad (66)$$

Note that $\phi \rightarrow x$ as $n \rightarrow 1$; that is, Eq. (66) reduces to the expression for the entropy of mixing for an ideal solution in the event that the solute molecule is no larger than the solvent molecule.

At this point we could proceed as in the previous derivation of an expression for B: Evaluate ΔG_m, $\Delta \mu_m$, and π by setting $\Delta H_m = 0$. Doing this, however, would sacrifice this model's ability to deal with different interaction energies between solvent and solute segments. Hence our next objective is to find an expression for ΔH_m in terms of the lattice model we have developed. Once this is done, we can combine ΔS_m and ΔH_m into an expression for ΔG_m and then proceed as above.

To do this we assign an energy of interaction w_{11} to a pair of solvent molecules, and w_{22} to a pair of polymer *segments*. The latter arises from the intermolecular forces between segments and not from the covalent bonds between them. In the same fashion we define w_{12} to be the energy of the solvent–segment interaction.

As noted above, a lattice site in the polymer solution has z nearest neighbors which may be occupied by either solvent molecules or solute segments. If ϕ_i is the volume fraction of the solution occupied by species i, then we assume that this fraction also applies to the z sites which adjoin the specific site under consideration. Hence any site is surrounded, on the average, by $z\phi_1$ solvent molecules and $z\phi_2$ chain segments. If the central site is occupied by a solute segment (component 2), then the inter-molecular forces between this segment and its neighbors contribute $z\phi_1 w_{12} + z\phi_2 w_{22}$ to the energy of the system. Since the lattice consists of N sites, of which $\phi_2 N$ are occupied by solute segments interacting in the manner just described, we write $\frac{1}{2} N\phi_2(z\phi_1 w_{12} + z\phi_2 w_{22})$ for the total interaction energy of all such segments. The factor $\frac{1}{2}$ enters the previous expression because each pair of interacting species is counted twice by this procedure. If the lattice is completely filled with polymer segments, then the interaction energy is simply given by $\frac{1}{2} z\phi_2 N w_{22}$.

Precisely the same series of steps gives the energy contribution by sites occupied by solvent molecules:

1. The contribution of interactions with its neighbors made by each solvent molecule is $z\phi_1 w_{11} + z\phi_2 w_{12}$.
2. The number of sites occupied by solvent molecules is $\phi_1 N$.
3. The total energy contribution by all solvent molecules is $\frac{1}{2}\phi_1 N(z\phi_1 w_{11} + z\phi_2 w_{12})$.
4. If the lattice is completely filled with solvent, then the interaction energy is given by $\frac{1}{2} z\phi_1 N w_{11}$.

Now consider the change in interaction energies that accompanies the mixing process $1 + 2 \rightarrow$ mixture. We must total the interactions of all chain segments and all solvent molecules to obtain the interaction energy of the mixture, and then subtract from this the interactions corresponding to pure solute and pure solvent. Assembling the required values from above, we write

$$\Delta H_m = \frac{1}{2} zN[\phi_2(\phi_2 w_{22} + \phi_1 w_{12}) + \phi_1(\phi_1 w_{11} + \phi_2 w_{12}) - \phi_1 w_{11} -$$
$$\phi_2 w_{22}] \tag{67}$$

Since $\phi_1 + \phi_2 = 1$, Eq. (67) can be rearranged as

$$\Delta H_m = -\frac{1}{2} zN\phi_1\phi_2(w_{11} + w_{22} - 2w_{12}) \tag{68}$$

In terms of pairwise interactions, the solution process can be represented

$$(1,1) + (2,2) \rightarrow 2(1,2) \tag{69}$$

and, following the usual thermodynamic notation, we can write Δw for the process as

$$\Delta w = 2w_{12} - w_{11} - w_{22} \tag{70}$$

Combining Eqs. (68) and (70) gives

$$\Delta H_m = \tfrac{1}{2} z N \phi_1 \phi_2 \, \Delta w = \tfrac{1}{2} z N_1 \phi_2 \, \Delta w \tag{71}$$

The forces between molecules which we measure by these w's are forces of attraction and, by convention, are represented by negative numbers. Thus when 1,2 attractions are stronger than 1,1 and 2,2 attractions, Δw and ΔH_m are both negative. Since a negative enthalpy of mixing makes a favorable contribution to a negative value for ΔG_m, this sign convention makes sense. Conversely, if the 1,1 and 2,2 attractions are stronger (more negative), then Δw and ΔH_m are positive. The case in which ΔH_m is zero is called athermal mixing; Eq. (70) shows that this corresponds to a situation in which solute–solute, solvent–solvent, and solute–solvent interactions are all equivalent in energy. This was assumed to be the case in the excluded volume model for solution nonideality discussed in the last section.

An additional development of Eq. (70) can be made by assuming that w_{12} is the geometric mean of w_{11} and w_{22}. It makes sense that the energy of interaction between unlike molecules is somehow related to the homogeneous interactions; and this manner of averaging the latter has advantages that will be evident presently. By assuming $w_{12} = (w_{11}w_{22})^{1/2}$, Eq. (70) becomes

$$\Delta w \propto w_{11} + w_{22} - 2(w_{11}w_{22})^{1/2} \propto (\sqrt{w_{11}} - \sqrt{w_{22}})^2 \tag{72}$$

This development cannot result in a negative value for Δw or ΔH_m and is therefore definitely inapplicable for systems in which the solute and solvent display some specific type of interaction, such as hydrogen bonding. However, when purely physical interactions are involved, Eq. (72) has proved to be quite useful.

The utility of this approach lies in the fact that Eq. (72) describes the mixing process in terms of homogeneous interactions and the latter are readily measured for pure liquids. The heat of vaporization, for example, is a liquid property which increases as the strength of intermolecular attractions increases. Rather than working with molar heats of vaporization, it is more convenient to divide the latter by the molar volume to define what is known as the cohesive energy density (CED) of a material. As the name implies, the CED measures the energy per unit volume which holds the molecules of a liquid together. As such, it is directly proportional to the w's of our discussion and can be evaluated from readily available data. Introducing the concept of cohesive energy density into Eq. (72) enables us to write

$$\Delta H_m \propto \Delta w \propto [\sqrt{(CED)_1} - \sqrt{(CED)_2}]^2 \tag{73}$$

Table 3.1 Values of the Square Root of the Cohesive Energy Density for Some Polymers and Low Molecular Weight Solvents

Solvent	$(CED)^{1/2}$ $[(cal\ cm^{-3})^{1/2}]$	Polymer	$(CED)^{1/2}$ $[(cal\ cm^{-3})^{1/2}]$
n-Decane	6.6	Poly(tetrafluoro-	6.2
Cyclohexane	8.2	ethylene) (Teflon)	
Toluene	8.9	Polyethylene	7.7–8.2
Acetone	9.9	Polystyrene	8.5–9.1
Cyclohexanol	11.4	Poly(methyl	9.1–9.5
Ethanol	12.7	methacrylate)	
Methanol	14.5	Polypropylene	9.2–9.4
Water	23.4	Poly(vinyl chloride)	9.7–9.9
		Poly(ethylene	10.7
		terephthalate)	
		Poly(acrylonitrile)	12.3–12.8

Source: H. Burrell, in J. Brandrup and E. H. Immergut (eds.), *Polymer Handbook*, 2nd ed., Wiley, New York, 1975.

Polymers decompose before they evaporate, so it appears that the concept of CED is not applicable to these materials. However, by finding a solvent with which a particular polymer mixes athermally, we can assign to the polymer the same CED as that solvent by Eq. (73). Thus cohesive energy densities for a number of polymers as well as low molecular weight solvents have been determined. Table 3.1 lists some representative examples of such data as $(CED)^{1/2}$.

Cohesive energy densities can be used on a limited basis to give quantitative meaning to the chemist's rule of thumb, "Like dissolves like." Specifically, the more alike a solvent and a polymer are in CED, the more nearly athermal their mixing will be. The more different the two are in this property, the more endothermic the mixing process will be. Remember that Eq. (72) makes no provision for exothermic mixing. In the next section we will see how such information might be used. Remember that Eq. (71) is not limited to endothermic situations. In the next section we will use the more general form of ΔH_m to consider the Flory–Huggins theory as it applies to the second virial coefficient.

3.8 The Second Virial Coefficient: Random Coils

Equations (65) and (71), respectively, give ΔS_m and ΔH_m according to the Flory–Huggins theory. From these components, ΔG_m can be assembled directly and by differentiation with respect to N_1, the Flory–Huggins expression for $\mu_1 - \mu_1^0$ may be obtained:

$$\mu_1 - \mu_1^0 = RT \ln \phi_1 + RT \left(1 - \frac{1}{n} \right) \phi_2 + \frac{1}{2} z \, \Delta w \, \phi_2^2 \tag{74}$$

It is conventional to let $\frac{1}{2} z \, \Delta w = \chi RT$, where χ is called the Flory–Huggins interaction parameter. Note that $\frac{1}{2} \Delta w$ is the energy change per 1,2 pair according to Eq. (69); therefore, with the coordination number z absorbed, the parameter χ measures this in units of RT. Finally, Eq. (21) establishes the connection between chemical potential and osmotic pressure; according to this equation,

$$- \pi \bar{V}_1 = RT \left[\ln \phi_1 + \left(1 - \frac{1}{n} \right) \phi_2 + \chi \phi_2^2 \right] \tag{75}$$

By expressing all volume fractions in terms of the solute, the first term on the right-hand side of Eq. (75) becomes $\ln(1 - \phi_2)$, which may be expanded (see Appendix A) as $-\phi_2 - \frac{1}{2} \phi_2^2$. With this modification, Eq. (75) becomes

$$\frac{\pi \bar{V}_1}{RT} = \phi_2 + \frac{1}{2} \phi_2^2 - \left(1 - \frac{1}{n} \right) \phi_2 - \chi \phi_2^2 \tag{76}$$

All that remains to be done to complete our derivation of the second virial coefficient in terms of the Flory–Huggins theory is convert volume fractions into practical concentration units. First of all, we can express the volume fraction of the solute in terms of partial molar volumes:

$$\phi_2 = \frac{n_1 \bar{V}_1}{n_1 \bar{V}_1 + n_2 \bar{V}_2} \simeq \frac{n_2}{n_1} \frac{\bar{V}_2}{\bar{V}_1} \tag{77}$$

where the approximate form applies to dilute solutions. If we recall Eqs. (24) and (32), Eq. (77) may be written

$$\phi_2 \simeq x_2 \frac{\bar{V}_2}{\bar{V}_1} = c_2 \frac{\bar{V}_2}{M_2} \tag{78}$$

Introducing practical concentration units to Eq. (78), the Flory–Huggins theory yields

$$\frac{\pi}{RTc} = \frac{\bar{V}_2}{M_2 n \bar{V}_1} + \frac{\frac{1}{2} - \chi}{\bar{V}_1} \left(\frac{\bar{V}_2}{M_2} \right)^2 c = \frac{1}{M_2} + \frac{\frac{1}{2} - \chi}{\bar{V}_1} \left(\frac{\bar{V}_2}{M_2} \right)^2 c \tag{79}$$

The last simplification is possible because the solute molecule is n times larger than a solvent molecule, and the same size relationship applies to the partial molar volumes. According to Eq. (79), the second virial coefficient is given by

$$B = \frac{\frac{1}{2} - \chi}{\bar{V}_1} \left(\frac{\bar{V}_2}{M} \right)^2 \tag{80}$$

Since χ measures Δw for the process described by Eq. (69), we see that Eq. (80) accomplishes what we set out to do, namely, relate the second virial coefficient to differences in the interaction energies between various pairs of molecules.

One of the first things to observe about Eq. (80) is the fact that it allows the second virial coefficient to be positive, negative, or zero, depending on whether χ is less than, greater than, or equal to $\frac{1}{2}$. Figure 3.4b reveals that positive and negative slopes are both observed in plots of reduced osmotic pressure versus concentration. We shall have more to say about this. For now it is enough to note that under conditions where $\chi = \frac{1}{2}$, $B = 0$ and solutions which are not too concentrated behave ideally. This is an advantage from the point of view of molecular weight determination, since a polymer–solvent system which meets this requirement satisfies the van't Hoff equation. This means that the molecular weight can be determined from a single solution without the need to do a series of experiments and extrapolate to infinite dilution.

The condition of $B = 0$ also marks the demarcation between good and poor solvent conditions. Positive values of the second virial coefficient characterize good solvents, and negative values poor solvents. This state of affairs—which is usually called the Θ condition—is very important in polymer chemistry. We shall encounter it again in our discussions of viscosity (Chap. 4) and light scattering (Chap. 5). Before examining the significance of the Θ condition any further, let us first consider the two states on either side of it.

A positive B value indicates a good solvent. According to Eq. (80), this is guaranteed for negative (or small positive) values of χ. Since χ is proportional to Δw and ΔH_m, a positive value for the second virial coefficient corresponds to an exothermic (or small endothermic) enthalpy of mixing. Conversely, a negative value for the second virial coefficient corresponds to a positive (or small negative) value of χ and an endothermic (or small exothermic) enthalpy of mixing. Classifying the solvent as good and poor on the basis of the slope of a plot of π/c versus c is therefore consistent with the contribution of ΔH_m to favorable mixing.

The Flory–Huggins expression for $\Delta \mu_m$, Eq. (74), can be related to other quantities besides osmotic pressure. One of the things that can be done is to calculate solubility limits and, therefore, miscibility diagrams for various polymer-solvent systems. Although we shall not pursue this in detail, it is of interest to note that this approach leads to the conclusion that $\chi = \frac{1}{2}$ is a critical value for this parameter; that is, the critical point on a miscibility diagram corresponds to $\chi = \frac{1}{2}$ for a polymer of infinite molecular weight [$n = \infty$ in Eq. (74)]. What is significant about this in the following: Suppose that we could somehow adjust the "goodness" of a particular solvent-polymer system, decreasing this quality from an initially good state. At $\chi = \frac{1}{2}$ the solvent would go from good to poor for a polymer of infinite molecular weight, and that particular molecular weight fraction would undergo phase separation. A polymer of somewhat lower molecular weight does not undergo phase separation until the value of χ is somewhat larger than $\frac{1}{2}$. The shorter the polymer chain, the more χ must exceed $\frac{1}{2}$ to reach the threshold of miscibility. Remember that $\chi > \frac{1}{2}$ corresponds to "poor" solvent conditions in terms of the second virial coefficient. The extension of these ideas to miscibility limits shows that systems of this sort are on their way to phase separation. It is only because the molecules have finite molecular weights that they remain in solution at all.

In the remarks above we considered adjusting solvent goodness as if it were an imaginary process. In fact, it can be carried out physically in two different ways for a particular polymer solution. One way is to lower the temperature of the system, another way is to dilute the initial system with a poor solvent. An application such as this is one place where the cohesive energy densities listed in Table 3.1 can be put to use. The utility of changes in solvent goodness lies in the possibility of fractionating a specimen with respect to molecular weight by such a variation. Since synthetic polymers are almost always highly polydisperse, the addition of a poor solvent to a solution of the polymer (or lowering its temperature) causes the highest molecular weight fraction to separate out of solution. The separated phase can be physically removed and the process repeated until a series of fractions are obtained. The same thing can be accomplished by temperature variations for a fixed polymer-solvent system.

Since the goodness of a polymer-solvent system can be adjusted by changing the temperature, it is desirable to recast Eq. (80) in a way that shows this effect explicitly. Toward this end we define the following identity to introduce a temperature variable into Eq. (80):

$$(\tfrac{1}{2} - \chi) = \psi(1 - \Theta/T) \tag{81}$$

We shall discuss this more fully below, but one thing to note immediately is that $T = \Theta$ describes the same state as is described by $\chi = \frac{1}{2}$, namely, the condition of $B = 0$. It is apparent that Θ is a temperature—variously known as the theta temperature or the Flory temperature. Introducing this parameter indicates why the $B = 0$ situation is called the Θ condition.

In order to justify the equivalence of the two sides of Eq. (81) consider the following steps:

1. The term $\frac{1}{2}$ on the left-hand side of Eq. (81) enters the Flory–Huggins theory as part of the series expansion of $\ln(1 - \phi_2)$ in the transition between Eqs. (75) and (76).

2. Thus $\frac{1}{2}R$ is an entropy contribution to the second virial coefficient.

3. Our expression for ΔS_m—on which this term is based—was derived by assuming purely random placement of polymer segments and solvent molecules. This may not be fully justified because of intermolecular forces which we did not consider in arriving at Eq. (66). We take advantage of this opportunity to allow for some bias in the placement of particles on the lattice and replace $\frac{1}{2}R$ with ΔS_ψ as the contribution of entropy to the second virial coefficient.

4. We recall from the transition between Eqs. (74) and (75) that χRT is the enthalpy contribution to the second virial coefficient. In the present context, we designate this ΔH_ψ.

5. By multiplying numerator and denominator of the terms by the same factors, the left-hand side of Eq. (81) can be transformed as follows:

$$\frac{1}{2} - \chi = \frac{\frac{1}{2}R}{R} - \frac{\chi RT}{RT} = \frac{\Delta S_\psi}{R} - \frac{\Delta H_\psi}{RT}$$

6. If $\Delta S_\psi/R$ is factored out of the last version, we obtain

$$\frac{\Delta S_\psi}{R}\left(1 - \frac{1}{T}\frac{\Delta H_\psi/R}{\Delta S_\psi/R}\right) = \psi\left(1 - \frac{1}{T}\frac{\Delta H_\psi/R}{\Delta S_\psi/R}\right)$$

where $\Delta S_\psi/R$ has been written ψ.

7. The ratio of ΔH_ψ to ΔS_ψ has Kelvin units; this quantity is called the theta temperature. Substituting Θ for this ratio gives the right-hand side of Eq. (81).

These manipulations may appear to add little except for needless complication to an interpretation of the second virial coefficient for random coils. Recall, however, that Eq. (81) allows the variation of solvent goodness caused by temperature changes to be described quantitatively. Thus the interaction parameter χ is used to describe how B changes when a polymer is dissolved in different solvents. By contrast, Θ is used to describe the variation in B when a given polymer–solvent system is examined at different temperatures. This has been done for the polystyrene-cyclohexane system at three different temperatures; the results are discussed in the following example.

Example 3.4   Values of the second virial coefficient along with some pertinent volumes are tabulated below for the polystyrene–cyclohexane system at three temperatures.

T (K)	$B \times 10^5$ (cm^3 g^{-2} mole)	$\bar{V}_1$ (cm^3 mole^{-1})	$\bar{V}_2/M_2$ (cm^3 g^{-1})
303	−4.55	109.5	0.930
313	4.45	110.9	0.935
323	9.01	112.3	0.940

Use these data to estimate Θ and ψ for this system. Does ΔS_ψ agree with the expected entropy contribution to the second virial coefficient?

Solution   The theta temperature is that value of T at which $B = 0$. It is apparent that B changes sign (i.e., passes through zero) about midway between 303 and 313 K. Equations (80) and (81) can be combined to give

$$B = \frac{(\bar{V}_2/M_2)^2}{\bar{V}_1}\,\psi\left(1 - \frac{\Theta}{T}\right)$$

which can be solved for ψ once Θ is known. Using $\Theta = 308$ K $= 35°$C, the following values of ψ can be calculated using the volumes provided:

T (K)	$1 - \Theta/T$	$\bar{V}_1 B(\bar{V}_2/M_2)^{-2}$	ψ
303	−0.017	−5.76 × 10^{-3}	0.339
313	0.016	5.65 × 10^{-3}	0.353
323	0.046	1.15 × 10^{-2}	0.249

For reasons we did not go into, it is correct to use the value of ψ interpolated to Θ conditions rather than, say, average the divergent values. Thus we estimate $\psi = 0.34$, or $\Delta S_\psi = 0.34R$. Since the Flory–Huggins theory predicts a contribution of $\frac{1}{2}R$ to the second virial coefficient, it seems that an additional entropy effect—given by $-0.16R$—must be included in order to account for the experimental B value. The fact that this "correction" to ΔS_m is negative implies that the entropy of mixing has been slightly overestimated by assuming random placement of the constituents on the lattice.

•

We have now looked at two models for the second virial coefficient of uncharged colloidal solutes. In Sect. 3.10 we will see that B depends on the magnitude of the particle charge for polyelectrolyte solutes.

3.9 Donnan Equilibrium and Electroneutrality

We have had no occasion as yet in this book to note that colloidal solutes may possess an electrical charge just like their low molecular weight counterparts. Chapters 12 and 13 are concerned with those properties of colloids which are direct consequences of the charge of the particles. For the present we shall introduce the idea of charged particles by examining the effect of the charge on the osmotic pressure of the system.

The charge on a colloidal particle may originate either from the dissociation of functional groups which are covalently bonded to the colloid or from the preferential adsorption of ions to the surface. The charge of a colloid cannot be regarded as a fixed quantity like molecular weight, but must be treated as a variable whose value depends on the nature and concentration of other components of the system. For example, proteins are positively charged at very low pH levels and negatively charged at very high pH levels; the point of electroneutrality varies from one protein to another. In the case of proteins, it is clearly the ionization of acidic and basic functional groups attached to the polypeptide chain that is primarily responsible for the charge characteristics of the molecules.

At this point we shall not concern ourselves any further with the origin of the charge of a colloidal system; rather, our attitude is that charge is one more characteristic that must be measured and understood in order to characterize certain systems fully.

Although it is not particularly difficult to formulate the thermodynamics of charged systems in perfectly general terms, the resulting notation is cumbersome. Instead of the completely general form, therefore, we shall consider a very specific case. The principal features will emerge clearly from this example; other situations may be readily derived by parallel arguments. The system we shall be concerned with consists of three components: Component 1 is the solvent, usually water; component 2 is the colloidal electrolyte; and component 3 is a low molecular weight uni-univalent electrolyte MX. We shall (arbitrarily) designate the colloidal electrolyte PX_z, consisting of a positively charged macroion having a valence number $+z$, paired with z X^- ions. It could be the negative ion of the colloidal electrolyte that is the macroion and the low molecular weight solute could have a different stoichiometry, but the essential features would remain the same.

In physical chemistry it is convenient to express concentrations as molalities and to use molality units to express the activity of the components. This is the convention we follow in this section. Accordingly, the standard state for a component consists of a solution in which that component has an activity of 1.0 mole (kg solvent)$^{-1}$.

vanaf deze paragraaf $7s$ $M = \left[\dfrac{mol}{kg\ solvent}\right]$

That specific situation we wish to consider is the osmotic equilibrium that develops in an apparatus which has a semipermeable membrane that is impermeable to the macroion only. That is, the membrane is assumed to be permeable not only to the solvent but also to both of the ions of the low molecular weight electrolyte, but not to the colloidal ion P^{z+}. At equilibrium the low molecular weight ions will be found on both sides of the membrane, but not in equal concentrations because of the presence of the macroions of one side of the membrane. We shall designate that side of the membrane which contains the macroions as the α phase and the solution from which the macroions are withheld as the β phase.

Equation (12) continues to describe the equilibrium condition; applying it to component 3 leads to the following:

$$\mu_{3,\alpha} = \mu_{3,\beta} \tag{82}$$

Substituting Eq. (13) for the β phase and Eq. (19) for the α phase which is under an osmotic pressure π yields

$$\mu_3^0 + RT \ln a_{3,\beta} = \mu_3^0 + RT \ln a_{3,\alpha} + \int_0^\pi \bar{V}_3 dp \tag{83}$$

At sufficiently low concentrations of the macroion the osmotic pressure term will be negligible compared with $RT \ln a_{3,\alpha}$, so Eq. (83) becomes

$$a_{3,\alpha} = a_{3,\beta} \tag{84}$$

It may be recalled from physical chemistry that the activity of a 1:1 electrolyte is given by the product of the activities of the positive and negative ions of the compounds; therefore

$$(a_{M,\alpha})(a_{X,\alpha}) = (a_{M,\beta})(a_{X,\beta}) \tag{85}$$

This expression describes what is known as the Donnan equilibrium. It does *not* say that the activity of M^+ and/or X^- is the same on both sides of the membrane, but that the ion activity product is constant on both sides of the membrane. In the sense that an ion product is involved, the Donnan equilibrium clearly resembles all other ionic equilibria.

Remembering that $a_+ = \gamma_+ m_+$ and $\gamma_\pm^2 = \gamma_+ \gamma_-$, where $\gamma_\pm$ is the mean ionic activity coefficient (appropriate to molality units), enables us to rewrite Eq. (85) as

$$(m_{M,\alpha})(m_{X,\alpha})\gamma_{\pm,\alpha}^2 = (m_{M,\beta})(m_{X,\beta})\gamma_{\pm,\beta}^2 \tag{86}$$

Of course, in the limit of infinite dilution $\gamma_\pm \to 1$. For the present we shall restrict our attention to sufficiently dilute solutions so that activity coefficients may be neglected and molalities may be used instead of activities. It might also be noted that in dilute solutions, where this simplification is apt to be valid, molality and molarity are almost equal.

Another factor which we have not yet taken into account is the requirement that both sides of the membrane be electrically neutral. For the α phase which contains the macroion, this condition is expressed by

$$zm_{P,\alpha} + m_{M,\alpha} = m_{X,\alpha} \tag{87}$$

In the β phase which contains only low molecular weight ions, electroneutrality requires

$$m_{M,\beta} = m_{X,\beta} \tag{88}$$

The significance of the Donnan equilibrium is probably best seen as follows. Combining Eqs. (86) and (88) yields

$$m_{M,\beta}^2 = m_{X,\beta}^2 = (m_{M,\alpha})(m_{X,\alpha}) \tag{89}$$

Next, we use Eq. (87) to substitute for either $m_{M,\alpha}$ or $m_{X,\alpha}$ in Eq. (89), obtaining the following quadratic equations for $m_{M,\alpha}$ and $m_{X,\alpha}$:

$$m_{M,\alpha}^2 + zm_P m_{M,\alpha} - m_{M,\beta}^2 = 0 \tag{90}$$

and

$$m_{X,\alpha}^2 + zm_P m_{X,\alpha} - m_{X,\beta}^2 = 0 \tag{91}$$

These expressions permit us to evaluate the concentration of the low molecular weight ions in the compartment with the macroions, the α phase, in terms of z and the concentration of ions in the other compartment.

The situation is most easily understood by considering a numerical example. Table 3.2 lists values of $m_{M,\alpha}$ and $m_{X,\alpha}$ calculated using Eqs. (90) and (91). These values have been determined for two different values of $m_{M,\beta} = m_{X,\beta}$: 10^{-3} and 10^{-2}. The parameter zm_P has been selected at six evenly spaced intervals between 10^{-3} and 10^{-2}. A solution containing 1 g of colloidal electrolyte of molecular weight 10^5 per 100 g of water, for example, would have a value of $m_P = 10^{-4}$; if the macroion carries a charge of $+10$, the parameter zm_P equals 10^{-3}. It is evident from Table 3.2 that the concentration of low molecular weight positive ions is larger in the β phase than in the α phase (which contains the macroions), and that the situation is reversed for the negative ions. The requirement of electroneutrality brings this about. To better show the uneven distribution of low molecular weight ions on the two sides of the membrane, Table 3.2 also lists the ratio of the concentrations on both sides of the membrane for both the positive and the negative ions.

Table 3.2 Values of $m_{M,\alpha}$ and $m_{X,\alpha}$ and the Ratios $(m_\alpha/m_\beta)_M$ and $(m_\alpha/m_\beta)_X$ for Two Values of m_β and a Range of Values of $zm_P{}^a$

zm_P	$m_{M,\beta} = m_{X,\beta} = 10^{-3}$				$m_{M,\beta} = m_{X,\beta} = 10^{-2}$			
	$m_{M,\alpha}$	$m_{X,\alpha}$	$(m_\alpha/m_\beta)_M$	$(m_\alpha/m_\beta)_X$	$m_{M,\alpha}$	$m_{X,\alpha}$	$(m_\alpha/m_\beta)_M$	$(m_\alpha/m_\beta)_X$
10^{-3}	6.18×10^{-4}	1.62×10^{-3}	0.62	1.62	9.51×10^{-3}	1.05×10^{-2}	0.95	1.05
2×10^{-3}	4.14×10^{-4}	2.41×10^{-3}	0.41	2.41	9.05×10^{-3}	1.11×10^{-2}	0.91	1.11
4×10^{-3}	2.36×10^{-4}	4.24×10^{-3}	0.24	4.24	8.20×10^{-3}	1.22×10^{-2}	0.82	1.22
6×10^{-3}	1.62×10^{-4}	6.16×10^{-3}	0.16	6.16	7.44×10^{-3}	1.34×10^{-2}	0.74	1.34
8×10^{-3}	1.23×10^{-4}	8.12×10^{-3}	0.12	8.12	6.77×10^{-3}	1.48×10^{-2}	0.67	1.48
10^{-2}	9.9×10^{-5}	1.01×10^{-2}	0.10	10.09	6.18×10^{-3}	1.62×10^{-2}	0.62	1.62

^{a}All concentrations are in moles per kilogram of solvent. The α phase contains the positive macroions.

Two conclusions may be drawn readily from an inspection of the results of Table 3.2. First, the uneven distribution of the simple ions becomes more pronounced as the quantity zm_P increases. The low molecular weight ions are free, after all, to pass through the membrane; it is only electroneutrality that holds them back. The more macroions present or the higher charge they carry, the more asymmetrically the simple electrolyte will be distributed. Table 3.2 also shows that the uneven distribution of low molecular weight electrolyte becomes less pronounced as the concentration of this electrolyte is increased. We return to this point in the next section, when we discuss the osmotic pressure of charged systems.

The combined effects of electroneutrality and the Donnan equilibrium permits us to evaluate the distribution of simple ions across a semipermeable membrane. If electrodes reversible to either the M^+ or the X^- ions were introduced to both sides of the membrane, there would be no potential difference between them; the system is at equilibrium and the ion activity is the same in both compartments. However, if calomel reference electrodes are also introduced into each compartment in addition to the reversible electrodes, then a potential difference will be observed between the two reference electrodes. This potential, called the membrane potential, reflects the fact that the membrane must be polarized because of the macroions on one side. It might be noted that polarized membranes abound in living systems, but the polarization there is thought to be primarily due to differences in ionic mobilities for different solutes rather than the sort of mechanism that we have been discussing. We shall return to a more detailed discussion of the electrochemistry of colloidal systems in Chap. 12.

3.10 The Osmotic Pressure of Charged Colloids

Now let us turn our attention to the osmotic pressure generated by the macroion in this system. Since we have already restricted ourselves to dilute solutions, it is adequate for our purposes to substitute into Eq. (35), making allowance for the fact that we have been expressing concentrations as molality in this section. The volume of 1 kg of solvent equals $1000V_1^0/M_1$, so Eq. (35) becomes

$$\pi \frac{1000V_1^0}{M_1} = mRT \tag{92}$$

where m is the molality of the solute responsible for the osmotic pressure and V_1^0 is the molar volume of the solvent. Since there are solute molecules on both sides of the membrane, the osmotic pressure will be due to the excess solute on the side of the membrane that carries the macroion (the α phase). Therefore we may replace m in Eq. (92) by

$$m = m_{P,\alpha} + m_{M,\alpha} + m_{X,\alpha} - m_{M,\beta} - m_{X,\beta} \tag{93}$$

Substituting from the electroneutrality equations (87) and (88) transforms Eq. (93) into

$$m = m_{P,\alpha} + m_{M,\alpha} + zm_{P,\alpha} - m_{M,\alpha} - 2m_{M,\beta} \tag{94}$$

This is further modified by the Donnan relationship, Eq. (86), also rewritten to include electroneutrality, to give

$$m = m_{P,\alpha}(1 + z) + 2m_{M,\alpha} - 2[m_{M,\alpha}(zm_{P,\alpha} + m_{M,\alpha})]^{1/2} \tag{95}$$

Combining Eqs. (92) and (95) yields the following for the osmotic pressure of the system:

$$\pi = \frac{M_1 RT}{1000V_1^0}\left[(1 + z)m_P + 2m_M - 2m_M\left(1 + \frac{zm_P}{m_M}\right)^{1/2}\right] \tag{96}$$

It should be noted that all concentrations have been expressed in terms of the molality of the solute in the compartment which contains the macroion, so the α subscript is no longer necessary.

Although Eq. (96) is rather awkward as written, several highly informative variations of it are obtained by considering different limiting situations. These are summarized in Table 3.3 for the cases in which $m_M = 0$, $m_M \gg m_P$, and $m_M > m_P$. In the table, the expressions for π which follow from Eq. (96) in each of these cases are written both in terms of the molality of the macroion m_P and with the concentration of the latter expressed in weight per unit volume c_2, specifically grams of colloid per liter of solution.

We shall not present the algebraic manipulations which lead to the various forms presented in Table 3.3; however, two relationships which are involved in generating these forms might be noted. First, the quantity $(1 + zm_P/m_M)^{1/2}$ in Eq. (96) may be approximated by the binomial series expansion (see Appendix A) to give

$$\left(1 + \frac{zm_P}{m_M}\right)^{1/2} = 1 + \frac{1}{2}\frac{zm_P}{m_M} - \frac{1}{8}\left(\frac{zm_P}{m_M}\right)^2 + \cdots \tag{97}$$

When $m_M \gg m_P$, only the first two terms are retained; when $m_M > m_P$, the first three are used. Second, the relationship between molality and grams per liter units is given by the following in dilute solutions:

$$m_P = \frac{1000V_1^0}{M_1 M_2}c_2 \tag{98}$$

Table 3.3 Special Cases of Eq. (96) for the Osmotic Pressure of a Charged Colloid in the Presence of a Low Molecular Weight Salt

	m_P(mole kg^{-1})	c_2(g liter^{-1})
$m_M = 0$	$\pi = \dfrac{M_1 RT}{1000 V_1^0}(1 + z)m_P$	$\pi = \dfrac{RT(1 + z)c_2}{M_2}$
$m_M \gg m_P$	$\pi = \dfrac{M_1 RT}{1000 V_1}m_P$	$\pi = \dfrac{RTc}{M_2}$
$m_M > m_P$	$\pi = \dfrac{M_1 RT}{1000 V_1}\left(m_P + \dfrac{z^2}{4}\dfrac{m_P^2}{m_M}\right)$	$\pi = \dfrac{RTc}{M_2} + \dfrac{z^2}{4}\dfrac{1000V_1^0 RTc^2}{M_1 M_2^2 m_M}$
		$\dfrac{\pi}{RTc} = \dfrac{1}{M_2} + \dfrac{z^2 1000c}{4\rho_1 M_2^2 m_{MX}}$

All the forms presented in Table 3.3 are readily obtained from Eq. (96) by incorporating Eq. (97) or (98) or both into Eq. (96). Now let us consider the physical significance of the resulting special cases.

The two extreme values of m_M, $m_M = 0$ and $m_M \gg m_P$ are especially interesting. The former corresponds to the case of no added salt (since the macroion is positive), and the latter to a large excess or "swamping" amount of added salt. Now suppose an osmotic pressure experiment were conducted on two solutions of the same colloid, assumed to have a fixed charge z, with the objective of determining the molecular weight of the colloid. Further suppose that the two determinations differ from one another in the sense that one corresponds to zero added salt, and the second to swamping electrolyte conditions. Finally, suppose the results are simply interpreted in terms of Eq. (35) to yield the molecular weight of the colloid. Comparing Eq. (35) with the results listed in Table 3.3 reveals that the correct molecular weight would be obtained for the charged colloid under swamping electrolyte conditions, but an apparent molecular weight less than the true weight by a factor $z + 1$ is obtained in the absence of salt.

How are we to understand this odd result? The answer is easy when we remember that osmotic pressure counts solute particles. The macroion cannot pass through the semipermeable membrane. In the absence of added salt, its counterions will not pass through the membrane either, since the electroneutrality of the solution must be maintained. Therefore the equilibrium pressure is that associated with $z + 1$ particles. Failure to consider the presence of the counterions will lead to the interpretation of a low molecular weight for the colloid. As we already saw, the presence of increasing amounts of salt leads to a leveling off of the ion concentrations on the two sides of the membrane. The effect of the charge on the macroion is essentially "swamped out" with increasing electrolyte.

One interesting aspect of the limiting case of swamping electrolyte is the fact that the conclusion is totally independent of the specific nature of the ions. This is a partial justification for an assumption that was implicitly made at an earlier point. In writing Eqs. (87) and (88), we assume that the only ions present are M^+, X^-, and P^{z+}. In aqueous solutions, however, H^+ and OH^- are always present also. The latter have clearly been assumed to be negligible in writing Eqs. (87) and (88). The swamping electrolyte concentration may always be chosen to justify neglecting these contributions.

Table 3.3 also includes an approximation for the case in which the concentration of the salt exceeds that of the colloid, but not to the swamping extent, $m_M > m_P$. Comparison of that case with the result given in Eq. (34) suggests that the contribution of charge to the second virial coefficient of the solution is given by

$$B = \frac{1000z^2}{4M_2^2 \rho_1 m_{MX}} \tag{99}$$

Strictly speaking, this contribution should be added to the excluded volume of the particles. The latter becomes more important as the concentration of the salt increases, a conclusion that may be seen by considering two facts:

1. Charged colloids behave as if they were uncharged under swamping electrolyte conditions.
2. The electrostatic contribution to B, Eq. (99), is inversely proportional to salt concentration.

This effect may be qualitatively understood as follows. In a charged system the colloid consists of the macroion *and* its low molecular weight counterions. The latter are, of course, distributed through a portion of the solution in the neighborhood of the macroion. Thus we visualize the colloidal ion as being surrounded by an ion atmosphere, the same sort of model that is invoked in the Debye–Hückel theory of electrolyte nonideality. The "excluded volume" that is required, therefore, includes both the volume of the colloidal particle *and* the volume of that part of the solution which contains the counterions. It is reasonable to expect the latter to be sufficiently larger than the former in dilute solutions so that the volume actually excluded by the colloidal particle can be neglected.

The precise distribution of counterions around a charged particle is the subject matter of Chap. 12 and, to a lesser extent, Chap. 13. In those chapters we shall see that the extent of the domain over which the ion atmosphere extends decreases as the electrolyte concentration increases.

We have already seen that the second virial coefficient may be determined experimentally from a plot of the reduced osmotic pressure versus concentration. Since all other quantities in Eq. (99) are measurable, the charge of a macroion may be determined from the second virial coefficient of a solution with a known amount of salt. As an illustration of the use of Eq. (99), we consider the data of Fig. 3.5 in the following example.

Example 3.5   In Example 3.3, we evaluated M and B for bovine serum albumin at pH = 5.37, where the molecule is known to be uncharged. Use the data in Fig. 3.5 to evaluate B and, from it, the charge of the molecule at pH = 7.00. The data in Fig. 3.5 was measured in 0.15 m NaCl.

Solution   The slope of the line at pH = 7.00 is 2.28×10^{-3} Torr kg^2 g^{-2}; we convert to SI units as in Example 3.3:

$$2.28 \times 10^{-3} \text{ Torr kg}^2 \text{ g}^{-2} \times \frac{1.01 \times 10^5 \text{ N m}^{-2}}{760 \text{ Torr}} \times \left(\frac{10^{-3} \text{ m}^3}{\text{kg}} \right)^2 \left(\frac{10^3 \text{ g}}{\text{kg}} \right)^2$$

$$= 0.303 \text{ N m}^4 \text{ kg}^{-2}$$

Division by RT gives B:

$$0.303/(8.314)(298) = 12.23 \times 10^{-5} \text{ m}^3 \text{ kg}^{-2} \text{ mole}$$

As noted above, it is the difference between this value and the B value for the uncharged molecule (Example 3.3) that is interpreted by Eq. (99). Also recall that M was determined to be 69,600 g mole^{-1} in the example cited. Therefore

$$(12.23 - 7.35)10^{-5} = \frac{1000z^2}{4M_2^2 \rho_1 m_{MX}}$$

$$z^2 =$$

$$\frac{4(4.88 \times 10^{-5} \text{ m}^3 \text{ kg}^{-2} \text{ mole})(69.6 \text{ kg mole}^{-1})^2(1.0 \text{ g cm}^{-3})(0.15 \text{ mole kg}^{-1})(100 \text{ cm m}^{-1})^3}{1000 \text{ g kg}^{-1}}$$

$$= 142$$

and $z = \pm 12$.

Since the pH is higher than that at which the molecule is uncharged, the albumin must be negatively charged. Hence we identify z as -12 in this case.

•

3.11 Dialysis

Substances with particles in the colloidal size range are often obtained in a form which contains low molecular weight impurities. For example, enzymes are separated from homogenized tissue samples by extraction in a buffer solution. The enzyme preparation, therefore, is "contaminated" by the components of the buffer. Likewise, synthetic high polymers generally contain unreacted monomer, initiator, and catalyst. There are many experiments in which traces of low molecular weight impurities would have no effect, as, for example, in sedimentation. This chapter has shown clearly, however, that the presence of low molecular weight solutes may have large effects in an osmotic pressure experiment if these substances are retained by the membrane, either directly or through the electroneutrality condition.

Experiments with semipermeable membranes not only point out the need for purification, but also suggest a means by which this may be accomplished. The procedure of dialysis is one technique for removing low molecular weight salts or nonelectrolytes from a colloid. The method consists merely of enclosing the colloid to be purified in a bag which is made of some semipermeable material. The sealed bag is then placed in a quantity of the solvent. The membrane must be permeable to the solvent and to any low molecular weight impurities which are present, but impermeable to the colloid. As a result of the semipermeability of the bag, the impurities will distribute themselves through both compartments. The outer portion is replaced often or even continuously, so the low molecular weight impurities are gradually flushed away.

Since the membrane is also permeable to the solvent, the latter simultaneously diffuses into the bag, diluting the colloid. Ample air space must be present in the bag to begin with, otherwise it will rupture owing to the pressure developed by the solvent imbibed. There is also a danger that the porosity of the membrane will increase if the bag is stretched as a result of internal pressure buildup. Cellophane tubing is most commonly used as the membrane material. It is sold in rolls for this purpose and may be cut to length and tied at the ends to make the required bags.

Purification by dialysis is a slow process. Its rate is increased, however, by increasing the surface area of the membrane, since that is where the exchange of solute between the two phases takes place. Stirring and frequent replacement of the solvent accelerate the process by maintaining the maximum gradient of concentration across the membrane. With ionic contaminants, the rate of dialysis may be enhanced by placing electrodes in the compartment surrounding the enclosed colloid and taking advantage of the migration of the ions in an electric field. This modification is known as electrodialysis.

Finally, it might be noted that colloids may be concentrated by a slight modification of the dialysis procedure. The liquid against which the colloid of interest is being dialyzed may itself be a concentrated colloid. With aqueous dispersions, for example, polyethylene oxide solutions may be used as the second colloid.

The second colloid is prepared at higher activity; therefore the solvent is drawn toward the more concentrated phase. This increases the concentration of the colloid of interest. Alternatively, the concentration increase may be accomplished by allowing the solvent to evaporate from the outer surface of the bag.

Just as with osmotic pressure, the membranes in dialysis must be carefully selected to be compatible with the system under study. Specifically, this amounts to impermeability with respect to the colloid(s) involved and permeability with respect to low molecular weight components.

3.12 Reverse Osmosis

As we saw in Sect. 3.2, samples of solution and solvent separated by a semipermeable membrane will be at equilibrium only when the solution is at a greater pressure than the solvent. This is the osmotic pressure. If the solution is under less pressure than the equilibrium osmotic pressure, solvent will flow from the pure phase into the solution. If, on the other hand, the solution is under a pressure greater than the equilibrium osmotic pressure, the pure solvent will flow in the reverse direction, from the solution to the solvent phase. In the latter case the semipermeable membrane functions like a filter which separates solvent from solute molecules. In fact, the process is referred to in the literature by the terms *hyperfiltration* and *ultrafiltration*, as well as *reverse osmosis*; however, the latter name is enjoying common usage these days.

As we have seen already, the property of membrane semipermeability applies to all sorts of systems. Likewise, reverse osmosis may be applied to a wide variety of systems. An application that has attracted a great deal of interest in recent years is the production of potable water from saline water. Since no phase transitions are involved, as, for example, in distillation, the method offers some prospect of economic feasibility in coastal regions.

Cellulose acetate seems to be the most thoroughly investigated of many possible membrane materials. Cellulose acetate membranes are capable of yielding 96–98% retention of NaCl, for example, and of delivering about 0.2 cm^3 s^{-1} atm^{-1} m^{-2}. This amounts to about 50 gal.

day^{-1} ft^{-2} at 100 atm. Note that in this application of osmometry, it is not the solute but the membrane that comprises the colloidal system. In this case, the solid portion of the membrane would be the continuous phase and the pores, necessarily small if the membrane is to be effective, the "dispersed" phase. According to this point of view, numerous aspects of membrane technology become part of the interests of colloid and surface chemists. Practical desalination by reverse osmosis depends on a membrane which (1) has enough contact area to process large volumes of solution and (2) is thin enough to do so rapidly, yet (3) sturdy enough to withstand these pressures. Since some of these points work in opposition to each other, the technical problem is one of optimization.

It is the rate of separation rather than the efficiency of salt retention that is the primary practical issue in the development of reverse osmosis desalination. In addition to a variety of other factors, the rate of reverse osmotic flow depends on the excess pressure across the membrane. Therefore the problem of rapid flow is tied into the technology of developing membranes capable of withstanding high pressures. The osmotic pressure of sea water at 25°C is about 25 atm. This means that no reverse osmosis will occur until the applied pressure exceeds this value. This corresponds to a water column about 840 ft high at this temperature.

Note that a solution more concentrated than the original one also results from the reverse osmosis process. This means that the method of reverse osmosis may also be used as a method for concentrating solutions. Fruit juices and radioactive wastes, for example, have been concentrated by this method.

References

1. A. W. Adamson, *A Textbook of Physical Chemistry*, 2nd ed., Academic, New York, 1979.
2. R. U. Bonnar, M. Dimbat, and F. H. Stross, *Number Average Molecular Weights*, Wiley, New York, 1958.
3. P. J. Flory, *Principles of Polymer Chemistry*, Cornell University Press, Ithaca, N.Y., 1953.
4. P. C. Hiemenz, *Polymer Chemistry: The Basic Concepts*, Marcel Dekker, New York, 1984.
5. H. Morawetz, *Macromolecules in Solution*, Wiley, New York, 1965.
6. E. G. Richards, *An Introduction to the Physical Properties of Large Molecules in Solution*, Cambridge University Press, Cambridge, 1980.
7. S. Souririjan, *Reverse Osmosis*, Academic, New York, 1970.
8. C. Tanford, *Physical Chemistry of Macromolecules*, Wiley, New York, 1961.

Problems

1. Criticize or defend the following proposition: As proof that no low molecular weight fractions of polymer have passed through the membrane in an osmotic pressure experiment, the following test may be performed. A quantity of "poor" solvent is added to an aliquot taken from the solvent side of the membrane. The absence of precipitate proves that no low molecular weight polymer passed through the membrane.

2. The osmotic pressure of solutions of a fractionated, atactic poly(isopropyl acrylate) solution was measured at 25°C with the following results*:

π (g cm^{-2})	1.39	2.46	4.20	6.52
$c \times 10^2$ (g cm^{-3})	0.47	0.69	1.05	1.36

   Prepare a plot of π/c versus c for these results and evaluate $(\pi/c)_0$. Calculate M and B for this system. Define what is meant by an "atactic" polymer and compare with syndiotactic and isotactic polymers. List reference(s) consulted for these definitions.

3. An important assumption made in truncating Eq. (34) is that the third virial coefficient is small. It is known (see Ref. 3) that the third virial coefficient depends strongly on the second so that it approaches zero in poor solvents even faster than B does. In fact, if Γ_2 is defined as the product BM_2, it is known that Γ_3 is approximately $0.25\Gamma_2^2$ in a good solvent; that is, Eq. (34) may be written with one additional term as

$$\frac{\pi}{RTc} = \frac{1}{M}(1 + \Gamma_2 c + 0.25\Gamma_2^2 c^2)$$

Describe how this result may be used to facilitate the evaluation of M and B in the event that a plot of reduced osmotic pressure versus c still contains too much curvature to permit a meaningful straight line to be drawn.

4. Osmotic pressures for aqueous solutions of n-dodecylhexaoxyethylene monoether, $C_{12}H_{25}$ $(OC_2H_4)_4OC_2H_5$, were measured at 25°C. At concentrations below 0.038 g liter^{-1} no osmotic pressure develops, indicating complete membrane permeability. Above this concentration a pressure develops, indicating the presence of impermeable species. In the following data† these pressures are reported for various $c - c_0$ values, where $c_0 = 0.038$ g liter^{-1}, the threshold for an osmotic effect:

π (cm)	4.90	6.53	7.62	10.58
$c - c_0$ (g liter^{-1})	29.72	38.12	43.90	58.46

---

*J. E. Mark, R. A. Wessling, and R. E. Hughes, *J. Phys. Chem.*, **70**:1895 (1966).

†D. Attwood, P. H. Elworthy, and S. B. Kayne, *J. Phys. Chem.*, **74**:3529 (1970).

Plot $\pi/(c - c_0)$ versus $c - c_0$ to evaluate the molecular weight and B for the species responsible for the osmotic pressure.

5. The data of the preceding problem may be interpreted by assuming the following model. Above 0.038 g liter^{-1}, solute molecules associate into aggregates. By comparing the M value obtained in Problem 4 with the molecular weight of the original ether, calculate the number of molecules in the aggregate according to this model. Assuming the colloidal particles are spherical, use the second virial coefficient evaluated in Problem 4 to estimate the molar volume and radius of the aggregates. Do the quantities calculated in this problem seem reasonably self-consistent?

6. The cohesive energy density of a low molecular weight liquid is given by the heat of vaporization of that liquid, expressed per unit volume. Verify the value given for the CED of acetone in Table 3.1 from the facts that $\Delta H_v = 30.2$ kJ mole^{-1} and $\rho = 0.792$ g cm^{-3}. Would it be better to use the change in internal energy upon vaporization ΔU_v rather than ΔH_v as a measure of intermolecular attraction? Does it make much difference quantitatively whether ΔU_v or ΔH_v is used?

7. The osmotic pressure of polystyrene (PS) solutions in toluene and methyl-ethyl ketone (MEK) was measured* at 25°C, and the results were analyzed to give B values of 4.59×10^{-4} and 1.39×10^{-4} cm^3 g^{-2} mole, respectively, for the two solutions. Use these results to criticize or defend the following proposition: According to Table 3.1, $\sqrt{\text{CED}} = 8.5$–9.1 for polystyrene. For toluene and MEK, $\sqrt{\text{CED}}$ equals 8.9 and 9.04, respectively. Since the MEK solution has a smaller B value than the toluene solution, it appears that the best value to use for $\sqrt{\text{CED}}$ for polystyrene is at the upper end of the range of values given and close to the value of MEK. In this way the quantity $[(\sqrt{\text{CED}})_{\text{MEK}} - (\sqrt{\text{CED}})_{\text{PS}}]^2$ will be smaller than the same quantity for polystyrene in toluene. This is consistent with the order of the B values.

8. The solvent activity in a solution of polybutadiene in benzene was determined by measuring† the vapor pressure p_1 of benzene over solutions containing various concentrations of polymer. A plot of $\ln(p_1/p_1^0) - \ln \phi_1 - (1 - 1/n)\phi_2$ versus ϕ_2^2—in which p_1^0 is the vapor pressure of the pure benzene—yields a straight line having an intercept of zero and a slope equal to 0.33. Evaluate the interaction parameter χ from this result. Is the Θ temperature above or below the experimental temperature? Explain.

9. Krigbaum and Geymer‡ measured the osmotic pressure of polystyrene in cyclohexane at several different temperatures. The following is a sample of their results for a single fraction of polymer:

*C. Bawn, R. Freeman, and A. Kamaliddin, *Trans. Faraday Soc.*, *46*:862 (1950).

†R. S. Jessup, *J. Res. Nat. Bur. Stand.*, *60*:47 (1958).

‡W. R. Krigbaum and D. O. Geymer, *J. Am. Chem. Soc.*, *81*:1859 (1959).

	$T = 24°C$		$T = 34°C$		$T = 44°C$
c (g cm^{-3})	$(\pi/RTc) \times 10^6$ (mole g^{-1})	c (g cm^{-3})	$(\pi/RTc) \times 10^6$ (mole g^{-1})	c (g cm^{-3})	$(\pi/RTc) \times 10^6$ (mole g^{-1})
0.0976	8.0	0.0081	13.3	0.0959	18.6
0.182	6.0	0.0201	14.2	0.178	28.1
0.259	8.7	0.0964	14.2	0.255	40.0
		0.180	18.7		
		0.257	26.2		

Plot all of these points on the same graph as π/RTc versus c. Although considerable nonideality exists, estimate the limiting slope at each temperature. Interpretation is assisted by realizing that each set of data approaches the same intercept as $c \to 0$. What is the approximate molecular weight of the polymer? How do the values of B estimated from the limiting slopes compare with the values given for the same system over an overlapping range of temperatures in Example 3.4?

10. Solutions of bovine serum albumin in 0.15 M NaCl were studied at other pH levels in addition to those shown in Fig. 3.5. The following data are examples of additional measurements (reference in legend of Fig. 3.5):

pH	m_P (g protein kg^{-1})	π (mm Hg)
6.19	57.71	21.48
6.64	56.17	21.40

Since the limiting value of π/m_P shown in Fig. 3.5 applies to these data also, it is possible to evaluate the second virial coefficient at these pH levels. Evaluate B and z, the effective protein charge, at the pH values shown. (*Note*: Slight variation in NaCl concentrations at these different pH levels should be taken into account for a more accurate determination of z.)

11. The osmotic pressure of salt-free (electrodialyzed) bovine serum albumin solutions was measured at pH = 5.37 (±?) (reference listed in Fig. 3.5). At this pH the net charge of the protein molecules is zero. The following data were obtained in different runs:

π (mm Hg)	4.26	7.44	8.14	12.97	11.31	19.62
m_P (g protein kg^{-1})	19.56	19.71	40.88	46.05	60.81	63.25

Discuss the reasons why it is so difficult to obtain meaningful osmotic pressure data in salt-free solutions. Consider specifically the reconciliation of the electrolyte-free aspect of the experiment with the accurate control of pH.

12. In reverse osmosis both solvent and solute diffuse because of gradients in their chemical potentials. For the solvent there is no gradient of chemical potential at an osmotic pressure of π; at applied pressures p greater than π, there is such a gradient which is proportional to the difference $p - \pi$. To a

first approximation the gradient of the solute chemical potential is independent of p and depends on the difference between concentrations on opposite sides of the membrane. This leads to the result that the fraction of solute retained varies as $[1 + \text{const.}/(p - \pi)]^{-1}$. Verify that the following data* for a reverse osmosis experiment with 0.1 M NaCl and a cellulose acetate membrane follow this relationship:

Applied p (atm)	10	13	20	38	51	75
Percent salt retained	63	79	88	94	95	97

(π is about 2.6 atm for 0.1 M NaCl.)

---

*Data of J. E. Breton, Jr., cited by H. K. Lonsdale, in *Desalination by Reverse Osmosis* (U. Merton, ed.), MIT Press, Cambridge, 1966.

4

THE VISCOSITY OF DILUTE DISPERSIONS

Imagine that your Tradesman drags behind his regular and respectable vertex, a parallelogram of twelve or thirteen inches in diagonal:—What are you to do with such a monster sticking fast in your house door?

[From Abbott's *Flatland*]

4.1 Introduction

The way liquids flow is one of their most obvious properties. We use a variety of terms in everyday and technical language to describe this property. Thus we speak of the "thickness" of cream, the "weight" of oil, and the "leveling" of paint. The science student will probably recognize that all these terms describe in one way or another the property known as the viscosity of the liquid. But substituting a technical term does little to make this somewhat elusive property more intelligible.

In this chapter we discuss viscosity in terms of several different models and assumptions. It is helpful to be aware of these differences from the outset. As an example, fluids are treated as continuous matter in discussing experimental viscometry; we use a statistical, particulate model in discussing polymer solutions. Another example is the development of different models for lyophobic and lyophilic systems. The Einstein theory of viscosity is based on a model of rigid spherical particles. As such, it is most appropriate for two-phase systems or solutions containing solutes with rigid structures. The Kirkwood–Riseman theory, by contrast, applies to flexible chains and is suitable for synthetic polymers in solution. A final example is the assumption—made through the bulk of the chapter—that flow has no effect on the size, shape, or structure of the dispersed units. In Sect. 4.11, we relax this

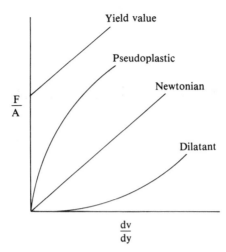

Figure 4.1   Illustration of the relationship between applied force per unit area and fluid velocity.

constraint and consider the viscous behavior of flocculated systems where the flow may disrupt the flocs.

Both the fluid mechanics and the statistical mechanics upon which some of the key theoretical results of the chapter are based are sufficiently complicated that we only sketch the highlights of these topics. We attempt to impart some physical plausibility to these theories, however, by using both force and energy perspectives in discussing the viscous resistance to flow.

4.2   The Coefficient of Viscosity

The coefficient of viscosity was introduced in Chap. 2, but this parameter is elusive enough to warrant further comment. In this section we shall examine the definition of the coefficient of viscosity—the viscosity, for short—of a fluid. This definition leads directly to a discussion of some experimental techniques for measuring viscosity; these will be taken up in the following sections.

Imagine two parallel plates of area A between which is sandwiched a liquid of viscosity η. If a force F parallel to the x direction is applied to one of these plates, it will move in the x direction as shown in Fig. 4.1. Our concern is the description of the velocity of the fluid enclosed between the two plates. In order to do this it is convenient to visualize the fluid as consisting of a set of layers stacked parallel to the boundary plates. At the

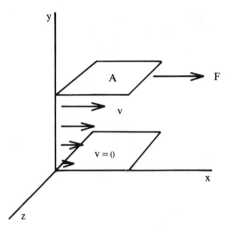

Figure 4.2 Comparison of Newtonian liquids with several forms of non-Newtonian behavior.

boundaries those layers which are in contact with the plates are assumed to possess the same velocities as the plates themselves; that is, $v = 0$ at the lower plate and equals the velocity of the moving plate at that surface. This is the non-slip condition that we described in Sect. 2.3. Intervening layers have intermediate velocities. This condition is known as laminar flow and is limited to low velocities. At higher velocities turbulence sets in, but we will not worry about this complication.

We can imagine within the fluid two layers separated by dy, over which distance the velocity changes by an amount dv. Therefore dv/dy defines a velocity gradient; Newton's law of viscosity states that F/A is proportional to dv/dy. The viscosity η of the sandwiched fluid is the factor of proportionality:

$$\frac{F}{A} = \eta \frac{dv}{dy} \tag{1}$$

Since $dv = dx/dt$ and dx/dy defines the shear acting on the sample, the velocity gradient may be written $d(\text{shear})/dt$ and is called the rate of shear. Note that dv/dy has units time^{-1}. Thus a dimensional statement of Eq. (1) gives mass length time^{-2} length^{-2} = η time^{-1}, which shows that η has dimensions mass length^{-1} time^{-1}, or kg m^{-1} s^{-1} in SI units. We observed in Chap. 2 that the cgs unit of viscosity, the poise, is widely used and that 10 P = 1 kg m^{-1} s^{-1}. Fluids which obey the form predicted by Eq. (1) are said to be Newtonian. Figure 4.2 is a sketch of F/A versus the velocity gradient for several different modes of behavior. For a Newtonian fluid this representation gives a straight line of zero intercept and slope equaling η. Non-Newtonian fluids generally show nonlinear plots; their "viscosity," the slope of the tangent to the curve at various points, is a function of the rate of shear. Most actual representations of experimental results display the data with the coordinates interchanged from the way they are shown in Fig. 4.2. In that case it is the cotangent of the angle which describes the slope of the line at any point that determines the true (if Newtonian) or apparent (if non-Newtonian) viscosity of the system. Figure 4.16 is an example of this alternative representation. For the present, Newtonian behavior is our concern. We will discuss some examples of non-Newtonian behavior in Sect. 4.11.

A second interpretation of η is as valid as Eq. (1) and perhaps more illuminating. To arrive at this alternative, we multiply both sides of Eq. (1) by dv/dy:

$$\frac{F}{A}\frac{dv}{dy} = \eta \left(\frac{dv}{dy}\right)^2 \tag{2}$$

Now consider the following points to interpret $(F/A)(dv/dy)$:

1. Write dv as dx/dt and regroup terms: $(f\,dx/A\,dy)/dt$.
2. A force times an increment of distance $F\,dx$ equals an increment of energy dE.
3. An area times an increment of distance $A\,dy$ equals an increment of volume dV.
4. Since the force under consideration measures viscous *resistance* to flow, the quantity dE/dV measures the energy dissipated per unit volume.
5. Dividing dE/dV by dt gives the rate of energy dissipation per unit volume.

Based on these ideas, Eq. (2) can be rewritten

$$\frac{dE/dV}{dt} = \eta \left(\frac{dv}{dy}\right)^2 \tag{3}$$

which shows that the volume rate of energy dissipation is proportional to the square of the velocity gradient with the viscosity of the fluid as the factor of proportionality.

Equations (1) and (3) are equivalent as definitions of the viscosity of a fluid. To convince ourselves that η as defined by these expressions does indeed measure resistance to flow, consider two liquids of widely different viscosity, say, water and molasses:

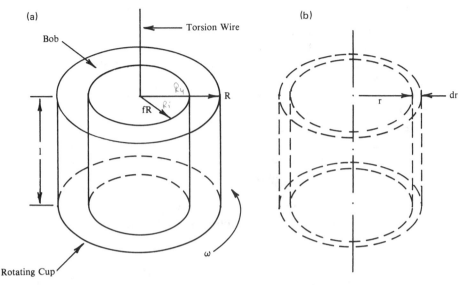

Figure 4.3   Schematic of a concentric cylinder viscometer: (a) geometry of a cup and bob and (b) volume element within a liquid gap.

1. Imagine an arrangement like that shown in Fig. 4.1 and consider the force that must be applied to sustain a velocity gradient of, say, 1 s^{-1}. The higher the viscosity of the fluid, the greater the force. Molasses would require more force than water.
2. Imagine the rate at which energy is dissipated in maintaining a unit velocity gradient. The higher the viscosity, the faster the energy dissipation per unit volume. Molasses would dissipate energy more rapidly than water.

In the next section we consider an experimental approach to viscosity. We generate the apparatus of interest by wrapping—in our imagination—the fluid in Fig. 4.1 into a closed ring around the z axis. The two rigid surfaces then describe concentric cylinders and the instrument is called a concentric cylinder viscometer.

4.3 Concentric Cylinder and Cone-and-Plate Viscometers

Imagine a viscous fluid enclosed in a gap between two concentric cylinders as shown in Fig. 4.3a. If one of the cylinders is caused to rotate, a viscous resistance to the rotation will be transmitted through the fluid to the nonrotating cylinder and produce a torque on the latter. An

$$\eta = \frac{I}{2\pi\,\omega_0}\,\frac{\Delta R}{R_4{}^3}$$

apparatus that is easily visualized—although not widely used—consists of a cup centered on a turntable with a bob concentrically suspended in it as shown in Fig. 4.3a. In addition to the characteristics of the apparatus, the torque on the suspending wire depends on the viscosity of the fluid— it would be greater for molasses than for water, all other things being equal. Such a torque can be measured, and our objective is to see how the viscosity of the fluid can be evaluated from such data.

We define r to be the distance variable in the radial direction, and ω to be the velocity of rotation with ω measured in radians second^{-1}. This means the velocity of a cylindrical layer of fluid a distance r from the axis of rotation is $r\omega$. There is a velocity gradient across the gap in this apparatus and this can be written $r(d\omega/dr)$. Next we represent a layer of fluid as a cylindrical shell of length l and thickness dr as shown in Fig. 4.3b and consider the viscous force acting on such a shell. By Eq. (1), this force is given by

$$F = \eta A \frac{dv}{dr} = \eta(2\pi rl)\left(r\frac{d\omega}{dr} \right) \tag{4}$$

where $2\pi rl$ is the area of the shell. Torque is given by the product of a force and the distance through which it operates, so this element of viscous force must be multiplied times r to give a torque T:

$$T = Fr = 2\pi\eta lr^3 \frac{d\omega}{dr} \tag{5}$$

When the apparatus begins to rotate, the fluid experiences an initial acceleration, but a stationary state is rapidly attained in which forces balance and acceleration is zero. Equation (5) gives a generalized expression for torque; under stationary state conditions it must be independent of r. If this were not the case, forces would be different in different parts of the fluid and acceleration would occur. Accordingly, we set the torque on this volume element equal to a constant:

$$2\pi\eta lr^3 \frac{d\omega}{dr} = \text{const.} \tag{6}$$

Now suppose the outer cylinder has a radius R and rotates with an experimental velocity ω_{ex}. Furthermore, assume the inner cylinder has a radius fR—where f is a fraction which is close to unity for small gaps— and is stationary. We can integrate Eq. (6) between these limits to obtain a value for the constant torque that characterizes the experiment:

$$2\pi\eta l \int_0^{\omega_{ex}} d\omega = \text{const.} \int_{fR}^{R} r^{-3}\, dr = -\frac{\text{const.}}{2}\left(\frac{1}{R^2} - \frac{1}{(fR)^2}\right) \qquad (7)$$

or

$$\text{const.} = 4\pi\eta l R^2 \omega_{ex} \frac{f^2}{1 - f^2} = FR \qquad (8)$$

where F is the force transmitted to the bob under stationary state conditions. Since everything in this expression except η is experimentally measurable, Eq. (8) can be used to evaluate viscosity. Since ω is the observed angular velocity, the subscript is no longer necessary. As noted above, an apparatus that is based on this geometrical arrangement is called a concentric cylinder viscometer. Just as the geometry of the concentric cylinder viscometer is based on Fig. 4.1, which serves to define η, Eq. (8) reduces to Newton's law of viscosity in the appropriate limit. This is examined in the following example.

Example 4.1 Examine Eq. (8) in the limit as $f \rightarrow 1$ to show that the relationship reduces to Eq. (1) under these conditions.

Solution The limit $f \rightarrow 1$ corresponds to a small gap between the two cylindrical walls. We are therefore justified in saying that the area of contact between the liquid and wall is $2\pi R l$; if we divide both sides of Eq. (8) by this area A, we obtain

$$\frac{F}{A} = 2\eta\omega\left(\frac{f^2}{1 - f^2}\right)$$

Since $1 - f^2 = (1 + f)(1 - f) \simeq 2(1 - f)$, this becomes

$$\frac{F}{A} = \frac{\eta\omega f^2}{1 - f}$$

Multiplying the numerator and denominator of this expression by R and letting $v_{max} = R\omega$ gives

$$\frac{F}{A} = \frac{\eta v_{max} f^2}{R(1 - f)} = \frac{\eta(v_{max} - 0)f^2}{R(1 - f)} = \eta f^2 \frac{\Delta v}{\Delta R}$$

since the velocity at the stationary inner wall is zero. In the limit we obtain

$$\lim_{f \to 1} \frac{F}{A} = \eta \frac{dv}{dr}$$

which is equivalent to Eq. (1), the defining equation for η. When the separation between the cylinders is negligible compared to their radius of curvature, the concentric cylindrical surfaces approximate the infinite parallel plates of the model used to define the coefficient of viscosity.

•

One of the appealing features of the concentric cylinder viscometer is the fact that the rate of shear can be varied by adjusting either the width of the gap or the angular velocity. This is a valuable capability in light of the kinds of non-Newtonian behavior that colloidal systems may display as shown in Fig. 4.2. Equation (6), however, reminds us that this capability must be viewed with caution. That relationship shows that it is $r^3(d\omega/dr) = r^2(dv/dr)$ that is constant across the gap, not dv/dr itself. Thus, even though we may change either the speed or the dimensions of the viscometer and thereby achieve different *average* rates of shear, the latter remains an average. The local velocity gradient varies from place to place within the fluid, although this variation is small for narrow gaps.

A variety of commercial viscometers are available which embody the essential features of the device we have analyzed here. In most commercial instruments it is the inner cylinder that rotates, even though there is a theoretical advantage to having the outer cylinder rotate rather than the inner one. The reason is that laminar flow is stable when the outer cylinder rotates, whereas centrifugal forces tend to induce turbulent flow in the reverse case. By rotating the inner member, however, interchangeable spindles can be used and the sturdiness and versatility of such a design offsets the theoretical advantage of the rotating cup arrangement. In actual practice, instruments may deviate from the concentric cylinder arrangement and their use to evaluate η depends on calibration with known standards rather than mathematical analysis of the instrument geometry. Figure 4.4 shows an example of a commercial viscometer which gives a viscosity value as digital output when the spindle is rotated in a beaker of the test fluid. This particular instrument can operate at eight different speeds with seven different spindles and therefore spans 56 different average values for dv/dr.

We conclude this section with a few remarks about the cone-and-plate type of viscometer, sketched schematically in Fig. 4.5. In this viscometer, the fluid is placed between a stationary plate and a cone which touches the plate at its apex. This apparatus also possess cylindrical symmetry, but this time in order to indicate a location within the fluid we must

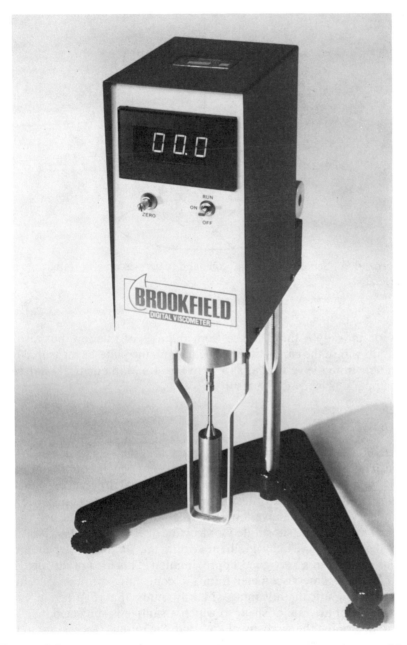

Figure 4.4   Brookfield Digital Viscometer, an instrument based on modified cylindrical geometry. (Photo courtesy of Brookfield Engineering Laboratories, Inc., Stoughton, Mass. 02072.)

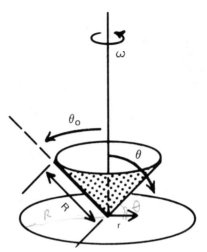

Figure 4.5 Schematic of a cone-and-plate viscometer (angle greatly exaggerated).

$$\eta = \frac{3 T \Delta\theta}{2\pi R^3 \omega_0}$$

specify not only r, the distance from the axis of rotation, but also the location within the gap between the cone and the plate, as measured by θ, the angle from the vertical as shown in Fig. 4.5. Mathematical analysis of this apparatus leads to the result

$$\frac{dv}{dr} \simeq \omega \, \frac{\cos \theta}{\cos \theta_0} \tag{9}$$

where θ_0 is the angle between the vertical and the wetted surface of the cone. This approximation holds for the case in which both θ and θ_0 are close to 90°. This condition is always met in actual practice, where the angle $90 - \theta_0$ is generally less than 5°. Some commercial concentric cylinder viscometers have interchangeable parts; this means that a conical rotor may be substituted for a cylindrical one. Equation (9) shows that the primary velocity gradient within the fluid is independent of radial position to a very good approximation. This feature distinguishes the cone-and-plate viscometer from the concentric cylinder viscometer and is a significant advantage in any study that requires accurate knowledge of the rate of shear to which a sample is subjected.

To determine the viscosity of the fluid, the torque T necessary to turn the cone at an angular velocity ω is measured. This torque depends on the viscosity of the fluid according to the equation

$$T = \frac{\frac{4}{3}\pi R^3 \eta \omega \sin \theta_0}{\cot \theta_0 + \frac{1}{2}\ln[(1 + \cos \theta_0)/(1 - \cos \theta_0)]\sin \theta_0} \qquad (10)$$

where R is the radius of the cone measured along the wetted surface. Like the concentric cylinder viscometer, this apparatus can be operated, in principle, at a variety of different angular velocities, and thus η can be studied at different rates of shear. Furthermore, this last quantity is reasonably constant throughout the fluid in the apparatus.

As noted previously, dilute colloidal systems display Newtonian behavior; that is, their apparent viscosity is independent of the rate of shear. Accordingly, the capability to measure η under conditions of variable shear is relatively superfluous in these systems. However, non-Newtonian behavior is commonplace in flocculated colloids.

Both the concentric cylinder and the cone-and-plate viscometers are widely used, especially in cases of non-Newtonian behavior. Liquids of low molecular weight and dilute solutions in such solvents generally display Newtonian behavior. In the latter case there is no advantage in being able to vary the shear rate, nor is it a disadvantage that this varies locally within the apparatus. Accordingly, a simpler viscometer is often used for such liquids. A capillary viscometer takes advantage of the fact that a velocity gradient exists whenever a fluid flows past a stationary wall. In the next section we examine the capillary viscometer and Poiseuille's law in terms of which it is analyzed.

4.4 The Poiseuille Equation and Capillary Viscometers

Figure 4.6 shows a portion of a cylindrical capillary of radius R and length l. Our interest is in the flow of a viscous liquid through such a capillary. This arrangement has the same cylindrical symmetry as the viscometers we discussed in the last section and, once again, it is convenient to focus attention on a cylindrical shell of fluid of radius r and thickness dr as shown in Fig. 4.6. Because of the non-slip condition at the wall of the capillary, the liquid shell adjacent to that wall has a velocity equal to zero. The velocity increases for shells of decreasing r so that a velocity gradient exists. Let us attempt to write an expression for v as a function of r, the radius of the cylindrical shell. Since stationary state conditions hold within the volume element, then any nonviscous forces must exactly balance in the viscous force.

The increment in viscous force acting on this element is the difference between the viscous forces on the outer and inner surfaces of the element, where each of these is given by Eq. (1), with $A = 2\pi r l$:

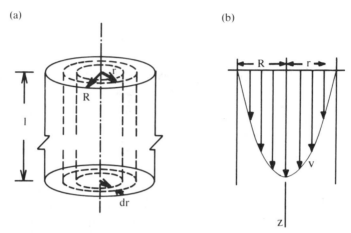

Figure 4.6 Flow in a cylindrical capillary: (a) a volume element in the flowing liquid and (b) parabolic flow profile.

$$\Delta F_{\text{vis}} = (F_{\text{vis}})_{\text{out}} - (F_{\text{vis}})_{\text{in}} = 2\pi(r + dr)\eta l \left(\frac{dv}{dr}\right)_{r+dr} - 2\pi r \eta l \left(\frac{dv}{dr}\right)_r \tag{11}$$

Next, we must relate $(dv/dr)_{r+dr}$ to $(dv/dr)_r$. The following expression accomplishes this, provided dr is small:

$$\left(\frac{dv}{dr}\right)_{r+dr} = \left(\frac{dv}{dr}\right)_r + \left(\frac{d^2v}{dr^2}\right) dr \tag{12}$$

If this result is substituted into Eq. (11), expanded, and only terms linear in dr retained, the expression becomes

$$\Delta F_{\text{vis}} = 2\pi\eta l \left(r\frac{d^2v}{dr^2} dr + \frac{dv}{dr} dr \right) = 2\pi\eta l \frac{d}{dr}\left(r\frac{dv}{dr} \right) \tag{13}$$

This force is counterbalanced by the increments in gravitational and pressure forces:

$$\Delta(F_g + F_{\text{press}}) = 2\pi l \rho g r \, dr + 2\pi \, \Delta p r \, dr \tag{14}$$

where the first term equals the weight of the shell and the second is the force on the shell if a pressure difference of Δp exists across the ends of the tube. Setting Eqs. (13) and (14) equal to each other gives

$$\eta \frac{d}{dr}\left(r \frac{dv}{dr} \right) = \left(\rho g + \frac{\Delta p}{l} \right) r \, dr \tag{15}$$

Integration of Eq. (15) yields

$$\eta r \frac{dv}{dr} = \frac{1}{2}\left(\rho g + \frac{\Delta p}{l} \right) r^2 \tag{16}$$

where the condition that $r(dv/dr)$ equals zero at $r = 0$ is used to evaluate the integration constant.

Equation (16) may be integrated again:

$$\eta \int dv = \frac{1}{2}\left(\rho g + \frac{\Delta p}{l} \right) \int r \, dr \tag{17}$$

Using the boundary condition that $v = 0$ at $r = R$ to evaluate the constant yields

$$v = \frac{(\rho g + \Delta p/l)(r^2 - R^2)}{4\eta} \tag{18}$$

This equation describes the velocity of a fluid element to be a parabolic function of its radial distance from the center of the tube as shown in Fig. 4.6b.

The rate of volume flow through the tube, V/t, equals the summation of the cross-sectional area of each shell multiplied by the velocity of that shell where the latter is given by Eq. (18):

$$\frac{V}{t} = \int_0^R \frac{(\rho g l + \Delta p)}{4\eta l} (r^2 - R^2) 2\pi r \, dr \tag{19}$$

or

$$\frac{V}{t} = \frac{(\rho g l + \Delta p)\pi R^4}{8\eta l} \tag{20}$$

Equation (20), known as Poiseuille's law, provides the basis for the most common technique for measuring the viscosity of a liquid or a dilute colloidal system, namely, the capillary viscometer.

Most capillary viscometers are designed with a relatively large bulb at both ends of the capillary, as shown in Fig. 4.7. A constant volume in the upper bulb is designated by two lines etched at either end of the bulb. The

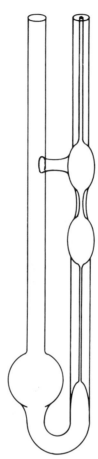

Figure 4.7   Schematic of a capillary viscometer.

viscometer is used by measuring the time required for the liquid level to drop from one line to the other as the fluid drains through the capillary. In such an apparatus the difference in height of the two liquid columns is relatively constant during the time required for flow. Generally, the only pressure difference across the liquid is due to the weight of the liquid. Under these conditions, Eq. (20) can be written

$$\eta = A\rho t \tag{21}$$

in which the constant A incorporates all the parameters which characterize the apparatus. Comparison of the flow times of two substances, one known (subscript 1) and one unknown (subscript 2), through the same apparatus provides an easy way to evaluate η for the unknown. In this case Eq. (21) becomes

$$\eta_2 = \frac{\rho_2}{\rho_1} \frac{t_2}{t_1} \eta_1 \tag{22}$$

For greater accuracy, an additional term may be added to Eq. (21) which corrects the latter for the fact that the Poiseuille equation does not apply exactly at the two ends of the capillary. With this correction for end effects, Eq. (21) becomes

$$\eta = A\rho t - B \frac{\rho}{t} \tag{23}$$

where $B = V/8\pi$. With this correction included, both A and B can be regarded as instrument constants and evaluated by calibrating the viscometer with two liquids of known density and viscosity. Commercial capillary viscometers are generally designed to make this end correction small, often negligible.

Example 4.2 Criticize or defend the following proposition: A set of capillary viscometers with different radii can be used in much the same way as a concentric cylinder viscometer with variable speed or gap width to conduct studies where the rate of shear is an independent variable.

Solution Equation (16) shows that the velocity gradient is not uniform in a capillary viscometer any more than it is in a concentric cylinder instrument. The rate of shear dv/dr is directly proportional to the radial distance from the axis of the cylinder. At the wall it has its maximum value which is proportional to R; at the center of the tube it equals zero. Some intermediate value, say, the average, might be used to characterize the gradient in a given instrument. This quantity will be different for capillaries of different radii. All of this is similar to the situation in concentric cylinder viscometers. The two types of viscometers differ in the following way, however. A much wider range of shear rates is attainable in concentric cylinder instruments with adjustable features than in capillaries. Because an average velocity gradient is used in describing these experiments, it is essential that a wide range of averages be spanned; otherwise, there is too much uncertainty in the individual values for meaningful results to be obtained. Since a wider range is possible with the concentric instrument, the latter is preferable when variable shear rates are to be investigated. As we saw in the last section, cone-and-plate viscometers are better yet.

•

4.5   The Equation of Motion

It is impossible to read much of the literature of viscosity without coming across some reference to the equation of motion. In the area of fluid mechanics this equation occupies a place like that of the Schrödinger equation in quantum mechanics. Like its counterpart, the equation of motion is a complicated partial differential equation, the analysis of which is a matter for experts. Our purpose in this section is not to solve the equation of motion for any problem, but merely to introduce the physics of the relationship. Actually, both the concentric cylinder and the capillary viscometers that we have already discussed are analyzed by the equation of motion, so we have already worked with this result without even realizing it. The equation of motion does in a general way what we did in a concrete way in the preceding discussions, namely, describe the velocity of a streamline within a flowing fluid as a function of location in the fluid. The equation of motion allows this to be considered as a function of both location *and* time and is thus useful in non-stationary-state problems as well.

As is the case with all differential equations, the boundary conditions of the problem are an important consideration, since they determine the "fit" of the solution. Many problems are set up to have a high level of symmetry and thereby simplify their boundary descriptions. This is the situation in the viscometers that we discussed above and which could be described by cylindrical symmetry. Note that the cone-and-plate viscometer—in which the angle from the axis of rotation had to be considered—is a case where we skipped the analysis and went straight for the final result, a complicated result at that. Because it is often solved for problems with symmetrical geometry, the equation of motion is frequently encountered in cylindrical and spherical coordinates, which complicates its appearance but simplifies its solution. We base the following discussion on rectangular coordinates, which may not be particularly convenient for problems of interest but which are easily visualized.

Figure 4.8 shows a small rectangular volume element $\Delta x \, \Delta y \, \Delta z$ located within a flowing fluid. The fluid passing through this element has a velocity v, which may be resolved into x, y, and z components— represented by v_x, v_y, and v_z, respectively—since, in general, v bears no special relationship to the coordinates. Note that v was in the axial direction in the cylindrical problems we discussed above; no such restriction operates here. Now let us examine the rate at which momentum accumulates in this volume element. To begin with, we recognize that the rate of change of momentum equals a force $\mathbf{F}$:

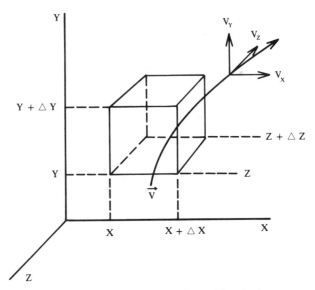

Figure 4.8 A volume element in a liquid flowing with velocity **v**.

$$\frac{d(m\mathbf{v})}{dt} = m\left(\frac{d\mathbf{v}}{dt}\right) = m\mathbf{a} = \mathbf{F} \tag{24}$$

Furthermore, the net rate of change of momentum *per unit volume* is given by $d(\rho\mathbf{v})/dt$. Therefore the net rate of change in momentum per unit volume equals the sum of all forces acting per unit volume. This last quantity is made up of three contributions—external forces such as gravity, pressure forces, and viscous forces—therefore we write

$$\frac{d(\rho\mathbf{v})}{dt} = \mathbf{F}_{ext} + \mathbf{F}_{press} + \mathbf{F}_{vis} \tag{25}$$

Treating density as a constant, the case for an incompressible fluid, enables us to write

$$\frac{d(\rho\mathbf{v})}{dt} = \rho\left(\frac{d\mathbf{v}}{dx}\frac{dx}{dt} + \frac{d\mathbf{v}}{dy}\frac{dy}{dt} + \frac{d\mathbf{v}}{dz}\frac{dz}{dt} + \frac{d\mathbf{v}}{dt}\right) \tag{26}$$

where we have used the chain rule for differentiation to expand $d\mathbf{v}/dt$ into the form shown.

The external, pressure, and viscous forces acting on the volume element can be developed more fully also. We will not go through the details of this development, but instead cite the final result followed by some explanatory remarks:

$$\rho \left(\frac{d\mathbf{v}}{dx} v_x + \frac{d\mathbf{v}}{dy} v_y + \frac{d\mathbf{v}}{dz} v_z + \frac{d\mathbf{v}}{dt} \right) = \rho g + \frac{dp}{dx} + \frac{dp}{dy} + \frac{dp}{dz}$$

$$+ \eta \left(\frac{d^2\mathbf{v}}{dx^2} + \frac{d^2\mathbf{v}}{dy^2} + \frac{d^2\mathbf{v}}{dz^2} \right) \tag{27}$$

Equation (27) is the equation of motion for an incompressible fluid. Remember, the solution to this differential equation for a set of specified boundary conditions gives a general expression for v as a function of $x, y, z,$ and t.

Now let us briefly consider the various terms in Eq. (27) with the idea of establishing their plausibility. Rewriting Eq. (27) in vector notation condenses it by combining terms with similar significance; in this notation Eq. (27) becomes

$$\rho(\mathbf{v} \cdot \nabla\mathbf{v}) + \rho \frac{\partial \mathbf{v}}{\partial t} = \rho g + \nabla p + \eta \nabla^2 \mathbf{v} \tag{28}$$

where $\nabla = \partial/\partial x + \partial/\partial y + \partial/\partial z$, $\nabla^2 = \partial^2/\partial x^2 + \partial^2/\partial y^2 + \partial^2/\partial z^2$, and the term in parentheses is a dot product. Starting on the right-hand side of this equation, we note the following:

1. The first term on the right describes the force of gravity acting on the volume element. In this development gravity is the only external force we consider. The gravitational force per unit volume is given by ρg, where g is the acceleration of gravity.
2. The second term on the right describes the force on the volume element due to pressure. The net force in the x direction due to pressure is given by $(p_{x+\Delta x} - p_x)\Delta y\, \Delta z$. Dividing through by $\Delta x\, \Delta y\, \Delta z$ gives the x component of force per unit volume as $[p_{x+\Delta x} - p_x]/\Delta x$, or dp/dx in the limit. Similar expressions apply to the y and z components of $\mathbf{F}_{press}$.
3. The third term on the right describes viscous forces. The viscous force acting on, say, the face of area $\Delta x\, \Delta y$ of the volume element is given by $\eta\, \Delta x\, \Delta y\, (dv/dz)_{z+\Delta z}$ on one side of the volume element and by $\eta\, \Delta x\, \Delta y (dv/dz)_z$ on the other side. The difference between the two is $(d^2v/dz^2)\Delta x\, \Delta y\, \Delta z$. Dividing through by the volume of the element and taking the limit gives the contribution to $\mathbf{F}_{vis}$ from the velocity gradient in the z direction. Similar terms apply to the x and y directions.

Next, we turn our attention to the terms on the left-hand side of Eq. (28):

1. The first term on the left is called the inertial term, and the second the viscous term. For low velocities the latter is more important than the former.
2. Under conditions where the inertial term can be neglected compared to the viscous, Eq. (28) becomes

$$\rho \frac{d\mathbf{v}}{dt} = \rho g + \nabla p + \eta \, \nabla^2 \mathbf{v} \tag{29}$$

This is called the Stokes approximation to the equation of motion.
3. Under stationary state conditions the velocity is independent of time and Eq. (29) becomes

$$\rho g + \nabla p + \eta \, \nabla^2 \mathbf{v} = 0 \tag{30}$$

This is called the Stokes-Navier equation.

We remarked at the beginning of this section that the equation of motion is the cornerstone of any discussion of fluid mechanics. When one considers the various coordinate systems in which it may be expressed, the vector identities which may transform it, or the approximations which may be used to simplify it, Eq. (27) takes on many forms, some of which are scarcely recognizable as the same relationship. The purpose of this section is to illustrate that—despite its complexity and variations—the equation of motion is really nothing more than a statement of Newton's second law!

Until now, we have been primarily concerned with the definition and measurement of viscosity, without regard to the nature of the system under consideration. Next we turn our attention to systems containing dispersed particles whose dimensions are in the colloidal size range. Viscosity measurements can be used to characterize both lyophobic and lyophilic systems; we shall discuss both in the order cited.

4.6 Einstein's Law of Viscosity: Theory

In 1906 A. Einstein (Nobel Prize, 1921) published his first derivation of an expression for the viscosity of a dilute dispersion of solid spheres. The initial theory contained errors that were corrected in a subsequent paper which appeared in 1911. It would be no mistake to infer from the historical existence of this error that the theory is complex. Therefore we will restrict our discussion to an abbreviated description of the assump-

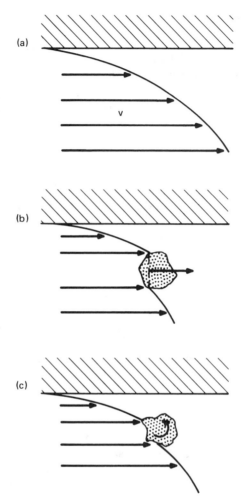

Figure 4.9 Schematic illustration of the flow pattern for (a) a pure liquid near a stationary wall and for a dispersion that contains (b) nonrotating and (c) rotating particles.

tions of the theory and some highlights of the derivation. Before examining the Einstein theory, let us qualitatively consider what effect the presence of dispersed particles is expected to have on the viscosity of a fluid.

The presence of a colloidal size particle in the liquid increases the viscosity because of the effect it has on the flow pattern. Two effects which readily come to mind are illustrated in Fig. 4.9. In Fig. 4.9a the velocity profile near a wall is shown for a pure liquid. The variation of velocity among layers is indicated by the arrows of different length. In Fig. 4.9b a nonrotating particle is pictured cutting across several layers in the flowing liquid. Since the particle does not rotate (by hypothesis), it must slow down the fluid so that the layers on opposite sides of the particle have the same velocity, that of the particle itself. The overall velocity gradient is thus reduced. Since the applied force is presumably the same in both Fig. 4.9a and b, the reduced velocity gradient must be offset by an increase in η. Alternatively, we might consider a particle which is induced to rotate by its position in the velocity gradient. Such a situation is shown in Fig. 4.9c. In this case some of the energy which would otherwise keep the liquid flowing is deposited in the particle, causing it to rotate. In both cases the presence of the particle increases the viscosity of the fluid.

The increase in viscosity due to dispersed particles is expected to increase with the concentration of the particles, a dependence we may tentatively describe in terms of a power series in concentration c:

$$\eta = A + Bc + Cc^2 + \cdots \qquad (31)$$

In this equation A, B, $C, \ldots$, are constants whose values are to be determined. This much is evident: As the concentration of a dispersion goes to zero, its viscosity must go to that of the continuous phase. Therefore $A = \eta_0$, the viscosity of the medium. Furthermore, the constants B, $C, \ldots$, might resonably be expected to depend on the size, shape, orientation, and so on, of the dispersed units.

The Einstein derivation is based on the hydrodynamic equations of the preceding section. As such, it is limited to those cases in which ρ and η for the fluid are constant and in which the flow velocity is low. Furthermore, the theory postulates an extremely dilute dispersion of rigid spheres with no slippage of the liquid at the surface of the spheres. Finally, the spheres are assumed to be large enough compared to the solvent molecules to permit us to regard the solvent as a continuum, but small enough compared to the dimensions of the viscometer to permit us to ignore wall effects. These size restrictions make this result applicable to particles in the colloidal size range.

Einstein considered a fluid in laminar flow through a dilute, random array of spherical particles. An obvious difficulty in applying these equations to the case in question is the enormous number of surfaces at which boundary conditions must be specified. This leads to the first reason why an infinitely dilute dispersion must be considered. If the particles are sufficiently far apart, each will modify the flow pattern in its environment as if it alone were present. This introduces two boundary conditions: no slippage at the surface and unperturbed flow at larger distances from the surface. Note the resemblance between this model and that of the Stokes' equation, Eq. (2.5). As in that derivation, two different expressions for the rate of energy dissipation in the dispersion are equated to give the final result. It is important to remember Eq. (3) in this context, since the latter establishes the coefficient of viscosity as the pertinent parameter in any discussion of the rate of energy dissipation per unit volume, $(dE/dV)/dt$.

Einstein was able to solve the equation of motion, first for the case of a single sphere present, to give $(dE/dV)/dt$ for a spherical volume of dispersion of radius r centered at the spherical particle. The distance r is so much larger than the radius R of the spherical particle that the disturbance of flow due to the particle has vanished at the surface of the hypothetical enclosing sphere. Einstein has shown the result of this integration to be

$$\left(\frac{dE/dV}{dt}\right)_{\text{sphere}} = K\eta_0(\tfrac{8}{3}\pi r^3 + \tfrac{4}{3}\pi R^3) \tag{32}$$

where η_0 is the viscosity of the solvent and K is a constant whose significance does not concern us, since it cancels out presently. Since $\tfrac{4}{3}\pi r^3$ represents the volume V of the system and $\tfrac{4}{3}\pi R^3$ represents the volume occupied by the spherical particle, we can rewrite this result:

$$\left(\frac{dE/dV}{dt}\right)_{\text{sphere}} = K\eta_0 V(2 + \phi) \tag{33}$$

where ϕ is the volume fraction occupied by the sphere. Examining the limit of Eq. (32) as $R \to 0$ permits us to evaluate the rate of deposition of energy in the same volume element in the absence of the sphere; that is, if solvent alone is present,

$$\left(\frac{dE/dV}{dt}\right)_{\text{solvent}} = 2K\eta_0 V \tag{34}$$

Next, we can use Eqs. (33) and (34) to consider the increment per sphere in the rate of energy deposition. Subtracting Eq. (34) from Eq. (33) and rearranging gives

$$\left(\frac{dE/dV}{dt}\right)_{\text{sphere}} - \left(\frac{dE/dV}{dt}\right)_{\text{solvent}} = \frac{\phi_i}{2}\left(\frac{dE/dV}{dt}\right)_{\text{solvent}} \tag{35}$$

in which the subscript i has been added to ϕ to remind us that this is the volume fraction occupied by a single sphere. In keeping with this notation, the left-hand side of Eq. (35) may be written $\Delta[dE/dV)/dt]_i$, the change due to one sphere in the rate of energy deposition. If the dispersion contains N spheres sufficiently far part for the effects of each to be independent of the others, we may write

$$\left(\frac{dE/dV}{dt}\right)_{\text{dispersion}} = \left(\frac{dE/dV}{dt}\right)_{\text{solvent}} + \sum_{i=1}^{N} \Delta\left(\frac{dE/dV}{dt}\right)_i$$

$$= \left(1 + \frac{N\phi_i}{2}\right)\left(\frac{dE/dV}{dt}\right)_{\text{solvent}} \tag{36}$$

Of course, the number of spheres times the volume fraction occupied per sphere equals the total volume fraction of the dispersed phase, ϕ. This fact combined with Eq. (34) transforms Eq. (36) into

$$\left(\frac{dE/dV}{dt}\right)_{\text{dispersion}} = 2K\eta_0 V\left(1 + \frac{\phi}{2}\right) \tag{37}$$

Einstein was able to derive another equation describing the rate of energy deposition per unit volume for a dispersion of spheres:

$$\left(\frac{dE/dV}{dt}\right)_{\text{dispersion}} = 2K\eta(1 - \phi)^2 V \tag{38}$$

in which η (without a subscript) is the measured viscosity *of the dispersion* and K has the same significance as in Eq. (37). Again in the derivation of this result, the particles are assumed to be far apart. Therefore, Eqs. (37) and (38), two expressions for the same quantity derived with comparable assumptions, may be equated:

$$2K\eta(1 - \phi)^2 V = 2K\eta_0 V\left(1 + \frac{\phi}{2}\right) \tag{39}$$

Rearranging this equation yields

$$\frac{\eta}{\eta_0} = \frac{1 + \phi/2}{(1 - \phi)^2} = \left(1 + \frac{\phi}{2}\right)(1 + \phi + \phi^2 + \cdots)^2 \tag{40}$$

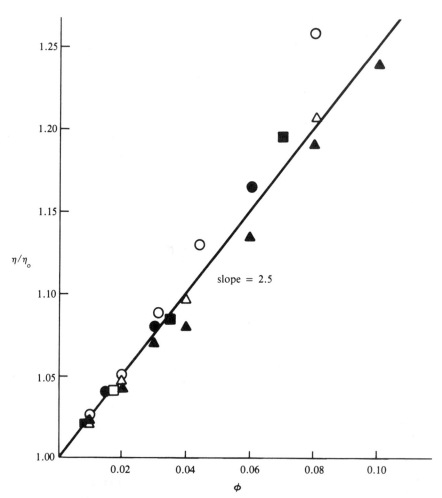

Figure 4.10 Experimental verification of Einstein's law of viscosity for spherical particles of several different sizes (squares, yeast particles, R = 2.5 μm; circles, fungus spores, R = 4.0 μm; triangles, glass spheres, R = 80 μm.). Open symbols represent measurements in concentric cylinder viscometers, and closed symbols measurements in capillary viscometers. [Data from F. Eirich, M. Bunzl, and H. Margaretha, *Kolloid Z.*, *74*:276 (1936).]

where only the leading terms in ϕ have been retained, since ϕ must be small to satisfy the requirement of large distances between spheres. If no term higher than first order in ϕ is retained, Eq. (40) becomes

$$\frac{\eta}{\eta_0} = 1 + 2.5\phi + \cdots \tag{41}$$

a result known as Einstein's law of viscosity. This derivation was not only an important accomplishment in its own right but also served as a model for many subsequent derivations. Before considering these, however, let us examine Einstein's law from an experimental point of view.

4.7 Einstein's Law of Viscosity: Experimental

The Einstein equation is one of those pleasant surprises that occasionally emerges from complex theories: a remarkably simple relationship between variables, in this case the viscosity of a dispersion and the volume fraction of the dispersed spheres. A great many restrictive assumptions are made in the course of the derivation of this result, but the major ones are (1) that the particles are solid spheres and (2) that their concentration is small.

These conditions are relatively easy to meet experimentally, so Eq. (41) has been tested in numerous studies. Figure 4.10 is an example of the sort of verification that has been obtained. In the work summarized in this figure, a variety of model systems were investigated, using both capillary and concentric cylinder viscometers. The solid line in the figure is the Einstein prediction; the agreement between theory and experiment is seen to be very good.

Note that the applicability of the Einstein equation seems equally good regardless of the size sphere used. Although the range of particle radii in Fig. 4.10 is relatively narrow, this conclusion has been verified for particles as small as individual molecules and as large as grains of sand. It might also be noted that experiments of this sort are carried out in mixed solvents or electrolyte solutions with equal the dispersed particles in density. Thus there is not tendency for the spheres to settle under the influence of gravity.

The simplicity of the Einstein equation makes it relatively easy to test, but also limits its usefulness rather sharply. With so few variables involved, the quantities we may evaluate by Eq. (41) are few in number. Viscosity is a measurable quantity from which we try to extract information about the dispersion. All that Eq. (41) offers directly in this area is the evaluation of ϕ from viscosity measurements, again provided ϕ is small and the particles are spheres.

The data in Fig. 4.10 are limited to concentrations below $\phi = 0.10$. We might ask what are the consequences of increasing the concentration to higher volume fractions of spherical particles? Equation (40) suggests that the range of applicability of Einstein's equation might be extended by retaining terms higher than first order in the power series; that is,

$$\frac{\eta}{\eta_0} = 1 + 2.5\phi + k_1\phi^2 + \cdots \tag{42}$$

might give a better fit to the data from dispersions whose concentrations are more than infinitely dilute. There are several points about this result which should be noted:

1. Equations (31) and (42) are identical in form. Equation (42) shows that volume fraction is the theoretically preferred unit of concentration as far as viscosity is concerned.
2. The theoretical evaluation of k_1 and higher coefficients in Eq. (42) requires more than merely retaining additional terms in Eq. (40). Einstein's derivation of Eqs. (32) and (38) is the origin of the restriction to very dilute systems.
3. A number of theoretical attempts have been made to evaluate k_1 by going back through the Einstein derivation and superimposing the effects of neighboring particles, rather than treating them as independent. A few of the theoretical values that have been obtained for k_1 from different models are 14.1, 12.6, and 7.35.

Rather than attempting to choose among these theoretical approaches, let us examine an empirical approach to the problem of deviations from the Einstein equation at high concentrations. Toward this end, it is convenient to rearrange Eq. (42) as follows:

$$\frac{(\eta/\eta_0 - 1)}{\phi} = 2.5 + k_1\phi + \cdots \tag{43}$$

This has the effect of reducing the order of each term on the right-hand side of the equation. Now if we plot the left-hand side of Eq. (43) versus ϕ, the result should be a straight line of slope k_1 and intercept 2.5, at least at low concentrations before still higher-order terms become important.

Figure 4.11 shows the viscosity of dispersions of glass spheres of radius 65 μm plotted in the manner suggested by Eq. (43). Several conclusions are evident from the data replotted in this way:

1. The Einstein coefficient, 2.5, is clearly the intercept toward which the quantity $(\eta/\eta_0 - 1)/\phi$ extrapolates as $\phi \to 0$.

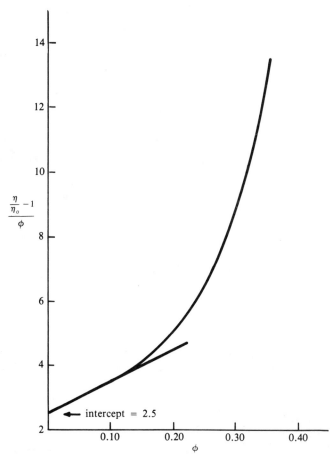

Figure 4.11 Reduced viscosity versus volume fraction for a dispersion of glass spheres ($R = 6.5 \times 10^{-3}$ cm) up to $\phi \simeq 0.40$. [Data from V. Vand, *J. Phys. Colloid Chem.*, *52*:300 (1948).]

2. The initial slope of Fig. 4.11 suggests k_1 is about 10.0 for these data, a reasonable value in the light of theoretical predictions.
3. At still higher volume fractions, η increases even more steeply than predicted by the two-term version of Eq. (43).

The results of experiments in viscometry are routinely reported in a variety of forms; the ones we have used in Figs. 4.10 and 4.11 are only two of the possibilities. Table 4.1 lists some of the functional forms in which

Table 4.1 Symbol, Common and IUPAC Names, and Limiting Values for a Variety of Forms Commonly Used to Present Viscosity Data

Functional form	Symbol	Common name	IUPAC name	$\lim_{c \to 0}$
—	η	Viscosity	—	η_0
η/η_0	η_r	Relative viscosity	Viscosity ratio	1
$\eta/\eta_0 - 1$	η_{sp}	Specific viscosity	—	0
$(\eta/\eta_0 - 1)/c$	η_{red}	Reduced viscosity	Viscosity number	$[\eta]$
$c^{-1} \ln(\eta/\eta_0)$	η_{inh}	Inherent viscosity	Logarithmic viscosity number	$[\eta]$
$\lim_{c \to 0} \eta_{red}$ or $\lim_{c \to 0} \eta_{inh}$	$[\eta]$	Intrinsic viscosity	Limiting viscosity number	—

data are often presented. Also listed are the symbols, common and IUPAC names, and the limiting values for these quantities as the concentration of the colloid goes to zero. In Table 4.1 the symbol c has been used to represent the concentration of the dispersed phase. For the present, we will continue to let ϕ be the unit of concentration.

The definition of the reduced viscosity is somewhat analogous to the reduced osmotic pressure [Eq. (3.34)] and the intrinsic viscosity is the limiting value of this quantity. Note that the reduced viscosity gives the relative increase in the viscosity of the solution over the solvent per unit of concentration. Since $[\eta]$ equals the limiting value of the reduced viscosity, $[\eta]$ has the significance of measuring the first increment of viscosity due to the dispersed particles. The intrinsic viscosity, therefore, is characteristic of the dispersed particles. The Einstein equation predicts that it should have a value of 2.5 for spherical particles. Note that $[\eta]$ may also be evaluated by extrapolating inherent viscosities to $\phi = 0$. The following example demonstrates the equivalence of these two procedures for evaluating $[\eta]$.

Example 4.3 Show that the inherent viscosity, $(1/\phi) \ln(\eta/\eta_0)$, reduces to $[\eta]$—defined as the limiting reduced viscosity—as $\phi \to 0$.

Solution By using volume fraction for c in Eq. (31), $\eta = A + B\phi + C\phi^2 +$ $\cdots$. Since $A = \eta_0$, $\eta/\eta_0 = 1 + B\phi + C\phi^2 + \cdots$. The intrinsic viscosity is the coefficient of the first-order concentration term; therefore this may be written $\eta/\eta_0 = 1 + [\eta]\phi + C\phi^2 + \cdots$. Replacing the argument of the logarithm in the definition of inherent viscosity by this expansion gives

$$\eta_{\text{inh}} = (1/\phi) \ln (1 + [\eta]\phi + C\phi^2 + \cdots)$$
$$= (1/\phi)([\eta]\phi + C\phi^2 + \cdots),$$

where the second form is obtained by using the series expansion (see Appendix A) for the logarithm. In the limit of $\phi \to 0$, this equals $[\eta]$, the limiting reduced viscosity.

•

The Einstein theory shows that volume fraction is the theoretically favored concentration unit in viscosity work, even though it is not a practical unit for unknown solutes. As was the case in the Flory–Huggins theory in Sect. 3.7, it is convenient to convert volume fractions into mass volume^{-1} concentration units for the colloidal solute. According to Eq. (3.78), $\phi = c(\bar{V}_2/M_2)$, where c has units mass volume^{-1} and $\bar{V}_2$ and M_2 are the partial molar volume and molecular weight, respectively, of the solute. In viscosity work volumes are often expressed in deciliters—a testimonial to the convenience of the 100-ml volumetric flask! In this case, $\bar{V}_2$ must be expressed in these units also. The reader is advised to be particularly attentive to the units of concentration in an actual problem, since the units of intrinsic viscosity are concentration^{-1}. With the substitution of Eq. (3.78), Eq. (42) becomes

$$\frac{\eta}{\eta_0} = 1 + 2.5 \frac{\bar{V}_2}{M_2} c + k_1 \left(\frac{\bar{V}_2}{M_2}\right)^2 c^2 + \cdots \tag{44}$$

or

$$\frac{1}{c}\left(\frac{\eta}{\eta_0} - 1\right) = 2.5 \frac{\bar{V}_2}{M_2} + k_1 \left(\frac{\bar{V}_2}{M_2}\right)^2 c + \cdots \tag{45}$$

Equation (45) shows that as long as balances, volumetric flasks, and viscometers are available, $[\eta]$ can be determined. All that is required is to measure viscosity at a series of concentrations, work up the data as $(1/c)(\eta/\eta_0 - 1)$, and extrapolate to $c = 0$. If the experimental value of $[\eta]$ turns out to be $2.5(\bar{V}_2/M_2)$, then the particles are shown to be unsolvated spheres. If $[\eta]$ differs from this value, the dispersed units deviate from the requirements of the Einstein model. In the next section we examine how such deviations can be interpreted for lyophobic colloids.

4.8 Deviations from the Einstein Model

The Einstein theory is based on a model of dilute, unsolvated spheres. In this section we examine the consequences on intrinsic viscosity of deviations from the Einstein model in each of the following areas:

1. Concentrations which are not very dilute
2. Particle swelling due to solvation
3. Nonspherical particles approximated as ellipsoids of revolution

We will consider each of these items as potential causes of deviation in the following paragraphs.

We have already indicated that the coefficient k_1 in Eq. (66) has been calculated for spheres by various theoretical models. While this coefficient is a measure of concentration effects, we shall not pursue its derivation. Instead, we qualitatively examine the effect of particle crowding as the origin of the positive deviations from the Einstein theory which inevitably set in at higher concentrations, as seen in Figs. 4.11 and 4.12.

To do this, we divide the spheres into a number of categories to each of which we assign an index number i. The number of spheres in each category is assumed to be small enough that the Einstein equation holds when the spheres are dispersed such that the ith set is at a volume fraction ϕ_i. Furthermore, we assume that the increment in viscosity that accompanies the addition of each successive increment is given by Eq. (41), that is,

$$d\eta = 2.5\eta \, d\phi_i \qquad (46)$$

In this case, η is the viscosity of the dispersion prior to the addition of the last fraction. It can be shown that $d\phi_i$ is also equal to $d\Phi/(1 - \Phi)$, where Φ is the *total* volume fraction of all spheres in the system. Therefore

$$d\eta = 2.5\eta \, \frac{d\Phi}{1 - \Phi} \qquad (47)$$

This result can be integrated, recalling that $\eta = \eta_0$ at $\Phi = 0$, to give

$$\frac{\eta}{\eta_0} = (1 - \Phi)^{-2.5} \qquad (48)$$

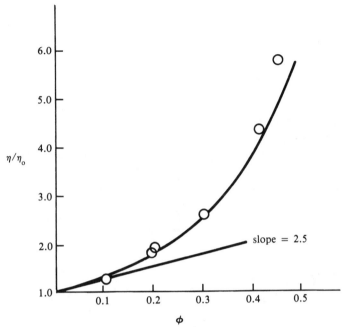

Figure 4.12 The effect of particle crowding on viscosity. The solid line is drawn according to Eq. (48); the points are experimental results. [Data from R. Roscoe, *Br. J. Appl. Phys., 3*:267 (1952).]

This result reduces to the Einstein equation as $\Phi \to 0$. The solid line in Fig. 4.12 was drawn according to Eq. (48). This result shows that the upward curvature displayed in Figs. 4.11 and 4.12 can be explained on the basis of particle crowding; unfortunately, this approach produces no new parameters to quantify the effect. It might be noted that the original derivation of this result was undertaken to explore the effects of polydispersity. Specifically, the different categories of spheres were assumed to be of different sizes. Since the Einstein result is independent of particle size, the latter is really irrelevant to the effect shown in Fig. 4.12.

Figure 2.3 suggested how either solvation or ellipticity could increase the effective size of a particle as far as its friction factor is concerned. A moment's reflection will convince us that this same conclusion is qualitatively true with respect to intrinsic viscosity also.

The solvation of a sphere swells its volume above the "dry" volume, and the latter is presumably what is used to evaluate ϕ. Therefore whatever factor describes the increase in particle volume due to solvation is absorbed into $[\eta]$. This is easily seen as follows. Suppose the mass of colloidal solute in a solution is converted to the volume of unsolvated of material using the "dry" density. In this way, mass volume^{-1} units are converted to volume volume^{-1} units, which we label ϕ_{dry}, since the unsolvated density was used in evaluating this quantity. If the particles are assumed to be uniformly solvated throughout the particle—as would be the case for, say, aqueous proteins—then the solvated particle exceeds the unsolvated particle in volume by the factor $1 + (m_{1,b}/m_2)[\rho_2/\rho_1]$ as shown by Eq. (2.38). Recall that in this expression, $m_{1,b}$ is the mass of bound solvent, m_2 is the mass of the solute particle, and ρ_i is the density of solvent or solute, as appropriate. Still assuming ϕ_{dry} has been used to evaluate $[\eta]$, we see

$$\frac{1}{\phi_{dry}}\left(\frac{\eta}{\eta_0} - 1\right) = 2.5\left(1 + \frac{m_{1,b}}{m_2}\left[\frac{\rho_2}{\rho_1}\right]\right) + \cdots \qquad (49)$$

or

$$\frac{1}{c}\left(\frac{\eta}{\eta_0} - 1\right) = 2.5\left(1 + \frac{m_{1,b}}{m_2}\left[\frac{\rho_2}{\rho_1}\right]\right)\left(\frac{1}{\rho_2}\right) + \cdots \qquad (50)$$

in practical units. The following example considers how intrinsic viscosity measurements can be used to reach conclusions about the state of solvation of colloidal particles.

Example 4.4 Suppose an aqueous solution of a spherical protein molecule shows an intrinsic viscosity of 3.36 cm^3 g^{-1}. Taking $\rho_2 = 1.34$ g cm^{-3} for the dry protein, estimate the extent of hydration of the protein.

Solution It is apparent from the units of $[\eta]$ that solute concentration has been expressed in g cm^{-3}. Dividing this concentration by the density of the unsolvated protein converts the concentration to "dry" volume fraction units. Since the concentration appears as a reciprocal in the definition of $[\eta]$, we must multiply the latter by ρ_2 to obtain $(1/\phi_{dry})(\eta/\eta_0 - 1)$. For this protein the latter is given by $(3.36)(1.34) = 4.50$. If the particles were unsolvated, this quantity would equal 2.5, since the molecules are stated to be spherical. Hence the ratio $4.50/2.50 = 1.80$ gives the volume expansion factor, which equals $1 + (m_{1,b}/m_2)[\rho_2/\rho_1]$. Therefore

(a)   (b)

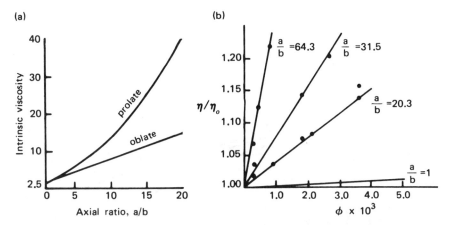

Figure 4.13   (a) Intrinsic viscosity as a function of the axial ratio a/b for oblate and prolate ellipsoids of revolution according to the Simha theory. [From Ref. 7; used with permission.] (b) Experimental values of relative viscosity versus volume fraction for tobacco mosaic virus particles of different a/b ratios. [M. A. Lauffer, *J. Am. Chem. Soc.*, *66*:1188 (1944).]

$m_{1,b}/m_2 = 0.80(1.00/1.84) = 0.60$. The intrinsic viscosity reveals the solvation of these particles to be 0.60 g H_2O (g protein)$^{-1}$.

•

We noted above that either solvation or ellipticity could cause the intrinsic viscosity to exceed the Einstein value. R. Simha and others have derived extensions of the Einstein equation for the case of ellipsoids of revolution. As we saw in Sect. 1.6, such particles are characterized by their axial ratio. If the particles are too large, they will adopt a preferred orientation in the flowing liquid. However, if they are small enough to be swept through all orientations by Brownian motion, then they will increase [η] more than a spherical particle of the same mass. Again, this is very reminiscent of the situation shown in Fig. 2.3.

Figure 4.13a shows plots of the intrinsic viscosity—in volume fraction units—as a function of axial ratio according to the Simha equation. Figure 4.13b shows some experimental results obtained for tobacco mosaic virus particles. These particles—an electron micrograph of which is shown in Fig. 1.9a—can be approximated as prolate ellipsoids. Intrinsic viscosities are given by the slopes of Fig. 4.13b, and the parameters on the curves are axial ratios determined by the Simha equation. Thus we see that particle asymmetry can also be quantified from intrinsic viscosity measurements for unsolvated particles.

(a)

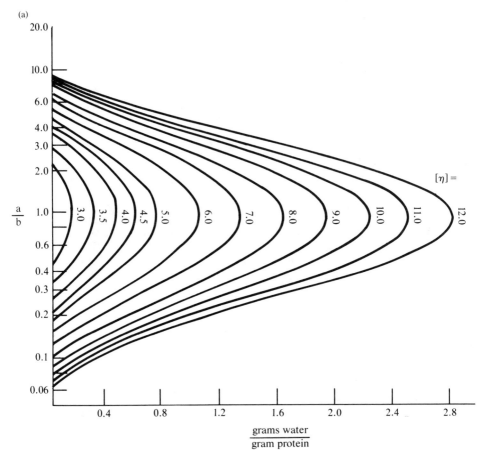

Figure 4.14 (a) Variation of the intrinsic viscosity of aqueous protein solutions with axial ratio a/b and extent of hydration $m_{1,b}/m_2$. [Redrawn from L. Oncley, *Ann. N.Y. Acad. Sci., 41*:121 (1941), used with permission.] (b) Superposition of the $[\eta] = 8.0$ contour from (a) and the $f/f_0 = 1.45$ contour from Fig. 2.8. The crossover unambiguously characterizes particles with respect to hydration and axial ratio.

Finally, we note that both solvation and ellipticity can occur together. The contours shown in Fig. 4.14a illustrate how various combinations of solvation and ellipticity are compatible with an experimental intrinsic viscosity. The particle considered in Example 4.4 has an intrinsic viscosity of 4.50 and was calculated to be hydrated to the extent of 0.60 g H_2O (g protein)$^{-1}$. The same value of $[\eta]$ is also compatible with

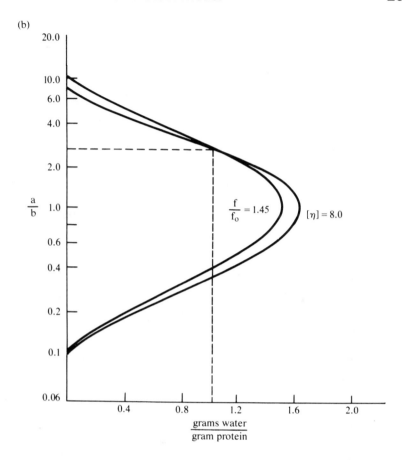

nonsolvated ellipsoids of axial ratios 4.0 and 0.22. Intermediate combinations of solvation and ellipticity can also be determined from Fig. 4.14a.

Contours like this are qualitatively the same sort of thing we obtain from sedimentation–diffusion experiments as shown in Fig. 2.8. Therefore let us consider the relationship between the two types of data. In general, exactly the same factors affect both the intrinsic viscosity and the friction factor ratio, but the functional dependencies are somewhat different. Figure 4.14b shows how a contour of f/f_0 selected from a sedimentation–diffusion study and an intrinsic viscosity contour selected on the basis of viscosity experiments might overlap. In this case the solvation–ellipticity combination is characterized unambiguously: $a/b = 2.5$ and $m_{1,b}/m_2 = 1.0$. Figure 4.14b shows the complementarity of viscosity and sedimentation–diffusion data.

The attentive reader will realize that we have strayed rather far from the hard spheres of the Einstein theory to find applications for it. It should also be appreciated, however, that the molecules we are discussing are proteins which—through disulfide bridges and hydrogen bonding— have fairly rigid structures. Therefore the application of the theory— amended to allow for solvation and ellipticity—is justified. This would not be the case for synthetic polymers, which are best described as random coils. For the latter a different formalism is employed. This is the topic of the next section.

4.9   The Staudinger-Mark-Houwink Equation

The viscosity of a polymer solution is one of its most distinctive properties. The spatial extension of the molecules is great enough so that the solute particles cut across velocity gradients and increase the viscosity in the manner suggested by Fig. 4.9. In this regard they are no different than the rigid spheres of the Einstein model. What is different for these molecules is the internal structure of the dispersed units, which are flexible and swollen with solvent. The viscosity of a polymer solution depends, therefore, on the polymer-solvent interactions, as well as on the properties of the polymer itself.

An important empirical generalization about the intrinsic viscosity of polymer solutions is given by

$$[\eta] = kM^a \tag{51}$$

in which k and a are experimentally determined constants. We shall call Eq. (51) the Staudinger-Mark-Houwink equation. H. Staudinger (Nobel Prize, 1953), one of the founders of modern polymer chemistry, originally proposed this relationship with $a = 1$. It was subsequently recognized that more commonly $0.5 < a < 0.8$. The constants k and a are called the Mark-Houwink coefficients; they depend on the temperature as well as the polymer-solvent system. Values of k and a are determined for a particular system by measuring the intrinsic viscosity of polymer fractions of known molecular weight. The following example illustrates how this is done.

Example 4.5   The molecular weights of various polycaprolactam preparations were determined by end group analysis (see Example 3.2); intrinsic viscosities of the various fractions in m-cresol were measured* at 25°C. The following values are representative of the results obtained:

---

*Data from H. K. Reimschussel and G. J. Dege, *J. Polym. Sci.*, 9:2343 (1971).

Table 4.2 Mark–Houwink Coefficients for Some Typical Polymer–Solvent Systems at the Indicated Temperatures

Polymer	Solvent	Temperature (°C)	$k \times 10^3$ $(cm^3\,g^{-1})$	a
Polypropylene	Cyclohexanone	92 = Θ	172	0.50
Poly(vinyl alcohol)	Water	80	94	0.56
Poly(oxy-ethylene)	CCl_4	25	62	0.64
Poly(methyl methacrylate)	Acetone	30	7.7	0.70
Polystyrene	Toluene	34	9.7	0.73
Natural rubber	Benzene	30	18.5	0.74
Poly(acrylo-nitrile)	Dimethyl formamide	20	17.7	0.78
Poly(vinyl chloride)	Tetrahydrofuran	20	3.63	0.92
Poly(ethylene terephthalate)	m-Cresol	25	0.77	0.95

Source: M. Kurata, M. Iwama, and K. Kamada, in *Polymer Handbook*, 2nd ed. J. Brandrup and E. H. Immergut (eds.), Wiley, New York, 1975.

$M \times 10^{-3}$	3.50	4.46	7.69	13.0	17.6	21.6	30.8
$[\eta]$ (dl g^{-1})	0.36	0.43	0.61	0.87	1.10	1.25	1.59

Determine the values of k and a which fit these data.

Solution By taking the logarithms of Eq. (51), a linear form is obtained: $\ln[\eta] = \ln k + a \ln M$. This could be plotted with a and $\ln k$ evaluated from the slope and intercept, respectively. Alternatively, a least-squares analysis of the data can be performed. When the latter is carried out using the logarithms of the above results, it is found that $a = 0.683$ and $\ln k = -6.593$, or $k = 1.37 \times 10^{-3}$, with $r = 0.9998$. The units of k are consistent with the concentration units dl g^{-1}. This result can be inverted to give M directly: $M = 1.51 \times 10^4 [\eta]^{1.46}$.

The practical significance of the result of this example lies in the great ease with which viscosity measurements can be made. Once the k and a values for an experimental system have been established by an appropriate calibration, molecular weights may readily be determined for unknowns measured under the same conditions. Extensive tabulations of Mark–Houwink coefficients are available, so the calibration is often unnecessary for well-characterized polymers (see Table 4.2).

If intrinsic viscosity is used to evaluate the molecular weight of a polydisperse sample, the molecular weight so obtained is an average value. Equation (1.30) defined the viscosity average, which is the kind of average obtained. We are now in a better position to see how this comes about.

It is apparent that the experimental (subscript ex) intrinsic viscosity depends on an average molecular weight M; it is the nature of this average that we want to establish. According to Eq. (51), we write

$$[\eta]_{ex} = k\overline{M}^{\,a} \tag{52}$$

The intrinsic viscosity contains a contribution of concentration which we note by writing $[\eta]_{ex} = (\eta_{sp})_{ex}/c_{ex}$, where it is understood that this is a limting value. If the sample is polydisperse, both $(\eta_{sp})_{ex}$ and c_{ex} are made up of contributions from molecules in different molecular weight classes which we designate by the subscript i; that is, we can write

$$(\eta_{sp})_{ex} = \sum_i (\eta_{sp})_i \tag{53}$$

and

$$c_{ex} = \sum_i c_i = \frac{1}{V} \sum_i m_i \tag{54}$$

where the second expression for c_{ex} recognizes that practical units are used in these experiments.

Each of the molecular weight fractions is expected to obey Eq. (51) independently; therefore

$$(\eta_{sp})_i = c_i[\eta]_i = c_i k M_i^a \tag{55}$$

Substituting Eqs. (53)–(55) into Eq. (52) yields

$$\frac{k\Sigma_i c_i M_i^a}{\Sigma_i c_i} = \frac{(k/V)\Sigma_i m_i M_i^a}{(1/V)\Sigma_i m_i} = k\overline{M}^{\,a} \tag{56}$$

which shows that

$$\overline{M}^a = \frac{\Sigma_i m_i M_i^a}{\Sigma_i m_i} \tag{57}$$

Since $m_i = n_i M_i$, Eq. (57) may be written

$$\overline{M} = \overline{M}_v = \left(\frac{\Sigma_i n_i M_i^{a+1}}{\Sigma_i n_i M_i} \right)^{1/a} \tag{58}$$

which is the form given in Chap. 1.

In this section we have looked at the Staudinger–Mark–Houwink equation as a purely empirical relationship which is useful for determining the molecular weight of unknown polymeric solutes. A considerable amount of work has been directed toward understanding the theoretical basis for this result. Although a detailed discussion would take us too far afield, we shall examine certain special cases of Mark–Houwink a values in the next section.

4.10 Polymer Chain Extension and Viscosity

Two special cases of the Staudinger–Mark–Houwink equation can be justified without much difficulty: $a = 1.0$ and $a = 0.5$. In our discussion we shall ignore all numerical coefficients and concentrate on the variables, particularly with respect to molecular weight dependence. This is sufficient to arrive at an understanding of the significance of the Mark–Houwink exponent. We shall examine two models for polymer chains in solution. These models picture the polymer chain as being either unwound, so that each segment experiences the flow streamlines, or tightly coiled, so that the flow leaves interior segments unaffected. The two models we discuss are known, respectively, as the free-draining and nondraining models. We shall see presently that these models—which clearly represent extremes of behavior—correspond to exponents of 1.0 and 0.5, respectively, in the Staudinger–Mark–Houwink equation. It is apparent that many polymer–solvent–temperature situations will result in chain coils of intermediate conformation; for these a is expected to take on values between the extremes we consider.

We begin by examining the free-draining model. Figure 4.15a shows a portion of a polymer chain situated in a velocity gradient. This will cause the molecule to rotate, thus converting some translational kinetic energy to rotational kinetic energy. This amounts to a dissipation of energy and thus connects with viscosity. Although not self-evident, it can be shown

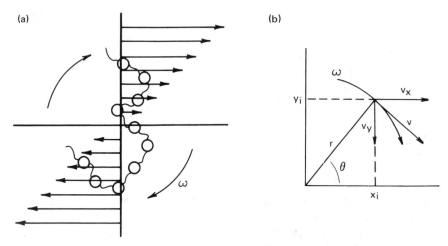

Figure 4.15 (a) Velocity gradient through the center of mass of a polymer chain. (b) Definition of coordinates for ith segment of rotating chain. [From Ref. 7; used with permission.]

that the angular velocity which is induced in the molecule in this situation is directly proportional to the velocity gradient. Furthermore, a chain segment a distance r from the center of mass of the molecule, say, segment i, acquires a velocity $v_i = r_i \omega$ from the rotation. Figure 4.15b defines the location of the ith segment relative to the center of mass.

Next, we consider the force of viscous resistance experienced by this segment as a result of moving through its surroundings with a velocity v. Equation (2.2) relates the force of viscous resistance to the velocity through the expression $F_v = fv$, where f is the friction factor. The friction factor of a chain segment plays the same role but is given the symbol ζ to emphasize that it applies to a segment rather than the entire molecule. Therefore, for the ith segment, we write

$$F_{v,i} = \zeta v_i \tag{59}$$

Both $F_{v,i}$ and v_i can be resolved into x and y components. Using the geometry defined by Fig. 4.15 b, we obtain

$$F_{v,i}^x = \zeta v_i \omega \sin \theta \tag{60}$$

and

$$F_{v,i}^4 = \zeta v_i \omega \cos \theta \tag{61}$$

Since force times distance equals energy, the rate at which energy is dissipated by viscous forces on segment i is given by $F_{v,i}^x v_x + F_{v,i}^y v_y$, which we identify as $(dE/dt)_i$. Substituting Eqs. (60) and (61) gives

$$\left(\frac{dE}{dt}\right)_i = \zeta r_i^2 \omega^2 (\sin^2 \theta + \cos^2 \theta) \propto \zeta r_i^2 \left(\frac{dv}{dy}\right)^2 \tag{62}$$

For the entire polymer chain (subscript p), this result must be totaled for the n segments of the chain to produce $(dE/dt)_p$:

$$\left(\frac{dE}{dt}\right)_p \propto \zeta \sum_{i=1}^{n} r_i^2 \left(\frac{dv}{dy}\right)^2 \tag{63}$$

Since r is the distance from the center of mass, we use Eq. (2.75) to replace $\Sigma_i r_i^2$ with $n\overline{R_g^2}$, where R_g is the radius of gyration of the molecule. Incorporating the latter, Eq. (63) becomes

$$\left(\frac{dE}{dt}\right)_p \propto \zeta n \overline{R_g^2} \left(\frac{dv}{dy}\right)^2 \tag{64}$$

To scale up this last result from a single molecule to all of the molecules in a volume of solution V, we multiply Eq. (64) by the particle concentration, given by c/M:

$$\frac{dE/dV}{dt} \propto \zeta n \overline{R_g^2} \frac{c}{M} \left(\frac{dv}{dy}\right)^2 \tag{65}$$

Comparing this result with Eq. (3) shows that the increment in viscosity caused by the free-draining chain is given by the coefficient of $(dv/dy)^2$ and may be written

$$\eta - \eta_0 \propto \zeta n \overline{R_g^2} \frac{c}{M} \propto \zeta n \overline{R_g^2} \frac{c}{n} \tag{66}$$

where the second form recognizes that M is proportional to the number of segments in the chain. Finally, Eq. (66) can be cast in the form of the intrinsic viscosity to yield

$$[\eta] = \frac{1}{c} \frac{\eta - \eta_0}{\eta_0} \propto \frac{\zeta \overline{R_g^2}}{\eta_0} \tag{67}$$

which shows that the intrinsic viscosity of a solution of freely draining chains is proportional to the square of the radius of gyration of those chains. The segmental friction factor has dimensions mass time^{-1}, and η_0 has dimensions mass length^{-1} time^{-1}; therefore the right-hand side of Eq. (67) has dimensions of length^3 mass^{-1}, which are the reciprocal concentration units appropriate for intrinsic viscosity.

Independent studies of the segmental friction factor reveal it to be essentially independent of the molecular weight of the polymer. Therefore for this model the molecular weight dependence of the intrinsic viscosity is the same as that of $\overline{R_g^2}$. According to Eq. (2.77), $\overline{R_g^2} \propto n \propto M$; therefore the only molecular weight dependence that survives out of all of this is a first-power dependence. We are thus led to conclude that the Mark–Houwink a coefficient equals unity for the case of free-draining chains. Since polymer chains are generally jumbled into coils, for which this may not be a good model, it is not surprising that experimental a values are usually less than this.

We discussed solvent goodness in Sect. 3.8. There are several aspects of that discussion which are pertinent here:

1. Solvent is squeezed out of the coil domain more and more as the solvent becomes poorer.
2. A change in solvent goodness can come about either from addition of a poorer solvent or from decreases in temperature.
3. Theta conditions correspond to a solvent so poor that precipitation would occur for a polymer of infinite molecular weight.
4. Theta conditions are identified experimentally as the situation where the second virial coefficient of osmotic pressure is zero.

In view of these considerations, it is not surprising that experimental a values vary systematically with decreasing solvent goodness. As the solvent goodness decreases, the chain becomes more tightly coiled so that flow streamlines penetrate the coil to a lesser extent. In an extreme situation we can imagine the coil so impermeable to the solvent flow that it behaves as a rigid sphere. Such a coil is said to be nondraining, since any solvent imbibed by the coil is essentially immobilized. There are two very important things to realize about this state of affairs. First, a coil in which polymer–polymer contacts are this highly favored sounds like an alternate description of theta conditions. Second, if the coil approaches a rigid sphere in behavior, Einstein's equation for viscosity becomes appropriate again.

J. G. Kirkwood and J. Riseman have developed a theory which allows for variable degrees of solvent drainage through the coil domain. We shall not go into this theory in any detail, except to note that it should reduce to Eq. (67) in the free-draining limit and to the Einstein equation in the nondraining limit. The Kirkwood–Riseman theory can be written in the form

$$[\eta] \propto \frac{(\overline{R_{g,0}^2})^{3/2}}{n} f(X) \tag{68}$$

where the subscript 0 on R_g indicates Θ conditions and $X = n\zeta/\eta_0(\overline{R_{g,0}^2})^{1/2}$. The function $f(X)$ in Eq. (68) is our concern. For our purposes, it is enough to note that $f(X)$ approaches a value which is proportional to $n\zeta/\eta_0(\overline{R_{g,0}^2})^{1/2}$ in the free-draining limit (we continue to ignore numerical constants) and approaches a constant value in the nondraining limit. If we substitute the free-draining limit into Eq. (68), we obtain

$$[\eta] \propto \frac{(\overline{R_{g,0}^2})^{3/2}}{n} \frac{n\zeta}{\eta_0(\overline{R_{g,0}^2})^{1/2}} \propto \frac{\zeta}{\eta_0} (\overline{R_{g,0}^2}) \tag{69}$$

The radius of gyration is expected to be different under theta and non-theta conditions, since the extent of coil swelling due to imbibed solvent changes with solvent goodness. We define a coil expansion factor α as follows:

$$\alpha = \frac{R_g}{R_{g,0}} \tag{70}$$

Although α is ordinarily greater than unity, fractional values are also possible. The range of fractional values is more limited, however, since the polymer tends to precipitate rather than squeeze out much more solvent under conditions poorer than theta conditions. Incorporating α into Eq. (69) gives

$$[\eta] \propto \frac{\zeta\alpha^2}{\eta_0} \overline{R_g^2} \tag{71}$$

which is equivalent to Eq. (67) as required.

Using the nondraining limit of the Kirkwood–Riseman theory gives

$$[\eta]_\Theta \propto \frac{(\overline{R_{g,0}^2})^{3/2}}{n} \tag{72}$$

Since Eq. (2.77) shows that $\overline{R_{g,0}^2} \propto n$, Eq. (72) becomes $[\eta] \propto n^{1/2} \propto M^{1/2}$. This important result shows that $a = 0.5$ is expected under Θ conditions. This expectation has been repeatedly verified by viscosity experiments under independently determined conditions. The subscript Θ has been attached to $[\eta]$ in Eq. (72) in recognition of this.

For high molecular weight polymers in good solvents, $[\eta]$ exceeds $[\eta]_\Theta$ because of coil expansion under nondraining conditions; that is, as more solvent enters the coil domain than would be present under Θ conditions, Eq. (72) continues to apply, with $\overline{R_g^2}$ replacing $\overline{R_{g,0}^2}$. Using Eq. (70) to quantify this expansion effect, we obtain

$$[\eta] \propto \frac{\alpha^3 (\overline{R_{g,0}^2})^{3/2}}{n} \tag{73}$$

Taking the ratio of Eq. (73) to Eq. (72) gives

$$\frac{[\eta]}{[\eta]_\Theta} = \alpha^3 \tag{74}$$

which shows how the effect of solvent uptake on the spatial extension of polymer coils can be quantitatively determined from intrinsic viscosity experiments under theta and non-theta conditions.

We remarked earlier that the Einstein equation might be pertinent in the case of a nondraining coil; Eq. (73) does not seem to bear this out. The Einstein theory in the form of Eq. (41) predicts that $\eta/\eta_0 - 1$ is proportional to the volume fraction of the dispersed particles. For the nondraining coils the volume fraction is proportional to the volume of each coil domain times the number concentration of the particles. The first of these factors is proportional to R_g^3, and the second to c/M. This leads to the prediction

$$\left(\frac{\eta}{\eta_0} - 1 \right) \propto \phi \propto R_g^3 \frac{c}{M} \propto \alpha^3 (\overline{R_{g,0}^2})^{3/2} \frac{c}{M} \tag{75}$$

which is equivalent to Eq. (72).

We have omitted a great deal of detail in this discussion of polymer viscosity. The interested reader will find some of the missing information supplied in Ref. 4. In particular, we have omitted all numerical coefficients, which limits us to ratios as far as computational capability is concerned. Numerical coefficients are available for Eq. (72), for example, and this allows coil dimensions to be evaluated from viscosity measurements. A general conclusion which unifies all of this section is that any factor which causes a polymer chain to be more extended in space—whether coil unfolding or swelling by solvent—tends to increase $[\eta]$. This is exactly what we expect in terms of the purely qualitative picture provided by Fig. 4.9. The following example illustrates this for some actual polymers.

Example 4.6   For cellulose triacetate in acetone at 25°C, $k = 8.97 \times 10^{-5}$ dl g^{-1} and $a = 0.9$. For polyisobutene in benzene at 24°C, $k = 1.07 \times 10^{-3}$

dl g^{-1} and a = 0.5. Calculate [η] for these systems if each of the respective polymers have M = 10^5. Comment on the correlation of the a values with the nature of the system in each case.

Solution   Intrinsic viscosities are calculated by direct substitution into the Staudinger–Mark–Houwink equation.
For cellulose triacetate: [η] = 8.97 × 10^{-5}(10^5)$^{0.9}$ = 2.84 dl g^{-1}.
For polyisobutene: [η] = 1.07 × 10^{-3}(10^5)$^{0.5}$ = 0.34 dl g^{-1}.
Even though the two polymers have the same molecular weight, the cellulose triacetate has an intrinsic viscosity more than eight times greater than the polyisobutene. Note that the Mark-Houwink a coefficient is primarily responsible for this; the intrinsic viscosities would be ranked oppositely if k were responsible.
These results are fully consistent with the nature of the systems involved. Cellulose triacetate repeat units are six-membered rings which carry three bulky acetate groups each. Internally, the rings are inflexible with respect to rotation and, because of the bulk of the acetate groups, are expected to be severely hindered in their rotation with respect to each other. As a consequence, such a molecule exists in solution in a highly extended form which approaches the predictions of the free-draining model in intrinsic viscosity. Polyisobutene, by contrast, is an aliphatic hydrocarbon for which low molecular weight aliphatic molecules are expected to be better solvents. Although benzene dissolves this polymer, it is a poorer solvent because of the polarizable pi electrons. It is therefore plausible that this system corresponds to theta conditions with an a value of 0.5.

•

We have restricted our discussion to Newtonian systems until now, except for a brief comparison of Newtonian and non-Newtonian behavior in connection with Fig. 4.2. This should not be interpreted to mean that colloidal systems do not display non-Newtonian behavior or that such behavior is unimportant. Neither of these is true. In the next section we outline some examples of non-Newtonian behavior, particularly in the viscous properties of flocculated colloids.

4.11   Non-Newtonian Behavior and Flocculation

We have already devoted a considerable amount of space to our discussion of viscosity without ever venturing beyond Newtonian systems. At least as much—probably more—could be said about non-

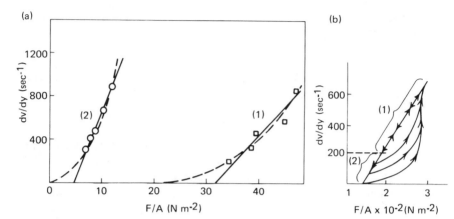

Figure 4.16 Rate of shear versus shearing stress for (a) 7% aqueous carbon black dispersions [data from A. I. Medalia and E. Hagopian, *Ind. Eng. Chem. Prod. Res. Dev., 3*:120 (1964)] and (b) 11% aqueous bentonite dispersion (pH = 8.7). The time of the cycle is 70 s. [Schematic based on data of W. F. Gabrysh, H. Eyring, P. Lin-Sen, and A. F. Gabrysh, *J. Am. Ceram. Soc., 46*:523 (1963).]

Newtonian systems. We shall not pursue this topic, however, beyond some brief remarks about the kinds of behavior that are observed and the areas where they find application. In particular, we shall discuss flocculated colloids as examples of systems where the apparent viscosity depends on the rate of shear. The existence of velocity gradients means that differences in flow velocity may extend over the dimensions of colloidal particles. If the dispersed particle is an aggregate of primary particles, a floc, then the shearing forces associated with viscosity which operate across the floc may disrupt or rearrange the aggregate. This corresponds to a dissipation of translational energy and hence contributes to the viscosity. It is clear that the extent to which this occurs may vary with the velocity gradient; this is one way in which the apparent viscosity may depend on the rate of shear.

Strictly speaking, *the* coefficient of viscosity is meaningful only for Newtonian fluids, in which case it is the slope of a plot of stress versus rate of shear, as shown in Fig. 4.2. For non-Newtonian fluids, such a plot is generally nonlinear, so the slope varies from point to point. In actual practice, the data are traditionally represented with the rate of shear (dv/dy) as the ordinate and the stress (F/A) as the abscissa, as shown in Fig. 4.16. In this case the apparent viscosity at some particular point is given by the *co*tangent of the angle which defines the slope at that particular point.

It is convenient to distinguish between three different types of behavior: (1) The apparent viscosity increases as the rate of shear increases, (2) the apparent viscosity decreases as the rate of shear increases, and (3) the apparent viscosity is very high—effectively infinite—until the applied force reaches a critical value, then it decreases abruptly and may be either constant or continue to decrease beyond this. The critical stress in this third case is called the yield value, since the system behaves as an elastic body for lesser stresses. The two regions of behavior in the third situation may merge more or less gradually in the vicinity of the yield value, giving rise to several different ways of defining this latter parameter.

These phenomena have different names, depending on whether they are observed under time-dependent or stationary state (time-independent) conditions. If the apparent viscosity increases with increasing rate of shear, the effect is called rheopexy, if time dependent, and dilatancy, if time independent. If the apparent viscosity decreases with increasing rate of shear, the effect is called thixotropy, if time dependent, and plasticity (with a yield value) or pseudoplasticity (without a yield value), if time independent.

As we shall see presently, thixotropic behavior is associated with flocs of asymmetrical particles. In certain instances quicksand apparently operates through a thixotropic mechanism. The struggles of the victim merely decrease the viscosity of the trap and worsen his plight. By contrast, wet beach sand is an example of a dilatant system in which the apparent viscosity increases with shear. Anyone who has wiggled his toes in the latter knows that under these conditions (low shear) the wet sand is very fluid. However, the same sand is hard to a firm footstep (high shear). In contrast to thixotropy, dilatancy is favored by symmetrical, nonflocculated particles. It is almost always observed at concentrations in the neighborhood of 40% dispersed particles. At this concentration and in the absence of disruptable flocs, the only way such a system can flow is by the gradual rolling of particles past one another. If the rate of shear is too great, this deformation is impossible.

Next let us see how thixotropic behavior relates to the phenomena of flocculation. The data shown in Fig. 4.16a were obtained for a 7% slurry of carbon black in water. Curve 1 shows the results obtained immediately after the dispersion is prepared in a high-speed blender. After additional mild agitation, the results shown in curve 2 are obtained. The latter are independent of further agitation. With shorter periods of mild agitation, a family of curves lying between 1 and 2 would be obtained.

The data presented in Fig. 4.16a are consistent with the following mechanism. The dispersion which emerges from the blender is fundamentally unstable with respect to flocculation and flocculates rapidly to form a volume-filling network throughout the continuous phase. Except for the size and structure of the "chains," the situation is comparable to a cross-linked polymer swollen by solvent. In both the liquid is essentially immobilized by the network of chains, and the system behaves as an elastic solid under low stress. The term *gel* is used to describe such systems, whether the dispersed particles are lyophilic or lyophobic.

As the force applied to the surface of the gel (see Fig. 4.1) is increased, however, a point is ultimately reached—the yield value—at which the network begins to break apart and the system begins to flow (curve 1). Increasing the rate of shear may result in further deflocculation, in which case the apparent viscosity would decrease further with increased shear. Highly asymmetrical particles can form volume-filling networks at low concentrations and are thus especially well suited to display these phenomena.

As the system is subjected to ongoing, low-level mechanical agitation, the network structure is rearranged to a dispersion of more compact flocs which display both a lower yield value and a lower apparent viscosity than the initial dispersion (curve 2). A certain amount of time is required for the dispersed units to acquire a size and structure which are compatible with the prevailing low level of agitation. This is why intermediate (not shown in Fig. 4.16a) cases are observed before the actual stationary state condition is obtained.

If the time for measurement of a stress/shear curve is short compared to the time required for rearrangement of the structure of the dispersed particles, then different results are obtained, depending on whether the rate of shear is increasing or decreasing. Figure 4.16b is an example of such hysteresis for 11% dispersions of bentonite (a montmorillonite clay with plate-shaped primary particles) in which the entire cycle is measured in 70 s. These data show the sensitivity of such experiments to the level of shear and to the time of observation. If the direction of the cycle is reversed along the descending branch of the curve, different results are obtained, depending on whether the reversal is done at a rate of shear above [region (1) in Fig. 4.16b] or below [region (2)] about 200 s^{-1}. This shows that the structure within the colloid builds up rapidly (compared to the cycle time of 70 s) at rates of shear below 200 s^{-1}, with negligible buildup at greater rates of shear.

These complicated observations are difficult to interpret in terms of fundamental interactions between particles; nevertheless, they have tremendous practical significance. For example, carbon black strengthens rubber against deformation and 3–4 kg of carbon black is introduced into every tire made for this reason. This accounts for an annual worldwide consumption of carbon black of over 4 billion kg for automobile tires alone. Likewise, dispersions of clay in oils are used as lubricating greases. Obviously, the viscosity of these substances under various conditions of shear is an important consideration. Printing inks, drilling muds (circulated around the shaft in well-drilling operations to cool the bit and flush away cuttings), paper coatings, paints, and innumerable industrial slurries may all be considered examples of areas in which these considerations are vitally important.

In a paint, for example, a controlled level of flocculation is important in both the actual application of the paint and its storage in the container during application. In the former, thixotropy (the word means "changing with touch") permits the paint to "thin" under the shearing influence of the paintbrush or spraygun. Once applied, it thickens, preventing the drip or sag of the paint on the surface. In addition, the time required for this yield value to develop should be sufficient to allow for the leveling of brush marks. Thixotropy is an important property of paint in the bucket as well as on the wall. The buildup of a yield value interferes with the sedimentation of the pigment and eliminates the need to stir the paint continuously to assure uniformity. The fact that these requirements are well met by commercial paints indicates the success of paint chemists in regulating thixotropy.

References

1. R. B. Bird, W. E. Stewart, and E. N. Lightfoot, *Transport Phenomena*, Wiley, New York, 1960.
2. E. J. Cohn and J. T. Edsall, *Proteins, Amino Acids and Peptides*, ACS Monograph, Hafner, New York, 1965.
3. A. Einstein, *Investigations on the Theory of the Brownian Movement*, Dover, New York, 1956.
4. P. J. Flory, *Principles of Polymer Chemistry*, Cornell University Press, Ithaca, N.Y., 1953.
5. H. L. Frisch and R. Simha, The Viscosity of Colloidal Suspensions and Macromolecular Solutions, in *Rheology*, Vol. 1 (F. R. Eirich, ed.), Academic, New York, 1956.
6. J. J. Hermans, *Flow Properties of Disperse Systems*, North-Holland, Amsterdam, 1953.
7. P. C. Hiemenz, *Polymer Chemistry: The Basic Concepts*, Marcel Dekker, New York, 1984.
8. E. G. Richards, *An Introduction to the Physical Properties of Large Molecules in Solution*, Cambridge University Press, Cambridge, 1980.
9. C. Tanford, *Physical Chemistry of Macromolecules*, Wiley, New York, 1961.

Problems

1. Gillespie and Wiley* used a cone-and-plate viscometer to measure F/A versus dv/dx for dispersions of silica and cross-linked polystyrene in dioctyl phthalate. At a volume fraction of 0.35 for both solids, the following results were obtained:

Silica	$F/A \times 10^{-3}$ (dyne cm^{-2})	2.2	1.4	1.0	0.50	0.25
	dv/dx (s^{-1})	500	325	235	125	60
Polystyrene	$F/A \times 10^{-3}$ (dyne cm^{-2})	1.6	0.80	0.55	0.25	
	dv/dx (s^{-1})	500	235	160	100	

Use these data to determine either η or the yield value for these dispersions, depending on whether or not the system is Newtonian. Are these results consistent with the fact that the axial ratio was nearer unity and the particle size distribution narrower for the polystyrene than the silica? Explain.

2. An aqueous polybutyl methacrylate latex ($\bar{r} = 200$ Å) has a viscosity of 50,500 cP at $\phi = 0.25$ and a viscosity of 36.7 cP at the same rate of shear when 1.71×10^{-5} g NaCl is added per gram of polymer.† Assuming that these charged particles must be surrounded by a layer of dissolved ions in solution, what conclusions can you draw about the dependence of the thickness of this layer of ions on the electrolyte content of the continuous phase?

3. A copolymer of vinylpyridine and methacrylic acid (62 and 38 mole %, respectively, in polymer) was studied in 90% methanol–10% water solution. The specific viscosity was measured‡ as a function of added NaOH or HCl and the following results were obtained:

η_{sp}	0.1	0.25	0.28	0.24	0.21	0.21	0.20
mEq added per gram	0	2	4	6	2	4	6
			HCl			NaOH	

Discuss these results in terms of the apparent effect of acid and base on the configuration of the polymer chain. Be sure your explanation is consistent with the chemical nature of the copolymer.

4. A dispersion of polydisperse spheres shows a relative viscosity of about 2.6 at a volume fraction of about 0.31.§ Compare this result with the predictions of Eq. (48).

---

*T. Gillespie and R. Wiley, *J. Phys. Chem.*, 66:1077 (1962).

†J. G. Brodnyan and E. L. Kelley, *J. Colloid Sci.*, 19:488 (1964).

‡T. Alfrey, Jr., and H. Morawitz, *J. Am. Chem. Soc.*, 74:436 (1952).

§ H. Eilers, *Kolloid Z.*, 97:313 (1941).

5. The viscosity of cross-linked polymethyl methacrylate spheres in benzene was measured* and found to be

ϕ	0.050	0.035	0.028	0.019	0.014	0.010
η/η_0	2.15	1.61	1.41	1.19	1.18	1.12

Calculate $[\eta]$ for these spheres. Is there evidence in these results that the polymer particles may be swollen by the solvent? Explain.

6. The viscosity of uniform, cross-linked polystyrene spheres of two different diameters was measured in benzyl alcohol at 30°C:†

$d = 0.382\ \mu m$	ϕ	0.013	0.030	0.059	0.075
	η_{sp}	0.036	0.086	0.178	0.233
$d = 0.433\ \mu m$	ϕ	0.02	0.04	0.08	
	η_{sp}	0.056	0.116	0.251	

Evaluate the intrinsic viscosity for each size of spherical particle and comment on the results in terms of the Einstein prediction that $[\eta]$ should be independent of particle size. Is the fact that benzyl alcohol is only a moderately good solvent for linear polystyrene consistent with the observed deviation between the experimental and theoretical values for $[\eta]$? Explain.

7. Criticize or defend the following proposition using the data of Problem 6 and the fact that the intrinsic viscosities of cross-linked polystyrene spheres in solvents such as benzene and CCl_4 (good solvents for linear polystyrene) lie in the range 5.8–7.5: Cross-linked polystyrene spheres swell by imbibing solvent, the effect being more extensive the better the solvent properties of the continuous phase for the non-cross-linked polymer.

8. Use the following data to evaluate the Mark–Houwink a and k constants for cellulose acetate in acetone at 25°C:‡

M_n	c (g dl^{-1})	η_{sp}
130,000	0.094	0.289
	0.273	0.99
	0.546	2.77
86,000	0.114	0.286
	0.351	1.10
	0.703	3.12
76,000	0.118	0.247
	0.353	0.89
	0.775	2.70

*A. Kose and S. Hachisu, *J. Colloid Interface Sci.*, 46:460
†Y. S. Papir and I. M. Krieger, *J. Colloid Interface Sci.*, 3
‡A. M. Sookne and M. Harris, *Ind. Eng. Chem.*, 37:475 (1945).

61,000	0.138	0.239
	0.275	0.52
	0.428	0.88
48,000	0.152	0.209
	0.303	0.45
	0.684	1.23

9. Various molecular weight fractions of cellulose nitrate were dissolved in acetone and the intrinsic viscosity was measured at 25°C*:

$M \times 10^{-3}$ (g mole^{-1})	77	89	273	360	400	640	846	1550	2510	2640
$[\eta]$(dl g^{-1})	1.23	1.45	3.54	5.50	6.50	10.6	14.9	30.3	31.0	36.3

Use these data to evaluate the constants k and a in the Staudinger–Mark–Houwink equation for this system.

10. The relative viscosity of solutions of cellulose nitrate in acetone was measured and extrapolated to zero rate of shear†:

η/η_0	1.45	1.53	1.67	1.89	2.31	3.41
c (g dl^{-1})	0.0151	0.0176	0.0212	0.0264	0.0352	0.0528

Use these data to evaluate the intrinsic viscosity for the polymer and use the a and k values from Problem 9 to calculate the molecular weight from the value of $[\eta]$.

11. The following intrinsic viscosity values of some high molecular weight polystyrene fractions have been reported‡:

$\overline{M} \times 10^{-6}$ (g mole^{-1})	$[\eta]$ (dl g^{-1})	
43.8	5.5	
27.4	4.4	Cyclohexane at 35.4°C
43.5	67.7	
26.8	36.5	Benzene at 40°C

Use these data to evaluate the constants in the Staudinger–Mark–Houwink equation. Are the values obtained consistent with the known facts that 35.4°C is the Flory (Θ) temperature for polystyrene in cyclohexane while benzene is a good solvent for polystyrene at 40°C?

12. Solutions of nylon-6,6 were studied in 90% formic acid solutions at 25°C and

---

*A. M. Holtzer, H. Benoit, and P. Doty, *J. Phys. Chem.,* *58*:624 (1954).
†A. M. Holtzer, H. Benoit, and P. Doty, *J. Phys. Chem.,* *58*:624 (1954).
‡D. McIntyre, L. J. Fetlers, and E. Slagowski, *Science,* *176*:1041 (1972).

the following data were obtained for two different molecular weight fractions*:

c (g dl^{-1})	0.744	0.527	0.368	0.164
η_{sp}/c (dl g^{-1})	0.485	0.477	0.478	0.450

and

c (g dl^{-1})	0.742	0.640	0.537	0.436	0.332	0.225	0.132	0.058
η_{sp}/c (dl g^{-1})	0.897	0.892	0.886	0.876	0.864	0.847	0.805	0.778

Calculate the intrinsic viscosity for these two polymers. The Mark–Houwink constants for this system are known to be $a = 0.72$ and $k = 11 \times 10^{-4}$ dl g^{-1}; calculate the molecular weights of the two nylon fractions.

13. The radius of gyration of polymer coils can be determined independently from light scattering. Fox and Flory† measured both R_g and $[\eta]$ for various molecular weight fractions of polystyrene in various solvents at several temperatures. The following results were obtained:

Solvent	T (°C)	$M \times 10^{-3}$ (g mole^{-1})	$[\eta]$ (dl g^{-1})	R_g (Å)
Methylethyl ketone	22	1760	1.65	437
	22	1620	1.61	414
	67	1620	1.50	400
	22	1320	1.40	367
	25	980	1.21	343
	22	940	1.17	306
	22	520	0.77	222
	25	318	0.60	194
	22	230	0.53	163
Dichloroethane	22	1780	2.60	576
	22	1620	2.78	545
	67	1620	2.83	529
	22	562	1.42	310
	22	520	1.38	278
Toluene	22	1620	3.45	527
	67	1620	3.42	523

Use these data to evaluate the factor of proportionality in Eq. (73). Does this factor seem reasonably constant?

---

*G. B. Taylor, *J. Am. Chem. Soc.*, 69:635 (1947).
†T. G. Fox, Jr., and P. J. Flory, *J. Am. Chem. Soc.*, 73:1915 (1951).

14. The intrinsic viscosity of a polystyrene solution in cyclohexane was measured* under theta conditions and found to be 0.078 dl g^{-1} for a polymer of $M = 8370$ g mole^{-1}. Use the average value of the factor of proportionality from the last problem to evaluate the radius of gyration of the polystyrene molecules in this solution. Calculate what this quantity is espected to be in terms of Eq. (2.77), using twice the degree of polymerization as the number of steps in the random walk (since there are two C—C bonds per segment) and taking $0.154(2)^{1/2} = 0.218$ nm as the value of l_0 (the C—C bond length corrected for tetrahedral bond angles). Briefly explain why the experimental value for R_g is larger than that calculated by Eq. (2.77).

---

*W. R. Krigbaum, L. Mandelkern, and P. J. Flory, *J. Polym. Sci.,* 9:381 (1952).

5

LIGHT SCATTERING

An interesting and oft-investigated question, "What is the origin of light?" and the solution of it has been repeatedly attempted, with no other result than to crowd our lunatic asylums with the would-be solvers. Hence, after fruitless attempts to suppress such investigations by making them liable to a heavy tax, the Legislature . . . absolutely prohibited them.

[From Abbott's *Flatland*]

5.1 Introduction

In Chapter 1 we described dark-field microscopy in which particles too small for direct microscopic observation could be detected against a dark field by horizontal illumination. Airborne dust or smoke particles show up in a beam of light in an otherwise dark room in the same way. In both cases the particles interact with the light which strikes them and deflect some of that light from its original direction. We speak of this light as being "scattered." A whole assortment of optical phenomena related to this are generally known as light scattering effects.

It turns out that the intensity of scattered light at any angle depends on the wavelength of the incident light, the size and shape of the scattering particles, and the optical properties of the scatterers, as well as the angle of observation. Furthermore, the functional relationship among these variables is known, at least for spherical particles and other geometries, under certain circumstances. By applying these relationships to light scattering experiments, information about the particles responsible for the scattering can be deduced. We shall see how the weight and a characteristic linear dimension of the particle may be determined for some systems from light scattering.

A general relationship for the intensity of light scattered by a spherical particle was derived by L. Lorenz in the latter part of the

223

nineteenth century and applied to colloids by G. Mie in 1908. Some quotations from two of the references at the end of this chapter will give an indication of the historical development of light scattering since Mie's complicated theory was presented. In a book published in 1956 [5], Stacey remarks that scattering patterns "have been tabulated in only a few cases because the computation is so laborious" [5, p. 56]. In his 1969 book [3], Kerker writes of the same patterns that so many have been published "that these can hardly be coped with in the usual tabular form, much less published in the normal way" [3, p. 77]. During the period of slightly more than a decade between these books, the computer arrived on the scene. Things haven't been the same since! Our concern in this chapter is not primarily with complicated systems which require elaborate calculations, although we shall briefly discuss several specific examples of such systems. Instead, we shall focus attention on systems where simplifying approximations to the general theory can be applied.

During the 50-year period between the publication of the Mie theory and the application of computers to its solution, much effort was devoted to finding approximations to the rigorous theory that would be easier to use. Lord Rayleigh's work (which was published in 1871, historically preceding the general theory) consists of an approximation which applies only to very small particles whose refractive index is not too large. Similarly, Debye's approximation (1909) applies to slightly larger particles than the Rayleigh approximation, but to a narrower range of refractive indices. The Rayleigh and Debye approximations have proved to be extremely valuable in the characterization of certain colloids, especially polymeric materials whose refractive index is not too much different from that of the solvent.

5.2 Electromagnetic Radiation

The phenomena with which we are concerned in this chapter are displayed by the entire spectrum of electromagnetic radiation. In applications to colloid chemistry, however, most work is done in the visible part of the spectrum—hence the common designation *light scattering*. Visible light shares a variety of parameters and descriptive relationships with other regions of the electromagnetic spectrum. The purpose of this section is to examine briefly some of the characteristics of electromagnetic radiation, particularly those that are needed for an understanding of light scattering.

Electromagnetic radiation consists of oscillating electrical (E) and magnetic (**H**) fields which are perpendicular to each other and perpendicular to the direction of propagation of the wave, as shown in Fig.

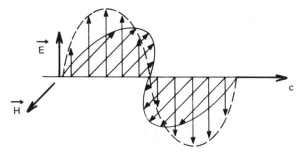

Figure 5.1 The relationship between the electric and magnetic fields and the direction of propagation of electromagnetic radiation.

5.1. Under vacuum the velocity of propagation of an electromagnetic wave c is about 3×10^8 m s^{-1} and is independent of the wavelength of the radiation. The frequency ν, wavelength λ_0, and velocity of the radiation are related through the familiar equation

$$c = \lambda_0 \nu \tag{1}$$

If the radiation is passing through a medium other than vacuum, its velocity and wavelength are both diminished by a factor n, the refractive index of the medium. Then Eq. (1) becomes

$$\frac{c}{n} = \frac{\lambda_0}{n} \nu \tag{2}$$

or

$$\upsilon = \lambda \nu \tag{3}$$

We shall use the symbols c and λ_0 to respectively refer to the velocity and wavelength under vacuum only. In this chapter the symbol λ, without a subscript, always refers to the wavelength of the radiation in the medium.

Our discussion of light scattering centers on the oscillations of the electric field. Before turning to the oscillating aspect of the field, let us first review a few points about electric fields per se. We begin by retreating to Coulomb's law, which states that the force between two charges q_1 and q_2 that are separated by a distance r is proportional to $q_1 q_2 / r^2$. In SI units the q's are measured in coulombs, r in meters, and F in Newtons; the proportionality factor in Coulomb's law must be dimensionally consistent. For charges under vacuum, it is traditional to write the proportionality factor as $\frac{1}{4}\pi\varepsilon_0$, where ε_0 is called the permittivity of vacuum. Thus Coulomb's law is written

$$F = \frac{1}{4\pi\varepsilon_0} \frac{q_1 q_2}{r^2} \tag{4}$$

to which the following numerical values apply:

1. The permittivity of vacuum $\varepsilon_0 = 8.854 \times 10^{-12}$ C^2 N^{-1} m^{-2}.
2. The proportionality constant in Eq. (4) $\frac{1}{4}\pi\varepsilon_0 = 8.988 \times 10^9$ N m^2 C^{-2}.
3. The SI units C^2 N^{-1} m^{-2} are equivalent to C^2 J^{-1} m^{-1} or kg^{-1} m^{-3} s^2.
4. The older literature uses cgs units in which the proportionality constant between F and $q_1 q_2 / r^2$ equals 1.00 dyne cm^2 (esu)$^{-2}$, where the esu is the electrostatic unit of charge which is defined to make this proportionality factor equal to unity.

If the charges are embedded in a medium, the electrical properties of the intervening molecules decrease the force from the value calculated by Eq. (4). The relative dielectric constant of the medium ε_r measures this effect quantitatively. In surroundings other than vacuum, the force between two charges is given by

$$F = \frac{1}{4\pi\varepsilon_0\varepsilon_r} \frac{q_1 q_2}{r^2} \tag{5}$$

It is important to remember several aspects of Eq. (5):

1. The relative dielectric constant ε_r is dimensionless. This quantity is also known as the relative permittivity.
2. The product $\varepsilon_0\varepsilon_r$ is sometimes written ε (without subscript); ε has the units of ε_0.
3. In cgs units ε_r also appears in the denominator of Coulomb's law.

Next, let us apply these ideas to the electric field. By definition, an electric field $\mathbf{E}$ describes the force experienced by a unit test charge $q_t = 1$. Thus, if we let one of the charges in Eq. (5) equal q_t, the following expression is obtained for the field associated with the remaining charge q:

$$\mathbf{E} = \frac{1}{4\pi\varepsilon_0} \frac{q}{r^2} \tag{6}$$

The dominant characteristic of the electrical and magnetic fields that comprise electromagnetic radiation is their periodically oscillating nature, a fact that enables us to describe them by the mathematics of waves. As far as light scattering is concerned, it is the electric field in which we are interested. The oscillating nature of an electric field propagating in the positive x direction is described by the equation

$$E = E_0 \cos \left[2\pi \left(vt - \frac{x}{\lambda} \right) \right] \qquad (7)$$

in which E_0 is the maximum amplitude of the field.

Since we have postulated the x direction as the direction of propagation, the electric field lies in the yz plane and may, in general, be resolved into y and z components, since E is a vector. Both the y and z components of the field are described by Eq. (7) when the latter is modified by the inclusion of phase angles. This is because the two components need not be in phase with each other. Accordingly, we write

$$E_y = E_{0y} \cos \left[2\pi \left(vt - \frac{x}{\lambda} + \delta_y \right) \right] \qquad (8)$$

and

$$E_z = E_{0z} \cos \left[2\pi \left(vt - \frac{x}{\lambda} + \delta_z \right) \right] \qquad (9)$$

in which the δ terms are the phase angles of the two components. In the most general case, these two equations mean that the electric field vector traces an ellipse in the yz plane. There are two special cases of note within this general situation. If the phase difference between the two components of the field, $\delta_y - \delta_z$, is zero or some integral multiple of π, the ellipse flattens to a line. If the phase difference is $\pi/2$ or any odd integral multiple of $\pi/2$ and the amplitudes of the two components are equal, the ellipse is rounded to a circle. In the former case we speak of the radiation as being plane polarized, and in the latter case as being circularly polarized.

Ordinary light is said to be unpolarized. This last term is somewhat unfortunate, because all light displays some form of polarization. In ordinary light, however, all forms of polarization are present, so the individual effects cancel out. The use of various filters makes it possible to conduct experiments with radiation which shows a unique state of polarization. Polaroid filters, for example, transmit plane-polarized light

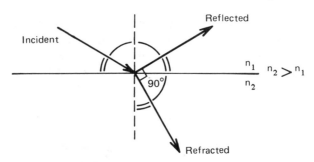

Figure 5.2 The relationship between the incident, reflected, and refracted beams of radiation at a plane surface.

only. In discussing light scattering, we are concerned primarily with unpolarized light and occasionally with plane-polarized light.

An interesting example of polarization arises in the study of light reflected from a surface. Suppose we consider a beam of light incident upon the planar surface of some material having a higher refractive index than the medium from which the beam approaches. At the surface some of the light will be refracted, and some will be reflected. Figure 5.2 illustrates this for the case in which the reflected and the refracted beams are separated by an angle of 90°. In this situation two very different results are obtained, depending on whether the incident light is linearly polarized in the plane of the figure or perpendicular to it. If the light is polarized in the plane of the figure, no light will be reflected at all. On the other hand, if it is initially polarized perpendicular to the plane of the figure, it will be reflected with the same polarization. If ordinary light—a mixture of the two types of light just mentioned—is used for the incident radiation, only one of the components contributes to the reflection. Furthermore, the reflected light is polarized perpendicular to the plane of the figure. Polaroid filters are used in photography and in sunglasses to reduce the glare of reflected light, since this light is polarized. This behavior is not observed uniquely when the angle between the two beams is 90°; rather, the intensity of the reflected beam varies continuously with the angle. At 90°, however, the polarization effect is most pronounced. We shall see that some scattering phenomena also show an angular dependence, as well as the fact that scattered light displays maximum polarization at 90°.

5.3 Oscillating Electric Charge

In this section we discuss the interaction between an electric field and a charge which is free to move with the field. Such a charge experiences a force which accelerates it with the field. If the field is oscillating, the acceleration of the charge will also oscillate. One of the basic results of classical electromagnetics is that the acceleration of a charge leads to the emission of radiation. It was the apparent violation of this requirement that led to the postulate of quantization in the Bohr theory of the hydrogen atom. However, we are concerned here with the classical result in which the charge does radiate. Our objective is to describe the emitted radiation some distance r from the emitter.

The radiation emitted by an oscillating charge may be described by its electric field vector, which is given by

$$E = \frac{q\mathbf{a}_p \sin \phi_z}{4\pi\varepsilon_0 c^2 r} \tag{10}$$

according to electromagnetic theory. In this equation q equals the magnitude of the charge, $\mathbf{a}_p$ is its periodic acceleration, and c is the velocity of light. The coordinate system defined by Fig. 5.3a will help describe this field. The origin of the coordinates is located at the emitting charge. The angles between the line of sight—along which r is meas-

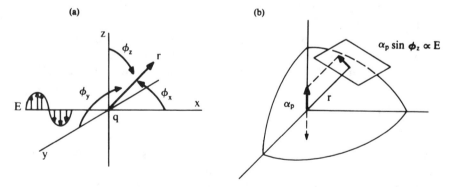

Figure 5.3 (a) The coordinates of an electric field E relative to an oscillating charge located at the origin. (b) Projection of the acceleration in the plane perpendicular to the line of sight.

ured—and the x, y, and z axes are designated ϕ_x, ϕ_y, and ϕ_z, respectively.

The oscillating charge behaves like an antenna and Eq. (10) describes the field of such an antenna, as long as r is large compared to the wavelength of the radiation which induces the oscillation. It should also be noted that the antenna to which Eq. (10) applies is aligned vertically (z axis) and is therefore "driven" by vertically polarized radiation.

We are using Eq. (10), presented without proof, as the point of departure for our discussion of light scattering. Therefore it is important that we find its predictions reasonable. First, let us consider the plausibility of the sin ϕ_z factor. This factor ranges between 0 and 1 as ϕ_z varies from 0 to $\pi/2$. This means that the maximum field will be observed at right angles to the oscillating charge and no field will be observed along the axis of the oscillation. It is the projection of the acceleration in the plane perpendicular to the line of sight that induces the field. The strength of the field is proportional to this projection at any location, as shown in Fig. 5.3b. This factor describes the entire angular dependence of the induced field produced by a vertical driving field. Since it depends on the angle ϕ_z alone, the induced field is seen to be symmetrical with respect to the z axis.

Next we note that the induced field varies inversely with r. It makes sense that the field should decrease as we get farther from the antenna, but the inverse first-power dependence may be unexpected, since we are more familiar with inverse-square laws. However, it is the energy or intensity of the light that varies according to an inverse-square law. In the next section we convert this expression for $\mathbf{E}$ to an expression for energy; the more familiar r^{-2} functionality will appear then.

Finally, it is sufficient for our purposes to think of the remaining factors in Eq. (10), $(4\pi\varepsilon_0 c^2)^{-1}$, as providing dimensional consistency to the expression. Taking a look at the SI units of the right-hand side of Eq. (10), we obtain

$$\frac{(C)(m\ s^{-2})}{(C^2\ N^{-1}\ m^{-2})(m\ s^{-1})^2(m)} = \frac{N}{C}$$

Newtons coulomb^{-1} are units of force per charge as required. Note that multiplication of both the numerator and denominator of this dimensional expression converts N C^{-1} to V m^{-1}, since N m = J and J/C = V. This latter set of units for $\mathbf{E}$ is particularly useful for describing the field between electrodes and, as such, will be encountered in Chapter 13, where we discuss electrokinetic phenomena.

Equation (10) describes the field emitted by an antenna which, in turn, is driven by another field. The oscillation of one field promotes the oscillation of a charge in the antenna, and this induces another electric field. The frequency is the same for all three.

This description of antennas may seem more appropriate to a discussion of radio or television waves. We must realize, however, that at the molecular level dipoles behave exactly like antennas. Since molecules are made up of charged parts, a dipole moment μ is induced by the electric field of the radiation in any material through which radiation passes. In this discussion the dipole moment equals the product of the effective charge displaced by the field and its distance of separation from the opposite charge. In SI, α has units C m. We shall consider isotropic materials characterized by a polarizability α. As the name implies, this property measures the ease with which charge separation—polarity—is induced in a molecule by an electric field. For isotropic substances the dipole moment and the field are related by the expression

$$\mu = \alpha E \tag{11}$$

Using Eq. (11) as the basis of a dimensional analysis of α shows the latter has SI units given by

$$\frac{C\ m}{N\ C^{-1}} = \frac{C^2\ m}{N}$$

The quantity $\alpha/4\pi\varepsilon_0$—which is the polarizability value used in cgs units—is informative. It is examined in the following example.

Example 5.1   Criticize or defend the following proposition on the basis of the units of $\alpha/4\pi\varepsilon_0$: The larger the volume of a particle, the easier it is to induce polarity in that particle.

Solution   The ratio $\alpha/4\pi\varepsilon_0$ has the units $(C^2\ m\ N^{-1})/(C^2\ m^{-2}\ N^{-1}) = m^3$, which are units of volume. In fact, polarizabilities of actual molecules are on the order of $10^{-29}\ m^3\ molecule^{-1} = 0.01\ nm^3\ molecule^{-1}$, which is the magnitude of molecular volumes. For individual atoms, we expect the polarizability to increase with atomic volume, since the outermost electrons are less tightly restrained by the nucleus in such cases. Extension of this principle to covalently bonded species must be done cautiously, however, since the bonding affects the overall picture. The accuracy of the proposition, then, depends on the nature of the "particle" under consideration. Even where the principle stated in the proposition is not literally true, it offers a convenient mnemonic for the definition of polarizability.

•

If we imagine the molecule to lie at the origin of a coordinate system so that $x = 0$, we may substitute Eq. (7) into (11) to obtain

$$\mu = \alpha E_0 \cos (2\pi \nu t) \qquad (12)$$

A dipole moment may be regarded as the product of the distance ξ between two charges and the magnitude of the charge q. A useful way of looking at Eq. (12) is to identify the charge as

$$q = \alpha E_0 \qquad (13)$$

and the separation of the charges as

$$\xi = \cos(2\pi\nu t) \qquad (14)$$

Then the periodic acceleration of the charge is given by

$$\mathbf{a}_p = \frac{d^2\xi}{dt^2} = -4\pi^2\nu^2 \cos(2\pi\nu t) \qquad (15)$$

Equations (13) and (15) may now be substituted into Eq. (10) to give

$$E = -\frac{\alpha E_0 4\pi^2\nu^2 \cos(2\pi\nu t)\sin \phi_z}{4\pi\varepsilon_0 c^2 r} \qquad (16)$$

For maximum generality, we must remember that the field is periodic in space as well as in time; hence the cosine factor in Eq. (16) is corrected by analogy with Eq. (7):

$$E = -\frac{\pi\nu^2\alpha E_0 \cos[2\pi(\nu t - r/\lambda)]\sin \phi_z}{\varepsilon_0 c^2 r} \qquad (17)$$

Equation (17) describes the induced field a distance r from the dipole.

5.4 Rayleigh Scattering

In this section we discuss the first of several light scattering theories to be considered in this chapter, Rayleigh scattering. We shall see presently that Rayleigh scattering applies only when the scattering centers are small in dimension compared to the wavelength of the radiation. As such, it is severely limited in its applicability to colloidal particles, at least when visible light is the radiation involved. Rayleigh scattering is the easiest of the scattering theories to understand, however, so it is a logical place to begin. Furthermore, we shall extend its applicability to larger particles in later sections by introducing suitable correction factors.

When a beam of radiation is incident upon a molecule, a certain fraction of that radiation will undergo the process described in the preceding section and be emitted by the dipole. Any light that does not interact this way will continue past the molecule along the original path. This undeviated or transmitted light will be attenuated compared to the incident light, since some of the original beam of light is scattered from its initial path. Note that this attenuation has nothing to do with absorption: The effect we are considering is a classical result and does not involve transitions between quantum states.

In order to evaluate the amount of light which is transmitted past a scattering center, it is necessary to add up the light scattered over all possible angles, since this is the light responsible for the attenuation. In saying that the total light equals the sum of the transmitted and the scattered light, we are merely stating that energy is conserved. In other words, the summation of which we speak must be carried out using the energy of the light rather than its electric field strength.

Light intensity is defined as the radiation energy falling on a unit of area in a unit of time. The ratio of intensities, therefore, is the same as the ratio of energies at a given location. Intensity is also proportional to the square of the electric field of the radiation. According to Eq. (17), therefore, the intensity i of light scattered at an angle ϕ_z and measured a distance r from the scattering is

$$i \propto \frac{\pi^2 \nu^4 \alpha^2 \mathbf{E}_0^2 \cos^2[2\pi(\nu t - r/\lambda)] \sin^2 \phi_z}{\varepsilon_0^2 c^4 r^2} \tag{18}$$

By analogy, the intensity I_0 of the incident light is

$$I_0 \propto \mathbf{E}_0^2 \cos^2 \left[2\pi \left(\nu t - \frac{r}{\lambda} \right) \right] \tag{19}$$

Since the proportionality factor is the same in both of these expressions, we may write the ratio of intensities as

$$\frac{i_v}{I_{0,v}} = \frac{\pi^2 \nu^4 \alpha^2 \sin^2 \phi_z}{\varepsilon_0^2 c^4 r^2} \tag{20}$$

Note that both the numerator and the denominator of the ratio have the subscript v. This indicates vertical polarization and reminds us that the entire development of the argument since Eq. (10) has been based on the assumption that the original radiation was polarized in the vertical plane. The scattered radiation therefore also possesses this polarization.

Now it is relatively easy to consider the scattering pattern that arises when the original radiation is polarized horizontally. In this case the electric field of the original radiation would lie in the xy plane of Fig. 5.3a. It is still the tangential projection of the dipole in which we are interested;

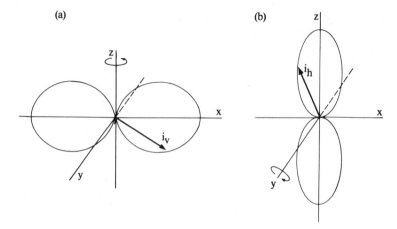

Figure 5.4 The intensity pattern for light scattered by a particle located at the origin. The incident and the scattered light are vertically polarized in (a) and horizontally polarized in (b). The length of the radius vector is proportional to the scattered intensity at each angle.

this would be given by the factor $\sin \phi_y$, where ϕ_y is the angle between the y axis and r as shown in Fig. 5.3a. Therefore the expression for $i_h/I_{0,h}$—the scattered intensity relative to the incident intensity when the latter is horizontally polarized—is identical to Eq. (20), except that $\sin^2 \phi_y$ is used instead of $\sin^2 \phi_z$.

The angular parts of these two scattered intensity functions are illustrated in Fig. 5.4. It should be recalled that the scatter pattern for i_v is symmetrical with respect to the z axis. The actual intensity envelope for the vertically polarized light is the figure of revolution that results from rotating Fig. 5.4a around the z axis. Likewise, the envelope for the horizontal component results from rotating Fig. 5.4b around the y axis.

If the original light is unpolarized, we may think of it as consisting of equal amounts of horizontally and vertically polarized light. In this case the intensity ratio equals

$$\frac{i}{I_{0,u}} = \frac{\frac{1}{2}(i_v + i_h)}{I_{0,u}}$$

$$= \frac{1}{2} \frac{\pi^2 \nu^4 \alpha^2}{\varepsilon_0^2 c^4 r^2} (\sin^2 \phi_y + \sin^2 \phi_z) \qquad (21)$$

where the subscript u reminds us that the incident light is unpolarized and the factor $\frac{1}{2}$ arises because two equal incident sources have been combined. In this case the scattered light consists of a horizontally polarized component (proportional to $\sin^2 \phi_y$) and a vertically polarized component (proportional to $\sin^2 \phi_z$). It is inconvenient to use the two different angles to describe the intensity pattern for light scattered from an initially unpolarized source. In the following example we examine how the two $\sin^2$ terms involving different angles can be replaced by a trigonometric term involving only one angle.

Example 5.2 Use Figure 5.3a to show that $\sin^2 \phi_y + \sin^2 \phi_z = 1 + \cos^2 \phi_x$. Verify that the angle ϕ is one half the tetrahedral angle when $\phi_x = \phi_y = \phi_z = \phi$.

Solution As shown in Fig. 5.3a, ϕ_x is the angle between the x axis and r. This means that $r \cos \phi$ is the projection of r onto the x, y, or z axis, depending on which angle is considered. It follows, therefore, that $r^2(\cos^2 \phi_x + \cos^2 \phi_y + \cos^2 \phi_z) = r^2$.

It is at this point that we can easily examine the situation in which all three angles are equal. In that case $3 \cos^2 \phi = 1$ or $\cos \phi = (\frac{1}{3})^{1/2}$ and $\phi = 54.75°$. Doubling this gives the tetrahedral angle.

Now we return to the relationship to be shown, substituting $1 - \sin^2 \phi$ for two of the $\cos^2 \phi$ terms. Upon rearrangement, this yields $\sin^2 \phi_y + \sin^2 \phi_z = 1 + \cos^2 \phi_x$, which was to be shown.

•

The result demonstrated in Example 5.2 may be substituted into Eq. (21) to give

$$\frac{i}{I_{0,u}} = \frac{\pi^2 \nu^4 \alpha^2}{2\varepsilon_0^2 c^4 r^2}(1 + \cos^2 \phi_x) \tag{22}$$

A plot of Eq. (22) is shown in Fig. 5.5. The intensity is clearly a maximum for $\phi_x = 0$ and π and falls to half this value at $\phi_x = \pi/2$. When $\phi_x = \pi/2$, the scattered light is totally polarized, since only one of the components contributes to the intensity at this angle. The envelope for the total intensity of scattered unpolarized light is generated by rotating the contour for total intensity around the x axis. Figure 5.5 also includes the contributions of the vertically and horizontally polarized components in the xz plane. Equation (22) and its precursors may be written in a slightly different form by incorporating Eq. (1) into them. Thus Eq. (22) becomes

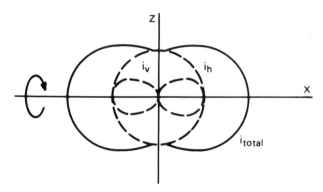

Figure 5.5 Projection in the xz plane of i_v, i_h, and their resultant i_{total} for light scattered by a molecule situated at the origin. The intensity of scattered unpolarized light at any angle ϕ_x is proportional to the length of the radius vector at this angle.

$$\frac{i}{I_{0,u}} = \frac{\pi^2\alpha^2}{2\varepsilon_0^2\lambda_0^4 r^2}(1 + \cos^2\phi_x) \tag{23}$$

This equation gives the intensity of the light scattered by a single molecule to the point described by the coordinates r and ϕ_x when the incident light is unpolarized.

Next we must consider how to evaluate the light scattered by an array of molecules. If the molecules in a scattering sample are far apart, as in a gas, then each may be regarded as an independent source of scattered light. Then we may multiply Eq. (23) by the number of molecules per unit volume to get an expression for the light scattered from that volume element. The number of molecules per unit volume is given by $N_A\rho/M$, where ρ and M are the density and molecular weight of the scattering gas, respectively, and N_A is Avogadro's number. Then Eq. (23) becomes

$$\frac{i_s}{I_{0,u}} = \frac{\pi^2 N_A\rho\alpha^2}{2\varepsilon_0^2 r^2\lambda_0^4 M}(1 + \cos^2\phi_x) \tag{24}$$

where i_s is the intensity of light scattered *per unit volume* (subscript s).

It is shown in physical chemistry that the polarizability and refractive index of dielectric particles are related by the expression

$$\alpha = \frac{3M\varepsilon_0}{N_A\rho}\frac{n^2-1}{n^2+2} \tag{25}$$

A particularly useful form of this relationship applies to material for which n is not much larger than unity, as would be the case for a nonabsorbing gas. In this case

$$\frac{n^2 - 1}{n^2 + 2} = \frac{(n + 1)(n - 1)}{n^2 + 2} \simeq \tfrac{2}{3}(n - 1) \tag{26}$$

and

$$\alpha \simeq \frac{2M\varepsilon_0}{\rho N_A}(n - 1) \tag{27}$$

Therefore Eq. (24) may be written

$$\frac{i_s}{I_{0,u}} = \frac{2\pi^2 M}{r^2 \lambda_0^4 N_A \rho}(n - 1)^2(1 + \cos^2 \phi_x) \tag{28}$$

The results we have considered in this section were derived by Lord Rayleigh in 1871. An interesting application of Eq. (28) is the explanation it offers for why the sky appears blue. This arises from the inverse fourth-power dependence on λ_0 for i_s. Suppose, for example, that two radiations are compared whose wavelengths differ by a factor of 2. Then the scattered intensity of the shorter wavelength will be 16 times as great as that of the longer wavelength. Although red and blue light do not differ by quite this much in wavelength, the blue component of white light, say, sunlight, is scattered very much more than the red. Accordingly, the sky overhead appears blue. At sunset, we see mostly transmitted light. Since the blue has been most extensively removed from this by scattering, the sky appears red at sunset. The following example describes the classic experiment in which the Rayleigh equation was used to determine an early value of Avogadro's number.

Example 5.3 The intensity of the light scattered by air was measured in the atmosphere at Mt. Wilson, California, in 1913. The following is a selection of the results obtained*:

λ_0 (nm)	350	360	371	384	397	413	431	452	475
$i_s/I_{0,u}$	0.459	0.423	0.377	0.338	0.285	0.245	0.213	0.174	0.147

Show that these results are in agreement with the predictions of the Rayleigh theory, given that $n - 1$ is constant at 2.97×10^{-4} over this range of wavelengths. Instead of $N_A\rho/M$, we could have multiplied Eq. (23) by N_0, the number of molecules per unit volume under appropriate conditions—taken to be STP in this example for simplicity—and used light scattering to measure this quantity. Under the conditions of this

---

*F. E. Fowle, *Astrophys. J.,* 40:435 (1914).

research, the numerical and geometrical factors work out to give $i_s/I_{0,u} = 0.1911/\lambda_0^4 N_0$ in SI units. Use a representative observation to evaluate N_0 and, from this, Avogadro's number.

Solution Since the refractive index term is independent of λ_0 over the range involved, Eq. (28) predicts that $i_s/I_{0,u}$ should vary with the inverse fourth power of the wavelength. To test this, prepare a plot of $\ln(i_s/I_{0,u})$ versus $\ln \lambda_0$ and examine the slope. Figure 5.6 shows this plot; the slope of the line shown is -4. It describes the data very well.

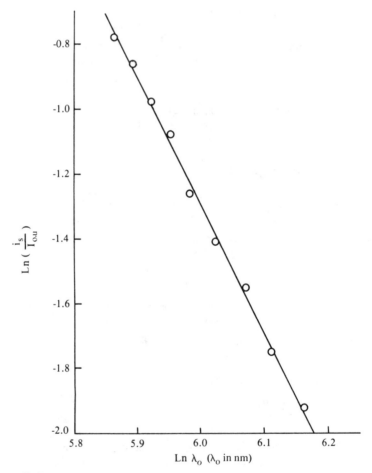

Figure 5.6 Verification of the inverse fourth-power dependence of scattering on wavelength for sunlight scattered by air (details and reference in Example 5.3).

Using the first data point to evaluate N_0, we obtain

$$N_0 = \frac{0.1911}{(0.459)(3.50 \times 10^{-7})^4} = 2.77 \times 10^{25} \text{ molecules m}^{-3}$$

Combining this with the gas law observation that a mole of (ideal) gas occupies 22.4 liters at STP gives

$$2.77 \times 10^{25} \frac{\text{molecules}}{\text{m}^3} \times \left(\frac{1 \text{ m}}{10 \text{ dm}}\right)^3 \times \frac{22.4 \text{ dm}^3 \text{ at STP}}{\text{mole}}$$

$$= 6.20 \times 10^{23} \text{ molecules mole}^{-1}$$

When N_0 is evaluated for each of the points and the average value used, $N_A = 5.98 \times 10^{23}$ molecules mole^{-1}, in good agreement with the accepted value.

•

Since Mt. Wilson is located within a 15-mile radius of the Los Angeles civic center, one wonders what sort of atmospheric light scattering results would be obtained at that location today! The Rayleigh theory does not apply when the scattering molecules are absorbing or when the atmosphere contains dust particles, water drops, or other particles whose dimensions are larger than ordinary gas molecules.

5.5 Rayleigh Scattering Applied to Solutions: Fluctuations

A crucial aspect of the transition from Eq. (23) to Eq. (24) is the requirement that the individual molecules are far enough apart to be treated as independent sources. This assumption is justified for gases, but in liquids the molecules are close enough together so that interference occurs between the waves emitted from different centers. As a matter of fact, there would be complete destructive interference of all scattered light in liquids if the molecules were randomly arranged and stationary. Nevertheless, pure liquids do scatter light. It is not the individual molecules which are the scattering centers in this case but, rather, the small domains of compression or rarefaction that arise from fluctuations.

We saw in Chapter 2 that molecular motion results in small fluctuations in density at the molecular level. Although the average density of a liquid is a constant equaling the experimental density, there will be small transient domains within it which have densities larger or smaller than the mean value.

Liquid solutions also scatter light by a similar mechanism. In the case of a solution the scattering may be traced to two sources: fluctuations in solvent density and fluctuations in solute concentration. The former are most easily handled empirically by subtracting a solvent "blank" correction from measurements of the intensity of light scattered from solutions. What we are concerned with in this section, then, is the remaining scattering, which is due to fluctuations in the solute concentration in the solution.

A fluctuation in the concentration of a small volume element δV of solution will result in a change in the properties of that volume element. We begin the analysis of this situation by defining δc and $\delta \alpha$ as the fluctuations in solution concentration and polarizability, respectively, in this element. The first thing that we recognize about these quantities is that their average values are zero, since both positive and negative fluctuations occur. Although their averages may be zero, the averages of their squares are not zero. Remember that similar situations were encountered in Sects. 2.11 and 2.12, where we discussed particle displacements due to diffusion and segment displacements in a random coil. In both of these cases it was by considering the average values of the square of the displacements that meaningful quantities could be obtained. Similarly, the average values of δc and $\delta \alpha$ are zero, but their mean square values will be different from zero. Equation (23) shows that the intensity of scattered light depends on the square of polarizability. We conclude, therefore, that the way to adapt this equation to the scattering by solutions is to replace α^2 in Eq. (23) by $\overline{\delta \alpha^2}$:

$$\frac{i}{I_0} = \frac{\pi^2 \overline{\delta \alpha^2}}{2\varepsilon_0^2 r^2 \lambda^4} (1 + \cos^2 \phi_x) \tag{29}$$

Note that the wavelength λ of light in the medium rather than the value under vacuum is used in this expression, since this is the light that reaches the scattering center.

Next we must consider how to extend this result to a unit volume of solution and how to relate $\delta \alpha$ to δc. The steps involved in these extensions are not difficult, but they are lengthy. Accordingly, we shall not develop the entire argument in detail. Instead, some of the key steps in the development along with a brief justification for each are summarized in Table 5.1. In this table each major substitution is presented along with variations of Eq. (29) which reflect the cumulative effects of all the substitutions. The first entry in the table, for example, converts Eq. (29) to an expression for the relative light scattered by a unit of volume i_s/I_0 by multiplying Eq. (29) by the number of volume elements in 1 cm^3: $1/\delta V$. Although the justifications in Table 5.1 are sketchy, they provide hints which will show the interested reader how to proceed in order to develop the required relationship in detail. Only the third entry in the table, the connection between $\overline{\delta c^2}$ and $\partial^2 G/\partial c^2$, requires a more elaborate proof than the qualitative justification supplied.

The cumulative result of these substitutions is to replace $\overline{\delta \alpha^2}$ in Eq. (29) by a number of other factors, all of which are experimentally measurable:

1. The refractive index gradient dn/dc. This is simply the local slope of a plot of the refractive index of a solution versus its concentration.
2. The concentration c of the solution. This is expressed in units of grams per volume.
3. The quantity $\partial\pi/\partial c$ is evaluated for an equilibrium solution. This is the significance of the subscript 0.

In Chap. 3 we developed expressions for the equilibrium osmotic pressure of a solution as a function of its concentration. In view of these substitutions, Eq. (29) becomes

$$\frac{i_s}{I_0} = \frac{2\pi^2[n(dn/dc)]^2 kTc}{r^2\lambda^4(\partial\pi/\partial c)_0}(1 + \cos^2 \phi_x) \tag{30}$$

Equation (3.34) may be written

$$\pi = RT\left(\frac{c}{M} + Bc^2\right) \tag{31}$$

Since Eq. (31) applies at equilibrium, we may evaluate $(\partial\pi/\partial c)_0$ from Eq. (31):

$$\left(\frac{\partial\pi}{\partial c}\right)_0 = RT\left(\frac{1}{M} + 2Bc\right) \tag{32}$$

Combining Eqs. (30) and (32) yields

$$\frac{i_s}{I_0} = \frac{2\pi^2[n(dn/dc)]^2 c}{N_A r^2\lambda^4(1/M + 2Bc)}(1 + \cos^2 \phi_x) \tag{33}$$

Before looking at the experimental aspects of light scattering, it is convenient to define several more quantities. First, a quantity known as the Rayleigh ratio R_θ is defined as

Table 5.1 Some Key Substitutions and Their Justifications for the Transformation of Eq. (29) to Eq. (30)

Substitution	Justification	Cumulative effect on Eq. (29)
$\dfrac{\overline{\delta\alpha^2}}{\text{volume}} = \dfrac{1}{\delta V}\,\overline{\dfrac{\delta\alpha^2}{\text{fluctuation}}}$	1. $\dfrac{1}{\delta V}$ domains of volume δV can fit into 1 cm³ of solution	$\dfrac{i_s}{I_0} = \dfrac{\pi^2 \overline{\delta\alpha^2}}{r^2 2\varepsilon_0^2 \lambda^4}\,\delta V\,(1+\cos^2\phi_x)$
$\overline{\delta\alpha^2} = \varepsilon_0^2\,\delta V^2\left(2n\dfrac{dn}{dc}\right)^2 \delta c^2$	1. δV replaces $\dfrac{M}{\rho N_A}$ in Eq. (25) 2. For $n \simeq 1$, $\alpha = \varepsilon_0\,\delta V(n^2 - 1)$ 3. $\delta\alpha = \varepsilon_0\,\delta V\,2n\,dn$ and $dn = \dfrac{dn}{dc}\,\delta c$	$\dfrac{i_s}{I_0} = \dfrac{2\pi^2 \delta V[n(dn/dc)]^2 \delta c^2}{r^2\lambda^4}$ $(1+\cos^2\phi_x)$
$\overline{\delta c^2} = \dfrac{kT}{(\partial^2 G/\partial c^2)_0}$ Subscript 0: evaluated at equilibrium	1. Taylor series (see Appendix A) expansion of G around equilibrium value: $G = G_0 + \left(\dfrac{\partial G}{\partial c}\right)_0 \delta c + \dfrac{1}{2}\left(\dfrac{\partial^2 G}{\partial c^2}\right)_0 \delta c^2$ 2. $\left(\dfrac{\partial G}{\partial c}\right)_0 = 0$ at equilibrium 3. $G - G_0 \simeq \dfrac{1}{2}kT$, since fluctuation is due to thermal energy ($\dfrac{1}{2}kT$ per degree of freedom)	$\dfrac{i_s}{I_0} = \dfrac{2\pi^2 \delta V[n(dn/dc)]^2 kT}{r^2\lambda^4(\partial^2 G/\partial c^2)_0}$ $\times (1+\cos^2\phi_x)$

$$\frac{i_s}{I_0} = \frac{2\pi^2 [n(dn/dc)]^2 kT\bar{V}_1 c}{r^2\lambda^4(-\partial\mu_1/\partial c)_0}$$
$$\times (1 + \cos^2\phi_x)$$

1. $dG = \mu_1 dn_1 + \mu_2 dn_2$

2. Since $\bar{V}_1 dn_1 = -\bar{V}_2 dn_2$,

$$dG = \left(\mu_2 - \frac{\bar{V}_2}{\bar{V}_1}\mu_1\right) dn_2$$

3. Since $M\, dn_2 = dc\,\delta V$,

$$\frac{dG}{dc} = \left(\mu_2 - \frac{\bar{V}_2}{\bar{V}_1}\mu_1\right) \frac{\delta V}{M}$$

4. By the Gibbs-Duhem equation,

$$\frac{\partial^2 G}{\partial c^2} = -\frac{\delta V}{M}\left(\frac{\bar{V}_2}{\bar{V}_1} + \frac{n_1}{n_2}\right) \frac{d\mu_1}{dc}$$

5. $c = \dfrac{n_2 M}{n_1 \bar{V}_1} + n_2\bar{V}_2$

1. By Eq. (3.21)

$$\frac{i_s}{I_0} = \frac{2\pi^2 [n(dn/dc)]^2 kTc}{r^2\lambda^4(\partial\pi/\partial c)_0}$$
$$\times (1 + \cos^2\phi_x)$$

$$\frac{\partial^2 G}{\partial c^2} = \frac{\delta V}{\bar{V}_1 c}\left(-\frac{\partial\mu_1}{\partial c}\right)_0$$

$$\left(\frac{\partial\mu_1}{\partial c}\right)_0 = -\bar{V}_1\left(\frac{\partial\pi}{\partial c}\right)_0$$

$$R_\theta = \frac{i_s r^2}{I_0(1 + \cos^2 \theta)} \tag{34}$$

where θ is the value of ϕ_x measured in the horizontal plane. According to Eq. (34), the Rayleigh ratio should be independent of both θ and r. An experimental verification of this is one way of testing the applicability of the Rayleigh theory to the experimental data. Next it is convenient to identify the numerical and optical constants in Eq. (33) as follows:

$$K = \frac{2\pi^2 n^2 (dn/dc)^2}{N_A \lambda^4} \tag{35}$$

With these changes in notation, Eq. (33) becomes

$$R_\theta = \frac{Kc}{1/M + 2Bc} \tag{36}$$

or

$$\frac{Kc}{R_\theta} = \frac{1}{M} + 2Bc \tag{37}$$

This suggests that a plot of Kc/R_θ versus c should be a straight line for which the intercept and slope have the following significance:

$$\text{intercept} = \frac{1}{M} \tag{38}$$

$$\text{slope} = 2B \tag{39}$$

Comparing Eqs. (38) and (39) with Eqs. (3.35) and (3.36) reveals that plots of π/RTc versus c and Kc/R_θ versus c have (1) identical intercepts, at least for monodisperse colloids (see Sect. 5.7 for a discussion of the average obtained for polydisperse systems), and (2) slopes differing by a factor of 2, with the light scattering results having the larger slope.

The Rayleigh ratio as defined by Eq. (34) has a precise meaning, yet it is a quantity which is somewhat difficult to visualize physically. After we have discussed the experimental aspects of light scattering, we shall see that R_θ is directly proportional to the turbidity of the solution, where turbidity is the same as the absorbance determined spectrophotometrically.

5.6 Experimental Aspects of Light Scattering

In order to determine M and B by means of Eq. (37), it is clear that all the other quantities in the equation must be measured. It is convenient to group these factors into two categories for the purposes of our discussion: optical and concentration terms. To begin with, concentration enters the light scattering expressions primarily through the equations for osmotic pressure. Hence the same conditions apply in this application as in osmometry. Specifically, light scattering should be measured under isothermal conditions. Although concentration units other than weight per unit volume (the units of c) may be used, few of the alternatives are as useful as these. In consulting the literature, however, one should be attentive to the possibility that various workers may use slightly different units for c.

All the remaining variables in Eq. (37) are optical in origin. The factors to be provided with numerical values are the refractive index of the solution and the refractive index gradient, dn/dc in K [Eq. (35)], and the Rayleigh ratio [Eq. (35)]. All these optical parameters are wavelength dependent; therefore each should be measured at the same wavelength. It is the value of this working wavelength that is used as the numerical substitution for λ in K.

The actual measurement of the refractive index of the solution poses no difficulty, but the evaluation of the refractive index gradient is more troublesome. The assumptions of the derivation of Eq. (33) restrict its applicability to dilute solutions. The refractive index of a dilute solution changes very gradually with concentration; hence a plot of n versus c, the slope of which equals dn/dc, will be nearly horizontal. Since the intensity ratio depends on the square of dn/dc, it is clear that successful interpretation of Eq. (33) depends on the accuracy with which this small quantity is evaluated. Measuring the absolute refractive indices of various solutions and determining dn/dc by difference or graphically would introduce an unacceptable error. A more precise method must be used to measure this quantity.

A differential refractometer is a device which specifically measures *differences* in refractive indices. By means of a differential refractometer the difference between the refractive index of a solution and solvent may be measured directly with the necessary precision. A variety of instrument designs are available for doing this. Most involve directing a light beam toward a two-compartment chamber, one portion of which contains the solvent while the other contains the solution. The light is deviated differently by each, and position-sensitive photodetectors measure the small differences in deflection and translate this into a refractive index difference. Differences as small as 10^{-7} refractive index units can be measured in this way. The search for detection methods suitable for liquid chromatography has contributed to the development of accurate techniques for measuring small changes in refractive index.

Note that we have also cited refractive index measurements in connection with size exclusion chromatography in Sect. 1.10 and with sediment-ation–diffusion studies in Chapter 2. A differential refractometer is an indispensible part of any laboratory where light scattering experiments are conducted.

Now let us consider the actual measurement of light intensity. A light scattering photometer differs from an ordinary spectrophotometer primarily in the fact that the photoelectric cell which measures the scattered light is mounted on an arm that permits it to be located at various angular positions relative to the sample. In commercial light scattering devices, the detector rests on a turntable, the center of which coincides with the center of the sample. Thus the angle ϕ_x is measured in the horizontal plane. From now on we shall use the symbol θ to signify the angle of observation in the horizontal plane. The angle θ is measured from the direction of the transmitted beam where $\theta = 0°$. The incident beam therefore strikes the sample at $\theta = 180°$. A schematic top view of a light scattering photometer is shown in Fig. 5.7a.

Figure 5.7b is a photograph of a light scattering photometer with one side removed to show its inner workings. The apparatus shown is the Brice-Phoenix Universal Scattering Photometer, Series 2000 (The Virtis Company, Gardiner, N.Y. 12525). Note that the arm holding the detector is located at about $\theta = -10°$.

The light scattering photometer includes a number of distinct components. First, there is the light source, usually a mercury vapor lamp. Next there are filters to provide monochromatic radiation. Most measurements are made at either the 436- or 546-nm line of mercury. The light must be carefully collimated so that the sample is, in fact, illuminated by a parallel beam as required by theory. Polarization filters are also included so that experiments with polarized light are possible. The intensity of the scattered light is low so that a sensitive photo-multiplier is required to amplify the signal. The output from the latter may be connected to a high-sensitivity galvanometer or to a recorder.

The inside of the scattering compartment must be painted black to prevent light reflected from the apparatus from entering the photo-multiplier. It is especially important to remove the transmitted light beam so a light trap is built into the scattering compartment at the $0°$ position.

The cells in which the scattering solutions are measured should have flat windows at the angle at which the scattering is measured. Cells with octagonal cross sections (actually, only half an octagon is used) are especially convenient, since they present flat faces at 0, 45, 90, 135, and $180°$. Cylindrical cells have been used, but they must be corrected for

(a)

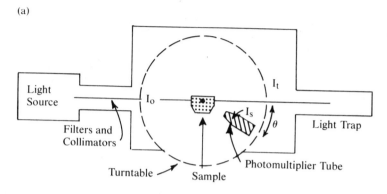

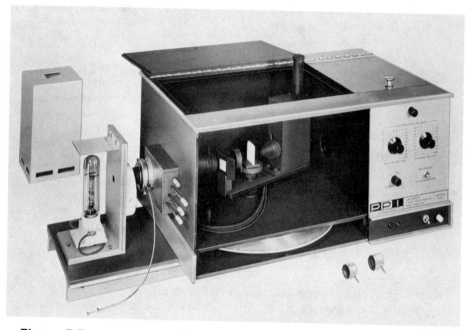

Figure 5.7 (a) Schematic top view of a light scattering photometer, showing the definition of θ. (b) Cutaway photograph of a light scattering photometer, the Brice-Phoenix Universal Scattering Photometer. (From The Virtis Company, Gardiner, N.Y., used with permission.)

reflections from the walls. Regardless of the cell geometry, it is imperative that the cells be clean, otherwise the scattering from a fingerprint may exceed that from the solute! The solvent must also be purified of all extraneous matter. Filtration through sintered glass or centrifugation is usually employed to remove any dust particles which would also invalidate the measurement.

The easiest way to calibrate a light scattering photometer is to use a suitable standard as a reference. Although polymer solutions and dispersions of colloidal silica have been used for this purpose, commercial photometers are equipped with opal glass reference standards.

Except for the movable photomultiplier tube, a light scattering photometer is very nearly identical to an ordinary spectrophotometer. The latter measures the ratio of the intensity of transmitted light to the intensity of incident light, I_t/I_0. The absorbance ε per unit optical path is defined in terms of this quantity as

$$\varepsilon = -\ln\left(\frac{I_t}{I_0}\right) \tag{40}$$

Now let us examine the relationship between absorbance and the intensity of scattered light. In a light scattering experiment with nonabsorbing materials, the intensity of the transmitted light equals the initial intensity minus the intensity of the light scattered *in all directions*, I_s:

$$I_t = I_0 - I_s \tag{41}$$

Combining Eqs. (40) and (41) leads to the result

$$\varepsilon = -\ln\left(\frac{I_0 - I_s}{I_0}\right) = -\ln\left(1 - \frac{I_s}{I_0}\right) \simeq \frac{I_s}{I_0} \tag{42}$$

where the approximation arises from retaining the first term of the series expansion of the logarithm (see Appendix A). The entire development of Sects. 5.4 and 5.5 is limited to dilute solutions and small n values. Therefore the approximation in Eq. (42) is applicable to the systems we have been discussing. When the light attenuation is due to scattering, the ratio I_s/I_0 is called the turbidity τ instead of absorbance.

The quantity I_s in Eq. (42) is not the same as the light scattered to a particular point (r, ϕ_x), but equals the summation of these contributions, totaled over all angles:

$$\frac{I_s}{I_0} = \sum_{\substack{\text{all} \\ \text{angles}}} \frac{i_s}{I_0} \tag{43}$$

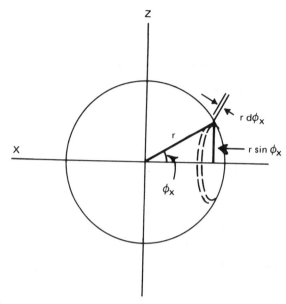

Figure 5.8 Definition of an element of area required for the summation over all angles of the intensity of scattered light.

This summation may be replaced by an integral as follows. An element of area on the surface of a sphere of radius r and making an angle ϕ_x with the horizontal is

$$dA = 2\pi r \sin \phi_x (r \, d\phi_x) \tag{44}$$

as shown by Fig. 5.8. Therefore the *total* scattered intensity ratio is given by

$$\frac{I_s}{I_0} = \int_0^\pi \left(\frac{i_s}{I_0}\right) 2\pi r^2 \sin \phi_x \, d\phi_x \tag{45}$$

Substituting Eqs. (33) and (35) into this expression yields

$$\frac{I_s}{I_0} = \int_0^\pi \frac{Kc(1 + \cos^2 \phi_x) 2\pi r^2 \sin \phi_x \, d\phi_x}{r^2(1/M + 2Bc)} \tag{46}$$

The factor r^2 cancels out of Eq. (46) and the value of the integral over ϕ_x is $\frac{8}{3}$. Therefore

$$\frac{I_s}{I_0} = \tau = \frac{16\pi Kc}{3(1/M + 2Bc)} \tag{47}$$

The parameter H is defined to equal the cluster of constants:

$$H = \frac{16\pi K}{3} = \frac{32\pi^3 n^2 (dn/dc)^2}{3N_A\lambda^4} \tag{48}$$

and in terms of this quantity, Eq. (47) becomes

$$\frac{Hc}{\tau} = \frac{1}{M} + 2Bc \tag{49}$$

The formal similarity between Eqs. (37) and (49) helps us understand somewhat better the physical significance of the Rayleigh ratio R_θ. It is directly proportional to the attenuation of the light per unit optical path, measured as absorbance, when the attenuation is due to scattering alone. In this case absorbance is more properly called turbidity.

Now let us consider some actual results from light scattering experiments on systems which satisfy the assumptions of the theory.

5.7 Results from Light Scattering Experiments: Weight Average Molecular Weights

In the preceding section we saw how turbidity measurements are made and how they may be analyzed to yield numerical values for some of the parameters of interest in colloid chemistry. Figure 5.9 shows a plot of Hc/τ versus c for three different fractions of polystyrene in methylethyl ketone. The measurements shown in the figure were made at 25°C and at a wavelength of 436 nm. The molecular weights for the three fractions may be determined from the intercepts of the lines as illustrated in the following example.

Example 5.4   Assuming that the same system of units is used throughout, what are the units for R_θ and τ and also for K and H? Verify that these lead to reciprocal molecular weight units for the clusters Hc/τ and Kc/R_θ. To what molecular weights do the three intercepts in the figure correspond? Do the three fractions behave as expected with respect to their second virial coefficients?

Solution   Since refractive index is dimensionless, dn/dc has reciprocal concentration units. In Eq. (30) the factors $[n(dn/dc)]^2 c/(d\pi/dc)$ have units pressure^{-1}, since the c's cancel. The product $r^2\lambda^4$ in Eq. (30) has units length^6 or volume^2; therefore the denominator of Eq. (30) has units

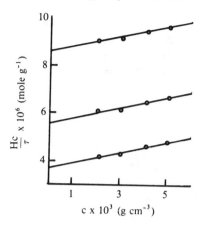

Figure 5.9 Plots of Hc/τ versus c for three different fractions of polystyrene in methylethyl ketone. [B. A. Brice, M. Halwer, and R. Speiser, *J. Opt. Soc. Am.*, 40:768 (1950), used with permission.]

pressure volume^2. The product kT in the numerator can also be expressed in pressure volume units. Therefore i_s/I_0 has units volume^{-1} and describes the intensity ratio per unit volume as described.

Equation (34) shows that the units of R_θ are those of $r^2(i_s/I_0)$. The units of the first factor are length^2, and those of the second length^{-3}. Therefore R_θ has units length^{-1}. By Eqs. (45) and (47), the units of τ are the same as R_θ. Each of these parameters measures the light attenuation per unit path length.

By Eq. (35), the units of K are concentration^{-2} length^{-4} = length^6 mass^{-2} length^{-4} = length^2 mass^{-2}. The fact that this quantity is divided by N_A places (mole^{-1})$^{-1}$ in the units. Equation (48) shows that H has the same units.

In terms of units, $Hc/\tau = Kc/R_\theta$ = (length^2 mass^{-2} mole)(mass length^{-3})/(length^{-1}) = mole mass^{-1}, reciprocal molecular weight units if mass is expressed in grams, as is the case in practical concentration units.

The three intercepts in Fig. 5.9 are 3.70×10^{-6}, 5.56×10^{-6}, and 8.62×10^{-6} mole g^{-1}. The reciprocals of these numbers give the molecular weights directly: 116,000, 180,000, and 270,000 g mole^{-1}.

The lines in Fig. 5.9 appear to be parallel and hence characterized by a single B value. This would be expected for a single polymer–solvent–temperature system.

•

At first glance it appears that light scattering experiments provide no information that is not already available from osmometry. Indeed, for monodisperse colloids this is true, at least for the experiments we have discussed until now. The apparent redundancy between osmotic pressure and light scattering results should not be interpreted to mean that the two procedures duplicate one another entirely. For one thing, light scattering is free from the limitations imposed on osmometry by the availability of a suitable semipermeable membrane. Furthermore, turbidity measurements do not require time for equilibration, and hence they may be used for systems which change with time in a manner that is not possible with osmometry. In addition to these practical considerations, there are other ways in which light scattering and osmometry differ. Some of these will become apparent only in subsequent sections where additional characteristics of light scattering are developed. Another important difference, however, arises in the type of average that is measured for polydisperse systems.

In Chapter 3 we saw that osmometry enables us to measure the number average molecular weight for a polydisperse colloid. In view of the way the osmotic pressure enters the development of Eqs. (37) and (49), it appears that the same type of average is obtained from turbidity experiments also. This is not the case. The following argument shows that light scattering measures the weight average molecular weight.

For a polydisperse system Eq. (49) relates the experimental concentration, the experimental turbidity, and the average molecular weight:

$$\frac{Hc_{exp}}{\tau_{exp}} = \frac{1}{\overline{M}} \tag{50}$$

It is sufficient to consider only the leading term of Eq. (49) in writing Eq. (50), since the molecular weight is evaluated from the intercept at infinite dilution. Likewise, we expect that Eq. (49) will also apply to each molecular weight fraction in the polydisperse system:

$$\frac{Hc_i}{\tau_i} = \frac{1}{M_i} \tag{51}$$

It is the relationship between the value of $\overline{M}$ and the distribution of M_i values that we wish to determine. To accomplish this we note that

$$c_{exp} = \sum_i c_i \tag{52}$$

and

$$\tau_{exp} = \sum_i \tau_i \tag{53}$$

Combining the last four equations gives

$$\overline{M} = \frac{\tau_{exp}}{Hc_{exp}} = \frac{\Sigma_i \tau_i}{H\Sigma_i c_i} = \frac{H\Sigma_i c_i M_i}{H\Sigma_i c_i} = \frac{\Sigma_i c_i M_i}{\Sigma_i c_i} \tag{54}$$

Now recalling that $c = n_i M_i / V$ enables us to write

$$\overline{M} = \frac{\Sigma_i n_i M_i^2}{\Sigma_i n_i M_i} = \overline{M}_w \tag{55}$$

Equation (55) corresponds to the weight average molecular weight $\overline{M}_w$ as defined by Eq. (1.26).

It will be recalled from Chapter 1 that the *number* of particles in a molecular weight class provides the weighting factor used to compute the number average molecular weight. The *weight* of particles in a class gives the weighting factor for the weight average molecular weight. For this reason the weight average is especially influenced by the larger particles in a distribution. Therefore the weight average molecular weight is always larger than the number average for a polydisperse system. As we saw in Chapter 1, the ratio of the two different molecular weights is a useful measure of the polydispersity of a sample. The authors of the research shown in Figure 5.9 also measured $\overline{M}_n$ for the same samples; the average value of $\overline{M}_w/\overline{M}_n$ for the three samples was 1.16. From polymerization theory this ratio is expected to be closer to 2 for polystyrene as synthesized, suggesting that these samples had been fractionated prior to molecular weight determination.

Thus we see that the redundancy between osmometry and light scattering is only an apparent effect for polydisperse systems. In fact the combination of the two analyses provides additional information about the characteristics of the system.

Equation (49) shows that the slope of a light scattering plot is twice the value of the slope of a comparable plot from osmometry. In addition to the factor of 2, there is a more subtle difference between the slopes arising from a difference in the two values of the second virial coefficient. Examination of Eq. (4.50) reveals that the second virial coefficient is inversely proportional to the square of the molecular weight of the solute. For polydisperse systems it is the average molecular weight that appears in this expression. Since the weight average molecular weight is larger than the number average, the second virial coefficient B will be somewhat smaller as determined by light scattering than by osmometry after the factor of 2 has been taken into account.

Aside from the difference just noted, the interpretation of the second virial coefficient in light scattering is exactly the same as that developed in Sects. 3.6 and 3.8. It should be noted, however, that Eq. (49) does not apply to charged systems. The reason for this lies in the fact that the charge of macroions is also a fluctuating quantity, and this must also be considered in developing a scattering theory for charged particles. The resulting analysis shows that it is a plot of Hc/τ versus $c^{1/2}$ which is linear in this case, with the limiting slope proportional to $\overline{z^2}$, the average value of the square of the charge. The slope is also predicted to be negative in this situation.

In concluding this section, it should be emphasized that the turbidity values which are plotted to interpret light scattering experiments are the *solution* turbidities, corrected for scattering by the solvent. Also, the entire theoretical development leading to Eq. (49) is based on the assumption that the scatterers are isotropic. In this case, unpolarized incident light will produce a scattered beam which is totally polarized at $\theta = 90°$, as may be seen from Fig. 5.5. When anisotropic particles are present, there is a depolarization of the light scattered at $90°$. The ratio of the horizontally to vertically polarized scattered light may be determined by inserting a Polaroid filter between the sample and the photomultiplier. From the measured value of this depolarization ratio, a correction factor (called the Cabannes factor) may be introduced to allow for anisotropy. In the sample with $M = 116,000$ in Fig. 5.9, for example, the ratio of the two different polarizations at $90°$ has a value of 0.013, for which the Cabannes factor equals 0.98. The turbidity should be multiplied by this factor to correct for the fact that the anisotropy enhances the amount of scattered light.

5.8 Extension to Larger Particles

In the remainder of this chapter we shall see that a good deal more information about scattering particles may be deduced, at least under some circumstances, from the study of the light scattered by a sample. In developing the Rayleigh theory and applying it to solutions, a definite model was postulated and certain variables emerged as factors which affect the intensity of the scattered light. Before we extend light scattering theory to more complex systems, it will be convenient to review the assumptions of the Rayleigh model:

1. The scattering centers are isotropic, dielectric, and nonabsorbing.
2. The scatterers have a refractive index which is not too large [see Eq. (26)].
3. The particles are small in dimension compared to the wavelength of the light.

This last assumption is fundamental to the theory and originates as early as Eq. (17), where it is assumed that the field which drives the oscillating dipole is the same throughout the scattering center. It is generally held that particles must have no dimension larger than about $\lambda/20$ for this assumption to apply.

Equation (49) suggests that light scattering is a technique which is ideally suited to the study of particles in the colloidal size range, since the turbidity increases with the molecular weight of the particle. However, the assumptions underlying the derivation of Eq. (49) impose a limitation. The Rayleigh theory shows that turbidity increases with molecular weight, at least until the particle is large enough to have some dimension exceeding about $\lambda/20$. In terms of the theory presented so far, we have no way of interpreting the light scattered by larger particles. The Debye theory which we examine in the following section will show us how to overcome this limitation.

The Rayleigh approximation shows that the intensity of scattered light depends on the wavelength of the light, the refractive index of the system (subject to the limitation already cited), the angle of observation, and the concentration of the solution (which is also restricted to dilute solutions). In the Rayleigh theory the size and shape of the scatterers (M and B) enter the picture through thermodynamic rather than optical considerations.

A fully developed theory of light scattering which allows all the variables, including particle size and shape, to take on a full range of values is extremely complex. Because of this complexity, many treatments such as the Rayleigh theory are approximations which apply only to a narrow range of values of the parameters. In the Debye approximation most of the preceding restrictions will continue to apply, except that the limitation on particle size will be relaxed considerably. At the same time, however, the stipulation of low values of the refractive index becomes even more stringent. As we shall see presently, the Debye approximation introduces some additional complexity to the theory of light scattering and trades off some range in refractive index for extra range in particle size. From this a positive dividend emerges: We shall be able to determine a characteristic linear dimension of the scattering particles without any assumptions about the shape of the particles. In the cases to which is applies, this information is definitely worth the price of a little additional complexity.

We shall omit most of the mathematical details in developing the additional theory; instead, we shall emphasize the major concepts of the theory, its range of applicability, and its plausibility in limiting cases.

5.9 The Debye Scattering Theory

The Rayleigh approximation is restricted to particles whose dimensions are small compared to the wavelength of light. Suppose we now relax this restriction in our model of a scattering particle, allowing the particles to take on dimensions comparable to λ. Under these circumstances different regions of the same particle will behave as scattering centers. Because the distances between these various scattering centers are of the same magnitude as the wavelength of light, there will be interference between the waves of light scattered from different parts of the same particle.

Figure 5.10 shows how this comes about. The electric field of the incident light beam is out of phase when it strikes two different portions of the scatterer, designated i and j. Although the figure shows i and j as polymer segments, they could be a pair of volume elements in any material. In fact, applications of the Debye theory to polymers are the most widely encountered. The light which is scattered from i and j is characterized by the same field which induces the oscillation at these locations. Therefore the light scattered from the two sites will be out of phase and display interference when observed at a large distance compared to Δx, say, along BB' in Fig. 5.10. As shown in the figure, this interference effect may be constructive or destructive, depending on the value of θ. We propose, therefore, that the Rayleigh ratio defined by Eq. (34) must be multiplied by a correction factor $P(\theta)$ to correct for the interference effects which were not considered previously. Several things may be anticipated about this factor:

1. As the dimensions of the particle become negligible compared to the

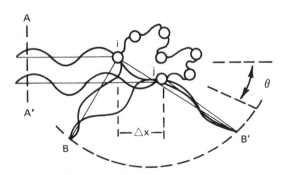

Figure 5.10 Interference of light rays scattered by segments i and j in a polymer chain. (From Ref. 1, used with permission.)

wavelength of light ($\Delta x \to 0$), $P(\theta) \to 1$—under these conditions, no correction is necessary.

2. The correction factor is a function of the angle of observation—as implied by the notation $P(\theta)$—since it is essentially an interference effect.

3. Multiplying the denominator of both sides of Eq. (37) by $P(\theta)$ corrects i_s (a theoretical quantity calculated without interference) for interference:

$$\frac{Kc}{i_s P(\theta) r^2 / I_0 (1 + \cos^2 \theta)} = \frac{1}{P(\theta)} \left(\frac{1}{M} + 2Bc \right) \tag{56}$$

4. The actual intensity of scattered light measured at P, i_P, equals the product i_s times $P(\theta)$. Therefore the experimental Rayleigh ratio obeys Eq. (37) modified as follows:

$$\frac{Kc}{R_\theta} = \frac{1}{P(\theta)} \left(\frac{1}{M} + 2Bc \right) \tag{57}$$

In view of these considerations, it is apparent that the way to extend the previous theories to larger particles is to evaluate $P(\theta)$. To do this in a general way, two complications must be introduced into Fig. 5.10:

1. Two scattering centers in a large particle are not simply displaced from one another in the x direction by an amount Δx—rather, the coordinates of one relative to the other must be described by a radial distance and two angles, for example, θ and ϕ.

2. A large particle does not consist of merely two scattering centers, but may be subdivided into several centers, the number of which increases with the size of the particle.

These two considerations must be incorporated into any general expression for $P(\theta)$.

We shall begin our summary of the derivation of $P(\theta)$, however, by considering only the pair of scattering centers shown in Fig. 5.10. To do this quantitatively, imagine that region i is at the origin of a coordinate system, so that the light which reaches j has to travel an additional distance Δx. The field at i and j is now represented by the following formulations of Eq. (7):

$$\mathbf{E}_i = \mathbf{E}_0 \cos(2\pi\nu t) \tag{58}$$

and

$$\mathbf{E}_j = \mathbf{E}_0 \cos\left[2\pi \left(vt - \frac{\Delta x}{\lambda} \right) \right] \tag{59}$$

The light scattered from each of these sites will be characterized by the same field as that which induces the oscillation. Therefore the light scattered from i and j will be out of phase by an amount $2\pi\,\Delta x/\lambda$.

Let us now consider the net scattered light which reaches a point P—located somewhere along BB'—a (large) distance r from the scatterer. The field at P is the sum of the fields emerging from i and j:

$$\mathbf{E}_P = \mathbf{E}_i + \mathbf{E}_j = \mathbf{E}_0 \left\{ \cos(2\pi vt) + \cos\left[2\pi \left(vt - \frac{\Delta x}{\lambda} \right) \right] \right\} \tag{60}$$

If we use the appropriate trigonometric formula for the sum of two cosines, this may be written

$$\mathbf{E}_P = \left[2\cos\left(\frac{\pi\,\Delta x}{\lambda} \right) \right] \mathbf{E}_0 \cos\left[2\pi \left(vt - \frac{\Delta x}{2\lambda} \right) \right] \tag{61}$$

This equation shows that the electric field of light scattered to P is not altered in frequency or wavelength, but that the amplitude is modified by the factor $2\cos\left(\pi\,\Delta x/\lambda\right)$. The intensity of light depends on the square of the field amplitude; therefore, with interference,

$$i_P \propto \left[2\cos\left(\frac{\pi\,\Delta x}{\lambda} \right) \right]^2 \mathbf{E}_0^2 \tag{62}$$

and without interference ($\Delta x \to 0$),

$$i_s \propto 2^2 \mathbf{E}_0^2 \tag{63}$$

Combining Eqs. (62) and (63) leads to

$$\frac{i_P}{i_s} = \cos^2\left(\frac{\pi\,\Delta x}{\lambda} \right) \tag{64}$$

In view of Eq. (56), the factor $\cos^2\left(\pi\,\Delta x/\lambda\right)$ must equal $P(\theta)$ for this simple case.

A fair amount of straightforward but tedious trigonometry is necessary to establish the relationship between Δx and r, ϕ_x, and θ for the general case of any orientation between i and j. The result of this analysis is the expression (given without proof)

$$\Delta x = 2r \cos \phi_x \sin \left(\frac{\theta}{2}\right) \tag{65}$$

where ϕ_x is defined the same as in Fig. 5.8 and θ continues to be the scattering angle measured in the horizontal plane.

It is not any specific value of ϕ_x in which we are interested, but in all possible values. This means that Eq. (65) is to be integrated over all values of ϕ_x, a procedure which leads to the result (given without proof)

$$\frac{i_P}{i_s} = 1 + \frac{\sin[(4\pi r/\lambda) \sin(\theta/2)]}{(4\pi r/\lambda) \sin(\theta/2)} \tag{66}$$

Equation (66) benefits considerably from some simplification in notation. Accordingly, we define

$$s = \frac{4\pi}{\lambda} \sin \left(\frac{\theta}{2}\right) \tag{67}$$

Using this notation permits us to rewrite Eq. (66) as

$$\frac{i_P}{i_s} = 1 + \frac{\sin(sr)}{sr} \tag{68}$$

The right-hand side of Eq. (68) correctly defines $P(\theta)$ for interference between *two* regions of a large particle. The next question we must consider is how this result may be applied to a particle which consists of N, not just two, scattering centers. If the composition of the particle is such that the scattering from one part of the particle has no effect (other than interference) on the scattering from another part, then the particle may be subdivided into N scattering elements and Eq. (68) can be applied to all possible pairs. If we define r_{ij} as the distance between the ith and jth members of the set of N scattering elements, then this argument leads to the result (given without proof)

$$\frac{i_P}{i_s} \propto \sum_i \sum_j \left(\frac{\sin(sr_{ij})}{sr_{ij}}\right) \tag{69}$$

Expression (69) is written as a proportionality rather than an equation because the summation requires that a normalization factor be introduced. The ratio i_P/i_s must equal unity when r_{ij} is small, since it explicitly corrects for interference effects which vanish under these circumstances. For small values of sr_{ij}, $\sin(sr_{ij})/sr_{ij}$ equals unity and $\Sigma_i \Sigma_j \sin(sr_{ij})/sr_{ij}$ equals N^2. Therefore normalization requires that we write

$$\frac{i_P}{i_s} = \frac{1}{N^2} \sum_i \sum_j \frac{\sin(sr_{ij})}{sr_{ij}} = P(\theta) \tag{70}$$

Note that the first term of Eq. (68) is implicitly present in Eq. (70) when r_{ij} equals zero. Equation (70) provides the general expression for $P(\theta)$ which we have sought. It is possible that the reader will recognize Eq. (70) from another context. It is exactly the same expression which describes the diffraction of x-rays by polyatomic molecules, the situation for which it was derived by Debye (Nobel Prize, 1936). An important insight emerges from this realization. All interference phenomena between electromagnetic radiation and matter follow the same mathematical laws. Interference becomes important when there is some characteristic distance D in the material under consideration which is of the same magnitude as the wavelength of the available radiation. The interference phenomena will be identical for identical values of D/λ, regardless of the separate values of D and λ. Thus x-ray diffraction is an experimental technique that uses x-rays for which $\lambda \simeq 0.1$ nm to measure interatomic distances which are on the order of nanometers. Likewise, we may use visible light for which $\lambda \simeq 500$ nm to measure particles whose dimensions are in the colloidal size range. The scattering of microwaves by atmospheric rain and snow particles is an example of still another situation in which interference phenomena may be observed by scaling the wavelength of the radiation to suit the particle sizes under investigation.

Before turning to the applications of the Debye approximation, we should elaborate more fully on a point that was glossed over. This is the assumption—made at the outset but explicated in going from Eq. (68) to Eq. (69)—that the scattering behavior of each scattering element is independent of what happens elsewhere in the particle. The approximation that the phase difference between scattered waves depends only on their location in the particle and is independent of any material property of the particle is valid as long as

$$\frac{2\pi D}{\lambda}(n-1) \ll 1 \tag{71}$$

It is the condition expressed by the inequality (71) that requires $n-1$ to become smaller and smaller as the theory is applied to progressively larger particles. For example, when $\theta = 10°$, the Debye approximation is good to within 10% for spheres whose radius is about 62, 37, and 25 times λ at $n = 1.1$, 1.2, and 1.3, respectively. As we shall see presently, the approximation applies to a somewhat wider range of R and n values at smaller angles of observation and to a narrower range at larger angles.

5.10 Zimm Plots

Although Eq. (70) may correct Eq. (57) for the turbidity of systems in which the scattering centers are not negligible in size compared to the wavelength of light, it is not a very promising looking result. A little more mathematical manipulation will correct this impression. Specifically, suppose we examine Eq. (67) in the case in which $\theta/2$ is small. This will make s small regardless of the value of r_{ij} so that $\sin(sr_{ij})$ in Eq. (70) may be expressed as a power series (see Appendix A):

$$P(\theta) = \frac{1}{N^2} \sum_i \sum_j \frac{\sin(sr_{ij})}{sr_{ij}} = \frac{1}{N^2} \sum_i \sum_j \frac{sr_{ij} - (sr_{ij})^3/3! + \cdots}{sr_{ij}} \quad (72)$$

For small values of s the expansion may be limited to the first two terms:

$$P(\theta) = \frac{1}{N^2} \sum_i \sum_j 1 - \frac{(sr_{ij})^2}{6} = 1 - \frac{s^2}{6N^2} \sum_i \sum_j r_{ij}^2 \quad (73)$$

It is actually $1/P(\theta)$ in which we are interested, so we may again take advantage of the fact that s is small to write Eq. (73) as

$$\frac{1}{P(\theta)} \simeq 1 + \frac{s^2}{6N^2} \sum_i \sum_j r_{ij}^2 \quad (74)$$

In spite of various simplifying approximations, Eq. (74) still does not appear particularly useful; the problem is the double summation of r_{ij}^2 terms. We have encountered summations like this before in discussing the radius of gyration in Sect. 2.12. A little manipulation of the results of that section will show us how to replace the summations in Eq. (74):

1. Equation (2.76) gives an expression for $(R_g^2)_j$, the square of the radius of gyration of a swarm of masses around the jth, assumed to be the center of mass.
2. The two summations in Eq. (2.76) can be consolidated into one—spanning all N of the mass elements—for the purposes of this discussion. Also, the product $P(r)r^2$ in Eq. (2.72) is equivalent to a specific value of r_{ij}^2.
3. In going from Eq. (2.72) to Eq. (2.73), we add together (i.e., integrate) N terms like those in Eq. (2.72) to allow for the fact that any of the N mass elements may play the role of the jth in the derivation. Doing this counts all mass elements twice, so a factor of $\frac{1}{2}$ must be inserted.

We summarize these observations by writing

$$\overline{R_g^2} = \frac{1}{2N^2} \sum_i \sum_j r_{ij}^2 \tag{75}$$

which may be substituted into Eq. (74) to give

$$\frac{1}{P(\theta)} = 1 + \tfrac{1}{3} R_g^2 s^2 \tag{76}$$

Equation (76) is valid in the limit of small values of $\theta/2$. Of course, if the scattering particle is not too large, the expansions of Eqs. (73) and (74) will be valid at larger θ, assisted by the fact that the values of r_{ij} will be small. This explains why the range of parameters to which the Debye equation applies depends on the angle of observation, becoming narrower for larger angles and broader at small angles.

Substitution of Eq. (76) into Eq. (57) yields

$$\frac{Kc}{R_\theta} = \left(\frac{1}{M} + 2Bc\right)\left[1 + \frac{16\pi^2 \overline{R_g^2}}{3\lambda^2}\sin^2\left(\frac{\theta}{2}\right)\right] \tag{77}$$

Let us consider Eq. (77) in three important limiting cases with the objective of developing a graphical technique for using Eq. (77):

1. In the limit of $\theta = 0$, Eq. (77) reduces to Eq. (37); that is, there is no interference effect in the scattered light.
2. In the limit of $c \to 0$, Kc/R_θ is proportional to $\sin^2(\theta/2)$.
3. If both c and θ equal zero, Kc/R_θ equals $1/M$.

These limits suggest how experimental data might be collected, plotted, extrapolated, and interpreted. The resulting graph is known as the Zimm plot, after its originator.

Equation (77) shows clearly that i should be measured as a function of both concentration *and* angle of observation in order to take full advantage of the Debye theory. The light scattering photometer described in Sect. 5.6 is designed with this capability, so this requirement introduces no new experimental difficulties. The data collected then consist of an array of i/I_0 values (no subscript is needed on i, since it now applies to small and large particles), measured over a range of c and θ values. By means of Eq. (34), the i/I_0 ratios are converted to R_θ values. The results are plotted with Kc/R_θ as the ordinate and $\sin^2(\theta/2) + c$ as the abscissa. Figure 5.11 shows this sort of plot for light scattering data collected from solutions of cellulose nitrate in acetone at 25°C. The measurements were made using 436-nm light of mercury. In the figure $\sin^2(\theta/2) + kc$ has been used for the abscissa. The numerical constant k—equal to 2000 in this case—spreads out the points and results in a more intelligible display. As

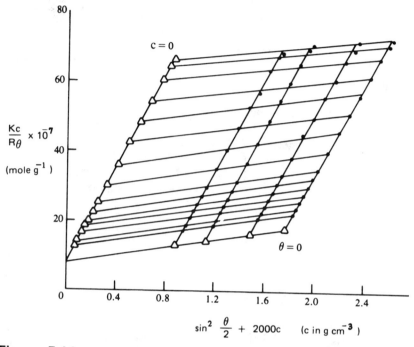

Figure 5.11 Experimental Zimm plot for cellulose nitrate in acetone. [Reprinted with permission from H. Benoit, A. M. Holtzer, and P. Doty, *J. Phys. Chem.*, 58:635 (1954), copyright by the American Chemical Society.]

with any scale factor in graphing, it is arrived at by trial and error. Each of the points in Fig. 5.11 corresponds to a particular pair of c, θ values. When the points measured at the same values of c and those measured at the same values of θ are connected, a grid of lines is obtained like that sketched in the figure. If all the experimental c and θ values are small enough for the theories to hold exactly, then the grid consists of two sets of parallel straight lines with different slopes. In general, however, the range of experimental c and θ values exceeds the range of validity of the theory, and the lines show some curvature.

The next step in the treatment of the data is the extrapolation of the curves drawn at constant θ to $c = 0$ and the extrapolation of those drawn at constant c to $\theta = 0$. This is done by placing a mark (the triangles in Fig. 5.11) on the smooth lines drawn through the experimental points at that value of the abscissa which corresponds to the value of that coordinate at the desired limit. For example, when the limit for $c = 0$ of the θ_2 line is located, it will lie on the θ_2 line and the value of the abscissa will be $\sin^2(\theta_2/2)$. Likewise, when the limit for $\theta = 0$ of the c_2 line is located, it will lie on the c_2 line and the value of the abscissa will be kc_2. The triangles in Fig. 5.11 were positioned on this basis. Note that the triangles describe two straight lines which possess a common intercept.

Let us now consider the interpretation of this graph. First, it should be remembered that interference effects vanish at $\theta = 0$. The line so labeled in Fig. 5.11 corresponds to values of Kc/R_θ at different values of c, all expressed at $\theta = 0$, where Eq. (37) is valid. According to that equation, the slope of this line equals $2B$ and the intercept equals $1/M$. This is the same result we obtained previously for small particles. By extrapolating to $\theta = 0$, the procedure now applies to larger particles as well. In formulas, then, we write for the $\theta = 0$ line

$$(\text{slope})_{\theta=0} = 2B \tag{78}$$

and

$$(\text{intercept})_{\theta=0} = \frac{1}{M} \tag{79}$$

The new feature of the Zimm plot is the second extrapolated line, corresponding to $c = 0$. This line connects values of Kc/R_θ measured at different values of θ and extrapolated to $c = 0$. Accordingly, it is described by Eq. (77) with $c = 0$; that is, theory predicts the $c = 0$ line to have an intercept of $1/M$ and a slope equal to $16\pi^2\overline{R_g^2}/3\lambda^2 M$. It is expected, then, that the two lines will extrapolate to a common intercept, $1/M$, and that the slope of the $c = 0$ line will be proportional to $\overline{R_g^2}$. Summarizing in formulas, we write for the $c = 0$ line

$$(\text{slope})_{c=0} = \frac{16\pi^2 R_g^2}{3\lambda^2 M} \tag{80}$$

and

$$(\text{intercept})_{c=0} = \frac{1}{M} \tag{81}$$

It follows, therefore, that the square of the radius of gyration equals

$$\overline{R_g^2} = \frac{3\lambda^2}{16\pi^2} \left(\frac{\text{slope}}{\text{intercept}} \right)_{c=0} \tag{82}$$

The Zimm plot shown in Fig. 5.11 is analyzed according to these relationships in the following example.

Example 5.5   The following are the intercept and slope values from Fig. 5.11:

Intercept = 7.87×10^{-7} mole g^{-1}
Slope $\theta = 0$ line = 5.70×10^{-7} mole g^{-1}
Slope $c = 0$ line = 6.78×10^{-6} mole g^{-1}

Evaluate M, B, and R_g for cellulose nitrate in acetone at 25°C from these data. Use 1.359 for the refractive index of acetone at this wavelength.

Solution   The molecular weight is the reciprocal of the common intercept and equals 1.27×10^6 g mole^{-1}.

The slope of the $\theta = 0$ line includes the factor k, which must have units concentration^{-1} for dimensional consistency. Hence the slope given must be multiplied by 2000 cm^3 g^{-1} to give the true slope to which the analysis applies. Since the latter equals $2B$ according to Eq. (78), we obtain

$$B = \tfrac{1}{2}(2000 \text{ cm}^3 \text{ g}^{-1})(5.70 \times 10^{-7} \text{ mole g}^{-1}) = 5.70 \times 10^{-4} \text{ cm}^3 \text{ mole g}^{-2}$$

The slope to intercept ratio for the $c = 0$ line equals $(6.78 \times 10^{-6})/(7.87 \times 10^{-7}) = 8.61$; this must be multiplied times $3\lambda^2/16\pi^2$ to obtain $\overline{R_g^2}$. Remember to convert $\lambda_0 = 436$ nm to the wavelength in acetone, which is the value used in this calculation: $436/1.359 = 321$ nm. Therefore $\overline{R_g^2} = 3(321)^2(8.61)/16\pi^2 = 1.69 \times 10^4$ nm^2 and $R_g = 130$ nm.

•

It is important to remember that the radius of gyration provides an unambiguous measure of a particle's extension in space. This quantity is evaluated from experimental data—as illustrated in the above example—with no assumptions as to particle geometry. If we happen to know the shape of a particle as well, R_g can be translated into a geometrical dimension of the particle. We shall examine this further in the next section.

5.11   The Dissymmetry Ratio

If the intensity of light scattered by a colloidal dispersion is measured as a function of c and θ, the Zimm method enables us to convert this information into several parameters which characterize the colloid: M, B, and R_g. In some situations this is more information than is actually needed. If spatial extension is the only information sought, a simpler method for evaluating the latter employs the so-called dissymmetry ratio.

In this technique, one measures i/I_0 at $\theta = 45°$ and $\theta = 135°$ for dispersions at several different concentrations. It should be noted that the factor $1 + \cos^2 \theta$ in Eq. (34) has the same value for these two angles of observation. Therefore any deviation of the ratio of the intensities, z, from unity must measures the ratio of the $P(\theta)$ values at these two angles:

$$z = \frac{i_{45°}}{i_{135°}} = \frac{P(45°)}{P(135°)} \tag{83}$$

For example, we may examine the approximate form for $P(\theta)$ provided by Eq. (76) to obtain

$$z = \frac{1 + (16\pi^2/3)(R_g/\lambda)^2 \sin 67.5}{1 + (16\pi^2/3)(R_g/\lambda)^2 \sin 22.5} \tag{84}$$

With this result—or its equivalent made with less restrictive approximations—values of z can be evaluated as a function of R_g/λ. From tables or plots of such calculated results, experimental dissymmetry ratios—extrapolated to $c = 0$ to eliminate the effects of solution nonideality—can be directly interpreted in terms of R_g.

We have noted previously that R_g is related to the geometrical dimensions of a body through expressions which are specific for the particle shape. Table 5.2 lists some of these relationships for shapes

Table 5.2 Relationships Between the Radius of Gyration and the Geometrical Dimensions of Some Bodies Having Shapes Pertinent to Colloid Chemistry

Geometry	Definition of parameters	Radius of gyration through the center of gravity
Random coil	$\overline{r^2}$: mean-square end-to-end distance ($\propto n$)	$\overline{R_g^2} = \dfrac{\overline{r^2}}{6}$
Sphere	R: radius of sphere	$R_g^2 = \tfrac{3}{5} R^2$
Thin rod	L: length of rod (approximation for prolate ellipsoids for which $a/b \gg 1$)	$R_g^2 = \dfrac{L^2}{12}$
Cylindrical disk	R: radius of disk (approximation for oblate ellipsoids for which $a/b \ll 1$)	$R_g^2 = \tfrac{1}{2} R^2$

Source: Ref. 1, with permission.

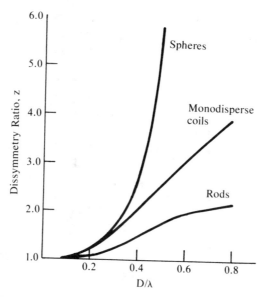

Figure 5.12 Values of the dissymmetry ratio z versus the size parameter D/λ for spheres, random coils, and rods. (Data from Ref. 5.)

which are pertinent for colloidal systems. For a selected shape, one of the tabulated relationships can be used to replace R_g in Eq. (84). What results is an expression which interprets z in terms of actual particle dimensions for the geometry chosen.

Figure 5.12 shows a plot of curves such as this drawn for spheres, coils, and rods. The characteristic dimension D used in the abscissa of the figure is R for a sphere, r_{rms} for a coil, and L for a rod; the pertinent wavelength is λ_0/n.

It will be noted from the ordinate values in Fig. 5.12 that the scattering is always larger in the forward direction for particles showing interference effects. This is one reason why the presence of dust particles raises such havoc in a scattering experiment on particles that lie in the Rayleigh region.

5.12 Large, Absorbing Particles: Scattering and Absorption Cross Sections

All the applications of light scattering that have been discussed so far have been restricted to very small particles and fairly small indices of refraction or to fairly small particles and very small indices of refraction. We have finally reached the point where it seems appropriate to relax all these restrictions and consider the scattering by a particle of arbitrary size and index of refraction.

The general problem is to solve Maxwell's electromagnetic equations both inside and outside a particle, where the indices of refraction are different in the two regions. The problem has been solved for both spherical and cylindrical particles; we shall limit our considerations to the former. In the most general case some provision must be made for the possibility that the particle absorbs as well as scatters light. This contingency is introduced by defining the refractive index of an absorbing material as a complex number $n - ik$, where $i = \sqrt{-1}$. For nonabsorbing materials $k = 0$. Both n and k are wavelength-dependent characteristics of the material; k obviously increases as the wavelength of an absorption peak is approached.

The idea of representing the refractive index of an absorbing material by a complex number may seem strange, so the following analysis will be helpful. Suppose we consider the passage of a beam of light through a layer (in the yz plane) of unspecified material of thickness Δx. If the layer contains a vacuum ($n = 1$), then the electric field transmitted through the layer will be given by Eq. (4):

$$\mathbf{E}_{n=1} = \mathbf{E}_0 \cos \left[2\pi \left(\nu t - \frac{x}{\lambda} \right) \right] \tag{85}$$

On the other hand, suppose the layer consists of a material of refractive index n. The light will now take an increment of time Δt longer to pass through the layer owing to the delaying effect of the medium. In this case the emerging field would be given by

$$\mathbf{E}_n = \mathbf{E}_0 \cos \left[2\pi \left(\nu(t + \Delta t) - \frac{x}{\lambda} \right) \right] \tag{86}$$

The delay may be related to the thickness of the layer and the refractive index as follows:

$$\Delta t = t_n - t_0 = \frac{\Delta x}{c/n} - \frac{\Delta x}{c} = (n - 1) \frac{\Delta x}{c} \tag{87}$$

This means Eq. (86) may be written

$$\mathbf{E}_n = \mathbf{E}_0 \cos \left[2\pi \left(\nu t + (n - 1) \frac{\Delta x}{\lambda} - \frac{x}{\lambda} \right) \right] \tag{88}$$

which shows that, compared with Eq. (85), the field experiences a phase shift in the medium.

A very important relationship involving complex numbers is

$$e^{i\theta} = \cos \theta + i \sin \theta \tag{89}$$

Cosine trigonometric functions, in other words, are given by the real part of the function $e^{i\theta}$. This means that Eqs. (85) and (88) may be written

$$\mathbf{E}_{n=1} = \mathbf{E}_0 \text{ Re} \exp \left[2\pi i \left(vt - \frac{x}{\lambda} \right) \right] \tag{90}$$

and

$$\mathbf{E}_n = \mathbf{E}_0 \text{ Re} \exp \left[2\pi i \left(vt + (n-1) \frac{\Delta x}{\lambda} - \frac{x}{\lambda} \right) \right] \tag{91}$$

where Re reminds us that it is only the real part of the complex number in which we are interested. Equation (91) may be written

$$\mathbf{E}_n = \mathbf{E}_0 \text{ Re} \exp \left[2\pi i \left(vt - \frac{x}{\lambda} \right) \right] \exp \left(2\pi i (n-1) \frac{\Delta x}{\lambda} \right) \tag{92}$$

$$= \mathbf{E}_{n=1} \exp \left(2\pi i (n-1) \frac{\Delta x}{\lambda} \right) \tag{92}$$

As we have seen repeatedly in this chapter, the intensity of light is proportional to the field amplitude squared; therefore we write

$$\frac{I_n}{I_{n=1}} = \exp \left(4\pi i (n-1) \frac{\Delta x}{\lambda} \right) \tag{93}$$

Now let us consider the case in which the refractive index of the layer is a complex number. In that case Eq. (93) becomes

$$\frac{I_n}{I_{n=1}} = \exp \left(4\pi i (n + ik - 1) \frac{\Delta x}{\lambda} \right) = \exp \left(4\pi i (n-1) \frac{\Delta x}{\lambda} \right)$$

$$\times \exp \left(-4\pi k \frac{\Delta x}{\lambda} \right) \tag{94}$$

The first factor of Eq. (94) is exactly what it would be if the material were nonabsorbing. It is modified, however, by a second term, $\exp(-4\pi k \, \Delta x/\lambda)$, which is real but contains the imaginary part of the index of refraction.

The Beer–Lambert equation is another formalism that might describe the arrangement we have been discussing. In the Beer–Lambert equation the intensity of the transmitted light I_t relative to the intensity of the incident light I_0 is given by Eq. (40), which may be written

$$\left(\frac{I_t}{I_0}\right)_{abs} = \exp(-\varepsilon \, \Delta x) \tag{95}$$

where ε is the absorbance of the material. Because of its usefulness in analytical chemistry, chemistry students are more likely to be familiar with this formalism than with that which leads to Eq. (94). Comparison of Eqs. (94) and (95) reveals that both describe the attenuation of light due to absorption in terms of a negative exponential which is proportional to the path length through the absorbing material. Since the two approaches describe the same situation in the same functional form, the two proportionality factors must also be equal. Therefore the imaginary part of the complex refractive index and the absorbance must be related as follows:

$$\varepsilon = \frac{4\pi k}{\lambda} \tag{96}$$

We observed in Sect. 5.6 that, for nonabsorbing systems, the turbidity is a concept analogous to the absorbance; that is,

$$\left(\frac{I_t}{I_0}\right)_{sca} = \exp(-\tau \, \Delta x) \tag{97}$$

Examination of Eqs. (95) and (97) shows how the two complement each other: The first describes absorption without scattering; the second describes scattering without absorption. For a system which displays these two optical effects simultaneously, the following composite relationship applies:

$$\left(\frac{I_t}{I_0}\right) = \exp[-(\varepsilon + \tau)\Delta x] \tag{98}$$

The experimental extinction in a system which displays these two effects equals the sum of ε plus τ.

The literature on this topic often factors absorbance and turbidity into the product of several terms. Both the nomenclature and notation which are used in this area vary from author to author, but the principal breakdown of both ε and τ goes as follows:

$$\varepsilon = N\pi R^2 Q_{abs} \tag{99}$$

and

$$\tau = N\pi R^2 Q_{sca} \tag{100}$$

where N is the number of dispersed particles per unit volume, R is their radius, and the Q terms are known as the efficiency factors for absorption and scattering. The quantity πR^2 in these equations gives the geometrical cross section of the dispersed spheres. This area times the efficiency factor defines a quantity known as the cross section C for absorption and scattering:

$$C_{abs} = \pi R^2 Q_{abs} \tag{101}$$

$$C_{sca} = \pi R^2 Q_{sca} \tag{102}$$

Dimensionally, ε and τ describe the attenuation of light per unit optical path and have units length^{-1}. The area πR^2 has units of length^2 per particle and N has units of particles length^{-3}. The efficiency factors are dimensionless. The cross sections have units of area per particle. These should not be taken literally as cross-sectional areas but, rather, as the "blocking power" of a particle as far as the transmission of incident light is concerned.

As stated previously, Maxwell's equations have been solved for spherical particles of arbitrary size and arbitrary refractive index. Furthermore, the refractive index may be complex, so the general theory applies to both nonabsorbing and absorbing particles. The solutions to Maxwell's equations are generally given in terms of the efficiency factors Q, where the magnitudes of the two efficiencies depend on the wavelength of light, the size of the particles, the real and imaginary parts of the refractive index, and the angle of observation:

$$Q = f(\lambda, R, n, k, \theta) \tag{103}$$

For nonscattering particles Q_{sca} is zero, and for nonabsorbing particles Q_{abs} is zero. The results of such an analysis are generally reported by giving values of Q_{sca} and Q_{abs} as a function of a size parameter α, where α is defined to be

$$\alpha = 2\pi \frac{R}{\lambda} \tag{104}$$

The calculations involved in computing these results are formidable and the results which were available prior to the widespread utilization of computers were very limited. As noted in the introduction to this chapter, the advent of the computer broadened the applicability of light scattering enormously.

It is difficult to make any generalizations about the variation of the efficiencies with α, since the functions are so complicated and vary greatly with the refractive index. About the best that can be done along these lines is the following: (1) At any given angle of observation, Q tends to be an oscillating function of α; (2) for nonabsorbing particles (n real), the amount of oscillation in the Q_{sca} versus α curve is more pronounced the larger n is—if the particles absorb, the amount of oscillation in the curves decreases with increasing k; (3) for any given value of α, Q varies with the angle of observation θ. The number of oscillations in this curve is greater for larger values of α and n. These generalizations are consistent with the approximations discussed earlier in the chapter: Small particles with low values of n are the simplest to describe.

In the following two sections we consider two specific systems which have been studied extensively, aqueous dispersions of colloidal gold (Sect. 5.13) and aqueous dispersions of colloidal sulfur (Sect. 5.14). These will illustrate some of the statements made previously, as well as show the sort of information that may be obtained from light scattering experiments in this very general case. No attempt has been made to be comprehensive in this presentation. Much more has been done with the systems to be discussed, and cross sections for many other systems have been calculated. The book by Kerker [3] contains a good bibliography of scattering functions published up to 1969.

5.13 The Mie Theory: Gold Sols

The preceding section indicated some of the complications that arise when the optical properties of dispersions are calculated without placing narrow limitations on the size and refractive index of the particles. The difficulties are still large but somewhat more manageable if only the refractive index is given full range, the particle dimensions being somewhat restricted. This is the situation that was treated by Mie in 1908.

Mie wrote the scattering and absorption cross sections as power series in the size parameter α, restricting the series to the first few terms. This truncation of the series restricts the Mie theory to particles with dimensions less than the wavelength of light but, unlike the Rayleigh and Debye approximations, applies to absorbing and nonabsorbing particles.

The following equations give some indication of the nature of these expansions:

$$Q_{abs} = A\alpha + B\alpha^3 + C\alpha^4 + \cdots \qquad (105)$$

and

$$Q_{sca} = D\alpha^4 + \cdots \qquad (106)$$

where the values of the coefficients A, B, C, and D are listed in Table 5.3. In this presentation, we have intentionally limited the series to include no terms higher than fourth order in α. Thus only the leading term in Q_{sca} is represented, even though the first three terms in Q_{abs} are included. The neglect of higher-order terms permits us to apply this discussion rigorously only to particles which are sufficiently small that terms in α^5 or higher would make negligible contribution. Even with only these terms retained, it is possible to draw several informative conclusions about Q_{abs} and Q_{sca}:

1. The absorption and scattering efficiencies do not show the same dependence on the particle size parameter α.
2. Both numerical coefficients of the complex refractive index, n and k, appear in both Q_{abs} and Q_{sca} (Table 5.3).
3. The coefficients A, B, and C equal zero and D reduces to Rayleigh's law if $k = 0$.
4. The efficiencies are functions of the dimensionless variable α alone.
5. For dispersions of uniform spheres, the entire wavelength dependence of the extinction is given by Eqs. (105) and (106).
6. The "wavelength dependence of the extinction" is simply the spectrum of the dispersion, which is therefore predicted theoretically by the general equations.

In the remainder of this section we shall see how the theoretical calculations of Mie account for the observed spectrum of colloidal gold. In the next section we shall consider the inverse problem for a simpler system: how to interpret the experimental spectrum of sulfur sols in terms of the size and concentration of the particles. Both of these example systems consist of relatively monodisperse particles. Polydispersity complicates the spectrum of a colloid, since the same Q value will occur at different λ values for spheres of different radii according to Eqs. (104)–(106).

Colloidal gold is of considerable historic importance in colloid chemistry, since many of the scientists who led the early development of the field conducted experiments on this system. Mie set out specifically to

Table 5.3 Values for the Constants A to D in Eqs. (105) and (106)

Coefficient	General case	Special case of $k = 0$
A	$\dfrac{24nk}{(n^2 + k^2)^2 + 4(n^2 - k^2) + 4}$	0
B	$\dfrac{4nk}{15} + \dfrac{20nk}{3[4(n^2 + k^2)^2 + 12(n^2 - k^2) + 9]}$ $+ \dfrac{4.8nk[7(n^2 + k^2)^2 + 4(n^2 - k^2 - 5)]^2}{[(n^2 + k^2)^2 + 4(n^2 - k^2) + 4]^2}$	0
C	$\dfrac{-192n^2k^2}{[(n^2 + k^2)^2 + 4(n^2 - k^2) + 4]^2}$	0
D	$\dfrac{\frac{8}{3}[(n^2 + k^2)^2 + n^2 - k^2 - 2]^2 + 36n^2k^2}{[(n^2 + k^2)^2 + 4(n^2 - k^2) + 4]^2}$	$\dfrac{\frac{8}{3}(n^2 - 1)^2}{(n^2 + 2)^2}$

Source: From R. B. Penndorf, *J. Opt. Soc. Am.*, *52*:896 (1962).

account for the brilliant colors displayed by sols of gold and other metals.

Chloroauric acid, $HAuCl_4$, is easily reduced to metallic gold by a wide variety of reducing agents. However, characteristics of the resulting gold are widely different for different reducing agents. Thus, if phosphorus is used, a polydisperse system containing very small particles forms rapidly. If the resulting colloid is used to seed a reaction in which hydrogen peroxide is the reducing agent, the following reduction takes place slowly without additional nucleation:

$$2AuCl_4^- + 3H_2O_2 \rightarrow 2Au + 8Cl^- + 6H^+ + 3O_2 \qquad \text{(A)}$$

The gold particles grow to a larger size by this process, with considerable sharpening of the particle size distribution. The particle size may be regulated to some extent by varying the amount of reagents used. Thus approximately monodisperse colloids of several different particle sizes may be prepared and compared. They are found to display different colors, depending on the particle size: The smaller particles produce a red dispersion; somewhat larger particles impart a blue color to the dispersion.

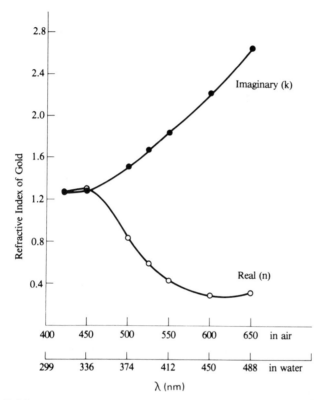

Figure 5.13   The real and imaginary parts of the complex refractive index of gold versus wavelength in air and in water. (Data from Ref. 7.)

In calculating efficiency factors for absorption and scattering, the wavelength dependence of both the real and imaginary parts of the refractive index must be considered.

Figure 5.13 shows the real and imaginary parts of the complex refractive index of gold plotted against the wavelength of light in air and in water. These values of the refractive index were used to calculate the values of C_{abs} and C_{sca} for three different sized spherical gold particles. The results are shown in Fig. 5.14 as a function of the wavelength in air. The figure also includes the sum of the two cross sections, C_{ext}, which describes the total extinction.

The magnitude and location of the maximum for each of the cross section curves are informative. It will be observed that the height and the location of the absorption curve in Fig. 5.14 change relatively little with

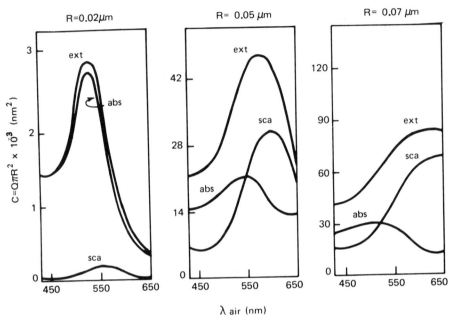

Figure 5.14 Scattering coefficients versus wavelength for spheres of colloidal gold having three different radii. Note the different scales for the ordinates in each figure. (Data from Ref. 7.)

particle size (note the different scales for the ordinates), as we might expect. The cross section for scattering, on the other hand, is practically negligible for the smallest of the particles and increases to be roughly 50 and 100% larger than absorption for the larger particles. Also noteworthy is the fact that the wavelength location of the scattering maximum shifts to longer wavelengths as the particle size increases. This is a consequence of the fact that the efficiency depends on α, rather than the separate values of R and λ. The wavelength at which maximum total extinction occurs lies in the green, yellow, and red portions of the spectrum, respectively, for the particles with $R = 0.02$, 0.05, and 0.07 μm. On the basis of color complementarity, these would appear red, violet, and blue, respectively. These are the hues displayed by actual dispersions. Thus the Mie theory not only accounts for the color of the dispersions but also shows how the color displayed may be used to characterize the particle size of the dispersed phase.

Note that the index of refraction of the continuous phase and that of the dispersed particles enter the evaluation of the various efficiencies.

The light which actually strikes the particles is used in the determination of Q. This light differs from that under vacuum by the refractive index of the medium. This effect enters the calculation of the Q values in that it is the ratio of the refractive index of the particle relative to that of the medium which determines the extinction.

5.14   Monodisperse Sulfur Sols: Higher-Order Tyndall Spectra

Another colloidal system whose light scattering characteristics have been widely studied is the so-called monodisperse sulfur sol. Although not actually monodisperse, the particle size distribution in this preparation is narrow enough to make it an ideal system for the study of optical phenomena.

The colloid is prepared by rapidly mixing dilute solutions of sodium thiosulfate and hydrochloric acid so that the final concentration of each is about 0.002 M. The following reaction then occurs so slowly that the sulfur precipitates only on those particles which nucleate first:

$$H^+ + S_2O_3^{2-} \rightarrow HSO_3^- + S \tag{B}$$

Slow growth on the original nuclei is how the narrow distribution of particle sizes is obtained, just as with the colloidal gold described in the preceding section. The formation of sulfur may be terminated at any time by adding I_2 to react with the remaining thiosulfate. These monodisperse sulfur sols have been studied extensively, notably by V. K. LaMer and co-workers.

Since these particles are nonabsorbing in the visible spectrum, the range of particle sizes which may be conveniently dealt with is broader than for absorbing particles such as gold.

Using the Mie theory, we can evaluate the scattering efficiency as a function of α for particles having a refractive index relative to the medium of 1.50, which describes the sulfur–water system.

Figure 5.15 shows a plot of Q versus α (on a double logarithmic scale) for this ratio of refractive indices. The same graph also shows the experimental spectrum (total extinction versus λ^{-1}, also double logarithmic) for a preparation which was allowed to grow for about 4.75 h before quenching. It will be observed that the theoretical and experimental curves agree very well in shape, but are displaced with respect to one another. By working with a prominent feature that appears on both curves, theoretical and experimental abscissa values can be related through Eq. (104) to give R for the dispersion. Likewise, theoretical and experimental ordinate values can be related through Eq. (100) to give N

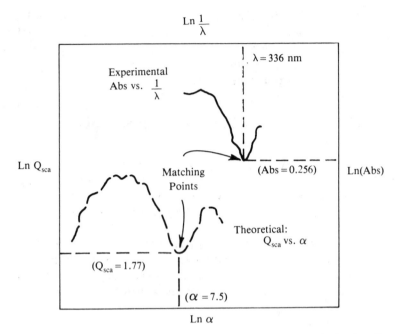

Figure 5.15 Analysis of the scattering spectrum for sulfur sols: broken line, Q versus α from theory; solid line, extinction versus $1/\lambda$ from experiment, both on double logarithmic scale. [Data from V. K. LaMer and M. D. Barnes, *J. Colloid Sci., 1*:71, 79 (1946).]

for the dispersion. The following example illustrates how this works using the data in Fig. 5.15.

Example 5.6   The minima on the theoretical and experimental curves in Fig. 5.15 are identified as matching points. The figure lists the absorbance and wavelength coordinates of the minimum for the experimental curve and the Q_{sca} and α coodinates of the minimum for the theoretical curve. Combine the corresponding pairs of experimental and theoretical parameters to evaluate R and N for this sulfur sol. The experimental absorbance values were determined in a 44.4-cm cell.

Solution   Since both the α and λ coordinates of the minima are known, Eq. (104) permits us to evaluate R: $R = \alpha\lambda/2\pi = (7.5)(336)/2\pi = 400$ nm.
    The experimental absorbance $\log_{10}(I_0/I)$ must first be converted into a value of τ using Eq. (97): $\tau = 2.303(\text{Abs})/\Delta x = 2.303(0.256)/0.444 = 1.33$ m^{-1}.

With both the τ and Q_{sca} coordinates of the minima known, Eq. (100) permits us to evaluate N (remember, the value of R has already been obtained): $N = \tau/\pi R^2 Q_{sca}$; $N = 1.33/\pi(400 \times 10^{-9})^2(1.77) = 1.49 \times 10^{12}$ $m^{-3} = 1.49 \times 10^6$ cm^{-3}.

•

By this type of analysis, it has been possible to show that the number of particles per unit volume does not change appreciably as R increases, thus substantiating the mechanism described to account for the mono-dispersity of sulfur sols.

This procedure or related variations have been applied to systems other than the sulfur sols described here. The success of the method clearly depends on having a system for which some outstanding feature of the spectrum (e.g., the minimum of Fig. 5.15) lies at a value of R/λ which is within the range of α values studied. Methods have been developed for generating additional functions from a measured spectrum which are systematically displaced from the original spectrum to assist the matching of theoretical and experimental curves.

Another interesting application of scattering theory consists of combining scattering functions with expressions for particle size distribution to generate theoretical spectra for polydisperse systems. Such spectra have been calculated for different values of the relative refractive index (particle/medium). In this manner the spectrum of a dispersion may be interpreted to yield estimates of both a mean and a standard deviation for the size of the dispersed particles. Excellent agreement with electron-microscopic results has been obtained by this method.

The monodisperse sulfur sols we have discussed herein are good examples of another light scattering phenomenon: the higher-order Tyndall spectrum. We observed in Sect. 5.12 that the scattering cross section is an irregularly oscillating function of θ, at least above a certain threshold value of α. Here it should be recalled that the complete theory reduces to the Rayleigh approximation for very small particles and to the Debye approximation for somewhat larger particles, provided the refractive index values are in the proper range. The full solution of the Mie theory provides quantitive information about the dependence of the efficiency factors on θ and λ. For uniform spheres over some range of refractive index and size, different colors of light will be scattered in different directions. The sulfur sols described here have the required properties to display this effect. Therefore, if a beam of white light is shown through a sample of the dispersion, various colors will be seen at different θ values. The resulting array of colors is known as the higher-order Tyndall spectrum (HOTS). Red and green bands are most evident in the sulfur sols, and the number of times these bands repeat increases with the size of the sulfur particles. Therefore the number and angular positions of the colored bands provide a unique characterization of the particle size. In the monodisperse sulfur sols, for example, particles having a radius of 0.30 μm are expected to show red bands at about 60, 100, and 140°. Particles with a radius of 0.40 μm, on the other hand, show red bands at about 42, 66, 105, 132, and 160°. Particle size determinations based on observations of this sort agree well with those determined from electron microscopy. These sulfur sols are quite easy to prepare, and it is interesting to observe the development of higher orders in the Tyndall spectrum as the thiosulfate decomposition reaction progresses.

It was once thought that the appearance of HOTS was evidence in itself for the presence of a monodisperse system. The argument was that one particle size would scatter, say, red light, at a particular angle, whereas another particle size would have the same α value and therefore the same scattering behavior for light of a complementary color. The resultant would be the obliteration of any distinct color: The scattered light would appear white. Although there may indeed be fortuitous cancellations of this sort at certain angles, it is also possible for certain bands to reinforce each other. In general, then, it is best to say that polydisperse systems may show HOTS, but in this case the angular distribution of bands is a characteristic of the particle size *distribution*. The angular location and number of bands as determined theoretically for uniform particles may not be used to interpret the HOTS of a polydisperse system correctly.

5.15 Wrap-Up: Chapters 1-5

Although they deal with a variety of different phenomena, Chaps. 1-5 have shared certain common features. In each chapter at least one of the major topics was the development of a theoretical framework and an experimental technique which would permit the determination of the molecular weight of particles in the colloidal size range. In most cases other facts concerning the nature of the system were also involved— either as part of the assumed model or as other parameters to be evaluated. Among the latter we have considered particle shape, linear dimension, solvation, and polydispersity.

It should be evident that not all the methods we have discussed are equally useful for all systems, nor are they equally convenient when more than one is applicable. Often it is only by combining the results from more than one technique that ambiguity about particle characterization can be removed, so the availability of more than one experimental method is a valuable asset.

Direct observation, whether by optical or electron microscopy, permits particle dimensions to be measured; with these, particle masses—the equivalent of molecular weights for such systems—may be calculated. The method is tedious, however, and subject to statistical complications and may be confused by artifacts of the preparative procedure.

Sedimentation studies are versatile inasmuch as they may be applied to particles as small as individual molecules (in an ultracentrifuge) and as large as grains of sand (in gravitational settling). In the latter case the assumption of spherical shape is needed to convert the results to molecular weights. For smaller particles, however, the combination of sedimentation and diffusion studies permits the unambiguous determination of particle mass.

Osmotic pressure gives the number average molecular weight of all particles restrained by a semipermeable membrane. If these are limited to the solute of interest, the molecular weight of that solute is measured with no assumptions about its shape. The procedure depends on the existence of a suitable membrane, however, and this may not always be as selective as might be desired.

Viscosity measurements are fast and through the Staudinger-Mark-Houwink equation permit the evaluation of molecular weight once suitable calibration data have been determined. For particles which combine unknown amounts of ellipticity and solvation, the viscosity cannot be interpreted unambiguously.

Light scattering is theoretically complicated, but not too difficult in practice. Rigorous solutions of scattering functions are becoming more and more readily available, so the range of this technique and the amount of information that may be obtained from it continue to expand.

In one way or another, all of the methods we have discussed in Chaps. 1-5 also provide some insight into the spatial extension of the particles. This property is measured directly by microscopy, is evaluated through the friction factor by diffusion, affects the excluded volume and hence the second virial coefficient in osmometry, and enters viscosity and light scattering through the radius of gyration.

In four of the next five chapters we shall be less concerned with the size and shape of dispersed particles than with the characteristics of their surfaces. Chapter 8 is the exception; in it we examine micelles and related structures. Except for Chapter 8, we are leaving colloid chemistry for a while and are entering the realm of surface chemistry. The two disciplines are closely related, however, since the specific surface area of any dispersed phase varies inversely with the dimensions of the dispersed particles. After this brief look at surface chemistry we shall return to a discussion of small particles dispersed in a continuous phase. We will then be in a position to understand how the behavior of the surfaces— particularly with respect to adsorption—affects the colloidal behavior of the dispersed particles.

References

1. P. C. Hiemenz, *Polymer Chemistry: The Basic Concepts*, Marcel Dekker, New York, 1984.
2. M. B. Huglin (ed.), *Light Scattering from Polymer Solutions*, Academic, New York, 1972.
3. M. Kerker, *The Scattering of Light and Other Electromagnetic Radiation*, Academic, New York, 1969.
4. D. McIntyre and F. Gormick (eds.), *Light Scattering from Dilute Polymer Solutions*, Gordon and Breach, New York, 1964.
5. K. A. Stacey, *Light Scattering in Physical Chemistry*, Butterworths, London, 1956.
6. C. Tanford, *Physical Chemistry of Macromolecules*, Wiley, New York, 1961.
7. H. C. Van de Hulst, *Light Scattering by Small Particles*, Wiley, New York, 1957.

Problems

1. The turbidity of solutions of sodium silicate containing SiO_2 and Na_2O in 3.75 ratio has been studied as a function of time.* When the total solute content is 0.02 g cm^{-3}, the following data are obtained:

Time (days)	0	1	3	15	45	88	166	199	351	455
$\tau \times 10^4$ (cm^{-1})		0.81	2.32	3.80	5.47	6.70	8.06	9.53	10.47	13.00	14.43

(During the same time the pH of the solution changes from 10.85 to 11.02.) Suggest a qualitative explanation for these observations in terms of the chemical behavior of silicates (be sure to include references to whatever sources you consult). Why is it important to the study of light scattering to realize that solutions such as these show a variation of turbidity with time?

---

*P. Debye and R. V. Hauman, *J. Phys. Chem.*, 65:5 (1961).

2. Equation (2.63) suggests that the quantity $\overline{\delta c^2}$ in Table 5.1 is evaluated by solving

$$\overline{\delta c^2} = \int_0^\infty (\delta c)^2 P(\delta c) d(\delta c)$$

Argue that the appropriate form for $P(\delta c)$ is

$$P(\delta c) = A \exp\left(- \frac{G - G_0}{kT} \right)$$

Use the value of $G - G_0$ for a fluctuation from Table 5.1 $[G - G_0 = \frac{1}{2}(\partial^2 G/\partial c^2)_0(\delta c)^2]$ to verify the value for $\overline{\delta c^2}$.

3. The refractive indices of NaCl solutions have been measured at 20°C as a function of concentration by means of a differential refractometer* with the following results:

$c \times 10^3$ (g NaCl cm^{-3})	3.749	5.468	7.498	7.920
$(n - n_0) \times 10^4$	6.73	9.83	13.18	14.06

Use these data to evaluate the quantity dn/dc for NaCl solutions in this concentration range. On the basis of these data, what is the apparent uncertainty introduced in a light scattering experiment through this quantity?

4. Above a certain concentration (the turbidity and concentration of which are taken to characterize the medium), the turbidity of sodium dodecyl sulfate (molecular weight = 288) solutions increases with concentration as if particles in the colloidal size range were present. Use the following data † to evaluate the apparent molecular weight of the species responsible for the scattering. For this system $H = 3.99 \times 10^{-6}$.

$c \times 10^3$ (g cm^{-3})	2.7	4.2	7.7	9.7	13.2	17.7	22.2
$\tau \times 10^4$ (cm^{-1})	1.10	1.29	1.71	1.98	2.02	2.14	2.33

Assuming the scattering centers are aggregates of sodium dodecyl sulfate molecules, estimate the number of these units in the aggregate.

5. The turbidity of Ludox (a colloidal silica manufactured by DuPont) has been studied as a function of concentration with the following results‡:

$c \times 10^2$ (g cm^{-3})	0.57	1.14	1.70	2.30

---

*P. Debye and R. V. Hauman, *J. Phys. Chem.*, 65:8 (1961).

† H. V. Tartar and A. L. M. Lelong, *J. Phys. Chem.*, 59:1185 (1956).

‡ G. Deželic and J. P. Kratohvil, *J. Phys. Chem.*, 66:1377 (1962).

$$\tau \times 10^2$$
(cm^{-1}) 1.56   2.97   4.25   5.36

Evaluate the molecular weight of the Ludox particles, using a value of $H = 4.08 \times 10^{-7}$ for the system. Calculate the characteristic diameter for these particles, assuming them to be uniform spheres of density 2.2 g cm^{-3}.

6. Criticize or defend the following proposition: The accompanying data for the turbidity of dodecylamine hydrochloride solutions* suggest that at concentrations exceeding about 0.003 g cm^{-3} the solute associates into aggregates of colloidal dimensions.

$c \times 10^3$ (g cm^{-3})   0.77   1.73   3.10   3.31   6.15   8.31

$\tau \times 10^4$ (cm^{-1})   0   0   0   0.95   1.91   2.55

For this system H is approximately 7.7×10^{-6}.

7. A. I. Krasna† has measured the turbidity of calf thymus DNA in aqueous solutions. The accompanying table gives $R_\theta \times 10^5$ (in cm^{-1}) for this system, measured at 546 nm:

$c \times 10^6$ (g cm^{-3})

θ (deg)	20.6	41.4	62.0	82.5
26	2.47	4.80	6.94	8.75
30	2.06	4.02	5.83	7.67
34	1.77	3.39	4.84	6.44
38	1.75	2.98	4.28	5.61
42	1.38	2.57	3.84	4.87
50	1.01	1.90	2.91	3.71
60	0.76	1.45	2.23	2.89

Prepare a Zimm plot of these results (using $K = 3.63 \times 10^{-7}$) and evaluate M, B, and the radius of gyration of the DNA in this preparation.

8. The following table gives $R_\theta/K \times 10^{-3}$ for different values of c and θ in the system polystyrene–decalin at 30°C, measured with the mercury 435.8-nm line‡:

$c \times 10^3$ (g cm^{-3})

θ (deg)	0.50	0.99	1.49
30	0.735	1.34	—
45	0.685	1.27	2.04
60	0.625	1.17	1.62
75	0.562	1.05	1.49
90	0.510	0.96	1.37
105	0.467	0.88	1.25

*P. Debye, *J. Phys. Chem.,* *53*:1 (1949).
†A. I. Krasna, *J. Colloid Interface Sci.,* *39*:632 (1972).
‡M. D. Lechner and G. V. Schulz, *J. Colloid Interface Sci.,* *39*:469 (1972).

Prepare a Zimm plot of these data and evaluate M, B, and the radius of gyration of the polymer under these conditions.

9. The following table shows the angular location of the green bands in the HOTS of various size spheres of relative refractive index 1.46 [3, p. 409]:

Number of green bands	Radius (μm)					
	0.2	0.3	0.4	0.5	0.6	0.8
First	25°	7.5°				
Second	140°	77.5°	57.5°	42.5°	32.5°	17.5°
Third		150°	95°	72.5°	60°	40°
Fourth			140°	117.5°	85°	62.5°
Fifth				115°	110°	87.5°
Sixth				(blue)	130°	110°
Seventh					157.5°	127.5°
Eighth						142.5°

Formulate a generalization correlating the observed number of green bands in the HOTS for this system with the approximate particle size. Briefly describe how this information could be used to "grow" a monodisperse sulfur sol whose dimension corresponds approximately to a predetermined size.

10. The spectra of monodisperse sulfur sols such as that shown in Fig. 5.15 have been observed after allowing the thiosulfate reduction to proceed for various lengths of time.* Each spectrum contains a recognizable minimum which may be identified with the minimum observed in the theoretical plot of K versus α. The coordinates of these minima for curves measured in a 44.4 cm cell at different times are as follows:

Age (hr)	λ_{min} (nm)	(Absorbance)$_{min}$
7.77	358	0.313
15.90	442	0.484
22.68	526	0.421

Evaluate the radius and concentration of the sulfur particles at each of these times and comment on the correlation between these results and the mechanism proposed for monodispersity in Sect. 5.14.

11. General solutions of light scattering equations generate oscillating curves when Q is plotted versus α for a particular value of the relative refractive index of the dispersed phase compared to the continuous phase. Such a curve can be described by the expression $Q = K\alpha^{-n}$, where $-n$ is the local value of the slope of the ln Q versus ln α curve at specific values of α.

---

*M. D. Barnes and V. K. LaMer, *J. Colloid Sci.*, *1*:79 (1946).

Suppose this parameter n is known as a function of α for an experimental system. Describe the kind of experimental data and the analysis required thereof to yield a size parameter for the dispersed particles. What are the limitations of this method? How does this method differ from the curve-matching techniques shown in Fig. 5.15?

12. Suppose your employer intends to develop a new laboratory to characterize particles in the colloidal size range. Your assignment is to prepare a list of the equipment which should be purchased for such a facility. A brief justification for each major item should be included along with a priority ranking based on the versatility of the method. Assume that your laboratory is already well stocked with such nonspecialized items as laboratory glassware, balances, and the like.

6

SURFACE TENSION AND CONTACT ANGLE
Application to Pure Substances

I don't think I said anything about the Third Dimension; and I am sure I did not say one word about "Upward, not Northward," for that would be nonsense, you know. How could a thing move Upward, and not Northward? Upward and not Northward! . . . How silly it is!

[From Abbott's *Flatland*]

6.1 Introduction

Everyone has had the experience of pouring more beverage into a cup or glass than that container could hold. In addition to the spills this causes, such an experience provides an opportunity to observe surface tension. Most liquids can be added to a vessel until the liquid surface bulges above the rim of the container. The liquid behaves as if it had a "skin" which prevents it—up to a point—from overflowing. Stated technically, a contractile force operates around the perimeter of the surface which tends to shrink the surface. This is what we mean when we talk about the surface tension of a liquid. All phase boundaries behave this way, not just liquid surfaces; however, the evidence for this is more apparent for deformable liquid surfaces.

The liquid "skin" described above is anchored to the solid walls of the container around the edges of the surface. The angle the liquid surface makes with the solid support is called the contact angle. The tendency of most liquids to climb walls—think of a meniscus in a capillary—is a manifestation of the existence of these angles.

Surface tension and contact angle are two different things, although they are closely related. Surface tension describes the interface between two phases, and contact angle describes the edge of the two-phase boundary where it ends at a third phase. Two phases must be specified to describe surface tension; three are needed to describe contact angle. We

287

shall designate solid, liquid, and vapor phases by S, L, and V, respectively; where two liquids are involved, we shall call them L_1 and L_2 or A and B.

In both this chapter and Chapter 7 we are primarily concerned with the equilibrium behavior of surfaces. In this chapter the emphasis is on the surfaces between "pure" phases; in Chapter 7 it is on phases for which concentration is a variable. Most of the phenomena we describe in this chapter continue to apply to solutions and carry over into the next chapter. The emphasis in Chapter 7 is on the effects of solutes on the behavior of surfaces; we do not worry about this in the present discussion. This explains the quotation marks around the word *pure* above. Equilibrium phases will be mutually saturated and hence not pure in the strict sense. On the other hand, concentration is not a variable in equilibrium phases.

6.2 Measuring Surface Tension and Contact Angle: Round One

As the name implies, surface tension γ is a force which operates on a surface and acts perpendicular and inward from the boundaries of the surface, tending to decrease the area of the interface. Figure 6.1a shows a simple apparatus based on this notion which might be used to measure surface tension. In fact, other methods are used, but the arrangement of Fig. 6.1a has the advantage of simplicity and serves to illustrate how the tension in the liquid surface is indeed measured by γ. The figure represents a loop of wire with one movable side upon which a film could be formed by dipping the frame into a liquid. The surface tension of a stretched film in the loop will cause the slide wire to move in the direction of decreasing film area unless an opposing force F is applied. In an actual apparatus, the friction of the slide wire might be sufficient for this. In an idealized, frictionless apparatus like that in Fig. 6.1a the force opposing γ could be measured. The force evidently operates along the entire edge of the film and will vary with the length l of the slide wire. Therefore it is the force per unit length of edge which is the intrinsic property of the liquid surface. Since the film in Fig. 6.1a has two sides, the surface tension as measured by this apparatus equals

$$\gamma = \frac{F}{2l} \tag{1}$$

Several points should be noted before proceeding any further:

1. Equation (1) defines the units of surface tension to be those of force

(a)   (b)

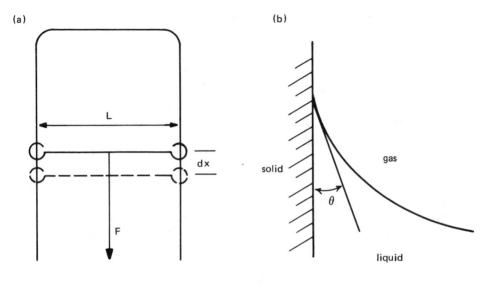

(c)

Figure 6.1   (a) A wire loop with a slide wire upon which a liquid film might be formed and stretched by an applied force *F*. (b) Profile of a three-phase (solid, liquid, gas) boundary which defines the contact angle θ. (c) The tilted plate method for measuring contact angles.

per length or N m^{-1} in SI or dynes cm^{-1} in the cgs system. We shall see presently that these are not the only units used for γ.

2. The apparatus shown in Fig. 6.1a resembles a two-dimensional cylinder/piston arrangement. With this similarity in mind, the suggestion that surface tension is analogous to a two-dimensional pressure seems plausible. With certain refinements, this notion will prove very useful in Chap. 7.

3. A gas in the frictionless, three-dimensional equivalent to the apparatus of the figure would tend to expand spontaneously. For a film, however, the direction of spontaneous change is contraction.

A quantity that is closely related to surface tension is the contact angle. The contact angle θ is defined as the angle (measured in the liquid) that is formed at the junction of three phases, for example, at the solid–liquid–gas junction, as shown in Fig. 6.1b. Although the surface tension is a property of the two phases which form the interface, θ requires that three phases be specified for its characterization.

In this section we take an initial look at the experimental determination of γ and θ. The methods we discuss show the complementarity between these two parameters while introducing some important phenomena. As we proceed through the chapter, some additional complications will be encountered. We shall return to the topic of measuring γ and θ in Sect. 6.9, where some refinements can be discussed with more meaning.

Contact angle seems like a more straightforward quantity to measure than surface tension. Junctions such as that illustrated in Fig. 6.1b are easy to observe and even photograph. This being the case, it seems easy enough to construct a tangent to the liquid surface at the point where it contacts the support and measure the enclosed angle. Actually, this is considerably easier to describe than to carry out. Although the foregoing is impractical as an experimental strategy, it does make the contact angle very accessible conceptually.

Figure 6.1c shows a variation of Fig. 6.1b which has been used extensively to measure contact angles. This technique, known as the tilted plate method, varies the angle of inclination between a smooth solid penetrating the liquid surface until a position is found in which the liquid makes a horizontal contact with the solid. The horizontal is easier to judge than the tangent to a curve, so this method is preferable to that based on direct measurement of a junction like Fig. 6.1b.

The situation shown in Fig. 6.1b is one in which surface tension–contact angle considerations pull a liquid upward in opposition to gravity. A mass of liquid is drawn up as if it were suspended by the surface from the supporting walls. At equilibrium the upward pull of the surface and the downward pull of gravity on the elevated mass must balance out. This elementary statement of force balance applies to two techniques whereby γ can be measured if θ is known: the Wilhelmy plate and capillary rise.

Figure 6.2a represents a thin vertical plate suspended at a liquid surface from the arm of a tared balance. For simplicity, the plate is

(a)   (b)

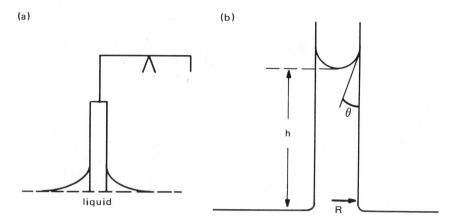

Figure 6.2 (a) The Wilhelmy plate method for measuring γ. (b) Schematic illustration of capillary rise in a cylindrical tube of radius R.

positioned so that the lower edge is in the same plane as the horizontal surface of the liquid away from the plate as shown in the figure. The manifestation of surface tension and contact angle in this situation is the entrainment of a meniscus around the perimeter of the suspended plate. Assuming the apparatus is balanced before the liquid surface is raised to the contact position, the imbalance that occurs on contact is due to the weight of the entrained meniscus. Since the meniscus is held up by the tension on the liquid surface, the weight measured by the apparatus can be analyzed to yield a value for γ.

The observed weight w of the meniscus must equal the upward force provided by the surface. This, in turn, equals the vertical component of γ—γ cos θ, where θ is the contact angle—times the perimeter of the plate, P, since surface tension is a force per unit length of surface edge. Therefore, at equilibrium,

$$w = P\gamma \cos \theta \qquad (2)$$

For a plate of rectangular cross section having length l and thickness t, $P = 2(l + t)$; these dimensions can be accurately measured. By suspending the plate from a sensitive balance, we can also measure w with considerable accuracy. The apparatus is called a Wilhelmy balance, and the technique the Wilhelmy plate method. Thus, if the contact angle is known from an independent determination by, say, the tilted plate, then γ can be evaluated by Eq. (2).

Strictly speaking, Eq. (2) allows the vertical component of surface tension to be measured. Since this equals $\gamma \cos \theta$, we are actually making a single measurement that involves two parameters. If γ were independently known, the Wilhelmy plate method could also be used to determine θ. Whether we seek to evaluate γ, θ, or both, two experiments are needed and these may not both involve the factor $\cos \theta$. In Sect. 6.9 we shall discuss a second type of measurement that can be made with the Wilhelmy apparatus which supplies a complementary observation so that both γ and θ can be determined on a single instrument.

Capillary rise is also a measure of the vertical component of surface tension, so it is an alternative to the Wilhelmy balance, and not its complement. Figure 6.2b shows how a capillary rise experiment is carried out. Conventionally, the height of a liquid column in a capillary above the reference level in a large dish is measured. It is important that the dish be large enough in diameter so that the reference level has a well-defined horizontal surface. In the capillary the liquid will have a curved meniscus and, as usual, it is the height of the bottom of the meniscus above the horizontal that is measured. It should be noted that capillary depression is also observed, as with mercury, for example. In this case the capillary "rise" is a negative quantity. Our objective is to relate the equilibrium liquid column height h to the surface tension of the liquid.

A simple—but incorrect—relationship between the height of capillary rise, capillary radius, contact angle, and surface tension is easily derived. At equilibrium the vertical component of the surface tension $2\pi R\gamma \cos \theta$ equals the weight of the liquid column, approximated as the weight of a cylinder of height h and radius R. This leads to the approximation

$$2\pi R\gamma \cos \theta \simeq \pi R^2 h \Delta \rho g \tag{3}$$

The difference in the density of the liquid and its surroundings is used in this formulation, since the latter could be a second liquid which has a buoyant effect on the liquid column. Rearranging Eq. (3), we obtain

$$Rh \cos \theta \simeq \frac{2\gamma}{\Delta \rho g} \tag{4}$$

where the cluster of constants on the right-hand side of the equation is called the capillary constant and is given the symbol a^2:

$$a^2 = \frac{2\gamma}{\Delta \rho g} \tag{5}$$

Since $a^2 \propto Rh$, a has units length. Note that the height to which a liquid climbs in a capillary (assuming $\theta < 90°$) increases as R decreases. As would be expected, h is larger for large γ and small $\Delta \rho$. By measuring h

for a capillary of known radius, Eq. (5) permits an approximate value of γ to be determined if θ is known.

The approximation which limits this analysis of capillary rise originates from neglecting the weight of the liquid in the "crown" of the curved meniscus. We shall see in Sect. 6.10 that the height of capillary rise can be related to surface tension without making this approximation, although the connection is somewhat unwieldy.

It is impossible to complete a discussion of the measurement of surface tension without saying something about the need for extreme cleanliness in any determination of γ. Any precision chemical measurement requires attention to this consideration, but surfaces are exceptionally sensitive to impurities. It is often noted that touching the surface of 100 cm^2 of water with a fingertip deposits enough contamination on the water to introduce a 10% error in the value of γ. Not only must all pieces of equipment be clean, but also the experiments must be performed within enclosures or in very clean environments to prevent outside contamination. In addition, both surface tension and contact angle should be measured under constant temperature conditions.

Both the Wilhelmy and capillary rise methods for determining γ have been based on the concept of surface tension as a force. While this point of view is useful for describing the experimental methods we have discussed, it is only one way of interpreting γ. An energetic interpretation is also possible which makes surface tension amenable to the powerful methods of thermodynamics.

6.3 Surface Tension as Surface Free Energy

Application of a force infinitesimally larger than the equilibrium force to the slide wire in Fig. 6.1a will displace the wire through a distance dx. The product of force and distance equals energy, in this case the energy spent in increasing the area of the film by the amount $dA = 2l\,dx$. Therefore the work done on the system is given by

$$\text{work} = F\,dx = \gamma 2l\,dx = \gamma\,dA \tag{6}$$

This supplies a second definition of surface tension: It equals the work per unit area required to produce new surface. In terms of this definition, the units of γ are energy per area—J m^{-2} in SI or erg cm^{-2} in the cgs system.

We see, therefore, that there are two equivalent interpretations of γ: force per unit length of boundary of the surface and energy per unit area of the surface. The dimensional equivalency of the two is evident if the numerator and denominator of force length^{-1} are multiplied by length.

Equation (6) relates γ to the work required to increase the area of a surface. From thermodynamics, it will be recalled that work is a path-dependent process: How much work is done depends on how it is done. Based on this realization, then, it seems desirable to examine Eq. (6) a little more fully. As already noted, there is a tendency for mobile surfaces to decrease spontaneously in area. Therefore it is convenient to shift our emphasis from work done *on* the system to work done *by* the system in such a reduction of area. If the quantity $\delta w'$ is defined to be the work done by the system when its area is changed, then Eq. (6) becomes

$$\delta w' = -\gamma dA \qquad (7)$$

According to Eq. (7), a decrease in area (dA negative) corresponds to work done *by* the system, whereas an increase in area requires work to be done on the system (dA positive and $\delta w'$ negative). This sign convention is consistent with the idea expressed in Chapter 1 that energy is stored in surfaces.

We are now in a position to relate the quantity $\delta w'$ to other thermodynamic variables. To do this a brief review of some basic thermodynamics is useful.

According to the first law, the change in the energy E of a system equals

$$dE = \delta q - \delta w \qquad (8)$$

in which δw is the work done *by* the system and δq is the heat absorbed *by* the system. The quantity δw is conveniently divided into a pressure-volume term and a non-pressure–volume term:

$$\delta w = \delta w_{pV} + \delta w_{\text{non-}pV} = p\, dV + \delta w_{\text{non-}pV} \qquad (9)$$

It will be recalled from physical chemistry that chemical work is the usual substitution for $\delta w_{\text{non-}pV}$. However, the work defined by Eq. (7) may also be classified as non-pressure–volume work.

The second law tells us that for reversible processes

$$\delta q_{\text{rev}} = T\, dS \qquad (10)$$

Substituting Eqs. (9) and (10) into Eq. (8), with the stipulation of reversibility as required by Eq. (10) enables us to write

$$dE_{\text{rev}} = T\, dS - p\, dV - \delta w_{\text{non-}pV} \qquad (11)$$

Next we recall the definition of the Gibbs free energy G:

$$G = H - TS = E + pV - TS \qquad (12)$$

which may be differentiated to give

$$dG = dE + p\ dV + V\ dp - T\ dS - S\ dT \tag{13}$$

Substituting Eq. (11) into (13) gives

$$dG_{\mathrm{rev}} = T\ dS - p\ dV - \delta w_{\mathrm{non}\text{-}pV} + p\ dV + V\ dp - T\ dS - S\ dT \tag{14}$$

This is a fundamental equation of physical chemistry because it enables us to assign a physical significance to G as defined by Eq. (12). Equation (14) shows that for a constant temperature, constant pressure, and reversible process

$$dG = -\delta w_{\mathrm{non}\text{-}pV} \tag{15}$$

that is, dG equals the maximum non-pressure–volume work derivable from such a process, since maximum work is associated with reversible processes.

We have already seen by Eq. (7) that changes in surface area entail non-pressure–volume work. Therefore we identify $\delta w'$ from Eq. (7) with $\delta w_{\mathrm{non}\text{-}pV}$ in Eq. (15) and write

$$dG = \gamma dA \tag{16}$$

Even better, in view of the stipulations made going from Eq. (14) to Eq. (15) we write

$$\gamma = \left(\frac{\partial G}{\partial A} \right)_{T,p} \tag{17}$$

Several things should be noted about Eq. (17). This relationship identifies the surface tension as the increment in Gibbs free energy per unit increment in area. The path-dependent variable $\delta w'$ is replaced by a state variable as a result of this analysis. Another notation that is often encountered which emphasizes the fact that γ is identical to the excess Gibbs free energy per unit area arising from the surface is to write it as G^s. The energy interpretation of γ, then, has been carried to the point where it has been identified with a specific thermodynamic function. Many of the general relationships which apply to G apply equally to γ. For example,

$$G^s = \gamma = H^s - TS^s \tag{18}$$

and

$$\left(\frac{\partial G^s}{\partial T} \right)_p = \left(\frac{\partial \gamma}{\partial T} \right)_p = -S^s \tag{19}$$

Table 6.1 Several Representative Values of γ, S^s, and H^s for a Variety of Liquids Near Room Temperature

Substance	γ at 20°C (mJ m^{-2})	$d\gamma/dT = -S^s$ (mJ m^{-2} K^{-1})	$H^s = \gamma - T(d\gamma/dT)$ (mJ m^{-2})
n-Hexane	18.4	−0.105	49.2
Ethyl ether	17.0	−0.116	51.0
n-Octane	21.8	−0.096	49.9
Carbon tetrachloride	26.9	−0.092	53.9
m-Xylene	28.9	−0.077	51.4
Toluene	28.5	−0.081	52.2
Benzene	29.0	−0.099	58.0
Chloroform	28.5	−0.135	68.3
1,2-Dichloroethane	32.2	−0.139	72.9
Carbon disulfide	32.3	−0.138	72.7
Water	72.8	−0.152	117.3
Mercury	484	−0.220	548

Source: Data from Ref. 7.

Equations (18) and (19) may be combined to give

$$\gamma = H^s + T\left(\frac{\partial \gamma}{\partial T}\right)_p \tag{20}$$

For water at 20°C, $\gamma = 7.28 \times 10^{-2}$ J m^{-2}; it is convenient to write this in millijoules, since 72.8 mJ m^{-2} is numerically equal to the cgs value. For water $d\gamma/dT = -0.152$ mJ m^{-2} deg^{-1}; therefore $H^s = 117$ mJ m^{-2} by Eq. (20). Additional values of G^s, S^s, and H^s for various substances are listed in Table 6.1.

6.4 The Laplace Equation

In Sect. 6.2 we discussed the Wilhelmy and capillary rise experiments as if the supported liquid were hanging from a surface skin. While this is a convenient device for generating formulas, it is not an adequate description of the physical situation. Now that we have established the connection between surfaces and thermodynamics, we can remedy this situation. The insight that is central to this development is the realization that a pressure difference operates across a curved interface. The pressure

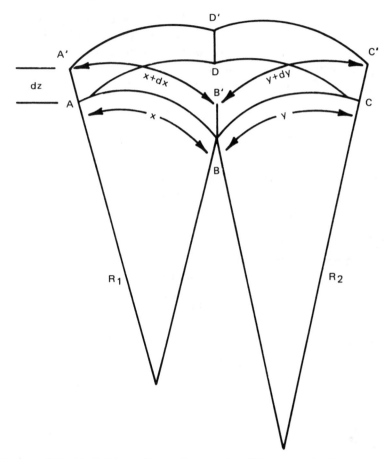

Figure 6.3 Definition of coordinates describing the displacement of an element of curved surface *ABCD* to *A'B'C'D'*.

difference is such that the greater pressure is on the concave side. Our objective in this section is to relate this pressure difference to the curvature of the surface.

Figure 6.3 shows a portion *ABCD* of a curved surface. The surface has been cut by two planes which are perpendicular to one another. Each of the planes therefore contains a portion of arc where it intersects the curved surface. In the figure the radii of curvature are designated R_1 and R_2, and the lengths are designated x and y, respectively, for these two intercepted arcs. Now suppose the curved surface is moved outward by a small amount dz to a new position $A'B'C'D'$. Since the corners of the surface continue to lie along extensions of the diverging radial lines, this move increases the arc lengths to $x + dx$ and $y + dy$. Obviously the area of surface must also increase. The work required to accomplish this must be supplied by a pressure difference Δp across the element of surface area.

The increase in area when the surface is displaced is given by

$$dA = (x + dx)(y + dy) - xy = x\,dy + y\,dx + dx\,dy \simeq x\,dy + y\,dx \tag{21}$$

where the approximation arises from neglecting second-order differential quantities. The increase in free energy associated with this increase in area is given by $\gamma\,dA$:

$$dG = \gamma(x\,dy + y\,dx) \tag{22}$$

If ordinary pressure–volume work is responsible for the expansion of this surface, then the work equals $\Delta p\,dV$, where dV is the volume swept by the moving surface. In terms of Fig. 6.3 this equals

$$dw = \Delta pxy\,dz \tag{23}$$

Setting Eqs. (22) and (23) equal to one another gives

$$\gamma(x\,dy + y\,dx) = \Delta pxy\,dz \tag{24}$$

By using similar triangles, we may set up the following proportions from Fig. 6.3:

$$\frac{x + dx}{R_1 + dz} = \frac{x}{R_1} \tag{25}$$

and

$$\frac{y + dy}{R_2 + dz} = \frac{y}{R_2} \tag{26}$$

which simplify to

$$\frac{dx}{x\,dz} = \frac{1}{R_1} \tag{27}$$

and

$$\frac{dy}{y\,dz} = \frac{1}{R_2} \tag{28}$$

Substituting Eqs. (27) and (28) into Eq. (24) enables us to write the relationship of Δp to R_1, R_2, and γ:

$$\Delta p = \gamma \left(\frac{1}{R_1} + \frac{1}{R_2} \right) \tag{29}$$

This expression is known as the Laplace equation and was derived in 1805.

Until now we have not been specific as to the location of the curved surface under consideration. Since this is the case, Eq. (29) is general and applies equally well to geometrical bodies whose radii of curvature are constant over the entire surface or to more intricate shapes for which the R's change from place to place on the surface. For the former category there are several special cases of Eq. (29) that are worthy of note:

1. For a spherical surface $R_1 = R_2 = R$; therefore

$$\Delta p = \frac{2\gamma}{R} \tag{30}$$

2. For a cylindrical surface $R_1 = \infty$; therefore

$$\Delta p = \frac{\gamma}{R_2} \tag{31}$$

3. For a planar surface $R_1 = R_2 = \infty$; therefore

$$\Delta p = 0 \tag{32}$$

It is also possible for a portion of a surface to be locally saddle shaped; in such a case the two radii of curvature lie on opposite sides of the surface and have different signs. It is possible for p to be zero in this situation also.

The Laplace equation applied specifically to spherical surfaces can be derived in a variety of ways. The following example considers an alternative derivation which points out the thermodynamic character of the result quite clearly.

Example 6.1 The Maxwell relations—of which Eq. (19) is an example—play an important role in thermodynamics. By including the term dA in the usual differential form for dG, show that $(\partial V/\partial A)_{p,T} = (\partial \gamma/\partial p)_{A,T}$. Evaluate $(\partial V/\partial A)_{p,T}$ assuming a spherical surface and, from this, derive the Laplace equation for this geometry.

Solution   The derivation of the Maxwell relations treats dG as an exact differential and expands the latter as

$$dG = \left(\frac{\partial G}{\partial p} \right)_T dp + \left(\frac{\partial G}{\partial T} \right)_p dT$$

The coefficients are then matched with their counterparts in an alternative expression for dG: $dG = V\, dp - S\, dT$.

When surface effects are included, these two expressions become

$$dG = \left(\frac{\partial G}{\partial p} \right)_{T,A} dp + \left(\frac{\partial G}{\partial T} \right)_{p,A} dT + \left(\frac{\partial G}{\partial A} \right)_{p,T} dA \quad \text{and}$$

$$dG = V\, dp - S\, dT + \gamma\, dA$$

or, at constant temperature,

$$dG = \left(\frac{\partial G}{\partial p} \right)_A dp + \left(\frac{\partial G}{\partial A} \right)_p dA \quad \text{and} \quad dG = V\, dp + \gamma\, dA$$

Since the order of differentiation is immaterial for exact differentials, it follows that

$$\frac{\partial}{\partial A} \left[\left(\frac{\partial G}{\partial p} \right)_A \right]_p = \frac{\partial}{\partial p} \left[\left(\frac{\partial G}{\partial A} \right)_p \right]_A \quad \text{or} \quad \left(\frac{\partial V}{\partial A} \right)_p = \left(\frac{\partial \gamma}{\partial p} \right)_A$$

So dV/dA can be written $(dV/dR)(dR/dA)$, which is easily evaluated for a sphere as $(4\pi R^2)(8\pi R)^{-1} = R/2$. Therefore $(\partial \gamma / \partial p)_{A,T} = R/2$ or $d\gamma = \frac{1}{2} R\, dp$. This result may be integrated to give $\gamma = \frac{1}{2} R\, \Delta p$, where the constant of integration has been set equal to zero, since there would be no pressure difference if $\gamma = 0$. This is the Laplace equation for spherical particles.

•

As noted above, it is possible that a different pair of radii of curvature apply at different locations on a surface. In this case the Laplace equation shows that Δp also varies with location. As is often true of pressures, it is convenient to define pressure variations relative to some reference plane. With this idea in mind, the horizontal surface in Fig. 6.2b can be taken as a reference level at which $\Delta p = 0$. Just under the meniscus in the capillary the pressure is less than it would be on the other side of the surface owing to the curvature of the surface. The fact that the pressure is less in the liquid in the capillary just under the curved surface than it is at the reference plane causes the liquid to rise in the capillary until the liquid column generates a compensating hydrostatic pressure. The capillary possesses an axis of symmetry; therefore at the bottom of the meniscus the radius of curvature is the same in the two perpendicular planes that include the axis. If we identify this radius of curvature by b, then the Laplace equation applied to the meniscus is $\Delta p 2\gamma/b$. Equating this to the hydrostatic pressure gives

$$\frac{2\gamma}{b} = \Delta\rho g h \tag{33}$$

or

$$bh = \frac{2\gamma}{\Delta\rho g} = a^2 \tag{34}$$

Comparison of this result with Eq. (5) shows the two relations to be of similar form, with the approximate form becoming exact under specific conditions. The condition under which the approximate result is exact is when the meniscus is a perfectly hemispherical depression at the liquid surface. When this is the case, $\theta = 0$ and $R_{cap} = R_{men} = b$. Since the meniscus usually only approximates a hemisphere, the use of the capillary radius in Eq. (5) remains an approximate relationship. Although Eq. (34) is exact, it is not particularly useful, since b is not readily measurable. Note that if b were available, Eq. (34) would permit the evaluation of γ without requiring that θ be known.

A variety of drop, bubble, and meniscus shapes have axial symmetry. As is the case in the capillary in Fig. 6.2b, p varies with z and, in general, the two radii of curvature may vary from position to position on the surface also. With these ideas in mind, the Laplace equation becomes

$$\Delta p(z) = \gamma[R_1^{-1}(x, y, z) + R_2^{-1}(x, y, z)] \tag{35}$$

In Eq. (35) we have added the notion that Δp, R_1^{-1}, and R_2^{-1} may be functions of location in space for any given surface.

The following expressions from analytical geometry are general functions for R_1^{-1} and R_2^{-1} for surfaces with an axis of symmetry:

$$R_1^{-1} = \frac{d^2z/dx^2}{[1 + (dz/dx)^2]^{3/2}} \tag{36}$$

and

$$R_2^{-1} = \frac{dz/dx}{x[1 + (dz/dx)^2]^{1/2}} \tag{37}$$

Substitution of Eqs. (36) and (37) into Eq. (35) generates a complicated differential equation whose solution relates the shape of an axially symmetrical interface to γ. In principle, then, Eq. (35) permits us to understand the shapes assumed by mobile interfaces and suggests that γ might be measurable through a study of these shapes. We shall not pursue this any further at this point, but shall return to the question of the shape of deformable surfaces in Sect. 6.10. In the next section we examine another consequence of the fact that curved surfaces experience an extra pressure because of the tension in the surface. We know from experience that many thermodynamic phenomena are pressure sensitive. Next we examine the effect of the increment in pressure small particles experience due to surface curvature on their thermodynamic properties.

6.5 The Kelvin Equation

In addition to capillarity, another important consequence of the pressure associated with surface curvature is the effect it has on the thermodynamic activity of substances. This is most readily understood for liquids, where the activity is measured by the vapor pressure of the liquid. Accordingly, suppose we consider the process of transferring molecules of a liquid from a bulk phase with a vast horizontal surface to a small spherical drop of radius r.

According to Eq. (32), no pressure difference exists across a plane surface: The pressure is simply p_0, the normal vapor pressure. However, a pressure difference given by Eq. (30) exists across a spherical surface. Therefore for liquid–vapor equilibrium at a spherical surface both the liquid and the vapor must be brought to the same pressure $p_0 + \Delta p$. If we assume the liquid to be incompressible and the vapor to be ideal, ΔG for the process of increasing the pressure from p_0 to $p_0 + \Delta p$ is given by the following:

1. For the liquid

$$\Delta G = \int_{p_0}^{p_0 + \Delta p} V_L \, dp = V_L \, \Delta p = \frac{2V_L \gamma}{r} \tag{38}$$

where V_L is the molar volume of the liquid.
2. For the vapor

$$\Delta G = RT \ln \left(\frac{p_0 + \Delta p}{p_0} \right) = RT \ln \left(\frac{p}{p_0} \right) \tag{39}$$

When liquid and vapor are at equilibrium, these two values of ΔG are equal:

$$RT \ln \left(\frac{p}{p_0} \right) = \frac{2V_L\gamma}{r} = \frac{2M\gamma}{\rho r} \tag{40}$$

since the volume per mole equals M/ρ, where M and ρ are the molecular weight and density of the liquid, respectively. In either of these forms, this expression is known as the Kelvin equation. The Kelvin equation enables us to evaluate the *actual* pressure above a spherical surface and not just the pressure difference across the interface, as was the case with the Laplace equation. In our derivation of the Kelvin equation, the radius is measured in the liquid. For a gas bubble in a liquid, the same equilibrium is involved, but the bubble radius is measured on the opposite side of the surface. As a consequence, a minus sign enters Eq. (37) when it is applied to bubbles. Since γ is on the order of millijoules while R is on the order of joules, Eq. (37) predicts that $\ln(p/p_0)$ is very small. It is important to realize, however, that γ/R is also divided by the radius of the spherical particle and therefore becomes more important as r decrease. For water at 20°C the Kelvin equation predicts values of p/p_0 equal to 1.0011, 1.0184, 1.1139, and 2.94 for drops of radius 10^{-6}, 10^{-7}, 10^{-8}, and 10^{-9} m, respectively. For bubbles of the same sizes in liquid water p/p_0 equals 0.9989, 0.9893, 0.8976, and 0.339. These calculations show that the effect of surface curvature, while relatively unimportant even for particles in the micrometer range, becomes appreciable for very small particles.

The Kelvin effect is not limited to spheres. For example, the "neck" of liquid between two supports is described by a paraboloid of revolution. In this case the radius of curvature of the concave surface is outside the liquid and therefore introduces a minus sign into Eq. (40). In addition, the factor of 2 is not required to describe the Kelvin effect for this geometry. The following example illustrates a test of the Kelvin equation based on this kind of liquid neck.

Example 6.2   Figure 6.4 shows a plot of experimental data which demonstrates the validity of the Kelvin effect. Necks of liquid cyclohexane were formed between mica surfaces at 20°C, and the radius of curvature was measured by interferometry. Vapor pressures were measured for surfaces with different curvature. Use these data to evaluate γ for cyclohexane. Comment on the significance of the fact that the linearity of Fig. 6.4 extends all the way to a p/p_0 value of 0.77.

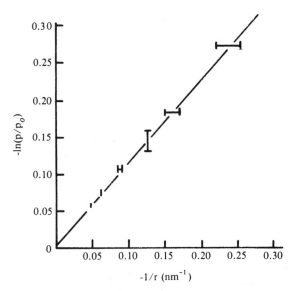

Figure 6.4 Verification of the Kelvin equation for cyclohexane "necks" of different curvature between mica surfaces (discussed in Example 6.2). [Reprinted from L. R. Fisher and J. N. Israelachvili, *Nature, 277*:548 (1979).]

Solution For the paraboloid of revolution, the Kelvin equation becomes $\ln(p/p_0) = -M\gamma/\rho RTr$. Therefore a plot of $\ln(p/p_0)$ versus $-1/r$ is expected to be a straight line of slope $M\gamma/\rho RT$. The slope of the line in Fig. 6.4 is 1.13×10^{-9} m. Equating this with $M\gamma/\rho RT$ and solving for γ, we obtain $\gamma = (1.13 \times 10^{-9})(779)(8.314)(293)/0.840 = 0.0255$ J m^{-2} = 25.5 mJ m^{-2}, which agrees with the value determined by conventional methods.

In addition to proving the Kelvin effect, the data in Fig. 6.4 show that the surface tension of cyclohexane may be regarded as constant down to a radius of curvature given by the reciprocal of the abscissa corresponding to $p/p_0 = 0.77$, which is 4 nm. It is remarkable that the same value of γ as determined on macroscopic surfaces continues to apply down to within an order of magnitude of molecular dimensions.

•

The Kelvin equation may also be applied to the equilibrium solubility of a solid in a liquid. In this case the ratio p/p_0 in Eq. (40) is replaced by the ratio a/a_0, where a_0 is the activity of dissolved solute in equilibrium with a flat surface and a is the analogous quantity for a spherical surface.

For an ionic compound having the general formula M_mX_n, the activity of a dilute solution is related to the molar solubility S as follows:

$$a = (mS)^m(nS)^n \qquad (41)$$

Therefore for a solid sphere

$$\frac{2M\gamma}{\rho r} = RT \ln\left(\frac{a}{a_0}\right) = (m + n)RT \ln\left(\frac{S}{S_0}\right) \qquad (42)$$

where S and S_0 are the solubilities of the spherical and flat particles, respectively. In principle, Eq. (42) provides a thermodynamically valid way to determine γ for an interface involving a solid. The thermodynamic approach makes it clear that curvature has an effect on activity for any curved surface. The surface free energy interpretation of γ is more plausible for solids than the surface tension interpretation, which is so useful for liquid surfaces. Either interpretation is valid in both cases, and there are situations in which both are useful. From solubility studies on particle of known size, γ_S can be determined by the method of Example 6.2.

Although the increase in solubility of small particles is unquestionably a real effect, using it quantitatively as a means of evaluating γ_{SL} is fraught with difficulties:

1. The difference in solubility between a small particle and a larger one will probably be less than 10%. Since a phase boundary exists at all, the solubility is probably low to begin with, so there may be some difficulty in determining the experimental solubilities accurately.
2. Solid particles are not likely to be uniform spheres, even if the sample is carefully fractionated; rather, they will be irregularly shaped and polydisperse, although the particle size distribution may be narrow. The smallest particles will have the largest effect on the solubility, but they may be the hardest to measure.
3. The radius of curvature of sharp points or protuberances on the particles has a larger effect on the solubility of irregular particles than the equivalent radius of the particles themselves.

The Kelvin equation helps explain an assortment of supersaturation phenomena. All of these—supercooled vapors, supersaturated solutions, supercooled melts—involve the onset of phase separation. In each case the difficulty is the nucleation of the new phase: Chemists are familiar with the use of seed crystals and the effectiveness of foreign nuclei to initiate the formation of the second phase.

The Kelvin equation shows that the ratio S/S_0 or p/p_0 increases rapidly as r decreases toward zero. Applying Eq. (40) rigorously (and incorrectly) down to $r = 0$ would imply infinite supersaturation and make the appearance of a new phase impossible. Equation (40) is derived on the basis of two phases already in existence. To arrive at an understanding of the emergence of a new phase, we must consider what is going on at the molecular level at the threshold of phase separation. A highly purified vapor, for example, may remain entirely as a gas even though its pressure exceeds the normal vapor pressure of the liquid for the temperature in question. At such a point the vapor state is thermodynamically unstable with respect to the formation of a liquid phase with flat surfaces. Whatever stability the vapor has is kinetic stability, arising from the high activation energy required to start the formation of the second phase. A crude kinetic picture of the processes occurring at the molecular level may be informative.

Although a supersaturated vapor is still a gas, it is a very nonideal gas indeed! Clusters of molecules are continually forming and disintegrating: embryonic nuclei of the new phase. Some of these clusters will be dimers, some trimers, and in general n-mers. Each will have its own characteristic radius r_n. An abbreviated derivation of the rate of formation of n-mers proceeds as follows. The rate law will contain both a frequency factor and a Boltzmann factor. The energy term in the latter may be estimated by Eq. (38) with V_L taken to be $\frac{4}{3}\pi r_n^3$. The frequency factor will involve the probability of additional molecules adding by collision; that is, it will depend on the surface area of the cluster $(4\pi r_n^2)$ and the frequency of collisions with a wall as given by kinetic molecular theory. Therefore the rate law may be approximated as

$$\text{rate} \simeq Z(4\pi r_n^2) \exp\left(-\frac{2\gamma}{r_n}\frac{4\pi r_n^3}{3kT}\right) = c_1 p r_n^2 \exp(-c_2 r_n^2) \qquad (43)$$

where Z is the collision frequency. Since the latter certainly increases with p, the pressure has been factored out of the second expression in Eq. (43), where c_1 and c_2 are constants.

Two aspects of Eq. (43) are especially informative. First, we observe that the pre-exponential term increases with increasing r_n, whereas the exponential term decreases. This means that there exists some critical radius for which the rate law shows a maximum. Clusters with this critical radius may be compared with reaction intermediates or transition states in ordinary chemical reactions. Those clusters which manage to overcome the energy barrier associated with this critical size are capable of further growth leading to the appearance of the new phase, and smaller clusters disintegrate. In addition, Eq. (43) also shows that the rate of cluster growth increases as the pressure increases. All clusters, including those of the critical size, form more rapidly at higher pressures.

The preceding considerations suggest that a point is ultimately reached in the course of increasing supersaturation where the liquefaction process becomes kinetically as well as thermodynamically favorable. It must be remembered here that the initial state of the system is one of instability, or, more correctly, metastability. Once liquefaction begins, the pressure drops until the radius–pressure combination that satisfies Eq. (40) is reached.

A great deal of work has been done on the kinetics of phase formation; the arguments presented here are intended merely to suggest the direction taken in more detailed treatments. Many aspects of nucleation are of extreme interest in colloid and surface chemistry. The monodisperse colloids described in Chap. 5 are formed by carefully controlling the formation of the solid phase. In the case of the monodisperse sulfur sols, for example, the decomposition of $S_2O_3^{2-}$ (Sect. 5.14) proceeds slowly to quite a high level of supersaturation (the solution must be free of foreign nuclei). Ultimately, clusters of sulfur atoms exceeding the critical size form, nucleate precipitation, and grow until the supersaturation is relaxed. Any additional sulfur that is formed beyond this point will deposit on these particles without forming a new "crop" of nuclei. Relatively monodisperse colloids formed by such condensation processes depend on the fact that supersaturation and therefore nucleation occur only once during the formation of the colloid. If the rate of crystallization were too slow, then a second stage of supersaturation and nucleation might be reached, and a polydisperse colloid would result.

Of course, nucleation may also be accomplished by seeding or adding externally formed nuclei. The monodisperse gold sols described in Sect. 5.13 are prepared by this method. The use of AgI crystals and other materials as nuclei in cloud seeding has also been studied extensively. This is especially interesting in view of possible applications to weather modification.

6.6 The Young Equation

In the preceding sections a variety of important related concepts have been introduced. Among these are the equivalence of surface tension and surface free energy, the applicability of these concepts to solids as well as liquids, and the notion of the contact angle. All these concepts are brought together in discussing the physical situation sketched in Fig. 6.5. Suppose a drop of liquid is placed on a perfectly smooth solid surface

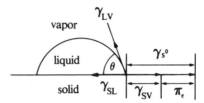

Figure 6.5 Components of interfacial tension needed to derive Young's equation.

and these phases are allowed to come to equilibrium with the surrounding vapor phase. Viewing the surface tensions as forces acting along the perimeter of the drop enables us to write immediately an equation which describes the equilibrium force balance:

$$\gamma_{LV} \cos \theta = \gamma_{SV} - \gamma_{SL} \qquad (44)$$

This result was qualitatively proposed by Thomas Young in 1805 and is generally known as Young's equation.

Young's equation is a plausible, widely used result, but its apparent simplicity is highly deceptive. The two terms which involve the interface between the solid and other phases cannot be measured independently, so experimental verification is difficult, although a variety of experiments have been directed along these lines. There are also a number of objections to Young's equation. These objections may be classified into two categories: those based on the noncompliance of the experimental system to the postulates of the derivation and those critical of the assumption of thermodynamic equilibrium in the solid. We shall discuss these two classes of objections separately.

Real solid surfaces may be quite different from the idealized one in this derivation. Actual solid surfaces are apt to be rough and even chemically heterogeneous. This statement is true on a fine scale, even for carefully prepared surfaces. We shall discuss these complications in more detail in Sect. 6.8. In Chapter 10 we will examine metal surfaces specifically to see the extreme conditions which must exist for these surfaces to be uniform down to an atomic scale. The model surface of the derivation is thus the exception rather than the rule for solid surfaces. In principle, both roughness and heterogeneity can be incorporated into Young's equation in the form of empirical corrections. For example, if a surface is rough, a correction factor r is traditionally introduced as a weighting factor for $\cos \theta$, where $r > 1$. The logic underlying this correction goes as follows. The factor $\cos \theta$ enters Eq. (44) by projecting

γ_{LV} onto the solid surface. If the solid surface is rough, a larger area will be "overshadowed" by the projection than if the surface were smooth. The roughness factor measures this effect. With the empirical correction factor for roughness included, Young's equation becomes

$$r\gamma_{LV} \cos \theta = \gamma_{SV} - \gamma_{SL} \qquad \qquad r = \text{correction factor.} \qquad (45)$$

A surface may also be chemically heterogeneous. Assuming, for simplicity, that the surface is divided into fractions f_1 and f_2 of chemical types 1 and 2, we may write

$$\gamma_{LV} \cos \theta = f_1(\gamma_{S_1V} - \gamma_{S_1L}) + f_2(\gamma_{S_2V} - \gamma_{S_2L}) \qquad (46)$$

where $f_1 + f_2 = 1$.

Both roughness and heterogeneity may be present in real surfaces. In such a case, the correction factors defined by Eqs. (45) and (46) are both present. Although such modifications adapt Young's equation to non-ideal surfaces, they introduce additional terms which are difficult to evaluate independently. Therefore the validity of Eq. (44) continues to be questioned.

More fundamental objections to Young's equation center on the issue of whether the surface is in a true state of thermodynamic equilibrium. In short, it may be argued that the liquid surface exerts a force perpendicular to the solid surface, $\gamma_{LV} \sin \theta$. On deformable solids a ridge is produced at the perimeter of a drop; on harder solids the stress is not sufficient to cause deformation of the surface. This is the heart of the objection. Is it correct to assume that a surface under this stress is thermodynamically the same as the idealized surface which is free from stress? Clearly, the troublesome stress component is absent only when $\theta = 0$, in which case the liquid spreads freely over the surface and Fig. 6.5 becomes meaningless.

In answer to this the following argument has been suggested, based on the fact that it is the difference $\gamma_{SV} - \gamma_{SL}$ which appears in Young's equation. Since the same solid is common to both terms, it is really only the local difference at the surface between an adjacent phase which is liquid and one which is vapor that is being measured. According to this point of view, the nonequilibrium state of the solid is immaterial: The same solid is involved at both interfaces.

Another approach to resolving this theoretical objection to Young's equation is to eliminate the difference $\gamma_{SV} - \gamma_{SL}$ from the equation entirely, replacing it by some equivalent quantity, thereby shifting attention away from the notion of solid surface tension. An example of this is the use of the heat of immersion to test Young's equation. We shall return to this presently.

In summary, then, Young's equation is still controversial despite the fact that it has been in existence since the beginning of the nineteenth century. The reader will appreciate that any relationship that has been around so long and has eluded definitive empirical verification has been the center of much research. Accordingly, the relationship is very widely encountered in the literature of surface chemistry.

To explore Young's equation still further, suppose we distinguish between γ_{SV} and γ_{S°, where the former describes the surface of a solid in equilibrium with the vapor of a liquid, and the latter a solid in equilibrium with its own vapor. Since Young's equation describes the three-phase equilibrium, it is proper to use γ_{SV} in Eq. (44). The question arises, however, what difference, if any, exists between these two γ's. In order to account for the difference between the two, we must introduce the notion of adsorption. In the present context adsorption describes the attachment of molecules from the vapor phase onto the solid surface. All of Chapter 9 is devoted to this topic, so it is inappropriate to go into much detail at this point. The extent of this attachment depends on the nature of the molecules in the vapor phase, the nature of the solid, and the temperature and the pressure.

For now we may anticipate a result from Chapter 7 to note that adsorption always leads to a decrease in γ. In the present context, therefore, we write

$$\gamma_{S^\circ} \geqslant \gamma_{SV} \tag{47}$$

We shall use the symbol π_e to signify the difference

$$\gamma_{S^\circ} - \gamma_{SV} = \pi_e \tag{48}$$

and call this quantity the equilibrium film pressure. The word *equilibrium* in this designation refers explicitly to the fact that the adsorbed molecules are in equilibrium with a drop of bulk liquid. The molecules adsorbed at an interface may be regarded as repelling one another or as rebounding off one another, thereby relieving some of the tension in the surface. This interpretation makes it sensible to call the reduction of γ due to adsorption, π_e, a "pressure." Note that π_e is a *two-dimensional* pressure, measuring the force exerted per unit length of perimeter (Newtons per meter) by the adsorbed molecules. We shall have a good deal more to say about this quantity in the following chapter. With these ideas in mind, Eqs. (44) and (48) may be combined to give

$$\gamma_{LV} \cos \theta = \gamma_{S^\circ} - \pi_e - \gamma_{SL} \tag{49}$$

Figure 6.5 shows the relationship between $\gamma_S°$, γ_{SV}, and π_e. It is apparent that the value of θ might be quite different between equilibrium and nonequilibrium situations, depending on the value of π_e.

There are several concepts which will assist us in anticipating the range of π_e values:

1. Spontaneously occurring processes are characterized by negative values of ΔG.
2. Surface tension is the surface excess free energy; therefore the lowering of γ with adsorption is consistent with the fact that adsorption occurs spontaneously.
3. Surfaces which initially possess the higher free energies have the most to gain in terms of decreasing the free energy of their surface by adsorption.
4. High-energy surfaces bind enough adsorbed molecules to make π_e significant. On the other hand, π_e is negligible for for a solid which possesses a low-energy surface.
5. A surface energy in the neighborhood of 100 mJ m^{-2} is generally considered the cutoff value between high- and low-energy surfaces.
6. Silica, glass, metals, metal oxides, metal sulfides, and inorganic salts are examples of high-energy surfaces. Most solid organic compounds, including polymers, have low-energy surfaces.
7. Hard and soft solids are generally classified as having high- and low-energy surfaces, respectively. This criterion must be used cautiously, since the mechanical properties of a solid depend on the concentration of defects and dislocations in the bulk.

Because of the simplification which results from $\pi_e = 0$ for low-energy surfaces, the latter are often chosen as model systems in fundamental research. Even when neglecting π_e is of questionabl validity, γ_{SV} and $\gamma_S°$ are often used interchangeably, for lack of suitable data. We shall see in Chapter 9—Eq. (9.7) for example—how π_e may be determined from experimental adsorption data. Otherwise, we shall generally assume $\pi_e = 0$.

Small quantities of heat are generally evolved when a dry solid is immersed in a liquid. This can be measured calorimetrically and is called the heat of immersion. The physical process with which this heat is associated may be represented by the following equation:

$$\text{dry solid} \xrightarrow{\text{liquid}} \text{wet solid} \tag{A}$$

Following the usual thermochemical convention, the heat of immersion ΔH_{im} may be written $H_{wet} - H_{dry}$, where these enthalpies are expressed per unit area. Since heat is released by this process, ΔH_{im} is negative. The "wet" surface clearly describes the SL interface. For the "dry" surface, we ignore π_e and describe the surface by the S° notation. Therefore the heat of immersion may also be written

$$\Delta H_{im} = H^s_{SL} - H^s_{S^\circ} \tag{50}$$

Next we recall Eq. (20), which gives us an expression for H^s:

$$H^s = \gamma - T \frac{\partial \gamma}{\partial T} \tag{51}$$

Applying Eq. (51) to the right-hand side of Eq. (44)—with $\gamma_{SV} = \gamma_{S^\circ}$—gives

$$\Delta H_{im} = \left(\gamma_{SL} - T \frac{d\gamma_{SL}}{dT} \right) - \left(\gamma_{S^\circ} - T \frac{d\gamma_{S^\circ}}{dT} \right) \tag{52}$$

Using $\gamma_{LV} \cos \theta$ as a replacement for $\gamma_{S^\circ} - \gamma_{SL}$, we obtain

$$-\Delta H_{im} = \gamma_{LV} \cos \theta - T \frac{d}{dT} (\gamma_{LV} \cos \theta) \tag{53}$$

Carrying out the indicated differentiation yields

$$-\Delta H_{im} = \gamma_{LV} \cos \theta - T \cos \theta \frac{d\gamma_{LV}}{dT} - T\gamma_{LV} \frac{d \cos \theta}{dT} \tag{54}$$

which shows how Young's equation may be tested by comparing experimental heats of immersion with calculated values. The following example gives an idea of the magnitude of some of these quantities.

Example 6.3   Estimate the heat of immersion for the system for which γ and θ are 22 mJ m^{-2} and 30°, respectively, at 20°C. The temperature variations of γ and $\cos \theta$ are -0.10 mJ m^{-2} K^{-1} and 0.0010 K^{-1}, respectively. These values are close to those observed for liquid alkanes on Teflon. Comment on the implications of this estimate for the ease or difficulty in measuring ΔH_{im}.

Solution   We can estimate ΔH_{im} by substitution into Eq. (54):

$$\Delta H_{im} = -\gamma_{LV} \cos \theta + T \cos \theta \frac{d\gamma_{LV}}{dT} + T\gamma_{LV} \frac{d \cos \theta}{dT} =$$

$$-(22)\cos 30 + 293 \cos 30 (-0.10) + 293(22)(0.0010) = -38 \text{ mJ m}^{-2}$$

To get an idea of the problems associated with this kind of experiment, we estimate the temperature change in the liquid as a result of absorbing this heat. Using 2.4 J g^{-1} K^{-1} for the heat capacity (the value for n-octane) and taking $T = 1.5$ K as an arbitrary but convenient temperature change, we calculate

$$\frac{2.4 \text{ J K}^{-1}}{g} \times 1.6 \text{ K} \times \frac{m^2}{0.038 \text{ J}} \simeq 100 \text{ m}^2 \text{ g}^{-1}$$

That is, for each gram of liquid about 100 m^2 of surface must be wet. This overestimates the temperature change, since it ignores the fact that the solid will absorb some heat. If the heat capacity of the solid is the same as the liquid, 100 m^2 of surface must be wet for each gram of solid–liquid mixture. Using a smaller temperature change would decrease the required area proportionately, but the fact remains that large areas are required to obtain measurable effects. Practically, this means work with powdered solids of small particle size.

•

Although heats of immersion are small, this quantity is measurable. For systems where both the heat of immersion and the necessary information about γ and θ have been measurable, the prediction of Eq. (54) has been verified. Figure 6.6 shows some experimental results for n-alkanes wetting Teflon (polytetrafluoroethylene) surfaces. The open circles were determined calorimetrically; the closed ones were calculated from Eq. (54). Even though the two sets of values diverge for alkanes larger than n-

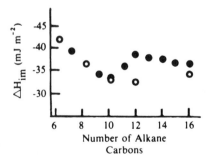

Figure 6.6 Heats of immersion determined by calorimetry and calculated by Eq. (54) for Teflon in various alkanes. [From A. W. Neumann, *Adv. Colloid Interface Sci.,* 4:105 (1974), used with permission.]

decane, the overall picture is quite acceptable. Incidentally, the value calculated in the example is close to the actual values, even though the numbers used in Example 6.3 were rounded off.

Several aspects of Fig. 6.6 and Example 6.3 deserve further comment:

1. Calorimetry requires large areas of interface, which virtually demands powdered solids. We have not considered (yet) the problem of measuring θ and $d\theta/dT$ for powders. Solids which are available as large specimens with smooth surfaces (suitable, say, for the tilted plate determination of θ) can be pulverized to increase their surface. The reverse process is often not possible for powders, so a method for measuring powders is important.
2. Results like those shown in Fig. 6.6 may be considered an experimental verification of Young's equation. In light of the first item, however, there may still be objections that the surfaces used for calorimetric studies and those used to study γ and θ are not identical even though they are nominally the same.
3. Accepting Eq. (54) and Young's equation, upon which it is based, suggests calorimetry as a method for measuring contact angles. At this time this is not practical, but the implication that contact angle is a thermodynamic property is a very important realization.

Although we established the thermodynamic significance of γ early in the chapter, θ has been allowed to drift along. Its role is clear when we think of surface tension as a force: We use θ to project γ in a specified direction. In thermodynamic terms, contact angle has been an outsider in our presentation. Young's equation is the remedy to this. Rewriting Eq. (44), we observe

$$\cos \theta = \frac{\gamma_{SV} - \gamma_{SL}}{\gamma_{LV}} \tag{55}$$

and $\cos \theta$ is fully described by various free energy terms. In view of this new-found (for us) significance, we shall return presently to some additional experimental methods for the determination of θ. First, however, we consider the notions of adhesion, cohesion, and spreading, which will enhance our ideas of γ and θ as thermodynamic quantities.

6.7 Adhesion, Cohesion, and Spreading

In this section we consider some hypothetical processes which provide us with additional ways of thinking about γ and θ. They are related to process (A) but are not studied calorimetrically as is the case with immersion. In defining adhesion, cohesion, and spreading, we designate the phases A and B without specifying their physical state. Their surface with each other is designated AB; their individual surfaces with their own vapor or air (we make no distinction here) are designated by either A or B. With this notation in mind, we consider the following processes as they affect a unit area:

1. Cohesion:

$$\text{no surface} \rightarrow 2\ A\ (\text{or } B)\ \text{surfaces} \tag{B}$$

2. Adhesion:

$$1\ AB\ \text{surface} \rightarrow 1\ A + 1\ B\ \text{surface} \tag{C}$$

3. Spreading (B on A):

$$1\ A\ \text{surface} \rightarrow 1\ AB + 1\ B\ \text{surface} \tag{D}$$

Schematic illustrations of these processes are shown in Fig. 6.7. Two things must be remembered about these sketches: One unit of surface is affected by the processes and the shape of the affected area is immaterial. It is understood that these are elements of volume and area which are portions of macroscopic samples. Our interest is in the free energy change accompanying each process.

In Fig. 6.7a—which applies to a pure liquid—the process consists of producing two new interfaces, each of unit cross section. Therefore for the separation process

$$\Delta G = 2\gamma_A = W_{AA} \tag{56}$$

The quantity W_{AA} is known as the work of cohesion, since it equals the work required to pull a column of liquid A apart. It measures the attraction between the molecules of the two portions. Recall the concept of *cohesive* energy density in terms of which we discussed interactions between pairs of identical molecules in Sect. 3.7. Interpreting γ as half the work of cohesion shows that surface tension measures the free energy change involved when molecules from the bulk of a sample are moved to the surface.

Now let us consider the value of ΔG for the separation of A and B as represented in (C) and Fig. 6.7b. Taking the difference between the final and the initial free energies for this process yields

$$\Delta G = W_{AB} = \gamma_{\text{final}} - \gamma_{\text{initial}} = \gamma_A + \gamma_B - \gamma_{AB} \tag{57}$$

This quantity is known as the work of adhesion and measures the attraction between the two different phases.

The work of adhesion between a solid and a liquid phase may be defined by analogy with Eq. (57):

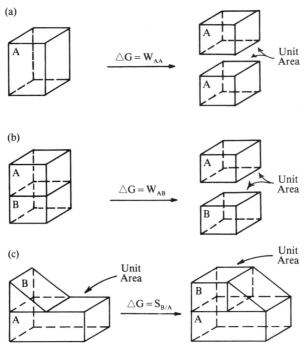

Figure 6.7 Schematic illustrations of the processes for which ΔG equals (a) the work of cohesion, (b) the work of adhesion, and (c) the work of spreading.

$$W_{SL} = \gamma_{S^\circ} + \gamma_{LV} - \gamma_{SL} \tag{58}$$

By means of Eq. (48), γ_{S° may be eliminated from this expression to give

$$W_{SL} = \gamma_{SV} + \pi_e + \gamma_{LV} - \gamma_{SL} \tag{59}$$

Finally, Young's equation may be used to eliminate the difference $\gamma_{SV} - \gamma_{SL}$:

$$W_{SL} = \gamma_{LV}(1 + \cos \theta) + \pi_e \tag{60}$$

Neglecting π_e as we did in the last section, Eq. (59) shows how γ and θ combined measure the work of adhesion between a solid and a liquid.

It is informative to apply Eq. (60) to low-energy surfaces for two extreme values of θ, $0°$ and $180°$, for which $\cos \theta$ is 1 and -1, respectively. For $\theta = 0°$, $W_{SL} = 2\gamma_{LV} = W_{AA}$; the work of solid–liquid adhesion is identical to the work of cohesion for the liquid. In this case interactions between solid and solid, liquid and liquid, and solid and liquid molecules are all equivalent. At the other extreme, where $\theta = 180°$, $W_{SL} = 0$. In this case the liquid is tangent to the solid; there is no interaction between the phases.

Lastly, if we taken the difference between the final and initial states for the process of spreading B over A (symbolized B/A) we obtain

$$\Delta G = \gamma_{AB} + \gamma_B - \gamma_A \tag{61}$$

As usual with free energies, a negative value for $\Delta G_{B/A}$ means that the process represented by (D) and shown in Fig. 6.7c occurs spontaneously. The negative of $\Delta G_{B/A}$ is called the spreading coefficient $S_{B/A}$; because of the sign change, a positive spreading coefficient means B spreads freely over A and wets it. The concept of wetting is very important in numerous applications: With lubricants and adhesives it is desirable, and in waterproofing undesirable. Since additives are frequently mixed into liquids to affect this property, it is appropriate to postpone any discussion of applications of the spreading coefficient until the next chapter.

By combining Eqs. (56), (57), and (61), we note that the spreading coefficient can also be written

$$S_{B/A} = W_{AB} - W_{BB} \tag{62}$$

If $W_{AB} > W_{BB}$, the A–B interaction is sufficiently strong to promote the wetting of A by B. This is the significance of a positive spreading coefficient. Conversely, no wetting occurs if $W_{BB} > W_{AB}$, since the work required to overcome the attraction between two B molecules is not compensated by the attraction between A and B. Thus a negative spreading coefficient means that B will not spread over A.

6.8   Contact Angles: Some Complications

In Sect. 6.2 we saw how the tilted plate method could be used to measure θ; we also noted that it could be determined by the Wilhelmy method if γ were measured independently. For that matter, the three-phase junction might be examined and θ determined by direct observation. These and many other methods—some of which we shall discuss later—have been used to measure θ. Even if the most careful experimental techniques are employed on carefully prepared surfaces, contact angle data are frequently confusing. The situation is best introduced by referring to Fig. 6.8, which shows a drop on a tilted plane. "Teardrop" shapes such as this are familiar to everyone: Just look at a raindrop on a window pane. The problem, of course, is that the contact angle is different at different points

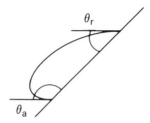

Figure 6.8 A drop on a tilted plane, showing advancing and receding contact angles. (From Ref. 6, used with permission.)

of contact with the support. It is conventional to call the larger value the advancing angle θ_a, and the smaller one the receding angle θ_r. The two may be quite different. The presence of contamination is definitely a contributing factor, but it is by no means the only one. Therefore, even with carefully purified materials, both advancing and receding contact angles should be measured.

All the techniques described here are easily conducted, so that both θ_a and θ_r may be observed. When the tilted plate method is used to evaluate the contact angle, θ_r values are obtained if the plate has been pulled out (emersion) from the liquid; θ_a results if the plate is pushed into the liquid (immersion). Likewise, both values of θ may be obtained from the Wilhelmy method, depending on whether the liquid is making an initial contact (θ_a) with the plate or is draining from it (θ_r).

A rather interesting example of the different between advancing and receding contact angles is obtained from the Wilhelmy plate method when the contact angle has a nonzero value. Suppose the Wilhelmy plate is allowed to dip beneath the horizontal liquid surface, as shown in Fig. 6.9a. In this case the weight of the meniscus as given by Eq. (2) will be decreased by a term w', the buoyant force on the submerged plate. The latter will clearly be proportional to the depth of immersion d. Therefore we write

$$w = \gamma P \cos \theta - w' = \gamma P \cos \theta - kd \qquad (63)$$

where k is a suitable proportionality constant. This equation shows that the apparent weight w of the meniscus should give a straight line when plotted against the depth of immersion and that the intercept should be proportional to $\cos \theta$. If a single value of θ applies to both the immersion and emersion steps, then the line shown in Fig. 6.9b would result. Because of the difference between θ_a and θ_r, a curve like that shown in Fig. 6.9c is obtained instead. When the immersion–emersion cycle is

(a)   (b)   (c)

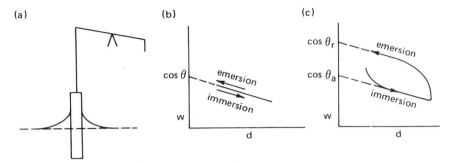

Figure 6.9 (a) Weighing a meniscus in a Wilhelmy plate experiment versus the depth of immersion of the plate. In (b) both the advancing and receding contact angles are equal; in (c) $\theta_a > \theta_r$.

repeated, a hysteresis loop is obtained which may be as reproducible as the analogous loops observed in magnetization–demagnetization cycles. The difference $\theta_a - \theta_r$ is called the hysteresis of a contact angle.

The general requirement for hysteresis is the existence of a large number of metastable states that differ slightly in energy and which are separated from each other by small energy barriers. The situation is shown schematically in Fig. 6.10. The equilibrium contact angle corresponds to the free energy minimum. However, systems may be "frozen" in metastable states of somewhat higher energy by lacking sufficient energy to overcome the energy barrier separating them from equilibrium. An interesting experimental observation is that advancing and receding values of θ converge to a common value when the surface is vibrated. Presumably, the mechanical energy imparted to the liquid by the vibration assists it in passing over the energy barrier and reaching

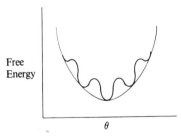

Figure 6.10 Schematic energy diagram for metastable states corresponding to different contact angles.

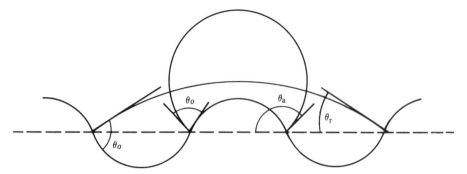

Figure 6.11 Cross section of a drop resting on a surface containing a set of concentric grooves. For both profiles the contact angles are microscopically identical although macroscopically different. (From Ref. 6, used with permission.)

equilibrium. In this sense maximum vibration rather than vibration-free conditions may appear to be the ideal conditions for measuring θ, but "vibration" is a difficult parameter to control reproducibly, so this concept is of little practical help.

Now let us briefly consider the origin of these metastable states. If we exclude the effect of impurities, the metastable states are generally attributed to either the roughness of the solid surface, its chemical heterogeneity, or both. Of course, a well-prepared laboratory sample will be fabricated and cleaned as effectively as possible to eliminate gross roughness and chemical heterogeneity. What we are talking about are microscopic irregularities that cannot be eliminated. In size and distribution, these will follow a random pattern on actual surfaces.

An informative model for contact angle hysteresis is obtained by postulating the surface to contain a set of concentric grooves upon which a drop rests. Figure 6.11 represents the profile of two different drops on such a surface. In both of the profiles shown the angles of contact between the liquid–vapor interface and the solid are identical, θ_0. With respect to the horizontal, however, two very different apparent contact angles are observed. The two extremes are identified as θ_a and θ_r in this model.

The two drop configurations in Fig. 6.11 differ in surface area and in the elevation of their centers of gravity. Thus they possess different energies. The change from one configuration to the other involves the distortion of the shape of the drop, which accounts for the energy barrier between the two configurations. Thus the model qualitatively accounts for the kind of metastable states shown in Fig. 6.10 which are required for hysteresis. According to this model, contact angle hysteresis arises when a three-phase boundary gets trapped in transit, lacking sufficient energy to surmount the energy barrier to a lower energy state. The teardrop profile of Fig. 6.8 corresponds to the situation in which one edge of the drop has one configuration while the other edge has the second configuration.

This model can also be applied to hysteresis which arises from chemical heterogeneity; however, this time the surface is assumed to be smooth and to contain concentric rings of different chemical composition and hence different θ's. Actual heterogeneity may arise from impurities concentrated at the surface, from crystal imperfections, or from differences in the properties of different crystal faces. The distribution of such heterogeneities on an actual surface will obviously be more complex than the model considers, but the qualitative features of hysteresis are explained by the model nevertheless. Johnson and Dettree [6] present additional details of model experiments of this sort.

If a surface consists of a patchwork array of high- and low-energy sites, it is generally assumed that the larger, advancing contact angle measures the contact angle that would be obtained for a smooth, homogeneous sample of the low-energy component:

$$\theta_a \simeq \theta_{\text{low E}} \tag{64}$$

Conversely, the receding angle is taken as a measure of θ appropriate to the high-energy component:

$$\theta_r \simeq \theta_{\text{high E}} \tag{65}$$

Both of these conclusions are consistent with Young's equation. All other things being equal in Eq. (55), the larger $\gamma_{SV} \simeq \gamma_{S}$° is, the larger $\cos \theta$ will be. Since the cosine increases as θ decreases, large values of γ_{S}° result in smaller contact angles, and vice versa. We noted previously that high-energy surfaces have more to gain by adsorption than low-energy surfaces. This helps explain why receding contact angles are less reproducible than advancing angles; adsorption by impurities affects this measurement the most. Correlations with surface roughness reveal that receding angles are also more sensitive to roughness than advancing angles. For these reasons θ_a is more reliable than θ_r as a characteristic of the solid. Obviously, the lower the hysteresis, the more we can be sure that this quantity truly represents the solid.

In addition to roughness and heterogeneity, time-dependent effects also influence contact angle hysteresis. Dynamic studies have been conducted with liquid fronts moving across the solid substrate at variable speeds. Even the extrapolation to zero rate is unsatisfactory, since the limiting dynamic values differ from equilibrium values. This suggests that insufficient mobility at the molecular level may also contribute to hysteresis. Viscous effects operating in the interface may contribute other forces besides those shown in Fig. 6.5 which influence the value of the contact angle. This type of effect is harder to analyze than roughness and heterogeneity, but future research may help clarify this third cause of hysteresis.

With this background, let us now return to experiments which yield values for γ and θ.

6.9 Measuring Surface Tension and Contact Angle: Round Two

We have now established that both γ and θ have thermodynamic significance and have seen that their values as well as their temperature coefficients are of interest. In addition, we have seen that the masurement of contact angles presents some complications of its own. All this adds up to a need for more reliable and more accurate methods for the measurement of these parameters than those presented in Sect. 6.2. One of the most powerful strategies for this involves a second measurement made with the Wilhelmy plate.

We saw in Sect. 6.2 that the Wilhelmy plate offers an accurate method for measuring $\gamma \cos \theta$. We thus have one experiment with two unknowns. The Wilhelmy balance measures the weight of the meniscus; in this section we examine the height to which the meniscus climbs on the same surface. We shall see that this distance—which may be accurately measured with a traveling microscope or cathetometer—also depends on γ and θ. The functional relationship between these parameters and the experimental variables is different from the case of the meniscus weight. Therefore we have two experiments with two unknowns which can be solved for γ and θ.

Since both of these measurements can be made with the same interface, difficulties arising from the pairing of mismatched data are eliminated. Of course, the complications associated with surface roughness and chemical heterogeneity persist, and the method applies only to solids that can be prepared as plates. Despite these limitations, this technique is the best general method available for determining γ and θ. In the next section we will describe some additional ways of obtaining γ values, and in Sect. 6.11 we shall discuss a method for measuring θ on powdered solids.

In Sect. 6.4 we discussed the pressure difference that exists across a curved surface. The Laplace equation, in the form provided by Eq. (35), gives a general description of this pressure difference Δp. Our objective is to apply this relationship to the meniscus formed by a liquid surface at a flat solid wall. The first thing to notice about this is that one of the R's in Eq. (35) becomes infinite, since the support is planar; hence this R^{-1} term disappears from the Laplace equation.

Figure 6.12a shows a cross-sectional view of this meniscus and defines some pertinent quantities for the experiment. The meniscus is formed by liquid A displacing B. The weight of the displaced B exerts a buoyant force on the meniscus; therefore the density difference must be used. The pressure difference is zero at the flat surface well removed from the wall. We define this to be the $z = 0$ plane. The meniscus makes contact with the wall at $z = h$. Our interest is in some general point on the surface having the coordinates (x,z). Just beneath the surface at this point the pressure is less by $\Delta p(x,z)$ than in the reference plane. The liquid is elevated at this point by an amount sufficient to produce a compensating hydrostatic pressure. The latter is given by $(\rho_A - \rho_B)gz$, which may be equated to the pertinent form of Eq. (35):

$$\Delta \rho g z = \gamma/R_1 \tag{66}$$

where R_1 is the radius of curvature in the plane of Fig. 6.12a. Substitution of Eq. (36) for $1/R_1$ generates a second-order differential equation whose solution gives $z = f(x)$. The mathematics are simplified considerably by using some trigonometric relations based on the tangent and the normal to the curve constructed at the point under consideration. Figure 6.12b shows the local radius of curvature of a general curve. The angle ϕ is the angle made by the extension of the normal with the z axis; ϕ also gives the angle between the tangent and the x axis. Inspection of Fig. 6.12b reveals

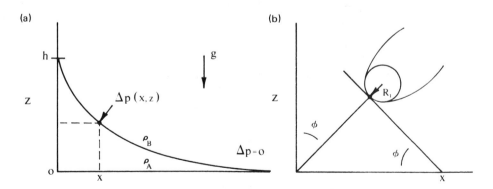

Figure 6.12 Definition of variables used to describe a meniscus formed by liquid A displacing B in terms of (a) x and z and (b) ϕ.

that the local slope of the tangent $dz/dx = -\tan \phi = \sin \phi/\cos \phi$. Substituting this ratio into Eq. (36) gives

$$\frac{1}{R_1} = \frac{d}{dx}\left(-\frac{\sin \phi}{\cos \phi}\right)\left[1 + \left(\frac{\sin \phi}{\cos \phi}\right)^2\right]^{3/2} \tag{67}$$

When the indicated differentiations are carried out and the result is simplified, Eq. (67) becomes

$$\frac{1}{R_1} = -\cos \phi \frac{d\phi}{dx} \tag{68}$$

which is the same as

$$\frac{1}{R_1} = -\frac{d \sin \phi}{dx} \tag{69}$$

Since $x \propto \sin \phi$ and $z \propto \cos \phi$, Eq. (69) is equivalent to

$$\frac{1}{R_1} = -\frac{d \cos \phi}{dz} \tag{70}$$

Substituting this result into Eq. (66) gives a relationship that is easily integrated:

$$\frac{\Delta\rho g}{\gamma}\int z \, dz = -\int d \cos \phi \tag{71}$$

Carrying out the indicated integration and evaluating the constant from the fact that $\phi = 0$ at $z = 0$ (i.e., the tangent is horizontal), we obtain

$$\cos \phi = 1 - \frac{\Delta\rho g}{2\gamma} z^2 \tag{72}$$

At the wall where $z = h$, ϕ is the complementary angle to θ, the contact angle. Therefore at the surface of the solid support

$$\cos \phi = \sin \theta \tag{73}$$

and

$$\sin \theta = 1 - \frac{\Delta\rho g}{2\gamma} z^2 = 1 - \left(\frac{z}{a}\right)^2 \tag{74}$$

where a^2 is given by Eq. (5). This result shows that θ may be evaluated by measuring the height to which the meniscus climbs against a wall, provided that γ is known. The following example will give us an indication of the magnitude of the effect to be determined.

Example 6.4   Calculate the height to which an n-octane surface will climb on a Teflon wall (this is the same system used in Example 6.3) if γ is 22 mJ m^{-2}, $\phi = 30°$, and $\rho = 0.70$ g cm^{-3}. Comment on the ease or difficulty of making this measurement.

Solution   The density of air is insignificant compared to the liquid; hence ρ can be used for $\Delta\rho$. Because the density is given in cgs units, it is convenient to use this system of units throughout; remember, millijoule meter^{-2} and erg cm^{-2} are identical. Substituting numerical values into Eq. (74) gives

$$\sin 30 = 1 - \frac{(0.70)(980)h^2}{2(22)}$$

from which $h^2 = 0.032$ and $h = 0.18$ cm.

Although h is not a large number, it is readily measurable by using a cathetometer or low-magnification traveling microscope to measure the vertical distance between the level surface of the liquid and the top of the meniscus. The sighting is done along the liquid surface through a window in the container holding the liquid. Therefore the junction of interest must be viewed through the meniscus at the window. Special care must be taken to establish the right reference plane. If the contact angle is very small, the top of the meniscus might be difficult to see, but good illumination should remedy this.

•

Equation (74) provides a second relationship in addition to Eq. (2) which expresses the connection between γ and θ and experimental quantities. The two simultaneous equations can be solved for γ and θ by the following steps:

1. Square Eq. (74):

$$\sin^2 \theta = \left(1 - \frac{\Delta\rho g h^2}{2\gamma} \right)^2$$

2. Square Eq. (2):

$$\cos^2 \theta = \left(\frac{w}{P\gamma} \right)^2$$

3. Recognize that $\sin^2\theta + \cos^2\theta$ equals unity:

$$1 = \left(\frac{w}{P\gamma}\right)^2 + \left(1 - \frac{\Delta\rho g h^2}{2\gamma}\right)^2$$

By multiplying out this last result and simplifying, we obtain

$$\gamma = \frac{\Delta\rho g h^2}{4} + \frac{w^2}{\Delta\rho\gamma h^2 P^2} \tag{75}$$

Substituting Eq. (75) into Eq. (2) gives

$$\cos\theta = \frac{4\,\Delta\rho g h^2 P w}{(\Delta\rho)^2 g^2 h^4 P^2 + 4w^2} \tag{76}$$

Equations (75) and (76) allow γ and θ to be evaluated in a perfectly straightforward way from experimental quantities measured on a single system with one apparatus. For systems which can be fabricated into the necessary plates, this is an excellent way for measuring these important parameters.

In the next section we examine the shapes of liquid surfaces possessing an axis of symmetry.

6.10   Drops, Bubbles, and Menisci

In the last section we saw that the Laplace equation describes the meniscus formed by a liquid surface at a flat wall. The situation is not essentially different—only more complicated—if the supporting wall is wrapped into a cylinder to generate a capillary meniscus with an axis of symmetry. As a matter of fact, any liquid interface with axial symmetry is described by the same mathematical formalism. For simplicity, we consider a liquid drop resting on a smooth horizontal surface as shown in Fig. 6.13. Such a drop, incidentally, is called a sessile (sitting) drop. In the absence of gravity, a drop like this would be spherical, since this geometry encloses the maximum volume within a minimum surface area. Any departure from a spherical shape increases the area and the surface free energy associated with it. A drop of liquid A in air—or, for that matter, any less dense medium B—is ordinarily acted upon by gravitational as well as surface forces. Gravity forces the drop to lower its center of mass and thus flatten. Flattening increases the surface area; the equilibrium shape depends on the balance of the two forces. The same situation holds for the shapes of bubbles and menisci.

With this physical picture in mind, let us consider how the Laplace equation can be written to describe the profile of a sessile drop. Figure

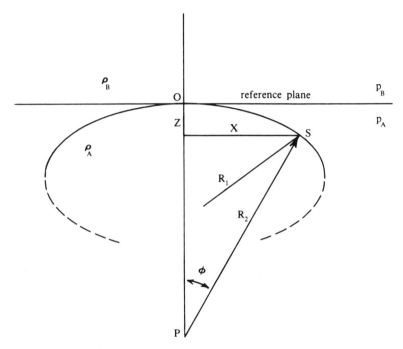

Figure 6.13 Definition of coordinates for describing surfaces with an axis of symmetry (*OP*).

6.13 represents the profile of a sessile drop; the actual surface is generated by rotating the profile around the axis of symmetry, the z axis in this representation.

In Fig. 6.13 the origin of the coordinate system O is situated at the apex of the surface. The two radii of curvature at point S are defined as follows. The one designated R_1 lies in the plane of the figure and describes the curvature of the profile shown. The radius of rotation of point S around the z axis equals x. The relationship between x and R_2 is given by

$$x = R_2 \sin \phi \qquad (77)$$

since R_2 is defined as a vector originating along the z axis at P and making an angle ϕ with the axis of symmetry as shown in the figure.

Because of the symmetry of the surface, both values of R must be equal at the apex of the drop. The value of the radius of curvature at this location is symbolized b. Therefore at the apex (subscript 0)

$$(\Delta p)_0 = \frac{2\gamma}{b} \tag{78}$$

according to Eq. (30).

Next let us calculate the pressure at point S. At S the value of Δp equals the difference between the pressure at S in each of the phases. These may be expressed relative to the pressure at the reference plane through the apex (subscript 0) as follows:

1. In phase A

$$p_A = (p_A)_0 + \rho_A gz \tag{79}$$

2. In phase B

$$p_B = (p_B)_0 + \rho_B gz \tag{80}$$

Therefore Δp at S equals

$$(\Delta p)_S = p_A - p_B = (p_A)_0 - (p_B)_0 + (\rho_A - \rho_B)gz = (\Delta p)_0 + \Delta\rho gz \tag{81}$$

where $\Delta\rho = \rho_A - \rho_B$. Now Eq. (78) may be substituted for Δp at the apex to give

$$(\Delta p)_S = \frac{2\gamma}{b} + \Delta\rho gz \tag{82}$$

The general form of the Laplace equation may be expressed in terms of the coordinates of Fig. 6.13 by combining Eqs. (35), (77), and (82):

$$\gamma\left(\frac{\sin\phi}{x} + \frac{1}{R_1}\right) = \frac{2\gamma}{b} + \Delta\rho gz \tag{83}$$

In Eq. (83) R_1^{-1} is given by Eq. (36). Expression (83) is known as the Bashforth–Adams equation. It is conventional to express this equation in dimensionless form by expressing all distances relative to the radius at the apex b:

$$\frac{\sin\phi}{x/b} + \frac{1}{R_1/b} = 2 + \frac{\Delta\rho gb^2}{\gamma}\frac{z}{b} = 2 + \beta\frac{z}{b} \tag{84}$$

The cluster of constants in Eq. (84) is defined by the symbol β:

$$\beta = \frac{\Delta\rho gb^2}{\gamma} \tag{85}$$

The Bashforth-Adams equation—the composite of Eqs. (36) and (84)—is a differential equation which may be solved numerically with β and ϕ as parameters. Bashforth and Adams solved this equation for a large number of β values between 0.125 and 100 by compiling values of x/b and z/b for $0° < \phi < 180°$. Their tabular results, calculated by hand before the days of computers, were published in 1883. Other workers subsequently extended these tables. A very useful compilation of these results is found in the chapter by Padday [9]. Table 6.2 shows a typical result from these tables for $\beta = 25$. The surface profiles sketched in Fig. 6.14 were drawn from these tabulated results for (a) $\beta = +10.0$ and (b) $\beta = -0.45$.

It is apparent from inspection of Fig. 6.14 that different values of β characterize different drop shapes. This is to be expected, since β varies with $\Delta\rho$ compared to γ. According to the convention established in Fig. 6.13, g and z are measured in the same direction. Therefore, when the Bashforth–Adams tables are applied to sessile drops, the sign of $\Delta\rho$ determines the sign of β. If $\rho_A > \rho_B$, β will be positive and the drop will be oblate in shape, since the weight of the fluid tends to flatten the surface. If $\rho_A < \rho_B$, a prolate drop is formed, since the larger buoyant force leads to a surface with much greater vertical elongation. In this case β is negative. A β value of zero corresponds to a spherical drop and, in a gravitational field, is expected only when $\Delta\rho = 0$. Positive values of β correspond to sessile drops of liquid in a gaseous environment. Negative β values correspond to sessile bubbles extending into a liquid. These statements imply that the drop is resting *on* a supporting surface. If, instead, the drop is suspended *from* a support (called pendant drops, or bubbles), g and z are measured in opposite directions and have different signs; that is, for a pendant drop z is measured upward from the apex, while g continues to define the downward direction. Because of this sign reversal, the liquid drop will be prolate ($\beta < 0$) shape, and the gas bubble oblate ($\beta > 0$).

The Bashforth–Adams tables provide an alternate way of evaluating γ by observing the profile of a sessile drop of the liquid under investigation. If, after all, the drop profiles of Fig. 6.14 can be drawn using β as a parameter, then it should also be possible to match an experimental drop profile with the β value that characterizes it. Equation (85) then relates γ to β and other measurable quantities. This method is claimed to have an accuracy of 0.1%, but it is slow and tedious, and hence not often the method of choice in practice.

In order to obtain the profile of one of these drops experimentally, it is best to photograph the silhouette of the surface. Then experimental and theoretical profiles are compared in an effort to identify the β value which

Table 6.2 x/b and z/b for $\beta = 25$ and $0° < \phi \leqslant 180°$

ϕ (deg)	x/b	z/b
5	0.08521	0.00368
10	0.16035	0.01348
15	0.22230	0.02712
20	0.27250	0.04288
25	0.31333	0.05974
30	0.34684	0.07713
35	0.37455	0.09475
40	0.39755	0.11236
45	0.41666	0.12985
50	0.43249	0.14711
55	0.44551	0.16405
60	0.45609	0.18063
65	0.46451	0.19678
70	0.47101	0.21246
75	0.47579	0.22761
80	0.47905	0.24221
85	0.48089	0.25626
90	0.48148	0.26966
95	0.48092	0.28243
100	0.47934	0.29435
105	0.47682	0.30594
110	0.47345	0.31665
115	0.46931	0.32662
120	0.46452	0.33585
125	0.45911	0.34433
130	0.45319	0.35204
135	0.44682	0.35901
140	0.44008	0.36519
145	0.43302	0.37061
150	0.42571	0.37526
155	0.41823	0.37915
160	0.41062	0.38231
165	0.40296	0.38472
170	0.39528	0.38643
175	0.38766	0.38744
180	0.38014	0.38776

Source: From F. Bashforth and J. C. Adams, *An Attempt to Test the Theory of Capillary Action*, Cambridge University Press, London, 1883.

(a)

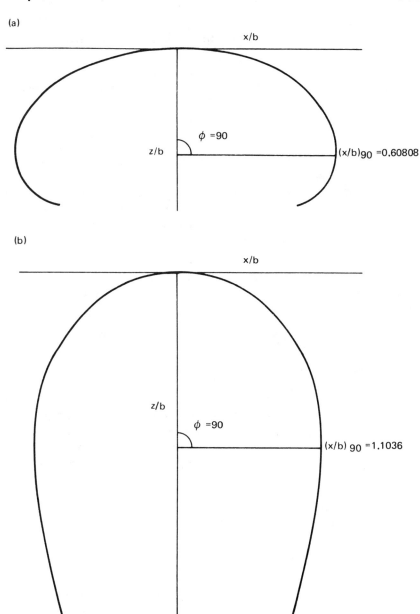

x/b

ϕ =90

z/b

$(x/b)_{90}$ =0.60808

(b)

x/b

z/b

ϕ =90

$(x/b)_{90}$ =1.1036

Figure 6.14 Sessile drop profiles drawn from results of Bashforth and Adams' tables: (a) β = 10.0 and (b) β = −0.45. (Data from Ref. 9.)

characterizes the experimental surface. With care, it is possible to interpolate between theoretical profiles and arrive at the β value which describes the surface under consideration. Equation (85) shows that knowledge of β alone is not sufficient to permit the evaluation of γ; b and Δρ must also be known. Evaluating the density difference poses no special difficulty. Let us next consider how b is measured.

Once β is known for a particular profile, the Bashforth–Adams tables may be used further to evaluate b:

1. For the appropriate β the value of x/b at ϕ = 90° is read from the tables. This gives the maximum radius of the drop in units of b.
2. From the photographic image of the drop this radius may be measured, since the magnification of the photograph is known.
3. Comparing the actual maximum radius with the value of $(x/b)_{90°}$ permits the evaluation of b.

Figure 6.14a may be used as a numerical example to illustrate this procedure. Suppose an actual experimental drop profile is matched with theoretical profiles and is shown to correspond to to a β value of 10.0. Then b is evaluated as follows:

1. The value of $(x/b)_{90°}$ for β = 10 is found to be 0.60808 from the tables.
2. Assume the radius of the actual drop is 5.00 mm at its widest point.
3. The first two items describe the same point; therefore b = 5.00/0.608 = 8.22 mm. This would be the radius at the apex of the drop shown in Fig. 6.14a if the maximum radius were 5.00 mm.
4. This numerical example may be completed by using the definition of β to evaluate γ for the liquid of Fig. 6.14a. Assuming Δρ to be +1.00 g cm^{-3} and taking g = 9.8 m s^{-2} gives

$$\gamma = \frac{\Delta\rho g b^2}{\beta} = \frac{(10^3)(9.8)(8.22 \times 10^{-3})^2}{10.0} = 0.0662 \text{ J m}^{-2}$$

$$= 66.2 \text{ mJ m}^{-2}$$

Several additional points might be noted about the use of the Bashforth–Adams tables to evaluate γ. If the interpolation is necessary to arrive at the proper β value, then interpolation will also be necessary to determine $(x/b)_{90°}$. This results in some loss of accuracy. With pendant drops or sessile bubbles (i.e., negative β values) it is difficult to measure the maximum radius, since the curvature is least along the equator of such drops (see Fig. 6.14b). The Bashforth-Adams tables have been rearranged to facilitate their use for pendant drops. The interested reader will find tables adapted for pendant drops in the chapter by Padday [9]. The pendant drop method utilizes an equilibrium drop attached to a support and should not be confused with the drop weight method, which involves drop detachment.

As a final application of these ideas, we consider the Bashforth–Adams tables in the context of a capillary rise experiment. Even though we have based our discussion of surface profiles on sessile drops, we should recognize that—inverted and cut off at their widest point by capillary walls—the profiles in Fig. 6.14 could equally well be those of menisci.

We have already seen by Eq. (34) that the height to which a liquid rises in a capillary is inversely proportional to the radius of curvature of the meniscus at its apex, with the capillary constant a^2 the proportionality constant. The radius at the apex in Eq. (34) is precisely the same b that appears in the value which characterizes the meniscus in the capillary.

For a liquid to make an angle of 0° (the usual situation in glass capillaries) with the supporting walls, the walls must be tangent to the profile of the surface at its widest point. Accordingly, $(x/b)_{90°}$ in the Bashforth–Adams tables must correspond to R/b. Since the radius of the capillary is measurable, this information permits the determination of b for a meniscus in which $\theta = 0°$. However, there is a catch. Use of the Bashforth–Adams tables depends on knowing the shape factor β. It is not feasible to match the profile of a meniscus with theoretical contours, so we must find a way of circumventing this problem. The following example suggests how this can be accomplished.

Example 6.5 Suppose a liquid for which $\theta = 0°$ and $\Delta\rho = 1.00$ g cm^{-3} shows a capillary rise of 4.00 mm in a capillary for which $R = 2.50$ mm. Outline a method of successive approximations whereby the sort of information provided by the Bashforth–Adams tables can be used to estimate β and b for a meniscus. Indicate how this approach can be used to yield exact values of γ from capillary rise experiments. Use the fact that $(x/b)_{90°} = 0.867$ for $\beta = 1.25$ and $(x/b)_{90°} = 0.846$ for $\beta = 1.44$ to illustrate the method.

Solution   As a first approximation (subscript 1) we assume the meniscus to be a hemisphere, in which case $b_1 \simeq R$ and $a_1 \simeq (Rh)^{1/2}$ by Eq. (4). From Eqs. (4) and (85), $\beta = 2(b/a)^2$; hence a first approximation to the value of β which characterizes this meniscus can be estimated as follows:

$$\beta_1 = 2[R/(Rh)^{1/2}]^2 = 2\{2.50/[(4.00)(2.50)]^{1/2}\}^2 = 1.25$$

For this β value, $(x/b)_{90°} = 0.867$; this value can be equated with R/b and used to make a second approximation for b; that is, $0.867 = 2.50/b_2$, from

Table 6.3   Use of the Bashforth-Adams Tables to Solve the Capillary Rise Problem by Successive Approximations (see Example 6.5)[a]

	Approximation number			
Parameter	First	Second	Third	Fourth
b (mm)	2.50	2.88	2.95	2.95
a (mm)	3.20	3.39	3.44	3.44
$\beta = 2(b/a)^2$	1.25	1.44	1.47	1.47
$(x/b)_{90°} = R/b_{i+1}$	0.867	0.846	0.847	—
R/a	0.78	0.736	0.727	0.728

[a]The β values are obtained by linear interpolation of the tabulated quantities.
Source: Data from Ref. 9.

which $b_2 = 2.88$ mm. From this value, a second approximation for a is obtained by Eq. (34): $a_2 = (b_2 h)^{1/2} = [(2.88)(4.00)]^{1/2} = 3.39$. From this pair of a and b values a second approximation for β can be determined: $\beta_2 = 2(2.88/3.39)^2 = 1.44$. Based on this estimate of β, a third round of approximations may be conducted. For $\beta = 1.44$, $(x/b)_{90°} = 0.846$. Therefore $0.846 = 2.50/b$ and $b_3 = 2.96$ mm. Also, $a_3 = [(2.96)(4.00)]^{1/2} = 3.44$ mm. From this, still further rounds of approximation could be conducted based on the next estimate of β ($\beta_3 = 1.47$).

From the best estimate of a obtained, a value of γ can be calculated by Eq. (5). In this case $a_3 = 3.39$ mm; hence $\gamma = \Delta\rho g a^2/2 = (10^3)(9.8)(3.39 \times 10^{-3})^2/2 = 56.3 \times 10^{-3}$ J m^{-2} = 56.3 mJ m^{-2}. This contrasts with the value 50.2 mJ m^{-2} that would have been obtained if Eq. (4)—that is, the first approximation to a—had been used to evaluate γ.

•

Table 6.3 summarizes Example 6.5 and extends the iterations until no further changes occur. The ultimate value of γ is 57.9 mJ m^{-2} when the process is carried out further.

Although the procedure just outlined in Example 6.5 shows the application of the Bashforth-Adams equation to the phenomenon of capillary rise, supplementary tables—compiled by Sugden and reproduced by Padday [9]—are somewhat more convenient to work with. The latter tabulate pairs of R/a, R/b ratios which satisfy the Bashforth-Adams equation. The values $R/a = 0.73$ and $R/b = 0.85$ in Table 6.3 are examples of a pair of points meeting this condition.

Several other methods for determining γ—notably, the maximum bubble pressure, the drop weight, and the DuNouy ring methods—all involve measurements on surfaces with axial symmetry. Although the

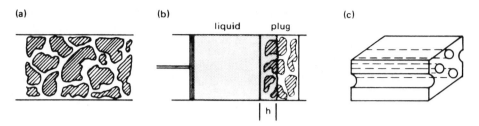

Figure 6.15 (a) Schematic view of pores through a plug of particles, (b) liquid intrusion into a plug under pressure and (c) an idealized plug of cylindrical pores.

Bashforth–Adams tables are pertinent to all of these, the data are generally tabulated in more practical forms which de-emphasize the surface profile.

6.11 Powders and Porosimetry

Any solid can be pulverized into particles of small size; not all can be fabricated into the smooth supports we have discussed until now in this chapter. This consideration alone—not to mention the many practical applications of powdered solids—encourages us to look for a relationship which describes the junction of a liquid interface with such solids. One thought might be to compress the powder into a pellet and treat the surface of the latter in the same way as we have treated other solid surfaces. On a fine scale the surface of such a pellet is rough, however, so hysteresis effects should be severe. Therefore we look for some alternative approach.

Instead of the exterior surface of a powder pellet, let us consider the network of irregular channels that permeate it—the hole instead of the doughnut! Figure 6.15 a represents a section through such a pellet for irregularly shaped particles. It is easy to imagine an arrangement such as that shown in Fig. 6.15b in which the powder pellet is positioned as a plug in a cylinder such that liquid can be forced through it by a movable piston. For now our main requirement for such an apparatus is that we are able to measure the applied pressure p needed to force the liquid into the plug. In a further development we will also be interested in measuring the depth h into the plug that the liquid penetrates under a pressure p.

Commercial instruments are available that carry out these operations under the control of a computer which also analyzes the results. Let us consider the penetration of the liquid into the pores of the plug to see the basis for this analysis.

Instead of the actual network of irregular channels, the interpretation of this experiment is based on a model which imagines the plug to consist of a bundle of cylindrical pores of radius R. The model is represented by Fig. 6.15c. The intrusion of the liquid into the cylindrical pores in response to the applied pressure follows the same mathematical description as the rise of a liquid in a capillary. In view of the approximate nature of the model, it is adequate to use the Laplace equation in the form given by Eq. (3) to describe this situation:

$$\frac{2\gamma_{LV} \cos \theta}{R} = p \tag{86}$$

In the capillary rise experiment $p = \Delta\rho g h$, while p is the applied pressure in this case.

Now suppose a liquid is pushed through the plug which forms a contact angle of zero with the walls of the channels. Using the subscript 0 to represent this situation, eq. (86) becomes

$$p_0 = \frac{2\gamma_{LV,0}}{R} \tag{87}$$

since $\cos \theta = 0$ in this case. Equation (87) can be used to evaluate R for the plug. The value of R thus obtained is a parameter which is characteristic of the experiment; however, it might be very different from any physical distance within the plug. It might be called the radius of an equivalent cylinder, although some researchers feel that this name attaches too much significance to the R value thus obtained. The term *tortuosity* is sometimes applied to this empirical R value. Whatever terminology is used, this evaluation of R amounts to a calibration with a known (with respect to θ) liquid so that R can be eliminated from Eq. (86) to give

$$\frac{\gamma_{LV} \cos \theta}{p_0 \gamma_{LV,0}} = p \tag{88}$$

or

$$\cos \theta = \frac{p}{p_0} \frac{\gamma_{LV,0}}{\gamma_{LV}} \tag{89}$$

Note that values of γ are assumed to be available when this equation is used. Since numerous methods are available for measuring γ, there is no loss of applicability in assuming this. Since all of the quantities on the right-hand side of Eq. (89) are measurable, this approach provides a method for determining θ for powdered solids. The method is not highly reliable, but is preferable to any technique based on the exterior surface of the plug as a liquid support.

Our approach until now has been to focus attention on the liquid-solid junction and eliminate the parameter which characterizes the plug. Numerous porous solids exist for which there may be considerable interest in knowing the pore dimensions. The foregoing analysis can also be applied to this problem by assuming that both γ and θ are known. For this application the liquid is chosen for convenience, and not because it is part of what is being measured. Mercury is generally used as the penetrating fluid, and the general technique is called mercury porosimetry. Contact angles of 130–140° are generally assumed, although it would clearly be preferable to use the θ value which is appropriate for the specific system.

When the pore structure of a solid is being investigated, it is desirable to learn as much as possible from porosimetry. In general, solids will possess a distribution of pore sizes, so the volume (or surface area) associated with pores of various radii is of interest. The significance of this sort of information is apparent for such solids as zeolite catalysts. These are aluminasilicates which can be synthesized with variable Si/Al ratios and variable pore sizes. Pore dimensions in the range of 0.8–2.0 nm can be obtained for solids with controllable acidity associated with Al^{3+} in the silica matrix. Zeolites are used on a large scale in the petroleum industry as catalysts for the cracking and isomerization of hydrocarbons. Part of the specificity of these catalysts arises from the pore size distribution, so porosimetry becomes a valuable tool for the characterization of these materials.

In an apparatus based on Fig. 6.15b, the volume of mercury forced into the pores of the solid can be measured as a function of the applied pressure. Equation (86) shows that higher pressures are required for smaller pores. Therefore incremental increases in p will result in the filling of pores of progressively smaller radii. The volume V that has intruded a porous solid at a pressure p gives the cumulative volume of all pores larger than the size associated with p. Since plots of V versus p give information about the cumulative pore distribution, it is the derivative of such data that measures the increment in pore volume associated with an increment in R. Written as a formula, $dV/dp \propto dV/(-dR)$, since V increases as R decreases.

To develop this last relation further, note that Eq. (86) can be written as $pR = \text{const.}$, since γ and θ are independent of pore structure. Therefore $p \, dR + R \, dp = 0$, or

$$\frac{dp}{dR} = -\frac{p}{R} \tag{90}$$

With Eq. (90), the slope of the experimental V–p plot can be developed as follows:

$$\frac{dV}{dR} = -\frac{dV}{dp}\frac{dp}{dR} = \frac{dV}{dp}\frac{p}{R} \tag{91}$$

Combining Eqs. (86) and (91) gives

$$\frac{dV}{dR} = \frac{dV}{dp}\frac{p^2}{2\gamma \cos \theta} \tag{92}$$

which shows how the local slope of the experimental data can be converted into the desired information about increments of volume for pores of different radius.

As noted above, much of the data manipulation of commercial porosimeters is computerized, so a pore size distribution is produced automatically by these instruments.

The rate at which a liquid penetrates a porous solid can also be measured and interpreted at the same level of approximation as the foregoing. The Poiseuille equation, Eq. (4.20), gives an expression for the rate at which a liquid flows through cylindrical pores of radius R under a pressure p. We shall neglect the gravitational contribution to the driving force, which means that the following is better suited to horizontal than to vertical arrangements. If we write the volume rate of flow $dV/dt = d(\pi R^2 h)/dt$ and simplify, Eq. (4.20) can be rewritten as

$$\frac{dh}{dt} = \frac{pR^2}{8\eta h} \tag{93}$$

where h is the depth of penetration of the intruding liquid and η is its viscosity.

If Eq. (86) is used to eliminate p, Eq. (93) becomes

$$\frac{dh}{dt} = \frac{2\gamma \cos \theta}{R}\frac{R^2}{8\eta h} = \frac{\gamma R \cos \theta}{4\eta h} \tag{94}$$

As with Eq. (86), R can be eliminated from Eq. (94) by calibration with a liquid for which $\theta = 0$. In that case Eq. (94) can be used to determine θ for powders or porous solids. Alternatively, a value of θ may be assumed and R can be evaluated from the rate of mercury intrusion. Equation (94) is called the Washburn equation.

Since advancing and receding contact angles are likely to be different in these experiments, mercury permeation curves are expected to be different, depending on whether the mercury is being pushed in or out of the plugs. This type of hysteresis is indeed observed. We shall encounter another type of hysteresis associated with pore filling in Sect. 9.10.

6.12   The Girifalco-Good-Fowkes Equation

Throughout this chapter we have dealt with surface tension from a phenomenological point of view almost exclusively. To chemists, however, descriptions from a molecular perspective are often more illuminating than descriptions of phenomena alone. In condensed phases, where interactions involve many molecules, rigorous derivations based on the cumulative behavior of individual molecules are extremely difficult. We shall not attempt to review any of the efforts directed along these lines for surface tension. Instead, we consider the various types of intermolecular forces which exist and interpret γ for any interface as the summation of contributions arising from the various types of interactions which operate in the materials forming the interface. To develop this idea, several broad categories of molecular interactions should be considered.

1. Hydrogen bonding: Hydrogen atoms serve as bridges linking together two atoms of high electronegativity. In the present context these atoms are in separate molecules so the molecules themselves are mutually "attracted" by these bonds.
2. Metallic bonding: A sea of mobile electrons shared by the atoms of a metal contributes to the attraction between metal atoms in bulk samples.
3. Permanent dipole interactions: Polar molecules have relatively positive and negative regions. Regions of opposite charge on different molecules result in an attraction between these molecules. Molecules must possess a permanent dipole moment to display this effect.
4. London forces: Deformable electron clouds in adjoining molecules distort one another, resulting in an instantaneous polarity with accompanying attraction between the molecules involved. The polarizability of a molecule (see Sect. 5.3) is a measure of its tendency to display this effect.

Although additional entries could be included in this list, the foregoing is sufficient for our purposes. The main thing to recognize about these interactions is that all but London forces are highly specific. London forces require only the presence of electrons and therefore operate between all molecules. The other types of interactions, by contrast, require some specific feature: the metallic state, high electronegativity, or certain molecular geometry. Whether any of the other types of interactions operate or not, London forces are always present.

In Chapter 11 we shall discuss in considerable detail the attractions between molecules arising from their polarity and polarizability. For now it is sufficient to recognize that London forces attract all molecules

together, regardless of their specific chemical nature. The other interactions itemized above operate only for systems possessing the requisite
special features. What this means for surface tension is best illustrated by
considering a specific surface. At the water–mercury interface, for
example, the metallic bonds of the mercury end at the interface.
Similarly, the hydrogen bonds of the water end at the surface. London
attraction exists between water and mercury particles, however, just as it
exists for water–water and mercury–mercury pairs. These various
London attractions may differ in magnitude as a result of different
molecular properties, but they share a common origin.

Next we apply these concepts to a molecule of A in two different
environments: in the interior of a bulk phase and near an interface
between two phases. A molecule in the interior of a bulk phase is
surrounded on all sides by a homogeneous molecular environment of A.
Any movement that would increase its separation from some neighbors
would automatically decrease the separation from others, so large
deviations from the equilibrium separation are improbable.

For a molecule at an interface between a condensed phase and the gas
phase the environment is quite asymmetrical. Movement toward the bulk
phase is impeded by the excluded volume of the A molecules. Movement
away from the bulk phase meets no resistance of this sort, although the
prevailing attraction between molecules in the condensed phase opposes
the A molecules from escaping the condensed phase altogether. Because
of this, the equilibrium separation between the molecules at the surface
will be larger than between those in the interior. The intermolecular
separation has been "stretched" in bringing a molecule to the surface.
The contractile force in the interface of the substance is simply the
restoring force attempting to return the molecules to their bulk spacing.
From an energetic point of view, the difference between the energy of the
bulk and surface minimum separations gives the work needed to bring
a molecule from the interior to the surface.

Next let us consider the situation in which a second condensed phase
B adjoins the reference phase A. The new consideration in this case is the
London component of attraction of the molecules in condensed phase B
for the A molecules in the interface. This A–B attraction partially
overcomes the A–A attraction which opposes the movement of an A
molecule to any interface. As a consequence, there is a difference in the
energy which must be expended to bring an A molecule to an interface
with a gas and to an interface with another condensed phase. We call this
difference ΔE^s.

It is argued that only the London component of intermolecular
attractions operates across the interface to decrease the work required to

bring a particle to the surface. Suppose we define ϕ as that fraction of the surface tension which is due to London forces. Next we use the sort of geometric mixing rule which was employed in Sect. 3.7 to estimate ΔE^s as follows:

$$\Delta E^s = \sqrt{(\phi_A\gamma_A)(\phi_B\gamma_B)} \tag{95}$$

The assignment of a geometric mean rather than an arithmetic mean or some other function of the two γ terms is justified primarily on the basis of the successful use of this type of averaging in the theory of nonelectrolyte solubility. Only the London component of γ is used, since it is the part of γ which crosses phase boundaries.

According to these ideas, the work required to bring an A molecule to the AB interface is given by

$$(\text{work})_A = \gamma_A - (\Delta E^s)_A = \gamma_A - \sqrt{(\phi_A\gamma_A)(\phi_B\gamma_B)} \tag{96}$$

when A and B are both in condensed phases. A similar expression applies to the work required to bring a B molecule to the interface. The total work of forming the AB interface is the sum of these contributions:

$$\gamma_{AB} = [\gamma_A - \sqrt{(\phi_A\gamma_A)(\phi_B\gamma_B)}] + [\gamma_B - \sqrt{(\phi_A\gamma_A)(\phi_B\gamma_B)}]$$
$$= \gamma_A + \gamma_B - 2\sqrt{(\phi_A\gamma_A)(\phi_B\gamma_B)} \tag{97}$$

Combining the two fractions into the parameter Φ, Eq. (97) becomes

$$\gamma_{AB} = \gamma_A + \gamma_B - 2\Phi(\gamma_A\gamma_B)^{1/2} \tag{98}$$

with $\Phi = (\phi_A\phi_B)^{1/2}$. The relationship given by Eq. (98) is called the Girifalco–Good equation. The parameter Φ may be viewed from two different points of view:

1. We may regard Φ as an empirical constant evaluated by fitting data to Eq. (98) or some other form of the basic relationship.
2. We may estimate Φ from first principles by using molecular parameters and relationships from Chapter 11.
3. With a few additional refinements, the two approaches are found to give satisfactory agreement.
4. The Φ values often lie in the range 0.5–1.0, which shows that, in general, half or more of the surface tension can be attributed to London forces.

Equation (98) may be combined with several other relationships of this chapter to generate expressions whereby the contributions of London forces can be investigated. For example, if one of the condensed phases is a solid and the other a liquid, Eqs. (98) and (49) may be combined to give

$$\gamma_L \cos \theta = \gamma_L - 2\Phi(\gamma_S\gamma_L)^{1/2} - \pi_e \tag{99}$$

Equation (99) can be solved for any one of the factors, depending on the data available. Neglecting π_e as usual (although we might solve for this also), we obtain the following:

1. If γ_S, γ_L, and θ are known,

$$\Phi_{SL} = \frac{\gamma_L(1 - \cos \theta)}{2(\gamma_S\gamma_L)^{1/2}}$$

2. If γ_L, γ_S, and Φ are known,

$$\cos \theta = -1 + 2\Phi \left(\frac{\gamma_S}{\gamma_L}\right)^{1/2}$$

3. If γ_L, θ, and Φ are known,

$$\gamma_S = \frac{\gamma_L(1 + \cos \theta)^2}{4\Phi^2}$$

The third item provides a valuable way for evaluating γ_S, a quantity which is otherwise difficult to determine.

Rather than pursue the fractions ϕ or their composite Φ any further, we turn next to an empirical approach due to Fowkes to estimate the product $\phi\gamma$ for various substances. We shall use the symbol γ^d to represent the London contributions to γ estimated by his method. The notation is derived from the fact that London forces are also called dispersion (superscript d) forces. In this notation Eq. (98) can be written

$$\gamma_{AB} = \gamma_A + \gamma_B - 2(\gamma_A^d\gamma_B^d)^{1/2} \tag{100}$$

In this version the relationship is called the Girifalco–Good–Fowkes equation. Although we shall use $\phi\gamma$ and γ^d interchangeably, it is important to recognize that γ^d values are determined by a particular strategy as illustrated in the following example.

Example 6.6   The following are the interfacial tensions for the various two-phase surfaces formed by n-octane (O), water (W), and mercury: for n-octane-water $\gamma = 50.8$ mJ m^{-2}, for n-octane-mercury $\gamma = 375$ mJ m^{-2}, and for water-mercury $\gamma = 426$ mJ m^{-2}. Assuming that only London forces operate between molecules of the hydrocarbon, use Eq. (98) to estimate γ^d for water and mercury. Do the values thus obtained make sense? Take γ values from Table 6.1 for the interfaces with air of these liquids.

Solution In general, Eq. (100) contains two unknowns: the dispersion components of γ for A and B. If γ^d for one of these is known, the other γ^d value can be calculated from experimental results. If it is assumed that $\gamma^d = \gamma$ for the hydrocarbon, then the experimental γ for an interface involving octane can be interpreted by Eq. (100) to give the other γ^d value. Thus $375 = 484 + 21.8 - 2(\gamma^d_{Hg} \times 21.8)^{1/2}$; therefore $\gamma^d_{Hg} = 196$ mJ m^{-2}. Also $50.8 = 72.8 + 21.8 - 2(\gamma^d_w \times 21.8)^{1/2}$; therefore $\gamma^d_w = 22.0$ mJ m^{-2}. The two γ^d values are 30 and 40%, respectively, of the γ's for water and mercury. These fractions are somewhat less than expected in terms of the preceding discussion, but are not totally out of line. A more meaningful test is to use these values to estimate γ for the water–mercury interface. Repeating the above procedure, we obtain

$$\gamma_{Hg-w} = 484 + 72.8 - 2[(196)(22)]^{1/2} = 425 \text{ mJ m}^{-2}$$

which compares quite favorably with the experimental value, 426 mJ m^{-2}.

•

Table 6.4 lists the data and calculated results for a number of hydrocarbons forming interfaces with water and mercury. The relative constancy of the calculated γ^d values adds to their plausibility.

The average values of γ^d for water and mercury are clearly successful in their ability to calculate γ correctly for the water–mercury interface. Individually, however, they are the averages of slightly divergent values measured for several different interfaces with hydrocarbons. In this sense the values of γ^d which we have considered are analogous to mean bond energies in physical chemistry. The latter, too, are averages obtained from a variety of compounds. Although mean bond energies are very useful, they are, by nature, insensitive to unique, specific effects. With both mean bond energies and values of γ^d, the user must be careful that no such special interactions are present, otherwise quite serious errors could arise. Furthermore, it is important to realize that any errors are perpetuated and compounded by this scheme for evaluating γ^d.

It is not difficult to apply the concept of the dispersion component of γ to solid surfaces. In doing this, it is necessary to treat high- and low-energy surfaces differently. We shall not consider solid interfaces in detail; our treatment is limited to the following observations:

1. For low-energy surfaces, $\pi_e \simeq 0$. Manipulation of Young's equation [Eq. (44)] generates a relationship that expresses γ^d_S in terms of θ and other experimental quantities.

Table 6.4 Experimental Values for γ_H, γ_{H-w}, and γ_{H-Hg}, With H = Hydrocarbon, at 20°C (all values in mJ m^{-2})[a]

Hydrocarbon	γ	Mercury (γ_{Hg} = 484)		Water (γ_W = 72.8)	
		γ_{H-Hg}	γ_{Hg}^d	γ_{H-w}	γ_W^d
n-Hexane	18.4	378	210	51.1	21.8
n-Heptane	18.4	—	—	50.2	22.6
n-Octane	21.8	375	199	50.8	22.0
n-Nonane	22.8	372	199	—	—
n-Decane	23.9	—	—	51.2	21.6
n-Tetradecane	25.6	—	—	52.2	20.8
Cyclohexane	25.5	—	—	50.2	22.7
Decalin	29.9	—	—	51.4	22.0
Benzene	28.85	363	194	—	—
Toluene	28.5	359	208	—	—
o-Xylene	30.1	359	200	—	—
m-Xylene	28.9	357	211	—	—
p-Xylene	28.4	361	203	—	—
n-Propylbenzene	29.0	363	194	—	—
n-Butylbenzene	29.2	363	193	—	—
Average			200±7		21.8±0.7

[a] Here γ_{Hg}^d and γ_W^d are calculated as in Example 6.6.
Source: Data from F. M. Fowkes, *Ind. Eng. Chem.*, 56:40 (1964).

2. For high-energy surfaces, $\pi_e > 0$ owing to adsorption. Relationships have been derived which express γ_S^d in terms of gas adsorption.

Values of γ_S^d which have been determined for high- and low-energy surfaces by these two methods are listed in Table 6.5.

Examination of the values of γ^d for high- and low-energy solids in Table 6.5 reveals several interesting points. To begin with, there is a slight overlap in the data inasmuch as polypropylene was studied by the gas adsorption technique, even though it might reasonably be expected to resemble paraffin wax or polyethylene in γ^d value. As the table shows, the value of γ^d for this substance lies in the same neighborhood as the values for these other compounds, even though very widely different procedures are used to arrive at the different values. The multiple values of γ^d listed for ,some high-energy solids correspond to evaluations based on the adsorption of different gases at widely different temperatures. For

Table 6.5 Values of γ_S^d (in mJ m^{-2}) for a Variety of Solids as Determined From Contact Angles and From Gas Adsorption

	γ_S^d values as determined from measurements of	
Material	θ	π_e
Dodecanoic acid on Pt	10.4, 13.1	—
Polyhexafluoropropylene	11.7,* 18.0	—
Polytetrafluoropropylene	19.5	—
n-C$_{36}$H$_{74}$	21.0	—
n-Octadecylamine on Pt	22.0,* 22.1	—
Paraffin wax	23.2,* 25.5	—
Polypropylene	—	26, 28.5
Polytrifluoromonochloro-		
ethylene (Kel F)	30.8	—
Nylon-6,6	33.6*	—
Polyethylene	31.3,* 35.0	—
Polyethyleneterephthalate	36.6*	—
Polystyrene	38.4,* 44.0	—
BaSO$_4$	—	76
Silica	—	78
Anatase (TiO$_2$)	—	89, 92, 141
Iron	—	89, 106, 108
Graphite	—	115, 120, 123, 132

Source: Most data from F. M. Fowkes, *Ind. Eng. Chem.*, *56*:40 (1964); those data labeled with an asterisk are from Ref. 7.

example, the γ^d values for TiO$_2$ are obtained from adsorption studies conducted with butane at 0°C, heptane at 25°C, and N$_2$ at -195°C. With these ideas in mind, the observed variation between ostensibly duplicate values becomes more acceptable. The evaluation of γ_S^d also depends on the accuracy of the value of γ^d for a substance other than the solid. Again we see that errors may be propagated in the analyses that led to the results presented in Table 6.5.

References

1. N. K. Adam, *The Physics and Chemistry of Surfaces*, Dover, New York, 1968.
2. A. W. Adamson, *Physical Chemistry of Surfaces*, 4th ed., Wiley, New York, 1982.
3. J. J. Bikerman, *Physical Surfaces*, Academic, New York, 1970.

4. J. T. Davies and E. K. Rideal, *Interfacial Phenomena*, Academic, New York, 1961.
5. R. J. Good, in *Surface and Colloid Science*, Vol. 11 (R. J. Good and R. R. Stromberg, eds.), Plenum, New York, 1979.
6. R. E. Johnson, Jr., and R. H. Dettree, in *Surface and Colloid Science*, Vol. 2 (E. Matijević, ed.), Wiley, New York, 1969.
7. D. H. Kaelbe, *Physical Chemistry of Adhesion*, Wiley, New York, 1971.
8. A. W. Neumann and R. J. Good, in *Surface and Colloid Science*, Vol. 11 (R. J. Good and R. R. Stromberg, eds.), Plenum, New York, 1979.
9. J. F. Padday, in *Surface and Colloid Science*, Vol. 1 (E. Matijević, ed.), Wiley, New York, 1969.
10. A. C. Zettlemoyer, in *Hydrophobic Surfaces* (F. M. Fowkes, ed.), Academic, New York, 1969.

Problems

1. The frictional force was measured on a strand of viscous rayon fiber moving through a wad of identical fibers as a function of the water content of the wadded fiber. It was found that the friction increased from 59 to 133 mN cm^{-1} when the water content decreased to the point at which capillary "necking" between the fibers occurred.* A model for this situation may be visualized by considering two parallel, tangent cylinders connected by a "neck" of water held in the neighborhood of the contact by capillary forces. Sketch the situation represented by this model and explain why the force of fiber–fiber attraction increases with decreasing water content. Use this model to discuss (a) the behavior of a wet paintbrush, (b) the practice of wetting the tip of a thread before threading a needle, and (c) the dewatering of cellulose fibers to form paper.

2. The accompanying data give experimental values of the capillary rise for various liquids†:

Liquid	$\Delta\rho$ (g cm^{-3})	h (cm)	R (cm)
Water	0.9972	1.4343	0.10099
Benzene	0.8775	1.5425	0.043135
$CHCl_3$	1.4869	1.921	0.01932

---

*J. Skelton, *Science, 190*:15 (1975).
†T. W. Richards and E. K. Carver, *J. Am. Chem. Soc.*, *43*:827 (1921).

Use values from the following table (interpolated from Ref. 9) to evaluate a^2 (and γ) by the successive approximation technique illustrated in Table 6.3:

β	0	0.02	0.04	0.06	0.08	0.10	0.12	0.14	0.16	0.18	0.20
$(x/b)_{90°}$	1.00	0.997	0.994	0.991	0.987	0.984	0.981	0.978	0.975	0.972	0.970

Compare the values of γ calculated by this procedure with those obtained by the approximation given by Eq. (4).

3. Finely dispersed sodium chloride particles were prepared, their specific area was measured, and their solubility in ethanol at 25°C was studied.* It was found that a preparation with a specific area of 4.25×10^5 cm^2 g^{-1} showed a supersaturation of 6.71%. Estimate the radius of the NaCl ($\rho = 2.17$ g cm^{-3}) particles, assuming uniform spheres. Calculate γ for the NaCl–alcohol interface from the solubility behavior of this sample.

4. Enüstün and Turkevich† prepared SrSO$_4$ ($\rho = 3.96$ g cm^{-3}) precipitates under conditions which resulted in different particle sizes. Particle sizes were characterized by electron microscopy and solubilities were determined at 25°C by a radiotracer technique. In the following data the supersaturation ratios are presented for different preparations, each of which is characterized by an average particle width and a minimum particle width:

x_{mean} (Å)	x_{min} (Å)	S/S_0
247	96	1.43
269	130	1.35
388	155	1.28
541	168	1.29
629	252	1.16
1260	378	1.10
1660	500	1.07

Which size parameter gives the best agreement with the Kelvin equation? Explain. Use the best fitting data to evaluate γ for the SrSO$_4$–H$_2$O interface.

5. Bartell and Osterhof‡ describe an experimental procedure for measuring the work of adhesion between liquids and solids. With carbon (lampblack) as the solid, the following values for the work of adhesion were obtained:

Liquid	Benzene	Toluene	CCl$_4$	CS$_2$	Ethyl ether	H$_2$O
W_{AB} (erg cm^{-2})	109.3	110.2	112.4	122.1	76.4	126.8

---

*F. Van Zeggeren and G. C. Benson, *Can. J. Chem.,* *35*:1150 (1957).
†B. V. Enüstün and J. Turkevich, *J. Am. Chem. Soc.,* *82*:4502 (1960).
‡F. E. Bartell and H. J. Osterhof, *J. Phys. Chem.,* *37*:543 (1933).

Use these data together with the surface tensions of the pure liquids from Table 6.1 to calculate the spreading coefficients for the various liquids on carbon black. Use your results to interpret the authors' observations: "About equal quantities of water and organic liquid were put into a test tube with a small amount of the finely divided solid and shaken. It was noted that the carbon went exclusively to the organic liquid phase."

6. The effect of mutual saturation on the L–V and L–L interfacial tensions is effectively illustrated by considering the spreading coefficient of one liquid on another using both the initial (unsaturated) and equilibrium values of γ. use the following data to calculate $S_{B/A}$ (equilibrium) and $S'_{B/A}$ (non-equilibrium)*:

	H_2O/air	H_2O/IAA	IAA/air	H_2O/air	H_2O/CS$_2$	CS$_2$/air
γ' (erg cm^{-2})	72.8	5.0	23.7	72.8	47.4	32.4
γ (erg cm^{-2})	25.9	5.0	23.6	70.3	48.4	31.8

Describe what happens when a drop of pure isoamyl alcohol (IAA) is placed on the surface of pure water. What happens with the passage of time? Repeat this description for the case of pure CS$_2$ on pure water.

7. Water drops were formed at the mercury–benzene interface by means of a syringe, and the contact angle (measured in the water) was recorded as a function of time. † For the interface between Hg and benzene saturated with water, γ was measured independently as a function of time. The following table summarizes these data (all measured at 25°C):

Time (h)	$\gamma_{Hg-benzene}$ (erg cm^{-2})	θ_{obs} (deg)
0.10	363.0	118
0.42	359.5	119
1.0	358.0	122
2.5	354.5	138
5.0	350.0	144
13	336.0	—
23	—	180

Using 379.5 and 34.0 erg cm^{-2}, respectively, as the values of γ for the mercury–water and benzene–water interfaces, compare the observed contact angles with the predictions of Young's equation. Comment on the fact that constant values are used for γ_{Hg-w} and $\gamma_{benzene-w}$.

---

*W. D. Harkins, *The Physical Chemistry of Surface Films*, Van Nostrand-Reinhold, Princeton, N.J., 1952.
†F. E. Bartell and C. W. Bjorkland, *J. Phys. Chem.*, *56*:453 (1952).

8. A cylindrical rod may be used instead of a rectangular plate in a slight variation of the Wilhelmy method. Derive an expression equivalent to Eq. (63) for a suspended solid of cylindrical geometry. Prepare semiquantitative plots analogous to Fig. 6.9c based on the following data, assuming cylindrical rods 1.0 mm in diameter at the air-water interface*:

Metal	ρ_s (g cm^{-3})	θ_a (deg)	θ_r (deg)
Au	19.3	70	40
Pt	21.5	63	28

In both cases the metal surfaces were carefully polished, washed, steamed, and then heated in an oven for 1 h at 100°C.

9. The vertical rod method of the preceding problem was used to study the contact angle of water at the gold-water-air junction at 25°C. The following data show how the value of θ (θ_a) depends on the prior history of the metal surface.† For a gold surface polished, washed, and heat-treated for 1 h at T°C, then allowed to stand in air for t h we obtain

T (°C)	100	200	300	400	500	600	600	600	600	600	600
t (h)	$<\frac{1}{4}$	$<\frac{1}{4}$	$<\frac{1}{4}$	$<\frac{1}{4}$	$<\frac{1}{4}$	$\frac{1}{4}$	1	5	10	24	120
θ_a (deg)	68	57	45	36	25	13	22	38	47	53	55

Calculate the work of adhesion between water and gold for each of these cases on the assumption that $\pi_e = 0$. Is the variation in W_{SL} consistent (qualitatively? quantitatively?) with the expected validity of the assumption concerning π_e?

10. The tendency of spilled mercury to disperse as small drops which roll freely on most surfaces is a well-known characteristic of this liquid. Discuss this behavior in terms of (a) the work of adhesion and the spreading coefficient for mercury on various substrates, (b) surface tension versus contact angle (measured in Hg) as causes of this behavior, and (c) the implications of the Kelvin equation on the health hazards associated with mercury spills.

11. Use the data of Table 6.2 to plot the profile of a drop with $\beta = 25$. Measure (in cm) the radius of the drop you have drawn at its widest point. By comparing this value with the value of $(x/b)_{90°}$ from the table, evaluate b (in cm) for the drop as you have drawn it. Suppose an actual drop is characterized by this value of β. If the actual radius at the widest point is 0.25 cm and $\Delta\rho = 0.50$ g cm^{-3}, what is γ for the interface of the drop?

12. Suppose the drop profile shown in Fig. 6.14b describes an actual drop for which the radius at the widest point equals 0.135 cm. Use the value of $(x/b)_{90°}$ from the figure to calculate γ for each of the following situations:

---

*F. E. Bartell, J. A. Culbertson, and M. A. Miller, *J. Phys. Chem.*, **40**:881 (1936).
†F. E. Bartell and M. A. Miller, *J. Phys. Chem.*, **40**:889 (1936).

System	$|\Delta\rho|$ (g cm^{-3})
(a) Oil in water	0.20
(b) Water in oil	0.20
(c) Oil in air	0.80
(d) Air in water	1.00
(e) Water in air	1.00

State whether the drop is pendant or sessile in each case.

13. Sometimes it is difficult to locate the bottom of the meniscus of a colorless solution in a buret. If this is the case, it is surely that much more difficult to try to estimate contact angles by direct observation of a meniscus. In view of this, criticize or defend the following proposition:

> Before attempting to read contact angles directly by viewing the surface through a low-power microscope, a new worker should practice with simulated drops.* By masking off the lower part of a sphere such as a ball bearing and looking at its silhouette, a practice junction is obtained. The angle estimated can be compared with the true value by calculating the latter from the height of the apex and the width of the base. In fact, values for actual drops can be determined by this method if the profile of the entire drop is visible in the microscope.

14. The following combinations of θ, γ_L, and p values were reported by Bartell and Whitney† for the wetting of silica plugs by various liquids:

Liquid	p (g cm^{-2})	γ_L (dyne cm^{-1})	θ
Nitrobenzene	125	25.3	41°25'
Chloroform	192	31.6	22°11'
Benzene	215	34.7	19°16'
Toluene	221	36.5	—
CCl$_4$	265	44.5	—
Hexane	333	51.0	—

Use the first three sets of data to determine a value of R for the plug; then use this value to determine θ for the remaining liquids against silica. Are large or small contact angles more sensitive to, say, a 5% error in R?

15. Drake‡ has reported the accompanying data for the porosimetry analysis of a catalyst preparation; the cumulative pore volume occupied by mercury is given for the applied pressures indicated:

---

*Recommended in Ref. 8.

†F. E. Bartell and C. E. Whitney, *J. Phys. Chem.*, *36*:3115 (1932).

‡L. C. Drake, *Ind. Eng. Chem.*, *41*:781 (1949).

Pressure (psi)	Volume (cm³ g⁻¹)	Pressure (psi)	Volume (cm³ g⁻¹)
1,000	0.082	20,000	0.228
1,500	0.115	25,000	0.249
2,000	0.132	30,000	0.276
3,000	0.150	35,000	0.307
4,000	0.160	40,000	0.336
6,000	0.177	45,000	0.358
10,000	0.190	50,000	0.363
15,000	0.213		

Plot these data and from the tangents to the curve estimate dV/dR at 2000, 10,000, 30,000, and 45,000 psi. To what radii do these pressure correspond? Use $\gamma = 484$ mJ m^{-2} and $\theta = 140°$ for these calculations.

16. By a suitable combination of Eqs. (49) and (97) show that

$$\gamma_S^d = \frac{\gamma_{LV}^2}{4\gamma_L^d}(1 + \cos\theta)^2$$

for low-energy surfaces. Use the following data (see Table 6.4 for reference) to evaluate either γ_S^d or γ_L^d as appropriate:

S	L	γ_{LV} (erg cm^{-2})	γ^d (erg cm^{-2})	θ (deg) in liquid
Dodecanoic acid on Pt	α-Bromo-naphthalene	44.6	10.4 for S	92
Kel F	α-Bromo-naphthalene	44.6	30.8 for S	48
Paraffin wax	Glycerol	63.4	36 for L	97
Paraffin wax	Fluorolube	20.2	13.5 for L	31

17. The equation derived in the preceding problem suggests that a plot of $\cos\theta$ (as ordinate) versus $\sqrt{\gamma_L^d/\gamma_L}$ (as abscissa) should be linear with a slope of $2\sqrt{\gamma_S^d}$ and an intercept of -1. Describe how this result can be used to evaluate γ_S^d when contact angle measurements are made on a particular solid with a variety of liquids for which γ_L and γ_L^d are known. Describe how the same graphing procedure can be used to evaluate $\sqrt{\gamma_L^d/\gamma_L}$ when contact angles are measured in a liquid of unknown γ_L^d on different solids of known γ_S^d. Use the data of the preceding problem to illustrate these two graphical interpretations. Include the additional datum that α-bromonaphthalene forms a contact angle of 58.5° with paraffin (see Table 6.4 for reference).

18. If γ_L and θ are measured for a homologous series of liquids on a given low-energy solid, a plot of $\cos\theta$ versus γ_L results in a straight line. Verify that this

is the case for the following data, determined for alkanes on Teflon at 20°C*:

Compound	γ_L (mJ m^{-2})	θ (deg)	Compound	γ_L (mJ m^{-2})	θ (deg)
Hexadecane	27.6	46	Nonane	22.9	32
Tetradecane	26.7	44	Hexane	21.8	26
Dodecane	25.4	42	Heptane	20.3	21
Undecane	24.7	39	Octane	18.4	12
Decane	23.9	35	Pentane	16.0	spreads

The intercept at $\gamma = 0$ is viewed as a kind of critical state for these systems, and the corresponding surface tension is represented by γ_c. What is the value of γ_c for Teflon?

19. Use Eq. (99) to show that γ_c as defined in the last problem is equal to γ^d for those systems where $\gamma_L = \gamma^d$. What else must be assumed to prove this?

---

*H. W. Fox and W. A. Zisman, *J. Colloid Sci.,* 5:514 (1950).

7
ADSORPTION FROM SOLUTION

You are living on a plane. What you style Flatland is the vast level surface of what I may call a fluid, on, or in, the top of which you and your countrymen move about, without rising above it or falling below it.

[From Abbott's *Flatland*]

7.1 Introduction

Until now, we have intentionally excluded solutes of variable concentration from our consideration of surfaces. In this chapter the effects of such solutes are our specific interest. We shall be especially concerned with a particular class of solutes which show dramatic effects on surface tension. These are said to be surface active and are often simply called surfactants. The primary emphasis is on the relationship between adsorption phenomena and surface tension–surface energy. In the process of describing experimental methods, results, and interpretations, however, a variety of related concepts will enter the picture.

In studying the material of this chapter it may be helpful to realize that the topics covered may be grouped in several different ways. Let us enumerate what these various ways of looking at the material are.

First, we may focus our attention on the solubility of the adsorbed species in one or both of the adjacent phases. In this way there is generated the study of two broad categories of phenomena: insoluble and soluble surface layers.

A second way of classifying the material of this chapter is on the basis of the experimental methods involved. For mobile interfaces surface tension is easily measured. For these it is easiest to examine the surface tension–adsorption relationship starting with surface tension data. When insoluble surface films are involved, we shall see how the difference in γ between a clean surface and one with an adsorbed film may be measured directly. For solid surfaces surface tension is not readily available from experiments. In this case adsorption may be measurable directly, and the relationship between adsorption and surface tension may be examined from the reverse perspective.

Third, the material of this chapter is a mix of descriptive and theoretical concepts. The most important descriptive observation is the existence of two-dimensional phases. The theoretical content of the chapter is mostly thermodynamic in origin. Three major results are the equations named after Gibbs, Langmuir, and Lippmann. We are mostly concerned with uncharged surfaces, except for a brief discussion of electrolyte adsorption at a polarizable mercury electrode.

Finally, the material of this chapter may be regarded as a mixture of fundamentals and applications. Although the entire book stresses principles, applications are considered from time to time as examples of more abstract ideas. This is also the intent of the sections on applications in this chapter. In addition, however, many applications of adsorption phenomena are the basis of large and important areas of technology. To omit mention of them would lead to a very incomplete picture of these fields. As it is, many important applications must be omitted for lack of space, and those which are mentioned are sketched in only a superficial way.

7.2 Spread Monolayers

Suppose a dilute solution is prepared from an aliphatic solvent and an organic solute RX, where R is a long-chain alkyl group and X is a polar group. Then a small amount of this solution is placed on a large volume of water with a horizontal surface. The components of this system were chosen because they are assumed to meet the following experimental criteria:

1. The solubility in water of both components of the organic phase is negligible at room temperature.
2. The likelihood of any complex being formed between the organic solvent and solute is exceedingly low.
3. The volatility of the organic solvent is high and that of the solute is low.

With these facts in mind, let us examine the fate of the drop of solution placed on the surface of water. The initial spreading coefficient [Eq. (6.61)] for the organic layer on water, $S_{O/W}$, is positive. This is primarily because γ_{OW} is unusually low and γ_W is high, even with an adsorbed layer of the organic solvent. After spreading, we allow sufficient time to elapse for all the solvent to evaporate from the spread layer. At this point the surface will contain a layer of the organic solute similar to that which would result from the spreading of a sessile drop of pure liquid solute or from the adsorption of vapors of the solute component from the gas phase. Using a solution with a volatile solvent to form such a layer is a very common technique and has the advantage of permitting very small amounts of solute to be quantitatively deposited on a surface.

After the solvent has evaporated, the nature of the remaining layer depends on the amount of solute deposited and the area available to it. It is convenient to distinguish between three situations in this regard. If the amount of added material and the area are such that the water surface is covered uniformly to a depth of one molecule with the solute, the resulting film is called a monolayer. On the other hand, submonolayer coverage and multilayer coverage result when the amount of added material per area is less or more, respectively, than that which produces the monolayer. In this chapter we shall be concerned mostly with degrees of coverage up to and including the monolayer. If a large excess of spread material (beyond the amount needed for monolayer coverage) is used, the excess collects into droplets of a bulk phase. The equilibrium situation is then identical to what would be produced by the spreading of a sessile drop of the solute material.

Films of the sort described here are called either spread monolayers (when the method of their preparation is stressed) or insoluble monolayers (when the chemical nature of the solute is emphasized). We shall use these terms interchangeably. Now let us examine some of the properties of the spread monolayer that we have described.

It was seen in the preceding chapter [e.g., Eq. (6.48)] that the presence of an adsorbed layer lowers the surface tension of an interface. The phenomenon is quite general, so we shall redefine π (no subscript) in the following symbols:

$$\pi = \gamma_0 - \gamma \tag{1}$$

where γ_0 refers to the surface of any phase in the absence of adsorption and γ refers to the same surface with an adsorbed layer. Specific subscripts are used only when the problem clearly involves more than one interface.

The spread monolayer just described may be discussed from two points of view. First, there are those aspects of the film that pertain explicitly to the chemical nature of the components: water and the organic solute. Second, there are certain properties of the monolayer that depend on physical variables such as temperature, area of the water surface, and number of molecules of RX present. Let us briefly discuss both of these viewpoints.

The organic solute RX is a prototype of an important array of surface-active materials. Many surface-active substances are composed of what are known as amphipathic molecules. This term simply means that the molecule consists of two parts, each of which has an affinity for a different phase. We shall be concerned mostly with surfaces in which one of the phases is aqueous, so the surfactants that we consider will contain polar or ionic groups, or "heads," and nonpolar organic residues, or "tails." In the compound RX, for example, R is an alkyl group, generally containing 10 or more carbon atoms. The literature of surface chemistry contains many references to these organic groups by both their IUPAC and their common names. Table 7.1 lists some of the more commonly encountered examples. In RX, the polar X group may be—OH,—COOH,

Table 7.1 IUPAC and Common Names for a Variety of Normal Saturated and Unsaturated Surface-Active Compounds

	Normal, saturated compounds		
	Carboxylic acids		Alcohols, amines, sulfates, etc.
n (number of C atoms)	IUPAC name	Common name	Common name
12	Dodecanoic	Lauric	Lauryl
14	Tetradecanoic	Myristic	Myristyl
16	Hexadecanoic	Palmitic	Cetyl
17	Heptadecanoic	Magaric	Heptadecyl
18	Octadecanoic	Stearic	Stearyl
20	Eicosanoic	Arachidic	Eicosyl, arachic
22	Docosanoic	Behenic	Docosyl

	Normal, unsaturated carboxylic acids	
	IUPAC name	Common name
18	cis-9-Octadecenoic	Oleic
18	cis,cis-6,9-Octadecenoic	Linoleic
18	cis,cis,cis-3,6,9-Octadecenoic	Linolenic
18	trans-9-Octadecenoic	Elaidic
22	cis-9-Docosenoic	Erucic
22	trans-9-Docosenoic	Brassidic

$-CN, -CONH_2$, or$-COOR'$ or an ionic group such as$-SO_3^-$$-OSO_3^-$, or$-NR_3^+$.

With the foregoing ideas in mind, one characteristic of the adsorbed monolayer becomes apparent: molecular orientation at surfaces. For a film of RX on water, the picture that emerges is one in which the polar groups are incorporated into the aqueous phase with the hydrocarbon part of the molecule oriented away from the water. Such details as the depth of immersion of the tail and the configuration of the alkyl group are best approached by considering how the properties of the monolayer depend on the physical variables.

Next let us consider some of the physical properties of the spread monolayer we have described. Equation (1) states that the surface tension of the covered surface will be less than that of pure water. It is quite clear, however, that the magnitude of γ must depend on both the amount of material adsorbed and the area over which it is distributed. The spreading technique already described enables us to control the quantity of solute added, but so far we have been vague about the area over which it spreads. Fortunately, once the material is deposited on the surface, it stays there—it has been specified to be insoluble and nonvolatile for precisely this reason. This means that some sort of barrier resting on the surface of the water may be used to "corral" the adsorbed molecules. Furthermore, moving such a barrier permits the area accessible to the surface film to be varied systematically. In the laboratory this adjustment of area is quite easy to do in principle. As we shall see later, the actual experiments must be performed with great care to prevent contamination.

Suppose that the initial film is spread on water which fills to the brim a shallow tray made of some inert material. Rods with low-energy surfaces may then be drawn across the water to adjust the area accessible to the molecules of the monolayer. Figure 7.1a indicates schematically how such an arrangement might appear. In practice, several barriers would be used, first to sweep the surface free of insoluble contaminants and then to confine the monolayer.

Next an experiment such as that shown in Fig. 7.1b could be conducted. The apparatus consists of a pair of Wilhelmy plates attached to two arms of a balance. One plate contacts the clean surface, and the other the surface with the monolayer. Note that the barrier separates the two portions of surface. The surface tension will be different in the two regions, and the weight (and volume) of the meniscus entrained by the plate will be larger for the clean surface because of its higher surface tension [Eq. (6.2)]. Ideally, the two plates are identical in weight, perimeter, and contact angle, although the last may be difficult to achieve

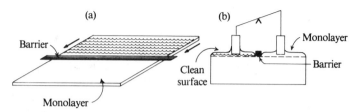

Figure 7.1 (a) Schematic illustration of a barrier delineating the area of a monolayer and (b) a Wilhelmy plate arrangement for measuring the difference in γ on opposite sides of barrier.

in practice. If these conditions are met, however, the additional weight needed to bring the apparatus to balance measures the difference in γ for the two surfaces. By Eq. (6.2), this is given by

$$w_{clean} - w_{film} = 2P \cos \theta(\gamma_0 - \gamma) \tag{2}$$

Of course, there is no necessity to measure both γ_0 and γ on the same apparatus; they may be determined independently by any of the methods of the preceding chapter. The experiment represented by Fig. 7.1b is intended mainly to emphasize that the surface tension of the two areas will be different. Furthermore, as the barrier is moved in such a way as to compress the area of the spread monolayer, the value of π will increase.

Although π and the area A of the surface vary inversely, the precise functional form by which they are related is more difficult to describe. For very large areas π and A show a simple inverse proportionality such as pressure and volume for an ideal gas. As the area is decreased, a more complex relationship is needed to connect these variables, just as the equation of state becomes more complex for nonideal gases and condensed phases. This analogy of π and A with p and V turns out to be a very profitable way of thinking about insoluble monolayers. For one thing, it suggests an alternative to the difference between two values of surface tension as a means of measuring π. In addition, the analogy to three-dimensional states suggests models for understanding monolayers. In succeeding sections each of these points is developed in greater detail.

Identification of area as the two-dimensional equivalent of volume is a straightforward geometrical concept. That π should be interpreted as the two-dimensional equivalent of pressure is not so evident, however, even though the notion was introduced without discussion in Sect. 6.6. Figure 7.2 will help to clarify this equivalency as well as suggest how to

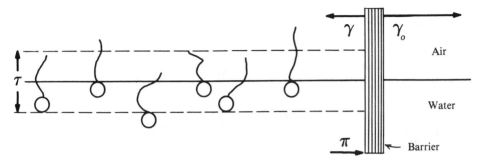

Figure 7.2 Schematic profile of the air–water interface at a barrier which separates a monolayer from the clean surface.

quantitatively compare two- and three-dimensional pressures. The figure sketches a possible profile of the air–water surface with an adsorbed layer of amphipathic molecules present. In general, we must allow for the fact that different configurations might exist among the adsorbed molecules; nevertheless, the surface layer has some mean thickness τ.

If the barrier represents the limit of the monolayer, then it is clear that the contractile force exerted by the surface is different on opposite sides of the barrier. Since γ is less than γ_0, it is as if the film were exerting a force on the barrier along the perimeter of the film equal to π. Force per unit length—the units of γ—is the two-dimensional equivalent of force per unit area, the units of pressure.

The surface layer does not have zero thickness, of course, even though it is conceptually convenient to think of it as two-dimensional matter. If we assume that the film pressure π extends over the entire thickness of the film, then it is an easy problem to convert the two-dimensional pressure to its three-dimensional equivalent. Taking 10 mN m^{-1} as a typical value for π and 1.0 nm as a typical value for τ enables us to write

$$p = \frac{\pi}{\tau} = \frac{10^{-2} \text{ N m}^{-1}}{10^{-9} \text{ m}} = 10^7 \text{ N m}^{-2} \tag{3}$$

or

$$p = 10^7 \text{ N m}^{-2} \times \frac{1 \text{ atm}}{1.013 \times 10^5 \text{ N m}^{-2}} \simeq 100 \text{ atm} \tag{4}$$

In view of this calculation, it is not too surprising that insoluble monolayers do not usually display a simple inverse proportionality

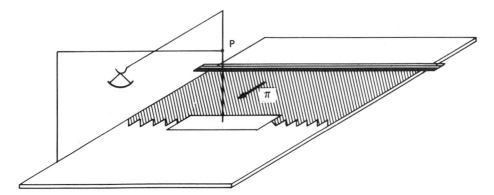

Figure 7.3   Schematic representation of a Langmuir film balance.

between π and A. At pressures this high, three-dimensional matter is not likely to obey the ideal gas law either.

7.3   The Langmuir Film Balance

The considerations of the preceding section suggest a second way to study spread monolayers. This technique involves measuring the film pressure directly rather than calculating it from surface tension differences by Eq. (1). Figure 7.3 is a schematic representation of an apparatus called the Langmuir film balance after I. Langmuir (Nobel Prize, 1932), a pioneering worker in this field. Its base is a shallow tray or trough of some inert material. As was the case in Fig. 7.1a, the surface must be swept by barriers both to clean the surface and to compress monolayers. In the Langmuir balance, however, one of the barriers is attached to a pivoted arm, arranged in such a way that a torque balance around point P can be measured, for example, by adding weights to the pan. As the figure shows, an insoluble monolayer may be confined to a portion of the surface adjacent to the pressure-sensing float. The shaded area in the figure corresponds to the area of the film. It is obviously adjustable by moving the other barrier. To prevent the film from leaking past the edges of the float, flexible barriers connect the ends of the float to the edges of the tray.

By means of this apparatus, it is possible to vary the area of a spread monolayer and measure the corresponding film pressure directly. Many different variations of the film balance have been built. Figure 7.4 is a photograph of a commercial apparatus in which the float is attached to a torsion wire which may be calibrated to read film pressures directly.

Figure 7.4 Photograph of a commercial film balance. (Courtesy of Central Scientific Company, Ill., used with permission.)

When the torsion wire is 0.010-in. steel, at $1°$ rotation of the wire corresponds to about 0.3 mN m^{-1} of pressure.

Although the Langmuir balance is quite simple conceptually, obtaining unambiguous results by this technique is far from simple. As we have done in discussing other experimental techniques in this book, we shall only touch upon those aspects of the method which will somehow contribute to our fundamental understanding of insoluble monolayers. Anyone considering experiments of this sort should consult more detailed discussions, such as the book by Gaines [4].

A convenient way to discuss the Langmuir balance is to examine the difficulties involved in measuring each of the two-dimensional state variables: π, A, T, and, of course, the number of moles n of material in the insoluble layer.

The float–torsion wire assembly is the pressure-sensing system in the film balance. The flexible barriers that connect the float to the edges of the tray must be considered part of this mechanism. As with gas pressure determinations, it is essential that the system be leakproof. For this reason the float and flexible barriers must always be hydrophobic, that is, not wetted by the aqueous substrate. If these surfaces were wet by water, the possibility of surfactant transferring to them—that is, a leak in the system—would be enhanced. Thin pieces of mica are commonly used as float material, and platinum ribbons or threads of silk or nylon are often used for flexible barriers. All of these are waxed to give them suitably hydrophobic characteristics. The sweeping barriers and the tray must also be hydrophobic for the same reasons. These are usually waxed metal, although Teflon is also quite popular because of its inertness. The barriers must make intimate contact with the edges of the tray, also to prevent leaks. Therefore both tray and barriers must be carefully machined to assure good contact. Plastics are generally unsuitable as barrier materials because they are too light to make good contacts.

To convert the measured torque into a two-dimensional pressure, it is necessary to know both the length of the float and the distance between the float and the torsion wire. The latter requires that the water level be controlled quite accurately. As far as the length of the float is concerned, the flexible connectors must be included in this figure. Since they are anchored at one end, only part of their length may be considered a part of the pressure-sensing system. Some approximations are required here, but if the length of the connector is small compared to the total length of the float, the error is negligible.

The float is effectively a two-dimensional manometer, and, like its open-ended counterpart, it measures the film pressure difference between the two sides of the float. This is another reason why it is imperative that no leakage occur past the float assembly: Leakage would increase the pressure on the reference side of the float. For the same reason, the side of the float opposite the monolayer must be carefully checked for any possible source of contamination, not just misplaced surfactant. One way of doing this is to slide a barrier toward the float from that side to verify that no displacement of the float occurs. In all aspects of film pressure measurement, the torque must be measured with sufficient sensitivity to yield meaningful results.

Measuring the area of the film is less troublesome. If the edges of the tray are parallel and the barriers perpendicular to them, the area of the rectangular surface is easily determined. The curvature of the surface at the hydrophobic boundaries introduces a small error, but—since the total area is of the order of magnitude of 10^{-2} m²—this is generally negligible.

The results obtained in π—A experiments may be sensitive to the rate at which the film area is changed. We shall not discuss the factors responsible for this but shall merely note that the same film pressures should be obtained on compression and expansion if true equilibrium values are being measured.

Temperature is an important variable in any equation of state. The experiments we are describing for isothermal; therefore it is important that both the water substrate and the adjoining vapor be thermostated.

Next let us consider those difficulties associated with the determination of the amount of material deposited on the surface. We have already noted that the method of depositing insoluble monolayers by spreading permits the accurate determination of n. Since the spreading technique requires solvent volatility, care must be exercised to prevent the stock solutions from changing concentration due to evaporation prior to their application to the surface. Also, precise microvolumetric methods must be used to dispense the solution on the aqueous surface, since the quantity used is small. The solvent (as well as the solute) must be free from contaminants. There is also the possibility that the solvent will extract spreadable contaminants from the waxed surfaces of the float, barriers, and tray. Some workers advocate addition and evaporation of one drop at a time to minimize this. Oily contaminants may also reach the water surface from the fingers and from the atmosphere. These last sources are particularly hard to control: Tests for reproducibility and blank compressions (i.e., moving the barrier toward the float on a "clean" surface) are the best evidence of their absence.

Not all solvents are equally suitable for spreading monolayers. The requirement that the solvent evaporate completely is self-evident. It has been suggested that if the organic solvent dissolves much water, the properties of the monolayer will be different from those in which no water is trapped. Verification that no artifacts are entering the observations from the solvent may be accomplished by conducting duplicate experiments with different solvents.

Until now we have concentrated on those difficulties in using the Langmuir balance which arise from the determination of π, A, T, and n. A few remarks are also in order about considerations that may affect the monolayer and which originate in the adjacent phases. We have already discussed contaminants originating in the gaseous phase. For some monolayer materials air oxidation may also be a problem. The aqueous substrate is the source of a wide assortment of contaminants in addition to spreadable oily matter. Ionic impurities, including those which affect the pH, are quite troublesome. The charge state of many amphipathic molecules, for example, amines and carboxylic acids, is obviously pH dependent. Salts or complexes formed between amphipathic molecules and ions in the aqueous phase will have different monolayer properties from those of the unreacted surfactant molecule.

7.4 Results of Film Balance Studies

The preceding section shows that is is possible to determine π–A isotherms for surfaces just as p–V isotherms may be measured for bulk

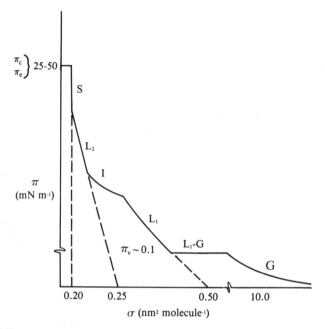

Figure 7.5 Composite two-dimensional pressure (π)–area (σ) isotherm which includes a wide assortment of monolayer phenomena. Note that the scale of the figure is not uniform so that all features may be included on one set of coordinates.

matter. The results that are obtained for surfaces are analogous to bulk observations also, although some caution must be expressed about an overly literal correlation between bulk and surface phenomena. We shall return to a discussion of these reservations later. There can be no doubt, however, that analogies with bulk behavior supply a familiar framework within which to consider π–A isotherms.

The curve sketched in Fig. 7.5, which is drawn with grossly distorted coordinates to encompass all features, contains several similarities to p–V isotherms. Not all the features shown here are always observed, nor are all known idiosyncrasies of π–A isotherms represented. The presence or absence of various features and their π–A coordinates vary with temperature for a particular amphipathic molecule and from one amphipathic substance to another. Lastly, there is some diversity in the terminology used to describe various monolayer phenomena. In short, Fig. 7.5 is a composite isotherm which will introduce and summarize a variety of observations.

In this section we shall discuss in turn the various two-dimensional phases and phase equilibria represented in Fig. 7.5, progressing from low values of π to high ones. The existence of these two-dimensional states and the properties they possess are presumably unfamiliar to most readers. Therefore it is important to keep the following ideas in mind in reading this section:

1. We are concerned with *two-dimensional matter* situated at the boundary between two bulk phases.
2. The properties of the two-dimensional phases are relatively independent of the properties of the bulk phases of the same material.
3. Many surface states are two-dimensional analogs of three-dimensional states. As with any analogy, however, there are points of similarity and points of difference between the surface and bulk states. For most of the states we discuss, we shall consider the phenomenological behavior as represented by Fig. 7.5 and also suggest a molecule interpretation in terms of Fig. 7.6.

If measurements can be made at sufficiently low pressures, all monolayers will display gaseous behavior, represented by region G in Fig. 7.5. The gaseous region is characterized by an asymptotic limit as π → 0. In the limit of very low film pressures, a two-dimensional equivalent to the ideal gas law applies:

$$\pi A = nRT \tag{5}$$

where R is the gas constant, usually in SI units. This is a convenient place to define another quantity, the area occupied per molecule in the interface, σ. Since R equals Avogadro's number times the Boltzmann constant and nN_A equals the total number of surface molecules, we may write

$$\pi \left(\frac{A}{nN_A} \right) = \pi\sigma = kT \tag{6}$$

As a model for this highly expanded state we may choose a situation like that shown in Fig. 7.6a, in which the hydrocarbon chain lies flat on the surface, blocking an area πl^2, where l is the length of the "tail." We may then use Eq. (6) to calculate the value of π corresponding to this area per molecule. At 25°C and with $l = 1.0$ nm, we obtain

$$\pi = \frac{kT}{\sigma} = \frac{(1.38 \times 10^{-23})(298)}{(3.14)(10^{-9})^2} = 1.31 \text{ mN m}^{-1} \tag{7}$$

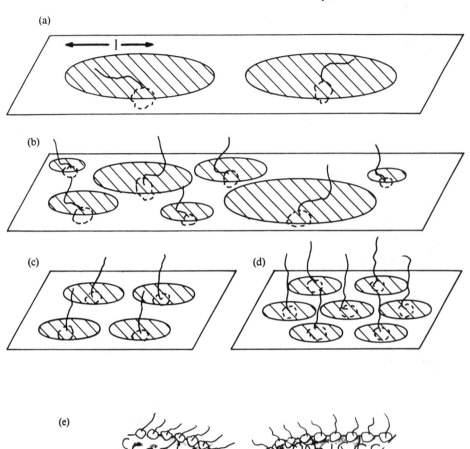

Figure 7.6   Schematic illustration showing by shading the effective area per molecule at various stages of monolayer compression: (a) gaseous state, (b) liquid expanded state, (c) liquid condensed state, and (d) solid state. In (e) the collapse of the film is illustrated.

Allowing only this area per molecule is equivalent to defining a distance of closest approach. Therefore ideal behavior is not expected until areas per molecule are larger than this, on the order of 10 nm² perhaps. Correspondingly lower film pressures will be involved also. If we recall that the pressure of this example is equivalent to a bulk gas pressure of about 13 atm, it is less surprising that such low film pressures are needed to observe gaseous monolayer behavior. The repulsion between particles in a charged monolayer increases the effective area these molecules occupy at the surface. The effect of this is to increase the pressure of charged films, making them more accessible to measurement. Since ionic surfactants are soluble, techniques other than the film balance must be used to study their π–σ isotherms.

Like its three-dimensional counterpart, Eq. (5) is a limiting law which means that deviations may be expected at higher pressures, lower temperatures, or with more strongly interacting molecules. Figure 7.7 is a plot of $\pi\sigma/kT$ versus π for several members of the carboxylic acid

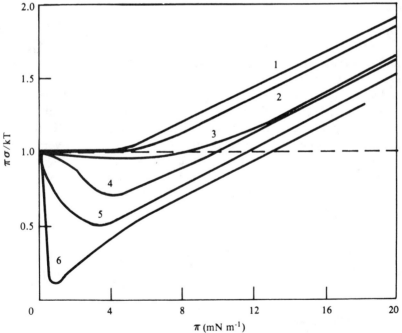

Figure 7.7   Plots of $\pi\sigma/kT$ versus π for n-alkyl carboxylic acids: (1) C_4, (2) C_5, (3) C_6, (4) C_8, (5) C_{10}, and (6) C_{12}. [Data from N. K. Adam, *Chem. Rev., 3*:172 (1926).]

homologous series. The film balance was used to collect the data only for the C_{12} acid. Shorter chain compounds are soluble and were investigated by surface tension measurements and interpreted by the Gibbs equation, which we discuss in Sect. 7.8. The main point to note about Fig. 7.7 is the strong resemblance it bears to similar plots for three-dimensional gases. Negative deviations occur at low pressures, becoming more pronounced as the length of the alkyl chain increases. As with gases, this may be attributed to attraction between molecules, an effect that increases with chain length. At higher pressures deviations tend to be positive. This is analogous to the excluded volume effect for gases, except that it becomes an excluded area in two dimensions. An application of this idea is taken up in the following example.

Example 7.1 The molecular weight of a gas may be determined by measuring the mass of the sample as well as a set of p, V, and T values for the same sample. Use this fact to criticize or defend the following proposition: The molecular weight of a solute in a monolayer may be calculated by the formula $M = mRT/\pi A$, where A/m gives the area mass^{-1} of the monolayer under a pressure π and at a temperature T.

Solution The formula given is the two-dimensional analog of the ideal gas law; therefore two conditions must be met to justify its use. First, the monolayer must be in the gaseous state and, second, the gas pressure must approach zero. The first point may appear trivial, since no one would apply $pV = nRT$ to a bulk liquid sample for which p, V, and T had been measured. The two-dimensional state of a monolayer is not directly perceptible, however, and π–σ data must be evaluated to verify that the monolayer is indeed in the G state. With π–A data measured over a range of (low) π's, Fig. 7.7 shows that the limiting value of $\pi A/RT$ is the number of moles in the sample. Dividing the mass of the sample by this value of n yields the correct molecular weight. The proposition needs to be qualified by adding "in the limit of $\pi \rightarrow 0$."

•

Next, let us return to Fig. 7.5, discussing the features labeled L_1 and L_1–G. The region L_1 is called the liquid state or, more commonly, the liquid expanded state to distinguish it from L_2, the liquid condensed state. The region L_1–G is a two-phase region in which the gaseous and liquid expanded state coexist in equilibrium.

The horizontal line of the L_1–G region in Fig. 7.5 is analogous in every way to the corresponding feature in bulk matter. At a given temperature there is a constant-pressure region over which a significant compression occurs. The film pressures at which L_1–G equilibrium occurs are known as film vapor pressures π_v. Like the gaseous state itself, the L_1–G equilibrium occurs at very low pressures. Tetradecanol, for example, has a two-dimensional vapor pressure of 1.1×10^{-4} N m^{-1} at 15°C.

It is important to remember the significance of π_v. It refers specifically to the equilibrium between two *surface* states. There is a danger of confusing π_v with the equilibrium spreading pressure π_e, introduced in Chap. 6. The latter is the pressure of the equilibrium film that exists in the presence of excess *bulk* material on the surface. It is the equilibrium spreading pressure that is involved in the modification of Young's equation [Eq. (6.49)], where a bulk phase is present on the substrate. For tetradecanol at 15°C the equilibrium spreading pressure is about 4.5×10^{-2} N m^{-1}, so π_e and π_v are very different from one another.

The temperature variation of π_v may be analyzed by a relationship analogous to the Clapeyron equation to yield the two-dimensional equivalent to the heat of vaporization. The numerical values obtained for this quantity more nearly resemble the bulk values for hydrocarbons than those for polar molecules. This suggests that most of the change in the surface transition involves the hydrocarbon tail of the molecule rather than the polar head.

Finally, note that the L_1–G equilibrium region disappears above a certain temperature which is the two-dimensional equivalent of the critical temperature for liquid–vapor equilibrium.

Because of the very low pressures at which they are observed, the surface states we have discussed so far are difficult to observe and relatively unimportant experimentally. They are easy to understand, however, so they serve well to introduce the idea of surface states. The liquid expanded state, L_1 in Fig. 7.5, is the first of several condensed states we shall discuss. Because it is bound on the low-pressure side by a two-phase region with a critical temperature, the L_1 state is easily compared to a bulk liquid state. Since gaseous behavior is observed only at very low pressures, it is easy to extrapolate the isotherm for the L_1 state to the $\pi = 0$ value as the dashed line in the figure shows. For amphipathic molecules with saturated unbranched R groups, this intercept is in the range 0.45–0.55 nm^2. We shall identify the limiting area per molecule (superscript zero) for this state (subscript) by the symbol $\sigma^0_{L_1}$. The presence of branched chains or double bonds—particularly in the cis configuration—increases the value of this limiting area.

The precise structural details of the liquid expanded state at the molecular level are not fully understood, but several generalizations do appear to be justified. The value of $\sigma^0_{L_1}$ is several times the actual cross-sectional area of the amphipathic molecule, with the latter oriented perpendicular to the surface. At the same time the area per molecule is considerably less than could be permitted if the entire tail were free to move in the surface. Figure 7.6b represents a model of the surface in the L_1 state at the molecular level. Here part of the hydrocarbon chain lies in the surface and some has been lifted out of the surface plane. That portion of the tail in the surface defines the effective area per molecule. Neither this area nor the length of the chain which is out of the surface will be the same for all molecules, so experimental values of $\sigma_{L_1}^0$ correspond to average values. Furthermore, there will be considerable lateral interaction between those segments which are not in contact with the substrate.

The compressibility of the L_1 state is expected to be far less than that of the gaseous state, but not yet incompressible, because the average area per molecule may be altered by squeezing additional CH_2 groups out of contact with the water.

With sufficient compression the isotherm of the L_1 state shows a sharp break to enter a situation variously known as the intermediate or transition state, indicated by I in Fig. 7.5. From the sharpness of the break in the isotherm, this situation might be initially regarded as another two-phase equilibrium. This is especially tempting in view of the fact that one finds at still greater compression one or more additional states which are far less compressible than the liquid expanded state. Region I is definitely not horizontal, however, as would be required for a first-order phase transition. As a matter of fact, there is scarcely a discontinuity in the isotherm at the most compressed extreme of the intermediate state. Figure 7.8 shows a set of experimental isotherms for tetradecanoic acid on a substrate of 0.01 M HCl which keeps the C_{14} acid un-ionized. These isotherms show region I particularly clearly and reveal that its characteristics vary quite regularly.

At present, no entirely satisfactory molecular interpretation of the I state exists. As its name implies, the I state is something between the L_1 and L_2 states. Perhaps some cooperative interaction between R groups causes these to cluster together with a vertical orientation in the manner of a transition state between the structures shown in Fig. 7.6b and c. If the I state is a transition state that somehow gets trapped in transit, it should be described as a metastable rather than a stable state. We saw in Sect. 6.8 that metastable states are associated with nonreversible behavior, and I states seem to be reversible. It would be informative to study the reversibility of the intermediate portion of a surface isotherm using different sweep rates to make sure that no kinetic features are being confused with stable two-dimensional states.

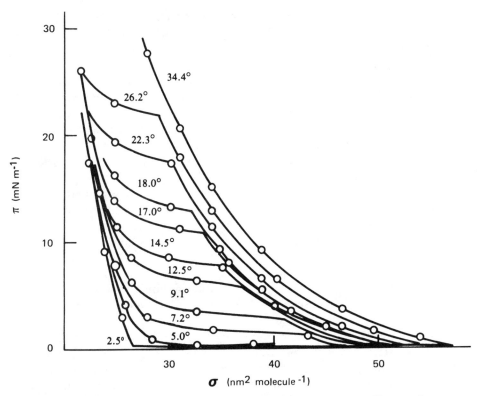

Figure 7.8 Plots of π versus σ for myristic acid on 0.01 M HCl at various temperatures. The L_1, I, and L_2 states are shown. [N. K. Adam and G. Jessop, *Proc. R. Soc., A112*:364 (1926).]

If the area of an insoluble monolayer is isothermally reduced still further, the compressibility eventually becomes very low. Because of the low compressibility, the states observed at these low values of σ are called condensed states. In general, the isotherm is essentially linear, although it may display a well-defined change in slope as π is increased as shown in Fig. 7.5. The (relatively) more expanded of these two linear portions is called the liquid condensed state L_2, and the less expanded is called the solid state S. It is clear from the low compressibility of these states that both the L_2 and S state are held together by strong intermolecular forces so as to be relatively independent of the film pressure. Because of the near linearity of these portions of the isotherm, it is easy to extrapolate both regions to their value at π = 0. The intercepts for the solid and liquid condensed regions, σ_S^0 and $\sigma_{L_2}^0$, respectively, differ only slightly. Values of $\sigma_{L_2}^0$ for alcohols are about 0.22 nm^2, and for carboxylic acids about 0.25 nm^2, more or less independent of the length of the hydrocarbon chain. The intercept σ_S^0 has a value of about 0.20 nm^2, independent of both the length of the chain and the nature of the head. The film pressures in the condensed states (L_2 or S) are of the same magnitude as the equilibrium spreading pressure for amphipathic molecules.

For the condensed L_2 and S states a molecular interpretation is again possible. In both the values of σ^0 are close to actual molecular cross sections when the molecules are oriented perpendicular to the surface. The difference between these two regions seems to involve the polar part of the molecule more than the hydrocarbon chain, which was more important for the more expanded states. The difference between σ_S^0 and $\sigma_{L_2}^0$ may involve a more efficient packing of the heads or the formation of fairly specific lateral interactions through hydrogen bonds, for example. Figure 7.6c and d presents models of how the L_2 and S states, respectively, may differ in packing efficiency. The values of σ^0 which are observed for monolayers of saturated n-alkyl compounds are only slightly larger than the close-packed cross sections obtained for these compounds in the bulk solid state by x-ray diffraction.

Additional compression eventually leads to the collapse of the film. The pressure at which this occurs, π_c, is somewhere in the vicinity of the equilibrium spreading pressure. Figure 7.6e represents schematically how this film collapse may occur. The mode of film buckling shown in Fig. 7.6e is not the only possibility: head-to-head as well as tail-to-tail configurations can be imagined. The second structure strongly resembles that of cell membranes, which we shall discuss in the next chapter.

Figure 7.6e represents the squeezing out of the surface of highly ordered aggregates which may quite plausibly be regarded as nuclei to bulk phase particles. If collapse marked the appearance of the amphipathic material in a bulk phase, then the collapse pressure and the equilibrium spreading pressure should be identical. Here a significant complication appears. The collapse pressure is highly sensitive to the rate at which the film is compressed. This indicates nonequilibrium conditions, showing that the two-dimensional solid phase resembles three-dimensional solids in this respect also, a difficulty in the attainment of thermodynamic equilibrium.

In summary, it must be emphasized again that there are wide variations in the properties of insoluble monolayers. Some of the phenomena reported in the literature are probably artifacts due to impurities or nonequilibrium conditions. Others are probably unique effects which apply only to a very specific system. In the descriptive material of this section both phenomenological and modelistic information were provided for various stages along the isotherm. At the very least, the models serve the pedagogical function of assisting the student in remembering an assortment of probably unfamiliar facts. At best, the models provide the basis for quantitatively understanding these phenomena. In the next section we shall take a more quantitative look at the model for the gaseous state.

7.5 Models for the Gaseous State of Monolayers

It is not difficult to propose and develop a model for the gaseous state of insoluble monolayers. The arguments parallel those developed in kinetic molecular theory for three-dimensional gases and lead to equally appealing results. The problem, however, is that many assumptions of the model are far less plausible for monolayers than for bulk gases. To see this, a brief review of the derivation seems necessary.

Suppose we imagine a single molecule bouncing back and forth across a surface between two restraining barriers. If we define the direction of this motion to be the x direction and the velocity of the molecule to be v_x, then the change in momentum at each collision (if we assume they are elastic) is

$$\frac{\Delta(\text{momentum})}{\text{collision}} = mv_x - (-mv_x) = 2mv_x \tag{8}$$

where m is the mass of the molecule. The time interval between two successive collisions at the *same* wall is given by

$$\frac{\text{elapsed time}}{\text{collision}} = \frac{2l}{v_x} \tag{9}$$

if the distance between barriers is l, since the molecule must cross the distance between the barriers twice before returning to the same spot. The force exerted by the molecule on impact equals the rate of change in momentum, which, in turn, equals the ratio of Eq. (8) to Eq. (9):

$$\frac{\Delta(\text{momentum})}{\Delta(\text{time})} = F_x = \frac{mv_x^2}{l} \tag{10}$$

This force is converted to two-dimensional pressure by dividing it by the length of the edge to which the force is applied. Assuming the accessible surface area to be a square means that the length of this edge is also l; the pressure contribution of this one collision equals

$$\pi = \frac{F_x}{l} = \frac{mv_x^2}{l^2} \tag{11}$$

The quantity l^2 in this equation clearly describes the area accessible to the molecule.

Since pressure is isotropic, we assume that the forces on the perpendicular barriers are identical. Therefore, if the surface contains N molecules, they behave as if $N/2$ were exerting a pressure given by Eq. (11) on the barriers perpendicular to the x direction, with the other $N/2$ exerting an identical pressure in the other direction. That is, for a surface containing N molecules

$$\pi A = \frac{N}{2} \overline{mv^2} = N(KE) \tag{12}$$

The average value of the square velocity has been used in Eq. (12) to allow for the fact that a distribution of molecular velocities exists. The nature of the averaging procedure to be used in this case is well established from physical chemistry. We also know from physical chemistry that the average kinetic energy per molecule (KE) per degree of freedom is

$$\frac{\overline{KE}}{\text{degree of freedom}} = \tfrac{1}{2} kT \tag{13}$$

Since the molecules on the surface have two translational degrees of freedom, Eqs. (12) and (13) may be combined to give

$$\pi \frac{A}{N} = \pi\sigma = kT \tag{14}$$

which is identical to Eq. (6).

The foregoing derivation is a straightforward two-dimensional analog of the three-dimensional case and leads to a result which describes the experimental facts. From a pragmatic point of view, it is a great success. One of the theoretical assumptions underlying Eq. (13), however, is that translational quantum states are sufficiently close together to justify treating them as continuous rather than discrete. This is unquestionably true for gases. For an amphipathic molecule whose polar head contacts—and interacts with—the aqueous substrate, it is somewhat harder to justify. We shall see presently that there is a totally different way of looking at Eq. (14) that is free from this objection.

We noted previously that the applicability of Eq. (14) to insoluble monolayers is severely restricted to very low values of π. Figure 7.7 shows that the deviations from Eq. (14) with increases in π are very similar to what is observed for nonideal gases. Specifically, the positive deviations associated with excluded volume effects in bulk gases and the negative deviations associated with intermolecular attractions are observed. It is tempting to try to correct Eq. (14) for these two causes of nonideality in a manner analogous to that used in the van der Waals equation:

$$\left(\pi + \frac{a}{\sigma^2} \right) (\sigma - b) = kT \tag{15}$$

where a and b are the two-dimensional analogs of the van der Waals constants. Note that b, the excluded area per molecule, is conceptually equivalent to σ^0, although which of the values ($\sigma^0_{L_1}$, $\sigma^0_{L_2}$, or σ^0_S) best fits the data cannot be predicted a priori.

The temperature at which the van der Waals equation goes from one having three real roots to one having one real root is generally identified with the critical temperature. In the same way, Eq. (15) may be considered to connect both the gaseous (G) and liquid expanded (L_1) states in monolayers, the transition between which also displays a critical point. Statistical mechanics shows, however, that the van der Waals a constant explicitly ignores orientation effects as contributing anything to the energy in gases; it is hard to imagine the properties of insoluble monolayers as being independent of orientation. Therefore any attempt to correct Eq. (14) in such a way as to extend its range encounters difficulties. Ultimately, all objections to two-dimensional equations of state seem to center on their neglect or unsatisfactory inclusion of the substrate.

An alternative way of looking at monolayers is to consider them as two-dimensional binary solutions rather than two-dimensional phases of a single component. The advantage of this approach is that it does acknowledge the presence of the substrate and the fact that the latter plays a role in the overall properties of the monolayer. Although quite an extensive body of thermodynamics applied to two-dimensional solutions has been developed, we shall consider only one aspect of this. We shall examine the film pressure as the two-dimensional equivalent of osmotic pressure. It will be recalled that, at least for low osmotic pressures, the relationship between π, V, n, and T is identical to the ideal gas law [Eq. (3.25)]. Perhaps the interpretation of film pressure in these terms is not too farfetched after all!

As we saw in Chap. 3, the heart of any osmotic pressure experiment is a semipermeable membrane which allows the solvent but not the solute to pass. The float of a Langmuir balance accomplishes this. That portion of surface with the monolayer is considered to be the two-dimensional solution; the clean surface is the two-dimensional solvent. The solvent can certainly pass from one region to another (remember that the mechanism of the partitioning has nothing to do with the equilibrium osmotic pressure) through the bulk substrate. However, the insoluble solute is restrained by the float.

For osmotic equilibrium the chemical potential of the solvent must be the same on both sides of the membrane. In the two-dimensional analog μ_1 must also be the same for the water on both sides of the float. The presence of the solute lowers the chemical potential of the solvent, but the excess pressure compensates for this. Therefore, by analogy with Eq. (3.19), we write

$$\mu_{1s}^0 = \mu_{1s}^0 + RT \ln a_{1s} + \int_0^\pi \bar{A}_1 dp \tag{16}$$

where the subscript s indicates the surface and the subscript 1 identifies the solvent. If the partial molal area of the solvent, $\bar{A}_1$, is assumed to be independent of π, Eq. (16) may be integrated to give

$$-RT \ln a_{1s} = \pi\bar{A}_1 \tag{17}$$

For ideal (dilute) solutions the activity is replaced by the mole fraction x_{1s}, which, for a two-component surface solution, equals $1 - x_{2s}$. With the customary expansion of the logarithm as a power series (see Appendix A), these substitutions yield

$$RTx_{2s} = \pi\bar{A}_1 \tag{18}$$

the two-dimensional equivalent of Eq. (3.23). Finally, it is necessary to relate the surface mole fraction and the molar area of the solvent to more familiar variables.

The surface mole fraction is entirely analogous to the bulk value of this quantity,

$$x_{2s} = \frac{n_{2s}}{n_{1s} + n_{2s}} = \frac{N_{2s}}{N_{1s} + N_{2s}} \tag{19}$$

where the n terms are the numbers of moles and the N terms are the numbers of molecules. For dilute surface solutions—that is, expanded monolayers—$N_{1s} \gg N_{2s}$; therefore

$$x_{2s} \simeq \frac{N_{2s}}{N_{1s}} \tag{20}$$

The total area of the surface may be written

$$A_T = n_{1s}\bar{A}_1 + n_{2s}\bar{A}_2 = N_{1s}\sigma_1^0 + N_{2s}\sigma_2^0 \tag{21}$$

Applying these various relationships to Eq. (18) leads to the following result for low film osmotic pressures:

$$\pi(A_T - N_{2s}\sigma_2^0) = n_{2s}RT = N_{2s}kT \tag{22}$$

Dividing through by N_{2s} to express the total area as area per solute molecule gives

$$\pi(\sigma - \sigma_2^0) = kT \tag{23}$$

Equation (23) obviously gives the two-dimensional ideal gas law when $\sigma > \sigma_2^0$, and with the σ_2^0 term included represents part of the correction included in Eq. (15). This model for surfaces is, of course, no more successful than the one-component gas model used in the kinetic approach; however, it does call attention to the role of the substrate as part of the entire picture of monolayers. We saw in Chap. 3 that solution nonideality may also be considered in osmotic equilibrium. Pursuing this approach still further results in the concept of phase separation to form two immiscible surface solutions.

Example 7.2   A monolayer of egg albumin was spread on a concentrated aqueous solution of ammonium sulfate and π-A data were collected* at 25°C. Use the two-dimensional van't Hoff equation to evaluate the molecular weight of the albumin if $(\pi/c')_0 = 5.54 \times 10^5$ erg g^{-1}. In this expression c' is the two-dimensional concentration in practical units, g cm^{-2}. How does this interpretation of π-A data compare with the interpretation given in Example 7.1?

Solution   According to the van't Hoff equation,

$$1/M = (1/RT)(\pi/c')_0 = (5.54 \times 10^5)/(8.314 \times 10^{-7})(298)$$

$$= 2.24 \times 10^{-5} \text{ mole g}^{-1}$$

therefore $M = 44,700$ g mole^{-1}. This is completely equivalent to the interpretation given in Example 7.1, since $\pi \to 0$ as $c' \to 0$. Therefore the same limiting value is obtained whether the limit is taken in terms of π or c'.

•

In summary, we see that insoluble monolayers may be viewed either as examples of two-dimensional phases of one component or as two-

---

*H. B. Bull, *J. Am. Chem. Soc.*, 67:4 (1945).

dimensional solutions with two components. The former model is somewhat simpler and is often adequate. The latter, although more complex, is more realistic. In spite of our interest in the monolayer, we must not neglect the fact that none of the monolayer phenomena would exist without the aqueous phase as the substrate.

7.6 Surface Viscosity

Our discussion of two-dimensional phases has drawn heavily on the analogy between bulk and surface behavior. This analogous behavior is not restricted to thermodynamic observations, but extends to other areas also. The viscosity of surface monolayers is an excellent example of this. To illustrate the parallel between bulk and surface viscosity, let us retrace some of the introductory notions of Chap. 4, restricting the flow to the surface region.

We begin by defining the surface coefficient of viscosity η^s. Figure 4.1 serves to define the bulk viscosity; for surfaces we ignore the area extending in the z direction and consider the force per unit edge of the surface, l. Assuming a velocity gradient dv/dy exists between two edges of an element of area, we can write the two-dimensional analog of Eq. (4.1) as

$$\frac{F}{l} = \eta^s \frac{dv}{dy} \tag{24}$$

This gives η^s units of mass time^{-1} in contrast to bulk viscosity, which has dimensions mass length^{-1} time^{-1}. In discussing Fig. 7.2 we noted that—from a chemical if not geometrical point of view—surfaces extend over a thickness τ. By analogy with Eq. (3), we might expect surface and bulk viscosities to be related by

$$\eta = \frac{\eta^s}{\tau} \tag{25}$$

Next let us consider how surface viscosities can be measured. A variety of methods for measuring η^s exist, including a method based on concentric rings, the two-dimensional equivalent of the concentric cylinder viscometer. We shall limit our discussion to the analog of the capillary viscometer. Figure 7.9a shows an arrangement whereby a monolayer can be pushed through a narrow channel by a moving barrier. Following a derivation which parallels the development of the Poiseuille equation [Eq. (4.20)], the area rate at which the monolayer emerges from the channel under an applied pressure $\gamma° - \gamma = \Delta\gamma$ can be obtained. As

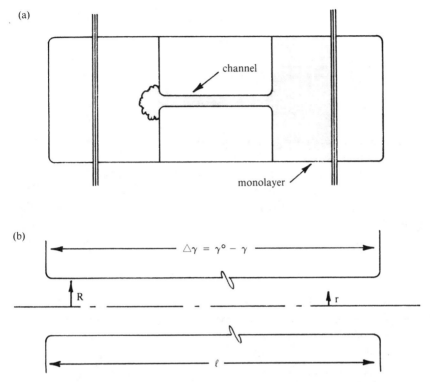

Figure 7.9 Schematic of a surface viscometer: (a) A monolayer is pushed through a narrow channel and (b) definition of variables for analysis.

in Sect. 4.4, we continue to describe the locations within the channel in terms of the distance r from its center line, with R the distance of the wall from the center line. The following steps are highlights of the derivation and parallel the presentation in Sect. 4.4:

1. Viscous and surface pressure forces balance under stationary state conditions along the edge of the film a distance r from the center line. In terms of Fig. 7.9b, this force balance is given by

$$\eta^s l \frac{dv}{dr} + \Delta \gamma r = 0 \tag{26}$$

2. Integrating Eq. (26) and using the non-slip condition at the wall ($v = 0$ at $r = R$) to evaluate the constant of integration yields

$$\frac{\eta^s l}{\Delta \gamma} v = \tfrac{1}{2}(R^2 - r^2) \tag{27}$$

3. The rate of emergence of the monolayer from the channel A/t is twice the integral over r of $dA/dt = v\, dr$:

$$\frac{A}{t} = \frac{2}{3} \frac{\Delta \gamma}{\eta^s l} R^3 \tag{28}$$

This is the two-dimensional equivalent of Poiseuille's equation. All of the other quantities besides η^s in Eq. (28) are measurable, so the latter can be evaluated by measuring the rate at which the monolayer flows through the channel. In practice, a second barrier is moved along in front of the advancing interface to maintain a constant film pressure for an insoluble monolayer.

In the two-dimensional gaseous state surface viscosities can be as low as 10^{-8} kg s^{-1}, while for condensed states values range from 10^{-7} to 10^{-5} kg s^{-1}. While it makes sense that the surface viscosity increases as we move from G to L to S states, the numbers themselves mean little to us. To remedy this we repeat the sort of calculation done for two-dimensional pressure and examine what Eq. (25) tells us about the equivalent bulk viscosity. As with Eq. (3), we assume $\tau = 1.0$ nm; therefore a surface viscosity of 10^{-7} kg s^{-1} is equivalent to a bulk viscosity of

$$\eta = \frac{\eta^s}{\tau} = \frac{10^{-7} \text{ kg s}^{-1}}{10^{-9} \text{ m}} = 10^2 \text{ kg s}^{-1} \text{ m}^{-1} = 10^3 \text{ P} \tag{29}$$

a value which suggests a consistency like that of butter for the molecules of the monolayer in condensed states. This is consistent with a high degree of lateral interaction between the vertically orientated chains. Once again, we see that a highly localized phenomenon translates into a dramatic effect when scaled up to macroscopic dimensions. In the next section we shall see some applications of monolayers or monolayer concepts that take advantage of the properties we have discussed.

Surface viscosities have been measured for soluble and insoluble monolayers, for charged and uncharged molecules, and at water–air and water–oil interfaces. We will not consider all these possibilites, but note instead that all share as a common feature the presence of polar head groups in the aqueous phase. In general, this means hydrogen bonding will occur between the amphipathic molecules at the surface and the substrate. A certain amount of water gets dragged along with the surface molecules as a consequence. This effect was not taken into account in the derivation of Eq. (28), although it has been investigated extensively. The effect of the entrained substrate may be incorporated into Eq. (28) by multiplying the latter by a correction factor of the form $(1 + 2R\eta/\pi\eta^s)^{-1}$. This shows that the resistance to the motion of the monolayer due to this effect increases as the ratio η/η^s increases. This factor is incorporated into Eq. (28) when η^s is determined experimentally by this method.

7.7 Applications of Monolayers and Monolayer Concepts

At the surface of water amphipathic molecules are oriented in such a way as to interact extensively, at least for ordinary surface concentrations. This results in the formation of the various two-dimensional condensed phases with the attendant effect on surface viscosity. In this section we consider some situations where monolayers or the concepts involved in their discussion find application. One area where monolayers have been successfully employed is the retardation of evaporation. Particularly in arid regions of the world, evaporation of water from lakes and reservoirs constitutes an enormous loss of a vital resource. Under some conditions the water level of such bodies may change a smuch as 1 ft per month due to evaporation. The usual unit for water reserves is the acre-foot, a volume of water covering an acre of surface to the depth of 1 ft. It equals about ⅓ million gal. for each acre of water surface.

Considerable research has been conducted both in the laboratory and in the field on the effectiveness of insoluble monolayers in reducing evaporation. An American Chemical Society Symposium in 1960 dealt exclusively with this topic; the proceedings of that symposium are given by LaMer [7].

Laboratory research in this area is conducted by suspending a porous box of desiccant very close to the surface of a film balance. The rate of water uptake is determined by weighing at various times. This way the retardation of evaporation may be measured as a function of film pressure and correlated with other properties of the monolayer determined by the same method. As might be expected, the resistance to evaporation which a monolayer provides is enhanced by those conditions which promote the most coherent films, most notably high film pressures and straight-chain compounds. To see how this is quantified, consider the following example.

Example 7.3   The rate of evaporation is quantified by a parameter called the transport resistance r. For water with octadecanol monolayers at surface pressures of 10, 20, 30, and 40 mN m^{-1}, r is about 1, 2, 3, and 4 s cm^{-1}, respectively. This resistance drops off rapidly at lower pressures approaching 2×10^{-3} s cm^{-1} for pure water. By considering the rate of water uptake as a diffusion problem, suggest how these r values are calculated from data collected in an experiment like that described above. Use the fact that $1/r$ is dimensionally equivalent to the diffusion coefficient D divided by a length.

Solution   To be collected by the desiccant, molecules evaporating from a surface of area A must diffuse across a gap of width Δx between the water surface and the desiccant. The gap contains the monolayer as well as the air space, so the diffusion coefficient used is an effective value rather than the actual D value for a homogeneous region. According to Eqs. (2.20) and (2.22), the rate at which the desiccant increases in weight is given by

$$\frac{dQ}{dt} = AD\,\frac{\Delta c}{\Delta x}$$

In this expression Δc describes the difference between the concentration of water vapor at the water surface and that at the desiccant surface: $c = c_W - c_{des}$. Since $c_{des} \ll c_W$, Δc may be replaced by c_W, which, in turn, equals pM/RT, where p and M are the vapor pressure and molecular weight, respectively, of water. With this substitution the expression for dQ/dt becomes $A(D/\Delta x)(pM/RT)$. The ratio $D/\Delta x$ has units length time^{-1}, so we identify it as the reciprocal of the transport resistance. Note that r increases as the effective value of Δx increases and the effective value of D decreases. Thus $1/r$ is the only unknown in the expression $dQ/dt = A(1/r)(pM/RT)$ and can be calculated from the measured rate of weight increase with different monolayers present.

•

To be acceptable for use in the field, the monolayer material must have the following properties:

1. It must spread easily, probably as bulk material, so a high value of π_e is desirable.
2. It must be self-healing, since surface ripples will disrupt the monolayer. This implies viscous rather than rigid monolayers.
3. It must be inexpensive, which means, effectively, capable of forming good films from naturally occurring mixtures.
4. It must be nontoxic and free from other deleterious effects on aquatic life.

Hexadecyl and octadecyl alcohol have been extensively studied and shown to be highly effective in evaporation retardation. Scattering powdered samples of commercial-grade alcohols by boat on lake surfaces or the continuous addition of alcohol slurries from floating dispensers are two of the methods that have been employed to apply these monolayers. Wind conditions and the activity of aquatic birds have a considerable effect on the stability of the monolayer and therefore on the rate at which the monolayer chemicals must be reapplied. Rates of application rarely exceed 0.5 lb acre^{-1} day^{-1}, however, so that the cost of the materials used is not excessive.

An indication of the effectiveness of such field treatment is seen in Fig. 7.10, which compares the amount of water lost by evaporation from two small adjacent lakes in Illinois. Treatment of North Lake (ordinate) with commercial hexadecanol was begun in late June 1957; untreated South Lake was the control (abscissa). Prior to treatment the evaporation losses from the two lakes were identical, as shown by the 45° line in the figure. After treatment was begun, however, the loss of water from North Lake fell considerably behind that from South Lake. By the end of the summer a difference of about 40% in the water loss was observed. This

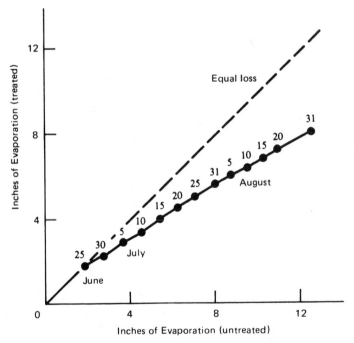

Figure 7.10 Comparison of the water level in two adjacent lakes during the summer, 1957. The ordinate shows the level in the lake with the monolayer; the abscissa is the level in the untreated lake. (From Ref. 7, redrawn with permission.)

was equivalent to about 7600 gal. of water saved per pound of hexadecanol used. For areas where water is scarce—the southwestern United
States, Israel, and western Australia, for example—such conservation of
water is highly valued, and research continues to look for methods to
improve the efficiency of this technique.

Surface viscosity also has observable effects on macroscopic bodies of
water. The calming effect on surface turbulence of pouring oil on the sea
has been known from antiquity. In terms of the concepts of this chapter,
the increase in surface viscosity produced by the film has a damping
effect on waves. Such damping has been studied both theoretically and in
laboratory situations; Ref. 3 contains some interesting photographs of the
ripples on a pond before and after the application of hexadecanol to the
surface.

Emulsions and foams are two other areas in which dynamic and
equilibrium film properties play a considerable role. Emulsions are
colloidal dispersions in which two immiscible liquids constitute the
dispersed and continuous phases. Water is almost always one of the
liquids, and amphipathic molecules are usually present as emulsifying
agents, components which impart some degree of durability to the
preparation. Although we have focused attention on the air–water surface
in this chapter, amphipatic molecules behave similarly at oil–water
surfaces as well. By their adsorption, such molecules lower the interfacial
tension and increase the interfacial viscosity. Emulsifying agents may
also be ionic compounds, in which case they impart a charge to the
surface, which, in turn, establishes an ion atmosphere of counter-ions in
the adjacent aqueous phase. These concepts affect the formation and
stability of emulsions in various ways:

1. Most emulsions are formed by some sort of comminution process in
    which large blobs of the dispersed phase are eventually ground down
    to small drops. This is a complex process, but basically consists of
    drops becoming elongated under shearing forces, necking, and
    finally separating into smaller drops. The adsorption of a surface
    film with the attendant lowering of γ and increase in η^s clearly enter
    the picture.
2. The first step in the "breaking" of an emulsion is the coming together
    of the individual drops. If water is the continuous phase and the
    emulsifier is ionic, then it is the ion atmospheres of the approaching
    particles that make the first contact. We shall see in Chapter 12 that
    the details of this first encounter can determine whether the drops
    aggregate into a floc or go their separate ways.

3. If droplets can aggregate into a single kinetic unit, they might also coalesce into a single geometrical unit. This involves rupture of the thin film of continuous phase that separates them in an aggregate. Again, surface tension and surface viscosity are certainly pertinent to the coalescence process.

The huge variety of emulsions used as food, medicinal, cosmetic, and other industrial products make these colloids important practical systems in which the surface monolayers exert considerable influence.

Foams are colloidal systems in which a gas is the dispersed phase. Although a whole range of concentrations is possible, we shall focus on those foams which consist of volume-filling, distorted polyhedra separated by liquid films. For aqueous foams the high area of air–water interface requires adsorption to lower the surface tension sufficiently to make the foam in the first place. Foams drain by losing liquid through the channels that occur at the junction of the planar film surfaces. Such surfaces meet with curved menisci between them. This means that the pressure is lower in the junction than in the flat faces of the film according to the Laplace equation, Eq. (6.29). As a consequence, liquid flows from the planar regions into these junctions—called plateau borders—through which the drainage occurs. As the film thickness of the continuous phase decreases, the probability of rupture due to thermal or mechanical fluctuations increases. Surface energetics and viscosity are important here also. As with emulsions, foams occur in many systems familiar to consumers, such as firefighting foams, whipped cream, shaving lather, and the head on a glass of beer!

Although we started out this chapter by discussing insoluble monolayers, it is evident that we have slipped into examples where soluble amphipathics are being considered. In the next section we examine the thermodynamics of adsorption from solution.

7.8 Adsorption from Solution and the Gibbs Equation

Until now we have discussed only insoluble monolayers. Although their behavior is complex, they have the conceptual simplicity of being localized in the interface. It has been noted, however, that even in the case of insoluble monolayers, the substrate should not be overlooked. The importance of the adjoining bulk phases is thrust into even more prominent view when soluble monolayers are discussed. In this case the adsorbed material has appreciable solubility in one or both of the bulk phases which define the interface.

Gibbs treated this situation as part of his investigations on phase equilibria. Suppose we consider two phases α and β in equilibrium with a surface s dividing them. For the system so constituted, we may write

$$G = G^\alpha + G^\beta + G^s \tag{30}$$

where the superscripts indicate the contribution from each category. For the bulk phases

$$G = E + pV - TS + \sum_i \mu_i n_i \tag{31}$$

where the chemical potential terms are summed for all components i. The volume term is replaced by an area term in the corresponding expression for G^s:

$$G^s = E^s + \gamma A - TS^s + \sum_i \mu_i n_i \tag{32}$$

Substituting Eqs. (31) and (32) into Eq. (30) and taking the total derivative yields

$$dG = \sum_{\alpha,\beta,s} \left(dE + p\,dV + V\,dp - T\,dS - S\,dT + \sum_i \mu_i\,dn_i \right.$$

$$\left. + \sum_i n_i\,d\mu_i \right) + A\,d\gamma + \gamma\,dA \tag{33}$$

For a reversible process

$$dE = \delta q - \delta w = \sum_{\alpha,\beta,s} dE = \sum_{\alpha,\beta,s} [T\,dS - (p\,dV + \delta w_{\text{non-}pV})] \tag{34}$$

Substituting this result into Eq. (33) gives

$$dG = \sum_{\alpha,\beta,s} \left(V\,dp - S\,dT + \sum_i \mu_i\,dn_i + \sum_i n_i\,d\mu_i - \delta w_{\text{non-}pV} \right)$$

$$+ A\,d\gamma + \gamma\,dA \tag{35}$$

As we saw in Chapter 6 [Eq. (6.16)], the quantity $\gamma\,dA$ may be equated to non-pressure–volume work when surface energy is being considered. With this consideration, Eq. (35) simplifies still further to become

$$dG = \sum_{\alpha,\beta,s} \left(V\,dp - S\,dT + \sum_i \mu_i\,dn_i + \sum_i n_i\,d\mu_i \right) + A\,d\gamma \tag{36}$$

Another well-known relationship from physical chemistry may be introduced at this point:

$$dG = V\,dp - S\,dT + \sum_i \mu_i\,dn_i \tag{37}$$

Applying Eq. (37) to the bulk phases and the surface and subtracting the result from Eq. (36) gives

$$\sum_i n_i^\alpha d\mu_i + \sum_i n_i^\beta d\mu_i + \sum_i n_i^s d\mu_i + A\,d\gamma = 0 \tag{38}$$

When only one phase is under consideration, only one of the terms in Eq. (30) is required, and only one of the bulk phase summations in Eq. (38) survives. The result in this case is the famous Gibbs–Duhem equation:

$$\sum_i n_i\,d\mu_i = 0 \tag{39}$$

It will be recalled from physical chemistry that this relationship permits the evaluation of the activity of one component from measurements made on the other in binary solutions.

By means of the Gibbs–Duhem equation, we may eliminate the terms in Eq. (38) which apply to bulk phases and write

$$\sum_i n_i^s d\mu_i + A\,d\gamma = 0 \tag{40}$$

This is the Gibbs adsorption equation which relates γ to the number of moles and the chemical potentials of the components in the interface. In subsequent developments we shall consider only two-component systems and shall identify the solvent (usually water) as component 1 and the solute as component 2. In terms of this stipulation, Eq. (40) becomes

$$n_1^s\,d\mu_1 + n_2^s\,d\mu_2 + A\,d\gamma = 0 \tag{41}$$

It is conventional to divide Eq. (41) through by A to give

$$d\gamma = -\frac{n_1^s}{A}\,d\mu_1 - \frac{n_2^s}{A}\,d\mu_2 \tag{42}$$

The quantity n_i^s/A is called the surface excess of component i and is given the symbol Γ_i:

$$\Gamma_i = \frac{n_i^s}{A} \tag{43}$$

In this notation, Eq. (42) becomes

$$-d\gamma = \Gamma_1\,d\mu_1 + \Gamma_2\,d\mu_2 \tag{44}$$

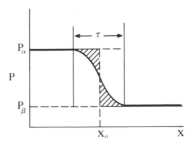

Figure 7.11 Variation of some general property P with perpendicular distance from the surface in the vicinity of an interface between two phases α and β.

Before proceeding any further, it is necessary to examine just what the concept of a surface excess means. To do this it is convenient to consider the changes which occur in some general property P as we move from phase α to phase β. The situation is represented schematically by Fig. 7.11, where x is the distance measured perpendicular to the interface. The scale of this figure is such that variations at the molecular level are shown. The interface is not a surface in the mathematical sense but, rather, a zone of thickness τ across which the properties of the system vary from values which characterize phase α to those characteristic of β. In spite of this, we generally do not assign any volume to the surface, but treat it as if the properties of α and β applied right up to some dividing plane situated at some specific value of x. What is this position x_0 at which we draw such a boundary?

Suppose the solid line in Fig. 7.11 represents the actual variation of property P. The squared-off extensions of the bulk values of this property represent the approximation made in assuming the surface to have zero thickness. Then the shaded area to the left of x_0 shows the amount by which the value of P for the system as a whole has been overestimated by extending P_α. Likewise, the shaded area to the right of x_0 shows how the extension of P_β leads to an underestimation of P for the system as a whole. In principle, the "surface" may be located at an x value such that these two areas compensate for one another; that is, x_0 may be chosen so that the two shaded areas in the figure are equal.

This is where the trouble begins! Generally speaking, the kind of profile sketched in Fig. 7.11 will be different for each property considered. Therefore we may choose x_0 to accomplish the compensation discussed herein for one property, but this same line will divide the profiles of other properties differently. The difference between the "overestimated" property and the "underestimated" one accounts for the "surface excess" of this property.

From the point of view of thermodynamics—which is oblivious to details at the molecular level—the dividing boundary may be placed at any value of x in the range τ. The actual placement of x_0 is governed by consideration of which properties of the system are most amenable to thermodynamic evaluation. More accurately, that property which is least convenient to handle mathematically may be eliminated by choosing x_0 so that the difficult quantity has a surface excess of zero.

For example, if the property in Fig. 7.11 were G and the dividing surface were placed so that the two shaded regions would be equal, then there would be no surface excess G: The last term in Eq. (30) would be zero. The Gibbs free energy is convenient to work with, however, so such a choice for x_0 would not be particularly helpful. Until now we have not had any reason to identify the surface of physical phases with any specific mathematical surface. We had not, that is, until Eq. (44) was reached. Now things are somewhat different.

Suppose the property represented in Fig. 7.11 is the number of moles of solvent per unit area in a slice of solution at some value of x. This quantity will clearly undergo a transition in the vicinity of an interface. We choose x_0 so that the shaded areas are equal when this is the quantity of interest. This placement of the dividing surface means

$$\Gamma_1 = 0 \tag{45}$$

With this situation, Eq. (44) becomes

$$d\gamma = -\Gamma_2\, d\mu_2 \tag{46}$$

The physical significance of Γ_2 is determined by the arbitrary placement of the mathematical surface that made $\Gamma_1 = 0$; that is, Γ_2 equals the algebraic difference between the "overestimated" and "underestimated" areas of the curve describing moles of solute when this curve is divided at a location x_0 which makes the surface excess of the solvent zero.

It is important to realize that the mathematical dividing surface just discussed is a *reference level* rather than an actual physical boundary. What is physically represented by this situation may be summarized as follows. Two portions of solution containing an identical number of moles of solvent are compared. One is from the surface region, and the other from the bulk solution. The number of moles of solute in the sample from the surface minus the number of moles of solute in the sample from the bulk gives the surface excess number of moles of solute, according to this convention. This quantity divided by the area of the surface equals Γ_2. To emphasize that the surface excess of component 1 has been chosen to be zero in this determination, the notation Γ_2^1 is generally used.

It should be evident from the foregoing discussion that the property defined to have zero surface excess may be chosen at will, the choice being governed by the experimental or mathematical features of the problem at hand. Choosing the surface excess number of moles of one component to be zero clearly simplifies Eq. (44). The same simplification could have been accomplished by defining the mathematical surface so that Γ_2 would be zero, a choice which would obviously de-emphasize the solute. If the total number of moles N, the total volume V, or the total weight W had been the property chosen to show a zero surface excess, then in each case both Γ_1 and Γ_2 (which would be identified as Γ^N, Γ^V, or Γ^W for these three conventions) would have nonzero values. Lastly, note that the surface "excess" is an algebraic quantity which may be either positive or negative, depending on the convention chosen for Γ. A variety of different experimental methods are encountered in the literature to measure "surface excess" quantities; one must be careful to understand clearly what conventions are used in the definition of these quantities.

Equation (46), one form of the Gibbs equation, is an important result because it supplies the connection between the surface excess of solute and the surface tension of an interface. For systems in which γ can be determined this measurement provides a method for evaluating the surface excess. It might be noted that the finite time required to establish equilibrium adsorption is why dynamic methods (e.g., drop detachment) are not favored for the determination of γ for solutions. At solid interfaces γ is not directly measurable; however, if the amount of adsorbed material can be determined, this may be related to the reduction of surface free energy through Eq. (46). To understand and apply this equation, therefore, it is imperative that the significance of Γ_2 be appreciated.

Now let us return to the development of Eq. (46). The chemical potential depends on the activity according to the equation

$$\mu_2 = \mu_2^0 + RT \ln a_2 \tag{47}$$

In applying these results to adsorption from solution, the activity equals the pressure or concentration multiplied by the activity coefficient f. Differentiation of Eq. (47) at constant temperature yields

$$d\mu_2 = RT \frac{da_2}{a_2} = RT \, d \ln(fc) \tag{48}$$

This relationship may also be applied to the adsorption of gases by replacing concentration by gas pressure and continuing to use the

The Gibbs Equation: Experimental Results

The Gibbs Equation: Experimental Results

appropriate activity coefficient. We shall return to the application of this result to the adsorption of gases in Chap. 9.

For adsorption from dilute solutions the activity coefficient approaches unity, in which case the combination of Eqs. (46) and (48) leads to the result

$$\Gamma_2^1 = -\frac{c}{RT}\left(\frac{d\gamma}{dc}\right)_T = -\frac{1}{RT}\left(\frac{d\gamma}{d\ln c}\right)_T \tag{49}$$

This form of the Gibbs equation shows that the slope of a plot of γ versus the logarithm of concentration (or activity if the solution is nonideal) measures the surface excess of the solute. It might also be noted that the choice of units for concentration is immaterial at this point.

7.9 The Gibbs Equation: Experimental Results

Surface tensions for the interface between air and aqueous solutions generally display one of the three forms indicated schematically in Fig. 7.12. The type of behavior indicated by curves 1 and 3 indicates positive adsorption of the solute. Since $d\gamma/dc$ and therefore $d\gamma/d\ln c$ are negative, Γ_2^1 must be positive. On the other hand, the positive slope for curve 2 indicates a negative surface excess, or a surface depletion of the solute. Note that the magnitude of negative adsorption is also less than that of positive adsorption.

Curve 1 in Fig. 7.12 is the type of behavior characteristic of most unionized organic compounds. Curve 2 is typical of inorganic electrolytes and highly hydrated organic compounds. The type of behavior indicated by curve 3 is shown by soluble amphipathic species, especially ionic ones. The break in curve 3 is typical of these compounds; however, this degree

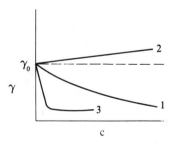

Figure 7.12 Three types of variation of γ with c for aqueous solutions: (1) simple organic solutes, (2) simple electrolytes, and (3) amphipathic solutes.

of sharpness is observed only for highly purified compounds. If impurities are present, the curve will display a slight dip at this point. All three of these curves correspond to relatively dilute solutions. At higher concentrations effects other than adsorption may lead to departures from these basic forms. We shall say a bit more about adsorption from binary solutions over the full range of compositions in Sect. 7.11.

For the limit as $c \rightarrow 0$, curve 1 may be presented by the equation of a straight line:

$$\gamma = \gamma_0 - mc \tag{50}$$

where m is the initial slope of the line. This is the same as

$$\pi = mc \tag{51}$$

From Eq. (50) $d\gamma/dc = -m$ and from Eq. (51) $c = \pi/m$; therefore Eq. (49) may be written

$$\Gamma_2^1 = \frac{\pi}{RT} \tag{52}$$

Recalling the definition of Γ_2^1 provided by Eq. (43), we see that Eq. (52) may also be written

$$\pi A = n_2^s RT \tag{53}$$

the two-dimensional ideal gas law again! Those carboxylic acids containing less than 12 carbons in the alkyl chain for which results were presented in Fig. 7.7 were investigated by this method. This same analysis also applies to the branch of curve 3 in Fig. 7.12 as $c \rightarrow 0$.

Curve 2 in Fig. 7.12 indicates a negative surface excess of simple electrolytes. This means that portions of solution from both the surface and bulk regions which contain the same number of moles of solvent will have more solute in the bulk region than at the surface. Obviously, the surface is enriched over the bulk in solvent, a fact that is easily understood when the hydration of the ions is considered. Water molecules interact extensively with ions, a fact that accounts in part for the excellent solvent properties of water for ionic compounds. To move an ion directly to the air–water interface would require considerable energy to partially dehydrate the ion. Accordingly, the first couple of molecular diameters into the solution will be a layer of essentially pure water, the ions being effectively excluded from this region. The surface tension is not that of pure water, but is increased slightly due to the small surface deficiency of solute. Other highly solvated solutes such as sucrose also show this effect.

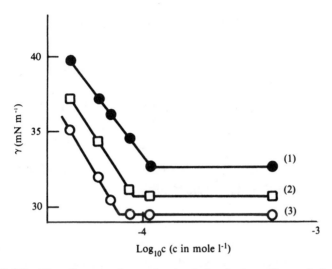

Figure 7.13 Plot of γ versus $\log_{10} c$ for the dodecyl ether of hexaethylene oxide at (1) 15°C, (2) 25°C, and (3) 35°C. [J. M. Corkill, J. F. Goodman, and R. H. Ottewill, *Trans. Faraday Soc.*, *57*:1927 (1961).]

Curve 3 in Fig. 7.12 applies primarily to amphipathic species. Most long-chain amphipathic molecules are insoluble unless the hydrophobic alkyl part of the molecule is offset by an ionic head or some other suitably polar head such as a polyethylene oxide chain, $-(CH_2CH_2O)_n-$. Like their insoluble counterparts, these substances form an oriented mono-layer even at low concentrations. Figure 7.13 shows some actual experimental plots of type 3 for the ether which consists of a dodecyl chain and a hexaethylene oxide chain ($n = 6$) in the general formula just given. The following example illustrates the application of the Gibbs equation to these data.

Example 7.4 The slope of the 25°C line in Fig. 7.13 on the low-concentration side of the break is about -16.7 mN m^{-1}. Calculate the surface excess and the area per molecule for the range of concentrations shown. How would Fig. 7.13 be different if accurate measurements could be made over several more decades of concentration in the direction of higher dilution? Could the data still be interpreted by Eq. (49) in this case?

Solution The surface excess is constant over this range of concen-trations as indicated by the linearity of Fig. 7.13. Equation (49) can be used directly to evaluate Γ_2^1. Since base 10 logarithms are used in the figure, we write

$$\Gamma_2^1 = -\frac{(-0.0167)}{(2.303)(8.314)(298)} = 2.93 \times 10^{-6} \text{ mole m}^{-2}$$

The reciprocal of this gives the area of surface occupied by a mole of adsorbed molecules. Division by Avogadro's number converts the reciprocal of Γ_2^1 into a value for σ:

$$\sigma = \frac{1 \text{ m}^2}{2.93 \times 10^{-6} \text{ mole}} \times \frac{1 \text{ mole}}{6.02 \times 10^{23} \text{ molecules}} \times \left(\frac{10^9 \text{ nm}}{1 \text{ m}}\right)^2$$

$$= 0.56 \text{ nm}^2$$

If accurate measurements could be made to increasingly lower concentrations, the surface excess would gradually decrease toward zero. This means that the lines in Fig. 7.13 must eventually curve until they show a slope of zero at infinite dilution. Curved lines on a semilogarithmic plot of γ versus c are interpreted by drawing tangents at the concentrations of interest and applying the Gibbs equation to the slopes of the tangents to give the corresponding surface excesses.

•

The polar heads of the solute molecules in Fig. 7.13 are much bulkier than those of the simple amphipathic molecules whose insoluble monolayers we discussed earlier. This is especially true when the hydration of the ether oxygens is considered. In view of this, the value calculated in Example 7.4 is probably about as small a value as these molecules can achieve. This suggests that the amphipathic molecules are in a highly condensed surface state at the concentrations investigated in Fig. 7.13.

The break in the curve 3 in Fig. 7.12 is characteristic of this type of plot for soluble amphipathic molecules. Note that it appears in the experimental curves of Fig. 7.13 also. The break is understood to indicate the threshold of micelle formation (see Sect. 1.3). We shall not discuss this phenomenon any further, since the next chapter is devoted entirely to micelles and related structures.

It is instructive to consider the effect of dissociation on the adsorption of amphipathic substances, since many of the compounds which behave according to curve 3 are electrolytes. We shall consider only the case of strong 1:1 electrolytes; for weak electrolytes the equilibrium constant for dissociation must be considered.

If an ionic solute is totally dissociated into positive and negative ions, then its activity is given by

$$a_{MR} = a_M a_R \simeq c_M c_R \tag{54}$$

where the subscripts M and R refer to the cation and amphipathic anion, respectively. Analogous results would be obtained if the cation were the amphipathic species. The approximation included in Eq. (54) applies to the case in which the activity coefficient equals unity. Substituting this result into Eq. (49) gives

$$\Gamma^1_{MR} = -\frac{1}{2RT}\left(\frac{d\gamma}{d\ln c}\right)_T \tag{55}$$

The assumption that no other electrolyte is present is implicit in this result. Now let us consider what happens when the system also contains a nonamphipathic electrolyte with a common ion to the surface-active electrolyte.

If a second electrolyte MX is present in addition to MR, then Eq. (44) must be written

$$-d\gamma = \Gamma^1_M \, d\mu_M + \Gamma^1_R \, d\mu_R + \Gamma^1_X \, d\mu_X \tag{56}$$

Now the condition of surface neutrality becomes

$$\Gamma^1_M = \Gamma^1_R + \Gamma^1_X \tag{57}$$

so Eq. (56) may be written

$$-d\gamma = \Gamma^1_R(d\mu_M + d\mu_R) + \Gamma^1_X(d\mu_M + d\mu_X) \tag{58}$$

This result may now be simplified by invoking some previous results. Recalling curve 2 from Fig. 7.12, we know that the surface excess of the X^- ions is likely to be a small negative number which we shall set equal to zero as a first approximation. With this approximation, Eq. (58) becomes

$$-d\gamma \simeq \Gamma^1_R(d\mu_M + d\mu_R) = \Gamma^1_R RT\left(\frac{dc_M}{c_M} + \frac{dc_R}{c_R}\right) \tag{59}$$

Now let us consider a small change in the concentration of MR while the concentration of MX remains constant and considerably greater than the total MR concentration. Under these conditions $dc_M = dc_R$ and $c_M \gg c_R$; therefore $d\ln c_M \ll d\ln c_R$, and Eq. (59) becomes

$$-d\gamma = \Gamma^1_R(RT \, d\ln c) \tag{60}$$

Equations (55) and (60) are thus seen to describe the adsorption of MR in the absence of electrolyte and in the presence of swamping amounts of electrolyte, respectively. It is clear from the difference between these two results that extreme care must be taken in the study of charged monolayers if the effect of the charge on the state of the monolayer is to be properly considered in the interpretation of experimental results.

The difference between Eqs. (55) and (60) may be qualitatively understood by comparing the results with the Donnan equilibrium discussed in Chap. 3. The amphipathic ions may be regarded as restrained at the interface by a hypothetical membrane which is of course permeable to simple ions. Both the Donnan equilibrium [Eq. (3.85)] and the electroneutrality condition [Eq. (3.87)] may be combined to give the distribution of simple ions between the bulk and surface regions. As we saw in Chap. 3 (e.g., see Table 3.2), the restrained species behaves more and more as if it were uncharged as the concentration of the simple electrolyte is increased. In Chap. 12 we shall examine the distribution of ions near a charged surface from a statistical rather than a phenomenological point of view.

We have noted previously that measuring γ as a function of concentration is a convenient means of determining the surface excess of a substance at a mobile interface. In view of the complications arising from charge considerations, the need for an independent method for measuring surface excess becomes apparent. Some elaborate techniques have been developed which involve skimming a thin layer off the surface of a solution and comparing its concentration with that of the bulk solution. A simpler method for verifying the Gibbs equation involves the use of isotopically labeled surfactants. If the isotope emits a low-energy β particle, the range of the β in water will be very low. Thus a detector placed just above the surface will count primarily those emissions originating from the surface region. Tritium (^{3}H), for example, emits a 0.0186-MeV β particle with a range in water of only about 17 μm, which means that only a negligible fraction of the β particles can travel farther than this in water. In fact, most are absorbed in an even shorter distance, so any ^{3}H β particles detected above an aqueous solution of tritiated surfactant probably originate within approximately 3 μm of the surface. The contribution of the bulk solution to the "background" of the former measurement is made using the same isotope in a compound which is known not to be adsorbed. By such studies the kinds of effects just described have been investigated and verified.

The surface-active substances we have discussed have been purified research-quality materials. In practical situations the cost of synthesizing

Table 7.2 Some Families of Commercial Surfactants and Specific Examples from Each

Name of series	General chemical nature	Specific designation ("___") and chemical nature of example	Example
Igepon " ___ "	Fatty acid amide of methyltaurine	"TN" R = palmityl	$RCON(CH_3)C_2H_4SO_3^- Na^+$
Aerosol " ___ "	Alkyl ester of sulfosuccinic acid	"OT" R = octyl	CH_2—COOR CH_2—COOR SO_3Na^+
Span " ___ "	Fatty acid esters of anhydrosorbitols	"60" R = stearyl	HO—CH—CHOH CH_2—O—CH—CHOH—COOR
Tween " ___ "	Fatty acid ester and ethylene oxide esters of anhydrosorbitols	"21" n = 4, R = lauryl	ROOCCH—CHOH CH_2—O—CH(OC₂H₄)$_n$OH
Triton " ___ "	Ethylene oxide ethers of alkyl benzene	"X-45" n = 5, R = octyl	R—⟨benzene ring⟩—$(OC_2H_4)_n$OH
Hyamine " ___ "	Alkylbenzyl dimethyl ammonium salts	"3500" R = C_{12}-C_{16}	R—⟨benzene ring⟩—$N^+(CH_3)_2Cl^-$ $CH_2\phi$

Source: Data from Surfactants, in *Encyclopedia of Chemical Technology*, Vol. 19, Wiley, New York, 1969.

and purifying such surfactants is prohibitive. The materials commercially used, therefore, are inevitably mixtures. Commercial surfactants originate, for example, from the esterification of sugars or the sulfonation of alkyl–aryl mixtures. Such mixtures are marketed under a bewildering variety of trade names, and often as members of number- or letter-designated series which correspond, roughly, to a set of homologs. Table 7.2 lists examples of several specific members of such series, along with a brief description of the general nature of the family to which they belong.

7.10   The Langmuir Equation: Theory

Throughout most of this chapter we have been concerned with adsorption at mobile surfaces. In these systems the surface excess may be determined directly from the experimentally accessible surface tension. At solid surfaces this experimental advantage is missing. All we can obtain from the Gibbs equation in reference to adsorption at solid surfaces is a thermodynamic explanation for the driving force underlying adsorption. Whatever information we require about the surface excess must be obtained from other sources.

If a dilute solution of a surface-active substance is brought in contact with a large adsorbing surface, then extensive adsorption will occur with an attendant reduction in the concentration of the solution. To meet the requirement of a large surface available for adsorption, the solid—which is called the adsorbent—must be finely subdivided. From the analytical data describing the concentration change in the solution as well as a knowledge of the total amount of solid and solution equilibrated, it is possible to determine the amount of solute adsorbed—which is called the adsorbate—per unit weight of adsorbing solid. If the specific area of the latter is known, then the results may be expressed as amount adsorbed per unit area. These studies are generally conducted at constant temperature, and the results—which relate the amount of material adsorbed to the equilibrium concentration of the solution—describe what is known as the adsorption isotherm.

One isotherm that is both easy to understand theoretically and widely applicable to experimental data is due to Langmuir and is known as the Langmuir isotherm. In Chap. 9, we shall see that the same function often describes the adsorption of gases at low pressures, with pressure substituted for concentration as the independent variable. We shall discuss the derivation of Langmuir's equation again in Chapter 9 specifically as it applies to gas adsorption. Now, however, adsorption from solution is our concern. In this section we consider only adsorption from dilute solutions. In the following section adsorption over the full range of binary solution concentrations is also mentioned.

Suppose we imagine a dilute solution in which both the solvent (component 1) and the solute (component 2) have molecules which occupy the same area when they are adsorbed on a surface. The adsorption of solute may then be schematically represented by the equation

adsorbed solvent + solute in solution → adsorbed solute

+ solvent in solution (61)

The equilibrium constant for this reaction may be written as

$$K' = \frac{a_2^s a_1^b}{a_1^s a_2^b} \tag{62}$$

where a stands for the activity of the species and the superscripts s and b signify surface and bulk values, respectively. Next let us assume that the two-dimensional surface solution is ideal, an assumption which enables us to replace the activity at the surface by the mole fraction at the surface x^s:

$$K' = \frac{x_2^s a_1^b}{x_1^s a_2^b} \tag{63}$$

Since the surface contains only two components, $x_1^s + x_2^s = 1$ and Eq. (63) becomes

$$K' = \frac{x_2^s a_1^b}{(1 - x_2^s)a_2^b} \tag{64}$$

Equation (64) may be rearranged to give

$$x_2^s = \frac{K'a_2^b/a_1^b}{K'a_2^b/a_1^b + 1} \tag{65}$$

In dilute solutions the activity of the solvent is essentially constant, so the ratio K'/a_1^b may be defined to equal a new constant K, in terms of which Eq. (65) becomes

$$x_2^s = \frac{Ka_2^b}{Ka_2^b + 1} \tag{66}$$

This is one form of the Langmuir adsorption isotherm.

An equivalent form of the Langmuir equation expressed in slightly different variables is obtained by multiplying both x_1^s and x_2^s in Eq. (63) by the total area of the surface A. We have already postulated that both the solvent and solute occupy equal areas on the surface. Therefore, $x_i^s A$ equals the fraction of the surface occupied by component i, θ_i. Since $\theta_1 + \theta_2 = 1$, the ratio θ_2/θ_1 rearranges the same as the ratio x_2^s/x_1^s to give

$$\theta_2 = \frac{Ka_2^b}{Ka_2^b + 1} \tag{67}$$

In this form the Langmuir equation shows how the fraction of surface adsorption sites occupied by solute increases as the solute activity in solution increases. From now on we shall drop the subscript 2 and the superscript b. Since Eq. (67) is written solely in terms of the solute, these designations are redundant.

Two limiting cases are of special interest:

1. At infinite dilution $a \to 0$ and Eq. (67) becomes

$$\theta = Ka \tag{68}$$

2. If $Ka \gg 1$, Eq. (67) becomes

$$\theta = 1 \tag{69}$$

Equation (68) shows that θ increases linearly with an initial slope that equals K. This slope will be larger the farther to the right the equilibrium represented by Eq. (61) lies. At higher concentrations Eq. (69) indicates that saturation of the surface with adsorbed solute is achieved. Figure 7.14a shows how these two limiting conditions affect the appearance of the isotherm.

Experimentally, one does not measure the fraction of sites containing adsorbed solute directly; instead, either the number of moles of solute adsorbed per unit weight of adsorbent, n_2^s/w, or the number of moles per unit area of adsorbent, n_2^s/A, is measured. These quantities are related by the equation

$$\frac{n_2^s}{A} A_{sp} = \frac{n_2^s}{w} \tag{70}$$

where A_{sp} is the specific area of the adsorbent (see Sect. 1.2). The fraction covered is related to these quantities as follows:

$$\theta = \frac{n_2^s}{A} N_A \sigma^0 = \frac{n_2^s N_A \sigma^0}{w A_{sp}} \tag{71}$$

where N_A is Avogadro's number and σ^0 is the area occupied per molecule. The level at which saturation adsorption occurs may be identified with

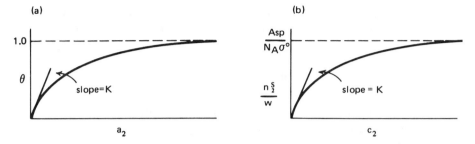

Figure 7.14 Schematic plots of the Langmuir equation showing the significance of the initial slope and the saturation value of the ordinate: (a) the fraction covered versus solute activity and (b) the number of moles of solute adsorbed per unit weight of adsorbent versus concentration.

$\theta = 1$. Therefore Eq. (71) shows the saturation values of the usual ordinates to be either

$$\left(\frac{n_2^s}{w}\right)_{sat} = \frac{A_{sp}}{N_A \sigma^0} \tag{72}$$

or

$$\left(\frac{n_2^s}{A}\right)_{sat} = \frac{1}{N_A \sigma^0} \tag{73}$$

Since the entire derivation of the Langmuir isotherm assumes dilute solutions, the concentration c of the solute rather than the activity is generally used in presenting experimental results. Figure 7.14b shows how actual experimental data might appear.

7.11 The Langmuir Equation: Application to Results

Many systems which definitely do not conform to the Langmuir assumptions—the adsorption of polymers, for example—nevertheless display experimental isotherms which resemble Fig. 7.14. Although these can be fitted to Eq. (67), the significance of the constants is dubious. Therefore the Langmuir equation is often written as

$$m\left(\frac{n_2^s}{w}\right) = \frac{(m/b)c}{(m/b)c + 1} \tag{74}$$

where m and m/b are regarded simply as empirical constants. A method for obtaining the numerical values for these constants from experimental data is easily seen by rearranging Eq. (74) to the form

$$\frac{c}{n_2^s/w} = mc + b \tag{75}$$

This form suggests that a plot of $c/(n_2^s/w)$ versus c will be a straight line of slope m and intercept b.

If the experimental system matches the model, then the values of m and b can be assigned a physical significance by comparing Eq. (74) with Eqs. (67) and (71):

$$m = \frac{N_A \sigma^0}{A_{sp}} \tag{76}$$

and

$$\frac{m}{b} = K \tag{77}$$

If the model does not apply, these constants are treated merely as empirical parameters which describe the adsorption isotherm.

When the model does apply, the experimental value of m permits A_{sp} to be evaluated if σ^0 is known, or σ^0 to be evaluated if A_{sp} is known. It is often difficult to decide what value of σ^0 best characterizes the adsorbed molecules at a solid surface. Sometimes, therefore, this method for determining A_{sp} is calibrated by measuring σ^0 for the adsorbed molecules on a solid of known area, rather than relying on some assumed model for molecular orientation and cross section.

The following example illustrates how adsorption data can be interpreted if the data conform to the Langmuir model.

Example 7.5   The moles of solute B adsorbed per gram of solid C were determined by measuring concentration changes in the solution. The accompanying results report the adsorption versus the concentration of the equilibrium solution:

c (mole B liter^{-1})	0.75	1.40	2.25	3.00	3.50	4.25
$n/w \times 10^4$ [mole B (g C)$^{-1}$]	6.00	8.00	9.57	10.0	10.4	10.8

Plot these data in the form suggested by Eq. (75) and evaluate the slope and intercept. If A_{sp} for solid C is known by independent study to be 325 m^2 g^{-1}, what is σ^0 for the adsorbate? Alternatively, suppose σ^0 for the adsorbate is known to be 0.25 nm^2 on this surface. What value of A_{sp} is consistent with the adsorption data?

Solution   The ratio $c/(n/w)$ is evaluated as required to test Eq. (75):

$c(n/w)^{-1} \times 10^{-3}$ (g C liter^{-1})   1.25   1.75   2.35   3.00   3.35   3.95

A plot of these values against the equilibrium concentrations is shown in Fig. 7.15. The slope and intercept of the line drawn are 769 g C (mole B)$^{-1}$ and 0.0700 g C liter^{-1}, respectively.

Equation (76) permits the interpretation of the slope m: If A_{sp} is known,

$$\sigma = \frac{mA_{sp}}{N_A} = \frac{(769)(325)(10^9)^2}{6.02 \times 10^{23}} = 0.42 \ nm^2$$

and if σ^0 is known,

$$A_{sp} = \frac{N_A\sigma^0}{m} = \frac{(6.02 \times 10^{23})(0.25)(10^{-9})}{769} = 196 \ m^2 \ g^{-1}$$

The value of K is given by the ratio m/b according to Eq. (77). For this (hypothetical) system, $K = (769)/(0.0700) = 1.10 \times 10^4$ liter (mole B)$^{-1}$. these reciprocal concentration units are appropriate for K of Eq. (61),

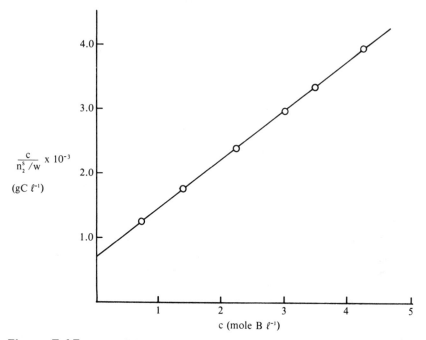

Figure 7.15   Plot of the Langmuir equation in the form given by Eq. (75) for data in Example 7.5.

since the activity of the bulk solvent has been absorbed into the definition
of K.

•

The method of Example 7.5 applied to the adsorption of benzene,
naphthalene, and anthracene on carbon black from heptane solutions
gives values of 0.42, 0.67, and 0.83 nm^2, respectively, for σ^0. The
progression of sizes indicates that the molecules lie flat on the surface of
the carbon.

It might also be noted that K' [Eq. (62)] may be related to ΔG^0 for the
adsorption process if the model applies to the experimental system.
Therefore, from studies of adsorption at different temperatures, values of
ΔH^0 and ΔS^0 may be determined for the process described by Eq. (61). It
must be emphasized that compliance with the form predicted by the
Langmuir isotherm is not a sensitive test of the model; therefore
interpretations of this kind must be used cautiously.

In summary, adsorption from dilute solutions frequently displays the
qualitative form required by the Langmuir equation. If this form is
observed, it may be quantitatively described by Eq. (75) in which m and b
are empirical constants. Sometimes there may be a justification for
further interpretation of these parameters in terms of the theoretical
model.

We should not be too surprised that the Langmuir equation often
yields only an empirical isotherm. There are several reasons why real
systems are likely to deviate from the theoretical model:

1. The adsorption process described by Eq. (61) is a complex one,
   involving several different kinds of interactions: solvent–solute,
   solvent–adsorbent, and solute–adsorbent.
2. Few solid surfaces are homogeneous at the molecular level.
3. Few monolayers are ideal.
4. Our interest often extends beyond the region of dilute concentrations.

We shall briefly comment on each of these limitations.

In discussing adsorption from solution, there is nothing that can be
done about the multiplicity of possible interactions, except possibly to
avoid systems in which highly specific interactions are to be expected. In
Chap. 9 we shall again discuss the Langmuir isotherm as it applies to the
adsorption of gases. In that case there are considerably fewer interactions
involved in the adsorption process, making the latter more amenable to
analysis.

The assumption of surface homogeneity is one that was not explicitly
stated in deriving the Langmuir equation. It is essential, however;

otherwise a different value of K would apply to Eq. (61) at various places on the surface. Attempts to deal with surface heterogeneity have been undertaken, but this enterprise seems more likely to be successful for gas adsorption rather than for adsorption from solution, because the variety of interactions which must be considered is less in the former than in the latter. There is an equation—known as the Freundlich isotherm—that may be derived by assuming a certain distribution function for sites having different ΔG^0 values for the process represented by Eq. (61) and assuming Langmuir adsorption at each type of site. The Freundlich isotherm is given by the expression

$$\theta = ac^{1/n} \tag{78}$$

in which a and n are constants with $n > 1$. This equation was in use long before the interpretation of a certain distribution of sites was assigned to it. Therefore it is best regarded as an empirical isotherm, the constants for which may be evaluated from the slope and intercept of a log–log plot of θ versus c. The Freundlich isotherm is no cure-all for surface heterogeneity: Its theoretical derivation depends on a highly specific distribution of site energies. In addition, the Langmuir equation gives adequate results in many cases where surface heterogeneity is known to be present. Note that with $n = 1$ the Freundlich isotherm is identical to the low-concentration limit of the Langmuir isotherm [Eq. (68)], and with $n = \infty$ to the high-concentration limit [Eq. (69)].

In the Langmuir derivation the adsorbed molecules are allowed to interact with the adsorbent but not with each other: The adsorbed layer is assumed to be ideal. This necessarily limits adsorption to a monolayer. Once the surface is covered with adsorbed molecules, it has no further influence on the system. The assumption that adsorption is limited to monolayer formation was explicitly made in writing Eqs. (72) and (73) for the saturation value of the ordinate. It is an experimental fact, however, that adsorption frequently proceeds to an extent that exceeds the monolayer capacity of the surface for any plausible molecular orientation at the surface. That is, if monolayer coverage is postulated, the apparent area per molecule is only a small fraction of any likely projected area of the actual molecules. In this case the assumption that adsorption is limited to the monolayer fails to apply. A model based on multilayer adsorption is indicated in this situation. This is easier to handle in the case of gas adsorption, so we shall defer until Chap. 9 a discussion of multilayer adsorption.

Next let us consider adsorption from solutions which are not infinitely dilute. Suppose, for example, that adsorption is studied over the full range of binary liquid concentrations. Figure 7.16 is an example of such results for the benzene–ethanol system adsorbed on carbon. At first

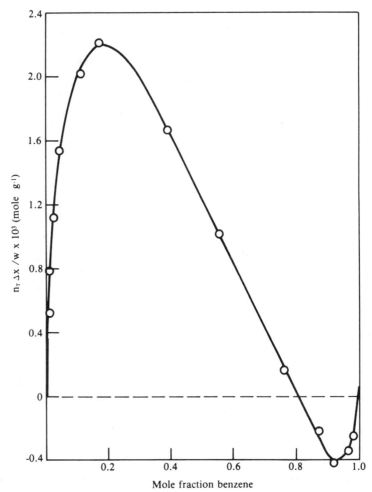

Figure 7.16 Adsorption on carbon from the ethanol–benzene system. The ordinate equals the total number of moles of solution times the change in solution mole fraction per unit weight of carbon. [Data from F. E. Bartell and C. K. Sloan, *J. Am. Chem. Soc., 51*:1643 (1929).]

these results appear quite bewildering, displaying maximum, minimum, and negative adsorption. Recall, however, that what is actually measured is an isotherm of concentration change. The observed change in concentration is then expressed as moles of solute adsorbed. In a totally different range of solution concentrations, the solvent rather than the solute may adsorb. The associated change in the solution would then be an increase in solute concentration or the apparent negative adsorption of solute. A curve like that shown in Fig. 7.16 should therefore be understood as a composite of two distinctly different isotherms. A good deal of work has been done with composite isotherms, particularly toward separating them into individual isotherms. A summary of this kind of research can be found in Ref. 5.

7.12 Applications of Adsorption from Solution

No discussion of adsorption from solution is anywhere near complete unless it includes some indication of the enormous practical applicability of this topic. As a matter of fact, the examples we shall briefly consider—detergency and flotation—encompass a wide variety of concepts from almost all areas of surface and colloid chemistry. We have chosen to stress principles rather than applications, however, so these subjects will receive an amount of attention which belies their actual importance.

It is impossible to do justice to the complex phenomena of detergency and flotation in a few paragraphs. All we can do is point out some of the ways in which the principles of colloid and surface chemistry apply in these areas. There are several ways in which detergency and flotation phenomena resemble one another:

1. Both terms give simple names to process involving many different steps. The more familiar of the two, detergency, may be defined as the process whereby some unwanted foreign matter is removed from a substrate by a combination of chemical treatment, temperature, and mechanical agitation. Flotation is the process by which a specific mineral component of an ore mixture is separated from other components (called "gangue") by being concentrated in the froth of an aerated slurry. Chemical additives and mechanical forces are involved here also.
2. In actual practice, both detergency and flotation deal with systems which are terribly difficult to idealize by any sort of model. In a laundering operation, for example, there will be present a variety of different fabric surfaces (cotton, polyester, etc.), different kinds of foreign matter (particulate, oily, etc.), and different chemical additives (detergent, inorganic phosphate, fluorescent whitening agents as well as the solvent, water). In flotation, all three states of matter—solid, liquid, and gas—are involved and each of these involves several chemical components. The ore is a complex mixture of minerals (assumed to be crushed to such an extent that each particle is a different phase), the air is a mixture of gases (including chemically reactive oxygen), and the liquid contains at least three deliberately added reagents (known as regulator, collector, and

frother), in addition to whatever dissolved minerals are present in the water.

3. A third point of resemblance between detergency and flotation (perhaps redundant in view of what has already been said) is that both have developed largely by empirical research with (partially satisfactory) explanations trailing far behind the actual practice.

With this much general background, let us now consider these two processes separately. In discussing detergency we must first examine the availability of the surfactant. Weak acid soaps form insoluble compounds with Ca^{2+}, for example, and are converted to insoluble molecular acids at low pH levels. One of the reasons for the addition of inorganic phosphate to laundry products is to prevent or minimize these reactions. In this discussion we shall assume that the impurity has not been imbibed into the interior of the fiber (soaking might help if it has) and that it is a semiliquid soiled spot rather than a solid contaminant with which we are dealing. One advantage of washing this type of soiled material at high temperatures is that the viscosity of the oily spot is lowered so that the shape of these drops is more readily altered.

The process of removing an oily drop from a solid substrate may be described in terms of the work of adhesion, given by Eq. (6.57). Applying this idea to the separation of oil (O) from solid (S) gives

$$W = \gamma_{ow} + \gamma_{sw} - \gamma_{os} \qquad (79)$$

where the subscript W describes the aqueous solution. For the separation to be spontaneous in the thermodynamic sense, this quantity (ΔG) must be negative. Positive surface excesses of surfactant molecules at the interface between the aqueous phase and the oil and/or the solid will lower γ for these surfaces. This change is a favorable one for the process of removing foreign matter.

In addition, the contact angle between an oil spot and a solid surface to be cleaned may be a contributing factor in detergency. For example, Fig. 7.17b and d illustrates schematically two different situations for an oily drop being lifted off a substrate by currents in the adjacent phase. The contact angles θ_1 between the drop and the substrate are assumed to be the same at "lift off" (b and d) as in the quiescent state (a and c). It is evident from the figure, however, the necking of the drop for $\theta_1 < 90°$ is likely to leave a residue, whereas $\theta_1 > 90°$ would lead to a clean detachment. Young's equation [Eq. (6.44)] may be applied to this situation:

$$\gamma_{ow} \cos \theta_1 = \gamma_{sw} - \gamma_{os} \qquad (80)$$

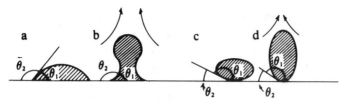

Figure 7.17 Schematic illustration of several configurations of three phases useful in the discussion of detergency and flotation. The shaded region represents the soiled spot in detergency and θ_1 is the relevant contact angle; the shaded region is an air bubble in flotation and θ_2 is the appropriate contact angle.

where θ_1 is measured in the oil drop as shown in Fig. 7.17. Equation (80) shows that $\theta_1 > 90°$ and $\theta_1 < 90°$ correspond to $\gamma_{sw} < \gamma_{os}$ and $\gamma_{sw} > \gamma_{os}$, respectively. Any adsorption at the solid–water interface will lower γ_{sw} and therefore be conducive to a contact angle which favors the complete "rollback" of the oily spot.

Once the dirty spot is removed from the substrate being laundered, it is important that it not be redeposited. Solubilization of the detached material in micelles of surfactant has been proposed as one mechanism which contributes to preventing the redeposition of foreign matter. Any process which promotes the stability of the detached dirt particles in the dispersed form will also facilitate this. We shall see in Chap. 12 how electrostatic effects promote colloidal stability. The adsorption of ions—especially amphipathic surfactant ions—onto the detached matter assists in blocking redeposition by stabilizing the dispersed particles. Materials such as carboxymethylcellulose are often added to washing preparations, since these molecules also adsorb on the detached dirt particles and interfere with their redeposition.

Now let us turn to a brief examination of flotation. Virtually all nonferrous metallic ores are concentrated by the flotation process. Sulfide ores have been studied particularly extensively, although the method has been used with oxides and carbonates as well as such nonmetallic materials as coal, graphite, sulfur, silica, and clay. Something on the order of a billion tons of ore a year are processed in this way.

We shall assume that the ore has been pulverized and mixed with water so that our involvement begins with a slurry known as "pulp". We shall consider, in turn, the chemical nature and the effects of each of the three broad classes of chemicals added in the flotation process.

The first class of chemical additives to be considered is the regulator, a compound which affects the adsorption of the collectors. Regulators, like catalysts, may be positive or negative in their role. For the case in which the collector adsorption is enhanced, the regulator is called an activator; when the effect is negative, it is called a depressant. Regulators are frequently compounds which control the pH and sequester metallic cations which would otherwise compete with the mineral particle surfaces for the surface-active collectors. The pH affects not only the availability of certain collectors, but also the charge of the mineral particles (see Sect. 12.2 for a discussion of potential-determining ions). This last consideration plays a role in determining whether the mineral particles will be dispersed as small units (easier to lift by flotation) or whether they will be flocculated. Ammonia, lime, and sources of CN^- and HS^- are commonly used as regulating agents.

Collectors are surface-active additives which adsorb onto the mineral surface and prepare the surface for attachment to an air bubble so that it will float to the surface. Therefore collectors must adsorb selectively if flotation is to result in any fractionation of the crude ore. In addition, the adsorbed collector must impart a hydrophobic character to the particle surface so that an air bubble will attach to the mineral, or vice versa.

Amphipathic substances such as we have discussed throughout this chapter are used as collectors. Alkyl compounds with C_8 to C_{18} chains are widely used with carboxylate, sulfate, or amine polar heads. For sulfide minerals sulfur-containing compounds such as mercaptans, monothiocarbonates, and dithiophosphates are used as collectors. The most important collectors for sulfides are xanthates, the general formula for which is

$$R-O-\underset{\underset{\displaystyle S-CH_3}{}}{\overset{\overset{\displaystyle S}{\|}}{C}}$$

In the collectors used R is generally in the C_2 to C_6 range. Xanthates are readily oxidized to dixanthogens, and the extent of this reaction may have a large effect on the efficiency of the collector.

The fundamental role of the collector is to produce a solid surface which is sufficiently hydrophobic so that it will attach to an air bubble when the pulp is aerated. Figure 7.17 may also be used to represent this situation, except that for flotation the shaded region is an air bubble. Since contact angles are measured in the liquid phase, the contact angle in the flotation case will be θ_2. For good bubble adhesion contact angles greater than 90° are preferred. Unlike the parallel situation in detergency, the adhesion of the bubble rather than its detachment is required for the success of the process.

Once again, we may use Young's equation to decide what adsorption situation is most conducive to values of $\theta_2 > 90°$:

$$\gamma_{WA} \cos \theta_2 = \gamma_{AS} - \gamma_{SW} \qquad (81)$$

From this, we see the optimum condition corresponds to $\gamma_{SW} \gg \gamma_{AS}$, or extensive adsorption at the air–solid surface and minimum adsorption at the solid–water interface. The hydrophobic nature of the collectors and their chemical affinity for specific solids promote this situation.

The formation of a large bubble which facilitates flotation requires a large area of attachment, or, more specifically, a large perimeter of attachment, since the three-phase contact boundary occurs along the perimeter. Increasing this perimeter is favored by a positive value of the spreading coefficient [Eq. (6.61)]. In the notation of this problem (air spreading on solid in water), the spreading coefficient equals

$$S_{A/S} = \gamma_{SW} - (\gamma_{AW} + \gamma_{SA}) \qquad (82)$$

The collector lowers γ_{SA}, an effect favorable to a positive spreading coefficient.

Finally, the frothing agents are intended to stabilize the mineral-laden foam at the surface of the aeration tank until it can be scooped off. Alkyl or aryl alcohols in the C_5 to C_{12} range are typical frothers. We have already seen how the long-chain members of this series form monolayers at the air-water surface. This lowers γ_{AW}, which is beneficial to the stability of the foam, and also favors a large contact angle (if $\theta_2 > 90°$), positive spreading, and large bubbles for flotation. Neither the collector nor the frother is adsorbed exclusively at the solid–air or the water–air surface where their respective effects would be greatest. To a certain extent these two classes of additives compete with each other for adsorption sites; therefore conditions under which each produces the maximum effect are difficult to achieve, so compromise conditions in which the net effect is optimized are sought.

Another aspect of the frothers used is the fact that they form fairly condensed and therefore relatively viscous slow-draining films. In addition to thermodynamic considerations, then, kinetic factors are also important in stabilizing the froth.

A number of additional applications of the ideas of this chapter could be profitably considered if space permitted. Included among these are adhesives, lubricants, waterproofing, and the recovery of oil from the pores of rocks. Like detergency and flotation, these topics involve a variety of surface and colloid phenomena. The interested reader will find an introduction to these fields in some of the references listed for this chapter, especially Refs. 2, 3, and 8.

7.13   Electrocapillarity

We conclude this chapter with a discussion of adsorption at the interface between mercury and a solution (usually aqueous) under the influence of an applied potential. The γ value for this interface is easily measured, and potential and electrolyte concentration can be studied as variables. The Nernst equation provides a familiar reminder that potentials can be dealt with by thermodynamics.

Moving a charge of q coulombs to a surface where the potential is ψ volts involves a quantity of work (in joules)

$$\delta w = q\psi \tag{83}$$

Since this is non-pressure–volume work, it can be identified with a change in Gibbs free energy. In addition, the charge q carried by n moles of ions having a relative charge of $\pm z$ is given by $q = \pm n N_A z e$, where e is the proton charge, 1.60×10^{-19} C. The product $N_A e$ is called the Faraday constant F and equals 96,480 C. In view of these ideas, we can write

$$\Delta G = \pm n N_A z e \psi = \pm z n F \psi \tag{84}$$

where the sign depends on whether the charge is moved with or against the potential.

Figure 7.18 illustrates an apparatus whereby the relationship between interfacial tension, applied potential, and electrolyte concentration can be investigated. Since mercury has a contact angle (measured in the mercury) which is greater than 90°, the mercury–solution interface is depressed (again in reference to the mercury). If an etched mark is placed on the capillary, it is possible to bring the mercury–solution meniscus to that mark by varying the height of the mercury reservoir. It turns out that this is quite sensitive to the potential E between the electrodes. The height of the capillary depression (a negative capillary "rise") can be readily converted to the interfacial tension through Eq. (6.4). We see, therefore, that γ for the Hg–solution interface depends on the electrical potential across the system. In addition, the detailed shape of the so-called electrocapillary curve—a plot of γ versus E—depends on the concentration and nature of the electrolyte present. Figure 7.19a and b shows examples of typical electrocapillary curves.

Several generalizations are evident from an inspection of Fig. 7.19:

1.   The general shape of the curves is roughly parabolic. The coordinates of the maximum in the electrocapillary curve depend on the

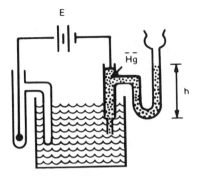

Figure 7.18 Schematic illustration of an apparatus to measure the electrocapillary effect.

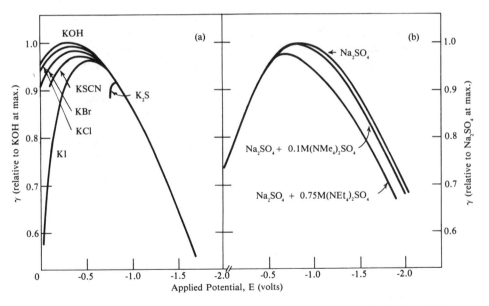

Figure 7.19 Typical electrocapillary curves: (a) Anions are adsorbed and (b) cations are adsorbed. (From Ref. 1, used with permission.)

electrolyte content of the system. Since γ decreases on both sides of the electrocapillary maximum and since reductions in γ are associated with adsorption, we conclude that adsorption increases as we move in either direction from the maximum; that is, the electrocapillary maximum seems to be a point of minimum adsorption.

2. The left-hand branch of the electrocapillary curves (also called the rising or ascending branch) is sensitive to the chemical nature of the anion present. In Fig. 7.19a, for example, different potassium salts are used as the electrolytes. Although the ascending branches of these curves differ, the descending branches (to the right of the maximum) lie on a common curve. Figure 7.19b shows that the right-hand branch of the curves is sensitive to the nature (and concentration) of the cation. These observations, coupled with the preceding remarks about adsorption, suggest that the anion is preferentially adsorbed on the ascending branch of the curve, the cation is preferentially adsorbed on the descending branch, and that neither ion is preferentially adsorbed at the maximum, although both may be adsorbed in equivalent amounts.

3. The charge carriers in the system—ions in the solution and electrons in the mercury—come to an equilibrium distribution which is consistent with the applied potential. To the left of the maximum the mercury surface is a positively charged anode (this branch is also called the anodic branch), toward which anions are attracted. Electrons in the mercury are repelled from the surface by the anions in the water, so the potential on the mercury must become progressively more positive (relative to the maximum) for increased anion adsorption to occur. To the right of the maximum the mercury is negative (the cathodic branch), attracts cations, and must become progressively more negative (relative to the maximum) for increased cation adsorption.

4. The potential at which the maximum occurs is different for different ions. In light of the third item, the differences in the voltage coordinates of the maxima must reflect differences in the chemical (as opposed to purely electrostatic) affinities of the ions for the interface. Nonionic solutes have also been investigated extensively, but we shall not go into this aspect of the subject.

Electrocapillary phenomena have been studied for a long time; the apparatus shown in Fig. 7.18 is essentially that used by G. Lippmann in 1875 in his comprehensive studies of electrocapillarity. We shall not examine either the experimental or the theoretical aspects of this system in great detail; however, an interpretation of the results that is more quantitative than that just outlined qualitatively is possible with relatively little additional effort.

As a starting point, it is convenient to return to the kind of argument that leads to the Gibbs–Duhem equation [Eq. (39)] for one-phase systems and to the Gibbs adsorption equation [Eq. (44)] for systems containing an interface. In the present context we are interested not only in the interface but also in possible charge effects at that interface. Accordingly, it is convenient to distinguish between charged and uncharged components in the system; we shall use the subscript i to identify the former and j for the latter. In this notation, Eq. (44) may be written

$$-d\gamma = \sum_i \Gamma_i \, d\bar{\mu}_i + \sum_j \Gamma_j \, d\mu_j \qquad \begin{array}{l} i = geladen \\ j = ongeladen \end{array} \qquad (85)$$

In this expression μ_j is the usual chemical potential for the uncharged species. The quantity $\bar{\mu}_i$ is called the *electro*chemical potential and is related to μ_i as follows:

$$\bar{\mu}_i = \mu_i + z_i e \psi \qquad (86)$$

The second term on the right-hand side of Eq. (86) arises from Eq. (84), since chemical potential is the partial molar free energy. In this expression z_i is the relative charge of the ith species and ψ is the potential of the surface.

In this discussion the mercury–solution interface is assumed to be completely polarizable; that is, there is no transfer of charge across the interface. Therefore each of the charged species (including electrons) occurs in only one of the phases. The requirement of complete polarizability means that the surface excess may be divided between that lying in the aqueous phase (W) and that in the mercury (Hg). In view of Eqs. (85) and (86), we write

$$-d\gamma = \sum_i (\Gamma_i \, d \, \mu_i)_w + \sum_i (z_i \Gamma_i e \psi)_w + \sum_i (\Gamma_i \, d\mu_i)_{Hg} + \sum_i (z_i \Gamma_i e \psi)_{Hg}$$

$$+ \sum_j \Gamma_j \, d\mu_j \qquad (87)$$

Since the system as a whole is electrically neutral, the charge density on each of the phases must be equal and opposite; that is,

$$\sum_i (z_i e \Gamma_i)_w = - \sum_i (z_i e \Gamma_i)_{Hg} \qquad (88)$$

Substituting Eq. (88) into Eq. (87) yields

$$-d\gamma = \sum_i (\Gamma_i \, d\mu_i)_\text{w} + \sum_i (\Gamma_i \, d\mu_i)_\text{Hg} + \sum_j \Gamma_j \, d\mu_j$$

$$+ \sum_i (z_i \Gamma_i e)_\text{w} (\psi_\text{w} - \psi_\text{Hg}) \tag{89}$$

The difference in potential between the two phases varies as dE, the externally applied potential difference. Therefore Eq. (89) may also be written

$$-d\gamma = \sum_i (\Gamma_i \, d\mu_i)_\text{w} + \sum_i (\Gamma_i \, d\mu_i)_\text{Hg} + \sum_j \Gamma_j \, d\mu_j + \sum_i (z_i \Gamma_i e)_\text{w} dE$$

$$\tag{90}$$

For a completely polarizable electrode the concentration of each of the components in both solutions is constant and therefore so is the chemical potential for each. Therefore we obtain

$$-\left(\frac{\partial \gamma}{\partial E} \right)_\mu = \sum_i (z_i e \Gamma_i)_\text{w} \tag{91}$$

This result is known as the Lippmann equation. The charge density on the right-hand side of this equation refers exclusively to the solution phase; therefore the subscript is no longer retained.

For monovalent ions, Eq. (91) is simply

$$\left(\frac{\partial \gamma}{\partial E} \right)_\mu = -e(\Gamma_+ - \Gamma_-) \tag{92}$$

When the surface excess of anions exceeds that of cations, $d\gamma/dE$ is positive, as is observed in Fig. 7.19a. Negative values of $d\gamma/dE$ correspond to larger surface excesses of cations, as shown by Fig. 7.19b. Finally, the condition $d\gamma/dE = 0$ corresponds to equal amounts of positive and negative adsorbed charge, that is, surface neutrality. Note that this is not the same as saying no ions are adsorbed. The slope of the electrocapillary curve measures the *difference* between the cation and anion surface excesses. This is explored further in the following example.

Example 7.6   In electrocapillary curves like those shown in Fig. 7.19a, the ascending branches cross for NaCl and $Ca(NO_3)_2$ solutions, with the steeper NaCl curve cutting across the flatter $Ca(NO_3)_2$ curve. Use this observation to criticize or defend the following proposition: Both solutes must show equal adsorption at the cross-over point, since the lowering of

γ depends on the surface excess. Since both show the same value for γ, both must have the same surface excess.

Solution   According to the Lippmann equation, it is not γ per se but, rather, $d\gamma/dE$ that is pertinent to this data. Even though the two interfaces show the same value of γ at the point in question, they do not have the same slopes. In fact, $(d\gamma/dE)_{\text{NaCl}} > (d\gamma/dE)_{\text{Ca(NO}_3)_2}$, which means, according to Eq. (91),

$$-e[(+1)\Gamma_{\text{Na}^+} + (-1)\Gamma_{\text{Cl}^-}] > -e[(+2)\Gamma_{\text{Ca}^{2+}} + (-1)\Gamma_{\text{NO}_3^-}]$$

or

$$(\Gamma_{\text{Cl}^-} - \Gamma_{\text{Na}^+}) > (\Gamma_{\text{NO}_3^-} - 2\Gamma_{\text{Ca}^{2+}})$$

This does not mean that $\Gamma_{\text{Cl}^-} > \Gamma_{\text{NO}_3^-}$ but, rather, that the surface excess of the anion compared to the cation is greater for NaCl than for Ca(NO$_3$)$_2$.

•

Another interpretation of the electrocapillary curve is easily obtained from Eq. (89). We wish to investigate the effect of changes in the concentration of the aqueous phase on the interfacial tension at constant applied potential. Several assumptions are made at this point to simplify the desired result. More comprehensive treatments of this subject may be consulted for additional details, for example, the chapter by Overbeek [9]. We assume that (1) the aqueous phase contains only 1:1 electrolyte, (2) the solution is sufficiently dilute to neglect activity coefficients, (3) the composition of the metallic phase (and therefore μ_i^{Hg}) is constant, (4) only the potential drop at the mercury–solution interface is affected by the composition of the solution, and (5) the Gibbs dividing surface can be located in such a way as to make the surface excess equal to zero for all uncharged components ($\Gamma_j = 0$). With these assumptions, Eq. (89) becomes

$$-d\gamma = RT(\Gamma_+ + \Gamma_-)_w d \ln c + e(\Gamma_+ - \Gamma_-)_w dE \tag{93}$$

where c is the concentration of the electrolyte. If we specify constant applied potential, Eq. (93) becomes

$$-\left(\frac{\partial \gamma}{\partial \ln c}\right)_E = RT(\Gamma_+ + \Gamma_-)_w \tag{94}$$

This result shows that the vertical displacements (at fixed potential) of the electrocapillary curve with changes in electrolyte concentration measure

418   **Adsorption from Solution**

the *sum* of the surface excesses at the solution surface. Curves such as those in Fig. 7.19b may be interpreted by this result. We have already seen that $\Gamma_+ = \Gamma_-$ at the electrocapillary maximum (where $E = E_{max}$); therefore

$$-\left(\frac{\partial\gamma}{\partial \ln c}\right)_{E_{max}} = 2RT\Gamma_{+,w} \tag{95}$$

A final result that can be extracted from the Lippmann equation is readily obtained by differentiating Eq. (91) with respect to E at constant μ:

$$-\left(\frac{\partial^2\gamma}{\partial E^2}\right)_\mu = \left(\frac{\partial(e\Sigma_i z_i\Gamma_i)}{\partial E}\right)_\mu = C \tag{96}$$

The quantity $(e\Sigma_i z_i\Gamma_i)$ gives the charge density of the surface, and d(charge density)$/dE$ is called the differential capacitance C of an interface. Although experimental values of surface capacitance provide valuable information about the distribution of charge at an interface, we shall not pursue this topic. Our interest in Eq. (96) is primarily in the suggestion it offers that the distribution of charge at an interface can be modeled as a capacitor. When we discuss the ion atmosphere at a charged surface in Chapter 12, we shall begin by using the parallel plate capacitor as a preliminary model.

References

1. N. K. Adam, *The Physics and Chemistry of Surfaces*, Dover, New York, 1968.
2. A. W. Adamson, *Physical Chemistry of Surfaces*, 4th ed., Wiley, New Yok, 1982.
3. J. T. Davies and E. K. Rideal, *Interfacial Phenomena*, Academic, New York, 1961.
4. G. L. Gaines, *Insoluble Monolayers at Liquid–Gas Interface*, Wiley, New York, 1966.
5. J. J. Kipling, *Adsorption from Solutions of Nonelectrolytes*, Academic, New York, 1965.
6. V. I. Klassen and V. A. Mokrousov, *An Introduction to the Theory of Flotation*, Butterworth's, London, 1963.
7. V. K. LaMer (ed.), *Retardation of Evaporation by Monolayers*, Academic, New York, 1962.
8. L. I. Osipow, *Surface Chemistry*, Van Nostrand-Reinhold, Princeton, N.J., 1962.

9. J. Th. G. Overbeek, in *Colloid Science*, Vol. 1 (H. R. Kruyt, ed.), Elsevier, Amsterdam, 1952.
10. A. M. Schwartz, The Physical Chemistry of Detergency, in *Surface and Colloid Science*, Vol. 5 (E. Matijević, ed.), Wiley, New York, 1972.

Problems

1. By analogy with the behavior of gases, monolayers are expected to expand to cover an entire surface. When the underlying liquid is flowing, the motion supplies a natural barrier to spreading at the edge of the monolayer. Use this concept to interpret the following bucolic scene*:

   On a calm day an observer who finds the right place on a stream or a river will see an unobtrusive yet startling phenomena, a line on the surface of the water. The line may lie still, or it may contort itself, one way and another, in response to eddies. Very likely he will think a spider thread has fallen onto the water and try to cut it with his canoe paddle. As the disturbance caused by the cutting fades, the line reappears, mended and whole.

   Use the concepts of this chapter to discuss the existence of the bulge or line at the edge of the monolayer.

2. In some general chemistry laboratory courses the following experiment is done to evaluate Avogadro's number. A watch glass is filled to the brim with water and then a solution of stearic acid in benzene is slowly deposited dropwise on the surface until such time that a drop is added which "will not spread out, but will instead form a thick, lens-shaped layer."† What is the name of the film pressure at the "end point" of this experiment? What would be a reasonable estimate for σ at this pressure? Estimate the number of 0.005-ml drops of stearic acid solution ($c = 0.200$ g liter^{-1}) needed to form a monolayer on a watch glass with a 14-cm diameter. Outline how data such as these could be interpreted to lead to a value for N_A. Discuss some of the sources of error in this experiment.

3. Isotherms of π versus σ at both 15 and 25°C were studied for oleic acid on a substrate of pH 2.0 using a high-sensitivity film balance.‡ The following results were obtained:

---

*C. W. McCutchen, *Science, 170*:61 (1970).
†J. B. Ifft and J. L. Roberts, Jr., *Frantz/Malm's Essentials of Chemistry in the Laboratory* (3rd ed.), Freeman, San Francisco, 1975.
‡R. E. Pagano and N. L. Gershfeld, *J. Colloid Interface Sci., 41*:311 (1972).

	π(dyne cm^{-1})	
σ(Å^2 molecule^{-1})	At 15°C	At 25°C
10,000	0.028	0.037
8,000	0.035	0.040
6,000	0.044	0.052
4,000	0.055	0.071
3,000	0.058	0.086
2,000	0.076	0.095
1,000	0.075	0.095
500	0.074	0.095

Prepare a plot of π versus σ from these data. What is the apparent significance of the discontinuity in the curves? What quantity can be evaluated from the temperature variation of this discontinuity? Estimate this quantity from the available data.

4. The accompanying data give σ values corresponding to different film pressures for monolayers of various lecithins spread on 0.1 M NaCl at 22°C*:

$$\sigma \ (\text{Å}^2 \ \text{molecule}^{-1})$$

π (dyne cm^{-1})	Dibehenoyl	Distearoyl	Dipalmitoyl	Dimyristoyl	Dicapryl
2	51.7	53.3	96.7	96.7	99.2
4	50.0	52.5	88.3	90.0	93.8
6	49.2	50.8	82.2	85.0	86.7
8	48.3	50.0	76.7	81.7	82.5
10	47.9	49.5	66.7	77.5	78.3
15	46.7	48.0	53.3	70.8	71.7
20	45.5	46.7	50.0	65.8	65.8
30	45.0	45.0	46.3	58.3	58.3
40	44.7	44.7	45.0	53.8	53.8

Plot π versus σ for these isotherms and label the apparent two-dimensional phase present for various parts of the curves. Write the structural formulas for each of the lecithins and discuss the features of the curves in terms of the structure of the molecules.

5. If gas densities can be measured as a function of pressure, the molecular weight of a gas may be calculated from the expression

$$M = RT \left(\frac{d}{p}\right)_{\lim p \to 0}$$

---

*M. C. Phillips and D. Chapman, *Biochim. Biophys. Acta, 163*:301 (1968).

Likewise, if surface concentrations (in weight per area) are measured as a function of π, the molecular weight of a solute which forms an insoluble monolayer may be determined. D. Romeo and H. L. Rosano* obtained the following data for a monolayer of acetyl lipopolysaccharide on 0.2 M NaCl at 20°C:

c (mg m^{-2})	0.06	0.09	0.11	0.14	0.17	0.23
π (millidyne cm^{-1})	10.3	16.4	20.4	25.9	34.3	50.0

What is the molecular weight of the acetyl lipopolysaccharide?

6. E. G. Cockbain† measured the interfacial tension of the water–decane surface at various concentrations of sodium dodecyl sulfate (NaDS). The experiments were done at 20°C both in the presence and absence of NaCl. Use the suitable form of the Gibbs equation in each case to calculate Γ_R^l and σ at γ values of 10 and 20 dyne cm^{-1} from the following data:

Pure H$_2$O		0.1 M NaCl (swamping)	
c_{NaDS} (mole liter^{-1})	γ (dyne cm^{-1})	c_{NaDS} (mole liter^{-1})	γ (dyne cm^{-1})
0.0079	8.5	0.0014	5.2
0.00694	10.8	0.000694	11.7
0.00521	15.3	0.000347	17.4
0.00347	20.8	0.000173	22.7
0.001735	28.3	0.0000867	27.5

Is the variation of σ with interfacial film pressure qualitatively consistent with the expected behavior of monolayers in general? Of charged monolayers in particular?

7. M. Blank (in Ref. 7) has reported the permeability (in cm^3 of gas s^{-1} cm^{-2} surface) of various spread monolayers to water vapor at 25°C. For several different RX-type compounds at different π values, the permeabilities are as follows:

R	X	π (dyne cm^{-1})	Permeability $\times 10^3$ (cm^3 s^{-1} cm^{-2})
C$_{16}$	OH	44	380
C$_{18}$	OH	44	300
C$_{17}$	COOH	24	430

Discuss the observed differences in permeability between (a) the two alcohols at the same film pressure and (b) the two 18-carbon surfactants at different pressures. In your comments include comparisons of the molecular structure of the surfactants and the efficiencies of these monolayers in retarding evaporation.

---

*D. Romeo and H. L. Rosano, *J. Colloid Interface Sci.*, *33*:84 (1970).
†E. G. Cockbain, *Trans. Faraday Soc.*, *50*:874 (1954).

8. The pendant drop method has been used to measure the interfacial tension at the surface between mercury and cyclohexane solutions of stearic acid at 30 and 50°C.* Interpret the accompanying data by means of the Gibbs equation to evaluate Γ^1 and σ for stearic acid at these two temperatures when the equilibrium bulk concentrations are 10^{-3}, 10^{-4}, and 10^{-5} M:

$T = 30°C$		$T = 50°C$	
γ (dyne cm^{-1})	c_{Eq} (mole liter^{-1})	γ (dyne cm^{-1})	c_{Eq} (mole liter^{-1})
362	4.8×10^{-6}	364	4.8×10^{-6}
355	8.5×10^{-6}	362	9.6×10^{-6}
334	6.6×10^{-5}	354	7.4×10^{-5}
307	2.7×10^{-4}	334	4.4×10^{-4}
286	1.0×10^{-3}	314	1.6×10^{-3}

Estimate the concentrations at which the stearic acid film at the interface reaches a condensed packing at the two temperatures.

9. A scintillation counter is used to measure tritium β particles adjacent to the surfaces of tritiated sodium dodecyl sulfate in 0.115 M aqueous NaCl solution and tritiated dodecanol in dodecanol. The former system is surface active and the latter is not, so the difference between the measured radioactivity above the two indicates the surface excess of sodium dodecyl sulfate. The number of counts per minute arising from the surface excess A_s is related to the surface excess in moles per square centimeter Γ^1 by the relationship $A_s = 4.7 \times 10^{12}\ \Gamma^1$.† Use the following data (25°C) to construct the adsorption isotherm for sodium dodecyl sulfate on 0.115 M NaCl:

	Activity $\times 10^{-3}$ (cpm)		
^{3}H μC (g solution)$^{-1}$	Tritiated nonsurfactant	Tritiated surfactant	Surfactant concentration $\times 10^3$ (mole kg^{-1})
2	—	1.9	0.17
4	—	2.1	0.34
6	—	2.3	0.50
8	0.50	2.5	0.67
10	—	2.6	0.84
15	0.95	2.9	1.20
20	—	3.2	1.65
30	1.85	3.8	2.45

Briefly outline how the proportionality constant between A_s and Γ^1 might be determined experimentally.

---

*D. S. Ambwani, R. A. Jao, and T. Fort, Jr., *J. Colloid Interface Sci.,* **42**:8 (1973).
†M. Muramatsu, K. Tajima, M. Iwahashi, and K. Nukina, *J. Colloid Interface Sci.,* **43**:499 (1973).

10. A quantity called the HLB (for hydrophile–lipophile balance) number has proved to be a useful way to match a surfactant to a particular application. For example, surfactants with HLB numbers in the range 4–6 produce water-in-oil emulsions; those in the range 7–9 are useful as wetting agents; those ranging between 8 and 18 produce oil-in-water emulsions; and those with values in the range 13–15 make good detergents. Use these considerations plus the following specific examples to formulate a generalization about the dependence of the HLB number on the molecular structure of the surfactant*:

Surfactant	HLB number
Sodium dodecyl sulfate	40
Sodium oleate	18
Tween 80 ($n = 20$, R = oleate)	15
Sorbitan monolaurate	8.6
Span 60	4.7
Sorbitan tristearate	2.1

11. The adsorption of straight-chain fatty acids from n-heptane on Fe_2O_3 has been studied.† In all cases studied the adsorption isotherms conform with the Langmuir equation. The following are values of the amount adsorbed at saturation:

Fatty acid	$\left(\dfrac{n_2^s}{w}\right)_{sat} \times 10^5$ (mole g^{-1})
Acetic	3.00
Propionic	2.36
n-Butyric	2.11
n-Hexanoic	1.78
n-Heptanoic	1.40
n-Octanoic	1.30
Lauric	1.04
Myristic	0.97
Palmitic	0.91
Magaric	0.82
Stearic	0.81

Calculate the area per molecule (σ) of each on the saturated surface if the Fe_2O_3 is known to have a specific surface of 3.45 m^2 g^{-1}. Do the adsorbed molecules appear to be in the same two-dimensional phase in each of these systems?

*Data from Ref. 8.
†T. Allen and R. M. Patel, *J. Colloid Interface Sci.,* *35*:647 (1971).

12. The adsorption of various aliphatic alcohols from benzene solutions onto silicic acid surfaces has been studied.* The experimental isotherms have an appearance consistent with the Langmuir isotherm. Both the initial slopes of an n/w versus c plot and the saturation value of n/w decrease in the order methanol > ethanol > propanol > butanol. Discuss this order in terms of the molecular structure of the alcohols and the physical significance of the initial slope and the saturation intercept. Which of these two quantities would you expect to be most sensitive to the structure of the adsorbed alcohol molecules? Explain.

13. The Michaelis–Menton equation is an important biochemical rate law. It relates the rate of the reaction v to a substrate concentration [S] in terms of two constants v_{max} and K_M:

$$v = \frac{v_{max}[S]}{K_M + [S]}$$

It will be noted that this equation follows the same functional form as the Langmuir equation, specifical $v \rightarrow v_{max}$ as [S] increases. The biochemical literature contains three different graphical procedures to evaluate the constants v_{max} and K_M from kinetic data:

Name of method	Plotted on ordinate	Plotted on abscissa
Lineweaver–Burk	$1/v$	$1/[S]$
Hanes	$[S]/v$	$[S]$
Eadie	v	$v/[S]$

Describe how the three equivalent variations of the Langmuir equation would be plotted. Give the interpretation of the slope and intercept in each case.

14. In laboratory tests of flotation, fluorite (CaF_2) particles (range of particle diameters, 0.074–0.147 mm) with oleic acid as collector were aerated under three different conditions. These conditions and the percent CaF_2 recovery after 10 min are as follows†:

Aeration conditions	Percent recovery after 10 min of aeration
Bubbles precipitated from solution	5
Bubbles produced by mechanical means	20
Combination of both means of bubble production	70

Suggest an interpretation for this variation in the efficiency of fluorite recovery.

---

*R. L. Hoffman, D. G. McConnell, G. R. List, and C. D. Evans, *Science, 157*:550 (1967).
†From Ref. 6.

15. D. C. Grahame* gives the following data for γ_{max} versus log c (corrected for activity) for the interface between mercury and aqueous KI at 18°C:

γ_{max} (dyne cm^{-1})	390	398	407	414	419	422
log c (c in mole liter^{-1})	0.5	0	−0.5	−1.0	−1.5	−2.0

Use these results to estimate Γ, the surface excess of KI at the electrocapillary maximum, for 1.0 and 0.1 M KI. Express your results as moles of KI adsorbed per square centimeter and as total charge (C) per square meter.

16. Show by the double integration of Eq. (96), that is $-(\partial^2\gamma/\partial E^2)_\mu = C$, that a parabolic relationship between $\gamma_{max} - \gamma$ and $E - E_{max}$ is expected if the capacitance of the double layer is constant. Use the equation you derive to test the assumed constancy of C for the interface between mercury and 1 M NaCl from the following data†

γ (dyne cm^{-1})	340	376	396	410	418	423^a	421
E (V)	−0.02	−0.08	−0.18	−0.28	−0.38	−0.56^a	−0.68
γ (dyne cm^{-1})	415	405	396	384	373	358	340
E (V)	−0.78	−0.88	−0.98	−1.08	−1.18	−1.28	−1.38

^{a}Maximum.

Double layer capacitance values are usually expressed as microfarads per square centimeter; remember that practical electrical units, including the farad, are consistent with SI units. Comment on these results in terms of anion and cation adsorption.

---

*D. C. Grahame, *Chem. Rev.*, *41*:441 (1947).

8

COLLOIDAL STRUCTURES IN SURFACTANT SOLUTIONS

I for my part have never known an Irregular who was not also what Nature evidently intended him to be— ... up to the limits of his power, a perpetrator of all manner of mischief.

[From Abbott's *Flatland*]

8.1 Introduction

In the last chapter we examined the tendency of surfactant molecules in aqueous solution to adsorb at a surface in the form of a monolayer. In this chapter we continue to study surfactant solutions, this time considering a few of the many possible modes of organization they can adopt within a bulk phase. Both chapters share an interest in amphipathic solutes: Chapter 7 focuses on surface activity, while the present chapter is concerned with structures in the colloidal size range formed by these molecules. In both, the head-to-head/tail-to-tail ordering of the amphipathic molecules is observed. The difference is that in one case the ordering occurs at the surface, under the influence of the surface, while it occurs in bulk in the present case. Quite an assortment of bulk surfactant structures are known, including ordinary and reverse micelles, liquid crystals, bilayers, vesicles, and microemulsions. This chapter is concerned primarily with micelles (Sects. 8.2 and 8.7) and microemulsions (Sects. 8.8 and 8.9), although some additional structures are mentioned in passing.

The colloidal structures we examine in this chapter are formed as a result of physical interactions among amphipathic molecules, rather than by covalent bonding. This sort of physical association has been recognized for a long time, although contemporary students may be relatively (or totally!) unaware of it. It is interesting to note that in the early days of polymer chemistry macromolecules were believed to be physically associated rather than covalently bonded structures. The birth of modern polymer chemistry can be traced to the acceptance of the covalent character of these substances. Associated structures do exist, however, and we shall see by the end of this chapter that their investigation is a very lively area of chemical research.

In the last couple of decades one important insight has triggered a tremendous upsurge of interest in surfactant structures. This is the recognition that these structures may mimic biological structures in some ways. Enzymes, for example, are protein molecules into which a reactant molecule somehow "fits" to form a reactive intermediate. The highly efficient and specific catalytic effect of enzymes makes their investigation one of the most fruitful areas of biochemical research. Likewise, cell membranes not only compartmentalize biological systems, but also play a variety of functions in the life of the cell. Surfactant structures can be used as model systems to mimic both enzymes and membranes. A whole new field of mimetic (exhibiting mimicry) chemistry has grown around this concept, and colloidal structures formed by surfactants are at the center of the entire subject.

8.2 Micelles: Some Experimental Observations

Curve 3 in Fig. 7.12 was presented as a typical illustration of the way surface tension γ varies with concentration for an amphipathic solute in aqueous solution. In discussing that figure, we noted that the break in the curve marks the concentration above which these solute molecules aggregate to form clusters known as micelles. If we represent the amphipathic species by S, then this clustering process can be described by the reaction

$$n\mathrm{S} \rightleftharpoons \mathrm{S}_n \tag{A}$$

in which S_n is the micelle with a degree of aggregation n. Reaction (A) is the first of a series of reactions of increasing complexity we shall use to represent the micellization process. Note that reaction (A) is reversible: Dilution shifts the equilibrium toward the monomeric surfactant. The surfactant species which undergo micellization are essentially the same amphipathic molecules whose adsorption behavior we discussed in the last chapter. They have the general formula RX, in which R is a hydrocarbon chain and X is a polar group. The hydrocarbon chains in R are ordinarily C_8 or greater, may be saturated or unsaturated, may be linear or branched, and may contain an aromatic ring.

The polar X group in the amphipathic molecule may be non-ionic or ionic. The nonionic X group most widely used in the study of surfactant structures are relatively short-chain polyoxyethylene moieties, $-(C_2H_4O)_x-$, in which x ranges from 3 or 4 to 20 or more. Several commercial surfactants of this sort (Tween and Triton) are listed in Table 7.2. The polyoxyethylene "heads" of these molecules are essentially short-chain polymers themselves and, as such, are generally polydisperse. The x values which characterize these preparations are average values and a distribution of chain lengths around the average is typical of commercial nonionic surfactants.

Among numerous possible ionic groups, sulfate ($-SO_4^-$), sulfonate ($-SO_3^-$), and carboxylate ($-CO_2^-$) are the most common anionic groups; various quaternary ammonium groups ($-NH_3^+$) are the most common X substituents in cationic surfactants. Zwitterionic surfactants which combine both positively and negatively charged groups have also been investigated. With ionic surfactants, the amphipathic ion is accompanied by a counter-ion. Although other possibilities exist, we shall consider only univalent counter-ions. In general, we shall write ionic surfactants M^+S^- and assume these are 100% dissociated into M^+ and S^-. A common surfactant of this type is sodium dodecyl sulfate, often written SDS, in which $M^+ = Na^+$ and $S^- = C_{12}H_{25}SO_4^-$.

The clustering of low molecular weight anionic surfactant molecules to form micelles can also be represented by the following reaction:

$$nS^- + mM^+ \rightleftharpoons (S_n^- M_m^+)^{z-} \qquad \text{(B)}$$

in which n is the degree of aggregation and the net charge z of the micelle is given by $z = n - m$. Note that the charge of the micelle can also be expressed as the fraction ionized, $\alpha = (n - m)/n$. This formalism is readily extended to micelles formed from cationic surfactants; for nonionics, m and z are zero.

Reaction (B) is clearly an extension of reaction (A), with the former admitting the possibility of counter-ion binding to the micelle. Additional refinements can be introduced into this reaction. The micelle still carries a net charge of $-z$, which means that $z\,M^+$ ions must be present in solution to assure electroneutrality. This may be included in the representation of micellization by writing

$$nS^- + mM^+ \rightleftharpoons (S_n^- M_m^+)^{z-} + zM^+ \qquad \text{(C)}$$

Unless noted otherwise, we shall use anionic micelles as prototypes; that is, we shall use reaction (B) to represent the formation of micelles.

The threshold concentration at which micellization begins is known as the critical micelle concentration (CMC). Surface tension is by no means the only property of the solution which displays a discontinuity when plotted against increasing concentration. Figure 8.1 shows schematically how several different experimental quantities vary with concentration. The figure idealizes things somewhat, since the various experimental quantities tend to "see" slightly different thesholds for micellization. The differences in CMC values by various methods of determination are not so diverse as to invalidate the notion of a "critical concentration." Below the breaks in the curves in the figure, anionic surfactants behave as expected for strong electrolytes; above the CMC, however, the behavior observed is consistent with the presence of particles in the colloidal size range. That this is true for the curves shown in Fig. 8.1 is seen by the following arguments:

1. To a first approximation, we know that π/c is proportional to $1/M$ [Eq. (3.34)]. This is roughly equivalent to saying that the slope of a plot of π versus c is proportional to $1/M$. The decrease in the slope of the osmotic pressure plot at the CMC indicates an increase in the average molecular weight of the solute at this point.

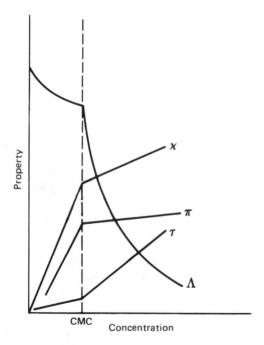

Figure 8.1 Schematic illustration of a variety of properties (κ, conductivity; τ, turbidity; Λ, equivalent conductivity; π, osmotic pressure) versus surfactant concentration. Note the change in behavior at the CMC.

2. To a first approximation Hc/τ is proportional to $1/M$ [Eq. (5.49)]. This means that the slope of a plot of τ versus c is roughly proportional to M. The break in the curve again corresponds to an increase in the molecular weight of the solute.

3. The conductivity κ of the solution decreases at the CMC owing to the lower mobility of the larger micelles. Dividing by concentration to convert to equivalent conductivity Γ leads to a sharp reduction in the latter quantity at the CMC.

These data can be used quantitatively as well as qualitatively to determine the molecular weight of the micelle. For example, the colloidal particles may be analyzed by light scattering by modifying Eq. (5.49) as follows:

$$\frac{H(c - c_{CMC})}{\tau - \tau_{CMC}} = \frac{1}{M} + 2B(c - c_{CMC}) \tag{1}$$

since the micelles form in a solution which already has a turbidity τ_{CMC} at the CMC. Turbidity data can then be plotted as in Fig. 5.9 to give the molecular weight and the second virial coefficient of the micelles. The latter can be analyzed further to give the net charge of the micelle if the amphipathic species are ionic as assumed. Most of the techniques we have considered in Chapters 2–5 along with numerous other experimental procedures have been used to characterized micelles. Of particular interest to us at this point are the average molecular weight M and the net charge z. Also of interest are the particle shape and hydration (available from sedimentation and diffusion measurements) and polydispersity (available from the combination of $\overline{M}_n$ and $\overline{M}_w$ determinations). For nonionic micelles the degree of association is given by the ratios of the molecular weights of the micelle and of the individual surfactant molecules. For ionic micelles the measured molecular weight includes the bound counter-ions, but n can be determined if z is known.

Table 8.1 lists experimental CMC values for several ionic surfactants in water and aqueous NaCl solutions, as well as n and α values measured at the CMC. The data in Table 8.1 supplemented by numerous additional studies allow us to make several generalizations about the state of aggregation and charge of micelles at the CMC:

1. All other things being equal, the aggregation number n increases as the length of the hydrocarbon chain increases.

2. The aggregation number for sodium dodecyl sulfate increases with increasing amounts of the indifferent electrolyte NaCl.

Table 8.1   Critical Micelle Concentration, Degree of Aggregation, and Effective Fraction Ionization for Several Surfactants With and Without Added Salt

Surfactant	Solution	Critical micelle concentration (mole liter^{-1})	Aggregation number n	Ratio of charge to aggregation number, z/n
Sodium dodecyl	Water	0.0081	80	0.18
sulfate	0.02 M NaCl	0.00382	94	0.14
	0.03 M NaCl	0.00309	100	0.13
	0.10 M NaCl	0.00139	112	0.12
	0.20 M NaCl	0.00083	118	0.14
	0.40 M NaCl	0.00052	126	0.13
Dodecylamine	Water	0.0131	56	0.14
hydrochloride	0.0157 M NaCl	0.0104	93	0.13
	0.0237 M NaCl	0.00925	101	0.12
	0.0460 M NaCl	0.00723	142	0.09
Decyl trimethyl	Water	0.0680	36	0.25
ammonium				
bromide	0.013 M NaCl	0.0634	38	0.26
Dodecyl trimethyl	Water	0.0153	50	0.21
ammonium				
bromide	0.013 M NaCl	0.0107	56	0.17
Tetradecyl				
trimethyl	Water	0.00302	75	0.14
ammonium				
bromide	0.013 M NaCl	0.00180	96	0.13

Source: J. N. Phillips, *Trans. Faraday Soc., 51*:561 (1955).

3. There is a general tendency for α to decrease owing to the addition of electrolyte. More extensive data show that α is generally in the range 0.1–0.4; a value of about 0.25 is a good choice for an "average" fraction of ionization.

4. As a further refinement of the previous item, additional studies suggest that the order of ion binding to negative micelles is $Cs^+ > Rb^+ > Na^+ > Li^+$, and for positive micelles $I^- > Br^- > Cl^-$; that is, larger ions that are more polarizable and which tend to be less hydrated bind more effectively.

5. For nonionics, increasing the polyoxyethylene chain length for a constant R group decreases n.

6. Those factors which increase n tend to lower the CMC.

7. An assortment of additional evidence suggests that at the CMC micelles are roughly spherical and relatively monodisperse.

The variables in terms of which micelles have been examined are by no means exhausted by this list. We shall encounter some additional experimental observations as we proceed. At this point, however, these results indicate that the tendency toward aggregation is increased by those variations which increase the hydrophobic character of the surfactant: increasing the hydrocarbon chain length, decreasing the polyoxyethylene chain length, or increasing counter-ion binding.

The last point we consider in this section is the question of whether micellization should be viewed in terms of chemical reaction equilibrium or phase equilibrium. If we think of micellization as a chemical reaction, then reaction (A) should surely be written as a sequence of stepwise additions:

$$2S \rightleftharpoons S_2 \overset{S}{\rightleftharpoons} S_3 \rightleftharpoons S_4 \rightleftharpoons \cdots S_n \qquad \text{(D)}$$

The law of mass action implies that a continuous distribution of species—monomers, dimers, trimers, etc.—should be present. Increasing the surfactant concentration should shift the equilibria in reaction (D) to the right, but the observed transition from monomers to n-mers with $n > 50$ is not expected. Experimentally, large micelles do form sharply at the CMC in aqueous solutions, and—to a good approximation—the concentration of monomeric surfactant in solution varies little above the CMC. This type of behavior is typical of phase equilibria; examples include the (constant) concentration of a saturated solution and the (constant) vapor pressure of a liquid. Energetically, some minimum value of n is apparently necessary before the exclusion of hydrophobic heads from the aqueous medium is effective. Once the solution is concentrated enough for aggregates with this critical n value to be formed, any additional surfactant added to the solution goes into the micelle.

We shall see in Sect. 8.7 that surfactants undergo aggregation in nonaqueous solvents also, but the degree of aggregation is very much less ($n < 10$) and the threshold for aggregation is far less sharp than in water. The mass action model for micellization seems preferable for non-aqueous systems.

In summary, whether a reaction equilibrium or a phase equilibrium approach is adopted depends on the size of the micelles formed. In aqueous systems the phase equilibrium model is generally used. In Sect. 8.4 we shall see that thermodynamic analyses based on either model merge as n increases. Since a degree of approximation is introduced by using the phase equilibrium model to describe micellization, micelles are sometimes called pseudophases.

An illustration of how both the reaction equilibrium and phase equilibrium models can be applied to micellization is provided by the following example:

Example 8.1   Research in which the CMC of an ionic surfactant M^+S^- is studied as a function of added salt, say, M^+X^-, shows that a plot of log CMC versus $\log([MX] + CMC)$ yields a straight line of slope $1 - \alpha$. Derive this relationship by considering reaction (B) as describing both a chemical equilibrium and a phase equilibrium.

Solution   As a chemical equilibrium, reaction (B) is described by an equilibrium constant K which may be written

$$K = \frac{[\text{micelle}]}{[S^-]^n [M^+]^m}$$

where the brackets signify molar concentrations of the indicated species, activity effects being neglected. Taking logs and rearranging gives

$$n \log[S^-] + m \log[M^+] = \log\left(\frac{[\text{micelle}]}{K}\right)$$

As a phase equilibrium, the concentration of surfactant in equilibrium with the micellar pseudophase equals the CMC regardless of the apparent concentration as long as micelles are present. Indicating the CMC value by a subscript, we write

$$\log[S^-]_{\text{CMC}} = \frac{1}{n} \log\left(\frac{[\text{micelle}]}{K}\right) - \frac{m}{n} \log[M^+]$$

At the CMC the total cation concentration equals the CMC value plus the concentration of the added 1:1 electrolyte: $[M^+] = [M^+]_{\text{salt}} + [M^+]_{\text{CMC}}$. Since $m/n = 1 - \alpha$, these results can be combined to give

$$\log[S^-]_{\text{CMC}} = \frac{1}{n} \log\left(\frac{[\text{micelle}]}{K}\right) - (1 - \alpha) \log([M^+]_{\text{salt}} + [M^+]_{\text{CMC}})$$

which is the desired result. The picture of micellar charge as developed here is in reasonable agreement with that determined by other methods.

•

8.3   The Structure of Micelles

In considering the structure of micelles, we continue to base our discussion on aqueous, anionic surfactant solutions as prototypes of amphipathic systems. Cationic micelles are structured no differently from anionics, and nonionics will be described parenthetically at appropriate places in the discussion. We shall summarize present thinking about the structure of micelles at surfactant concentrations equal to or only slightly above the CMC. We shall see that in nonaqueous systems (Sect. 8.7) and in concentrated aqueous systems (Sect. 8.5), the surfactant molecules are organized quite differently from the structure we describe here.

We saw in the last section that at the CMC surfactant molecules cluster into roughly spherical aggregates containing 50–100 amphipathic units. For purposes of orientation, let us take an imaginary journey radially outward from the center of a micelle into its aqueous surroundings, identifying some distinctly different regions for subsequent discussion. This first inventory is incomplete; we shall add refinements as we proceed. Moving outward from the center of the micelle, we find the following:

1. A central core that is predominantly hydrocarbon. The expulsion of the hydrophobic tails of the surfactant molecules from the polar medium is an important driving force behind micellization. The amphipathic molecules aggregate with their hydrocarbon tails pointing together toward the center of the sphere and their polar heads in the water at its surface. For the surfactant this mode of organization competes with monolayer adsorption, which it somewhat resembles.
2. In ionic micelles the hydrocarbon core is surrounded by a shell that more nearly resembles a concentrated electrolyte solution. This consists of ionic surfactant heads and bound counter-ions in a region called the Stern layer (see Sect. 12.10). Water is also present in this region, both as free molecules and as water of hydration.
3. In typical nonionic micelles the shell surrounding the hydrocarbon core also resembles a concentrated aqueous solution, this time a solution of polyoxyethylene. The ether oxygens in these chains are heavily hydrated and the chains are jumbled into coils to the extent that their length and hydration allow.
4. Beyond the Stern layer, the remaining z counter-ions exist in solution. These ions experience two kinds of force: an electrostatic attraction drawing them toward the micelle and thermal jostling,

which tends to disperse them. The equilibrium resultant of these opposing forces is a diffuse ion atmosphere, the second half of a double layer of charge at the surface of the colloid. Chapter 12 provides a more detailed look at the diffuse part of the double layer.

As the first step in refining our ideas about the various regions in and around micelles, it is important to realize that in reality these domains are very different from the static well-delineated regions described. The various steps represented by reaction (D) take place very rapidly, so the micelle exists in a state of dynamic equilibrium. Furthermore, by simple rotations around carbon–carbon bonds in the hydrocarbon chains, the spatial extension of that chain varies, either pulling the head deeper into the core or allowing it to protrude into the Stern layer. This, coupled with the comings and goings of ions and water molecules, makes the surface of the micelle quite turbulent at the molecular level. These considerations alone argue that the regions cited above are not sharply defined on a molecular scale. We shall see presently that uncertainties as to the extent of water penetration into the core also blur the distinction between these regions.

Now let us retrace our way, in reverse order, through the various regions of the micelle enumerated above.

The diffuse part of the double layer is of little concern to us at this point. Chapters 12 and 13 explore in detail various models and phenomena associated with the ion atmosphere. At present it is sufficient for us to note that the extension in space of the ion atmosphere may be considerable, decreasing as the electrolyte content of the solution increases. As micelles approach one another in solution, the diffuse parts of their respective double layers make the first contact. This is the origin of part of the nonideality of the micellar dispersion and is reflected in the second virial coefficient B as measured by osmometry or light scattering. It is through this connection that z can be evaluated from experimental B values.

Ionic micelles will migrate in an electric field and the ion atmosphere of the colloidal particle is dragged along with it. Interpretation of micellar mobility (conductivity experiments) must take this into account. The same is true, however, of the mobility of simple ions, but the situation is more involved here, since the micelle and the ion atmosphere have comparable dimensions. We shall see in Chapter 13 how particle and double layer dimensions affect the interpretation of mobility experiments.

In the second item above, the presence of bound and free water molecules was noted. Both bound ions and ionic surfactant groups are hydrated to about the same extent in the micelle as would be observed for the independent ions. The dehydration of these ionic species is an endothermic process and this would contribute significantly to the ΔH of micellization if ion dehydration occurred. In the next section we shall discuss the thermodynamics of micellization, but it can be noted for now that there is no evidence of a dehydration contribution to the ΔH of micelle formation. The extent of micellar hydration can be estimated from viscosity data, assuming spherical particles. From this approach, sodium dodecyl sulfate micelles are estimated to be hydrated to about 39% by weight. For this system, 24 wt% hydration can be explained by assuming that sulfate groups and sodium ions are hydrated with 1 and 4 water molecules, respectively, at $\alpha = 0.3$. Although charge solvation accounts for a considerable part of the hydration of a micelle, these estimates show that an additional 15 wt% hydration remains unaccounted for.

One important point to recognize about the Stern layer in ionic micelles is that the bound counter-ions help overcome the electrostatic repulsion between the charged heads of the surfactant molecules. For non-ionics no such repulsion exists. It is incorrect to think that ionic micelles form and *then* adsorb counter-ions. The Stern layer is part of the micelle, and the energetics of its formation are part of the thermodynamics of micellization.

Next let us return to the hydrocarbon core of the micelle. Of particular interest are the geometrical constraints imposed on the micellar structure by the length of the hydrocarbon tail and the location of the yet unexplained water of hydration. As a first approximation, the core of the micelle can be viewed as a drop of liquid hydrocarbon, the maximum radius of which equals the length of the fully extended hydrocarbon tail. To get an idea of the size of this region, let us consider a numerical example.

Example 8.2 Calculate the radius, volume, and surface area of the core of a micelle formed by the aggregation of dodecyl groups. The fully extended chain has a zigzag structure with angles of 109.5° and the carbon–carbon bond length is 0.154 nm. The van der Waals radius of the terminal methyl group equals 0.21 nm, and 0.06 nm may be taken as half the length of the bond to the polar head.

Solution The radius of the spherical core equals the length of the fully extended hydrocarbon tail. The 12 carbon atoms are connected by 11 bonds, each of length 0.154 nm. What must be added together, however, are the projections of these bond lengths along the direction of the chain. The distance between every other carbon in the fully extended chain—the base of a triangle opposite the tetrahedral angle—is given by the law of cosines:

$$a = (b^2 + c^2 - 2bc \cos \theta)^{1/2} = [(2)(0.154)^2(1 - \cos 109.5)]^{1/2} = 0.252 \text{ nm}$$

Half of this is the projection of each bond along the chain length; the sum of these projections is $(11)(0.252/2) = 1.39$ nm. Adding the contributions of the two ends to this gives the radius of the sphere: $1.39 + 0.21 + 0.06 = 1.66$ nm.

The volume and surface area of the core are now readily calculated:

$$V = \tfrac{4}{3}\pi R^3 = \tfrac{4}{3}\pi (1.66)^3 = 19.1 \text{ nm}^3$$

and

$$A = 4\pi R^2 = 4\pi (1.66)^2 = 34.5 \text{ nm}^2$$

•

Dividing the surface area calculated in the example by typical aggregation numbers gives the area per head group at the surface of the core. Using n values of 50 and 100 gives 0.69 and 0.35 nm^2 per group, respectively, quite plausible numbers in view of the behavior of surfactants in monolayers.

The alkyl tails of surfactant molecules are not lines without thickness which can radiate outward from a common center in unlimited number. Instead, the chains themselves occupy a certain volume, represented by the bars in Fig. 8.2. In this figure the circles represent head groups and the dotted region is water. Figure 8.2a shows schematically that close packing of the head groups requires unacceptable overlapping of the chain ends if the radius of the core is equal to the length of the fully extended chain. Figure 8.2b continues to use bars for the radius of the sphere, but fans them out in such a way as to avoid overlapping. The hole in the center is an artifact of this pictorial representation, but the wedges which allow water to penetrate deeply into the core is a real feature of this model. The representation in Fig. 8.2c differs from the preceding models as follows:

1. Some surfactant molecules protrude out from the surface of the core farther than others, thereby alleviating crowding at the center of the micelle.

2. Chain flexibility is explicitly acknowledged, allowing some chains to twist and bend in such a way as to fill wedges which would otherwise contain water.
3. These modifications, taken separately or combined, overcome objections to the hypothetical structures shown in Fig. 8.2a and b. The latter, incidentally, are whimsically known as reef and fjord structures, respectively.
4. Because of the protrusion of portions of the hydrocarbon chains into the Stern layer, the core acquires a rough surface. This model seems like a reasonable "snapshot" of what we have already described as a rapidly changing surface region.

One consequence of roughness at the surface of the micellar core is an increased contact between water and hydrocarbon. Figure 8.2b seems unrealistic, because the water–hydrocarbon contact is scarcely less than in the bulk solution, a situation which apparently undermines an important part of the driving force for micellization. Figure 8.2c minimizes this effect without eliminating it. At the same time it allows for some water entrapment, which accounts for that part of the micellar hydration which was unexplained by the hydration of ions and charged groups.

The rough water–hydrocarbon surface of the core introduced in Fig. 8.2c suggests that the core of the micelle should really be considered as two distinct regions: an inner core which is essentially water free and a hydrated shell between the inner core and the polar heads. This partly aqueous shell is sometimes called the palisade layer. The extent to which the hydrocarbon chains protrude into the water is problematic, but we can get an idea of the volume of the palisade layer as follows. Suppose a section of chain three methylenes long defines the thickness of the

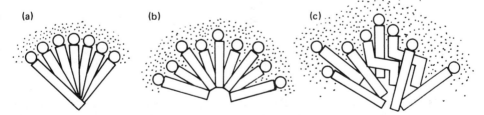

Figure 8.2 Three schematic representations of aqueous micelles: (a) tails overlap at the center, (b) water penetrates core, and (c) chain protrusion and bending correct deficiencies of (a) and (b).

palisade layer. For a dodecyl chain this corresponds to the outer $\frac{1}{12}$ of the radius, meaning that the radius of the inner, anhydrous core is only $\frac{3}{4}$ of what we have been using. Cubing this fraction shows that $(\frac{3}{4})^3 = 0.42$ is the fraction of the original core that is anhydrous, while 58% is hydrated. Other fractions can be calculated by assuming other extents of protrusion. What is significant about this calculation is the fact that hydration of a relatively small fraction of the chain results in the presence of water in a significantly larger fraction of the volume of the micelle. After these idealized calculations, we should ask: Is there any experimental evidence for the presence of water in micelles?

We shall see later in this chapter that there is a great deal of interest in molecules that are "guests" in micelles. In the present context water molecules are the guests and their abundance and location have been and continue to be the basis for considerable research and controversy. We shall merely describe one set of pertinent experiments and their interpretation.

Figure 8.3a shows a portion of the ultraviolet spectrum of benzene in three media: (1) heptane, (2) water, and (3) 0.4 M aqueous sodium dodecyl sulfate. It is not the prominent peaks in these spectra that interest us but, rather, the small bands located 3.6 nm on the long-wavelength side of the major features. This band is absent in benzene vapor, but is present with variable intensity in solutions. Accordingly, it is described as a solvent-induced band whose intensity depends on the polarity of the solovent. This latter property is shown in Fig. 8.3b, in which the characteristics of spectra measured in different reference liquids and liquid mixtures are plotted. The abscissa in Fig. 8.3b is an index of solvent polarity, specifically, the molar concentration of —OH groups in the reference liquids relative to the concentration of such groups in water, namely, 55.5 mole liter^{-1}. Thus abscissa values of 0.0 and 1.0 correspond to hydrocarbon and water, respectively, as solvents; intermediate values describe solvents of intermediate polarity. The ordinate in Fig. 8.3b gives the ratio of the intensity of the solvent-induced band relative to the intensity of the major peak at 3.6 nm shorter wavelength. With this as a calibration curve, the behavior of the benzene in the sodium dodecyl sulfate micelles can be interpreted in terms of the micellar microenvironment of the benzene molecules. In this case the microenvironment is about 62% of the way between hydrocarbon and water. Figure 8.3b also shows the relative dielectric constant of the reference liquids plotted against the same polarity index. The spectrum of the benzene suggests that the benzene molecules are located in a microenvironment of relative dielectric constant 46 (ε_r values for alkanes and water are about 2 and 78, respectively; $\varepsilon_r = 43$ for glycerol).

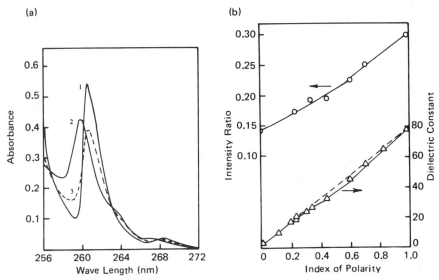

Figure 8.3 (a) A portion of the ultraviolet spectrum of benzene in (1) heptane, (2) water, and (3) 0.4 M sodium dodecyl sulfate. (b) Ratio of the intensity of the solvent-induced peak to that of the major peak for benzene versus the index of solvent polarity. The relative dielectric constant is also shown versus the index of polarity. (Reprinted with permission from P. Mukerjee, J. R. Cardinal, and N. R. Desai in Ref. 5.)

This experiment introduces the use of a probe molecule to explore the microenvironment within a micelle. The results in this case show the environment to be quite "wet," but this observation alone does not tell us where either the benzene or the water is. Any one of the models in Fig. 8.2 has regions of water–hydrocarbon contact. We shall examine some additional data in Sect. 8.5 that support the palisade layer as the location of benzene in these micelles.

8.4 The CMC and the Thermodynamics of Micellization

In this section we shall consider the thermodynamics of micellization from two points of view: the law of mass action and phase equilibrium. This will reveal the equivalency of the two approaches and the conditions under which this equivalence applies. In addition, we define the thermodynamic standard state, which must be understood if derived parameters are to be meaningful.

In the mass action approach we use reaction (B) as a prototype for the process of micellization. The equilibrium constant for this reaction is given by

$$K = \frac{a_{\text{micelle}}}{a_S^n \, a_M^m} \tag{2}$$

in which the a's are the activities of the indicated species. The well-known thermodynamic result $\Delta G^0 = -RT\text{n } K$ can be applied to Eq. (2) to give ΔG^0 for reaction (B), that is, the ΔG^0 value for micelle formation:

$$\Delta G^0 = -RT(\ln a_{\text{mic}} - n \ln a_S - m \ln a_M) \tag{3}$$

Dividing both sides of Eq. (3) by n expresses this free energy change per mole of surfactant; we shall label this ΔG^0_{mic}. At the CMC, $a_M \simeq a_S = a_{\text{CMC}}$, so per mole of surfactant Eq. (3) becomes

$$\Delta G^0_{\text{mic}} = RT\left[\left(1 + \frac{m}{n} \right) \ln a_{\text{CMC}} - \frac{1}{n} \ln a_{\text{mic}} \right] \tag{4}$$

Since n is large, the second term on the right is smaller than the first at the CMC and can be neglected. Introducing this approximation yields

$$\Delta G^0_{\text{mic}} \simeq RT \left(1 + \frac{m}{n} \right) \ln a_{\text{CMC}} \tag{5}$$

Critical micelle concentrations occur in dilute solutions, so activity may be replaced by the concentration of the surfactant at the CMC:

$$\Delta G^0_{\text{mic}} = RT \left(1 + \frac{m}{n} \right) \ln c_{\text{CMC}} \tag{6}$$

Equation (6) can be used to evaluate ΔG^0_{mic} from readily available CMC values. Note that setting $m = 0$ for ionic micelles is equivalent to reverting from reaction (B) to (A) for a description of micellization. The ΔG^0_{mic} value calculated in this case describes the contribution of the surfactant alone without including the contribution of counter-ion binding. Since $m = 0$ for non-ionics, the surfactant contribution alone is useful when comparisons between ionic and nonionic micelles are desired.

It is apparent that CMC values can be expressed in a variety of different concentration units. The measured value of c_{CMC} and hence of ΔG^0_{mic} for a particular system depends on the units chosen, so some uniformity must be established. The issue is ultimately a question of defining the standard state to which the superscript on ΔG^0_{mic} refers. When mole fractions are used for concentrations, ΔG^0_{mic} directly measures the free energy difference per mole between surfactant molecules in micelles and in water. To see how this comes about, it is instructive to examine reaction (A)—this focuses attention on the surfactant and ignores bound counter-ions—from the point of view of a phase equilibrium. The thermodynamic criterion for a phase equilibrium is that the chemical potential of the surfactant (subscript S) be the same in the micelle (superscript mic) and in water (superscript W): $\mu_S^{mic} = \mu_S^W$. In general, $\mu_i = \mu_i^0 + RT \ln a_i$, in which μ_i^0 is the standard state for the chemical potential. We shall write the activity as the product of the mole fraction and an appropriate activity coefficient γ_i. (Unfortunately, this is the same symbol that is used for surface tension; however, it should be clear from the context which is meant.) We recognize that some of the surfactant is in the water and some is in the micelle and use the labels W and mic to differentiate between the two. If x is the mole fraction surfactant, then

$$\mu_S^W = \mu_S^{0,W} + RT \ln x_W + RT \ln \gamma_W \qquad (7)$$

In this equation the standard state corresponds to the state that results from letting $\gamma_W \to 1$ and $x_W \to 1$, in which case $\mu_S^W = \mu_S^{0,W}$. Letting $\gamma_W \to 1$ is equivalent to saying that the surfactant behaves ideally, and letting $x_W \to 1$ is equivalent to having "pure" surfactant possessing the kind of interactions it has when surrounded by water. Physically, this corresponds to an infinitely dilute solution of surfactant in water. Using the primed symbol to represent the chemical potential of surfactant in micelles *per mole of micelles*, we write

$$(\mu_S^{mic})' = (\mu_S^{0,mic})' + RT \ln \left(\frac{x_{mic}}{n} \right) + RT \ln \gamma_{mic} \qquad (8)$$

The mole fraction of micelles is given by x_{mic}/n for particles of aggregation n. By the same logic as used above, $(\mu_S^{0,mic})'$ describes the surfactant in a standard state of pure micelle. We can write Eq. (8) per surfactant molecule by dividing it by n:

$$\mu_S^{mic} = \mu_S^{0,mic} + \frac{RT}{n} \ln \left(\frac{x_{mic}}{n} \right) + \frac{RT}{n} \ln \gamma_{mic} \qquad (9)$$

The condition for phase equilibrium is given by equating Eqs. (7) and (9):

$$\mu_S^{0,\text{mic}} + \frac{RT}{n} \ln\left(\frac{x_{\text{mic}}}{n}\right) + \frac{RT}{n} \ln \gamma_{\text{mic}} = \mu_S^{0,\text{W}} + RT \ln x_\text{W} + RT \ln \gamma_\text{W}$$

$$(10)$$

This last result is seen to be equivalent to Eq. (6) with $m = 0$ under the following conditions:

1. Both activity coefficients are set equal to unity, causing the $\ln \gamma$ terms in Eq. (10) to go to zero.
2. For large n

$$\frac{RT}{n} \ln\left(\frac{x_{\text{mic}}}{n}\right) \ll RT \ln x_\text{W}$$

so the smaller term can be neglected.
3. Since the concentration of surfactant in water is essentially constant and equal to the CMC value after micellization, $x_\text{W} = c_{\text{CMC}}$.
4. This shows that $\mu_S^{0,\text{mic}} - \mu_S^{0,\text{W}}$ is identical to ΔG_{mic}^0, since it describes the difference per mole between the free energy of surfactant in a micelle and in water.

Although there are some aspects of micellization that we have not taken into account in this analysis—the fact that n actually has a distribution of values rather than a single value, for example—the foregoing shows that CMC values expressed as mole fractions provide an experimentally accessible way to determine the free energy change accompanying the aggregation of surfactant molecules in water. For computational purposes, remember Eq. (3.24), which states that $x_2 \simeq n_2/n_1$ for dilute solutions. This means that CMC values expressed in molarity units, [CMC], can be converted to mole fractions by dividing [CMC] by the molar concentration of the solvent, [solvent]: $x_2 \simeq$ [CMC]/[solvent]; for water, [solvent] = 55.5 mole liter^{-1}.

The Gibbs–Helmholtz equation provides another familiar thermodynamic relationship which is useful in the present context:

$$\Delta H_{\text{mic}}^0 = \left(\frac{\partial(\Delta G^0/T)}{\partial(1/T)}\right)_p = -T^2\left(\frac{\partial(\Delta G^0/T)}{T^2}\right)_p$$

$$(11)$$

Using Eq. (6) as the expression for ΔG^0, Eq. (11) becomes

$$\Delta H_{\text{mic}}^0 = R\left(1 + \frac{m}{n}\right)\left(\frac{\partial \ln c_{\text{CMC}}}{\partial(1/T)}\right)_p = -RT^2\left(1 + \frac{m}{n}\right)\left(\frac{\partial \ln c_{\text{CMC}}}{\partial T}\right)_p$$

$$(12)$$

Table 8.2 Some Thermodynamic Properties for the Micellization Process at or Near 25°C for Various Surfactants

Surfactant	ΔG^0_{mic} (kJ mole^{-1})	ΔH^0_{mic} (kJ mole^{-1})	ΔS^0_{mic} (J K^{-1} mole^{-1})
Dodecyl pyridinium bromide	−21.0	−4.06	+56.9
Sodium dodecyl sulfate[a]	−21.9	+2.51	+81.9
N-Dodecyl-N,N-dimethyl glycine	−25.6	−5.86	+64.9
Polyoxyethylene(6) decanol	−27.3	+15.1	+142
N,N-dimethyl dodecyl amine oxide	−25.4	+7.11	+109

[a]Calculated in Example 8.3.
Source: Data from Ref. 3.

The first version of Eq. (12) suggests that a plot of ln c_{CMC} versus $1/T$ is a straight line of slope $\Delta H^0_{mic}/R(1 + m/n)$ if ΔH^0_{mic} and m/n are independent of T. Backtracking to Eq. (4) or (10) shows that n must also be constant with respect to temperature for this to be valid. In fact, n increases with temperature for polyoxyethylene non-ionics; any temperature dependence of n is generally assumed to be absent in ionic systems. Even if the ln c_{CMC} versus $1/T$ plot is nonlinear, the second form of Eq. (12) allows a value of ΔH^0_{mic} to be evaluated from the slope of a tangent to a plot of ln c_{CMC} versus T at a particular temperature. Once ΔG^0_{mic} and ΔH^0_{mic} are known, the entropy of micellization is readily obtained from $\Delta G = \Delta H - T \Delta S$. The following example illustrates the use of these relationships.

Example 8.3 At 25°C the CMC for sodium dodecyl sulfate is 8.1×10^{-3} M. An analysis of the temperature variation of the CMC according to Eq. (12) (with $m = 0$) gives $\Delta H^0_{mic} = 2.51$ kJ mole^{-1}. Evaluate ΔG^0_{mic} and ΔS^0_{mic} from these data. Omit counter-ion binding by taking $m = 0$.

Solution Convert molarity to mole fraction concentration units:

$$x = \frac{8.1 \times 10^{-3} \text{ m liter}^{-1}}{55.5 \text{ m liter}^{-1}} = 1.46 \times 10^{-4}$$

Then $\Delta G^0_{mic} = (8.314)(298) \ln(1.46 \times 10^{-4}) = -21.9$ kJ mole^{-1};

$$\Delta S^0_{mic} = \frac{\Delta H^0_{mic} - \Delta G^0_{mic}}{T} = \frac{2.51 - (-21.9)}{298} = 81.9 \text{ J K}^{-1} \text{ mole}^{-1}$$

Note that the quantity $T \Delta S^0_{mic}$ represents about 90% of the value of ΔG^0_{mic}.

•

Some values of ΔG^0_{mic}, ΔH^0_{mic}, and ΔS^0_{mic} as determined by this method are listed in Table 8.2. Several observations can be made concerning these results:

1. By taking $m = 0$, we have focused on the aggregation of surfactant only, making it easier to compare ionic and non-ionic micelles.
2. The ΔG^0_{mic} values are negative, indicating spontaneous micellization.
3. The ΔH^0_{mic} values are both positive and negative. Values of ΔH^0_{mic} calculated by this method generally show poor agreement with those determined calorimetrically, at least for ionic surfactants.
4. The ΔS^0_{mic} values are positive and make a far larger contribution to ΔG^0_{mic} than ΔH^0_{mic}.

Despite controversies and uncertainties as to the best values for ΔH^0_{mic}, there is no question that the principal driving force behind the aggregation of surfactant molecules is a large, positive value of ΔS^0_{mic}. We shall return to this presently.

The variation of ΔG^0_{mic} with changing molecular parameters has been extensively investigated. It is an experimental fact that $\ln c_{CMC}$ varies linearly with the number of carbon atoms in the alkyl chain of the surfactant molecules, the CMC decreasing with increasing chain length. This variation is readily explained by breaking ΔG^0_{mic} into contributions from the terminal methyl group (subscript CH$_3$), chain methylene groups (subscript CH$_2$), and the polar head group (subscript PH):

$$\Delta G^0_{mic} = \Delta G_{CH_3} + \nu \, \Delta G_{CH_2} + \Delta G_{PH} \tag{13}$$

In Eq. (13) ν is one less than the number of carbons in the alkyl chain. If we assume that neither ΔG_{CH_3} nor ΔG_{CH_2} is affected by the length of the tail, then Eq. (13) can be combined with Eq. (6) to give

$$\ln c_{CMC} = \frac{\nu \, \Delta G_{CH_2}}{RT} + \text{const.} \tag{14}$$

which is the equation of a straight line ($\ln c_{CMC}$ versus ν) and explains the observed behavior. The slope of such a plot for an assortment of n-alkyl ionic surfactants averages about -0.69, which corresponds to a ΔG_{CH_2} of -1.72 kJ mole^{-1} at $25°$C. Since 0.693 equals ln 2, it follows that the addition of a methylene group to a hydrocarbon chain decreases the CMC of these surfactants by almost exactly a factor of 2. In the presence of added indifferent electrolyte, $\Delta G_{CH_2} \simeq -2.9$ kJ mole^{-1}, the aggregation being facilitated by the additional ions. With this type of data, it is possible to back-calculate the contribution to ΔG_{mic}^0 of various polar head groups and, with these, to estimate the ΔG_{mic}^0 and CMC values of other surfactants as the sum of group contributions. Extensive tabulations of these contributions are available.

We conclude this section with a brief discussion of the relatively large, positive values of ΔS_{mic}^0, which we have seen are primarily responsible for the spontaneous formation of micelles. At first glance it may be surprising that ΔS for reaction (A) is positive; after all, the number of independent kinetic units decreases in this representation of the micellization process. Since such a decrease results in a negative ΔS value, it is apparent that reaction (A) is incomplete as a description of micelle formation. What is not shown in reaction (A) is the aqueous medium and what happens to the water as micelles form. The water must experience an increase in entropy to account for the observed positive values for ΔS_{mic}^0.

The key to understanding this entropy increase is the extensive hydrogen bonding that occurs in water. To a first approximation, the water molecule has a tetrahedral shape with the oxygen atom at the center, hydrogen atoms at two of the apex positions, and two lone pairs of electrons at the remaining apex positions. Hydrogen bonds form between the hydrogen atoms on one molecule and the oxygen lone pairs on another, effectively building a loose network of tetrahedra bound at the corners. Because of thermal fluctuations, various parts of this network continuously break and reform in liquid water, but at equilibrium a high average level of hydrogen bonding prevails. Many properties of water are due to this hydrogen bonding.

Next suppose a hydrocarbon moiety such as the tail of a surfactant is imbedded in the water. Since water forms no hydrogen bonds with the alkyl group, the latter merely occupies a hole in the liquid water structure. Our first thought might be that such cavities require the the breaking of hydrogen bonds and that their formation should involve a positive enthalpy contribution. As we have seen, a positive ΔS_{mic}^0 is the dominant driving force behind surfactant clustering. We are therefore forced to conclude that water molecules at the surface of the cavity regenerate the hydrogen-bonded water network, as if the hydrocarbon chains were nucleation sites for network formation. As a result, water molecules become more ordered around the hydrocarbon with an attendant decrease in entropy. The standard state we have used means that the various Δ's measure the difference between surfactant in micelles and surfactant in water. Removing the surfactant from the water and placing it in a micellar environment allows the cavity to revert to the structure of pure liquid water. The highly organized cavity walls return to normal, hydrogen-bonded liquid with an increase in entropy. Incidentally, enhanced hydrogen bonding at the walls of the cavity largely compensates for the breaking of hydrogen bonds to form the cavity, so the enthalpy change is slight.

This explanation for the entropy-dominated association of surfactant molecules is called the hydrophobic effect or, less precisely, hydrophobic bonding. Note that relatively little is said of any direct "affinity" between the associating species. It is more accurate to say that they are expelled from the water and—as far as the water is concerned—the effect is primarily entropic. The same hydrophobic effect is responsible for the adsorption behavior of amphipathic molecules and plays an important role in stabilizing a variety of other structures formed by surfactants in aqueous solutions.

8.5 Solubilization

Above the CMC a number of solutes which would normally be insoluble or only slightly soluble in water dissolve extensively in surfactant solutions. The process is called solubilization, the substance dissolved, the solubilizate, and—in this context—the surfactant the solubilizer. The resultant is a thermodynamically stable, isotropic solution in which the solubilizate is somehow taken up by micelles, since the enhancement of solubility begins at the CMC. This observation, in fact, provides one method for determining the CMC of a surfactant; it must be used cautiously, however, since the solubilizate may change the CMC from what would be observed in its absence. There is an upper limit to the amount of material that can be solubilized in a given surfactant solution; beyond this limit the excess solubilizate displays normal phase separation.

There is considerable interest in establishing the location within a micelle of the solubilized component. As we have seen, the environment changes from polar water to nonpolar hydrocarbon as we move radially toward the center of a micelle. While the detailed structure of the various zones is disputed, there is no doubt that this gradient of polarity exists. Accordingly, any experimental property that is sensitive to molecular

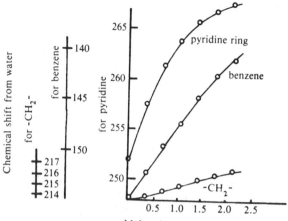

Figure 8.4 Chemical shifts of protons in benzene, pyridium rings, and methylene groups versus the ratio of moles of solubilized benzene to moles of hexadecyl pyridinium chloride. [Reprinted from J. C. Eriksson, *Acta Chem. Scand.*, 17:1478 (1963).]

environment can be used to monitor the whereabouts of the solubilizate in the micelle. Spectroscopic measurements are ideally suited for determining the microenvironment of solubilizate molecules. This is the same principle used in Sect. 8.3, where the ultraviolet spectrum of solubilized benzene was used to explore the solvation of micelles. Here we take the hydration for granted and use similar methods to locate the solubilizate.

In addition to ultraviolet spectroscopy, resonance methods—such as electron spin resonance and nuclear magnetic resonance (NMR)—have been widely used to establish the site of solubilization. We shall briefly consider the use of proton NMR for such an investigation. Like the electron, the proton may have spin quantum numbers of $\pm\frac{1}{2}$, resulting in two different nuclear quantum states. In a magnetic field these states differ in energy, the difference corresponding to the nuclear magnetic moment aligning with or against the magnetic field. If energy of the appropriate frequency is supplied, transitions from one state to the other are possible and the radiation is absorbed. The absorbed frequency measures the separation of the two states by the familiar formula $\Delta E = h\nu$, where h is Planck's constant. The energy separation of the affected states depends on the strength of the applied magnetic field as modified by the local environment. Thus even protons in different parts of a molecule absorb at slightly different frequencies, since the magnetic field each experiences is somewhat different from the applied field owing to the shielding effect of nearby electrons. Furthermore, neighboring protons cause a peak in an absorption spectrum to be split into a multiplet with characteristic relative intensities. This aids the identification of peaks in an NMR spectrum. In addition to one part of a molecule influencing the NMR spectrum of another part, the medium in which the molecule is embedded also has an effect. Therefore the NMR spectrum of a solubilized molecule has the potential not only to reveal the location of a solubilized molecule in a micelle but also to give information about its orientation.

Figure 8.4 shows some data measured for benzene solubilized in hexadecyl pyridinium chloride. The abscissa in the figure shows the extent of solubilization expressed as moles of solubilizate per mole of surfactant. The ordinate values show shifts in the resonance frequencies from water taken as an internal standard. Shifts of peaks arising from protons in the pyridinium ring, in benzene, and in methylene groups in the alkyl tail are shown versus the extent of solubilization.

The following points summarize these observations:

1. All peaks are shifted toward higher fields because of the diamagnetic effect of the solubilized benzene.
2. The slope of the shift versus extent of solubilization curve is much steeper for the protons on the benzene and pyridinium rings than for protons in methylenes in the tail. In fact, the first two increase in roughly parallel fashion.
3. Since the charged pyridinium ring must be at the surface, the benzene which parallels it in chemical shift must have a similar location.
4. The far less sensitive response of the chemical shift of the methylene protons to the extent of solubilization suggests that the core of the micelle is mainly composed of alkyl chains and is relatively unaffected by the benzene.

In another related study (using hexadecyltrimethyl ammonium bromide micelles), isopropyl benzene was solubilized and the chemical shifts of aromatic and alkyl protons were observed. The results suggest that the isopropyl benzene molecules are oriented such that the isopropyl groups are buried more deeply in the core of the micelle while the benzene ring is in the more hydrated palisade layer. This plus the conclusion of item 3 is consistent with the description presented in Sect. 8.3 which located the benzene in a relatively polar portion of the micelle.

The extent of solubilization and the location of solubilizate in the micelle are related to one another. In the most hydrophobic inner core of the micelle there simply is not as much room for the solubilizate compared to the hydrated palisade layer. In the case of polyoxyethylene non-ionics, the core is surrounded by a mantle of aqueous hydrophilic chains, and solubilization may occur in both the core and the mantle. The relative amount of solubilization in these two regions of the nonionic micelle depends on the polarity of the solubilizate. Nonionics appear relatively more hydrophobic at higher temperatures, presumably owing to an equilibrium shift which favors dehydration of the ether oxygens. At some elevated temperature called the cloud point, solutions of these surfactants undergo phase separation. As the cloud point is approached, the solubilization of nonpolar solubilizates increases, probably because of an increase in the aggregation number of the micelles. For polar solubilizates solubilization decreases owing to dehydration of the polyoxyethylene chains accompanied by their coiling more tightly. These observations demonstrate that nonpolar compounds are solubilized in the core of the micelle while polar solubilizates are located in the mantle. Both of the temperature effects cited here are consistent with variations in the space available for the solubilized molecules in the micelle.

We saw in the last section how the removal of a surfactant molecule from water into a micellar environment has a negative ΔG^0 value. Qualitatively, it is not surprising that other organic solutes—the solubilizates—should decrease their free energy by entering micelles. Next we briefly consider how this general notion is quantified.

Using the same logic as we applied to micellization in Sect. 8.3, we can write the following for the solubilizate (component 3):

$$\mu_3^{0,mic} - \mu_3^{0,W} = RT \ln \left(\frac{a_3^{mic}}{a_3^W} \right) = RT \ln \left(\frac{x_3^{mic} \gamma_3^{mic}}{x_3^W \gamma_3^W} \right) \tag{15}$$

Since solubility in water for many solubilizates is low, the aqueous phase may be treated as an ideal solution, that is, $\gamma_3^W = 1$. As we have seen, however, the micellar phase is neither ideal nor simple, with γ_3^{mic} varying from place to place within the micelle. Aside from noting that solubilization occurs spontaneously, which makes $\Delta \mu^0$ negative, we shall not pursue this approach to solubilization any further.

Until now we have taken a rather narrow view of solubilization, having considered only the uptake of solubilizate starting with aqueous solutions of surfactant. This three-component system can clearly take on a much wider range of proportions than we have discussed so far. A ternary phase diagram describes the full range of possibilities. For surfactant systems these diagrams can be quite complex, even though we shall limit ourselves to relatively simple cases. We shall begin with a brief review of the method for reading a ternary phase diagram:

1. The apex points of an equilateral triangle are identified with the three pure components, say, A, B, and C.
2. Each of the sides represents one of the possible binary combinations, say, AB, BC, and CA. The fraction of the distance along the side of the triangular diagram measures the fractional composition of the binary mixture.
3. The perpendicular distance from any point within the triangle to one of the bases is proportional to the amount in the mixture of the component opposite the base in question. Since the sum of the three perpendiculars from any point to the sides equals the height of the triangle, expressing these lengths as fractions of the height permits any point in the triangle to be described as being some fraction toward A, some fraction toward B, and some fraction toward C. These same fractions can be used to describe either the weight fraction or mole fraction compositions of a ternary mixture.
4. Since ternary phase diagrams are drawn at constant temperature and pressure, the phase rule allows one, two, or three phases to be present. In two-phase regions tie lines connect points having the compositions of the equilibrium phases. Three-phase regions have triangular shapes; the coordinates of the corners of the triangular zone give the compositions of the equilibrium phases.

Readers desiring a more detailed review of ternary phase diagrams will find the topic discussed in most physical chemistry textbooks.

Figure 8.5 shows the ternary phase diagram for water, hexanoic acid, and sodium dodecyl sulfate at 25°C. Seven different areas are shown in the figure, which has been used to describe the solubilization of polar dirt by surfactant solutions in detergency applications. The following comments refer to these seven different regions and explain the labeling used in Fig. 8.5:

1. Two liquids: This is a two-phase region in which two liquid solutions—each containing three components—are in equilibrium. The two solutions could exist as distinct layers (e.g., in a separatory funnel) or as an emulsion in which droplets of one phase are dispersed in the second.
2. L_1 and L_2: Each of these phases is a homogeneous, isotropic liquid solution containing three components. The sort of system we have focused on in this section is the L_1 region in which the acid is

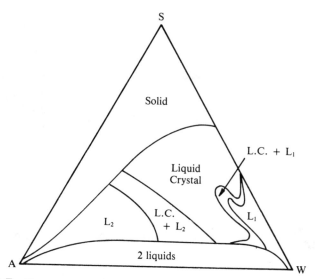

Figure 8.5 Ternary phase diagram for water (W)–hexanoic acid (A)–sodium dodecyl sulfate (S) at 25°C. See text for a description of the various regions. [Reprinted from A. S. C. Lawrence, *Chem. Ind.*, 1764 (1961).]

solubilized in aqueous micelles. In L_2 the reverse is true: The acid is the solvent and water is solubilized in reverse micelles. We shall have more to say about the latter in Sect. 8.7.

3. Liquid crystal: As the name implies, this is an ordered yet fluid phase in which water, surfactant, and solubilizate combine to form anisotropic, organized structures. These are called lyotropic meso-morphic phases, as opposed to thermotropic mesomorphs, which form when certain organic crystals are heated.

4. L_1 or L_2 plus liquid crystal: Each of these is a two-phase region in which the liquid crystalline phase exists in equilibrium with one of the isotropic liquid phases. Tie lines must be determined experi-mentally to give the precise compositions of the phases in equili-brium.

5. Solid: Eventually the system becomes saturated with surfactant and solid sodium dodecyl sulfate precipitates out.

The phase diagram shown in Fig. 8.5 is simpler than many other ternary phase diagrams inasmuch as only one liquid crystal phase exists and there are no three-phase triangles. By contrast, the water–decanol–

(a) (b) (c)

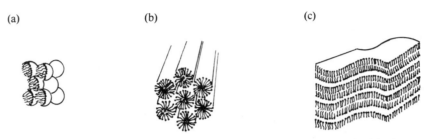

Figure 8.6 Schematic representations of surfactant structures in (a) viscous isotropic, (b) middle, and (c) neat liquid crystal phases.

sodium caprylate phase diagram at 20°C shows 5 different liquid crystal phases and 10 triangular regions in each of which three different sets of phases are in equilibrium.

Liquid crystals are a fascinating topic of study in their own right, but we shall limit our discussion to a brief description of the ordering in some of the possible structures. In all cases the amphipathic molecules are oriented in such a way as to minimize the contact between water and the alkyl chains. Whether the polar head points out or not depends on which component dominates the continuous phase; the minor component is solublized inside the micellar structures.

At relatively low concentrations of surfactant, the micelles are essentially the spherical structures we have discussed until now in this chapter. As the amount of surfactant and the extent of solubilization increase, these spheres become distorted into prolate or oblate ellipsoids and, eventually, into cylindrical rods or lamellar disks. Figure 8.6 schematically shows (a) spherical, (b) cylindrical, and (c) lamellar micelle structures. The structures shown in the three parts of the figure are called (a) the viscous isotropic phase, (b) the middle phase, and (c) the neat phase. Again, we emphasize that the orientation of the amphipathic molecules in these structures depends on the nature of the continuous and the solubilized components.

The "crystallinity" of liquid crystal phases refers to the large assortment of ways these micellar structures can be organized within a bulk phase. For example, spherical micelles of either type can form an ordered bulk phase by arranging themselves into either face-centered or body-centered cubic arrangements. Likewise, rodlike micelles of either type can be arranged into square or hexagonal packing. When all of the shape and packing possibilities are taken into account, it is no wonder that surfactant phase diagrams are complicated!

8.6 Catalysis by Micelles

For about four decades—between 1920 and 1960—the study of micelles was conducted mostly by researchers who would be classified as colloid scientists. Since about 1960 there has been an immense upswing of interest in micelles, with biochemists, organic chemists, and inorganic chemists joining the ranks of those engaged in micellar research. The discovery that micellar media can affect the rates of chemical reactions has triggered this surge of interest, and the notion that micelles might serve as models for enzymes has further stimulated research in this area. Studies have been conducted with a variety of objectives, including elucidation of reaction mechanisms, clarification of enzyme catalysis, investigations of micelles themselves, and applications to synthetic chemistry.

Enzymes and micelles resemble each other both with respect to structure (e.g., globular proteins and spherical aggregates) and catalytic activity. Probably the most common form of enzyme catalysis follows the mechanism known in biochemistry as Michaelis–Menton kinetics. In this the rate of the reaction increases with increasing substrate concentration, eventually leveling off. According to this mechanism, enzyme E and substrate A first react reversibly to form a complex EA which then dissociates to form product P and regenerate the enzyme:

$$E + A \underset{k_{-1}}{\overset{k_1}{\rightleftharpoons}} EA \overset{k_2}{\rightarrow} P + E \tag{E}$$

The various k's are the rate constants for the specific reactions shown. Standard kinetic analysis of this mechanism predicts that the rate of product formation is given by

$$\text{rate} = \frac{k_1 k_2 [E]_0 [A]}{k_2 + k_{-1} + k_1 [A]} = \frac{k_2 [E]_0 [A]}{K_M + [A]} \tag{16}$$

in which $[E]_0$ is the total concentration of enzyme and $K_M = (k_1 + k_{-1})/k_2$ is called the Michaelis constant. Note that as $[A]$ increases, K_M can be neglected in the denominator, allowing cancellation of $[A]$ and explaining the plateau in rate. The Michaelis constant inversely measures the affinity between enzyme and substrate: A small value of K_M means the enzyme binds the substrate tightly.

Figure 8.7 illustrates that micellar catalysts product similar effects. The chemical reaction under investigation in the figure is that between the crystal violet carbonium ion and the hydroxide ion to form the corresponding alcohol:

$$(CH_3)_2N - C6H4 - \overset{N(CH_3)_2}{\underset{N(CH_3)_2}{C^+}} + OH^- \longrightarrow (CH_3)_2N - C6H4 - \overset{N(CH_3)_2}{\underset{N(CH_3)_2}{C}} - OH \qquad (F)$$

In Fig. 8.7 pseudo-first-order rate constants at 30°C for the rate of this reaction in 0.003 M NaOH are plotted versus the concentration of various alkyl trimethyl ammonium bromides. Several things should be noted about these data:

1. The catalytic role of the surfactant begins more or less sharply at the CMC. The C_{10} and C_{12} surfactants also show catalytic activity above their respective CMCs at higher concentrations.
2. The enhancement of rate qualitatively follows Michaelis–Menton kinetics, with both the initial slope and the final plateau increasing with increasing length of the alkyl tails of the surfactant.
3. The initial (starting at the CMC) slope of these curves is inversely proportional to K_M and increases with increasing binding effectiveness between the micelle and the substrate (since K_M itself inversely measures affinity).

These results can be rationalized by picturing the crystal violet carbonium ion as solubilized and oriented in the micelle, followed by attack by the aqueous hydroxide ion. The catalytic effect of the micellar solution has a twofold origin: a concentration effect and an effect on the transition state of the reaction. For the system shown in Fig. 8.7 the micelles are positively charged and the OH^- reactant is concentrated in the Stern layer of such micelles. In addition, the cationic crystal violet should be less stable in the cationic micelles than the zwitterionic transition state. Thus the transition state is electrostatically favored in the micelle compared to a nonsolubilized species. That electrostatic as well as hydrophobic effects are involved in the enhancement of the rate of reaction (F) is evident from the fact that the relative rates with and without surfactant are 241/17 for cationic micelles and 1/17 for anionic micelles. It has also been observed that other anions inhibit the catalysis of reaction (F), presumably by competing with OH^- for adsorption sites in the Stern layer.

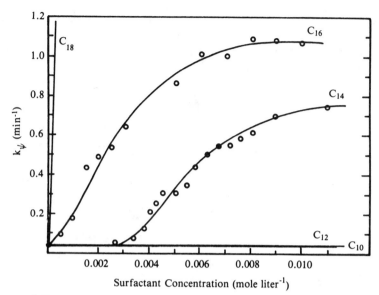

Figure 8.7 Experimental rate constants for reaction (F) versus concentration for alkyl trimethyl ammonium bromides with the indicated chain length. [Reprinted with permission from J. Albrizzio, J. Archila, T. Rodulfo, and E. H. Cordes, *J. Org. Chem.*, 37:871 (1972), copyright 1972 by the American Chemical Society.]

Having looked at an example of micellar catalysis, let us next consider how such results are analyzed quantitatively. By analogy with reaction (E), we visualize the micelle M and the substrate A entering a solubilization equilibrium characterized by an equilibrium constant K:

$$\text{M} + \text{A} \underset{k_0}{\overset{K \quad k_m}{\rightleftharpoons}} \text{MA} \rightarrow \text{P} \tag{G}$$

The solubilized substrate MA is the analog of the complex ES in reaction (E), and the product P is formed in the micelle with the rate constant k_m. The product can also form from the substrate without involving the micelle; k_0 is the rate constant for the latter process. The experimental (subscript ψ) rate constant for reaction (G) is then a weighted sum of the two constants k_m and k_0:

$$k_\psi = f_0 k_0 + f_m k_m = f_0 k_0 + (1 - f_0) k_m \tag{17}$$

where f_0 and f_m are the fractions of substrate in the bulk solution and solubilized in the micelles, respectively. The second version of Eq. (17) arises from the recognition that $f_0 + f_m = 1$. The equilibrium constant in reaction (G) can be written as

$$K = \frac{[MA]}{[M][A]} = \frac{f_m[A]_0}{[M]f_0[A]_0} = \frac{f_m}{(1 - f_m)[M]} \tag{18}$$

where $[A]_0$ is the total substrate concentration. Solving Eq. (18) for $f_0 = (1 + K[M])^{-1}$ and substituting into Eq. (17) gives

$$k_\psi = \frac{k_0 + Kk_m[M]}{1 + K[M]} \tag{19}$$

Subtracting k_0 from both sides of Eq. (19) yields

$$k_\psi - k_0 = \frac{k_0 + Kk_m[M]}{1 + K[M]} - k_0 = \frac{Kk_m[M] - Kk_0[M]}{1 + K[M]} \tag{20}$$

which, on inversion, gives

$$\frac{1}{k_\psi - k_0} = \frac{1}{k_m - k_0} + \frac{1}{(k_m - k_0)K[M]} \tag{21}$$

Finally, the concentration of micelles can be eliminated from Eq. (21) by noting that $[M]$ is given by the number of moles of surfactant in excess of the CMC value divided by the degree of aggregation of the micelle, or

$$[M] = \frac{c - c_{CMC}}{n} \tag{22}$$

where c is the total concentration of surfactant. Substituting Eq. (22) into Eq. (21) gives

$$\frac{1}{k_\psi - k_0} = \frac{1}{k_m - k_0} + \frac{n}{K(k_m - k_0)} \frac{1}{c - c_{CMC}} \tag{23}$$

Recall that k_ψ and k_0 are rate constants with and without surfactant for the reaction in question and that c and c_{CMC} are surfactant concentrations in the reaction mixture and at the CMC, respectively. Therefore k_ψ, k_0, c, and c_{CMC} are experimentally accessible. Equation (23) predicts that a plot of $(k_\psi - k_0)^{-1}$ versus $(c - c_{CMC})^{-1}$ is a straight line whose slope and intercept have the following significance:

$$\text{intercept} = \frac{1}{k_m - k_0} \tag{24}$$

$$\text{slope} = \frac{n}{K(k_m - k_0)} \qquad (25)$$

$$\text{slope/intercept} = n/K \qquad (26)$$

Since k_0 is known, Eq. (24) allows k_m to be evaluated. Likewise, the equilibrium constant for the binding of the substrate to the micelle can be evaluated from Eq. (26) if n is known from a separate experiment. This method of analysis of catalyzed reactions is called a Lineweaver-Burke plot after the corresponding technique in biochemistry. The following example illustrates the use of these relationships.

Example 8.4 Using $k_0 = 0.050$ min^{-1} and $c_{CMC} = 1.0 \times 10^{-3}$ M, estimate k_m and K/n for the C_{16} data in Fig. 8.7. Because of the scatter in the data, take points at regular intervals from the drawn curve in the figure as the basis for this estimate.

Solution The following points are read from the C_{16} curve in Fig. 8.7:

$c_{CMC} \times 10^3$ (M)	1.5	2.0	2.5	3.0	4.0	5.0	6.0	7.0	8.0	9.0
k_ψ (min^{-1})	0.25	0.40	0.52	0.66	0.80	0.90	1.0	1.1	1.1	1.1

Calculate $(c - c_{CMC})^{-1}$ and $(k_\psi - k_0)^{-1}$ using $c_{CMC} = 1.0 \times 10^{-3}$ M and $k_0 = 0.05$ min^{-1}:

$(c - c_{CMC})$ $\times 10^3$ (M)	$k_\Psi - k_0$ (min^{-1})	$(c - c_{CMC})^{-1}$ (liter mole^{-1})	$(k_\Psi - k_0)^{-1}$ (min)
0.5	0.20	2000	5.0
1.0	0.35	1000	2.9
1.5	0.47	670	2.1
2.0	0.61	500	1.6
3.0	0.75	330	1.3
4.0	0.85	250	1.2
5.0	0.95	200	1.1
6.0	1.05	170	0.95
7.0	1.05	140	0.95
8.0	1.05	130	0.95

These data produce a reasonably linear Lineweaver-Burke plot of slope 2.2×10^{-3} min mole liter^{-1} and intercept 0.62 min.

According to Eq. (24), $k_m - k_0 = \text{intercept}^{-1} = 1/0.62 = 1.61$; therefore $k_m = 1.7$ min^{-1}, a value 34 times larger than the estimated k_0 value.

According to Eq. (26), K/n = intercept/slope = $0.62/2.2 \times 10^{-3}$ = 280 liter mole^{-1}.

•

The derivation of Eq. (19) involves a number of assumptions, some of which are questionable in many cases. For example, in the above derivation, we assume the following:

1. The association between substrate and micelle shown in reaction (G) follows 1:1 stoichiometry; that is, only one substrate molecule is taken up per micelle. If this is not the case, n is not the actual aggregation number of the micelle but, rather, the number of surfactant molecules per solubilizate molecule.
2. The substrate does not complex with monomeric surfactant and rate enhancement begins sharply at the CMC. Reactant molecules like other solubilizates may change the CMC of a surfactant and premicellar clusters could influence the rate.
3. There are no competitive processes that inhibit the reaction.

In discussing reaction (F), we remarked that other anions are observed to compete with OH$^-$ in the Stern layer. This sort of electrolyte inhibition is widely observed and the dependence of the inhibition on both the size and charge of the ions generally corresponds to expectations. For example, in the base-catalyzed hydrolysis of carboxylic esters in the cationic micelles, anions inhibit the reaction in the order $NO_3^- > Br^- > Cl^- > F^-$. For acid-catalyzed ester hydrolysis in anionic micelles, the cation of added salt inhibits the reaction in the order $R_4N^+ > Cs^+ > Rb^+ > Na^+ > Li^+$.

Inhibition may be incorporated into the mechanism of micellar catalysis in the same way it is handled in enzyme kinetics. Representing the inhibitor by I, we can revise reaction (G) as follows:

$$M + A \underset{k_0}{\overset{K \quad\quad k_m}{\rightleftharpoons}} MA \rightarrow P$$
$$+$$
$$I$$
$$\updownarrow K_I$$
$$MI \qquad\qquad\qquad\qquad\qquad\qquad (H)$$

where K_I is the binding constant for the inhibitor. Assuming that solubilization of one inhibitor molecule into the micelle blocks the addition of substrate leads to

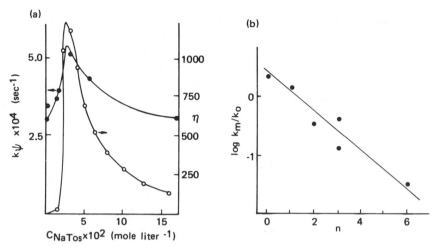

Figure 8.8   (a) The micelle catalyzed rate constant for reaction (I) versus the concentration of sodium tosylate. The viscosity of the system versus sodium tosylate concentration is also shown [reprinted with permission from C. A. Bunton, in *Reaction Kinetics in Micelles* (E. H. Cordes, ed.), Plenum, New York (1973)]. (b) Ratio of k_m to k_0 for reaction (J) versus the number of carbon atoms in the acyl group [reprinted from A. K. Yatsimirski, K. Martinek, and I. V. Berezin, *Tetrahedron, 27*:2855 (1971)].

$$k_\Psi = \frac{k_0 + k_0 K_I[I] + k_m K[M]}{1 + K_I[I] + K[M]} \qquad (27)$$

which reduces to Eq. (19) when [I] or K_I equal zero. Equation (27) is also an oversimplification, but, nevertheless, it satisfactorily accounts for many observations.

Figure 8.8a is an example of some data in which the effect of added salt is more than a competitive ion-binding phenomenon. The reaction involved is the decarboxylation of 6-nitrobenzisoxazole-3-carboxylate, catalyzed by hexadecyl trimethyl ammonium bromide micelles:

$$\text{(I)}$$

In the presence of sodium tosylate (NaTos), both the rate constant k_Ψ and the viscosity of the solution show maxima at the same electrolyte concentration. The dramatic variation in η shown in Fig. 8.8a suggests that the sodium tosylate alters the shape of the micelles, first producing rodlike structures which subsequently break up into more compact structures. The complicated phase diagrams of surfactants make this a plausible explanation. Effects such as this clearly complicate the picture, not only of inhibition but also of micellar catalysis in general.

Equilibrium constants for the binding between substrates and micelles—reaction (G)—generally range from 10^3 to 10^6 for hydrophobic organic substrates. Furthermore, they are expected to increase as the hydrophobic character of the substrate increases. Figure 8.8b shows that this effect sometimes overshoots optimum solubilization. The figure shows, on a logarithmic scale, the ratio k_m/k_0 versus the number of carbon atoms in acyl groups for the following reaction:

$$(\text{J})$$

The reaction is catalyzed by anionic micelles. In this case increasing the carbon content of the acyl group inhibits the reaction. Apparently, increasingly hydrophobic substrates are buried more and more deeply in the micelle such that they are inaccessible to the aqueous aldoxime anion.

Exploring the parallel between catalysis by enzymes and micelles has been beneficial to both areas of research but has also revealed two major differences between the two systems. First, the effectiveness of micelles in enhancing the rates of reactions is generally less than that of enzymes; for micelles the effect rarely exceeds 100-fold. Second, micellar catalysts are less specific than enzymes. The lock-and-key model for enzyme activity implies specific binding sites in a protein molecule whose configuration is maintained by covalent bonds. The corresponding features in micelles are zones of a particular polarity in rapidly changing equilibrium structures. Although the parallels between enzymes and micellar catalysts will undoubtedly continue to be profitably explored, it seems likely that future developments of micellar catalysts will probably emphasize synthetic chemistry.

8.7 Reverse Micelles

Examination of Fig. 8.5—taken as a prototype of ternary phase diagrams involving surfactants—shows L_1 and L_2 phases occupying roughly symmetrical regions in the aqueous and organic corners of the diagram, although the sizes and shapes of these two areas are very different. The interpretation offered in Sect. 8.5 for these regions is that the minor component is solubilized in surfactant micelles which are dispersed in the major component as a solvent. Until now we have focused attention on aqueous micelles, emphasizing their relatively nonpolar cores—where nonpolar molecules are solubilized—and their polar surfaces. In the L_2 region of Fig. 8.5, solubilized micelles are also present, but with the orientation of the surfactant molecules reversed. In nonaqueous media, amphipathic molecules cluster with their polar heads together in the micellar core and their tails in the organic continuous phase. Water is solubilized in the core of these structures which are known as reverse micelles, terminology that emphasizes their difference from aqueous micelles.

Far more work has been done on aqueous than on nonaqueous micelles, partly because the number of amphipathic species that dissolve in nonpolar solvents is considerably less than water-soluble surfactants. Aerosol OT, Triton X-100, and various Spans—the structures of which are given in Table 7.2—are examples of substances which have been widely used in the study of reverse micelles. Fundamental understanding of these structures is most likely to come from research involving pure, well-defined chemicals. Commercial products—petroleum sulfonates and the like—are widely studied from an applications point of view, but the interpretation of such experiments is often clouded by the uncertain effect of impurities. Even though we have approached this discussion from the perspective of water-containing systems, the behavior of anhydrous systems is obviously of interest. Through solvation of polar groups and hydrogen bonding, water can promote aggregation in nonpolar media; complete removal of water as an impurity in both the solvent and the surfactant can be very difficult. Since two-component systems are simpler than those with three components, let us consider some anhydrous solutions of surfactant in nonaqueous solvents.

Surfactant aggregation in an anhydrous, nonpolar medium differs in several important respects from aggregation in water. The most apparent of these differences is that the hydrophobic effect plays no role in the formation of reverse micelles. The amphipathic species are relatively passive in aqueous micellization, being squeezed out of solution by the water. By contrast, surfactant molecules play an active role in the formation of reverse micelles which are held together by specific interactions between head groups in the micellar core. The differences in the solubility parameters (Sect. 3.7) of the hydrocarbon tail of the surfactant and the solvent have also been examined as contributing to reverse micelle formation. It is interesting to note that for the micellization of potassium benzene sulfonate in heptane, $\Delta H = -79.5$ kJ mole^{-1} and $\Delta S = -62.8$ J K^{-1} mole^{-1}. In contrast to aqueous systems, spontaneous micellization is largely due to a large negative enthalpy change with an unfavorable entropy change opposing micellization. Another striking difference between aqueous and anhydrous, nonaqueous systems is the size of the aggregates that are first formed. As we have seen, $n \gtrsim 50$ for aqueous micelles, while for many reverse micelles $n \lesssim 10$. A corollary of the small size of nonaqueous micelles and closely related to the matter of size is the blurring of the CMC and the breakdown of the phase model for micellization. Instead, the stepwise buildup of small clusters as suggested by reaction (D) is probably a better way of describing micellization in anhydrous systems. When the clusters are extremely small, the whole picture of a polar core shielded from a nonaqueous medium by a mantle of tail groups breaks down. It is sometimes argued that the "reverse micelle" terminology is an inappropriate comparison to aqueous micelles. Since water can be solubilized by these micelles, causing an increase in n, the reverse micelle model and vocabulary do seem useful for ternary systems.

Proton NMR has been used to measure both the onset of micellization and the tendency toward solubilization in nonaqueous surfactant solutions. In an NMR spectrum the observed displacement of a resonance frequency—the chemical shift, δ—depends on the environment of the molecule. Also, protons in the methylene groups of surfactants absorb at a different point in the spectrum than, say, aromatic protons, so a surfactant can be readily monitored in, say, a benzene solution. If δ for a proton in a surfactant molecule is plotted against the concentration of the amphipathic species, a graph showing two linear regions of different slope results. In contrast with the properties of aqueous systems (Fig. 8.1), the transition between the two linear regions is gradual, but extrapolation to their point of intersection defines the CMC. If a solubilized molecule has an identifiable feature in an NMR spectrum, then solubilization can also be monitored by NMR. The following example explores this further.

Example 8.5 Let δ be the chemical shift produced by the proton of a solubilized molecule, and differentiate between the experimental shift (subscript Ψ) and the δ's produced by molecules in bulk (subscript 0) or micellar (subscript m) environments. Arguing by analogy with the derivation of Eq. (19) show that

$$\delta_\Psi = \frac{\delta_0 + \delta_m K[M]}{1 + K[M]}$$

where K is the binding constant for the interaction of the solubilizate and the micelle. Suggest how experimental NMR data can be interpreted to give a quantitative value for K.

Solution Incorporation of solubilizate B into micelle M can be represented as

$$B + M \overset{K}{\rightleftharpoons} MB$$

for which K is the equilibrium constant. Define f_0 and f_m as the fraction of the solubilizate in the bulk solution and in the micelle, respectively. It follows by analogy with Eq. (17) that $\delta_\Psi = f_0\delta_0 + f_m\delta_m = f_0\delta_0 + (1 - f_0)\delta_m$, since $f_m + f_0 = 1$.

By analogy with Eq. (18),

$$K = \frac{f_m[B]_0}{[M]f_0[B]_0} = \frac{1 - f_0}{f_0[M]}$$

with $[B]_0$ = total solubilizate concentration.

Finally,

$$\delta_\Psi = \frac{\delta_0 + \delta_m K[M]}{1 + K[M]}$$

with $[M] = (c - c_{CMC})/n$ by analogy with Eq. (19).

This can be rearranged by analogy with Eq. (23) to give

$$\frac{1}{\delta_\Psi - \delta_0} = \frac{1}{\delta_m - \delta_0} + \frac{n}{K(\delta_m - \delta_0)} \frac{1}{c - c_{CMC}}$$

If δ is measured for different concentrations of surfactant and δ_0 and c_{CMC} are known, then this result can be plotted in a linear form and K/n evaluated from the intercept/slope ratio [compare Eq. (26)].

•

Some of the amphipathic species that have been used in the investigation of reverse micelles are capable of dissociation under suitable conditions. Metal carboxylates, alkyl aryl sulfonates, sulfosuccinates, and alkyl ammonium salts are examples of compounds with a high degree of ionic character. Coulomb's law describes the force between two charges q_1 and q_2 separated by a distance r in a medium of relative dielectric constant ε_r:

$$F_{\text{Coul}} \propto \frac{q_1 q_2}{\varepsilon_r r} \tag{28}$$

Electrolyte dissociation is expected to decrease as this force increases. This dissociation—as measured by pK for the dissociation process— should be inversely proportional to ε_r when a given electrolyte is studied in a series of solvents of various polarities. Since r in Eq. (28) is the sum of anion and cation radii, the preceding statement is true only if the state of solvation of the ions remains the same as the solvent varies. Alkali metal salts of dinonyl naphthalene sulfonic acid show decreasing dissociation as the solvent changes from ethanol ($\varepsilon_r = 24.3$) to acetone ($\varepsilon_r = 20.7$) to ethyl acetate ($\varepsilon_r = 6.02$). The quantitative magnitude of the difference between solvents is not the same for all cations, however, indicating that solvation of the cation also enters the picture. In solvents of very low polarity—like hydrocarbons ($\varepsilon_r = 2\text{--}3$)—dissociation is negligible.

The situation with respect to dissociation makes the small core of the reverse micelle a unique microenvironment, approaching the properties of ionic crystals, but at the same time readily accessible to solubilizates and reactants. This is why the interior of the reverse micelle is so effective in solubilizing water and explains why anhydrous systems are so difficult to obtain. Under nearly anhydrous conditions, as few as one solubilized water molecule per reverse micelle might be obtained. This degree of solubilization corresponds to 10 mole of solute per 0.018 kg water, or about 550 molal—an intriguing microenvironment for the investigation of catalytic effects! In addition, for amphipathic molecules with, say, weakly basic polar groups like sulfonates, cations with various strengths as Lewis acids, can be used as counter-ions.

One investigation of the catalytic activity of such a system used the reverse micelles formed in decane by sulfosuccinates with various cations. This medium was used to study the reaction

$$\tag{K}$$

in which the benzylchloroformate reactant is basic and should therefore react strongly with acidic cations. The results were qualitatively similar to those shown in Fig. 8.7 and could be analyzed similarly. The rate constants were evaluated at different temperatures, and activation

Table 8.3 Activation Energies and Arrhenius
Pre-exponential Factors for Reaction (K)
Catalyzed by Reverse Micelles Containing the
Indicated Cation

Cation	E_a (kJ mole^{-1})	ln A
Na$^+$	129	25.3
Al^{3+}	148	32.2
Ce^{3+}	84	11.7
Zn^{2+}	267	71.4

Source: F. M. Fowkes, D. Z. Becher, M. Marmo, C.
Silebi, and C. C. Chao in Ref. 5.

energies were determined by a standard Arrhenius plot of ln k versus $1/T$.
Table 8.3 shows the kinetic parameters so determined when the cations in
the reverse micelle were Na$^+$, Al^{3+}, Ce^{3+}, and Zn^{2+}. Since the slope in this
kind of plot is proportional to the activation energy and since the latter is
very different for various cations, the lines in the Arrhenius plots can
cross at accessible temperatures. This makes the relative effectiveness of
the various cations in catalyzing reaction (K) a matter of temperature.

Use of reverse micelles in synthetic chemistry to improve the rate and
the yield of reactions seems likely to be a fruitful area of research in the
future. In addition to catalysis, several other applications of reverse
micelles can be cited. Just as nonpolar dirt is solubilized in aqueous
micelles, so, too, polar dirt that would be unaffected by nonpolar solvents
may be solubilized into reverse micelles. This plays an important role in
the dry-cleaning of clothing. Motor oils are also formulated to contain
reverse micelles to solubilize oxidation products in the oil which might
be corrosive to engine parts.

8.8 Microemulsions

The term *microemulsion* was coined in 1958 to describe a fairly specific
class of colloidal systems. Before we discuss these, a brief discussion of
the broader term *emulsion* seems in order. If two immiscible liquids are
shaken together, they will ordinarily separate rapidly into two distinct
layers which can be divided in, say, a separatory funnel. Although any
immiscible liquids might be considered, in this discussion we shall refer
to the two as oil (abbreviation O) and water (abbreviation W). Next,
instead of merely shaking the two liquids together, suppose we add a
surfactant—often called an emulsifying agent in this context—and then

vigorously mix the components in a blender or homogenizer of some sort. The milling together of the constituents causes one to be dispersed (the inner phase) in the other (the continuous phase): An emulsion is produced. There can be a great many variations in this procedure—in the nature of the components, their proportions, the milling process, the temperature, and so on—however, a few broad generalizations are possible:

1. The dispersed particles are spheres of great polydispersity. As noted in Sect. 1.12, this is generally the case in dispersions prepared by comminution such as this.

2. The average particle size is at the upper end of the colloidal size range (on the order of micrometers) and the particles are usually visible in a light microscope. We shall describe these as "coarse emulsions" when we want to emphasize their size range. Because the particles are relatively large and polydisperse, coarse emulsions look white when examined visually.

3. Emulsions are two-phase systems and—because of the free energy associated with the oil–water interface—are thermodynamically unstable with respect to separation into oil and water layers.

4. Oil may be the dispersed phase and water the continuous phase—designated an O/W emulsion—or water may be dispersed in oil (W/O). The form obtained depends on the specifics of the system, including the temperature. Compatibility with either oil or water upon dilution is an easy way of establishing which phase is continuous.

5. The surfactant is adsorbed at the oil–water interface in the oriented fashion of monolayers. Judging from monolayer studies at the air–water interface, saturating the surface with surfactant lowers the surface tension γ by 25–50 mN m^{-1} (Fig. 7.5).

6. If the surfactant is ionic and imparts a charge to the interface, then the dispersed particle will be surrounded by an ion atmosphere. We shall see in Chap. 12 how an ion atmosphere surrounding a particle may slow down the rate at which such particles come together. This is one of the ways by which an emulsion may achieve some degree of kinetic stability.

Many of these concepts were introduced in Chapter 1 and seem fairly straightforward as abstract propositions. Things are now always so clear in concrete instances however. The situation of microemulsions is a case in point.

Historically, the term *microemulsion* was applied to systems prepared by emulsifying an oil in aqueous surfactant and then adding a fourth component, called a cosurfactant, generally an alcohol of intermediate chain length. Benzene, water, potassium oleate, and hexanol might be the components of a typical microemulsion formulation. What is observed experimentally is that the usual milky emulsion becomes transparent upon addition of the alcohol. Light scattering and an assortment of other techniques reveal that the resulting system consists of either O/W or W/O dispersions with particles having diameters in the 10- to 100-nm size range. Whether oil or water is continuous, the extent of uptake of the other component may be appreciable. In summary, the following differences between microemulsions and coarse emulsions should be noted:

1. Microemulsions contain particles at least an order of magnitude smaller than those in coarse emulsions.
2. Microemulsions are clear, and coarse emulsions cloudy.
3. Microemulsions form spontaneously; coarse emulsions ordinarily require vigorous stirring.
4. Microemulsions are stable with respect to separation into their components; coarse emulsions may have a degree of kinetic stability but ultimately separate.

The term *microemulsion* seems quite firmly established as the name for the sort of system described above. Still, there has been and continues to be a great deal of controversy as to the exact nature of these systems and the suitability of this vocabulary. The word *emulsion* implies the presence of two phases with an interfacial free energy associated with the phase boundary. Their small size makes the specific area large for microemulsions with a large free energy contribution from γ. Both the spontaneous formation of microemulsions and their stability with respect to separation are hard to reconcile with these considerations. It has even been suggested that the mixed film of surfactant and cosurfactant make the interfacial free energy negative. Alternatively, the increase in overall free energy with decreasing particle size may be offset by a favorable and hence negative $T \, \Delta S$ term in which ΔS describes the entropy of mixing the microemulsion particles with molecules of the dispersion medium. Since the number of microemulsion particles increases with decreasing particle size, the $T \, \Delta S$ term becomes more favorable with decreasing size. This idea is hard to apply quantitatively because of uncertainty as to the value—or meaning—of γ in these systems.

A totally different way of looking at microemulsions—and one that connects this topic with previous sections of the chapter—is to view them as complicated examples of micellar solubilization. From this perspective, there is no problem with spontaneous formation or stability with

Figure 8.9   Schematic progression from micelle (at left) to emulsion droplet (at right). Various degrees of solubilization, including microemulsions, lie between the two extremes.

respect to separation. Furthermore, ordinary and reverse micelles provide the basis for both O/W and W/O microemulsions. From the micellar point of view, it is the phase diagram for the four-component system rather than γ that holds the key to understanding microemulsions.

The difference in perspective between the emulsion and micellar points of view is suggested by Fig. 8.9. The small shaded circle on the left represents a micelle with little or no solubilization. From left to right, the "particles" increase in size owing to increasing solubilization. The circle on the right represents an emulsion particle: an oil drop with a monolayer of surfactant on the surface. An actual continuum of states such as that suggested in Fig. 8.9 is not physically attainable—most emulsions are prepared by comminution rather than condensation—but microemulsions *do* lie between the extremes. The conflicting schools of thought concerning microemulsions arise from the difference in perspective: One side looks from the emulsion point of view, and the other from the micellar point of view. *Swollen micelles* is another term for microemulsions; this terminology clearly reflects the different perspective of its originators.

We saw in Sect. 8.5 that phase diagrams are an effective way of representing the complex behavior of surfactant systems. Let us take a look at microemulsions in terms of phase diagrams. It turns out that nonionic surfactants form microemulsions at certain temperatures without requiring cosurfactants. Since only three components are present, these have somewhat simpler phase diagrams; this kind of system offers a convenient place to begin. Figure 8.10 is a composite of both experimental and schematic, interpretive portions. The rectangular diagram shows the experimental behavior of the system water–cyclohexane–polyoxyethylene-(8.6)-nonyl phenol ether. All systems studied contained 5% surfactant with variable proportions of cyclohexane and water. Temperature was the experimental variable and the nature of the phases present was recorded as the temperature was changed. The presence of different equilibrium phases is represented by different areas on the rectangular diagram. The latter is divided diagonally by a pair of lines

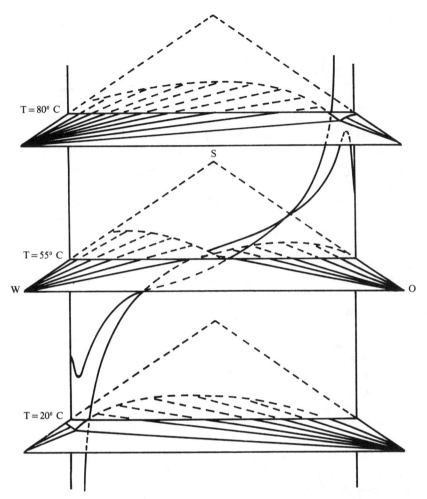

Figure 8.10 Rectangular figure shows the phase diagram for water-cyclo-hexane (at 5% surfactant) versus temperature. Superimposed ternary phase diagrams offer an interpretation of the phases present. (Reprinted with permission from M. L. Robbins in Ref. 5.)

that trace a ribbonlike pattern. This pair of lines cross twice so the ribbon is divided into three areas. These, plus the regions above and below the ribbon, define five different phase situations encountered at various temperatures and compositions.

The idealized ternary phase diagrams that have been superimposed on the experimental plot in Fig. 8.10 help us understand the five different regions on the rectangular diagram. The stacked triangles represent the oil–water–surfactant (abbreviation S) phase diagrams at different (constant) temperatures. Since the experimental data were collected at 5% surfactant, the rectangular plot slices through the triangles 5% of the distance toward S from the O–W base of the triangle. The following example offers practice in reading triangular phase diagrams and helps identify the phases present in the five different regions of the rectangular composition–temperature diagram.

Example 8.6 At each of the three temperatures for which a ternary phase diagram is provided, describe the phases present when the diagram is crossed from the W–S side to the O–S side in the slice at 5% S. Describe the homogeneous phases in terms of the apparent solubilization.

Solution Recall that on ternary diagrams the radial lines are tie lines representing two-phase regions. The triangular regions are three-phase regions and the remaining area toward the S apex of the triangles is a homogeneous phase.

At 20°C: Moving from left to right, we cross from a one-phase region [homogeneous micellar (O/W)] into a two-phase region [oil plus homogeneous micellar (O/W)].

At 55°C: Moving from left to right, we cross from a two-phase region [water plus homogeneous micellar (O/W)] into a three-phase region [oil plus water plus homogeneous micellar (?)]. Next we enter a different two-phase region [oil plus homogeneous micellar (W/O)].

At 80°C: Moving from left to right, we cross from a two-phase region [water plus homogeneous micellar (W/O)] to a one-phase region [homogeneous micellar (W/O)].

•

Since the micellar phase changes from water continuous to oil continuous with increasing temperature, it is an intriguing question how to describe the micelles in the three-phase region that exists at intermediate temperatures.

In this study it is the homogeneous micellar phases that comprise the microemulsions. For this 5% surfactant system it is only over a relatively

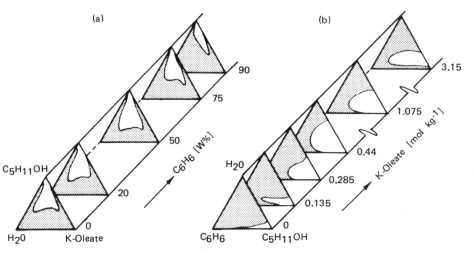

Figure 8.11 Two representations of a portion of the phase diagram for the water–benzene–potassium oleate–pentanol system. The unshaded regions represent homogenous solutions. [(a) Reprinted with permission from S. Friberg and I. Burasczeska, *Prog. Colloid Polym. Sci., 63*:1 (1978); (b) reprinted with permission from C. U. Herrmann, U. Wurz, and M. Kahlweit in Ref. 6.]

narrow range of temperatures that is is possible to have fairly extensive solubilization of either oil or water in the micellar solutions.

It could be argued that the system described in Fig. 8.10 is so different from those to which the name *microemulsion* was first applied as to make the figure irrelevant to the present discussion. However, one of the ways to represent a four-component phase diagram (p and T constant) is to use a trigonal prism in which one of the components—rather than T— varies along the rectangular faces. Figure 8.11a is a partial phase diagram for the system water–benzene–potassium oleate–pentanol. In this figure homogeneous micellar solutions are represented by unshaded areas in the individual triangles. These micellar systems are equivalent to the homogeneous areas in the ternary phase diagrams in Fig. 8.10. At equilibrium, up to 50 wt% benzene can be incorporated into the system with only minor variations in the maximum water uptake. The resulting four-component microemulsion is the same as that produced by first emulsifying the oil and water with potassium oleate and then adding pentanol. This is true because we are considering equilibrium phase diagrams. As with all thermodynamic conclusions, the diagram tells nothing about the rate at which equilibrium is achieved.

Figure 8.11b shows the same data as Fig. 8.11a, replotted with the variables interchanged. Note how the homogeneous area expands in size with increasing potassium oleate concentration up to 0.44 mole kg^{-1}. At still higher concentrations of potassium oleate the homogeneous area shrinks as other surfactant phases compete for the components.

8.9 Applications of Microemulsions

Systems in which one liquid phase is finely dispersed in another under the stabilizing influence of one or more additional components find applications in countless areas. As consumers, we encounter many of these every day. Floor waxes, shaving lotions, beverage concentrates, pesticide preparations, cold creams, and pharmaceutical products are a few of the more common examples. In recent years a great deal of research in this area has been directed toward the problem of tertiary oil recovery.

First of all, the recovery of oil from natural reservoirs occurs in three stages. During the primary recovery stage the pressure of natural gases in the reservoir pushes the oil out. When the gas pressure is no longer adequate, water is pumped into the reservoir to force the oil out. This is called water flooding and represents the second stage of oil recovery. Primary and secondary oil recovery leave about 70% of the total oil in place, much of it trapped in the pore structure of the reservoir by capillary and viscous forces. It is estimated that in U.S. oil fields alone 300 billion barrels of oil are not recoverable by primary or secondary processes and that, of this, 25 billion to 60 billion barrels are potentially recoverable through some tertiary process.

Numerous methods have been explored to recover at least some of this vast resource. Injection of oil-miscible fluids, gases under high pressure, and steam—either separately or in combination—have all been tried with various degrees of success. This is where microemulsions enter the picture. Under optimum conditions an aqueous surfactant solution—which may also contain cosurfactants, electrolytes, polymers, and so on—injected into an oil reservoir has the potential to solubilize the oil, effectively dispersing it as a microemulsion.

Any attempt to represent the trapped oil by a manageable model is bound to be an oversimplification. To see qualitatively, however, how capillary forces trap the oil and how surfactant solutions offer a potential for freeing it, imagine a cylindrical pore containing a slug of oil. Furthermore, assume the oil is in contact with water and that the interface is hemispherical. This assumption about the shape of the interface makes the water–oil-rock contact angle zero and is equivalent to neglecting θ (cos 0° = 1.0). Although a primitive picture of oil in a rocky reservoir, this model can be described by a single size parameter r, the radius of the pore and the radius of curvature of the interface. The Laplace equation [Eq. (6.30)] may then be used to describe the pressure across the oil–water interface, a pressure that must be exceeded to displace the oil. According to the Laplace equation, $\Delta p \propto \gamma/r$ and, since r is small, Δp will be large unless r is offset by a small value of γ. Using r values that are sensible for geological structures, it has been estimated that $\Delta p \simeq 500$ psi/ft if γ is on the order of 10 mN m^{-1}. This kind of pressure drop is unattainable under field conditions. Working backward and taking Δp as an attainable 1–2 psi/ft, the Laplace equation shows that γ must be less than about 0.1 mN m^{-1}, preferably closer to 10^{-3} mN m^{-1} for effective oil displacement.

Any surfactant adsorption will lower the oil–water interfacial tension, but these calculations show that effective oil recovery depends on virtually eliminating γ. That microemulsion formulations are pertinent to this may be seen by reexamining Fig. 8.9. Whether we look at microemulsions from the emulsion or the micellar perspective, we conclude that the oil–water interfacial free energy must be very low in these systems. From the emulsion perspective, we are led to this conclusion from the spontaneous formation and stability of microemulsions. From a micellar point of view, a "pseudophase" is close to an embryo phase and, as such, has no meaningful γ value.

It is apparent that extrapolating laboratory studies on microemulsions to oil recovery is a formidable task. While laboratory research is conducted with pure solutes, distilled water, and at constant temperature, these are meaningless in the field where the following applies:

1. The oil itself is a complex mixture containing surface-active components.
2. Commercial petroleum sulfonates, a mixture of compounds, are the most widely used surfactants.
3. Groundwater contains dissolved minerals and, in practice, brine is used as the aqueous component.

In addition, viscosity considerations may be as important or more important than capillarity; fortunately, microemulsions also have relatively low viscosities!

Despite the obvious difficulty of the tertiary oil recovery problem, this is a major area of surfactant research, since the potential rewards for success are very great.

A second area of microemulsion application is in the synthesis of certain polymers. The process is called emulsion polymerization, a

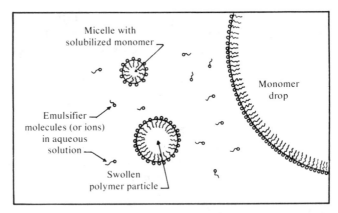

Figure 8.12 Schematic representation of the distribution of surfactant and monomer in an emulsion polymerization. [From J. W. Vanderhoff, E. B. Bradford, H. L. Tarkowski, J. B. Shaffer, and R. M. Wiley, *Adv. Chem., 34*:32 (1962).]

misnomer, since micelles rather than emulsion drops are the site of the polymerization reaction. Because of the commercial importance of polymers, this process has been extensively researched and is quite well understood. We shall only consider some highlights of the process.

Emulsion polymerization is applicable only to monomers which are relatively insoluble in water, such as styrene. A coarse emulsion of monomer in aqueous surfactant is prepared with a water-soluble initiator, say, H_2O_2 in the solution. The surfactant concentration is above the CMC, so surfactant molecules are present as monomers, micelles, and emulsifiers at the oil–water interface. Even an insoluble liquid like styrene dissolves in water to some extent. Therefore the monomer is present in coarse emulsion drops, solubilized in micelles, and as dissolved molecules in water. A schematic illustration of the distribution of surfactant, monomer, and polymer in an emulsion polymerization process is shown in Fig. 8.12.

The H_2O_2 molecules undergo thermal decomposition to form hydroxyl free radicals $\cdot OH$ which initiate the polymerization. The overall reaction for the polymerization of styrene can be represented as

$$C{=}C + \cdot OH \longrightarrow HO{-}C{-}C\cdot \xrightarrow{\text{sty-}\atop\text{rene}} {-}\!\!\left(\!C{-}C\!\right)_{\!x}$$

(L)

When x is large, the uniqueness of the end group(s) can be ignored; familiar polystyrene is the product.

In emulsion polymerization the first step in reaction (L) takes places in water between dissolved monomer and initiator fragments. The resulting free radical is solubilized in micelles, where it quickly reacts with solubilized monomer to form polymer. The low concentration of monomer in the aqueous phase prevents this from occurring to any appreciable extent in the water, although, by diffusion, there continues to be a flux of monomer from emulsion drops into micelles. Likewise, any polymerization that occurs in the coarse emulsion drops themselves is insignificant because of the much greater numerical abundance of micelles than the emulsified styrene droplets. The polymer chains grow in micelles until the process is terminated by reaction with another radical. Thus polymer growth is either propagating or terminating in micelles at any time; therefore half the micelles in a reaction mixture contain growing chains under stationary state conditions. Both the rate of polymerization and the average molecular weight of the polymer depend on the surfactant concentration—via the concentration of micelles—in emulsion polymerization, while this has no effect on polymerizations conducted in nonmicellar solutions.

The aqueous polymer dispersion that results from emulsion polymerization is called a latex. In applications the polymer may be separated or the latex may be used directly as in paints and floor coatings.

As the conversion to polymer proceeds, the micelles become progressively more and more swollen by the polymer–monomer mixture. As with other microemulsions, it eventually becomes problematic as to whether the resulting dispersed particles should be called micelles or swollen polymer particles with adsorbed surfactant.

8.10 Biological Membranes

It is neither feasible nor appropriate in a book like this to give a detailed presentation of biological membranes which compartmentalize living matter and perform numerous cell functions as well. However, because of the impetus to the study of surfactants that the membrane-mimetic properties of surfactant structures has provided, it would be a mistake to exclude some mention of membranes from this chapter. We have already noted in connection with Fig. 7.6e that a monolayer may collapse into a bilayer which leaves the surfactant in a tail-to-tail configuration. This is exactly the arrangement of molecules in the lipid portion of a cell membrane. Protein molecules are arrayed in or on this lipid bilayer.

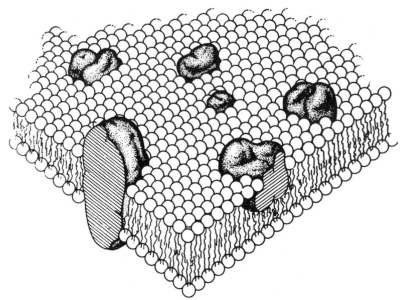

Figure 8.13 Schematic representation of a biological membrane. The amphipathic phospholipid molecules form a bilayer with protein molecules embedded therein. [Reprinted with permission from S. J. Singer and G. L. Nicolson, *Science, 175*:720 (1972), copyright 1972 by the American Association for the Advancement of Science.]

Figure 8.13 is a sketch of one possible relationship between the lipid bilayer and the membrane proteins. Molecules are free to move laterally in these membranes; hence the structure pictured in Fig. 8.13 is called the fluid mosaic model of a cell membrane.

Cell membrane lipids are natural surfactants and display most of the properties of synthetic surfactants. The principal difference between these molcules and the surfactants that we discussed earlier in the chapter is that lipids contain two hydrocarbon tails per molecule. Table 8.4 shows the general structural formula of these cell membrane lipids and the names and formulas for some specific polar head substituents. The alkyl groups in these molecules are usually in the C_{16}–C_{24} size range and may be either saturated or unsaturated.

Much of our understanding of the chemical aspects of cell membranes has been derived from model systems based on surfactants, especially membrane lipids. In this section we shall be primarily concerned with the use of monolayers, bilayers, and especially black lipid membranes and vesicles as cell membrane models.

Table 8.4   Names and Structures of Some Typical Phospholipid Surfactants

$$
\begin{array}{l}
\quad\quad O \\
\quad\quad \parallel \\
R-C-O-CH_2 \\
\quad\quad\quad\quad | \\
R-C-O-CH \\
\parallel \quad\quad\quad\;\; | \\
O \quad\quad\quad\;\; | \quad\quad\quad O \\
\quad\quad\quad\;\;\; | \quad\quad\quad \parallel \\
\quad\quad\quad CH_2-O-P-X \\
\quad\quad\quad\quad\quad\quad\quad | \\
\quad\quad\quad\quad\quad\quad\quad O\text{-}
\end{array}
$$

Name	X
Phosphatidic acid	$-OH$
Phosphatidylcholine	$-OCH_2CH_2\overset{+}{N}(CH_3)_3$
Phosphatidylethanolamine	$OCH_2CH_2NH_2$

Phosphatidylserine	$-OCH_2CCHNH_2$
	          |
	         COOH

Phosphatidylthreonine	$-OCH-CH-NH_2$
	        |
	     CH_3COOH

Phosphatidylglycerol	$-OCH_2CHCH_2OH$
	        |
	      OH

Fundamental membrane research has benefited greatly from the study of monolayers. One of the most important discoveries from this sort of research is the very existence of two-dimensional phases and phase transitions. Generally studies of the sort that can be carried out with monolayers and bilayers cannot be directly extended to living cells, but some exceptional cases have shown that the extrapolation is valid. For example, it is known from monolayer studies that the presence of unsaturated hydrocarbon chains in lipid monolayers prevents some phase transitions from occurring as the temperature is lowered. Certain mutants of *Escherichia coli* are unable to synthesize fatty acids and hence can be manipulated through the compounds they are provided as nutrients. Abnormal levels of saturated hydrocarbon can thereby be introduced into the cell membranes of these organisms. Under appropriate conditions the same phase transitions occur in cell membranes as in model monolayers; still the organisms survive. This is probably because patches of different two-dimensional phases segregate on the cell walls, with the unchanged regions carrying out normal biological functions.

Studies with monolayers and bilayers as models show that modifications of chain conformation can alter the effective thickness of the hydrocarbon portion of a bilayer; this, in turn, can have profound effects on its permeability. Suppose, for example, that a two-dimensional phase transition involved cooperatively inducing just one kink per molecule in otherwise fully extended chains. Such a kink shortens the extension of the chain by 0.127 nm, or thins a bilayer by 0.25 nm. We saw in Example 8.2 that this distance is about the length projected by two carbon–carbon bonds in the direction of the chain. Therefore cooperative induction of such kinks has the same effect on the thickness of the hydrocarbon part of the bilayer as removing two carbon atoms from the tail groups.

Even closer to cell membranes than monolayers and bilayers are organized surfactant structures called black lipid membranes, abbreviated BLMs. Their formation is very much like that of an ordinary soap bubble, except that different phases are involved. In a bubble a thin film of water—stabilized by surfactants—separates two air masses. In BLMs an organic solution of lipid forms a thin film between two portions of aqueous solution. As the film drains and thins, it first shows interference colors but eventually appears black when it reaches bilayer thickness. The actual thickness of the BLM can be monitored optically as a function of experimental conditions. Since these films are relatively unstable, they are generally small in area and may be formed by simply brushing the lipid solution across a pinhole in a partition separating two portions of aqueous solution. Because they are binary lipid films sandwiched in water, BLMs are excellent models for membranes in terms of electrical and permeability properties. Membrane potential, conductivity, and capacitance measurements have all been made on BLMs and the effects on these quantities of electrolytes, proteins, and organic additives have all been studied. For example, 2,4-dinitrophenol, a known hydrogen ion carrier, has been observed to significantly decrease the resistance between electrodes on opposite sides of a BLM. Similarly, high applied voltages thin the BLM, with a simultaneous reversible increase in its permeability and conductance. The rate of permeation of water through BLMs is greater than that of nonpolar compounds; however, ions permeate BLMs more slowly than biological membranes.

The final surfactant structures we consider as models for biological membranes are vesicles. These are spherical or ellipsoidal particles formed by enclosing a volume of aqueous solution in a surfactant bilayer. When phospholipids are the surfactant, these are also known as liposomes. Vesicles may be formed from synthetic surfactants as well. Depending on the conditions of preparation, vesicle diameters may range from 20 nm to 10 μm, and they may contain one or more enclosed compartments. A multicompartment vesicle has an onionlike structure with concentric bilayer surfaces enclosing smaller vesicles in larger aqueous compartments.

Phospholipid vesicles form spontaneously when distilled water is swirled with dried phospholipids. This method of preparation results in a highly polydisperse array of multicompartment vesicles of various shapes. Extrusion through polymeric membranes decreases both the size and polydispersity of the vesicles. Ultrasonic agitation is the most widely used method for converting the lipid dispersion into single-compartment vesicles of small size.

Phase transitions, electrical properties, and permeability have all been investigated for vesicle bilayers. Especially important are the molecular dynamics of tranverse motion through the bilayer. Use of isotopic labels shows that lipid molecules can flip-flop from one surface to another in the bilayer. Techniques have been developed for incorporating protein and other molecules into liposomes, and it has been observed that proteins facilitate lipid flip-flop in the bilayer. Liposomes shrink and swell osmotically as the activity of water in the surrounding aqueous phase is changed by additives. Depending on the surface phase state, the addition of cholesterol to the liposome may decrease the water permeability of the bilayer in vesicles. Transport of ions through the vesicle walls is thought to occur either through channels or via carriers. Since these surfaces are orders of magnitude more permeable to H^+ and OH^- compared to other univalent ions, transmission through the vesicle wall by a hydrogen-bonded network of water molecules is proposed just as the high mobility of these ions in water is explained.

Only a few of the observed properties of vesicles have been enumerated here. The examples cited are sufficient, however, to illustrate how these "synthetic cells" are ideally suited as models for research into the structure and functioning of cell membranes.

References

1. P. H. Elworthy, A. T. Florence, and C. B. Macfarlane, *Solubilization by Surface Active Agents*, Chapman and Hall, London, 1968.
2. J. H. Fendler, *Membrane Mimetic Chemistry*, Wiley, New York, 1982.
3. J. H. Fendler and E. J. Fendler, *Catalysis in Micellar and Macromolecular Systems*, Academic, New York, 1975.

4. A. S. Kertes and H. Gutmann, in *Surface and Colloid Science*, Vol. 8 (E. Matijević, ed.), Wiley, New York, 1976.
5. K. L. Mittal (ed.), *Micellization, Solubilization and Microemulsions*, Vols. 1 and 2, Plenum, New York, 1976.
6. K. L. Mittal (ed.), *Solution Chemistry of Surfactants*, Vols. 1 and 2, Plenum, New York, 1979.
7. L. M. Prince (ed.), *Microemulsions*, Academic, New York, 1977.
8. M. J. Rosen, *Surfactants and Interfacial Phenomena*, Wiley, New York, 1978.
9. C. Tanford, *The Hydrophobic Effect: The Formation of Micelles and Biological Membranes*, 2nd ed., Wiley, New York, 1980.

Problems

1. Use the data for sodium dodecyl sulfate in Table 8.1 to test the equation developed in Example 8.1 and to evaluate α for these micelles. Criticize or defend the following proposition: The α values in the table are more accurate than that evaluated here because they are based on individual light scattering experiments and reflect the variation of α with changing salt concentration; by contrast, the α value calculated here is an average value based on the assumption that m/n and hence α is independent of salt concentration.

2. Shinoda* examined the variation of the CMC of various potassium alkyl malonates, $RCH(COOK)_2$, upon the addition of univalent salts. For $R = C_8$, C_{12}, C_{14}, and C_{16}, the log–log plots of CMC versus counter-ion concentration produce parallel straight lines of slope -1.12. Criticize or defend the following proposition: According to the analysis presented in Example 8.1, the slope of this type of plot equals $-(1 - \alpha)$, meaning that $\alpha = -0.12$; this negative fraction apparently means that the micelle binds an excess of counter-ions and has the opposite charge from that expected.

3. The spectra of substituted pyridinium iodides are characterized by charge transfer bands involving the interaction of pyridinium and iodide ions. Mukerjee and Ray† showed that this band is shifted about 90 nm toward the red for dodecyl pyridinium iodide, which forms micelles, compared to methyl pyridinium iodide, which does not. They measured λ_{max} for the micelles in mixed solvents of variable relative dielectric constant and obtained the following results:

λ_{max} (nm)	281	284	288	298
ε_r	43	38	33	24

Estimate the effective dielectric constant at the surface of the micelle from the fact that λ_{max} occurs at 286 nm for dodecyl pyridinium iodide micelles in

---

*K. Shinoda, *J. Phys. Chem.*, *59*:432 (1955).
†P. Mukerjee and A. Ray, *J. Phys. Chem.*, *70*:2144 (1966).

water. In light of the value estimated in Sect. 8.3 for the dielectric constant in the vicinity of solubilized benzene, does it seem likely that the value of ε_r for bulk water applies in the Stern layer?

4. Ionescu et al.* measured the CMC of hexadecyl trimethyl ammonium bromide in water–dimethyl sulfoxide (DMSO) mixtures at 25 and 40°C:

	$CMC \times 10^3$ (mole liter^{-1})	
Mole fraction of DMSO	$T = 25°C$	$T = 40°C$
0.0	0.92	1.00
0.027	1.48	1.51
0.060	2.24	2.51
0.098	3.60	3.98
0.144	5.62	6.30
0.201	8.91	10.00
0.275	14.00	22.00
0.366	None	None

Use these data to evaluate ΔG^0_{mic} for hexadecyl trimethyl ammonium bromide and to estimate (remember, only two temperatures were measured) ΔH^0_{mic} and ΔS^0_{mic}. Discuss the values obtained in light of infrared and NMR experiments which indicate formation of the stoichiometric compound DMSO · 2H$_2$O) at x_{DMSO} = 0.33.

5. Both adsorption from solution and micellization occur as a result of the hydrophobic effect. To test the correspondence between these two effects, Rosen† assembled ΔG^0 values for adsorption at the air–water interface and for micellization of a numer of linear and branched surfactants. The following is a selection of these data:

		ΔG^0 (kJ mole^{-1})	
Compound	Temperature (°C)	Adsorption	Micellization
n-C$_9$SO$_4$Na	25	−22.4	−17.3
n-C$_{14}$SO$_4$Na	25	−30.0	−25.2
p-n-C$_8\phi$SO$_3$Na	70	−27.7	−23.4
p-n-C$_{14}\phi$SO$_3$Na	70	−38.5	−34.1
n-C$_{10}$SO$_4$Na	50^a	−27.8	−19.8
n-C$_{16}$SO$_4$Na	50^a	−38.6	−30.4
p-n-C$_{12}\phi$SO$_3$Na	75	−35.6	−32.4
p-C$_6$CHCH$_2\phi$SO$_3$Na C$_4$	75	−34.8	−28.7
p-(C—C—)$_4\phi$SO$_3$Na C	75	−34.4	−27.5

^{a}Measured at the hexane–water interface.

---

*L. G. Ionescu, T. Tokuhiro, B. J. Czerniawski, and E. S. Smith in Ref. 6.

†M. J. Rosen, in Ref. 6.

Use these data to criticize or defend the following propositions: (a) The CMC is a good indicator of a surfactant's adsorption effectiveness, since the ΔG^0 values for adsorption and micellization both show parallel changes with increasing chain length. (b) The dodecyl benzene sulfonates have some of the most favorable ΔG^0 values among the data shown; this shows that branching has no adverse effect on either adsorption or micellization.

6. Danbrow and Rhodes* used potentiometric titration data to determine the distribution of benzoic acid (HB) between water and non-ionic micelles. The commercial surfactant used has the average formula $C_{16}(OC_2H_4)_{24}$. They obtained the following results in 4% surfactant solutions:

$[HB]_0$ (mmole liter^{-1}) 4.055   8.804   13.63   18.18   24.30   28.21

HB_m (mmole)   0.1859   0.3497   0.5067   0.6660   0.7790   0.9460

These authors consider two postulates: (a) If the benzoic acid is solubilized in the micellar core, then dimerization should occur in which case $HB_m/[HB]^2 = const$. (b) If the solubilization occurs at the micellar surface, then surface saturation analogous to Langmuir adsorption should occur. Test each postulate with the data provided and decide which gives the better fit.

7. Tokiwa and Aigami† used proton NMR to study the solubilization of benzyl alcohol, 2-phenyl ethanol, and 3-phenyl propanol in sodium dodecyl sulfate micelles. The upfield chemical shifts of the aromatic protons in the micelle from their location in water was measured as a function of solubilization; it tends to increase with solubilization much like the results in Fig. 8.4. The following are some of the results obtained:

Compound	Chemical shift at 0.05 mole of solubilizate (100 g surfactant solution)$^{-1}$ (cps)	$\dfrac{(alcohol)_m}{(alcohol)_w}$	$\dfrac{mole\ alcohol}{mole\ surfactant}$
ϕCH_2OH	3.0	2.28	4.48
ϕCH_2CH_2OH	8.5	4.35	4.08
$\phi CH_2CH_2CH_2OH$	17	11.8	2.81

Criticize or defend the following proposition: The more carbon atoms there are in the alcohols, the more hydrophobic these compounds becoe and the more enriched the micelles become relative to the aqueous phase; the magnitude of the chemical shift increases as the extent of solubilization in the micelles increases owing to the diamagnetic effect of the phenyl groups.

---

*M. Danbrow and C. T. Rhodes, *J. Chem. Soc.*, 6166 (1964).

†F. Tokiwa and K. Aigami, *Kolloid Z. Z. Polym.*, 246:688 (1971).

8. Duynstee and Grunwald* present some experimental data for reaction (F) in the presence of hexadecyl trimethyl ammonium bromide (CTABr, C = cetyl) and sodium dodecyl sulfate (NaLS, L = lauryl). Sodium hydroxide was the source of OH^- in all cases. A pseudo-first-order rate constant of 2.40×10^{-2} s^{-1} is observed for k_{CTABr}. Use the following absorbance data to evaluate k_{NaLS} for this reaction:

t (min)	0	4	7	10	13	17	25	35	41	51
Absorbance	0.734	0.652	0.618	0.577	0.537	0.489	0.408	0.327	0.293	0.244

	60	68	85	124	⋯
	0.211	0.189	0.160	0.129	0.115

Recall that for first-order kinetics a plot of ln(fraction unreacted) versus time has a slope $-k$. Also note that the reaction reaches an equilibrium characterized by an absorbance 0.115; the data must be corrected for this. For both the anionic and cationic micelles, qualitatively sketch, emphasizing the charge state, the micelle, the solubilized substrate, and the approaching OH^- reactant. Indicate how these pictures are consistent with the experimental rate constants.

9. In the same research described in the last problem, the authors† also examined the rate of reaction (F) in pure water and in NaCl solution to guarantee that the ionic surfactants were not displaying an electrolyte activity effect as is ordinarily observed with ion combination reactions. For 10^{-5} M crystal violet and 0.01 M NaOH, they observed the following:

Solution	Pure water	0.01 M NaCl	0.01 M CTABr
$k \times 10^4$ (s^{-1})	17.1	16.4	240

Use these data and the Debye–Hückel theory of electrolyte nonideality to criticize or defend the following proposition: Indifferent electrolytes always inhibit the rates of ion combination reactions, because the activity coefficients are fractions. The data for CTABr show an enhancement of rate so this cannot be due to an activity effect. In these data the k's for pure water and aqueous NaCl are essentially identical, so no activity effects operate in the absence of micelles either.

10. Bunton and Robinson‡ studied the effect of sodium dodecyl sulfate micelles on the rate of the reaction between OH^- and 2,4-dinitrochlorobenzene. These negative micelles have an inhibiting effect on the reaction, yet the kinetic data can be analyzed according to Eq. (19). Use the following data and the authors' CMC value of 0.0064 M to estimate K/n, where K is the binding constant between the dodecyl sulfate micelles and the 2,4-dinitrochlorobenzene:

---

*E. F. J. Duynstee and E. Grunwald, *J. Am. Chem. Soc.*, *81*:4542 (1959).
†*Ibid.*
‡C. A. Bunton and L. Robinson, *J. Am. Chem. Soc.*, *90*:5972 (1968).

$c_{NaC_{12}SO_4} \times 10^2$ (M)	0	1.41	1.85	2.80	3.90	5.60
$k \times 10^5$ (liter mole^{-1} s^{-1})	14.2	9.4	7.7	6.2	5.1	3.1

From the increased solubility of 2,4-dinitrochlorobenzene in sodium dodecyl sulfate solutions (without NaOH), the authors of this research estimate a K/n of 44. Criticize or defend the following proposition: The presence of OH$^-$ could change the n value for sodium dodecyl sulfate micelles, so exact agreement between K/n values determined by the two methods is not necessarily expected.

11. The reaction of the last problem was studied at two different temperatures and, from the temperature dependence of the rate constants, the authors* determined $\Delta H\ddagger$ and $\Delta S\ddagger$, the enthalpy and entropy of activation, respectively. The following values of these parameters were obtained in pure water and in 0.01 M sodium dodecyl sulfate (NaLS) and 0.01 M hexadecyl trimethyl ammonium bromide (CTABr):

Solvent	$\Delta H\ddagger$ (kcal mole^{-1})	$\Delta S\ddagger$ (cal K^{-1} mole^{-1})
Water	21.3	−5.2
CTABr	16.4	−13.4
NaLS	21.3	−5.6

Use these values to criticize or defend the following proposition: The less endothermic value for $\Delta H\ddagger$ in CTABr means the product molecules must be more readily expelled from these micelles, making the enthalpy contribution more favorable to the reaction in this case. The solubilized substrate has a higher entropy, so the decrease in entropy for the micellar reaction is larger. The last problem shows that the reaction occurs about twice as fast in water as in 0.01 M NaLS. The rate in water determines the kinetic parameters in the latter case.

12. Chemistry students are certainly familiar with the regular tetrahedron (methane, sp^3 hybrids, etc.) but may not have considered this geometry as the basis for a quaternary phase diagram. McCarthy† discusses these as extensions of triangular phase diagrams: Each of the four faces of a regular tetrahedron is the ternary phase diagram that results as the concentration of component X in a system goes to zero. The opposite apex represents pure component X, and planes slicing through the tetrahedron parallel to any face have a constant percentage of X, the magnitude of which depends on their placement. Resketch one of the versions of Fig. 8.11 using this kind of tetrahedral representation. This should be done qualitatively and on a large enough scale to separate the various slices. Criticize or defend the following proposition: Both the tetrahedral and prismatic quaternary diagrams show

---

*C. A. Bunton and L. Robinson, *J. Am. Chem. Soc.*, *90*:5972 (1968).
†P. McCarthy, *J. Chem. Educ.*, *60*:922 (1983).

essentially the same thing; in the tetrahedral diagram the advantage of having all four ternary diagrams contained therein is offset by the shrinking size of the slices as an apex is approached.

13. Proteins are polyamids formed from amino acids having the formula H_2N—CHR—COOH; therefore different R groups occur along the polymer backbone according to the amino acid sequence in the protein. Write structural formulas for the R groups in the following amino acids:

Alanine (Ala)*	Glycine (Gly)	Serine (Ser)*
Arginine (Arg)	Isoleucine (Ile)*	Threonine (Thr)
Asparagine (Asn)	Leucine (Leu)*	Tyrosine (Tyr)*
Aspartic acid (Asp)	Lysine (Lys)	Valine (Val)*
Glutamine (Gln)	Phenylalanine (Phe)*	

Kaiser and Kezdy† have suggested that helical structures for certain amino acid sequences in proteins may be stabilized by a hydrophobic interaction with cell membrane lipids. In the list above R groups marked with an asterisk may be considered hydrophobic. Construct a model for residues 1–22 of human growth hormone-releasing factor in the following way. Roll a piece of paper into a cylinder and sketch a helix on the surface. Evenly space the names of five consecutive amino acids along each turn of the helix. On a second cylinder mark off portions of helix two turns long and carefully enter seven amino acids along each two-turn length. The models with 5 and 3.5 amino acid residues per turn are approximations of the π and α helical structures shown in protein crystals. Which of the two appears more effective in concentrating hydrophobic groups alone one edge of the helix? The first 22 amino acids in this protein occur in the following order:

1.	Tyr	12.	Lys
2.	Ala	13.	Val
3.	Asp	14.	Leu
4.	Ala	15.	Gly
5.	Ile	16.	Gln
6.	Phe	17.	Leu
7.	Thr	18.	Ser
8.	Asn	19.	Ala
9.	Ser	20.	Arg
10.	Tyr	21.	Lys
11.	Arg	22.	Leu

14. On the basis of the model produced in the previous problem, criticize or defend the following proposition: A roughly 20-residue α-helix has a length approximately equal to the thickness of a membrane bilayer. A bundle of three, four, or more of these helices—with hydrophobic and hydrophilic

---

† E. T. Kaiser and F. J. Kezdy, *Science, 223*:249 (1984).

sides—could orient themselves in such a way as to form a hydrophilic channel through the membrane. While the individual helices may be stable, their aggregation into this sort of channel involves bringing several membrane proteins together and is therefore entropically unfavorable.

9
PHYSICAL ADSORPTION AT THE GAS-SOLID INTERFACE

When I cut through your plane as I am now doing, I make your plane a section which you, very rightly, call a Circle. For even a sphere—which is my proper name in my own country—if he manifest himself at all to an inhabitant of Flatland—must needs manifest himself as a Circle.

[from Abbott's *Flatland*]

9.1 Introduction

Adsorption at the solid–gas interface is traditionally subdivided into two broad classes: chemisorption and physical adsorption. As the name implies, the former comes very close to the formation of chemical bonds between the adsorbent and the adsorbate. Two consequences of this are that the associated heat effects are comparable to those which accompany ordinary chemical reactions and that the process is not always reversible. It is possible, for example, to absorb (chemisorb) oxygen on carbon and desorb CO or CO_2. In physical adsorption, on the other hand, the energy effects are comparable to those which accompany physical changes such as liquefaction and are completely reversible for nonporous solids. Physical adsorption is the easier of the two types of adsorption and provides much background needed for an understanding of chemisorption. In this chapter we limit our discussion of the solid–gas interface to cases in which only physical adsorption occurs; chemisorption is discussed briefly in the next chapter.

There are several different ways in which the topics pertaining to physical adsorption can be subdivided. We shall be primarily concerned with nonporous solids, briefly discussing porous materials only in Sect. 9.10. It is convenient to divide the extent of adsorption into three categories, submonolayer, monolayer, and multilayer, and we shall discuss them in this order. The thermodynamics of adsorption may be developed around experimental isotherms or around calorimetric data. In this presentation we shall follow the former approach. Adsorption isotherms may be derived from a consideration of two-dimensional equations of state, from partition functions by statistical thermodynamics, or from kinetic arguments. Even though these methods are not fundamentally different, they differ in ease of visualization. We shall consider examples of each method. Finally, much of this chapter will lead up to and be developed around the determination of specific areas by gas adsorption. Low-temperature N_2 adsorption and the Brunauer–Emmett–Teller method of analysis are so widely used for this purpose that these topics will receive special attention.

9.2 Experimental and Theoretical Adsorption Isotherms: An Overview

Adsorption experiments are conducted at constant temperature, and an empirical or theoretical representation of the amount adsorbed as a function of the equilibrium gas pressure is called an adsorption isotherm. Adsorption isotherms are studied for a variety of reasons, some of which focus on the adsorbate while others are more concerned with the solid adsorbent. In Chap. 7 we saw that adsorbed molecules can be described as existing in an assortment of two-dimensional states. Although the discussion in that chapter was concerned with adsorption at liquid surfaces, there is no reason to doubt that similar two-dimensional states describe adsorption at solid surfaces also. Adsorption also provides some information about solid surfaces. The total area accessible to adsorption for a particular mass of solid—the specific area A_{sp}—is the most widely encountered result that is determined from adsorption studies. The energy of adsorbate–adsorbent interaction is also of considerable interest.

We saw in Sect. 6.8 that solid surfaces are notoriously heterogeneous, particularly with respect to roughness and chemical composition. From the point of view of specific area determination, roughness is not too troublesome, since the adsorbed gas molecules can generally cover the hills and valleys of the surface with ease. Pores with very small dimensions pose more of a problem. For now, we assume that such pores are absent; we shall take up the question of adsorption on porous solids in Sect. 9.10. Chemical heterogeneity affects the energetics of adsorption. For simplicity, we often assume that the surface is characterized by a single adsorption energy. Actually, a distribution of surface sites with differing adsorption energies may be present, and some indication of this may be extracted from adsorption data. As an approach to surface characterization, adsorption studies are indirect and give average rather than specific descriptions. We shall see in the next chapter that solid surfaces can be probed more directly for information on a molecular scale. For high specific area solids, however, gas adsorption is the method of choice for quantifying this feature.

Adsorption studies are conducted in a vacuum apparatus from which all gases can be removed prior to the addition of the adsorbate being studied. After the solid is introduced into the sample tube and the latter is attached to the vacuum line, it is generally pretreated by some sort of degassing procedure. This is a combination of heating and pumping to ensure the removal of physically adsorbed contaminants. Consideration must be given to the possibility of changing the solid when the temperature of degassing is selected. A heat treatment which is too vigorous may result in changes in any chemisorbed layer, which—from our point of view, at least—amounts to a change in the adsorbent itself. If extensive enough, such changes may alter the surface area of a solid. Even less drastic changes are sufficient to alter the adsorption energy of a surface.

The range of pressures over which adsorption studies may be conducted is—in principle—from zero to p_0, the saturation pressure or the normal vapor pressure of the material at the temperature of the experiment. At the low-pressure end of this range adsorption will be slight, so the determination of the isotherm involves measuring small differences in pressure at low pressures. This is not easy to do experimentally, although relatively modern low-pressure techniques have greatly extended this region. As the pressure approaches p_0, adsorption often increases rapidly as if anticipating phase separation by occurring as multilayer adsorption in which most of the adsorbed molecules behave as if they were in the bulk liquid state. As a matter of fact, if the solid is porous, the vapor may actually condense in the small pores at $p < p_0$.

Figure 9.1 is a sketch of an apparatus which can be used to determine the equilibrium extent of gas adsorption as a function of pressure. We shall outline how such an experiment is conducted at ambient temperature, even though adsorption studies are frequently conducted at low temperatures, particularly when determination of A_{sp} is the objective of the experiment. A known mass of adsorbent is introduced into the sample tube and degassed as described above. Then the following set of pressure/volume readings are made, described here in terms of Fig. 9.1.

1. The sample tube and gas buret are evacuated and then a non-adsorbing gas—frequently helium—is introduced into the gas buret. The latter is graduated with respect to volume and also serves as one

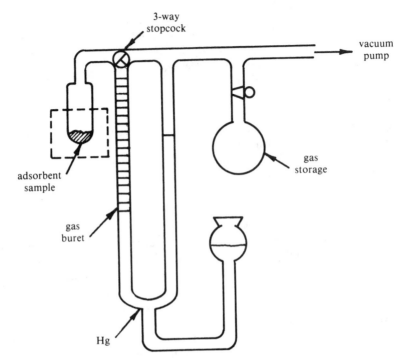

Figure 9.1 Schematic illustration of a gas adsorption apparatus.

leg of a manometer so that both the volume and pressure of gas in the buret can be measured. Ambient temperature is assumed to apply throughout.

2. The three-way stopcock is opened to connect the gas buret with the sample tube. The new pressure and volume are read. From this, the volume of the dead space—that volume beyond the three-way stopcock that is not occupied by sample—can be determined.

3. The nonadsorbed gas is pumped out and replaced by the adsorbate. Its volume, pressure, and temperature are measured, and from these the number of moles of gas introduced (initially) into the apparatus can be determined.

4. The three-way stopcock is opened to connect the buret and sample tube, and volume and pressure are measured again. Since the dead

space is known, the (final) number of moles of gas can be calculated. The difference between the initial and final number of moles gives the number of moles adsorbed.
5. This amount of adsorbed gas is in equilibrium with bulk gas at the pressure read in step 4.

These steps describe the determination of a single point on an adsorption isotherm. By adjusting the mercury level, we can increase the pressure of the equilibrium gas with more adsorption occurring. Steps 3 and 4 are thus repeated until the full isotherm is mapped. The following example illustrates numerically how a point on the isotherm is established.

Example 9.1   The following pressure-volume (p-V) data were collected at a temperature of 22°C. The V's are volumes in the gas buret and the numerical subscripts refer to the steps itemized above.

With helium, $p_1 = 21.71$ Torr, $V_1 = 12.90$ cm^3, $p_2 = 16.50$ Torr, $V_2 = 10.90$ cm^3. With adsorbate, $p_3 = 12.85$ Torr, $V_3 = 13.70$ cm^3, $p_4 = 3.24$ Torr, $V_4 = 5.00$ cm^3. What is the volume of the dead space? How many moles are adsorbed at the final equilibrium pressure, 3.24 Torr?

Solution Successive applications of the ideal gas law allow us to calculate the desired quantities.

The total volume to which the gas has access after the stopcock is opened is the sum of the buret volume and the dead space V_d. Therefore $V_d + V_2 = p_1 V_1/p_2 = (21.71)(13.70)/(16.50) = 16.97$, or $V_d = 6.07$ cm^3. It is convenient to use $R = 62,360$ cm^3 Torr K^{-1} mole^{-1} as the value of the gas constant in these calculations. The initial moles of adsorbate are given by $n_i = p_3 V_3/RT = (12.85)(13.70)/(62360)(295) = 9.57 \times 10^{-6}$. After adsorption equilibrium is established, $n_f = (3.24)(5.00 + 6.07)/(62360)(295) = 1.95 \times 10^{-6}$. The difference $n_i - n_f = (9.57 - 1.95) \times 10^{-6} = 7.62 \times 10^{-6}$ mole gives the amount adsorbed at an equilibrium pressure of 3.24 Torr.

•

If the adsorption isotherm is to be determined at some temperature other than room temperature—liquid nitrogen temperature, for example—the sample tube is placed in a suitable thermostat. This is indicated by the dotted line in Fig. 9.1. In this case two sets of readings are made with the nonadsorbed gas; one at room temperature and one with the thermostat in place. In this way the partitioning of the dead space between the two temperature regions can be determined. Several

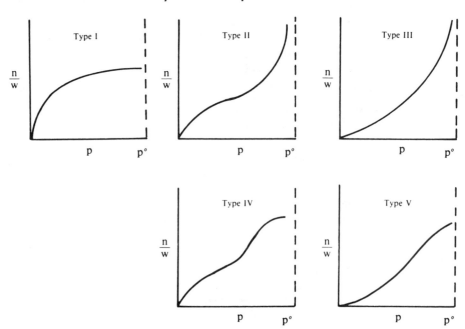

Figure 9.2 Qualitative shapes of the five general types of gas adsorption isotherms (see text for a discussion of their physical significance).

additional considerations should be cited which are important in actual practice:

1. The gases used for adsorption must be of high purity.
2. The gas buret should itself be thermostated unless the laboratory has very good temperature control.
3. The nonideality for the adsorbate at low temperatures must be taken into account unless the volume of the low-temperature dead space is minimized.
4. Sufficient time must be allowed for equilibrium to be established. A way of checking this experimentally is to lower the pressure and observe that desorption follows the same curve as adsorption.

Experimental gas adsorption isotherms are traditionally classified into one of the five types shown in Fig. 9.2. The type I isotherm is reminiscent of Fig. 7.14, the Langmuir isotherm. The plateau is interpreted as indicating monolayer coverage. We shall see that this type of behavior implies a sufficiently specific interaction between adsorbate and adsorbent to be more typical of chemisorption than physical adsorption. Type II adsorption, by contrast, is widely observed with physical adsorption and is interpreted to mean multilayer adsorption. We shall see in Sect. 9.7 that type III behavior occurs when the heat of liquefaction is more than the heat of adsorption. Types IV and V are analogs of types II and III, except for the leveling off which occurs at pressures below p_0. These cases are associated with porous solids where the adsorbate condenses in the small pores at $p < p_0$.

If the specific area of the solid is the information sought, it is the amount of adsorption at monolayer coverage that must be measured. This is readily available in type I adsorption, but requires considerable interpretation when multilayer adsorption takes place. Once determined, however, the number of molecules required to saturate a surface times the area occupied per molecule gives the surface area of a sample. This divided by the mass of the sample gives A_{sp}. In addition, once the adsorption at monolayer coverage is identified, all other extents of adsorption can be expressed as fractions or multiples of the monolayer.

This makes a convenient point of contact with theory, since models for adsorption inevitably subdivide the surface into an array of adsorption sites which gradually fill as the pressure increases. θ is defined as the fraction of sites filled, then $\theta = 1$ corresponds to the monolayer, with $\theta < 1$ or $\theta > 1$ at submonolayer and multilayer coverages, respectively. Theoretical isotherms predict how θ varies with p in terms of some particular model for adsorption. It turns out that a set of experimental points can often be fitted by more than one theoretical isotherm, at least over part of the range of the data; that is, theoretical isotherms are not highly sensitive to the model upon which they are based. A comparison between theory and experiment with respect to the temperature dependence of adsorption is somewhat more discriminating than the isotherms themselves.

Since it is relatively easy to fit experimental adsorption data to a theoretical equation, there is some controversy as to what constitutes a satisfactory description of adsorption. From a practical point of view, any theory that permits the amount of material adsorbed to be related to the specific surface area of the adsorbent and which correctly predicts how this adsorption varies with temperature may be regarded as a success. From a theoretical point of view, what is desired is to describe adsorption in terms of molecular properties, particularly in terms of an equation of state for the adsorbed material, where the latter is regarded as a two-dimensional state of matter.

As we shall see in the course of the chapter, these two approaches frequently clash. The adsorption isotherm of Brunauer, Emmett, and Teller (BET), which is discussed in Sect. 9.6, is an excellent example of this. The model upon which the BET isotherm is based has been criticized by a great many theoreticians. At the same time, the isotherm itself has become virtually the standard equation for determining specific areas from gas adsorption data. Ross [8] summarizes the situation effectively by comparing it to a master chef who concocts a palatable dish out of an old shoe. For some, the end result is what matters: a palatable dish. For others, the starting material dominates their opinions: the old shoe. To present a relatively accurate picture of the current state of affairs in this area, it is necessary to present both of these viewpoints. We shall attempt not to take too one-sided a position but to give some indication of each.

9.3 Two-Dimensional Equations of State and Isotherms

To see how the equation of state of two-dimensional matter and the adsorption isotherm are related, we return to the Gibbs equation [Eq. (7.46)]:

$$-d\gamma = \Gamma_2 \, d\mu_2 \qquad (1)$$

Since we are concerned here with adsorption from the gas phase, the chemical potential may be related to the pressure of the gas by

$$\mu_2 = \mu_2^0 + RT \ln fp \qquad (2)$$

where f is the activity coefficient [see Eqs. (7.47) and (7.48)]. In this discussion we assume that the gas behaves ideally, although in analyzing experimental results it may be necessary to include the correction required by the nonideality of the gas. Combining Eqs. (1) and (2) leads to

$$-d\gamma = RT\Gamma_2 \, d \ln p \qquad (3)$$

In the present context it is convenient to use Eq. (7.43) to eliminate Γ_2 from Eq. (3) and express the surface excess as the number of moles adsorbed per unit area:

$$-d\gamma = \frac{nRT}{A} d \ln p \qquad (4)$$

Next we substitute the product of sample weight times specific area for A in this equation to obtain

$$-d\gamma = \frac{RT}{A_{sp}} \frac{n}{w} d \ln p \tag{5}$$

For a given adsorbent and an isothermal experiment, T and A_{sp} are constants. The ratio n/w is the equilibrium amount adsorbed which will be a function of p. Therefore Eq. (5) may be integrated as follows:

$$-\int d\gamma = \frac{RT}{A_{sp}} \int \frac{n}{w} d \ln p \tag{6}$$

The constant of integration may be evaluated by recognizing that n/w goes to zero as p approaches zero. Under these circumstances $\gamma \rightarrow \gamma_0$; therefore Eq. (6) becomes

$$\gamma_0 - \gamma = \pi = \frac{RT}{A_{sp}} \int_0^p \frac{n}{w} d \ln p \tag{7}$$

If the experimental isotherm (n/w as a function of p) is known, then Eq. (7) may be integrated either analytically or graphically to give the two-dimensional pressure as a function of coverage. This relationship therefore establishes the connection between the two- and three-dimensional pressures which characterize the surface and bulk phases. This is how adsorption data could be used to determine the film pressure in equilibrium with a drop of bulk liquid on a solid surface as discussed in Sect. 6.6.

As an illustration of the kind of information obtainable from Eq. (7), suppose we consider the situation in which the equilibrium adsorption of a gas is described by the isotherm

$$\frac{n}{w} = mp \tag{8}$$

where m is a constant. Equations (7) and (8) may be combined to give

$$\pi = \frac{RT}{A_{sp}} \int_0^p mp \frac{dp}{p} = \frac{RT}{A_{sp}} mp = \frac{N_A k T n}{A_{sp} w} \tag{9}$$

Since the quantity $A_{sp} w / n N_A$ equals σ, Eq. (9) may be written

$$\pi\sigma = kT \tag{10}$$

the two-dimensional ideal gas law for the surface phase!

The adsorption isotherm—Eq. (8)—which is associated with this surface equation of state is called the Henry law limit, in analogy with the

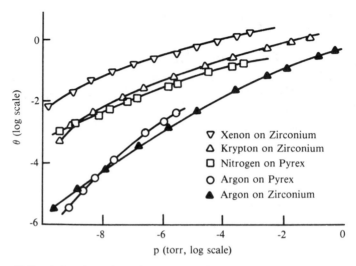

Figure 9.3 A log–log plot of θ versus p for xenon, krypton, and argon on zirconium and nitrogen and argon on Pyrex. (Reprinted from Ref. 6, by courtesy of Marcel Dekker, Inc.)

equation which describes the vapor pressure of dilute solutions. The constant m, then, is the adsorptions equivalent of the Henry law constant. When adsorption is described by the Henry law limit, the adsorbed state behaves like a two-dimensional ideal gas.

Equation (8) may also be written

$$\theta = m'p \tag{11}$$

if the specific area of the adsorbent and the cross-sectional area of the adsorbate are known [Eq. (7.71)]. We see, therefore, that compliance with Henry's law implies that a log–log plot of θ versus p yields a straight line of unit slope. Figure 9.3 shows some experimental results for adsorption at 77.4 K plotted in this way for pressures down to 10^{-10} Torr. Note that even at these low pressures the Henry law limit is not yet reached (it would give a 45° line in Fig. 9.3), although argon on Pyrex appears to be approaching this limiting behavior. Note, further, that any errors introduced in evaluating θ would vertically shift the curves but would not charge their slopes. We may conclude, therefore, that—just as with monolayers on aqueous substrates—the compliance of an adsorbed layer with the ideal gas law is a form of behavior which is extremely difficult to observe. The implication of this is that significant departures from two-dimensional ideality already set in at very low surface coverage.

In general, there are two different types of interactions in which adsorbed molecules may participate. There is the interaction between the adsorbed molecules and the adsorbent and the interaction between the adsorbed molecules themselves. In the Henry law region, we have seen that the adsorbed layer behaves ideally. This is to be expected in view of the low surface concentration (θ very small) of adsorbate. The adsorbed molecules definitely do interact with the adsorbent, however, and at very low coverage the interaction energy might be very sensitive to surface heterogeneity. Any "hot spots" on the surface would adsorb first, less energetic patches next, then the normal surface sites. In Sects. 9.5 and 9.9 we shall see how isotherms measured at several different temperatures may be interpreted to yield information on the energy of adsorption.

The second type of interaction possible for adsorbed molecules is direct adsorbate–adsorbate interaction. Interactions of this sort are expected to lead to deviations from ideality in the two-dimensional phase just as they lead to deviations from ideal behavior for bulk gases. In this case surface equations of state which are analogous to those applied to nonideal bulk gases are suggested for the adsorbed molecules. The simplest of these allows for an excluded area correction [see Eq. (7.23)]:

$$\pi(\sigma - \sigma^0) = kT \tag{12}$$

Let us now consider the kind of adsorption isotherm associated with this two-dimensional equation of state.

Equation (12) may be used as a starting point in the evaluation of the isotherm from the equation of state. From this equation, $d\pi$ is given by

$$d\pi = -kT(\sigma - \sigma^0)^{-2}\, d\sigma \tag{13}$$

Next, we recall that according to Eq. (7)

$$d\pi = -d\gamma \tag{14}$$

and that

$$\sigma = \frac{A_{sp}w}{nN_A} \tag{15}$$

Therefore Eq. (5) may be written

$$\frac{\sigma\, d\pi}{kT} = d \ln p \tag{16}$$

Combining Eqs. (13) and (16) gives

$$-\frac{\sigma \, d\sigma}{(\sigma - \sigma^0)^2} = d \ln p \tag{17}$$

Suppose we divide the numerator and denominator of the left-hand side of this equation by σ^0 and let $\sigma/\sigma^0 = x$ for the time being. Then Eq. (17) becomes

$$-\frac{x \, dx}{(x - 1)^2} = d \ln p \tag{18}$$

in which the left-hand side is a standard integral. Evaluating the integral leads to the result

$$-\ln(x - 1) + \frac{1}{x - 1} = \ln p + C \tag{19}$$

The quantity we have called x may also be recognized as $1/\theta$:

$$x = \frac{\sigma}{\sigma^0} = \frac{1}{\theta} \tag{20}$$

Therefore Eq. (19) may be written

$$\ln\left(\frac{\theta}{1 - \theta}\right) + \frac{\theta}{1 - \theta} = \ln p + C \tag{21}$$

Next we evaluate the integration constant C. We know that $\theta \to 0$ as $p \to 0$. Therefore, at first glance, we are tempted to equate C to zero. It must be remembered, however, that Henry's law must apply as $p \to 0$. This condition is met if we let $C = \ln m'$ in which m' is defined by Eq. (11):

$$\ln\left(\frac{\theta}{1 - \theta}\right) + \frac{\theta}{1 - \theta} = \ln p + \ln m' \tag{22}$$

This may be readily verified by examining the limit of Eq. (22) as $\theta \to 0$.

Equation (22) may also be written

$$m'p = \frac{\theta}{1 - \theta} \exp\left(\frac{\theta}{1 - \theta}\right) \tag{23}$$

by taking the antilog of both sides of the equation. It is interesting to look at this form of the isotherm in the limit of small values of θ but still above

Henry's limit. In this case the exponential term approaches unity and Eq. (23) becomes

$$m'p = \frac{\theta}{1 - \theta} \tag{24}$$

or

$$\theta = \frac{m'p}{1 + m'p} \tag{25}$$

which is identical in form to the Langmuir equation [see Eq. (7.67)]. Equation (23) also reveals that for $m'p \gg 1$, θ approaches unity as an upper limit. Thus at both the upper and lower limits, Eq. (23) gives the same results as the Langmuir equation. At intermediate values the two functions differ slightly, but it would probably be difficult to distinguish between them in fitting experimental data.

Several points might be noted in summarizing the results of this section:

1. In principle, it is possible to correlate an adsorption isotherm and a two-dimensional equation of state by working from either direction; that is, we may start with an experimental isotherm and develop the associated equation of state [as in going from Eq. (8) to Eq. (10)] or we may proceed from the equation of state to the isotherm [as in the derivation between Eqs. (12) and (23)].
2. Relatively small increases in complexity for the equation of state result in considerably more complex equations for the adsorption isotherms. The gross feaures of the more complex isotherms are also given by simpler isotherms. This means that it is very difficult to choose among various isotherms in terms of the goodness of fit to experimental data. Therefore it is difficult to conclude from an experimental isotherm what the two-dimensional surface phases are like.
3. Two-dimensional equations of state are a useful source of isotherms, however, even though the test of the isotherm must be made in terms of some criterion other than an ability to describe adsorption. For example, the ability of an isotherm to predict the temperature dependence of adsorption or the specific area of an adsorbent is a more sensitive test of an isotherm than merely describing the way n/w increases with p.

In the next section we shall briefly describe some additional isotherms which have been generated by consideration of two-dimensional equations of state.

9.4   Other Isotherms from Surface Equations of State

Still greater nonideality in the two-dimensional equation of state might be represented by the van der Waals analog:

$$\left(\pi + \frac{a}{\sigma^2} \right)(\sigma - b) = kT \tag{26}$$

in which the b factor and σ^0 from Eq. (12) are identical in principle. What makes this expression especially interesting is the fact that there exists a temperature above which there is only one real root to Eq. (26) and below which some values of π correspond to three values of θ. The situation is shown schematically in Fig. 9.4a. With bulk gases and also insoluble monolayers on water, the three-root region is identified with a region of two-phase equilibrium. Is there any evidence for this type of phase equilibrium in the two-dimensional layer of adsorbed gas on a solid substrate? Figure 9.4b shows several data for the adsorption of krypton on specially treated graphite over a range of temperatures from about 77 to 91 K. These plots are adsorption isotherms, not π–σ diagrams, but nevertheless there is quite clear evidence of a two-phase region with a critical temperature at about 86 K. The bulk critical temperature for Kr is 210 K, so the two-dimensional value is about 41% of the three-dimensional one.

It is possible to develop a mathematical isotherm corresponding to the van der Waals equation for the two-dimensional state, following the procedure used in developing Eq. (23). The calculation is tedious, however, and yields no new insights into the nature of the adsorbed state. Therefore we shall not go through the derivation in detail. The final result is given by the equation

$$m'p = \frac{\theta}{1 - \theta} \exp \left(\frac{\theta}{1 - \theta} - \frac{2a\theta}{bkT} \right) \tag{27}$$

Note that when a equals zero, Eq. (27) becomes identical to Eq. (23). A few additional assumptions lead to the prediction that the two-dimensional critical temperature should be one-half the value of the three-dimensional critical temperature according to this model. As just noted,

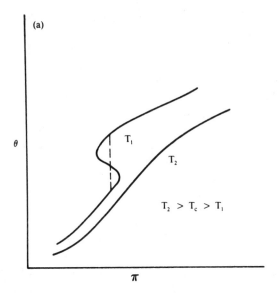

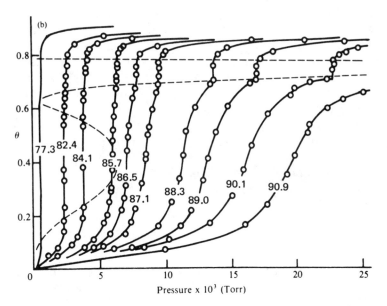

Figure 9.4   (a) Schematic illustration of π-σ isotherms in the vicinity of a two-dimensional critical temperature. (b) Experimental data for the adsorption of krypton on exfoliated graphite showing similar features. [Data from A. Thomy and X. Duval, *J. Chim. Phys.*, 67:1101 (1970).]

the two T_c values for Kr are close to this prediction, and the corresponding values for argon are 65 and 151 K.

There is another feature in Fig. 9.4b that deserves additional comment. It is the existence of a second set of vertical segments in the isotherms at values of θ in the range of 0.65–0.78. This suggests a second phase equilibrium in the two-dimensional matter, perhaps something analogous to the transition between the L_1 and L_2 states of monolayers on water or to a liquid–solid transition. Indeed, beyond the second two-phase region the slope of the isotherm changes sharply in a way which corresponds to a much less compressible surface state.

Before exploring the consequences of this feature, it is first necessary to make certain that the result is not just an artifact arising from surface heterogeneities. We noted that the graphite substrate for which the data of Fig. 9.4b were collected was "specially treated." A few details of this treatment should be mentioned. Graphite and anhydrous $FeCl_3$ are introduced into opposite ends of a Pyrex tube; the evacuated tube is then sealed and brought to about 300°C. A slight temperature gradient is maintained between opposite ends of the tube so the $FeCl_3$ distills to the graphite, where it reacts to form an interlamellar compound. Rapidly heating the latter to 900°C or higher results in the expulsion of the $FeCl_3$ and the attendant delamination of the graphite. The specific surface area of the final product—known as exfoliated graphite—is roughly two orders of magnitude greater than that of the initial graphite sample. The newly exposed graphite planes constitute a remarkably uniform solid adsorbent if all traces of $FeCl_3$ are removed. There is some evidence that the same vertical segments are observed in isotherms measured on the parent graphite as are observed in the exfoliated graphite. This argues that these features are not artifacts of the delamination process.

With these ideas in mind, we may consider another equation of state for the two-dimensional matter. The relatively incompressible surface state (e.g., above θ ≃ 0.70 in Fig. 9.4b) may be approximately described by an equation of the type

$$\pi = -m\sigma + b \tag{28}$$

by analogy with the (approximately) linear π–σ isotherms for insoluble monolayers. The parameters m and b are merely empirical constants. From Eq. (28) it follows that

$$d\pi = -m \, d\sigma \tag{29}$$

which may be substituted into Eq. (16) to yield

$$- \frac{m\sigma \, d\sigma}{kT} = d \ln p \tag{30}$$

This expression is readily integrated to give

$$- \frac{m\sigma^2}{2kT} = \ln p + C \tag{31}$$

We shall not bother to evaluate the constant of integration in this equation, instead we note that $\sigma = A_{sp}(w/n)$. Therefore Eq. (31) may be written

$$\ln p = - \frac{m}{2kT} A_{sp}^2 \left(\frac{n}{w} \right)^{-2} + C \tag{32}$$

This is one form of an adsorption isotherm known as the Harkins–Jura equation. It suggests that a plot of $\ln p$ versus $(n/w)^{-2}$ should be a straight line whose slope is proportional to the square of the specific area. In experiments with solids of known area the linearity predicted by Eq. (32) has been observed. Furthermore, the proportionality constant relating the observed slopes to A_{sp}^2 is independent of the nature of the adsorbent to a first approximation. Thus solids for which A_{sp} is known may be used to "calibrate" this method for a particular adsorbed species; then the specific area of an unknown may be determined using the same adsorbate.

The Harkins–Jura isotherm therefore introduces the possibility of determining specific areas by gas adsorption studies. This is probably the most important practical information to be derived from any study of the physical adsorption of gases. The Harkins–Jura equation is only one of many isotherms that permit the evaluation of A_{sp}. Although it may give satisfactory values for A_{sp}, Eq. (32) leaves something to be desired at the molecular level.

We have seen that the condensed state which characterizes the surface in the Harkins–Jura model appears only as $\theta \to 1$. In most instances of physical adsorption, however, no saturation limit of adsorption appears. As $p \to p_0$ the amount of material adsorbed increases asymptotically. Multilayer adsorption is the only reasonable model for this observation. Except for special cases, the neighborhood of $\theta = 1$ is obscured by the onset of multilayer adsorption. In summary, there is a mismatch between the situation described by the Harkins–Jura model and that suggested by macroscopic observations. We noted earlier that many isotherms are insensitive to the assumptions of their derivation. In line with that observation is the fact that the Harkins–Jura equation does fit a fairly wide range of experimental data and gives reasonable values of specific surfaces despite these objections. This is one example of an "old shoemaster chef" situation; we shall see presently that it is only one of several such cases.

Our approach until now has been to discuss adsorption isotherms on the basis of the equation of state of the corresponding two-dimensional matter. This procedure is easy to visualize and establishes a parallel with adsorption on liquid surfaces; however, it is not the only way to proceed. In the following section we consider the use of statistical themodynamics in the derivation of adsorption isotherms and shall examine some other approaches in subsequent sections.

9.5 Partition Functions and Isotherms

The partition function is the central feature of statistical thermodynamics. From the partition function the various thermodynamic variables such as entropy, enthalpy, and free energy may be evaluated. It is also possible, in principle, to deduce the equation of state for a system from the partition function.

It should be apparent—since an adsorption isotherm can be derived from a two-dimensional equation of state—that an isotherm can also be derived from the partition function, since the equation of state is implicitly contained in the latter. The use of partition functions is very general, but it is also rather abstract, and the mathematical difficulties are often formidable (note the cautious *in principle* in the preceding paragraph). We shall not attempt any comprehensive discussion of the adsorption isotherms which have been derived by the methods of statistical thermodynamics; instead, we shall derive only the Langmuir equation for adsorption from the gas phase by this method. The interested reader will find other examples of this approach discussed by Broeckhoff and van Dongen [2]. Statistical mechanics does not rely on pictorial models, but proceeds instead from mathematical statements about the energy states of the system. It turns out, however, that this freedom from more mechanical models helps reveal the underlying physical phenomena much more clearly.

A brief review of the statistical thermodynamics of ideal (bulk) gases will help us get started. In addition to reviewing some relevant physical chemistry, it will supply us with some expressions which may be useful, since the two-dimensional ideal gas law applies to adsorbed molecules as a limiting case.

The partition function is defined by the equation

$$Q = \sum_i g_i \exp\left(-\frac{\varepsilon_i}{kT}\right) \tag{33}$$

in which g_i and ε_i represent the degeneracy and the energy, respectively, of the ith state. An important property of a partition function is its factorability into contributions arising from translation and internal degrees of freedom:

$$Q = Q_{trans}Q_{int} \tag{34}$$

The translational portion of the partition function is relatively easy to evaluate for ideal gases. Substituting its value into Eq. (34) gives

$$Q = \frac{1}{N!}(Q_{trans}Q_{int})^N = \left[V\left(\frac{2\pi mkT}{h^2}\right)^{3/2}\right]^N \left(\frac{e}{N}Q_{int}\right)^N \tag{35}$$

for N molecules of mass m in a volume V at temperature T.

The easiest quantity to evaluate from this expression is the Helmholtz free energy A:

$$A = -kT \ln Q \tag{36}$$

A variety of other thermodynamic functions may be evaluated from this. For example, the chemical potential—the quantity equalized in equilibrium calculations—is

$$\mu_i = \left(\frac{\partial A}{\partial N_i}\right)_T = -kT\frac{\partial \ln Q}{\partial N_i} \tag{37}$$

Also, to calculate the equation of state, we recall

$$p = -\left(\frac{\partial A}{\partial V}\right)_T \tag{38}$$

If we apply Eqs. (36) and (38) to Eq. (35), for example, we get the ideal gas law:

$$p = kT\left[\left(\frac{\partial \ln [Q(V)]}{\partial V}\right)_T + \left(\frac{\partial (const.)}{\partial V}\right)_T\right] = kT\frac{\partial \ln (V^N)}{\partial V} = \frac{NkT}{V} \tag{39}$$

The approach of statistical thermodynamics to the derivation of an adsorption isotherm goes as follows. First, suitable partition functions describing the bulk and surface phases are devised. The former is usually assumed to be that of an ideal gas. From the latter, the equation of state of the two-dimensional matter may be determined if desired, although this quantity ceases to be essential. The relationships just given are used to evaluate the chemical potential of the adsorbate in both the bulk and the surface. Equating the surface and bulk chemical potentials provides the equilibrium isotherm.

We shall apply this method to the derivation of the Langmuir isotherm both to illustrate the method and to see the assumed nature of the surface energy states upon which it is based.

The Langmuir isotherm is based on the assumption of localized adsorption. This means that an adsorbed molecule has such a high statistical preference for a certain surface site as to possess a negligible translational entropy in the adsorbed state. Localized adsorption is thus seen to be very plausible for chemisorption, in which the adsorbed molecules and the adsorbent interact quite specifically. For nonspecific physical adsorption a nonlocalized or mobile layer seems to be a more plausible picture. We have already discussed the Henry law type of isotherm and the ideal gas equation of state that are associated with the simplest type of mobile adsorption.

The adsorption sites on the surface are assumed to be uniform and to bind the adsorbate with an energy ε per molecule or E per mole; that is, the potential energy of a molecule in the gaseous state is zero, and in the adsorbed state it is $-\varepsilon$. Note that this adsorption energy is a characteristic of the interaction between the adsorbed molecules and the adsorbent. As such, it is the same not only for all parts of the surface but also for all degrees of surface coverage. This is equivalent to saying that the adsorbed molecules do not interact with each other.

The surface is assumed to consist of S adsorption sites. Suppose we consider the case in which N of the sites are occupied, that is, when N molecules are adsorbed. To write the partition function, Q, for the surface molecules, we must ask how these molecules differ from those in the gas phase (superscript g). Some of the internal degrees of freedom may be modified by the adsorption (Q_{int}^{s}), but the most notable difference will be in the translational degrees of freedom. From three equivalent translational degrees of freedom, the adsorbed molecule goes to two highly restrained translational degrees of freedom (remember the adsorption is localized) and one vibrational degree of freedom normal to the surface(s):

$$Q^g_{\text{trans, 3d}} \rightarrow Q^s_{\text{trans,2d}} Q^s_{\text{vib}} \tag{40}$$

With these ideas in mind, we may assemble the partition function of the surface molecules as follows:

$$Q^s = g_N \{ Q^s_{\text{trans,2d}} Q^s_{\text{vib}} Q^s_{\text{int}} \exp[-(-\varepsilon)] \}^N \tag{41}$$

In this expression, the degeneracy factor g_N represents the number of ways the N molecules may be placed on S sites. The latter is given by the combinatorial formula (see Sect. 2.9):

$$g_N = \frac{S!}{N!(S-N)!} \tag{42}$$

Combining Eqs. (41) and (42) gives the following expression for the partition function of the adsorbed molecules:

$$Q^s = \frac{S!}{N!(S-N)!} (Q^s_{\text{trans,2d}} Q^s_{\text{vib}} Q^s_{\text{int}})^N \exp\left(\frac{N\varepsilon}{kT}\right) \tag{43}$$

Application of Eq. (36) to Eq. (43) gives the Helmholtz free energy of the adsorbed molecules:

$$A^s = -kT\left(\ln S! - \ln N! - \ln(S-N)! + \frac{N\varepsilon}{kT} \right.$$

$$\left. + N \ln (Q^s_{\text{trans,2d}} Q^s_{\text{vib}} Q^s_{\text{int}}) \right) \tag{44}$$

Since N and S are large, the factorials may be expanded by Sterling's approximation ($\ln x! \simeq x \ln x - x$) to give

$$A^s = -kT\left(S \ln S - N \ln N - (S-N)\ln(S-N) + \frac{N\varepsilon}{kT} \right.$$

$$\left. + N \ln (Q^s_{\text{trans,2d}} Q^s_{\text{vib}} Q^s_{\text{int}}) \right) \tag{45}$$

The chemical potential of the adsorbed molecules is given, according to Eq. (37), by differentiating this quantity with respect to N:

$$\mu^s = kT\left[\ln\left(\frac{N}{S-N}\right) - \frac{\varepsilon}{kT} - \ln (Q^s_{\text{trans,2d}} Q^s_{\text{vib}} Q^s_{\text{int}}) \right] \tag{46}$$

The condition of equilibrium between the adsorbed molecules and molecules in the gas state requires that the chemical potential for the adsorbed species be the same in both the gas phase and the adsorbed state:

$$\mu^s = \mu^g \tag{47}$$

Applying Eqs. (36) and (37) to Eq. (35) shows the chemical potential for an ideal gas to be

$$\mu^g = -kT \ln\left(\frac{kT}{p}\right)\left(\frac{2\pi m kT}{h^2}\right)^{3/2} Q_{int}^g \tag{48}$$

since $V = RT/p$.

Equating Eqs. (46) and (48) gives

$$\frac{N}{S - N} = \frac{p}{kT}\left(\frac{h^2}{2\pi m kT}\right)^{3/2} \frac{Q_{trans,2d}^s Q_{vib}^s Q_{int}^s}{Q_{int}^g} \exp\left(\frac{\varepsilon}{kT}\right) \tag{49}$$

We may group the following terms together to define a new quantity K:

$$K = \frac{1}{kT}\left(\frac{h^2}{2\pi m kT}\right)^{3/2} \frac{Q_{trans,2d}^s Q_{vib}^s Q_{int}^s}{Q_{int}^g} \exp\left(\frac{\varepsilon}{kT}\right) \tag{50}$$

in terms of which Eq. (49) becomes

$$\frac{N}{S - N} = Kp \tag{51}$$

Now if we divide both the numerator and denominator of the left-hand side by S and recognize that $N/S = \theta$, we obtain

$$\frac{\theta}{1 - \theta} = Kp \tag{52}$$

or

$$\theta = \frac{Kp}{1 + Kp} \tag{53}$$

two forms of the Langmuir adsorption isotherm [see Eq. (7.67)].

The quantity K defined by Eq. (50) may easily be expanded somewhat further. The two-dimensional partition function may be written by analogy with its three-dimensional counterpart, Eq. (35). To do this, V is replaced by the area accessible to the adsorbed molecule σ and the exponent $\frac{2}{2}$ (= 1) rather than $\frac{3}{2}$ is used, since two rather than three degrees of freedom are involved. Therefore we write

$$Q^s_{trans,2d} = \sigma \frac{2\pi m k T}{h^2} \tag{54}$$

If the energy separating the vibrational quantum states is small, the partition function for vibration is approximately given by

$$Q^s_{vib} = \frac{kT}{\varepsilon_{vib}} \tag{55}$$

For physical adsorption the approximation involved here is expected to be valid. Finally, the energy of vibration may be replaced by $h\nu$, where ν is the frequency with which the adsorbed molecules vibrate against the adsorbent:

$$Q^s_{vib} = \frac{kT}{h\nu} \tag{56}$$

Combining Eqs. (54) and (56) with Eq. (50) gives

$$K = \frac{\sigma}{\nu}(2\pi m k T)^{-1/2} \frac{Q^s_{int}}{Q^g_{int}} \exp\left(\frac{\varepsilon}{kT}\right) \tag{57}$$

It should be noted that this quantity has the units length^2 force^{-1}, reciprocal pressure units, as required by Eq. (53).

We saw by Eq. (7.75) how to rearrange the Langmuir equation into a form which permits graphical evaluation of the parameters. For the adsorption of gases this becomes

$$\frac{p}{n/w} = mp + b \tag{58}$$

and predicts a straight line when $p/(n/w)$ is plotted versus p, with

$$\text{slope} = m = \frac{N_A \sigma}{A_{sp}} \tag{59}$$

and

$$\frac{\text{slope}}{\text{intercept}} = \frac{m}{b} = K \tag{60}$$

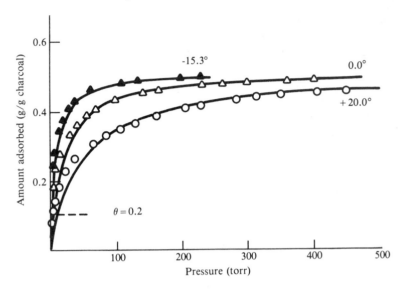

Figure 9.5 Plot showing how the amount of ethyl chloride adsorbed on charcoal (in g g^{-1}) varies with pressure at -15.3, 0, and 20°C. [Data from F. Goldman and M. Polanyi, *Z. Phys. Chem., 132*:321 (1928).]

Figure 9.5 shows some data for the adsorption of ethyl chloride on charcoal. Since these data were collected at different temperatures, the ratio p/p_0 is used as the independent variable in fitting the data to Eq. (58). That is, to compare the adsorption at different temperatures, the pressure is expressed as a fraction of the equilibrium vapor pressure at that temperature. The data for each of the three temperatures in Fig. 9.5 give quite good straight lines when plotted according to the linear form of the Langmuir equation [Eq. (58)]; the interpretation of this analysis is taken up in the following example.

Example 9.2 Slope and intercept values for the linearized plots of the data in Fig. 9.5 are as follows:

Temperature (°C)	Slope (g g^{-1})	Intercept (g g^{-1})
-15.3	1.9	0.040
0.0	2.0	0.047
20.0	2.0	0.088

Note that since p/p_0 is used in the linearization, the abscissa is dimensionless and the slope and intercept have the same units. Figure 9.5 suggests that the data at -15.3 and $0.0°C$ converge to a common saturation value, but this is less clear for the data at $20°C$. Do the linear plots clarify this situation? From the temperature dependence of the K's for these data, estimate the adsorption energy in this system.

Solution The slopes of these lines are identical (within experimental error), and the combination of Eqs. (59) and (7.72) shows that the reciprocal of the slope gives the adsorption at saturation. We conclude, therefore, that all three isotherms converge to the same saturation level of adsorption, namely, $(2.0)^{-1} = 0.50$ g g^{-1}.

The ratio of the slope to intercept values gives K for the adsorption at the three temperatures: 48, 43, and 23 at -15.3, 0.0, and $20.0°C$, respectively. Multiplying Eq. (57) through by $T^{1/2}$ and taking logarithms gives ln $(T^{1/2}K = (\varepsilon/k)(1/T) + \text{const.} = (E/R)(1/T) + \text{const.}$ While the quality of the data and the number of points scarcely justify the interpretation, a value for E/R of about 1700 can be estimated, suggesting an apparent value for E of about 14 kJ mole^{-1}. The word *apparent* is included until the effect of using p/p_0 instead of p in the linearization of the Langmuir equation is clarified.

•

As noted previously, the ability to predict correctly the temperature dependence of adsorption is a more stringent test of an isotherm than mere correlation of adsorption data. For this reason an independent measure of the energy of adsorption is clearly desirable. We shall return to this in Sect. 9.10.

Until now, we have focused our attention on those adsorption isotherms which show a saturation limit, an effect usually associated with monolayer coverage. We have seen two ways of arriving at equations which describe such adsorption: from the two-dimensional equation of state via the Gibbs equation or from the partition function via statistical thermodynamics. Next we shall turn our attention to multilayer adsorption. At the same time we shall introduce a third method for the derivation of isotherms, a kinetic approach.

9.6 Multilayer Adsorption: The BET Equation

As noted previously, the range of pressures over which gas adsorption studies are conducted extends from zero to the normal vapor pressure of the adsorbed species, p_0. An adsorbed layer on a small particle may readily be seen as a potential nucleation center for phase separation at p_0. Thus at the upper limit of the pressure range, adsorption and liquefaction appear to converge. At very low pressures it is plausible to restrict the adsorbed molecules to a monolayer. At the upper limit, however, the imminence of liquefaction suggests that the adsorbed molecules may be more than one layer thick. There is a good deal of evidence supporting the idea that multilayer adsorption is a very common form of physical adsorption on nonporous solids. In this section we shall be primarily concerned with an adsorption isotherm derived by S. Brunauer, P. H. Emmett, and E. Teller in 1938; the theory and final equation are invariably known by the initials of the authors: BET.

The BET isotherm has subsequently been derived by a variety of methods, but we shall follow the approach of the original derivation: a kinetic description of the equilibrium state. Like the Langmuir isotherm, the BET theory begins with the assumption of localized adsorption. There is no limitation as to the number of layers of molecules which may be adsorbed, however; hence there is no saturation of the surface with increasing pressure. In general, the derivation assumes that the rates of adsorption and desorption from each layer are equal at equilibrium and that adsorption or desorption can occur from a particular layer only if that layer is exposed, that is, provided no additional adsorbed layers are stacked on top of it.

Suppose we consider a surface consisting of a total of S adsorption sites, S_1 of which are filled with adsorbed molecules to a depth of one molecule. Thus the number of bare sites S_0 is $S - S_1$. These two surface situations are defined to be in equilibrium when the rate at which molecules attach to bare spots is the same as the rate at which they escape from monolayer regions. The rate of the adsorption process is proportional to both the pressure and the number of available sites:

$$R_a = k_a p S_0 \tag{61}$$

The rate of desorption is proportional to the number of sites occupied:

$$R_{d,1} = k_{d,1} S_1 \tag{62}$$

The subscript 1 on R_d and k_d in Eq. (62) reminds us that these refer to layer 1 only. At equilibrium the rates given by Eq. (61) and (62) are equal:

$$k_a p S_0 = k_{d,1} S_1 \tag{63}$$

It is profitable to interrupt the general BET development at this point to consider Eq. (63) in the special case in which the adsorption is restricted to a monolayer. In this case

$$S_0 + S_1 = S \tag{64}$$

a result which may be substituted into Eq. (63) to give the Langmuir equation:

$$\frac{k_a}{k_{d,1}} p = \frac{S_1}{S - S_1} = \frac{\theta}{1 - \theta} \tag{65}$$

Before returning to the case of multilayer adsorption, it is worthwhile to examine the expected form of the two rate constants k_a and $k_{d,1}$. The rate constant for desorption consists of a frequency factor and a Boltzmann factor. The former may be assumed to be proportional to a frequency ν, and the energy in the Boltzmann factor may be identified with the energy of interaction between the adsorbate and the adsorbent:

$$k_{d,1} \propto \nu \exp\left(-\frac{\varepsilon}{kT}\right) \tag{66}$$

From kinetic molecular theory, the number of collisions per unit area per unit time, Z, between gas molecules and a wall equals

$$Z = (2\pi mkT)^{-1/2} p \tag{67}$$

If this is multiplied by a surface area σ, the result is the rate of surface collisions. The coefficient of p may then be taken as proportional to k_a:

$$k_a \propto (2\pi mkT)^{-1/2} \sigma \tag{68}$$

The ratio $k_a/k_{d,1}$ may be set equal to the Langmuir K. Therefore

$$K = \frac{k_a}{k_{d,1}} \propto (2\pi mkT)^{-1/2} \frac{\sigma}{\nu} \exp\left(\frac{\varepsilon}{kT}\right) \tag{69}$$

a result that is identical to Eq. (57) as far as translational factors are concerned.

This digression back to the Langmuir result provides no new information about the latter, but it does remind us of the common starting assumptions of both the Langmuir and BET derivations. This kinetic derivation of the Langmuir equation is especially convenient as a means for obtaining an isotherm for the adsorption of two gases. This is illustrated in the following example.

Example 9.3   Assume that two gases A and B individually follow the Langmuir isotherm in their adsorption on a particular solid. Use the logic that results in Eq. (65) to derive an expression for the fraction of surface sites covered by one of the species when a mixture of the two gases is allowed to come to adsorption equilibrium with that solid.

Solution   The rate of desorption of A is given by Eq. (62) as $R_{d,A} = k_{d,A}S_A$, and the rate of adsorption of A by Eq. (61). Allowing for the fact that both A and B occupy surface sites, we have $R_{a,A} = k_{a,A}p_A S_0 = k_{a,A}p_A$ $(S - S_A - S_B)$. In this expression p_A is the partial pressure of A. At equilibrium the rates of adsorption and desorption are equal; therefore $k_{d,A}S_A = k_{a,A}p_A(S - S_A - S_B)$. Defining $\theta_i = S_i/S$ and $K_i = k_{a,i}/k_{d,i}$, we can write this last result as $\theta_A = K_A p_A(1 - \theta_A - \theta_B)$. A similar expression can be written for θ_B: $\theta_B = K_B p_B(1 - \theta_A - \theta_B)$.

Next the ratio of θ_A/θ_B can be used to eliminate one of the θ's from the above expressions. Since $\theta_A/\theta_B = K_A p_A/K_B p_B$, $\theta_B = (K_B p_B/K_A p_A)\,\theta_A$. Using this to eliminate θ_B, we obtain $K_A p_A = \theta_A(1 + K_A p_A + K_B p_B)$, which rearranges to $\theta_A = K_A p_A/(1 + K_A p_A + K_B p_B)$.

A similar expression is obtained for θ_B: $\theta_B = K_B p_B/(1 + K_A p_A + K_B p_B)$.

The fraction of surface sites covered by one gas or the other is given by $\theta_A + \theta_B = K_A p_A + K_B p_B/(1 + K_A p_A + K_B p_B)$. Note that the expressions for θ_A, θ_B, and $\theta_A + \theta_B$ all reduce to simple Langmuir expressions when one of the p or K values is zero.

•

Now let us return to the case of multilayer adsorption.

If we allow the possibility that a surface site may be covered to a depth of more than one molecule, then the following modifications are required. First, we define S_i to be the number of sites covered to a depth of i molecules. Second, Eq. (64) is modified to

$$S = S_0 + S_1 + S_2 + \cdots = \sum_{i=0}^{n} S_i \tag{70}$$

where the summation includes all thicknesses of coverage from zero to n, the maximum.

Now let us consider the composition of the second layer ($i = 2$). By analogy with Eq. (63), the rate at which molecules adsorb to form layer 2 is proportional to p and S_1. For simplicity, the proportionality constant is assumed to be the same as that for adsorption on the bare surface. Therefore we write

$$R_a = k_a p S_1 \tag{71}$$

In a similar fashion, we assume the rate of desorption from the second layer to be proportional to S_2:

$$R_{d,2} = k_{d,2} S_2 \tag{72}$$

The constant $k_{d,2}$ is assumed to be of the same form as that given by Eq. (66) for the first layer, *with one important modification*. While desorption from the first layer involves detaching a molecule from the adsorbent as a substrate, desorption from the second layer involves detachment from another adsorbed molecule of the same kind. The adsorption energy ε was used in the Boltzmann factor in the first case; the energy of vaporization ε_v is a more suitable value to use for the analogous quantity in the second case. We shall assume the frequency factor to be unchanged and write

$$k_{d,2} = v \exp\left(- \frac{\varepsilon v}{kT} \right) \tag{73}$$

At equilibrium, the rate of adsorption and the rate of desorption from the second layer are also equal; therefore

$$k_a S_1 p = k_{d,2} S_2 \tag{74}$$

As a matter of fact, the same expression also applies to the third, fourth, . . . , ith levels. As a first approximation, the "activation energy" for desorption is the same for all layers after the first. This leads to the generalization

$$k_a S_{i-1} p = k_{d,i} S_i \tag{75}$$

for $2 \leqslant i < n$.

Equation (75) enables us to relate S_i to S_{i-1} just as Eq. (63) relates S_1 to S_0. Therefore we may write

$$S_i = \left(\frac{k_a}{k_{d,i}} p \right)^{i-1} S_{i-(i-1)} = \left(\frac{k_a}{k_{d,i}} p \right)^{i-1} S_1 = \left(\frac{k_a}{k_{d,i}} p \right)^{i-1} \frac{k_a}{k_{d,1}} p^i S_0 \tag{76}$$

Since k_a is assumed to be the same for each layer and $k_{d,i}$ differs from k_d only in the value of the exponential energy, this may be written

$$S_i = \frac{k_a^i p^i S_0}{v^i [\exp(-\varepsilon_v/kT)]^{i-1} \exp(-\varepsilon/kT)} \tag{77}$$

Multiplying the numerator and denominator of the right-hand side of this equation by $\exp(-\varepsilon_v/kT)$ yields

$$S_i = \frac{k_a^i p^i S_0}{[v \exp(-\varepsilon_v/kT)]^i} \frac{\exp(-\varepsilon_v/kT)}{\exp(-\varepsilon/kT)} \tag{78}$$

which may be written

$$S_i = x^i c S_0 \tag{79}$$

with c and x defined as follows:

$$x = \frac{k_a p}{v \exp(-\varepsilon_v/kT)} = \frac{k_a}{k_{d,i>2}} p \tag{80}$$

and

$$c = \exp\left(\frac{\varepsilon - \varepsilon_v}{kT}\right) \tag{81}$$

Equation (79) may now be substituted into Eq. (70) to give

$$S = S_0 + S_1 + S_2 + \cdots = S_0 + \sum_{i=1}^{n} x_i c S_0 \tag{82}$$

Under some circumstances there may be a reason to restrict adsorption to a finite number of layers, that is, assign some specific value to n. In general, however, $n \rightarrow \infty$ as $p \rightarrow p_0$ is usually taken as the upper limit for this summation.

At this point we may define two other quantities in terms of the variables involved in Eq. (78): the total volume of gas adsorbed and the volume adsorbed at monolayer coverage, V and V_m, respectively. The total volume is obviously the sum of the volume held in each type of site V_i, where the latter is proportional to iS_i:

$$V = \sum_{i=1}^{n} V_i \propto \sum_{i=1}^{n} i S_i \tag{83}$$

The volume adsorbed at monolayer coverage is simply proportional to the total number of sites irrespective of the depth to which they are covered:

$$V_m \propto S \propto S_0 + \sum_{i=1}^{n} S_i \tag{84}$$

Note that in writing Eq. (84) it is not assumed that the monolayer is completely filled before other layers are formed. On the contrary, the picture allows for the coexistence of all types of patches, with adsorbed molecules stacked to various depths on each. Therefore V_m equals the volume of gas which would be adsorbed *if* a monolayer were formed. There is no implication that a filled monolayer is a step along the way toward multilayer formation.

Taking the ratio of Eq. (83) to Eq. (84) eliminates the unspecified proportionality constant and gives

$$\frac{V}{V_m} = \frac{\Sigma_i i S_i}{S_0 + \Sigma_i S_i} = \frac{c \, \Sigma_i i x^i}{1 + c \, \Sigma_i x^i} \tag{85}$$

The ratio V/V_m may be identified with θ, which may have values greater than unity in the case of multilayer adsorption. All that remains to be done to complete the BET derivation is to evaluate the summations in Eq. (85).

To assist in the evaluation of the summations, suppose we consider the quantity $x(1 - x)^{-1}$. The factor in parentheses may be expanded as a power series (see Appendix A) to give

$$x(1 - x)^{-1} = x(1 + x + x^2 + \cdots) = \sum_{i=1}^{n} x^i \tag{86}$$

Next we consider the quantity $x(d/dx)(\Sigma_i x^i)$. Carrying out the indicated differentiation gives

$$x \frac{d}{dx} \sum_i x^i = x \sum_i i x^{i-1} = \sum_i i x^i \tag{87}$$

Equation (86) may now be used as a substitution for $\Sigma_i x^i$ in Eq. (87) to yield

$$\sum_i i x^i = x \frac{d}{dx} [x(1 - x)^{-1}] = \frac{x}{(1 - x)^2} \tag{88}$$

Substituting Eqs. (86) and (88) into Eq. (85) gives

$$\frac{V}{V_m} = \frac{cx(1-x)^{-2}}{1+cx(1-x)^{-1}} \tag{89}$$

This last result may be simplified further to become

$$\frac{V}{V_m} = \frac{cx}{(1-x)[1+(c-1)x]} \tag{90}$$

Equation (90) is the result generally defined as the BET equation.

The condition that $V \rightarrow \infty$ is seen to correspond to $x = 1$ by Eq. (90). Recalling the definition of x given by Eq. (80) and that $V \rightarrow \infty$ as $p \rightarrow p_0$ permits us to write

$$1 = \frac{k_a}{k_{d,i>2}} p_0 \tag{91}$$

Therefore Eq. (80) becomes

$$x = \frac{p}{p_0} \tag{92}$$

Thus the independent variable in the BET theory is the pressure relative to the saturation pressure. Therefore the BET equation describes the volume of gas adsorbed at different values of p/p_0 in terms of two parameters V_m and c. Furthermore, the model supplies a physical interpretation to these two parameters.

9.7 Testing the BET Equation

The easiest way to evaluate the BET constants is to rearrange Eq. (90) into the following linear form:

$$\frac{1}{V}\frac{x}{1-x} = \frac{c-1}{cV_m}x + \frac{1}{cV_m} \tag{93}$$

This form suggests that a plot with $(1/V)[x/(1-x)]$ as the ordinate and x as the abscissa should yield a straight line, the slope and intercept of which have the following significance:

$$\text{slope} = m = \frac{c-1}{cV_m} \tag{94}$$

and

$$\text{intercept} = b = \frac{1}{cV_m} \tag{95}$$

Equations (94) and (95) may be solved to supply values of V_m and c from the experimental results:

$$V_m = \frac{1}{m + b} \tag{96}$$

and

$$c = \frac{m}{b} + 1 \tag{97}$$

In the following few paragraphs we shall first examine some of the general features of gas adsorption as predicted by the BET theory. Next we shall consider how well the theory actually fits experimental data. The use of experimental V_m values in the evaluation of A_{sp} is taken up in the following section.

Figure 9.6 enables us to observe some of the general features of the BET isotherm. The most apparent aspect concerning all the curves in this figure is their rapid increase as $p/p_0 \rightarrow 1$. It is also apparent, however, that the shape of the curve is sensitive to the value of c, especially for low values of c. Note particularly that the BET equation encompasses both type II and type III isotherms (Fig. 9.2). For values of c equal to 2 or less, the curves show no inflection point, whereas the inflection becomes increasingly pronounced as c increases above 2. In view of the wide diversity of curve shapes that are consistent with the BET equation and the relative insensitivity of adsorption data to the model underlying a particular equation, we might expect that the BET equation will fit experimental data rather successfully.

Figure 9.7a is a plot of some actual experimental data showing the volume of N_2—expressed as cubic centimeters at STP per gram—adsorbed by a sample of nonporous silica at 77 K. Figure 9.7b shows these same results plotted according to the linear form of the BET equation, given by Eq. (93). The BET equation is seen to fit the adsorption data in the range $0.05 \gtrsim p/p_0 \gtrsim 0.30$. From the values of the slope (0.0257 g cm^{-3} at STP) and intercept (2.85×10^{-4} g cm^{-3} at STP), Eqs. (96) and (97) may be used to evaluate V_m and c for this system:

$$V_m = \frac{1}{(257 + 2.85) \times 10^{-4}} = 38.5 \text{ cm}^3 \text{ g}^{-1} \text{ at STP} \tag{98}$$

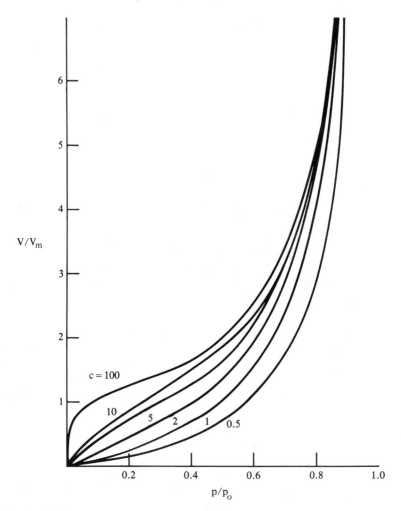

Figure 9.6   Plots of V/V_m versus p/p_0 for several values of the parameter c, calculated according to the BET theory by Eq. (90).

$$c = \frac{257 \times 10^{-4}}{2.85 \times 10^{-4}} + 1 = 91.2 \tag{99}$$

The range of pressures over which the linear form of the BET equation fits the experimental data in Fig. 9.7 is fairly typical for a variety of gases and adsorbents. At relative pressures below the range of fit, the BET equation underestimates the actual adsorption, whereas above

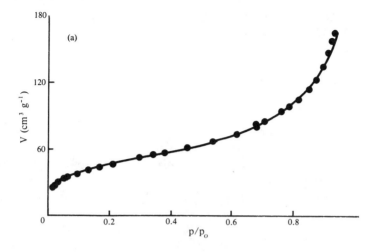

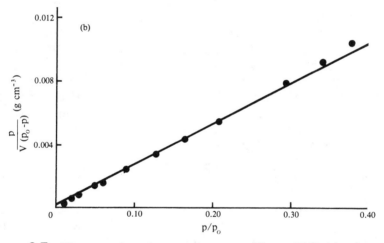

Figure 9.7 Nitrogen adsorption on nonporous silica at 77 K; (a) volume per gram (in cm³ at STP) versus p/p_0 and (b) according to the linear form of the BET equation [Eq. (93)]. [Data from D. H. Everett, G. D. Parfitt, K. S. W. Sing, and R. Wilson, *J. Appl. Chem. Biotechnol.*, 24:199 (1974).]

$p/p_0 = 0.35$, the equation overestimates it. This range of p/p_0 values encompasses the region in which $V = V_m$ for $2 \gtrsim c \gtrsim 500$. In applying the BET equation to surface area determination, this is an important region of fit.

One way to appreciate the significance of V_m is to consider Eq. (90) under the following limiting conditions. If c is large and x is small, then the factors $c - 1$ and $1 - x$ in Eq. (90) become c and 1, respectively. In that case the BET equation becomes

$$\frac{V}{V_m} = \theta = \frac{cx}{1 + cx} \tag{100}$$

which is the same form predicted by the Langmuir equation. We have already seen how the saturation value of the Langmuir equation may be related to the specific area of the adsorbent [see Eq. (59)]. Equation (101) shows the equivalence of $\theta = 1$ and $V = V_m$.

Inspection of the curves in Fig. 9.6 reveals a sharpening of the "knee" in the curve near $V/V_m = 1$ as c increases. One gets the impression in such a case that the volume adsorbed at monolayer coverage is almost recognizable, being just slightly obscured by the onset of multilayer adsorption.

At smaller values of c the "knee" of the curve—or point B as it is often called—is much more definitely buried in the multilayer region. Another way to see this is to note that the fraction of surface covered when $V = V_m$ is 0.24, 0.50, 0.91, and 0.97 for values of c equal to 0.1, 1.0, 100, and 1000, respectively; that is, V indeed approaches monolayer coverage as c increases. Some critics of the BET theory are reluctant to accept Eq. (96) literally as a measure of the saturation capacity of the surface. They prefer instead to regard $(m + b)^{-1}$ as a precise way of locating point B. According to this interpretation, point B represents a degree of coverage at which the affinity of the adsorbed molecules for the adsorbent is changing most rapidly. Conceptually, this may be identified with the completion of the monolayer, without the quantitative restrictions assumed by the BET model.

Basically there are three criteria against which the success of the BET theory may be evaluated: its ability to "fit" adsorption data, correct prediction of the temperature dependence of adsorption, and correct evaluation of specific area. We have already commented on the data-fitting capabilities of the equation. In the following section we shall discuss specific area determination exclusively. Therefore we shall conclude this section with a few comments on the ability of the BET theory to predict the temperature dependence of adsorption.

The parameter c describes the temperature dependence of adsorption. In the derivation already presented, certain simplifying assumptions were made as to the constancy of k_a and v_i in each of the layers. Various modifications of this assumption might be made, but they would involve minor temperature effects at best. To a first approximation, then, one should be able to evaluate $\varepsilon - \varepsilon_v$ from a knowledge of experimental c values, or vice versa. Proceeding in the first manner leads to values of $\varepsilon - \varepsilon_v$ which are too low, perhaps half their expected value. When we consider the region of fit on which the evaluation of c is based, it is not difficult to see why values of c are too low. The data are not fitted to the earliest stages of adsorption which are associated with larger values of ε; hence c is underestimated. The definition of c allows us to conclude some unfinished business pertaining to Example 9.2. This is examined in the following example.

Example 9.4   The BET analysis uses p/p_0 rather than p as a variable just as we used this pressure ratio to compare Langmuir adsorption at different temperatures in Example 9.2. What corrections, if any, are needed in the "apparent adsorption energy" of about 14 kJ mole^{-1} as calculated in Example 9.2?

Solution   The answer to this is obtained by comparing Eqs. (53) and (98). When p/p_0 is used instead of p, we obtain c, not K, from the linearized Langmuir equation via Eq. (60). The exponential energy term evaluated from the analysis of ln K versus $1/T$ is seen by Eq. (81) to be the difference between the adsorption energy and the energy of vaporization of the adsorbate. For ethyl chloride $E_v = 23$ kJ mole^{-1}; therefore $E = 23 + 14 = 37$ kJ mole^{-1}; this is the actual adsorption energy for the ethyl chloride-charcoal system discussed in Example 9.2.

•

Note also that small values of c—which account for type III adsorption isotherms—result when $\varepsilon_v > \varepsilon$ in Eq. (81). This was anticipated in our remarks about Fig. 9.2.

The final criterion for judging the success of a theoretical isotherm is its ability to measure specific surface areas. Since the BET equation has become a standard in this regard, we shall devote an entire section to the discussion of this topic.

9.8   Specific Surface Area: The BET Method

With monolayer adsorption, we saw how the saturation limit could be related to the specific surface area of the adsorbent. The BET equation permits us to extract from multilayer adsorption data [by means of Eq. (96)] that volume of adsorbed gas which would saturate the surface if the adsorption were limited to a monolayer. Therefore V_m may be interpreted in the same manner that the limiting value of the ordinate is handled in the case of monolayer adsorption. Since it is traditional to express both V and V_m in cubic centimeters at STP per gram, we write [see Eq. (7.72)]

$$V_m = \left(\frac{n}{w}\right)_{sat} (22{,}414 \text{ cm}^3 \text{ mole}^{-1}) = \frac{A_{sp}(22{,}414)}{N_A \sigma^0} \qquad (101)$$

Note that it is assumed that V_m has been expressed on a "per gram" basis in writing Eq. (101). If this is not the case, the area of the *actual sample* rather than its specific area is given by Eq. (101). If the area occupied per molecule on the surface is known, the specific surface may be evaluated:

$$A_{sp} = \frac{V_m N_A \sigma^0}{22{,}414} \qquad (102)$$

This last quantity is something which may be determined by independent methods; therefore the BET theory may be tested at this point. Two ways of proceeding come to mind. First, it is clear that the same value of A_{sp} must be obtained for a particular adsorbent, regardless of the nature of the adsorbate used. Studies of this sort lead to quite consistent values of A_{sp}, provided the adsorbed species all have access to the same surface. On a porous surface, for example, large adsorbate molecules may not be able to enter small cavities which are accessible to smaller molecules.

If the particles are approximately spherical, the specific surface is given by Eq. (1.2). BET surface areas have been found to agree quite reasonably with values calculated from electron micrographs. Of course, proper statistical consideration of the polydispersity of the sample must be included. Also, if the surface is too rough, the gas adsorption area will be higher than a value based on the assumption of spherical particles.

In the preceding discussion, it has been assumed that a value of σ^0 is known unambiguously. This quantity is obviously the "yardstick" by which moles of adsorbed gas are converted to areas. Any error in this quantity will invalidate the determination of A_{sp}. What is generally done is to assume the adsorbed material has the same density on the surface that it has in the bulk liquid at the same temperature and to assume the molecules are close packed on the surface. In view of the earlier discussion of surface phases, this is seen to be a somewhat risky procedure. A safer way to proceed would be to evaluate σ^0 for a particular adsorbate from independent measurements of V_m and A_{sp}.

In view of the difficulty in translating measured gas adsorption into absolute specific surface areas, it is not surprising that self-consistency is

Table 9.1 Values of A_{sp} and c as Determined in 10 Different Laboratories by the BET Method for the Same Silica Sample Shown in Fig. 9.7

Laboratory	A_{sp} (m^2 g^{-1})	c
A	166.4	92
B	162.8	101
C	174.0	100
D	148.5	166
E	173.5	70
F	166.0	98
G	167.9	91
H	143.6	113
I	169.5	62
J	161.7	122
Average	163.4	102
Standard deviation	10.0 (6%)	29 (28%)

Source: From D. H. Everett, G. D. Parfitt, K. S. W. Sing, and R. Wilson, *J. Appl. Chem. Biotechnol., 24*:199 (1974).

often normative in this matter. In this sense, at least, an area of 16.2 Å^2 for nitrogen has become something of a standard. Values for other common adsorbed species may be found in the works of Adamson [1], Broeckhoff and van Dongen [2], and Kantro et al. [7]. It is probably not surprising that polar molecules such as water display values of σ^0 which are sensitive to the nature of the substrate. Using nitrogen adsorption to evaluate A_{sp} and then using the latter to evaluate σ^0 for water has led to values of 12.5, 10.4, and 11.4Å^2 for amorphous silica, calcium hydroxide, and calcium silicate hydrate, respectively.

In spite of a variety of objections to the BET theory, V_m values from N$_2$ adsorption studies have become a very common means for determining specific surface areas. As a matter of fact, a IUPAC commission was organized in 1969 to study N$_2$ adsorption with the objective of preparing reference standards for surface area determinations. In this project four silicas and four carbon blacks of different particle size were investigated independently in 13 different laboratories. Nitrogen adsorption data were analyzed by the BET method over the best fit region, and values of V_m were converted to A_{sp} using 16.2 Å^2 as the value of σ^0. Table 9.1 shows the values of A_{sp} and c obtained in the 10 laboratories that studied the silica of Fig. 9.7. The average value of the specific surface from these determinations is 163.4 ($\pm$6%) m^2 g^{-1}. This value agrees with our analysis of Fig. 9.7 obtained by combining Eqs. (98) and (102):

$$A_{sp} = \frac{(38.5)(6.02 \times 10^{23})(16.2 \times 10^{-20})}{22,400} = 168 \text{ m}^2 \text{ g}^{-1} \tag{103}$$

Samples of this material as well as others investigated are now available for calibration purposes as gas adsorption standards. Details may be found in the reference cited for the data of Table 9.1.

At first glance, it may seem surprising that the BET method is as successful as it is in the evaluation of A_{sp}. After all, the values of c are not in particularly good agreement with expected values, nor, conversely, are $\varepsilon - \varepsilon_v$ values calculated from experimental c's in good agreement with expectations. Probably a significant part of the discrepancy between theory and experiment with respect to c arises from the heterogeneity of the surface. The BET equation—like the Langmuir equation to which it reduces in certain limits—assumes that a single adsorption energy applies to all surface sites. As we saw in Sect. 6.8, a distribution of surface energy states is not uncommon with solids. In this regard, both the Langmuir and BET models are unrealistic. However, sites with high adsorption energies are apt to be covered by adsorbate molecules first. If these represent a relatively small fraction of the surface sites, this effect gets masked fairly quickly as the first layer fills. In addition, lateral interactions between adsorbed molecules start at zero and increase as the surface begins to fill. Even these interactions tend to level off after a certain degree of coverage is achieved. Therefore, even though a single adsorption energy is unrealistic, there are compensating effects which tend to minimize the variation.

With these ideas in mind, it is not difficult to understand that a log–log plot of V/V_m versus p/p_0 gives a common curve when nitrogen adsorption is studied on a variety of nonporous solids. Another way to say this is that log–log plots of V versus p/p_0 would describe a family of curves of the same shape but displaced vertically from one another by an amount equal to log V_m. The only significant contributor to the differences in adsorption for $p/p_0 \gtrsim \frac{1}{3}$ is the difference in V_m for the solids involved. The BET method provides a relatively easy way to evaluate this quantity.

Although we have often mentioned the adsorption energy in this chapter, we have not yet discussed any procedure by which this can be measured quantitatively, except as some sort of an average quantity. Since most solid surfaces are heterogeneous, it is desirable to be able to examine adsorption energy in a more discriminating way, for example, as a function of coverage or pretreatment. In the following section we shall see how this can be done.

9.9 Heats of Adsorption

As we have seen, an adsorption isotherm is one way of describing the thermodynamics of gas adsorption. However, it is by no means the only way. Calorimetric measurements can be made for the process of adsorption and thermodynamic parameters may be evaluated from the results. To discuss all of these in detail would require another chapter. Rather than develop all the theoretical and experimental aspects of this subject, therefore, it seems preferable to continue focusing on adsorption isotherms, extracting as much thermodynamic insight from this topic as possible. Within this context, results from adsorption calorimetry may be cited for comparison, without a full development of this latter topic.

The approach we shall follow is essentially that used to derive the Clapeyron equation. Suppose we consider an infinitesimal temperature change for a system in which adsorbed gas and unadsorbed gas are in equilibrium. The criterion for equilibrium is that the free energy of both the adsorbed (subscript s) and unadsorbed (subscript g) gas change in the same way:

$$dG_s = dG_g \tag{104}$$

The following equations may be written for these two quantities if the temperature change is assumed to cause no change in the amount of adsorbed material:

$$dG_g = -S_g \, dT + V_g \, dp \tag{105}$$

and

$$dG_s = -S_s \, dT + V_s \, dp \tag{106}$$

Substituting Eqs. (105) and (106) into Eq. (104) and rearranging gives

$$\left(\frac{\partial p}{\partial T} \right)_{n_s} = \frac{S_g - S_s}{V_g - V_s} \tag{107}$$

In writing this last result, it has been explicitly noted that the number of moles of adsorbed gas n_s is constant. If the process under consideration is carried out reversibly, $S_g - S_s$ may be replaced by q_{st}/T, where q_{st} is known as the isosteric (the same coverage) heat of adsorption:

$$S_g - S_s = \frac{q_{st}}{T} \tag{108}$$

Combining Eqs. (107) and (108) leads to the result

$$\left(\frac{\partial p}{\partial T} \right)_{n_s} = \frac{q_{st}}{T(V_g - V_s)} \tag{109}$$

Equation (109) may be integrated if the following assumptions are made: (1) $V_g \gg V_s$, so that V_s may be neglected, (2) the gas behaves ideally so that the substitution $V_g = RT/p$ may be used, and (3) q_{st} is independent of T. With these assumptions, Eq. (109) integrates to

$$\ln \left(\frac{p_1}{p_2} \right) = -\frac{q_{st}}{R} \left(\frac{1}{T_1} - \frac{1}{T_2} \right) \tag{110}$$

Equation (110) shows that the isosteric heat of adsorption is evaluated by comparing the equilibrium pressure at different temperatures for samples showing the same amount of surface coverage. The data of Fig. 9.5 may be used as an example to see how this relationship is applied.

For an arbitrarily chosen extent of adsorption, a horizontal line such as the dashed line in Fig. 9.5 may be drawn which cuts the various isotherms at different pressures. The pressure coordinates of these intersections can be read off the plot. According to Eq. (110), a graph of $\ln p$ versus $1/T$ should be linear with a slope of $-q_{st}/R$. From Fig. 9.5, for example, when the adsorption is 0.10 g ethyl chloride (g charcoal)$^{-1}$ (which corresponds to $\theta = 0.2$) the equilibrium pressures are 0.20, 0.63, and 2.40 Torr at -15.3, 0, and 20°C, respectively. When plotted in the manner just described, these data yield a line of slope -5330 K. Multiplication by R gives $q_{st} = 44.3$ kJ mole^{-1} as the isosteric heat of adsorption for this system at $\theta = 0.2$. Table 9.2 lists values of q_{st} for different θ values as calculated from the data in Fig. 9.5.

Table 9.2 is based on the same data that were analyzed according to the Langmuir equation in Sect. 9.5. Examples 9.2 and 9.4 show that these data are consistent with an adsorption energy of about 37 kJ mole^{-1} according to the Langmuir interpretation.

In presenting this result initially, the remark was made that an independent determination of the adsorption energy would be desirable. Although the present reinterpretation of the same data is not exactly an "independent" determination, it does extract an energy quantity from the experimental results which is free of any assumed model for the mode of adsorption. Accordingly, it is informative to compare the two inter-

Table 9.2 Values of the Isosteric Heat of Adsorption at Different Values of θ for the Data Shown in Fig. 9.5 as Evaluated by Eq. (110).

θ	q_{st} (kJ mole^{-1})
0.06	56.9
0.08	47.3
0.10	46.4
0.20	44.4
0.30	41.4
0.40	40.2
0.50	40.2
0.60	41.0
0.70	38.1
0.80	37.2

pretations. The Langmuir model assumes that a single energy applies to all adsorption sites. Therefore any data which are analyzed according to this model cannot yield more than one energy. The isosteric heat of adsorption, on the other hand, is evaluated at different degrees of surface coverage. The data in Table 9.2 show that this quantity definitely varies with coverage, tending to level off as θ → 1. This is consistent with the picture of the more active "hot spots" being covered first. The average energy which the Langmuir analysis yields is approximately the same as the energy toward which q_{st} converges.

There are several additional thermal quantities besides q_{st} and E which may be generically called "heats of adsorption." One of these is the integral heat of adsorption Q_n, which is related to the isosteric heat of adsorption as follows:

$$Q_n = \int_0^n q_{st}\, dn \tag{111}$$

It applies to the process in which n moles of adsorbate are transferred from the bulk gas to the surface, starting from a bare surface. The integral heat of adsorption is determined from data such as those contained in Table 9.2 by graphical integration.

In addition, there are a number of different calorimetric methods to determine heats of adsorption. For example, we may distinguish between isothermal and adiabatic heats, depending on the type of calorimeter

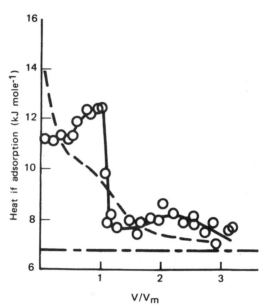

Figure 9.8 Calorimetric heats of adsorption as a function of coverage for argon on carbon black at 78 K. The dashed line represents untreated black; the solid line is after graphitization at 2000°C. The horizontal line is the heat of vaporization of argon. [Reprinted from R. A. Beebe and D. M. Young, *J. Phys. Chem.*, 58:93 (1954), copyright by the American Chemical Society.]

involved. Of course, thermodynamic relationships exist between these various quantities. We shall not pursue these topics, but the reader should be aware of the differences and seek precise definitions if the need arises.

The data shown in Fig. 9.8 indicate both the kind of data that may be obtained by direct calorimetric study of gas adsorption and some evidence of the effect of preheating on the properties of surfaces. The figure shows the calorimetric heat of adsorption of argon on carbon black. The broken line indicates the behavior of the untreated black, and the solid line is the "same" adsorbent after heating at 2000°C in an inert atmosphere, a process known as graphitization. The horizontal line indicates the heat of vaporization of argon.

There are several interesting aspects of this figure which are quite generally observed.

1. The untreated carbon black shows the effect of surface heterogeneity, an effect which becomes smeared out as the coverage increases. The

Table 9.3 Effect of Traces of
Adsorbed Water on the Heat of
Immersion of TiO_2 in Benzene

Amount of water adsorbed (mmol kg^{-1})	Heat of immersion (mJ m^{-2})
0.0	150
2.0	250
4.0	320
10.0	450
17.0	506
Pure H_2O	520

Source: Reprinted with permission from
G. E. Boyd and W. D. Harkins, *J. Am. Chem.
Soc., 64*:1195 (1942), copyright by the Amer-
ican Chemical Society.

surface of the untreated black contains a certain amount of oxygen in
a variety of functional groups (e.g., ether, carbonyl, hydroxyl, and
carboxyl). Graphitization results in both the reduction of these
oxygen-containing groups and the sharpening of both the basal and
prismatic crystallographic planes. Electron micrographs of carbon
black before and after graphitization are shown in Fig. 1.5.

2. After graphitization, the most notable feature is the sharp discon-
    tinuity at monolayer coverage. Beyond the monolayer the heat of
    adsorption is close to the heat of vaporization (as required by the
    BET theory) but does show some influence of the surface as well. At
    coverage below $V/V_m = 1$ the heat of adsorption increases with
    increasing coverage, probably due to lateral interactions between the
    adsorbed molecules.

Figure 9.8 is an extreme example of the effect of heating on the
properties of an adsorbent. Degassing prior to measuring an isotherm is
done under far less severe conditions; nevertheless, these conditions
should always be reported when adsorption studies are conducted
because of the possibility of surface modification upon heat treatment.

Another calorimetric technique for measuring the heat of adsorption
consists of comparing the heat of immersion (see Sect. 6.6) of bare solid
with that of a solid which has been pre-equilibrated with vapor to some
level of coverage. Table 9.3 summarizes some results of this sort. The

experiment consisted of measuring the heats of immersion of anatase (TiO_2) in benzene with the indicated amount of water vapor preadsorbed on the solid. Small quantities of adsorbed water increase the heat of immersion more than threefold so that it approaches the value for water itself. Most laboratory samples will be contaminated with adsorbed water. Unless the material has been carefully pretreated, the actual nature of the surface may be quite different from what would be expected nominally.

9.10 Capillary Condensation and Adsorption Hysteresis

High specific area solids of the type that are studied by gas adsorption consist of small particles for which the radius of an equivalent sphere is given by Eq. (1.2). This figure may be a fairly reasonable measure of a characteristic linear dimension even of irregularly shaped particles, provided the gas adsorption is restricted to the exterior surface of the particles. Any cracks or pores in the solid particles will expose additional adsorbing surfaces and increase the total specific area of the material. Because of this complication, we have specified nonporous solids at several places in this chapter. The particles of high specific area solids are small to begin with, so it follows immediately that any pores in these solids will necessarily have very small dimensions.

In Sect. 6.11 we discussed the pressure required to force a liquid—most commonly mercury—into the pores of a solid. Our emphasis in that section was on the pores between powder particles when the latter are tightly pressed into a plug. In this section we are not concerned with these interstitial pores, since powders are not densely packed in adsorption studies; instead, our interest is in the pores within the particles themselves. The difference is a matter of emphasis, however, and both liquid intrusion and gas adsorption complement one another for the study of porous solids.

We shall restrict our attention to only one aspect of the adsorption behavior of porous solids: the hysteresis they display in their adsorption isotherms. A schematic illustration of the phenomenon is shown in Fig. 9.9. Although the region enclosed by the hysteresis loop may have a variety of shapes, in all cases there are two quantities of adsorbed material for each equilibrium pressure in the hysteresis range. That branch of the loop which corresponds to adsorption (increasing pressure) inevitably displays less adsorption at any given pressure than the desorption branch (decreasing pressure). In many cases hysteresis loops such as this are reproducible, although they are not reversible in the

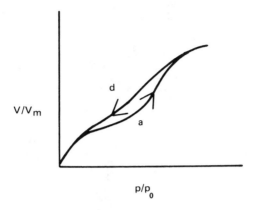

Figure 9.9   A schematic illustration of hysteresis between the adsorption (a) and desorption (d) branches of an experimental isotherm.

thermodynamic sense. Since irreversible processes are involved, the substitution of q_{st}/T for ΔS in the derivation of Eq. (110) is not valid. One can go through the motions of evaluating $\partial p/\partial T$ for a system which displays hysteresis, but the "apparent q_{st} values" so obtained (one for each branch) are not easily related to calorimetric adsorption energies.

Adsorption hysteresis is often associated with porous solids, so we must examine porosity for an understanding of the origin of this effect. As a first approximation, we may imagine a pore to be a cylindrical capillary of radius r. As just noted, r will be very small. The surface of any liquid condensed in this capillary will be described by a radius of curvature which is related to r. According to the Laplace equation [Eq. (6.29)], the pressure difference across a curved interface increases as the radius of curvature decreases. This means that vapor will condense in small capillaries at pressures less than the normal vapor pressure p_0 which is defined for flat surfaces. The condensation of vapors in small capillaries is an equilibrium phenomenon, however, so capillary condensation in itself does not account for hysteresis. It does point out the fact that a liquid–vapor surface is also involved in the adsorption on porous solids for $p < p_0$. In fact, the leveling off of the adsorption in type IV and type V isotherms before p_0 is reached is the result of liquid condensation in small pores at these pressures.

For simplicity, let us assume that the liquid condensed in a pore has a surface which is part of a sphere of radius R, with $R > r$. For a spherical surface we may use the Kelvin equation [Eq. (6.40)] to calculate p/p_0:

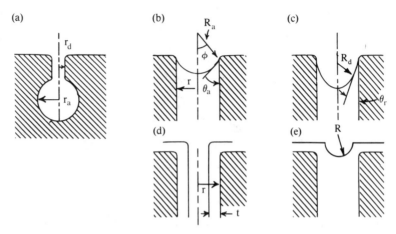

Figure 9.10   Five models for capillary condensation. The radius of the pore equals r, the radius of curvature of the spherical meniscus is R, and t is the thickness of the adsorbed layer. The subscripts a and d refer to adsorption and desorption.

$$N_A kT \ln \left(\frac{p}{p_0} \right) = - \frac{2M\gamma}{\rho R} \qquad (112)$$

A minus sign has been introduced in the Kelvin equation because the radius is measured outside the liquid in this application, whereas it was inside the liquid in the derivation of Chap. 6. In hysteresis, adsorption occurs at relative pressures that are higher than those for desorption. According to Eq. (112), it is as if adsorption–condensation took place in larger pores than desortion–evaporation. Since the pore dimensions are presumably constant, we must seek some mechanism consistent with this observation to explain hysteresis.

Figure 9.10 contains sketches for several different models of pores which will be useful in our discussion of capillary condensation. Figure 9.10a is the simplest, attributing the entire effect just described to variations in pore radius with the depth of the pore. That is, when liquid first begins to condense in the pore, the larger radius r_a determines the pressure at which the adsorption–condensation occurs. Once the pore has been filled and the desorption-evaporation branch is being studied, the smaller radius r_d determines the equilibrium pressure. Although bottlenecked pores of this sort may exist in some cases, this model seems far too specialized to account for the widespread occurrence of hysteresis.

Figure 9.10b and c represents another model, based on contact angle hysteresis. These sketches represent the surface of liquid in a capillary during the adsorption and desorption stages of hysteresis, respectively. In Fig. 9.10b the capillary is filling; in Fig. 9.10c it is emptying. Accordingly, the advancing and receding values of the contact angle apply to adsorption and desorption, respectively. The radius R of the spherical surface and the radius r of the capillary are related through the contact angle θ and its complementary angle ϕ (see Fig. 9.10):

$$r = R \sin \phi = R \cos \theta \qquad (113)$$

Substituting Eq. (113) into Eq. (112) gives

$$N_A kT \ln \left(\frac{p}{p_0} \right) = - \frac{2M\gamma \cos \theta}{\rho r} \qquad (114)$$

For a pore of constant radius the equilibrium pressure decreases as $\cos \theta$ increases or as θ decreases. It will be recalled from Chap. 6 that advancing contact angles are larger than receding ones. Therefore this model is consistent with the observation that desorption–evaporation occurs at lower relative pressures than adsorption–condensation. The only objection to this explanation of adsorption hysteresis is that it makes no reference whatsoever to adsorption!

Figure 9.10 d and e illustrates another model for adsorption hysteresis which considers multilayer adsorption explicity. During adsorption the capillary is viewed as a cylinder of radius $r - t$, where t is the thickness of the adsorbed layer at that pressure. This is represented by Fig. 9.10d. For such a surface the Kelvin equation becomes

$$N_A T \ln \left(\frac{p}{p_0} \right) = - \frac{M\gamma}{\rho(r - t)} \qquad (115)$$

since one of the radii of curvature is infinite. As multilayer adsorption proceeds, however, the thickness of the adsorbed layer equals r. Once the pore is thus filled, its surface becomes a meniscus which may be treated by Eq. (114) as a portion of a sphere, as shown in Fig. 9.10e. In the event that $\theta = 0$, Eq. (112) applies with $R = r$. Comparison of adsorption and desorption in this situation is particularly easy if $t \ll r$ in Eq. (115). In that case, the right-hand side of Eq. (115) is proportional to $1/r$, and comparison with Eq. (112) reveals that

$$\left(\frac{p}{p_0} \right)_d^{1/2} = \left(\frac{p}{p_0} \right)_a \qquad (116)$$

Since both of these ratios are less than unity, $p_a > p_d$. This model is qualitatively consistent with the observed hysteresis but is difficult to apply quantitatively because it neglects differences between the adsorbed material on the surface and that in the bulk liquid.

This third mechanism for adsorption hysteresis clearly imposes some sort of upper limit to the formation of a multilayer: the condition in which $t = r$. Suppose we assume that this limiting thickness occurs after n layers have been deposited. In that case the summations involved in the BET derivation may not be carried out for $1 < i < \infty$, but must be truncated at $i = n$. This leads to a modification of the BET equation to the following form:

$$\frac{V}{V_m} = \frac{cx[1 - (n + 1)x^n + nx^{n+1}]}{(1 - x)[1 + (c - 1)x - cx^{n+1}]} \qquad (117)$$

where $x = p/p_0$ and V_m and c have the same meaning as before. In this form the BET theory results in a three-parameter equation; the maximum numbers of layers n must be specified. The three-parameter equation is virtually indistinguishable from the two-parameter equation for $n > 3$ in the range of applicability of the BET theory. For values of n less than this, however, it extends the range of relative pressures over which the BET theory applies.

It is always a good idea to examine the plausibility of a relationship like Eq. (117)—a result that is presented without proof—by checking whether it reduces to known expressions under suitable limits. This is considered in the following example.

Example 9.5   Examine Eq. (117) in the limits of $n \to \infty$ and $n = 1$ in which cases it should reduce to the BET and Langmuir equations, respectively.

Solution   The case of $n \to \infty$ is easy, since the n's always appear as exponents of $x = p/p_0$, which is always less than unity. Raising any fraction to a power results in an even smaller fraction. In the limit of $n \to \infty$, $x^n \to 0$. With the extra terms in Eq. (117) thus eliminated, we obtain Eq. (90), the BET equation.

The case of $n = 1$ is best considered by making this substitution and then multiplying out the various factors. Doing this, we obtain $V/V_m = (cx + 2cx^2 + cx^3)/(1 - 2x + x^2 + cx - 2cx^2 + cx^3)$. The right-hand side of this equation can be refactored as follows:

$$V/V_m = cx(1 - x)^2/[(1 - x)^2 + cx(1 - x)^2] = cx(1 - x)^2/(1 - cx)$$
$$\times (1 - x)^2 = cx/(1 + cx)$$

This last form is identical to Eq. (100), the Langmuir equation.

•

An even better fit of the data with a more logical justification is obtained by truncating only the numerator of Eq. (85) after n terms and allowing the denominator to take on values to $i = \infty$. Only in the latter case is the total surface of the adsorbent represented. This leads to

$$\frac{V}{V_m} = \frac{cx(1 - x^n)}{(1 - x)[1 + (c - 1)x]} \tag{118}$$

an even simpler result than Eq. (117).

Gas adsorption data may be analyzed for the distribution of pore sizes. What is generally done is to interpret one branch of the isotherm and use an appropriate equation to calculate the effective pore radius at a given pressure. The amount of material adsorbed or desorbed for each increment or decrement in pressure measures the volume of pores with that effective radius.

References

1. A. W. Adamson, *Physical Chemistry of Surfaces*, 4th ed., Wiley, New York, 1982.
2. J. C. P. Broeckhoff and R. H. van Dongen, Mobility and Adsorption on Homogeneous Surfaces, in *Physical and Chemical Aspects of Adsorbents and Catalysts* (B. G. Linsen, ed.), Academic, New York, 1970.
3. S. Brunauer, L. E. Copeland, and D. L. Kantro, The Langmuir and BET Theories, in *The Solid–Gas Interface*, Vol. 1 (E. A. Flood, ed.), Marcel Dekker, New York, 1967.
4. J. H. de Boer, *The Dynamical Character of Adsorption*, 2nd ed., Oxford University Press, London, 1968.
5. S. J. Gregg, *The Surface Chemistry of Solids*, 2nd ed., Chapman and Hall, London, 1961.
6. J. P. Hobson, Physical Adsorption at Extremely Low Pressures, in *The Solid–Gas Interface*, Vol. 1 (E. A. Flood, ed.), Marcel Dekker, New York, 1967.
7. D. L. Kantro, S. Brunauer, and L. E. Copeland, BET Surface Areas—Methods and Interpretations, in *The Solid–Gas Interface*, Vol. 1 (E. A. Flood, ed.), Marcel Dekker, New York, 1967.
8. S. Ross, Monolayer Adsorption on Crystalline Surfaces, in *Progress in Surface and Membrane Science*, Vol. 4 (J. F. Danelli, M. D. Rosenberg, and D. A. Cadenhead, eds.), Academic, New York, 1971.

Problems

1. An isotherm which is not too difficult to derive by the methods of statistical mechanics assumes an adsorbed layer which obeys the two-dimensional

analog of the van der Waals equation. The result of such a derivation is the equation [see Eq. (27)]

$$\ln p = \ln\left(\frac{\theta}{1-\theta}\right) + \frac{\theta}{1-\theta} - \frac{2a}{bkT}\theta + \ln K$$

where a, b, and K are constants, the first two being the two-dimensional van der Waals constants. Like its three-dimensional counterpart, this equation predicts that at the critical point for the two-dimensional matter

$$\left(\frac{\partial \ln p}{\partial \theta}\right)_{T_c} = 0 \quad \text{and} \quad \left(\frac{\partial^2 \ln p}{\partial \theta^2}\right)_{T_c} = 0$$

From the second of these derivatives, evaluate θ_c as predicted by this model. Use this value of θ_c and the first of these derivatives to evaluate the relationship between T_c and the two-dimensional a and b constants. How does this result compare with the three-dimensional case? The van der Waals b constant is four times the volume of a hard-sphere molecule. What is the relationship between the two dimensional b value and the area of a hard-disk molecule?

2. Use the linear form of the Langmuir equation to evaluate $(n/w)_{sat}$ and K for the adsorption of pentane on carbon black from the higher pressure values in the following data. Use the ratio p/p_0 rather than p only to normalize pressures relative to the equilibrium vapor pressure of pentane at different temperatures.* (All pressures are in Torr.)

T (°C)	−63.7		0		5.24		20.5	
p_0	3.48		187.5		235.6		445.1	
	p	g C_5 g^{-1}	p	g C_5 g^{-1}	p	g C_5 g^{-1}	p	g C_5 g^{-1}
	0.024	0.2827	0.0533	0.1062	3.8	0.2299	0.284	0.1061
	0.028	0.2904	0.1405	0.1322	7.2	0.2535	0.675	0.1317
	0.067	0.3162	0.203	0.1427	19.6	0.2939	0.950	0.1421
	0.103	0.3276	0.890	0.1908	36.7	0.3147	3.62	0.1884
	0.233	0.3459	2.5	0.2305	53.6	0.3231	9.7	0.2262
	0.288	0.3507	5.0	0.2506	88.5	0.3313	15.9	0.2469
	0.671	0.3581	14.6	0.2964	199.9	0.3418	62.9	0.3002
	1.46	0.3647	29.2	0.3184			90.4	0.3107
			45.4	0.3272			155.2	0.3204
			80.7	0.3353			328.7	0.3321
			161.0	0.3433			428.7	0.3372

3. Examine the temperature variation of K values from the preceding problem by means of the procedure given in Example 9.2. (To decrease computational

*M. Polanyi and F. Goldmann, Z. Phys. Chem., 132:321 (1928).

effort, various members of the class may be assigned different temperatures to analyze in Problem 2. The K values may then be pooled for this problem.)

4. The accompanying data give the volume of N_2 at STP adsorbed on colloidal silica at the temperature of liquid nitrogen as a function of the ratio p/p_0.* Plot these results according to the linear form of the BET equation. Evaluate c, V_m, and A_{sp} from these results, using 16.2 $Å^2$ as the value for σ^0.

V at STP (cm^3 g^{-1})	p/p_0	V at STP (cm^3 g^{-1})	p/p_0
44	0.008	117	0.558
52	0.025	122	0.592
57	0.034	130	0.633
61	0.067	148	0.692
64	0.075	165	0.733
65	0.083	194	0.775
70	0.142	204	0.792
77	0.183	248	0.825
78	0.208	296	0.850
85	0.275		
90	0.333		
96	0.375		
100	0.425		
109	0.505		

5. The following data give the volume at STP of nitrogen and argon adsorbed on the same nonporous silica at $-196°C$†:

	V at STP (cm^3 g^{-1})	
p/p_0	Nitrogen	Argon
0.05	34	23
0.10	38	29
0.15	43	32
0.20	46	38
0.25	48	41
0.30	51	43
0.35	54	45
0.40	58	50
0.45	58	54
0.50	61	55
0.60	68	62
0.70	77	69
0.80	89	79
0.90	118	93

---

*D. H. Everett, G. D. Parfitt, K. S. W. Sing, and R. Wilson, *J. Appl. Chem. Biotechnol.*, 24:199 (1974).

†D. A. Payne, K. S. W. Sing, and D. H. Turk, *J. Colloid Interface Sci.*, 43:287 (1973).

Using 16.2 Å^2 as the N_2 cross section, calculate A_{sp} for the silica by the BET method. What value of σ^0 is required to give the same BET area for the argon data?

6. Use the accompanying data* to criticize or defend the following proposition: Self-consistent A_{sp} values for nonporous solids are obtained at 77 and 90 K by using values of σ^0 for N_2 equaling 16.2 and 17.0 Å^2, respectively. These are consistent with the density of liquid N_2 at these two temperatures. For the same self-consistency in A_{sp} using O_2 as the adsorbate, σ^0 values of 14.3 and 15.4 Å^2 must be used at these two temperatures. This suggests that O_2 is somewhat more loosely packed on the surface than in the liquid state at 90 K compared to 77 K.

	N_2		O_2	
T (K)	ρ (g cm^{-3})	V_{sp} (cm^3 g^{-1})	ρ (g cm^{-3})	V_{sp} (cm^3 g^{-1})
77	0.808	1.238	1.204	0.831
90	0.751	1.332	1.14	0.877

7. Use the data from Problem 2 to estimate the isosteric heat of adsorption of pentane on carbon black at $\theta \simeq 0.3, 0.6$, and 0.9. Under what conditions would greater variation of q_{st} be expected? What prevents these conditions from being examined in this problem? How does q_{st} compare with the energy of adsorption for this system as determined in Problem 3? How does the difference between q_{st} and E_{ads} compare with ΔH_v for pentane?

8. Colloidal carbon was formed by the slow pyrolysis of polyvinylidene chloride. Surface oxidation occurs during subsequent storage in air. By heating at elevated temperatures, the following gases are evolved† (heats of immersion were also measured after the different thermal pretreatments):

	Cumulative amount desorbed (mmol g^{-1})			Heat of immersion (cal g^{-1})
T (°C)	CO	CO$_2$	H$_2$	
100	0.03	0.02	—	11.0
200	0.05	0.14	—	9.8
300	0.15	0.33	—	9.2
400	0.35	0.44	—	8.7
500	0.63	0.51	—	8.2
600	0.95	0.55	—	7.7

---

*K. M. Hanna, I. Odler, S. Brunauer, J. Hagymassy, and E. E. Bodor, *J. Colloid Interface Sci.,* 45:27 (1973); also see Osipow in Ref. 8.

†S. S. Barton, M. J. B. Evans, and B. H. Harrison, *J. Colloid Interface Sci.,* 45:542 (1973).

	Cumulative amount desorbed (mmol g^{-1})			Heat of immersion (cal g^{-1})
T (°C)	CO	CO$_2$	H$_2$	
700	1.30	0.55	—	7.0
800	1.58	0.55	—	6.2
900	1.80	0.55	0.03	6.0
1000	1.823	0.556	0.269	5.9

Using 7.9 and 9.1 Å^2 as the areas of oxide desorbing as CO and CO$_2$, respectively, estimate the area (in m^2 g^{-1}) occupied by each of these oxide types. If the specific area of the carbon is 1100 m^2 g^{-1}, what percentage of the surface is covered with oxide? Discuss the variation of the heat of immersion with the removal of surface oxides.

9. Samples of rutile TiO$_2$ were outgassed at elevated temperatures: In some instances this was followed by subsequent exposure to O$_2$ at 150°C. The following results were obtained by the BET analysis of N$_2$ adsorption after various preliminary treatments*:

Pretreatment	A_{sp} (m^2 g^{-1})	c
150°C + O$_2$	11.0	200
210°C + O$_2$	11.7	100
200°C, no O$_2$	11.9	520
240°C, no O$_2$	11.3	370
470°C + O$_2$	11.7	450
560°C + O$_2$	11.9	1210

Criticize or defend the following proposition: After the initial removal of physically adsorbed water, partial dehydroxylation of the surface occurs with increasing temperature. In the absence of O$_2$, Ti^{3+} cations are present which interact with N$_2$ in pretty much the same way as the isolated hydroxyl groups on the surface. At still higher temperatures surface diffusion permits the hydroxyl groups to migrate into patches, which show stronger interactions with N$_2$. The specific surface is not essentially changed throughout treatment.

---

*G. D. Parfitt, D. Urwin, and T. J. Wiseman, *J. Colloid Interface Sci.*, *36*:217 (1971).

10
METAL SURFACES
Microscopy, Spectroscopy, and Diffractometry

I saw before me a vast multitude of small Straight Lines ... interspersed with other Beings still smaller and of the nature of lustrous points—all moving to and fro.

[From Abbott's *Flatland*]

10.1 Introduction

We discovered in the last chapter that certain things can be learned about solid surfaces by the study of gas adsorption. The latter, being a thermodynamic observation, is insensitive to details at the atomic level and can only yield a gross characterization of the surface. Properties such as the specific surface area and the presence or absence of pores can be determined, since the average surface and not atomic details are involved. The existence of a distribution of surface energy sites can also be inferred from adsorption data, but the method falls short when it comes to specifics about this distribution. Observations on an atomic scale are needed to learn more about the details of the surface structure. Such observations comprise the subject matter of this chapter.

A great assortment of high-technology (and high cost!) methods exist for examining surfaces on a fine scale, and new methods are being developed continuously. Rather than attempt to catalog all of these and say a few things about each, it seems preferable to single out a few of the more important techniques and develop them in a bit more detail. Accordingly, in this chapter we shall discuss scanning electron microscopy, Auger electron spectroscopy, low-energy electron diffraction, and field ionization microscopy as fairly typical "modern" methods for studying surfaces. Although the names of these experimental approaches are descriptive of what the technique involves, they are cumbersome and

545

it is standard practice in this field to use acronyms to name experimental methods. Thus the procedures we shall discuss will become familiar as SEM, AES, LEED, and FIM. Since there are numerous other techniques that we shall not discuss but which also tend to be known by their initials, the literature on this subject contains what Adamson [1] calls "a veritable alphabet soup of designations."

The techniques we discuss will permit us to characterize solid surfaces in a variety of ways: pictorally, chemically, and crystallographically. All of these methods have in common the fact that they explore the surface with probe particles—generally electrons—which leave the surface carrying certain information about that surface with them. The experiments must therefore generate the probes, allow them to interact with the surface, collect them, and interpret the information they contain. The presence of gas molecules interferes with these electron beams, and so the techniques we discuss are conducted under vacuum, although the effects of adsorbed molecules are detected in some instances.

Just as an assortment of experimental techniques exist, so too are there numerous types of solid surfaces that might be examined. Not all methods can be used for all surfaces, and the techniques we shall consider are particularly well suited for the study of metal surfaces. Therefore we shall also limit our examples to metal surfaces. Although this obviously excludes a number of interesting and important solid surfaces, it also includes some topics of great significance, including friction and corrosion in engineering materials, semiconductors for electronic devices, and catalytic surfaces for many chemical reactions.

10.2 Clean Surfaces: Ultra-High Vacuum

In the last chapter we considered what could be learned about solid surfaces from the study of gas adsorption. Although the focus of our attention was the solid surface, the experimental variable was gas pressure. In this chapter we look at solid surfaces more directly. Our strategy is to begin by examining clean surfaces and then surfaces with a controlled amount of adsorption. We saw in connection with the discussion of Fig. 9.1 that measurable gas adsorption occurs even at gas pressures as low as 10^{-10} Torr. As a matter of fact, the two-dimensional density of the adsorbed molecules is not low enough to conform to the two-dimensional ideal gas law even when the pressure is on the order of 10^{-10} Torr. A question of considerable practical importance, then, is how low the pressure must be for an initially clean surface to remain that way for a reasonable period of time. The foregoing reference to adsorption cites equilibrium data which is not useful for answering questions of rate.

Instead, we must turn to the kinetic molecular theory of gases for an estimate of the frequency with which molecules collide with a solid surface. We shall not be misled, however, if we anticipate that this pressure is low. The following example is a numerical examination of gas collisions with walls.

Example 10.1 Assuming 10^{19} atoms per square meter as a reasonable estimate of the density of atoms at a solid surface, estimate the time that elapses between collisions of gas molecules at 10^{-6} Torr and 25°C with surface atoms. Use the kinetic molecular theory result that relates collision frequency to gas pressure through the relationship $Z = \frac{1}{4}\bar{v}N/V$, where the mean velocity of the molecules $\bar{v} = (8RT/\pi M)^{1/2}$ and N/V is the number density of molecules in the gas phase and equals pN_A/RT. Repeat the calculation at 10^{-8} and 10^{-10} Torr.

Solution If 10^{19} collisions occurred per square meter per second, each surface atom would be hit an average of once each second. Therefore we must divide 10^{19} by Z at the required pressures to obtain the elapsed time. Assuming a molecular weight of 30 g mole^{-1} for the gas, we obtain

$$(8RT/\pi M)^{1/2} = [8(8.314 \text{ J K}^{-1} \text{ mole}^{-1})(298)/\pi(0.028 \text{ kg mole}^{-1})]^{1/2}$$

$$= 475 \text{ m s}^{-1}$$

Converting the gas pressure to SI units gives for $p = 10^{-6}$ Torr

$$p = 10^{-6} \text{ Torr} \times \frac{1.013 \times 10^5 \text{ N m}^{-2}}{760 \text{ Torr}} = 1.33 \times 10^{-4} \text{ N m}^{-2}$$

and

$$\frac{N}{V} = \frac{(1.33 \times 10^{-4} \text{ N m}^{-2})(6.02 \times 10^{23} \text{ molecules mole}^{-1})}{(8.314 \text{ J K}^{-1} \text{ mole}^{-1})(298 \text{ K})}$$

$$= 3.24 \times 10^{16} \text{ molecules m}^{-3}$$

Therefore

$$Z = \frac{1}{4}\bar{v}N = (475)(3.24 \times 10^{16})/4 = 3.85 \times 10^{18} \text{ collisions m}^{-2} \text{ s}^{-1}$$

and

elapsed time $= 10^{19}/3.85 \times 10^{18} = 2.60$ s

Since N/V and Z are directly proportional to p and the elapsed time is inversely proportional to p, the values of these quantities at other pressures can be written by inspection. At $p = 10^{-8}$ Torr, $N/V =$

3.24×10^{14} molecules m^{-2}, $Z = 3.85 \times 10^{16}$ m^{-2} s^{-1}, and time = 260 s = 4.3 min; at $p = 10^{-10}$ Torr, time = 2.6×10^4 s = 7.2 hr.

•

This calculation shows that until gas pressure is lowered below, say, 10^{-8} Torr, the rate of surface bombardment is too great for the surface to remain clean long enough for an experiment to be carried out. Pressures on the order of 10^{-6} Torr are achieved in ordinary vacuum lines with diffusion and mechanical pumping. Any pressure below this is considered ultra-high vacuum (UHV); today devices are commercially available which routinely achieve pressures in the range 10^{-9}–10^{-12} Torr. As we consider the problems of UHV technology, we shall also be introduced to some of the topics and vocabulary we shall encounter elsewhere in this chapter.

First of all, it should be noted that the types of spectroscopic analyses of surfaces with which we are concerned in this chapter involve bombarding the clean surfaces with projectiles which may be electrons, ions, or photons. The first two of these must be able to travel from the emitting source to the surface under investigation and, from there, to the detector without colliding with any gas molecules. This means that the mean free path of these projectile particles should be on the order of about 0.1 m. We can turn once again to kinetic molecular theory for the relationship that shows the mean free path to be inversely proportional to pressure. Therefore the high vacuum needed to assure clean surfaces is also required for the sample examination as well.

In UHV work stainless steel and aluminosilicate glass are the preferred materials for apparatus construction. Gold or some other malleable metal is used to form leak-resistant valves. Extreme care is needed to remove enough ambient gas to achieve ultra-high vacuum. Ordinary diffusion pumps are used as forepumps to remove most of the bulk gas. Of course, trapping is critical here so that vapors from the diffusion pump do not contaminate the UHV system. Bulk gases are only the beginnings of our concern, however. Desorption of any gas molecules adsorbed on the surfaces of the apparatus is equivalent to a leak in this kind of vacuum system. In UHV work desorption is the analog of a pinhole leak in an ordinary vacuum line. To remove adsorbed gases the working volume of a UHV system is baked for 10–20 hr at 300–500°C before use, a procedure called outgassing. The results of this can be complicated. For example, when water vapor is deliberately introduced, it is found that H_2 is the principal gas present in the apparatus, with carbon monoxide also appearing. Apparently what has happened is that water reacts with carbonaceous material present as contamination under the catalytic influence of the surface to produce these products. It should be pointed out that these results were obtained in work with surfaces that would be regarded as "clean" by all ordinary standards. We shall see that contamination of surfaces by carbon-containing molecules is a very widespread phenomenon. The catalytic properties of solid surfaces is another topic about which we shall have more to say later in this chapter.

An assortment of different pumping arrangements are employed to achieve UHV, including ion pumps, cryopumps, and getter pumps. With each of these the pumped gas is stored within the system rather than being removed as is the case with diffusion pumps. This means that caution must be exercised so that the stored gas is not re-released into the system during the course of an experiment. Ion pumping is achieved by ionizing with electrons the gas to be removed and then collecting these ions at a metallic cathode, where they remain adsorbed. Getter pumps selectively adsorb or dissolve the gas without using ionization to collect the molecules, and cryopumps use low temperatures to promote adsorption. These three modes of pumping are thus the highly specific reverse of outgassing procedures: Adsorption at a specific site (the pump) is promoted to lower the pressure.

It is not enough merely to reach a low pressure; it must also be possible to measure it. Ionization gauges are almost always used for this purpose. In these the residual gas is ionized, collected at an electrode, and the resulting current measured. The current varies linearly with the gas pressure down to about 10^{-11} Torr. If the ions are separated by mass—making the gauge a mass spectrometer—then the partial pressures of various gases in the vacuum chamber can be determined.

Next let us consider the preparation of the sample itself. Above we referred to the desirability of keeping a surface clean; nothing was said about the problem of obtaining such a surface in the first place. The process begins before the sample is introduced into the vacuum system with preliminary mechanical polishing. Depending on the coarseness of the abrasive used, such treatment disturbs the surface to a depth of 10^2–10^4 nm. A variety of subsequent treatments are possible once the sample is installed in the vacuum chamber. We mention specifically ion or electron bombardment, since these methods of sample preparation correspond to to the categories of spectroscopic probes we shall discuss in later sections. These bombardment methods constitute a kind of molecular sandblasting which causes surface damage and produces debris, the extent and nature of which depend on the bombarding particles, the surface, and so on. Bombarding the surface by an accelerated beam of gaseous ions sputters off surface atoms and thus removes layers damaged in earlier stages of preparation. Electron bombardment does not damage the solid substrate but does remove adsorbed molecules. A pulsed laser is also capable both of removing adsorbates and of damaging the surface, depending on the energy and duration of the pulses. Under optimum conditions desorption can be achieved with minimal surface damage.

Two additional methods for preparing clean surfaces consist of generating the surfaces under high vacuum to begin with. Using remote control manipulators to crush a sample under vacuum exposes fresh surface to an environment in which adsorption equilibrium is very slow. This technique produces a heterogeneous array of crystal faces, however. Far more suitable for the specific and localized examination that spectroscopic methods offer is crystal cleavage under vacuum. Again by use of remote manipulators, a sample may be cleaved along a predetermined crystal plane to expose a fresh surface of known character. Surface repair through annealing may be necessary after this relatively violent procedure. The need for annealing also means that UHV chambers must possess the capability for heating samples in a controlled and measurable way. Sample preparation by cleavage plus annealing gives the best defined surfaces for subsequent examination. In Sect. 10.6 we shall review how such surfaces are described in terms of crystallographic notation and atom packing.

We conclude this section with the observation that the preparation of a sample for study along the lines presented in this section is a costly matter. We have not even considered the acquisition of data and the expense involved in that! High costs are typical of surface characterization by spectroscopy, and this consideration makes these experiments considerably less accessible than most of the other experimental procedures discussed in this book.

10.3   Scanning Electron Microscopy

In Sect. 1.5 we saw that the transmission electron microscope is an instrument which is ideally suited for the determination of the size and shape of colloidal particles. This suitability is the consequence of the remarkable resolving power of an electron beam in producing an image. Equation (1.5) argues that the theoretical resolving power is directly proportional to the wavelength of the photon or particle responsible for the formation of the image. The word *theoretical* creeps into the previous sentence because loss of resolution can arise from extraneous causes such as vibrations, sample preparation, voltage drift, and stray fields. In Chap. 1 we discussed transmission electron microscopy, whose capabilities we did not explore beyond their ability to form silhouette images of the specimens under examination. Relief features on a particle surface are only indirectly accessible from transmission electron microscopy, since the image is highly sensitive to sample thickness. A second version of electron microscopy also exists in which the sample is scanned by a highly focused beam of electrons. The instrument, known as the scanning electron microscope (SEM), is ideally suited for the study of surface topography.

In the SEM a beam of electrons from a tungsten filament is focused by a system of magnetic lenses on an area of surface 5–15 nm in diameter in an evacuated chamber. There is nothing unique in this description. We have already seen that surfaces may be cleaned by such bombardment; likewise, Auger electron spectroscopy (see Sects. 10.4 and 10.5) and low-energy electron diffraction (see Sects. 10.7 and 10.8) are two additional techniques we shall consider in which a beam of electrons interacts with a solid surface and the results are interpreted in terms of surface properties. There is an underlying similarity in these and many other experimental procedures in surface science, but important differences also exist. Differences arise from the energy and geometry of the initial bombardment, from the mode of observation, and from the interpretation in terms of surface characteristics. Although the SEM itself can be used in a variety of ways, we shall concentrate on its ability to produce images of surface relief.

There are several modes of interaction between the incident electron beam—the primary electrons—and the atoms that comprise the sample. The first of these is elastic scattering, in which the primary electrons are reflected from the topmost layer of atoms in the surface without any energy interchange. Alternatively, the energy of the incident electrons may be dissipated in the sample by knocking electrons out of the atoms of the specimen such that a cascade of ions and electrons trace the path of the primary electron through the sample. Since the energy of the primary electron is very much greater than the energy with which electrons are bound to atoms, quite a number of ionizations may be induced by a single primary electron. The electrons released as a result of this ionization—the secondary electrons—possess a rather broad distribution of energies, considerably lower than the primary electrons. Figure 10.1 schematically shows the relationship between the primary and secondary electrons in terms of relative energy and breadth of energy distribution. In the figure the number of electrons having a particular energy is plotted as a function of energy. It will be observed that there is some fine structure in this distribution of electron energies, but this is of no concern to us at this point. It might be noted that the unit of choice for

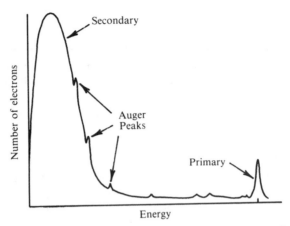

Figure 10.1 Schematic illustration of the number of electrons ejected from a surface as a function of their energy.

reporting particle energies in this sort of work is the electon volt (eV), the quantity of energy associated with moving an electron through a potential difference of 1 V. Since the product of coulombs and volts is joules, the conversion factor between electron volts and joules is simply 1 eV = 1.602×10^{-19} J. When multiplied by Avogadro's number for conversion to a molar basis, the corresponding energy is the familiar factor 96,485 J mole^{-1}. The primary electrons are accelerated to an energy of about 30 keV in typical SEM operation.

In discussing SEM, our interest is in the broad peak of secondary electrons, some of which are sufficiently energetic to escape from the sample surface. It is these that are used to form an image in what is known as the emissive mode of SEM operation. Some secondary electrons recombine with ions, and the resultant reshuffling of electrons through the various quantum states releases photons which are the basis for the cathodoluminescent mode of SEM operation. Our discussion will be limited to the emissive mode.

The secondary electrons emitted from the sample are collected at a positive electrode and the resulting signal is amplified through a photomultiplier circuit to modulate the display on a cathode ray tube (CRT). The key to SEM operation is the synchronized scanning of the specimen surface by the primary electron beam and of the viewing screen by the modulated beam of the CRT. The synchronous scanning of the sample and the CRT by their respective electron beams is accomplished by an electronic unit called the scan generator. A schematic illustration of

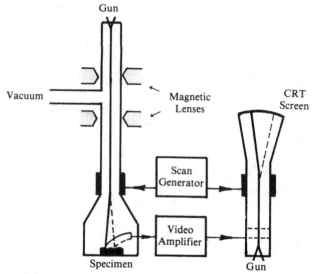

Figure 10.2 Schematic diagram showing how a scan generator synchronizes electron beams sweeping both the specimen and CRT display in a scanning electron microscope.

the relationship between these two scanning operations is shown in Fig. 10.2. The pattern of scan is called the raster and is usually a line-by-line (up to 1000 lines) sweep of a square area. The magnification is given by the ratio of the edge length of the image to that of the sample. Often more than one CRT display is included to allow both visual and photographic observation. These CRTs may be coated with phosphors of different persistence, and the scan rates may be different, depending on the mode of observation. Because the actual time of bombardment for any given surface is short, the damage produced by the primary electron beam may be kept to a minimum and the flexibility offered by different scan rates under different modes of observation optimizes this.

The contrast between one area and another in the image produced by a SEM is the result of differences in the signal produced by the respective areas of the specimen. Backscattered primary electrons as well as secondary electrons are involved in image production, although some discrimination between these may be accomplished by controlling the voltage at the collector electrode. Backscattered primary electrons have much greater energies than secondaries and therefore may originate from a much larger volume of the sample than the latter. For example, a 15-keV primary electron probes a volume on the order of micrometers in

(a) **Figure 10.3** Photomicrographs produced by the SEM: (a) low-carbon steel broken at $-196°C$ (reprinted with permission from Y. Kanai and K. Uchibori, *Scanning Electron Microscopy*, SEM, Inc., AMF O'Hare, Chicago, 1969) and (b) silicon from an integrated circuit after ion sputtering (reprinted with permission from S. I. Ingrey, *Scanning Electron Microscopy*, SEM, Inc., AMF O'Hare, Chicago, 1982).

dimensions, while the secondary electrons—whose energies are generally less than 50 eV—can only penetrate a 5–10 nm thickness of the solid. This is an instance of actual resolving power being determined by the interactions of the electrons with the solid rather than by the wavelength of the electron alone.

The contrast in an image may arise from chemical or physical differences in the sample. Although the SEM can be used directly for surface chemical analysis, we shall not consider this. It is informative to

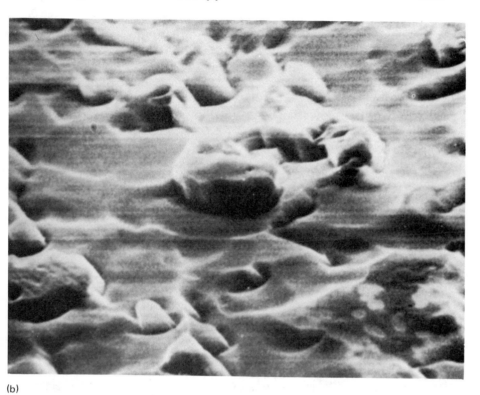

(b)

note, however, that images produced by backscattered primaries are more sensitive to chemical composition, and those produced by secondaries to surface topography. As noted above, the potential at the collector discriminates between the two; thus some tuning may be done to optimize the image sought.

Surface roughness produces variations in the trajectory of the emitted electrons: Surfaces facing the collector appear brighter than those facing in other directions, since the collector subtends a larger part of the electron beam in the former case. Secondary electrons are lower in energy and have a wider assortment of trajectories; therefore they are highly sensitive to the potential at the collector. By contrast, the more energetic backscattered primaries are relatively unaffected by this potential.

Differences in contrast in crystalline samples may also arise from a mechanism known as electron channeling. In this, secondary emission varies with the angle of incidence of the primary beam and reflects different electron distributions along various planes in the sample. This is a weak effect and is not observed unless conditions for its occurrence are optimized. In principle, channeling can yield information about crystal lattice parameters, but we shall turn to low-energy electron diffraction in Sect. 10.7 for this kind of information.

Figure 10.3 shows photomicrographs of two metal surfaces prepared by the SEM. The surface shown in Fig. 10.3a is a low-carbon (0.22% C) steel fractured by cleavage at $-196°C$. The portions of the photographic field which show undulation correspond to fracture along grain boundaries, while the flat-looking areas correspond to grain cleavage. In this study steels of various carbon contents were examined on fracture. It is observed that cleavage fracture surfaces are less clearly defined as the carbon content of the steel increases. At higher temperatures ductile fracture occurs, resulting in surfaces of totally different appearance in photomicrographs. This method of examining the relief of fractured surfaces is called fractography; it is the depth of field and high resolution of the SEM that make these studies feasible.

Figure 10.3b shows an SEM micrograph of a silicon surface from an integrated circuit. In the semiconductor device a thin film of conducting aluminum is layered on the silicon, with a layer of insulating SiO_2 sandwiched in between. The surface in the figure is that of silicon after the aluminum contact pad was removed by the ion-sputtering technique. The texturing of the surface probably reflects grain boundaries which face the sputtering beam at different angles and therefore erode at different rates. We shall encounter this same aluminum–SiO_2–silicon boundary in Sect. 10.5, where it is examined by Auger electron spectroscopy. The latter is the subject of the next section.

10.4 Auger Electron Spectroscopy

In discussing Fig. 10.1, we ignored the fine structure in the secondary peak of the distribution of electron energies. The small "blips" appear at specific energies and are all but buried by the broad peak due to secondary scattering. Their sharpness suggests that these peaks are due to a quantum phenomenon; in fact, this phenomenon was described by P. Auger in the early 1920s, and the electrons with these discrete energies are called Auger electrons. Auger observed this effect when a noble gas was irradiated with x-rays in a cloud chamber. In some instances he detected two sets of tracks—one long, one short—emanating from a single source. This was interpreted to mean that two electrons—one more energetic than the other—were being ejected from the atom as a result of the irradiation. The K, L, M designation of electron shells is the system generally used to describe this behavior. In this notation, the Auger effect may be understood in terms of the following sequence of events:

1. The incident radiation causes an electron to be ejected from an inner shell, say, the K shell, in the atom being bombarded.
2. An electron from an outer shell, say, the L shell, drops to fill the K shell vacancy. A quantity of energy equal to the difference between the energy of these two states, $E_K - E_L$, must now be disposed of.
3. X-ray fluorescence is one way for the atom to eliminate the excess energy; a photon of characteristic frequency may be emitted, where ν is given by

$$E_K - E_L = h\nu \tag{1}$$

4. Competing with the mechanism described in item 3 is the radiationless de-excitation of the atom, in which another L shell electron is ionized. It is ejected with an energy equal to the difference between its ionization energy and the original energy to be dissipated:

$$(E_K - E_{L_1}) - E_{L_2} = E_{\text{Auger}} \tag{2}$$

Note that subscripts have been added to the L-shell electrons to distinguish between them.

The electrons emitted as part of this de-excitation process are the Auger electrons, and since their energy depends on the energy levels of the parent atom, their appearance is therefore a "fingerprint" indicating the presence of the parent atom.

There are several additional points that should be made concerning Auger electrons. First, transition combinations other than KLL are possible; LMM and MNN combinations are encountered for heavier atoms. Figure 10.4 shows Auger electron energies for about the first half of the elements in the periodic table and indicates the states involved in the transition. A second thing to note is that more combinations are possible in the three-electron Auger process for higher-order transitions, which makes the resulting spectra more complex. Since x-ray fluorescence and Auger emission are competitive processes for energy dissipation, the probability of one occurring increases as the probability of the other decreases. X-ray emission is more probable for heavier elements, so the yield of Auger electrons is greater for lighter elements. Note that there is a lower limit, however, since hydrogen and helium have no L-shell electrons and do not show Auger emission; lithium with only one L-shell electron shows an Auger peak through a cooperative mechanism involving more than one atom.

The potential for using Auger electrons in analytical chemistry becomes apparent when we recognize that in this phenomenon we have a signal which is characteristic of an element; thus the location of a peak

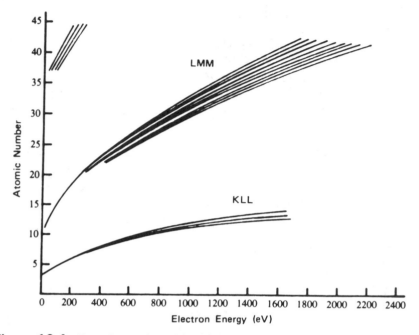

Figure 10.4   Experimental energies of Auger electrons for atoms of different atomic number through the first half of the periodic table. Various families of emissions are classified as to the shells involved. (Data from P. W. Palmberg, G. W. Riach, R. E. Weber, and N. C. MacDonald, *Handbook of Auger Electron Spectroscopy*, Physical Electronics Industries, Edina, Minn., 1972.)

offers the possibility of qualitative analysis, and the area under the peak the possibility of quantitative analysis. Thus we arrive at the possibility of Auger electron spectroscopy (AES). One question that comes to mind, however, is why use this relatively exotic phenomenon when an assortment of other types of spectroscopy are already available? While other types of spectroscopy are indeed useful and a variety of methods complement one another, AES is widely used for surface analysis. What makes AES so suitable for this sort of application is the shallow escape depth of Auger electrons from solid surfaces. Only those electrons originating within two or three atomic layers of the surface are able to escape without suffering the sort of inelastic interactions that result in the broad secondary peak.

It is apparent that there is a close similarity between SEM and AES. In both cases a surface is bombarded with electrons under vacuum and the resulting shower of emitted electrons is collected. In SEM this signal is used to produce an image; in AES analysis of the emitted electrons with respect to energy is necessary. Of several possible energy analyzers, the cylindrical mirror analyzer is one of the most widely used. Commercial scanning AES instruments are available that use this type of analyzer and which scan the surface with a beam size as low as 25 nm in diameter. These instruments are frequently designed in such a way as to allow more than one method of analysis to be used. Several techniques which are widely used along with AES are known by acronyms of their names: UPS, XPS, and SIMS. These spectroscopic methods are briefly summarized in the following:

1. Ultraviolet photoelectron spectroscopy (UPS): Photoelectrons are ejected and analyzed to provide information about the valence electron structure of surface atoms. The kinetic energy of the ejected electrons is the difference between their binding energy and the ionizing source, namely, low-energy ultraviolet radiation.
2. X-ray photoelectric spectroscopy (XPS): This is similar to UPS, except that x-rays are the source of the ionization and the properties of inner shell electrons are probed by this method.
3. Secondary-ion mass spectroscopy (SIMS): A surface is bombarded with ions with energies on the order of 1 keV and the ions ejected are analyzed with respect to mass. The method reveals clusters of atoms which are bonded together at the surface.

In addition, it should be noted that AES is frequently used in conjunction with low-energy electron diffraction (LEED). We shall return to this last technique in Sect. 10.7, but we shall not discuss UPS, XPS, or SIMS any further.

One obstacle in the development of AES was the unfavorable signal-to-noise ratio of the Auger peaks located as they are in the same part of the electron spectrum as the secondaries. In the late 1960s a means for overcoming this difficulty became available when instrumentation was developed which permitted the electronic differentiation of the electron energy distribution. It is easy to understand how differentiation of a distribution like that shown in Fig. 10.1 leads to a sharpening of the Auger peaks. The secondary peak which partially obscures them is so broad that it changes relatively slowly in the vicinity of an Auger peak. Electronic differentiation yields a signal which measures the slope of the original signal; hence a relatively constant value is produced for the secondaries. This becomes the baseline for the AES. Against this background the Auger electron peak first increases, then goes through a maximum, and finally decreases again. In the differentiated version this corresponds to a positive signal passing through zero and then going negative before returning to the baseline. Most Auger spectra are presented in this manner; the ordinate in this type of representation is frequently labeled dN/dE, reflecting the fact that it is literally the derivative of a plot such as Fig. 10.1.

Figure 10.5 shows Auger electron peaks for silicon displayed in both the differential and integral modes. Something that should be noted in these spectra is the fact that the peak location and shape are sensitive to the chemical environment of the atom. In this case the larger the difference in electronegativity between the bonded atoms, the lower the energy at which the Auger peak is found. In differential spectra it is the peak-to-peak height that is taken as a quantitative measure of the amount of material present, just as the area under a peak is used in the case of integrated data. If the Auger peak is symmetrical, the peak-to-peak height

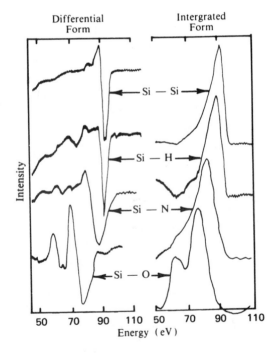

Figure 10.5 Auger electron peaks for pure silicon and silicon in different chemical environments. The peaks are shown in both differential and integrated form. [Reprinted with permission from H. H. Madden, *J. Vacuum Sci. Technol.*, *18*:677 (1981).]

is directly related to the integral of the peak. The latter method of quantification is more accurate, since it can be applied even when the shape of the peak differs between a sample and a calibration standard as is the case for the data in Fig. 10.5. The Si–O situation in Fig. 10.5 is clearly better suited to quantification by area because of the low-energy peak which would not be included if the peak-to-peak height were used. Most contemporary Auger spectrometers include dedicated micro-computers so that background stripping may be used to flatten the baseline. Spectrum stripping is clearly desirable when the Auger spectra of several elements overlap.

Quantitative use of Auger spectra ultimately involves comparing the Auger signal of an unknown sample with a standard. Although samples of pure elements might be used as standards, it is more suitable to use known materials whose composition is similar to the test specimen. There are a variety of reasons for this, including the shape variation of the Auger peak which we have already noted. In addition, the density of the material affects the escape distance and the backscattering of the electrons. Of course, any comparison between unknowns and standards must be done under identical instrument conditions.

Now that we have seen the physical and instrumental basis for AES, let us consider some of the features of solid surfaces that are revealed by this type of spectroscopy. This is the topic of the next section.

10.5 AES Applied to Solid Surfaces

One of the first observations that we note is that supposedly clean surfaces are very often contaminated with carbonaceous material. In most instances this originates in the laboratory environment and collects at the surface through adsorption. If the sample has been prepared outside of the vacuum chamber, this layer may be many molecules thick; even surfaces cleaved in the UHV system may be contaminated by submonolayer quantities of adsorbed carbon-containing molecules owing to the ubiquitous nature of these substances. Alternatively, trace impurities in a solid specimen may collect by diffusion at grain boundaries within the solid.

The correlation between embrittlement of solids and the accumulation of impurities at grain boundaries is one of the first problems that was investigated by AES. In these studies samples are fractured in the UHV chamber and the fracture surface is examined by AES for evidence of surface enrichment of trace impurities. Some early research along these lines—motivated, perhaps, by the extreme brittleness of the tungsten filaments used in electron guns—examined whether the fracture surfaces of tungsten contained trace amounts of oxygen atoms concentrated at grain boundaries. Instead of oxygen, it was found that the grain boundaries were populated to a significant extent by phosphorous atoms forming a layer a few atoms thick. These data suggest that phosphorus may be responsible for the embrittlement of tungsten, a surprising result, since no phosphorus was detected in the bulk material. This is another example—reminiscent of monolayer phenomena discussed in Chap. 7— of a situation in which an amount of material that may be negligible in terms of bulk phases has a profound effect at a surface. Similar studies on steel containing chromium, nickel, and antimony showed enrichment of antimony in the grain boundaries where fracture occurs. Concentrations of chromium and nickel were also higher at these boundaries than in bulk, so the possibility of their contributing to embrittlement cannot be ruled out.

Another interesting application of AES and other surface spectroscopies is the study of the surface composition of alloys. It has been observed, for example, that binary metal alloys display quite a different composition from that of the bulk phase. The Gibbs equation, Eq. (7.44), provides a thermodynamic basis for this. Furthermore, studies of monolayers on liquids and low-pressure gas adsorption on solids prepare us to think in terms of two-dimensional phases with properties which parallel yet differ from those of bulk phases. Since it is the surface of a sample that makes contact with the surrounding environment, it is evident that something like the corrosion behavior of an alloy may be more sensitive to surface than bulk composition. In this type of work ratios of the Auger peak-to-peak heights for two different elements are taken to equal the atom ratio in the surface zone probed by the Auger method. Figure 10.6 shows the result of one such study conducted on the silver–gold system. In the figure the mole fraction of gold in the surface is plotted along the ordinate, and the mole fraction of gold in the bulk alloy along the abscissa. The 45° line corresponds to a situation in which both surface and bulk have identical compositions. The experimental points reveal that the surface is actually enriched in silver. Incidentally, some of the data points shown were determined by ion scattering spectrometry; the fact that these and the points determined by AES agree corroborates the validity of the approach.

Figure 10.6 strongly reminds us of certain phase diagrams for liquid–vapor and solid–liquid systems. These latter phase diagrams are figures in which the pressure is held constant and temperature is plotted on the ordinate and mole fraction on the abscissa and two lines enclose a lens-shaped region in which two phases—liquid and vapor or solid and liquid, depending on the figure—are present. The following example pursues the

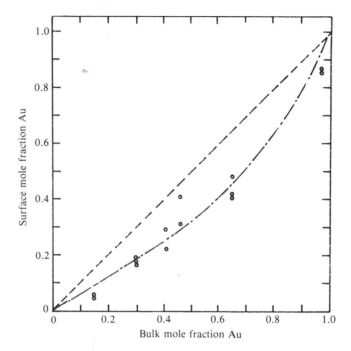

Figure 10.6 Mole fraction of gold in the surface and bulk phases of a silver-gold alloy. (Data from Ref. 9.)

resemblance between this type of bulk phase diagram and a surface composition diagram like that shown in Fig. 10.6.

Example 10.2 It has been suggested that the temperature versus composition phase diagram for the solid–liquid equilibrium of binary systems may be useful to predict which component will be in excess at a solid surface of the same binary system. Indicate how this sort of prediction might be made on the basis of the bulk phase diagram. Suggest why solid–liquid rather than liquid–vapor diagrams are chosen for this purpose.

Solution A slice across a solid-liquid phase diagram at some constant temperature will establish a tie line between the solid and liquid phases in equilibrium at the temperature chosen. The composition coordinates of the ends of the tie line will give the composition of the two equilibrium phases. If the liquid phase is richer in, say, A than B, compared to the

Table 10.1 Identity of the Metallic Element Concentrated at the Surface for Various Binary Alloys with a Brief Description of the Phase Diagram of the Alloy

Alloy	Concentrated component	Phase diagram
Ag–Pd	Ag	Simple
Ag–Au	Ag	Simple
Au–Pd	Au	Simple
Ni–Pd	Pd	Simple
Cu–Ni	Cu	Miscibility gap
Au–Ni	Au	Miscibility gap
Au–Pt	Au	Miscibility gap
Au–In	In	Complex
Al–Cu	Al	Complex
Pt–Sn	Sn	Complex
Fe–Sn	Sn	Complex
Au–Sn	Sn	Complex

Source: Data collected by Somorjai [9].

solid, this suggests that the surface of a solid A–B alloy might also be richer in A than B, compared to the bulk. Conversely, if the bulk liquid is richer in B than A, the surface is also predicted to be richer in B than A compared to the bulk.

It is more appropriate to use solid–liquid phase diagrams for this sort of prediction than liquid–vapor, since interatomic distances and hence atomic interactions are more similar when surface and bulk phases are compared to condensed solid and liquid phases than to liquid and low-density vapor phases. When we make this comparison, the surface is assumed to be more similar to the bulk liquid, since surface molecules are generally more disordered and are surrounded by a smaller number of neighboring atoms than is the case for bulk atoms.

•

The method suggested in this example for predicting which component tends to concentrate in the surface has been found to give satisfactory agreement with experiment, even in situations in which the bulk phase diagrams are fairly complicated. Table 10.1 lists the component which shows enrichment for several binary alloys, including several for which the phase diagram shows some complexity.

Probably no other area has utilized the capabilities of AES as extensively as the semiconductor industry. The explosive growth of microelectronics has created a high demand for ultrapure semiconducting materials. Surface cleanliness is crucial in the semiconductor industry, and the use of AES to monitor for possible contamination has made the latter an indispensible tool in the manufacture and processing of semiconductor devices. In this context the combination of AES and ion sputtering to give a depth profile of a surface is particularly important.

We have seen that it is the shallow escape depth of Auger electrons that makes AES so valuable as an analytical tool for surface science. We also saw in Sect. 10.2 that sputtering a surface strips away layer after layer of surface atoms. When these two techniques are combined, the Auger spectrum of successive layers may be determined to give a picture somewhat reminiscent of Fig. 7.11 showing how the chemical composition varies with distance away from the surface. Argon ions are most widely used for sputtering, and a focused ion beam is accelerated to energies as high as 5 keV for bombarding the surface. Alternating periods of AES measurement and sputtering allows the profile of chemical composition with depth to be monitored.

Figure 10.7 is an example of such a depth profile measured on a faulty integrated circuit. The surface shown in the figure was obtained in the following way. "Windows" were etched into the SiO_2 layer on the surface of ultra-high-purity silicon. Next, an aluminum film was deposited to make electrical contact with the silicon. This step sometimes results in faulty devices with unacceptably high resistances at the point of contact. Accordingly, the depth profile study was undertaken in an attempt to discover the source of the problem. The ordinate in Fig. 10.7 is a percentage scale and the abscissa is the sputtering time. In independent studies the depth of surface erosion was determined as a function of time, and the regions blocked out along the top of the figure indicate the approximate depth of the various zones from the initial surface. Starting at zero time of sputtering—that is, at the initial surface—let us follow along the profile shown in Fig. 10.7:

1. The most abundant element at the surface is carbon, an observation consistent with the earlier generalization about the nearly universal contamination of surfaces by carbonaceous materials. This is probably hydrocarbon matter; remember, hydrogen is not detectable by AES.
2. Oxygen is the second most abundant species near the surface. Both carbon and oxygen concentrations drop off with increasing penetration of the surface, with the carbon disappearing somewhat sooner than the oxygen. Most of the oxygen is presumably present in the

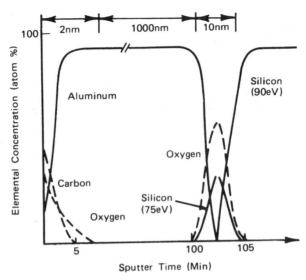

Figure 10.7 Depth profile of failed integrated circuit surface showing the presence of SiO_2 at the junction of aluminum and silicon. (Reprinted with permission from S. I. Ingrey, *Scanning Electron Microscopy*, SEM Inc., AMF O'Hare, Chicago, 1982.)

form of water and surface oxide. By the time a depth of about 2 nm is reached, the last of the oxide has been sputtered away.

3. There next follows a long plateau in which pure aluminum is recorded. This might be a good time to note that sputtering is a random process and does not cleanly remove one layer at a time. Instead, a sampling of several atomic layers is explored by AES, so the figures are average compositions over a few adjoining layers.

4. Next we encounter a region some 10 nm thick in which both silicon and oxygen are detected. Note that the silicon peak monitored here is the 75-eV peak which, according to Fig. 10.5, is the location of Si in the presence of oxygen. This indicates a layer of SiO_2 at this depth.

5. Finally, we enter a region of pure silicon, now monitored by a 90-eV peak (see Fig. 10.5 again)

This depth profile makes it clear why the resistance of the aluminum–silicon junction was unacceptably high: Not all of the SiO_2 initially covering the Si was removed when the window was etched for the aluminum. Numerous other examples of this sort could be cited, but this should suffice to give an indication of the capabilities of this analytical technique.

In Sect. 10.7 we shall take up the use of low-energy electron diffraction to determine the crystal structure of the surface layer and related properties. Before turning to this, however, it will be helpful to review some crystallographic concepts pertaining to bulk solids. This is the topic of the next section.

10.6 Some Crystallographic Concepts

Right along with colloid and surface chemistry, crystallography is one of those topics that is frequently slighted in general and physical chemistry courses. To correct for this possibility or at least to review some little-used information, we shall summarize a few useful ideas from crystallography in this section. Readers desiring a more detailed presentation will find the subject treated in most physical chemistry texts. Our main objective is to review the notion of crystallographic planes and their designation, since much surface spectroscopy is conducted on specified crystal faces. While doing this, however, we shall encounter some additional facts pertaining to the orderly packing of atoms in crystals.

A crystal is an orderly array of atoms or molecules but, rather than focusing attention on these material units, it is helpful to consider some geometrical constructs that characterize its structure. It is possible to describe the geometry of a crystal in terms of what is called a unit cell: a parallel-piped of some characteristic shape that generates the crystal structure when a three-dimensional array of these cells is considered. We then speak of the point lattice defined by the intersections of the unit cells upon translation through space. Note that the edge lengths and angles between faces in these unit cells have not been specified; when the former are equal and the latter are equal 90°, the unit cell is a cube. There are seven different unit cell shapes, including triclinic, hexagonal, trigonal, and so on, and all of them satisfy the very important requirement of being space filling upon translation. The lattice points need not be occupied by atoms but, when the positions of atoms within the unit cells are considered, it has been shown that only 14 possible lattice types exist. A cubic unit cell, for example, forms a primitive cubic structure when it contains an atom at each corner; body-centered and face-centered cubic structures retain the geometry of the primitive structure but contain additional atoms in the positions indicated. In two dimensions the equivalent of a unit cell is called a unit mesh, and a net is the two-dimensional equivalent of a lattice. Only four different two-dimensional unit meshes are possible.

Figure 10.8a shows a set of lattice points in one plane of a cubic structure. If this is the xy plane, identical layers would be stacked in the z

direction to generate the crystal. Each of the dots in the figure, then, is the end view of a row of points in the z direction. Several sets of parallel lines connecting various lattice points have been drawn in the figure. These may be regarded as the edges of planes slicing through the crystal. Much of our discussion in this section is based on cubic unit cells for simplicity, but the results are quite general. It also turns out that many metals crystallize with cubic unit cells.

The sets of numbers labeling the lines (i.e., planes) in Fig. 10.8 are called the Miller indices—represented hkl—for that plane. The easiest way to remember the significance of these indices is to note that h, k, and l count the number of planes (of type hkl) crossed in moving from one lattice point to the next in the x, y, and z directions, respectively, for cubic cells. In general, it is the axis system of the unit cell that is used in this description rather than the mutually perpendicular x, y, and z axes. Thus the 110 plane in Fig. 10.8a slices through the crystal in such a way that one of these planes is crossed in moving from one lattice point to the next in the x direction and one is crossed in moving from point to point in the y direction. In the z direction the movement is parallel to the plane, not through it, so the Miller index is zero in the z direction for this plane. Any plane in which one Miller index is zero is parallel to one of the axes of the crystal; likewise, if two Miller indices are zero, the plane includes two of the crystal axes. Figure 10.8b shows the 100, 110, and 111 planes drawn through a cube.

A more general definition of the Miller indices is that the hkl plane intersects the A axis at a/h, the B axis at b/k, and the C axis at c/z, where a, b, and c are the lengths of the unit cell along the A, B, and C axes of the crystal. A plane that includes an axis intersects that axis at ∞; hence its index in that direction is zero.

Before we leave Fig. 10.8a, there are two additional points to be made concerning the spacing and atom density in the various planes. The first thing to note for the simple array of points in Fig. 10.8a is that the actual spacing of the lattice points is exactly the same in all directions, the length of the cube edge D. Each of the various sets of planes that can be drawn through this lattice has its own characteristic spacing, say, d_{hkl} for the hkl plane. What makes this significant is that in diffraction studies x-rays are reflected off the various crystallographic planes and constructive interference occurs when the spacing between the diffracting surfaces d_{hkl} and the angle of incidence satisfy the Bragg equation,

$$n\lambda = 2d_{hkl} \sin \theta \tag{3}$$

where λ is the wavelength of the x-rays and n is the order of the diffraction, $n = 1, 2, 3,\ldots$. The Bragg equation is discussed in connection

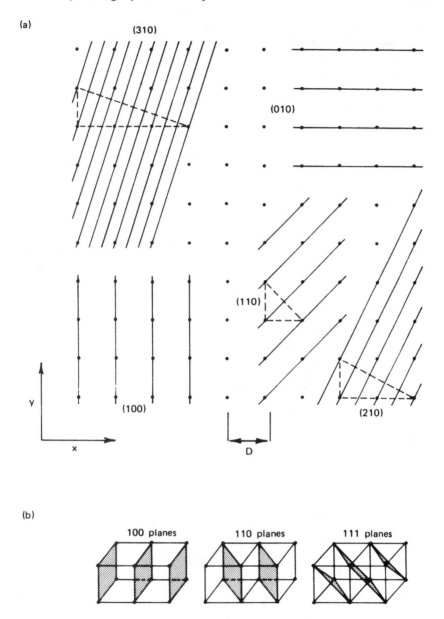

Figure 10.8 (a) Some crystallographic planes through a simple cubic lattice in which the atomic spacing is *D*. The numbers in parentheses are the Miller indices of the planes shown. (b) The 100, 110, and 111 planes through a simple cubic lattice.

with microscopy in Sect. 1.5. The point here is this: First-order ($n = 1$) diffraction with monochromatic x-rays (λ = const.) results in reinforcement at different angles for different reflecting planes, even when the actual lattice is characterized by a single distance D. Our interest is in the relationship between D and the spacing d_{hkl} for planes with those indices. This is explored numerically in the following example for the planes in Fig. 10.8a.

Example 10.3   For each of the sets of planes shown in Fig. 10.8a, a right triangle has been drawn in which the hypotenuse is perpendicular to the planes. Calculate the distance between the planes from geometry and compare the results with distances calculated by the formula $d_{hkl} = D/(h^2 + k^2 + l^2)^{1/2}$.

Solution   For the 100 and 010 planes, the spacing is D by inspection. For these planes, $(h^2 + k^2 + l^2)^{1/2}$ equals unity, so the formula gives the correct spacing for these planes. Because of the simple geometry involved, the separation is readily calculated by the Pythagorean theorem for the remaining planes. The length of the hypotenuse must be divided by the number of planes crossed to give a spacing which may then be compared with the distance calculated by the formula given. Calculated results are listed below:

Plane	Length of hypotenuse	Number of planes crossed	Distance of separation	$\dfrac{D}{(h^2 + k^2 + l^2)^{1/2}}$
110	$(D^2 + D^2)^{1/2} = \sqrt{2}D$	2	$\sqrt{2}D/2$	$D/\sqrt{2}$
210	$(4D^2 + D^2)^{1/2} = \sqrt{5}D$	5	$\sqrt{5}D/5$	$D/\sqrt{5}$
310	$(9D^2 + D^2)^{1/2} = \sqrt{10}D$	10	$\sqrt{10}D/10$	$D/\sqrt{10}$

These calculations show that the distance of separation between planes is simply related to the characteristic spacing of the lattice and the Miller indices of the planes. By continued application of the Pythagorean theorem, the generality of this result for planes that do not include the axes can be demonstrated.

•

All of the lines in the x-ray diffraction pattern of a cubic crystal satisfy the Bragg equation with distances that are simple fractions of the characteristic distance D.

By examining the diffraction pattern of a cubic crystal, it is possible to to evaluate D, to determine the values of h, k, and l responsible for the observed pattern, and to decide whether the pattern is compatible with unit cells which are primitive cubic, face-centered cubic, or body-centered cubic. This is easy to see by imagining some variations on the primitive cubic structure shown in Fig. 10.8b. For example, placing an atom in the center of each face of the cube converts the primitive structure to a face-centered cubic and promotes the 200 planes (among others) to the status of reflecting planes instead of the 100 planes which characterize the primitive structure. Similarly, 200 planes are characteristic of body-centered structures. It is 110 planes that pass through the central atom in body-centered cubic crystals, however, and 220 planes that include the face atoms of the face-centered cubic structures. This shows that different sets of planes (hence different hkl values and different diffraction patterns) characterize primitive, face-centered, and body-centered cubic structures. Among various other differences that exist among their diffraction patterns, evidence of the following planes in diffraction patterns enables us to decide whether a cubic crystal is primitive, face centered, or body centered:

1. Primitive cubic: 100 and 110
2. Face-centered cubic: 200 and 220
3. Body-centered cubic: 200 and 110

A second result that is evident from an inspection of Fig. 10.8a is the fact that planes with lower Miller indices have higher atomic densities. Those planes characterized by high Miller indices strike out across the lattice at acute angles and cross several planes of atoms before actually intercepting an atom. This makes the average distance between atoms greater in planes which have high Miller indices. The surface tension is less in planes of low Miller index than in planes of higher index, where the interatomic spacing is larger. If it were possible to deform the former surface into the latter, work would have to be done against the attractive forces between atoms to increase this separation. We may think of this work as an increment in surface free energy in the planes of high index which is not present when the Miller indices are lower. Surfaces of lower γ and hence lower Miller index are thermodynamically more stable. Surface studies conducted on the specific face of a crystal generally involve the 100, 110, or 111 plane for this reason.

A primitive cubic structure is generated by packing spheres in the three-dimensional equivalent of Fig. 10.8a. The resulting crystal is a very open structure with 52% of the total volume occupied by the spherical

(a)

(b)

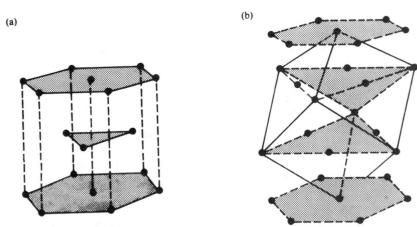

Figure 10.9 Hexagonal close packing of spheres: (a) ABABAB pattern and (b) ABCABC pattern. The cube in part (b) shows that hexagonally close-packed planes are 111 planes in a face-centered cubic structure.

particles. More efficient packing is obtained by considering a hexagonally close-packed arrangement of spheres rather than a cubic array as the basis for a crystal structure.

In a hexagonally close-packed layer of spheres each particle is surrounded by six neighboring spheres in the same plane, thereby accounting for the hexagonal symmetry of the base plane. Another hexagonally close-packed layer may be deposited on the first by nesting the second layer in the depressions of the first layer. The atom that we regarded as central in the first layer thus acquires three more neighbors in the second layer, each nesting in alternating cusps of the first layer. If the first layer had actually been deposited on still another layer, the central atom in a hexagonal array would be nested in a cluster of three spheres in the underlying plane so that it would have a total of 12 nearest neighbors.

As noted above, it is alternating depressions in the hexagonal array that are filled when a second layer is deposited. This introduces two possibilities for the third layer. If the atoms in the third layer are positioned directly above those in the first layer as shown in Fig. 10.9a, the resulting structure is called the hexagonally close-packed structure. Successive layers follow the pattern ABABAB ... in this structure. Alternatively, the third layer could be placed in the *other* set of three cusps in the second layer, generating the pattern ABCABCABC ... shown in

Fig. 10.9b. Both of the structures shown in Fig. 10.9 are based on hexagonally close-packed layers and both fill space with 74% efficiency, the most efficient packing possible for uniform spheres. These structures have different symmetries, however, because the third layer is positioned differently in the two cases.

In Fig. 10.9b, a cube has been superimposed on the hexagonally close-packed layers with the ABC pattern of layering. This cube shows that the close-packed layers are 111 planes in the cubic structure. In addition, it is seen that this corresponds to a face-centered cubic arrangement in the unit cell. This face-centered arrangement is also called cubic close packing.

Body-centered cubic structures are not based on hexagonally close-packed layers. Accordingly, the packing efficiency is less; only 68% of the space is filled in body-centered cubic structures.

With spherical atoms as their packing units, it is not surprising that many metals form crystals of the type we have considered here. For some metals more than one crystalline phase may be stable, depending on the conditions of pressure and temperature. Keeping this in mind, we note that the following structures are formed by the metals indicated:

1. Hexagonally close packed: Be, Ca, Cd, Co, Cr, Mg, Ti, and Zn
2. Cubic close packed: Al, Ag, Au, Ca, Cu, Ni, Pb, and Pt
3. Body-centered cubic: Ba, Cr, Cs, Fe, K, Li, Mo, Na, and W

The ideas of this section make it clear why the 111 plane of Pt (plane of high atomic density in a face-centered cubic crystal and hence thermodynamically stable) and the 110 plane of W are examples of crystal faces that have been extensively studied.

10.7 Low-Energy Electron Diffraction

Just as AES yields information regarding the chemical composition of surfaces, low-energy electron diffraction (LEED) provides insight into their physical structure, particularly crystallographic features as might be expected from a diffraction technique. In LEED a beam of low-energy electrons rather than x-rays are used to form the diffraction pattern, but otherwise many of the concepts, relationships, and vocabulary are based on x-ray diffraction. Accordingly, our discussion of LEED includes a review of pertinent aspects of this topic. Since diffractometry can get quite involved, we shall tailor our review to those subjects most helpful in getting us started and leave more advanced concepts for further study.

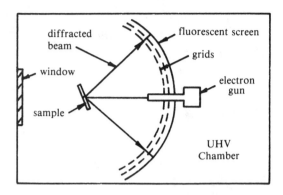

Figure 10.10   Schematic illustration of a LEED apparatus.

The de Broglie concept of wave–particle duality enables us to calculate the wavelength of an electron. According to the de Broglie equation,

$$\lambda = \frac{h}{p} = \frac{h}{[2m(KE)]^{1/2}} \tag{4}$$

in which h is Planck's constant, and the momentum of the electron, $p = [2m(KE)]^{1/2}$, can be calculated in terms of its mass and the accelerating voltage which accounts for its kinetic energy. A straightforward application of this formula enables us to calculate that a voltage of 150 V gives electrons a wavelength of 0.1 nm, the order of magnitude of interatomic spacings in crystals:

$$\lambda = \frac{6.63 \times 10^{-34} \text{ J s}}{[2(9.11 \times 10^{-31} \text{ kg})(1.6 \times 10^{-19} \text{ C})(150 \text{ V})]^{1/2}} = 10^{-10} \text{ m} = 0.10 \text{ nm}$$

Thus the electrons used in LEED are usually in the range 10–300 eV, lower in energy by one to two orders of magnitude than those used in SEM and AES. C. J. Davisson (Nobel Prize, 1937), working with a series of collaborators, is generally credited with the first application of electrons to diffraction studies.

What makes LEED of particular interest in the study of surfaces is the fact that low-energy electrons such as these are only able to penetrate and re-emerge by diffraction from a few atomic thicknesses of the target material. Even at that, the diffracted beam acquires a background of secondary electrons picked up through the types of inelastic interactions we have discussed earlier in this chapter. Figure 10.10 is a schematic of a LEED apparatus. Note that several grids are interposed between the specimen and the viewing screen. One of these is adjusted in potential so that only electrons of the original energy—those elastically diffracted—are allowed to pass; secondary electrons of lower energy are collected. The diffracted beam is accelerated by a final, positively charged grid to produce an image on the fluorescent screen. The experimental aspects of LEED resemble the other techniques we have discussed in this chapter in several ways:

1. UHV is required both because electron beams are involved and to assure surface cleanliness.
2. Ability to cleave samples and then clean, anneal, and align the surfaces under vacuum is essential.
3. Various supplementary experiments are often conducted along with LEED, including AES and high-energy electron diffraction (HEED), in which the higher energy of the electron is offset by a lower angle of incidence upon the surface.

Although UHV is required for LEED measurement, there is considerable interest in applying this technique to surfaces which carry adsorbed species. In view of our discussion of adsorption equilibrium in the last chapter, there is a difficulty here, since adsorbed molecules imply an equilibrium gas phase. One way around this problem is to study surfaces at a sufficiently low coverage that the equilibrium gas pressure is compatible with the LEED technique. When higher pressures are desired, the surface is first equilibrated, then the excess gas is pumped out, and the surfaces before and after adsorption are compared through LEED. Chemisorption is better suited for study by this method than physical adsorption because the adsorbed layer remains intact when the equilibrium gas is removed.

Generally, LEED experiments are conducted on specified faces of single crystals. When this is done, the diffraction pattern produced consists of a series of spots whose location, shape, and intensity can be interpreted in terms of the surface structure. We shall focus attention on what can be learned from the location and shape of the spots, since the study of intensity is beyond the scope of this book. It is generally assumed that the surface examined by LEED is an extension of an already-known bulk crystal structure. The correctness of this assumption can be tested and results are often expressed in terms of modifications of the three-dimensional structure at the surface. Before we turn to the LEED patterns shown in Fig. 10.12, we must first figure out how they are read.

The Bragg equation is the central relationship in diffractometry; we considered the standard geometrical derivation of this equation in Sect. 1.5 [Eq. (1.5)]. Using the notation of the last section and writing the spacing of the diffracting planes d_{hkl}, the Bragg condition for the diffraction of particles of wavelength λ is given by

$$2d_{hkl} \sin \theta = \lambda \tag{5}$$

for first-order diffraction. For higher orders of the diffraction the right hand-side of Eq. (5) should be multiplied by the integer characterizing the order. Because of its importance and to introduce another important notion—the idea of reciprocal space—a slightly different approach to the Bragg equation will be useful. Although the present discussion involves vector notation and some very elementary vector operations, the reader should not be intimidated: We shall not pursue this potentially complex topic far enough to be troublesome!

In Fig. 10.11a we represent the diffraction of electrons from a family of lattice planes by the vectors $\bar{k}_0$ and $\bar{k}$ representing the incident and diffracted beams, respectively. We define these vectors to be equal to each other in magnitude—following standard practice, the magnitude is represented $|k| = |k_0|$—and to have a magnitude equal to the reciprocal of the wavelength of the diffracted electrons:

$$|k| = |k_0| = \tfrac{1}{\lambda} \tag{6}$$

The fact that the incident and diffracted beams are represented by vectors of equal length is appropriate, since the diffraction is the result of an elastic reflection off the lattice planes. The diffraction planes are characterized by the Miller indices hkl and are separated by a distance d_{hkl}. The angle of incidence of the primary beam is θ and the angle between the vectors is 2θ, as seen from the figure. Finally, by a

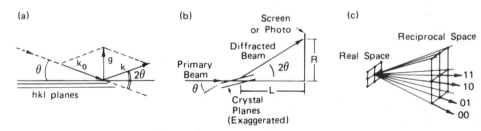

Figure 10.11 (a) Vector diagram showing the Bragg condition for diffraction from planes of Miller index hkl. (b) Interaction of a diffracted beam with a photographic plate for small angles of incidence. (c) Illustration of the indexing of points in reciprocal space relative to the primary beam, labeled 00.

parallelogram construction, we define another vector $\bar{g}$; it is apparent from the figure that $\bar{g}$ is given by the vector equation

$$\bar{g} = \bar{k} - \bar{k}_0 \tag{7}$$

Note that the diffracting surface is the origin of vector $\bar{g}$; we shall see presently that following $\bar{g}$ takes us into a different kind of space, something called reciprocal space.

In order to appreciate the significance of the vector $\bar{g}$, it is useful to carry out the following operations:

1. Multiply both sides of Eq. (7) by $\bar{g}$, using the dot product:

$$\bar{g} \cdot \bar{g} = \bar{g} \cdot (\bar{k} - \bar{k}_0) = (\bar{k} - \bar{k}_0) \cdot (\bar{k} - \bar{k}_0) \tag{8}$$

    The second version of this equation makes use of the definition of $\bar{g}$ provided by Eq. (7).

2. Since the dot product of a difference is formed by term-by-term multiplication, Eq. (8) becomes

$$\bar{g} \cdot \bar{g} = \bar{k} \cdot \bar{k} + \bar{k}_0 \cdot \bar{k}_0 - 2\bar{k} \cdot \bar{k}_0 \tag{9}$$

3. Since the dot product of two vectors is the product of their magnitudes times the cosine of the angle between the two, Eq. (9) may be written

$$|g|^2 = |k|^2 + |k_0|^2 - 2|k||k_0| \cos 2\theta \tag{10}$$

4. Substituting Eq. (6) into Eq. (10) yields

$$|g|^2 = 2(\tfrac{1}{\lambda})^2(1 - \cos 2\theta) \tag{11}$$

5. Next we use the trigonometric relationship $2 \sin^2 \theta = 1 - \cos 2\theta$ to write

$$|g|^2 = 4(\tfrac{1}{\lambda})^2 \sin^2 \theta \tag{12}$$

    or

$$|g| = 2(\tfrac{1}{\lambda}) \sin \theta \tag{13}$$

Equation (13) becomes identical to the Bragg equation, Eq. (5) for first-order diffraction, if the magnitude of $\bar{g}$ is identical to d_{hkl}^{-1}, the reciprocal of the spacing of the diffracting planes.

Although Eq. (13) is a plausibility argument rather than a general proof, it does serve to introduce us to the idea of reciprocal space. In Fig. 10.11a the diffracting planes and the incident and diffracted beams exist in real space. Vector $\bar{g}$ is a mathematical construct, and following it leads us into the world of reciprocal space; the distance that we travel into this space following $\bar{g}$ is a measure of $1/d_{hkl}$. The inverse relationship between distance in reciprocal space and actual physical distance is the reason the former is called *reciprocal* space. What makes this significant for LEED is the fact that the photographic image of the LEED pattern may be regarded as a map of this reciprocal space. In practice, there may be more than one surface structure compatible with a LEED pattern, and intensity considerations as well as spot locations must be considered to prove a structure. We shall ignore this complication.

The foregoing discussion is rather absract, but we can obtain a "feel" for its significance by considering Fig. 10.11b. This figure represents the same situation shown in Fig. 10.11a except that the path of the diffracted beam is followed to the surface, where the LEED pattern is observed. The diffracting planes are shown in greatly exaggerated size in this version and the angle of incidence is purposely drawn to be exceptionally small. Small angles like these are more typical of HEED than LEED, but the argument that follows will help clarify the idea that the LEED pattern is a map of reciprocal space. This is true even though the approximations we make are fairly bad for an actual LEED experiment.

In Fig. 10.11b the diffracted beam produces a spot on the photographic plate a distance R from the point where the primary beam strikes. From trigonometry, it is evident that $\tan 2\theta = R/L$. For the small angles considered here $\tan x \simeq x$; therefore $2\theta \simeq R/L$, or $\theta \simeq R/2L$. For small angles, $\sin x \simeq x$ also; therefore $\sin \theta \simeq R/2L$ for the situation shown in the figure. Combining this small-angle approximation with Eq. (5) to describe the Bragg condition for first-order diffraction, we obtain

$$\lambda = d_{hkl} \frac{R}{L} \qquad\qquad (14)$$

Since d_{hkl} and λ are constant, Eq. (14) predicts that R is inversely proportional to d_{hkl}, where R is the distance of the diffraction spot from the spot produced by the primary. Example 10.3 demonstrated that d_{hkl} is largest for planes of low Miller indices; since R varies inversely with d_{hkl}, it follows that spots nearest the primary spot are due to low Miller index planes. Likewise, more distant spots are due to planes of higher index. There is a reciprocal relationship between the location of the spot on the photographic plate and the separation of the planes responsible for the spot.

Finally, we note that R is measured on the photographic plate and that the plate fails to be perpendicular to the diffracting planes by the small angle θ. The closer to zero this angle is, the nearer to perpendicular are the viewing and diffracting planes. Under these same conditions the approximations involved in Eq. (14) also improve. Under these conditions there is a convergence between R, a distance on a photograph, and $\bar{g}$, a vector in reciprocal space. This is true because $\bar{g}$ is also perpendicular to the diffracting planes and has a length which measures d_{hkl}^{-1}. While this comparison breaks down for larger angles, it does illustrate the statement made earlier that the LEED pattern pictures the diffraction pattern in reciprocal space.

Since every point in reciprocal space is associated with a set of planes, it is standard to identify the points by the same index numbers as their originating planes. In LEED it is the two-dimensional array of surface atoms that is responsible for the diffraction, rather than the full three-dimensional array. Accordingly, the diffracting surfaces require only two indices. We shall use the term *net* to describe the two-dimensional array, just as we describe the three-dimensional analog by the term *lattice*. Figure 10.8a was drawn originally to represent a slice through a bulk crystal so that the lines shown are the edges of the diffraction planes. In LEED the two-dimensional section shown in that figure is the net of interest and the lines are the actual diffracting surfaces. Only two indices are needed to describe these. Note that the third Miller index in Fig. 10.8a is redundant, since all of the planes shown are perpendicular to the plane of the figure.

Figure 10.11c shows how the points in reciprocal space are identified by a pair of indices hk. The primary beam is labeled 00 and the length of $\bar{g}$ (or R in the approximation described above) is measured outward from 00. For a rectangular array of points like that shown in Fig. 10.11c, we note that $d_{11} = (d_{10}^2 + d_{01}^2)^{1/2}$. If $d_{10} = d_{01} = D$, this is the same result that we demonstrated in Example 10.3.

Table 10.2 Relationship Between the Spacing d_{hk} and the Side Lengths D and D' in the Two-Dimensional Unit Cells of Square, Rectangular, and Oblique Parallelogram Nets.

Geometry	Relationship
Square $(D = D')$	$d_{hk}^{-2} = (h^2 + k^2)/D^2$
Rectangle $(D \neq D')$	$d_{hk}^{-2} = (h/D)^2 + (k/D')^2$
Oblique, general $(D \neq D'; \alpha \neq 90°)^a$	$d_{hk}^{-2} = (h^2/D^2 \sin^2 \alpha) + (k^2/D'^2 \sin^2 \alpha)$ $- (2hk \cos \alpha/DD' \sin^2 \alpha)$
Oblique, hexagonal $(D = D'; \alpha = 120°)^a$	$d_{hk}^{-2} = 4(h^2 + hk + k^2)/3D^2$

[a]The angle between sides is α in the oblique case and equals 120° for the hexagonal element.

Just as there are seven basic shapes giving rise to 14 unit cells in three-dimensional lattices, it is possible to show that there are only three area-filling, two-dimensional shapes: square, rectangular, and oblique parallelogram. Table 10.2 shows how the inter-row spacing d_{hk} is related to the Miller indices and the side lengths a and b for these different shapes of two-dimensional unit cells.

Now that the theory and vocabulary of LEED are established, let us look at some simple LEED patterns and see what they tell us about solid surfaces.

10.8 LEED Applied to Metal Surfaces

Figure 10.12 is a set of photographs of LEED patterns on clean metal surfaces and surfaces with adsorbed gases. Several variables are in effect to produce the LEED patterns shown:

1. Parts (a) and (d) are the 100 surface of tungsten and the 111 surface of platinum, respectively. The symmetry of these patterns characterizes the cubic and hexagonal packing of the crystal faces; the spacing of the spots depends on the size of the atoms involved.
2. LEED patterns could be used to distinguish between different crystal faces.
3. Parts (a) and (d) are clean surfaces, while parts (b) and (c) are surfaces with adsorbed species present.
4. LEED patterns could be used to distinguish between clean surfaces and those with adsorption, at least in some cases.
5. Part (b) is the tungsten 100 face with adsorbed oxygen, while part (c) is the same surface with adsorbed hydrogen.
6. LEED patterns could be used to distinguish between adsorbed species, at least in some cases.

In most instances there are much simpler ways to determine the orientation of a crystal, the presence or absence of adsorbate molecules, and their nature than to use LEED. The unique power of LEED is its ability to measure order at a surface. Therefore we may state the following:

7. LEED enables us to determine the structure of the solid surface and to compare this with the bulk structure.
8. LEED allows us to observe and measure the order that exists in some adsorbed monolayers.

Item 7 is discussed somewhat further in the following example.

Example 10.4 If we accept that a LEED pattern has the same symmetry as the net of surface atoms responsible for its formation, what additional information is needed to complete the comparison between bulk and surface structures for the tungsten surface shown in Fig. 10.12a?

Solution We assume the crystal structure of the bulk metal, including the dimensions of the unit cell, are known. Comparison, then, depends on determining this quantity for the surface.

1. Taking photographic magnification into account, measure the distance of the diffraction spots from the origin, the spot produced by the primary beam.

2. These distances are multiples of the crystallographic spacing d_{hk}. The latter may be calculated by evaluating a magnification factor from the experimental geometry in the manner suggested by Fig. 10.11b.

3. The d_{hk} values so calculated are expressed in units of λ, so the wavelength of the electrons must be known to obtain absolute values for these spacings.

4. To convert d_{hk} values to the dimensions of the two-dimensional unit cell by a formula from Table 10.2, the spots in the LEED pattern must be indexed.

5. Since the pattern is a simple square in Fig. 10.12a and since the spots nearest the origin have the lowest indices, the pattern can be indexed just as rectilinear graph paper might be marked off. Negative indices are possible; these are written above the number affected. Fig. 10.13a shows how the points in Fig. 10.12a are indexed.

From the index numbers and the experimental parameters, the dimensions of the two-dimensional unit cell can be determined.

•

For the 100 face of tungsten, written W(100), it has been found that the characteristic dimension of the unit cell in the two-dimensional lattice is 6% less at the surface than in the bulk crystal, although the packing geometry remains the same in both. The change in separation without a change in symmetry for surface atoms is called relaxation and is widely observed, particularly in crystal faces with relatively low packing efficiency. For example, the contraction of the Al(110) and Mo(100) surfaces are 5–15 and 11–12%, respectively. By contrast, the high atomic density 111 surfaces of silver and platinum show no contraction and the Fe(111) surface contracts by 1.5%. This phenomenon may be explained as an attempt by surface atoms to compensate for their lower coordination

(a)

Figure 10.12 LEED patterns for clean metal surfaces and surfaces with adsorption: (a) clean W(100) surface, (b) W(100) with adsorbed oxygen, (c) W(100) with adsorbed hydrogen, and (d) clean Pt(111). [(a)–(c) reprinted with permission from Ref. 3; (d) reprinted with permission from Ref. 9.

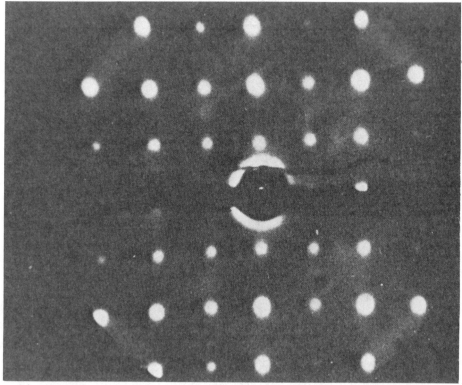

(b)

number by pulling closer to their neighbors for more effective bonding. In some instances it is observed that clean surfaces have totally different structures from what would be expected on the basis of bulk structures; this is called surface reconstruction.

One of the most interesting and important applications of LEED is in the study of submonolayer adsorption. Figure 10.12b and c is an example of this, and we shall discuss this figure in detail. First, let us consider the circumstances in which we might expect LEED to be sensitive to adsorbed species. From the perspective of approaching gas molecules, the surface of a single crystal presents an ordered array of adsorption sites. We might anticipate that this order is carried over into the adsorbed layer, at least at low coverage, where crowding and lateral interactions are not complications. Note also that this implies highly specific adsorption; this is another reason why LEED is especially valuable in the study of

(c)

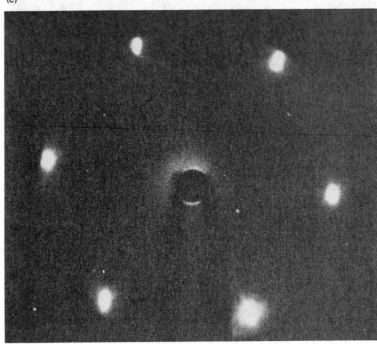

(d)

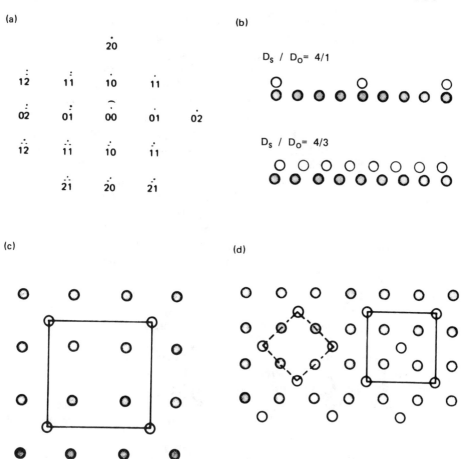

(a)

20

12   11   10   11

02   01   00   01   02

12   11   10   11

21   20   21

(b)

D_S / D_O= 4/1

D_S / D_O= 4/3

(c)

(d)

Figure 10.13 (a) Indexing of the LEED spots of W(100) pattern in Fig. 10.12a as described in Example 10.4. (b) Side view of coherent structures formed by adsorbed (open circles) species on metal surface. (c) Unit mesh for W(100) with adsorbed oxygen (open circles). (d) Unit mesh for W(100) with adsorbed hydrogen (open circles).

chemisorption. Since LEED responds to surface order, randomly adsorbed molecules such as those held by weak physical adsorption might go undetected. It was suggested at the beginning of this section that the LEED pattern of a clean surface might be taken as evidence that no adsorption had occurred. Actually, it may only indicate the absence of order in an adsorbed layer. A supplementary chemical analysis such as an AES experiment is a more definitive proof of surface cleanliness.

Figure 10.13b schematically represents a cross-sectional view of the surface of a solid and represents the topmost layer of atoms by shaded circles. The open circles represent molecules in an ordered pattern on the solid substrate. Since the adsorbed molecules are ordered, their structure on the surface is characterized by what is called a superlattice or supernet. Suppose we define D_0 as the characteristic spacing of the substrate and D_s the equivalent quantity for the supernet. Then the two arrangements in Fig. 10.13b are described the ratios $D_s/D_0 = 4/1$ and $D_s/D_0 = 4/3$. Since a LEED pattern is a picture in reciprocal space, it follows that the adsorbed layer with $D_s/D_0 = 4/1$ should produce spots whose separation is $\frac{1}{4}$ that of the substrate. Likewise, for the case where $D_s/D_0 = 4/3$, a pattern of spots whose separation is $\frac{3}{4}$ that of the substrate is predicted. Thus, if the substrate produces spots at, say, 0 and 1, extra spots would be expected at $\frac{1}{4}$, $\frac{2}{4}$, and $\frac{3}{4}$ for the $D_s/D_0 = 4$ case, and at $\frac{3}{4}$, $\frac{6}{4}$, and $\frac{9}{4}$ when $D_s/D_0 = 4/3$. The cases illustrated here are called coincident structures, since the two patterns coincide periodically. When there is no correlation between two structures, the latter are said to be incoherent.

With this background we are now in a position to make sense out of the LEED pattern shown in Fig. 10.12b. The substrate is the same W(100) surface shown in Fig. 10.12a, but in Fig. 10.12b the surface carries adsorbed oxygen. In the LEED pattern additional spots appear midway between the spots produced by the clean tungsten. The extra spot at the $\frac{1}{2}$ position means that $D_s/D_0 = 2$ for the oxygen in this experiment; that is, the dimension of the oxygen unit mesh is twice that of the unit mesh of the substrate. Figure 10.13c is a schematic illustration of a possible supernet that is consistent with this description. It should be pointed out that spot location alone does not tell us where in the unit mesh of the supernet the adsorbed oxygens are located, only that they display a periodic structure with two times the repeat distance of the solid. The relative positions of the adsorbate and substrate atoms within their respective nets may be calculated from intensity data.

This example of a supernet makes it evident that an assortment of these supernets are possible with different values of D_s/D_0. This suggests that a system of designation is needed to distinguish between the various possibilities. One such system uses the notation p($n \times m$) in which the letter p (for primitive) indicates that the adsorbed net has the same primitive unit mesh as the substrate. In this system of notation n and m are integers, not necessarily the same, which express the expansion of the supermesh dimensions relative to the substrate along the mesh axes. Since LEED patterns of this type portray specific crystal faces under definite adsorption conditions, they are labeled to identify both the solid and adsorbate structures. Thus Fig. 10.12b is a photograph of the W(100) − p(2 × 2) − oxygen LEED pattern. This system of labeling is known as the Wood notation.

Figure 10.12c is another example of a LEED pattern of the W(100) surface carrying an adsorbate, this time hydrogen. By their greater brightness, the spots originating from the substrate can be identified, and it is seen that the extra spots caused by the adsorbate once again lie midway between the tungsten spots, but this time with a different orientation than the array produced by the substrate. The midpoint positioning of these extra spots suggests that this adsorbate also shows a 2 × 2 supernet, but one that is rotated by 45° compared to the substrate net. A schematic illustration of the two nets is shown in Fig. 10.13d, in which the substrate atoms are represented by filled circles and the adsorbed species by open circles. The dashed square in the figure shows the unit mesh of the supernet and emphasizes its 45° orientation relative to the substrate. The unit mesh dimensions in the dashed square are the same as those of the substrate, so the distances seem to be wrong even though the symmetry is correct. The solid square in the figure shows that there is a second way of looking at the identical supernet: as a 2 × 2 enlargement of the substrate net but possessing a centered rather than primitive packing. This is a preferable way of describing the supernet, since it accounts for both the symmetry and the separation of the unit mesh. These two descriptions do not involve different supernets, but merely different ways of looking at the same net.

Since the version represented by the solid square in Fig. 10.13d best characterizes the adsorption, it is described as a c(2 × 2) net, the c reminding us that this is a centered structure. The full description of the LEED pattern in Fig. 10.12c in the Wood notation is therefore written W(100) − c(2 × 2) − hydrogen.

The solid squares in Fig. 10.13c and d represent the unit mesh of the adsorbed layers of oxygen and hydrogen, respectively, on the W(100) surface. Both are seen to have the square symmetry of the underlying tungsten surface, but with oxygen showing a primitive net, and hydrogen a centered structure. An additional detail about the surface with the adsorbed hydrogen is that the c(2 × 2) structure shown in Figs. 10.12c and 10.13d persists until the surface coverage is half a monolayer. This seems too neat for an accidental circumstance. On the other hand, if we assume the H_2 is dissociated and the adsorbed species are hydrogen atoms, the same amount of adsorbed material is enough to form a monolayer. The chemisorbed hydrogen atoms form a surface layer characterized by a square, centered unit mesh. In Wood notation the surface layer is $W(100) - c(2 \times 2) - H$, indicating the fact that atomic hydrogen rather than H_2 is the adsorbate. Oxygen is also adsorbed in the monatomic state on the $W(100)$ surface.

In the adsorption studies we have discussed, the expansion of the unit mesh is the same in both directions, but this need not be the case. Examples in which the expansion along different axes of the mesh varies are $p(4 \times 2) - O$ for the adsorption of O_2 on $Mo(111)$, $p(3 \times 15) - O$ for O_2 on $Pt(111)$, $c(4 \times 2) - S$ for H_2S on $Au(100)$, and $c(9 \times 5) - CO$ for CO on $W(110)$. Somorjai has assembled extensive tables of this sort of information in Ref. 9. Note that many but not all adsorbates are dissociated on the metal surfaces. Finally, it is not necessary for the supernet and the substrate to show the same angles between sides of their respective meshes. The Wood notation does not apply in these cases, but alternative notation exists which we shall not pursue. The following example deals with a surface at which the adsorbate and the substrate display different nets.

Example 10.5   The unit mesh of the $Pt(111)$ surface is a parallelogram with $D = D' = 0.277$ nm and having angles of 60 and 120°. This surface adsorbs n-butane with a unit mesh which is a parallelogram having $D = 0.480$ nm and $D' = 0.733$ nm and angles of 71 and 109°. Show that these two nets come into periodic register if the short side of the supermesh coincides with the diagonal of the substrate mesh.

Solution   Use the law of cosines to determine the length of the two diagonals of the substrate unit mesh:

$$L^2 = D^2 + D'^2 - 2DD' \cos 60° = [2(0.277)^2 - 2(0.277)^2(0.500)]^{1/2}$$

$$= 0.277 \text{ nm}$$

$$L'^2 = [2(0.277)^2 - 2(0.277)^2(-0.500)]^{1/2} = 0.480 \text{ nm}$$

Since the short side of the supernet mesh has $D = 0.480$ nm, this mesh registers with the substrate along the diagonal of the latter. Since the long diagonal of the substrate and the short side of the super mesh register, this common line makes a convenient reference line to test the register in the perpendicular direction. Half the length of the short diagonal, $0.277/2 = 0.139$ nm, measures the perpendicular distance from this base to the corner atom. Also, the long side of the supernet mesh makes an angle of $71°$ with the base. Therefore the long side of the supernet mesh projects a length $0.733 \sin 71 = 0.693$ nm onto the short diagonal of the substrate. The ratio $0.693/0.139 = 5.00$ shows that the edge of the supermesh and the diagonal of the substrate net will coincide again at the fifth row of atoms perpendicular to the diagonal that serves as the base for this calculation. Even though the two parallelograms have different side lengths, angles, and orientations, they do become coincident at periodic intervals.

In Sect. 10.10 we shall see some additional examples of LEED applied to the study of metallic catalysts.

10.9 Field Ionization Microscopy

Another technique for surface imaging—one which allows individual atoms to be seen—is the field ionization microscope (FIM). The remarkable resolution that permits this is accomplished by a combination of geometrical and physical elements in the experimental design. Let us begin our discussion of the FIM by examining an oversimplified but pertinent geometry problem. Consider two lines radiating from a common center and including the angle θ. These lines will subtend arcs of different lengths on concentric spherical shells of different radii. If we represent the radius of the smaller sphere by R_s and the radius of the larger sphere by R_L, then the ratio of subtended arc lengths L on the two surfaces is given by

$$\theta = \frac{L_s}{R_s} = \frac{L_L}{R_L} \tag{15}$$

Thus the ratio of the arc lengths L_L/L_s is the same as the radius ratio R_L/R_s. If the latter equals, say, 10^6, then points on the surface of the smaller sphere which are separated by 1 nm will be 1 mm apart on the larger surface.

Magnifications on the order of 10^6 are accomplished in the FIM by using essentially this concept. The metal surface to be imaged is fabricated from fine wire or vapor-grown whiskers. These are etched to sharp points whose radius is on the order of 10^{-7} m, and the point is annealed to smooth the surface. Although the resulting tip is polyhedral rather than spherical, this corresponds approximately to the smaller sphere in the geometrical arrangement described above. In the FIM the equivalent of the larger sphere is a phosphor-coated screen. The radial lines that result in the point-by-point mapping of the smaller surface on the larger are the trajectories of imaging particles.

In the FIM the sharp tip is maintained at a positive potential relative to the screen. Because of the small radius of curvature of the tip, the strength of the electrical field at the point is enormous. This field may be shown to equal V/R_s, where V is the potential difference between the tip and the screen. In the experimental apparatus this potential is variable over a range of about 5–30 kV. If we take $V = 10^4$ V and $R_s = 10^{-7}$ m as typical, the field at the tip is 10^{11} V/m, or 10 V/Å. Relatively few substances are able to withstand fields like this without damage; this restricts the variety of surfaces that can be studied by the FIM. For this reason refractory metals such as tungsten, tantalum, iridium, and rhenium have been most widely studied by this method.

Fields of this strength are able to strip electrons from gas molecules passing over the surface—hence the name *field ionization* for this technique. The resulting positive ions of the gas are drawn radially to the negative screen, where their impact produces an image on the phosphor.

The forerunner of the FIM was a similar arrangement, called the field emission microscope (FEM). In this variation the sharp point is at a negative potential and electrons are emitted from the metal by the field. The electrons are collected at the screen to produce the image. The resolution in the field emission microscope is about 2.5 nm, while this quantity is an order of magnitude better for the FIM. In both cases it is the velocity component of the imaging particles tangential to the surface that limits the resolution, rather than geometrical considerations. Both the FEM and the FIM were invented by E. W. Müller, in whose laboratory a great deal of research pertaining to these experimental methods has originated.

In the FIM the imaging gas is usually helium, although a number of other gases have been used. The potential needed to produce field ionization differs from gas to gas, so it is possible to work at lower, less damaging fields by using more readily ionized gases. Argon, for example, requires about half the field as helium to produce an image of comparable brightness. The imaging gas is introduced into a previously evacuated chamber which holds the specimen. The working gas pressure is also chosen to optimize two opposing pressure effects. The pressure must be high enough to provide sufficient collisions with the surface to yield a bright image. At the same time the pressure must be low enough so that the resolution is not destroyed by collisions between the ion beams and the gas molecules. Pressures on the order of 10^{-3} Torr are commonly used in the FIM, and the optimization of image brightness and resolution is assisted by cooling the tip, which both increases the gas density around the surface and decreases the translational velocity of the gas molecules.

There are two aspects of the FIM which differ from the simple geometrical picture presented above. Neither the point nor the screen are spherical surfaces. The simplified geometry is adequate to account for the magnification achieved by the FIM, but not to interpret the pattern of spots that make up the actual image. To do the latter we must examine the polyhedral shape of the point and consider its projection onto the flat screen that is actually used in the FIM.

As a result of specimen annealing, the crystal structure of the metal is revealed at the tip in the manner indicated schematically in Fig. 10.14. On an atomic level the tip is an array of crystal faces of low Miller index—remember from Sect. 10.6 that these are the thermodynamically favored surfaces—with stepped transitions between them. The local field is inversely proportional to the local radius of curvature and is therefore greatest at the steps which show up most brightly on the screen. The screen is a plane rather than a spherical surface, and a planar slice through a sphere is a circle. Hence the image that is projected onto the screen is a set of concentric circular rings each centered on the projection of a low-index pole as shown in Fig. 10.14. The resulting pattern of rings is characteristic of the crystal type and the indices of the pole planes, and from the observed pattern the various planes in the image can be indexed.

Figure 10.15 is a photograph of the FIM pattern of a platinum tip imaged at 26 kV and at 21 K. From the symmetry of the pattern, the axis of the projection is identified to be a 001 pole plane. The symmetrical distribution of 111 planes around this plane is labeled in the figure. The FIM image has the symmetry of a cut gem stone, except that here the facets are projections of crystallographic planes on an atomic level. Since the bright spots are formed by individual protruding atoms, the field ionization micrograph photographs surface topography on an atomic scale.

The number of rings around a pole projection can be counted in high-resolution FIM images and, through the geometry shown in Fig. 10.14, can be identified with the number of steps in the transition between indexed planes. If S is defined as the height of a step and θ as the angle between faces as defined in Fig. 10.14, then it is clear from the figure that the local radius of curvature R is related to the number of rings n by the relationship

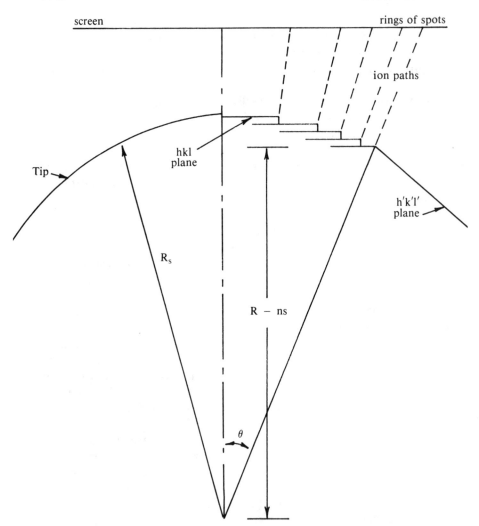

Figure 10.14 Schematic illustration of the metal tip in the FIM. The left half shows the tip as a spherical surface, and the right half as an array of crystallographic planes with stepped transitions between them.

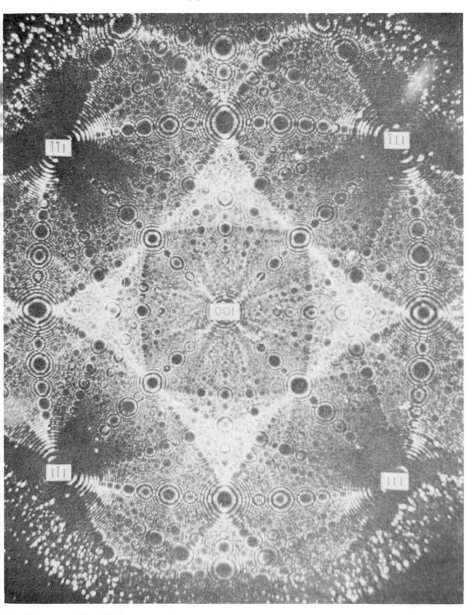

Figure 10.15 Field ionization micrograph of platinum surface. (Reprinted with permission from Ref. 7.)

$$R - nS = R \cos \theta \qquad (16)$$

Both the step height and the angle between planes can be evaluated from the geometry of the crystals. For cubic crystals, for example, S is related to the unit cell dimension D and the Miller indices of the pole plane by

$$S = \frac{D}{\delta(h^2 + k^2 + l^2)^{1/2}} \qquad (17)$$

where δ is either 1 or 2, depending on the specific values of h, k, and l. For the 001 plane of platinum shown in Fig. 10.15, $\delta = 2$ and $S = D/2$. Similarly, the angle between a specified pair of crystallographic planes may be determined for a particular crystal type. The 001 and 111 planes in the face-centered cubic structure of platinum form an angle of $55°44'$ (half the tetrahedral angle) between their projections. Thus for the stepped transition between the 001 and 111 planes in Fig. 10.15 the effective radius of curvature can be evaluated by substitution into Eq. (17): $R/nS = 2R/nD = 1/(1 - \cos \theta)$, or $R/D = n/2(1 - \cos 55°44') = 1.19n$. Therefore, by counting the number of rings (and knowing the unit cell dimension of platinum), the effective tip radius in any region of the surface may be evaluated from the indexed field ionization micrograph.

Since the field is greatest where the surface has the smallest radius of curvature, a phenomenon known as field evaporation may be employed to assure atomically perfect facets at the FIM tip. As the name implies, field evaporation is the stripping of atoms from the surface—as positive ions—under the influence of the field. Field evaporation limits the number of surfaces that can be imaged by the FIM because of the surface damage it produces. At the same time, however, application of a high potential before the imaging gas is admitted is highly effective in smoothing the surface. Likewise, alternating surface-vaporizing potentials with imaging potentials produces a series of photomicrographs which represent a depth profile of the surface. Studies of this sort are particularly useful in the study of grain boundaries, impurity occlusion, and other types of crystal defects.

Because of the large amount of detail in a field ionization micrograph, an ingenious method has been devised to make slight changes—such as before and after the stripping of an atomic layer—more apparent. This is done by superpositioning projected images—say, before and after stripping—using different colors of light for each image. Thus, if red light is used to project a "before" picture, and green an "after," those atoms which remain unchanged will appear yellow in the composite by color additivity. In the combined image a red spot indicates a site which has become vacant and a green spot denotes a site which was previously unoccupied. This ability to pinpoint an atom on a particular crystal face and follow its movement with time makes the FIM an ideal method for the study of surface diffusion.

We saw in Chap. 2 that diffusion is described by random walk statistics; Eq. (2.65) shows how the diffusion coefficient may be evaluated from a series of experiments in which the displacement x of a particle is measured during a period of time t. The FIM allows these quantities to be measured directly for surface atoms on preselected crystallographic planes. The experiment consists of the following steps:

1. A surface is perfected by field evaporation and is brought to the desired temperature.
2. A scattering of atoms is deposited on the smooth surface by adsorption. The adsorbed species are often atoms of the same metal as the substrate; these are generated by heating a filament of the metal near the surface.
3. The imaging gas is introduced and the imaging potential is turned on so that a photographic record of the surface protuberances can be made.
4. With the ionizing field removed, the specimen is held a time t during which surface migration occurs.
5. The field is increased and another image is photographed.
6. By the color comparator method described above, the new location of the adsorbed atoms can be determined, as well as the displacement x from their previous location.
7. This information is recorded for a number of atoms in similar crystallographic locations and the results are averaged as required by random walk theory.

With this sort of data, Eq. (2.65) enables us to calculate the surface diffusion coefficient D:

$$D = \frac{\bar{x}^2}{2t} \tag{18}$$

For tungsten atoms on the surface of tungsten, diffusion coefficients of 3×10^{-6}, 1×10^{-7}, and 2×10^{-11} m^2 s^{-1} were obtained for the 110, 321, and 211 planes, respectively. These data show clearly that the underlying crystal structure influences the mobility of surface atoms.

Since the FIM is designed so that the temperature of the metal tip can be controlled, this feature can also be incorporated into surface diffusion studies. If the diffusion coefficient for a particular surface is studied at a series of temperatures, then the activation energy for diffusion E^* may be determined by an Arrhenius analysis. An application of this notion is taken up in the following example.

Example 10.6   The activation energy for the diffusion of oxygen on the 100 surface of tungsten has been found to equal 95.0 kJ mole^{-1}.† Calculate the ratio of the root-mean-square distances traveled by oxygen at 298 K compared to 273 K on the W(100) surface.

Solution   By Eq. (18) the ratio of the rms distances is equal to the ratio of the square root of diffusion coefficients at the two temperatures. This last ratio equals the ratio of the Arrhenius equation at the two temperatures:

$$\frac{x_{\mathrm{rms,298}}}{x_{\mathrm{rms,273}}} = \left(\frac{\exp(-E^*/298R)}{\exp(-E^*/273R)}\right)^{1/2} = \left\{\exp\left[-\frac{E^*}{R}\left(\frac{1}{298} - \frac{1}{273}\right)\right]\right\}^{1/2}$$

or

$$\ln\frac{x_{\mathrm{rms,298}}}{x_{\mathrm{rms,273}}} = -\frac{E^*}{2R}\left(\frac{1}{298} - \frac{1}{273}\right) = -\frac{95,000}{2(8.314)}\left(\frac{1}{298} - \frac{1}{273}\right)$$

Thus $\ln(x_{\mathrm{rms,\,298}}/x_{\mathrm{rms,273}}) = 1.76$ and $x_{\mathrm{rms,298}}/x_{\mathrm{rms,273}} = 5.79$. In the same period of time the root-mean-square distance traveled by the oxygen is 5.79 times greater at 25°C than at 0°C.

•

Studies of the sort cited in Example 10.6 reveal that the activation energy for diffusion is only a fraction of the energy of adsorption. For the oxygen-W(100) system as an example, this fraction is about 20% the adsorption energy.

Now that we have considered several of the experimental techniques that may be used to study metal surfaces, let us look at the surfaces themselves in the very important application of surface catalysis.

10.10   Catalytic Properties of Metal Surfaces

The topic of heterogeneous catalysis is where the study of surfaces and the study of catalysts meet. It has been recognized for a long time that

---

†Data from R. Gomer, *Discuss. Faraday Soc.,* 28:23 (1959).

Table 10.3 Examples of Some Metal Catalysts and the Reactions They Catalyze

Metal	Reaction
Cobalt	Fischer–Tropsch synthesis of hydrocarbons from CO and H_2
Iron	Haber synthesis of ammonia from N_2 and H_2
Platinum	Hydrogenation of vegetable oils
Platinum–palladium	Oxidation of hydrocarbons and CO and reduction of NO_x in automobile exhaust
Platinum–rhenium and platinum–tin	Reforming alkanes to aromatic hydrocarbons
Platinum–rhodium	Oxidation of NH_3 to HNO_3

heterogeneous catalysts owe their activity to the properties of their surfaces. Until fairly recently, however, catalyst particle size and hence specific area were the major surface parameters that could be varied and monitored in the study of the surface effects. Much of the interest in characterizing high-area solids through physical adsorption that we discussed in the last chapter originates from this application. With the advent of the sort of techniques we have discussed in this chapter, the role of surface science in the study of these catalysts is greatly expanded.

A wide variety of solid surfaces are used as catalysts in an even wider assortment of industrial processes. In view of the contents of this chapter, we shall limit our discussion to metal catalysts. While these represent only a fraction of all catalytic systems, they do include a number of industrially important examples. Table 10.3 lists some metals used as commercial catalysts and indicates briefly the types of reactions for which they are employed. In this section we shall emphasize those features of metal surfaces which are most closely related to the specific techniques we have examined. In particular, the effect on catalytic activity of the chemical and crystallographic properties of metal surfaces will be explored. There is a vast difference between the microenvironment of the catalyst surface as examined by these atomic techniques and the overall surface that influences commercial processes. Until these modern techniques became available, however, catalyst preparation was mostly a matter of trial and error; we have now entered an era in which the science has a chance to catch up with the technology. It seems fairly safe to predict that a greatly increased understanding of heterogeneous catalysis will emerge as modern surface chemistry matures.

In our discussion of micellar catalysts in Chapter 8, we noted that effective catalysts have two features: the ability to accelerate the rate of a reaction and the ability to do so selectively. Chemistry students are familiar with the general notion that catalysts modify the mechanistic path of a reaction in such a way as to lower the activation energy and make the conversion of reactants to products more probable. One of the easiest places to see this is in reactions of diatomic gas molecules. In the gas phase the mechanism for the reaction of hydrogen and oxygen to form water involves the following steps, among others:

$$H_2 + O_2 \rightarrow HO_2 \cdot \ + H \cdot \qquad\qquad\qquad (A)$$

$$H_2 + HO_2 \cdot \ \rightarrow HO \cdot \ + H_2O \qquad\qquad\qquad (B)$$

$$H_2 + HO \cdot \ \rightarrow H \cdot \ + H_2O \qquad\qquad\qquad (C)$$

How much simpler things would be if these were monatomic gases and there were no need for all the juggling between intermediate species to dispose of unused molecular fragments! We saw in our discussion of LEED that molecules of this sort are chemisorbed at metal surfaces in the dissociated state. The combination of chemisorbed hydrogen and oxygen atoms to form water clearly follows a different mechanism than in the gas phase. The fact that the reaction occurs rapidly in the presence of platinum and not at all when the reactants are mixed without the metal shows that the activation energy has been lowered tremendously by this modification.

A similar reaction of great commercial importance is the synthesis of ammonia from the diatomic elements. The catalysts that are used commercially in this reaction are mixtures of iron and iron oxide with the oxides of potassium and aluminum. Indicating the adsorbed species by the subscript (a), the mechanism for the surface-catalyzed synthesis of ammonia is thought to involve the steps

$$N_{2\,(g)} \rightleftharpoons N_{2\,(a)} \qquad\qquad\qquad (D)$$

$$H_{2\,(g)} \rightleftharpoons H_{2\,(a)} \qquad\qquad\qquad (E)$$

$$N_{2\,(a)} \rightleftharpoons 2N_{(a)} \qquad\qquad\qquad (F)$$

$$H_{2\,(a)} \rightleftharpoons 2H_{(a)} \qquad\qquad\qquad (G)$$

$$N_{(a)} + H_{(a)} \rightleftharpoons NH_{(a)} \rightleftharpoons NH_{2\,(a)} \rightleftharpoons NH_{3\,(a)} \qquad\qquad\qquad (H)$$

$$NH_{3\,(a)} \rightarrow NH_{3\,(g)} \qquad\qquad\qquad (I)$$

The reaction between chemisorbed atoms dispenses with the problem of disrupting the triply bonded N_2 molecule in the gas phase, but reaction (F) may also be more complicated than what is shown here.

The foregoing are equilibrium reactions, and their successful exploitation requires that they be carried out under conditions where the equilibrium favors the product. Specifically, this requires that the adsorbed species in reactions (D)–(I) not be held so tightly on the catalyst surfaces so as to inhibit the reaction. On the other hand, strong interaction between adsorbate and catalyst is important to break the bonds in the reactant species. Optimization involves finding a compromise between scission and residence time on the surface. Although we are especially interested in metal surfaces, those constituents known as promoters in catalyst mixtures are also important. It is known, for example, that the potassium in the catalyst used for the ammonia synthesis shifts equilibrium (F) to the right and also increases the rate of reaction (D) by lowering the activation energy for the latter from 12.5 kJ mole^{-1} to about zero.

In addition to affecting reaction energetics, other atoms at metal surfaces also influence the selectivity of the catalyst. Platinum catalysts are excellent examples of this, since the platinum surface is capable of catalyzing a number of hydrocarbon reactions, including hydrogenation, dehydrogenation, isomerization, ring opening, and dehydrocyclization. Which of these processes is most favored by a particular catalyst is sensitive to the presence of foreign atoms at the platinum surface. Auger electron spectroscopy and other surface spectroscopies are clearly important techniques for the study of these effects. It is known, for example, that a partial monolayer of oxygen on platinum enhances reactions involving scission of the carbon–carbon bond and inhibits dehydrogenation reactions. Conversely, gold on platinum blocks C—C scission without affecting dehydrogenation or isomerization. Sulfur, which often poisons catalysts, increases the selectivity of Pt/Re catalysts.

At low pressures hydrocarbon reactions on Pt are controlled by the clean metal surface. At higher pressures a carbonaceous layer controls the selectivity of the catalyst. Since the second condition describes how catalysts are generally used, the carbonaceous layer becomes an intrinsic part of the catalyst. At low temperatures hydrocarbons are physically adsorbed on Pt surfaces; at high temperatures a graphitic coating poisons the catalyst. Under intermediate conditions of temperature, various hydrocarbon fragments are present on the surface which determine its reactivity. It is noteworthy that the hydrogen atoms in these surface fragments are readily available for reaction as evidenced by isotope exchange studies with deuterium. Carbonaceous matter, oxygen, and other atoms are thought to affect catalyst activity in several ways. For example, the electronic structure of the metal may be altered, surface restructuring may occur, and specific surface sites may be blocked.

The crystallographic character of surfaces has been shown to be of great importance in determining their catalytic behavior. Particularly interesting are those surfaces which are designed with well-characterzed "roughness." To see how this is accomplished, consider slicing through a face-centered cubic platinum crystal at a small angle relative to, say, the 111 plane. This would result in a surface of high Miller index, low atomic density, and high surface free energy according to the notions presented in Sect. 10.6. Such a slice can be stabilized, however, by forming terraces of 111 planes separated by steps of, say, 100 planes. If the angle of the cut is small relative to the 111 plane, the steps are only one atom high; the width of the terrace depends on the angle of the cut, with narrower terraces resulting from steeper cuts. Application of crystallographic concepts shows that a slice through a face-centered cubic crystal along a plane of Miller indices 7,5,5 is equivalent to a series of 111 terraces five atoms wide separated by 100 steps one atom high.

The slice through a bulk crystal can differ from *both* the 111 plane *and* the 100 plane by small angles. This produces a kink in the face of the step. By an extension of the analysis that leads to step characterization, these kinks can also be characterized. For example, a plane of Miller indices 10,8,7 has 111 terraces seven atoms wide, 110 steps one atom high, and kinks of 100 orientation every two atoms. Because of the greater thermodynamic stability of the planes of low Miller index, these surfaces of ordered roughness are stable and can be prepared and studied. Since it is sensitive to periodicity over a domain about 20 nm in diameter, LEED "sees" the pattern associated with terraces of various widths and may be used to characterize these surfaces. Satisfactory LEED patterns do not require absolute uniformity of terrace width but may be obtained with experimental surfaces which display a distribution of widths. Another experimental technique which may be used to monitor terraced surfaces is FIM.

The idea that catalyst surfaces possess a distribution of sites of different energies has been around since the 1920s, but it has not been possible until fairly recently to show that adsorption sites on terraces, steps, and kinks differ in energy. For example, hydrogen shows stronger bonding to steps and kinks on platinum than on the 111 terraces. In addition, the activation energy for H_2 dissociation is about zero on the step face and about 8.4 kJ mole^{-1} on the terrace plane. In addition,

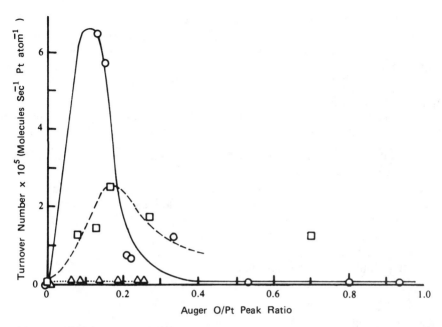

Figure 10.16   Number of molecules of cyclohexene converted to cyclohexane per second per Pt atom versus oxygen content of platinum surface (expressed as the ratio of O to Pt Auger peak heights). The data shown are for kinked (circles), stepped (squares), and terraced (triangles) platinum surfaces. (Reprinted with permission from Ref. 9.)

carbon monoxide is adsorbed with dissociation on the kinks of Pt, but in the molecular form on the steps and terraces.

The behavior of catalyst surfaces with respect to selectivity and poisoning by foreign atoms is now known to depend on the location of these atoms with respect to the terrace, step, and kink structure of the surface. By using the well-characterized faces of single crystals as catalytic surfaces, the effects of these variables have been examined for some reactions. Fig. 10.16 is an example of the results of such a study. The hydrogenation of cyclohexene to cyclohexane is the reaction involved and the figure shows how the ordered roughness and oxygen content of the platinum surface affect this reaction at 150°C. In Fig. 10.16 the ordinate gives the number of molecules converted to product per second per Pt atom—the turnover number of the catalyst—and the abscissa gives the oxygen content of the surface. The latter is measured by determining the ratio of the oxygen to platinum Auger peak heights for the various surfaces. The fraction of surface coverage is about double the Auger ratio.

The three different lines in the figure correspond to platinum surfaces of different Miller indices, corresponding to different degrees of ordered roughness. The data points marked by circles were measured on a 11,9,8 surface which consisted of terraces, steps, and kinks; the squares on a 7,5,5 surface which had terraces and steps; and the triangles on 111, which corresponds to terrace sites only.

The experiments described in Fig. 10.16 were conducted under low-pressure conditions so the following generalizations apply:

1. In the absence of oxygen all three surfaces show undetectable catalytic activity.
2. The Pt(111) surface continues to show undetectable activity as the oxygen content increases, although the activity passes through a maximum for both the stepped and kinked surfaces with increasing oxygen content.
3. Compared at their respective maxima, the kinked surface is more active than the stepped surface.
4. The oxygen content at the maximum is slightly higher for the stepped surface than for the kinked surface.

These observations suggest that maximum catalytic activity for this reaction occurs at a kinked surface with about one-third of the surface covered by oxygen. For a competitive reaction, the dehydrogenation of cyclohexene to benzene, the kinked surface is also the most active. In this case, however, the optimum oxygen content is at a coverage of about one-half compared to one-third for the hydrogenation reaction on which Fig. 10.16 is based. Differences such as these show that the selectivity of various catalysts is traceable to the chemical and physical nature of the surface. Even though the reasons for this behavior are not yet understood, these types of data hold out the promise that catalysts may someday be custom designed for maximum effectiveness. Polymer and pharmaceutical chemistry offer many examples of molecules which are designed with a specific function in mind. The methods and results of this chapter suggest that this may become possible for catalysts in the not too distance future.

10.11 Wrap-Up: Chapters 6–10

In Chapters 6, 7, 9, and 10 attention has been focused on the properties of the surfaces which separate phases. This contrasts with the first five chapters, which emphasized finely subdivided particles as kinetic units. Chapter 8 also falls into the latter category, but the surfactant molecules that are involved connects this material with topics which pertain to surfaces. The connection between these two blocks of chapters is simply this: The properties of surfaces play an increasingly important role as the size of the dispersed particles decreases.

Many of the topics we have discussed in Chaps. 6–10 appeared in more than one place. Some of these recurring themes might be mentioned again by way of summary:

1. We saw that there is a general thermodynamic equivalency among all types of interfaces, even though there are very real differences in the way they are handled experimentally. The idea, for example, of treating γ as a force is useful when discussing the shapes assumed by liquid surfaces; treating it as a free energy is more satisfying for solids. Likewise, it is easier to measure γ than Γ at liquid surfaces, whereas for rigid surfaces the reverse is true.

2. We saw that surface excess concentrations lower the free energy of an interface. This not only provides us with an insight as to the driving force responsible for adsorption but also shows us how to study adsorption by means of surface tension or vice versa.

3. We also encountered the notion of two-dimensional states of matter. This point of view permits us to visualize many phenomena in terms of the more familiar properties of bulk matter.

4. We saw that solid surfaces are likely to be rough and chemically heterogeneous, both of which factors make them energetically heterogeneous as well. Thus measurements on solids as well as liquid surfaces are especially susceptible to errors arising from contamination.

5. Modern UHV techniques allow both roughness and chemical heterogeneity to be examined on a nearly atomic scale. Their effects on such applications as metal catalysts is beginning to be appreciated.

A number of important topics have received minimal coverage in this book. This is partly because of space limitations and partly because of the level of the text. Chromatography, molecular sieves, and liquid crystals are examples of topics which could logically have been developed as part of a discussion of adsorption, porosity, and ordered structures, respectively. The fundamental principles we have covered are pertinent to these topics and should provide some necessary background to the reader who wishes to pursue them.

Two topics from among these chapters will reappear in the next block of chapters. In Chapter 11 we will see how surface tension is related to van der Waals forces and, via the latter, pertinent to flocculation phenomena. We have already noted that ions adsorbed at surfaces and ionic micelles have ion atmospheres associated with them. The structure of this ion atmosphere and its consequences with respect to flocculation and electrokinetics are taken up in Chaps. 12 and 13.

In the first five chapters of this book we were concerned with dispersed particles almost exclusively. Chapters 6–10 involved surfaces (almost) exclusively, without regard to the state of subdivision of the phases per se. In the remaining three chapters properties originating in the surface region which affect the behavior of dispersions are discussed. This will complete the merger of surface and colloidal considerations.

References

1. A. W. Adamson, *Physical Chemistry of Surfaces*, 4th ed., Wiley, New York, 1982.
2. D. K. Bowen and C. R. Hall, *Microscopy of Materials*, Wiley, New York, 1975.
3. P. J. Estrup, in *Modern Diffraction and Imaging Techniques in Materials Science* (S. Amelincks, R. Gevers, G. Remaut, and J. Van Landuyt, eds.), North Holland, Amsterdam, 1970.
4. D. Finnello and H. L. Marcus, in *Electron and Positron Spectroscopies in Materials Science and Engineering* (O. Buck, J. K. Tien, and H. L. Markus, eds.), Academic, New York, 1979.
5. R. Gomer, *Field Emission and Field Ionization*, Harvard University Press, Cambridge, Mass., 1961.
6. J. P. Hobson, in *Surface and Colloid Science*, Vol. 11 (R. J. Good and R. R. Stromberg, eds.), Plenum, New York, 1979.
7. E. W. Müller and T. T. Tsong, *Field Ion Microscopy—Principles and Applications*, American Elsevier, New York, 1960.
8. E. N. Sickafus and H. P. Bonzel, in *Progress in Surface and Membrane Science*, Vol. 4 (J. F. Danielli, M. D. Rosenberg, and D. A. Cadenhead, eds.), Academic, New York, 1971.
9. G. A. Somorjai, *Chemistry in Two Dimensions: Surfaces*, Cornell University Press, Ithaca, N.Y., 1981.
10. M. A. Van Hove and S. Y. Tong, *Surface Crystallography by LEED*, Springer-Verlag, Berlin, 1979.

Problems

1. Alloys of copper and bismuth containing 0.02% (by weight) Bi were annealed to yield grain domains of diameter 0.2–0.3 mm. Samples of the alloy were then fractured in UHV and the bismuth content normal to the surface was determined by AES after argon ion sputtering.* The Auger peak-to-peak

---

*B. D. Powell and H. Mykura, *Acta Metallogr.*, *21*:1151 (1973).

height, expressed as a percent of the copper peak-to-peak height, is given here as a function of the distance below the surface:

Depth (Å)	0	2	4	6	8	10	15	21
Bi Auger (%)	18.5	10.8	7.7	5.0	3.7	3.1	1.2	0.6

There is variation from surface to surface in the extent of Bi enrichment, but when both faces of the fracture surface are compared, they match in Bi content. Taking 10^{19} atoms m^{-2} as a density of surface atoms, compare the Bi content of this alloy with the amount of Bi needed to cover the grain domains of Cu ($\rho = 8.92$ g cm^{-3}) with a monolayer of Bi. Criticize or defend the following proposition: Monolayer coverage is not indicated by the depth profile or the even distribution of the Bi on both sides of the fracture surface.

2. Fracture surfaces of the copper–bismuth alloys described in the last problem were studied by AES and the SEM.* Surfaces in which the Bi Auger peak was 3% and 20% of the Cu were photographed with the SEM. One photomicrograph shows a smooth, undulating surface while the other is relatively rough and pockmarked. Match the SEM micrographs described with the Auger characterization of their surfaces. Explain.

3. Madden† has tabulated shifts from the Auger peak of pure vanadium for vanadium in the indicated compounds. The following data include both the *LMM* and *LMV* (*V* = valence-level electrons) transitions:

Compound	$VO_{0.92}$	V_2O_3	VO_2	V_2O_5
Shift in *LMV* (eV)	+0.3	+0.35	−0.35	−4.9
Shift in *LMM* (eV)	−0.65	−1.4	−2.3	−2.0

Do these displacements decrease the usefulness of AES as a method for elemental analysis? Does it make sense that the range of shifts is narrower (and more irregular) for the *LMM* transitions than for the *LMV* transitions? Explain.

4. Bouwman et al.‡ broke alloy rods of silver-gold, silver-tin, and gold-tin under UHV. They measured the Auger peak height ratio for the metals involved on freshly cleaved surfaces as a function of the atom ratio in the alloys. With constant instrument parameters, they found the Auger peak ratio to vary linearly with the atom ratio, suggesting that the former can be used to determine the surface composition of unknowns when the readings are calibrated by this technique. The observed ratio of the Auger signals for Ag–Au, Sn–Ag, and Sn–Au were 11.7, 2.15, and 21, respectively. Use these

---

*B. D. Powell and H. Mykura, *Acta Metallogr., 21*:1151 (1973).

† H. H. Madden, *J. Vacuum Sci. Technol., 18*:677 (1981).

‡ R. Bouwman, L. H. Toneman, and A. A. Holscher, *Vacuum, 23*:163 (1972).

observations to criticize or defend the following proposition: Surface segregation may occur at grain boundaries, and the alloy rods are most likely to break at grain boundaries; therefore the composition of the reference surface is not necessarily the same as the bulk composition. Since exact atom ratios at the surface are not known, this method is unsuitable as a calibration technique.

5. Auger spectra were measured on successive layers as SiO_2 was stripped off the surface of metallic silicon by sputtering. Initially a prominent Auger peak appeared at 74.2 eV; ultimately it appears at 90.3 eV. Interpret these peaks in terms of Fig. 10.4. Helms et al.* constructed a synthetic spectrum by adding together the spectra of Si and SiO_2, normalized to show equal maxima, and subtracted this synthetic spectrum from an experimental spectrum measured at the midpoint in a depth profile of the SiO_2–Si interface. This difference spectrum shows a distinct maximum at 83.3 eV. Careful analysis of nonsputtered oxide coatings also reveal evidence of a peak at 83.3 eV. Use these results to criticize or defend the following proposition: On an atomic scale, sputtering is a violent bombardment technique which may knock oxygen atoms from SiO_2 into Si, effectively scrambling the atoms at the interface and producing a peak at a location intermediate between pure Si and Si in SiO_2.

6. LEED diagrams reveal that O_2 adsorbs on a Pd(110) surface through a series of ordered structures which interchange reversibly with increasing surface coverage obtained by varying the pressure and temperature:

clean Pd(110) $\rightleftharpoons$ (1 × 3) $\rightleftharpoons$ (1 × 2) $\rightleftharpoons$ c(2 × 4) $\rightleftharpoons$ c(2 × 6)

The equilibrium pressure–temperature coordinates of the transitions between one LEED pattern and another were measured by Ertl and Rau† and were found to obey the two-dimensional Clausius–Clapeyron equation. When $\log_{10} p$ is plotted versus T^{-1}, straight lines of slope -1.75×10^4, -1.68×10^4, -1.35×10^4, and -1.05×10^4 K, respectively, are obtained for the four transitions above. Use these data to evaluate ΔH for each of the phase transitions of the adsorbed oxygen layer. Criticize or defend the following proposition: Since the only difference between these structures is the extent of oxygen coverage, the values of ΔH may be viewed as isosteric heats of adsorption for the surface at their respective degrees of coverage.

7. The object of this problem is to demonstrate that only certain shapes are possible for the unit mesh of a surface net. The following steps outline the proof: (a) Draw two parallel lines; one is the x axis, the other is at y = const. (b) Draw a transverse cutting across the two lines and making an angle θ with the x axis. (c) Cutting the x axis at another location, draw another transverse, this time making an angle $-\theta$ with the axis. The two transverse

*C. R. Helms, Y. E. Strausser, and W. E. Spicer, *Appl. Phys. Lett.*, *33*:767 (1978).

†G. Ertl and P. Rau, *Surface Sci.*, *15*:443 (1969).

lines should be mirror reflections of each other. (d) Imagine a surface atom at each of the four intersections of these lines. If the unit mesh is to be a regular polygon, then three of the lines must be equal to each other and equal to the interatomic spacing d. Make any necessary adjustments in the drawing so that the three shorter lengths—say, the base and the two transverses—are equal. (e) The longer side of the drawing has a length $d + 2d \cos \theta$, since each of the transverse lines projects a length $d \cos \theta$ in the x direction. For this array of atoms to come into periodic register, the ratio $(d + 2d \cos \theta)/d$ must be an integer. What values of n and θ satisfy this condition and what shapes do the corresponding meshes possess?

8. Firment and Somorjai* showed that the C_4–C_8 n-alkanes adsorb on the Pt(111) surface in ordered monolayers for which the unit mesh is a parallelogram with the following dimensions:

Hydrocarbon	D (Å)	D' (Å)	Degree
n-Butane	4.80	7.33	71
n-Pentane	4.80	16.79	74
n-Hexane	4.80	9.99	75.6
n-Heptane	4.80	22.29	78
n-Octane	4.80	12.69	79.5
Pt(111)	2.77	2.77	60

(Note that n-butane is the same system described in Example 10.6.) Look up the formula for the area of a parallelogram and calculate the area per mole, assuming one molecule per unit mesh. Compare these area with the Pt(111) unit mesh for which the dimensions are also given. Do the data make any more sense if it is assumed that some of these alkanes form surface structures with two molecules per unit mesh? Prepare a plot of the area per molecule versus the number of carbon atoms in the chain. Criticize or defend the following proposition: The amount of free area per unit mesh in these packings is equivalent to one Pt mesh.

9. The authors* of the research cited in the last problem point out that the width of the alkane molecule in the bulk crystals is 4.79, 4.78, and 4.70 Å for n-octane, n-heptane, and n-hexane, respectively. This corresponds almost exactly to one of the dimensions of the unit mesh of these molecules on the Pt(111) surface. The data tabulated in the last problem show that the longer dimensions of these cells tend to increase with the chain length. To establish whether this is a quantitative correlation, plot D' versus the number of methylenes N in the alkane and determine the slope and intercept of the resulting graph. This gives an equation of the type $D' = mN_{CH_2} + b$, where m is the length contribution per methylene and $b/2$ is the length contribution

---

per terminal methyl group. Compare these lengths with the dimensions of a fully extended hydrocarbon chain as presented in Example 8.2. What does this reveal about the configuration of these molecules on the Pt(111) surface?

10. Draw a circle and imagine it to be a section through the center of a sphere in reciprocal space (i.e., all distances have units length^{-1}). Assign the radius of the sphere a value λ^{-1} and, since this is the defined length of $\bar{k}$ and $\bar{k}_0$, draw in these vectors. Place $\bar{k}_0$ so that it points toward the 12 o'clock position, and $\bar{k}$ so that it points toward 2 o'clock. (This sphere in reciprocal space is called an Ewald sphere.) What is the origin of the vector $\bar{g}$ in terms of this drawing? Where is the head of vector $\bar{g}$? What is the diffraction angle θ? This sphere in reciprocal space makes contact with real space at the crystal. In terms of the drawing, where is the crystal located? Where is a LEED spot? Use the law of cosines to show that the length of $\bar{g}$ must be some multiple of d_{hkl}^{-1} if the Bragg condition is met by this construction. Criticize or defend the following proposition: As the angle of incidence decreases toward zero, $\bar{g}$ comes closer to tangency with the sphere; in this case the the interpretation of reciprocal space provided by Eq. (14) applies.

11. Experimental values of the fields $\bar{E}$ required to evaporate metals in the FIM are listed below for several elements. Theory predicts that these fields are related to the atomic radius r and the ionization energy I of the atoms and the sublimation energy λ and the work function ϕ of the metals by the expression $\bar{E} = (\lambda + I_n - n\phi - 3.6n^2/r)/nr$ for a process in which n is either 1 or 2, depending on the number of electrons removed from the evaporated atoms. Use the accompanying data* to evaluate $\bar{E}$ for both $n = 1$ and $n = 2$ (use $I_1 + I_2$ in this case) and compare the calculated and experimental values:

Metal	$\bar{E}_{exp}$ (V/Å)	λ (eV)	I_1 (eV)	$I_1 + I_2$ (eV)	ϕ (eV)	r (Å)
Si	3.00	4.90	8.15	24.49	4.80	1.17
Ti	2.50	4.85	6.82	20.39	4.17	1.32
Fe	3.60	4.13	7.87	24.05	4.17	1.17
Ni	3.60	4.36	7.63	25.78	5.01	1.15
Cu	3.00	3.50	7.72	28.01	4.55	1.18
W	5.70	8.67	7.98	25.68	4.52	1.30
Pt	4.75	5.62	9.00	27.56	5.32	1.30

Criticize or defend the following proposition: In the cases of Fe, Ni, and Pt the discrepancy between theory and experiment is too large to permit a value of n to be selected. For Cu and Au, $n = 2$, while for Si, Ti, and W, $n = 1$. Any residual mismatch between theory and experiment may be attributed to surface effects which cause bulk values of the parameters used in the calculation to be inappropriate for this purpose.

---

*From Ref. 7.

12. Listed below are the fields typically used* to produce images in the FIM with the indicated gases:

Gas	H_2	He	Ne	A	Kr
$\bar{E}$ (V/nm)	23	45	37	23	19

Compare these with the fields, tabulated in the last problem, which cause metal evaporation. Indicate which gases can be used to image which of the metals listed.

13. McAllister and Hansen† studied the decomposition of NH_3 on tungsten 100, 110, and 111 crystal faces in the temperature range 800–970 K. Some representative rates at the indicated NH_3 pressures in the neighborhood of 850 K are presented below for the indicated crystal faces.

Rate $\times$ 10^{-15} (molecules H_2 cm^{-2} s^{-1})

Pressure (μm)	On W(111)	On W(100)	On W(110)
2.83	1.6	0.7	0.2
5.20	2.2	0.8	0.3
8.00	2.6	0.9	0.4
11.2	3.1	1.0	0.4
14.7	3.5	1.1	0.5
18.5	3.9	1.1	0.5
22.6	4.4	1.2	0.6
31.6	5.3	1.4	0.7
58.1	7.7	1.9	1.1

Show that these data follow the expression Rate $= A + Bp^{2/3}_{NH_3}$. This suggests that the rate is the sum of two contributions: those arising from mechanisms which are zero order and $\frac{2}{3}$ order. The A mechanism is thought to have $2WN \rightarrow W_2N + \frac{1}{2}N_2$ as the rate-determining step. The B mechanism involves WN, $W_2N_3H_2$, WNH, and W_2N, with WN the dominant species. Quantitatively compare W(111), W(100) and W(110) in their ability to catalyze mechanism A and in their ability to catalyze mechanism B.

14. In describing the experimental procedure used to collect the data of the last problem, McAllister and Hansen† note that the crystal faces they examined were cut to within 1.5° of the nominal orientation. Criticize or defend the following proposition: We saw in Sect. 10.10 that the Pt(111) face readily forms steps and kinks when cut just slightly off the crystallographic plane. The steps and kinks are 100 and 110 planes; this should also be possible for W. Since the present study shows that the 100 and 110 faces have lower catalytic activity than the 111 face, we can rule out the possibility that steps and kinks arising from off-angle cutting are responsible for the enhanced activity of the 111 face.

---

*E. W. Müller, *Science, 149*:591 (1965).
†J. McAllister and R. S. Hansen, *J. Chem. Phys., 59*:414 (1973).

11

VAN DER WAALS ATTRACTION
AND FLOCCULATION

Already the difficulties of avoiding a collision in a crowd are enough to tax the sagacity of even a well-educated Square; but if no one could calculate the Regularity of a single figure... all would be chaos and confusion, and the slightest panic would cause serious injuries.

[From Abbott's *Flatland*]

11.1 Introduction

This chapter is concerned with two major topics in surface and colloid chemistry: van der Waals attraction between particles and one of its principal manifestations, flocculation. One of the most important things to bear in mind in studying this material is that each of these topics has ramifications that extend far beyond our discussion here. Van der Waals interactions, for example, contribute to the nonideality of gases and, closer to home, gas adsorption. We shall also see how these forces are related to surface tension, thereby connecting this material with the contents of Chapter 6. Furthermore, flocculation can sometimes be explained without explicitly invoking van der Waals forces at all. This is discussed in Sect. 11.11. In Chapter 12 we shall see how ion atmospheres near charged surfaces can override van der Waals attraction and prevent flocculation. In spite of these things, it is convenient and realistic to link these two topics in a single chapter.

In this chapter we make use of potential energy curves to describe the interaction of approaching particles. Analogous curves are used in ordinary chemistry to describe bonding and activation energies in reaction kinetics. It will become apparent as we proceed that we use such curves for the same purposes in describing the "reaction" between dispersed particles. We shall see that van der Waals forces scale up from atomic distances to colloidal distances undiminished. This means that the attractive influence of a particle can extend considerable distances from its surface. An important way of protecting a dispersion against flocculation is to arrange for some mechanism of repulsion between particles to be present. To be effective, these repulsive forces must extend over distances comparable to the range of van der Waals attraction.

We shall see in the next chapter that certain electrostatic forces are ideally suited to oppose van der Waals attraction between colloidal particles. This is not the only mechanism for protection against flocculation, however, and we shall discuss polymer adsorption on solid particles in this chapter as another mechanism for stabilization against flocculation.

11.2 Flocculation and Potential Energy Curves

We introduced the concept of flocculation in Sect. 1.4 as that process whereby two (or more) dispersed particles (primary or otherwise) cluster together to form an aggregate in which the individual units retain their identity but lose their kinetic independence. The fact that the primary particles are held together in these aggregates is evidence of the existence of attractive forces between the particles. The fact that some dispersions are stable with respect to the flocculation process is evidence that other forces which compete against attraction are also operative. In any specific system it is the relative magnitude of the attractive and repulsive forces between the particles that governs their flocculation behavior.

Figure 11.1a is a schematic illustration of the kinds of interactions described above. The figure shows the potential energy of the interacting particles as a function of the distance of separation between them. It turns out that potential energy is a more useful quantity to deal with than force, although the latter is given by the local slope of one of the curves in an illustration like Fig. 11.1a. The figure shows two potential energy curves, one corresponding to attraction and the other to repulsion. By convention, the potential energy associated with repulsion is defined to be positive, while the attraction is negative. This convention allows us to speak of the "height" of energy "barriers" and the "depth" of energy minima.

An important aspect of Fig. 11.1a is the fact that both the attraction and repulsion vary with the distance of separation between the bodies involved. Regardless of the specific shapes of the curves, both modes of interaction become weaker as the separation becomes larger. At sufficiently large distances the particles exert no influence on each other. The curves are deliberately interrupted at very small separations in view of the possibility of highly specific interactions at small separations. Our

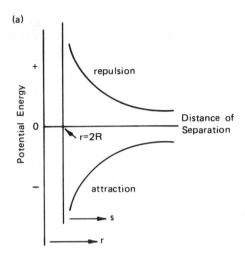

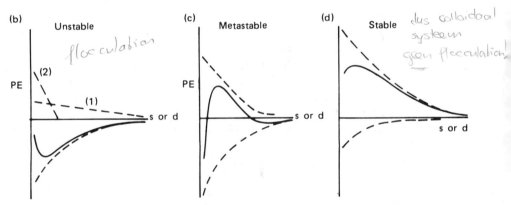

Figure 11.1 Potential energy curves for the interaction of two colloidal particles. Negative values correspond to attraction, and positive values to repulsion. (a) Definition of variables. (b) Repulsion less than attraction in magnitude and/or range. (c) Repulsion and attraction comparable in magnitude and range. (d) Attraction less than repulsion.

interest is in interactions of a more general type. In the schematic illustration shown the two components vary in roughly the same way with distance, but this effective cancellation is only one of an assortment of possible attraction–repulsion combinations. One of our objectives in this section is to consider some of the combinations that are of interest.

The attractions and repulsions between dispersed particles originate in a variety of ways. In this chapter we consider three types of intermolecular force which result in attraction between colloidal particles. Collectively, these are known as van der Waals forces, and we shall see presently that they are scaled-up versions of the same intermolecular attractions that contribute to the nonideality and ultimately liquefaction of gases. In view of Fig. 11.1a, we shall be interested in both the magnitude and the distance dependence of these attractive forces.

Our discussion of the repulsion between particles requires a more drawn out development. In this chapter we discuss the effects of adsorbed polymers as a source of repulsion; in Chapter 12 we consider ion atmosphere effects as the origin of repulsion. In these cases also, we are interested in both the magnitude and the distance dependence of these interactions.

Whether it is attraction or repulsion that we are discussing, it is important to realize that the interaction between a pair of colloidal particles involves some fundamental physical phenomena and some geometrical considerations. The precise shapes of the curves in a plot like Fig. 11.1a depend on both. For the quantitative interpretation of experimental results, it is important that both the physics and the geometry of the theory accurately (or at least approximately) describe the experimental system. From the point of view of pedagogy, however, we are more interested in physical phenomena than in particle geometry. For this reason we idealize dispersed particles as spheres and often discuss interactions between facing planar surfaces of bulk specimens of matter.

The distance of separation axis in Fig. 11.1a is defined in terms of the geometry of the interacting bodies. For spherical particles we define r to be the variable that describes the separation of their centers. For spheres of (equal) radius R the separation of their surfaces along the line of centers is given by $r-2R$, which we define as s. These are illustrated in Fig. 11.1a. We define the distance of surface separation of two blocks of material by the symbol d.

There is some analogy between the flocculation process and a polymerization reaction. Analogy implies difference and similarity, and it is important to be aware of both in making this comparison:

1. The primary particles and the floc are the analogs of the monomer and the polymer, respectively. Combination is possible between any elements of a pair: two primaries, a primary and a floc, or two flocs.

2. Polymerization involves covalent bond formation, whereas the net potential energy profile in curves like those shown in Fig. 11.1 describes the physical attraction that holds the floc together.
3. Flocculation and polymerization can both be described in thermodynamic and kinetic terms. In the thermodynamic formulation the initial and final states must be clearly understood and the transition between them is immaterial. In kinetics the path and any energy barriers along the way are important.

The attraction and repulsion curves which describe the interaction of a pair of colloidal particles are usually not as evenly matched as those shown in Fig. 11.1a. Next let us consider some other possible combinations of interest. In Fig. 11.1b–d the individual attraction and repulsion curves are shown as dashed lines, and the resultant of the two is indicated by a solid line. It is the net interaction that governs the behavior, and the components have been adjusted to span an assortment of possible behaviors.

In Fig. 11.1b the attraction is much stronger than the repulsion at large distances of separation. It is easy to imagine this occurring as a result of two different situations with respect to the repulsion component. The repulsion might be relatively small at all distances, as indicated by curve 1 in Fig. 11.1b, or it may drop off over a much shorter range of distance than the attraction, as shown by curve 2. In either case the attraction dominates at the larger separations, and the particles can achieve a lower state of energy by maintaining the distance of separation corresponding to the potential energy minimum. Stated somewhat differently, it would take an energy corresponding to the depth of the minimum to disrupt a floc that had formed as a result of the particles adopting this equilibrium separation. If we identify the potential energy axis with free energy, it is apparent that the initial dispersed state is unstable with respect to the final flocculated state in this case.

In Fig. 11.1d the relative magnitudes of attraction and repulsion are just the opposite of those in Fig. 11.1b. In Fig. 11.1d the attraction predominates. In this case the separated particles are lower in energy than the floc and stability with respect to flocculation is indicated.

Figure 11.1c is an intermediate situation in which attraction and repulsion each have regions of dominance. In this case there is a shallow minimum at large separations, a maximum at somewhat smaller separations, and a deep minimum at small separations. These minima are known as the secondary and primary minima, respectively, in the order cited. The actual depths, or heights, of the minima and maximum depend on the particulars of the components. Except for admitting the possibility of its existence, we shall not consider the secondary minimum any further. What is of particular interest to us is the fact that a minimum in energy results from flocculation in the primary minimum; however, access to that minimum first requires that an energy barrier be overcome. Such a system may be viewed as metastable, possessing a degree of kinetic stability even though it lacks thermodynamic stability: that is, flocculation is predicted, but it is expected to occur slowly. In this sense the barrier serves as an obstacle along the path to flocculation; its height is analogous to the activation energy in ordinary reaction chemistry. This last observation suggests that the kinetics of flocculation may offer some clue as to the height of the maximum in this intermediate situation.

Potential energy curves of the type shown in Fig. 11.1 are thus seen to be useful constructs for understanding and describing flocculation phenomena. In this chapter and the next we shall discuss the origins of the attraction and repulsion components. Our strategy is "divide and conquer." We shall examine various contributions to the total picture one at a time. It must be remembered, however, that the net or resultant curve is what governs the observed behavior. If we have accurately accounted for one contribution but misjudged another, the end result is wrong. A hazard associated with this kind of analysis is the danger of over-specializing in one area at the expense of another. This is especially true in a textbook such as this, where the full picture is developed over two chapters.

In the next section we consider the molecular origins of attractions between colloidal particles.

11.3 Molecular Interactions and Power Laws

To understand the origin of the attraction between colloidal particles, it is necessary to back off a bit and consider the interactions between individual molecules. Macroscopic interactions—as we shall call the interactions between colloids, since these particles are large compared to atomic dimensions—are the summation of the pairwise interactions of the constituent molecules in the individual particles. Therefore we begin by examining the interactions between a pair of isolated molecules.

Our primary interest in this section is to discuss the functional form which relates potential energy to the distance of separation x for various types of interactions. For many interactions an inverse power dependence on the separation describes the potential energy. Several examples of this are shown in Table 11.1. The main point to be observed for now is that the value of the exponent in the inverse power dependence on the separation differs widely for the various types of interactions. An immediate consequence of this is that the range of the interactions is quite different also.

It is those functions with an inverse sixth-power dependence on the separation that are our main concern in this chapter. Those power laws with exponents greater or less than 6 are included in Table 11.1 mainly to emphasize the point that many types of interactions exist and that these are governed by different relationships. The interactions listed are by no means complete: Interactions of quadrupoles, octapoles, and so on might also be included, as well as those due to magnetic moments; however, all of these are less important than the interactions listed. Let us now examine Table 11.1 in greater detail.

The first three entries in Table 11.1 include Coulomb's law and two results which follow directly from it by treating stationary dipoles as a pair of charges and adding all pairwise interactions. What is important to note about these results is that the sign may be positive or negative— corresponding to repulsion or attraction—depending on the charge of ions or the orientation of the dipoles, or both.

By contrast, those results which involve an inverse sixth-power law are always negative; that is, attraction always results from interactions of the following types:

1. Permanent dipole–induced dipole interaction (Debye equation)
2. Permanent dipole–permanent dipole interaction (Keesom equation)
3. Induced dipole–induced dipole interaction (London equation)

The inverse seventh-power law is a special case of the induced dipole–induced dipole interaction which applies to the case of large separations. This set of attractive interactions is collectively known as van der Waals attraction. In the following section we shall discuss in greater detail the significance and the origin of the van der Waals attractions listed in Table 11.1. In this section it is only the exponent in the power law that we are considering.

The last entry in Table 11.1 is the least well defined of those listed. This is of little importance to us, however, since our interest is in attraction, and the final entry always corresponds to repulsion. The reader may recall that so-called hard sphere models for molecules involve a potential energy of repulsion which sets in and rises vertically when the distance of closest approach of the centers equals the diameter of the spheres. A more realistic potential energy function would have a finite though steep slope. An inverse power law with an exponent in the range 9–15 meets this requirement. For reasons of mathematical

Table 11.1　Partial List of Interactions Between Pairs of Isolated Ions and/or Molecules, with a Listing of Functions Which Describe the Potential Energy Versus Separation, Along With the Appropriate Proportionality Constants

Description	Φ	Definitions and restrictions	Attributed to	Value of n in $\Phi \propto x^{-n}$
Ion 1–ion 2	$\dfrac{(ze)_1(ze)_2}{4\pi\varepsilon_0 x}$	z = valence, e = electron charge under vacuum—otherwise ε_r in denominator (sign depends on the z value)	Coulomb	1
Ion 1–permanent dipole 2	$\dfrac{(ze)_1\mu_2\cos\theta}{4\pi\varepsilon_0 x^2}$	μ = dipole moment, θ = angle between line of centers and axis of dipole; length of dipole small compared to x (sign depends on z and orientation)	(Coulomb)	2
Permanent dipole 1–permanent dipole 2	$\dfrac{(\text{const.})\mu_1\mu_2}{4\pi\varepsilon_0 x^3}$	Numerical constant (including sign) depends on orientation: const. = $\sqrt{2}$ for average over all orientations; const. = $+2$ for parallel and -2 for antiparallel alignment	(Coulomb)	3

Interaction	Formula	Notes	Name	Exponent
Permanent dipole 1–induced dipole 2	$\dfrac{(\alpha_{0,1}\mu_2^2 + \alpha_{0,2}\mu_1^2)}{(4\pi\varepsilon_0)^2 x^6}$	α_0 = polarizability (always negative)	Debye	6
Permanent dipole 1–permanent dipole 2	$-\dfrac{2}{3}\dfrac{\mu_1^2\mu_2^2}{(4\pi\varepsilon_0)^2 k_B T}\,x^6$	Free rotation of dipoles (always negative)	Keesom	6
Induced dipole 1–induced dipole 2	$-\dfrac{3h}{2}\dfrac{\nu_1\nu_2}{\nu_1+\nu_2}\dfrac{\alpha_{0,1}\alpha_{0,2}}{(4\pi\varepsilon_0)^2}\,x^6$	ν = characteristic vibrational frequency of electrons (always negative)	London	6
Induced dipole 1–induced dipole 2 (retarded)	$-\dfrac{23}{8\pi^2}\,hc\,\dfrac{\alpha_{0,1}\alpha_{0,2}}{(4\pi\varepsilon_0)^2 x^7}$	h = Planck's constant, c = speed of light; applies if $x > c/\nu$ (always negative)	Casimir and Polder	7
Repulsion	$+\dfrac{\xi}{x^{12}}$	Exponent in range 9–15; 12 mathematically convenient (always positive)		12

convenience, an inverse 12th-power dependence on the separation is frequently postulated for the repulsion between molecules.

In general, the combined effects of van der Waals attraction and interparticle repulsion may be represented by the equation

$$\Phi = \xi x^{-12} - \beta x^{-6} \tag{1}$$

in which the constant β has been used to represent the various constants in the Debye, Keesom, and London equations. Since the two terms on the right-hand side of Eq. (1) correspond to opposing tendencies, the total potential energy function will display a minimum, the coordinates of which will describe an equilibrium situation. By differentiating Eq. (1) with respect to x and setting the result equal to zero, the coordinates of the minimum can be evaluated. These are readily found to be

$$x_m = \left(\frac{2\xi}{\beta}\right)^{1/6} \tag{2}$$

and

$$\Phi_m = -\frac{\beta}{2}x_m^{-6} = -\xi x_m^{-12} \tag{3}$$

where the subscript reminds us that these are values at the minimum (equilibrium).

Equation (3) can be used to eliminate β and ξ from Eq. (1) to obtain

$$\Phi = -\Phi_m\left[\left(\frac{x}{x_m}\right)^{-12} - 2\left(\frac{x}{x_m}\right)^{-6}\right] \tag{4}$$

Equations (1) and (4) or other variations of the 6–12 power law are often called the Lennard–Jones potential. The numerical values of the constants in the Lennard–Jones potential may be obtained from studies of the compressibility of condensed phases, the virial coefficients of gases, and by other methods. A summary of these methods and other expressions for the molecular interaction energy can be found in the book by Moelwyn-Hughes [5].

Figure 11.2 is a plot of the Lennard–Jones function for methane, for which $\xi = 6.2 \times 10^{-134}$ J m^{12} and $\beta = 2.3 \times 10^{-77}$ J m^6 ($x_m = 0.42$ nm and $\Phi_m = -2.1 \times 10^{-21}$ J).

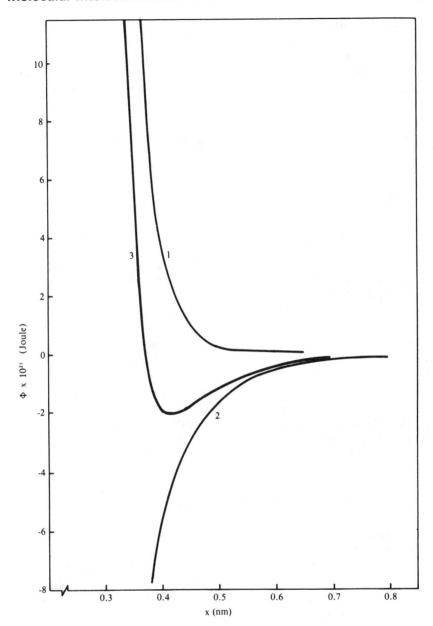

Figure 11.2 Potential energy versus distance of separation for two methane molecules. (1) repulsion according to an inverse 12th-power law, (2) attraction according to an inverse 6th-power law, and (3) resultant of curves 1 and 2.

Curves of this sort occur in many places in physical chemistry and it is important to realize that they are the resultant of two contributions: a very short-range repulsion and a relatively long-range attraction. It is the latter with which we are primarily concerned, so we shall turn next to an examination of the origins of these inverse sixth-power attractions.

11.4 Molecular Origins of Van der Waals Attractions

In Table 11.1 we saw that all random dipole-dipole interactions follow the inverse sixth-power law, except the so-called retarded van der Waals attraction, which varies with the inverse seventh power of the separation. In this section we shall examine briefly the physical basis of the three different inverse sixth-power laws which describe intermolecular attractions. Space limitations prevent us from deriving the Debye, Keesom, and London expressions in detail. More complete derivations may be found in many physical chemistry textbooks, such as that by Moelwyn-Hughes [5]. The abbreviated discussion we present should be sufficient, however, to indicate the connection between these attractions and molecular parameters.

Interactions between dipoles, whether permanent or induced, are the result of the electric field produced by one dipole (subscript 1) acting on the second dipole (subscript 2). Therefore the first factor to consider in discussing such interactions is the field $\bar{E}$ produced by a dipole and measured a distance x from the dipole, where x is large compared to the length of the dipole. Since we are dealing with the forces between charges, we must be attentive to the matter of units as was the case in the similar discussion in Sect. 5.2. The field has units charge length^{-2}, but in SI this requires that we divide the charge by $4\pi\varepsilon_0 = 1.112 \times 10^{-10}$ J^{-1} C^2 m^{-1}, in which case field has units N C^{-1} (i.e., force per unit charge) or V m^{-1} (i.e. potential gradient). The factor $4\pi\varepsilon_0$ is not required when cgs units are used; and, since the factor 4π may cancel, some of these relationships look different when written for other systems of units.

The field is a vector quantity and may be resolved into components as shown in Fig. 11.3. We define θ to be the angle between the axis of the dipole and the line which connects the point under consideration with the center of the dipole and along which x is measured. The field may be resolved into the following components:

1. Parallel to the line of centers:

$$\bar{E}_{\parallel} = -\frac{2\mu_1 x^{-3}}{4\pi\varepsilon_0}\cos\theta \qquad (5)$$

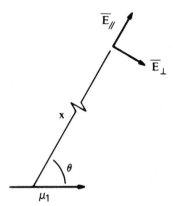

Figure 11.3 The electric field a distance x from a dipole. The field is resolved into components parallel and perpendicular to the line of centers along which x is measured.

2. Perpendicular to the line of centers:

$$\bar{E}_\perp = -\frac{\mu_1 x^{-3}}{4\pi\varepsilon_0} \sin \theta \qquad (6)$$

The total field is the square root of the sum of the squares of Eqs. (5) and (6), or

$$\bar{E} = \frac{\mu_1 x^{-3}}{4\pi\varepsilon_0} (\sin^2 \theta + 4 \cos^2 \theta)^{1/2} = \frac{\mu_1 x^{-3}}{4\pi\varepsilon_0} (1 + 3 \cos^2 \theta)^{1/2} \qquad (7)$$

Now let us consider the effect of such a field on a particle with no permanent dipole of its own. The field will induce a dipole in the second molecule; the magnitude of the induced dipole moment μ_2 is proportional to the field with the polarizability of the second molecule, α_2, the proportionality constant [see Eq. (5.11)]:

$$\mu_2 = \alpha_2 \bar{E} \qquad (8)$$

The potential energy of the second dipole due to this field is given by $-\mu_2 \bar{E}$ or $-\alpha_2 \bar{E}^2$. To this must be added the energy necessary to induce the dipole, $\frac{1}{2}\alpha_2 \bar{E}^2$, since the second is not a permanent dipole. Therefore the total potential energy Φ of the second dipole is

$$\Phi_2 = -\tfrac{1}{2}\alpha_2 \bar{E}^2 \qquad (9)$$

Substituting Eq. (7) into this expression and averaging over all orientations yields

$$\Phi_2 \doteq - \frac{\alpha_2 \mu_1^2}{(4\pi\varepsilon_0)^2} x^{-6} \tag{10}$$

The second dipole acts on the original dipole in a similar fashion, giving a second contribution to the interaction energy which is identical to Eq. (10) except that the subscripts are interchanged. The total potential energy of attraction is the sum of these two contributions:

$$\Phi_D = - \frac{(\alpha_2 \mu_1^2 + \alpha_1 \mu_2^2)}{(4\pi\varepsilon_0)^2} x^{-6} \tag{11}$$

This is the Debye equation (subscript D) in Table 11.1 which describes the attraction between a permanent dipole and an induced dipole.

If this argument is applied to two permanent dipoles, the polarizability may be regarded as the sum of two contributions: one which is independent of the presence of the permanent dipole, α_0, and a second which is the average effect of the rotation of the molecules in the electric field. The molar polarization P of a substance is given by

$$P = \frac{N_A}{3\varepsilon_0} \left(\alpha_0 + \frac{1}{3} \frac{\mu^2}{k_B T} \right) \tag{12}$$

The second term inside the parentheses thus gives the average value of the orientation contribution to the polarizability. Note that the magnitude of the permanent dipole contribution is expressed relative to thermal energy (k_B is the Boltzmann constant), since increased thermal jostling tends to scramble the permanent dipoles. It is the latter that is used as a substitution for α in Eq. (11) for the interaction of two permanent dipoles:

$$\Phi_K = - \frac{2}{3} \frac{\mu_1^2 \mu_2^2}{k_B T (4\pi\varepsilon_0)^2} x^{-6} \tag{13}$$

This is the Keesom equation (subscript K) from Table 11.1; it applies to the interaction of two permanent dipoles.

Finally, we turn our attention to the third contribution to van der Waals attraction, London (or dispersion) forces between a pair of induced dipoles. It will be noted that (at least) one permanent dipole is needed for the preceding sources of attraction to be operative. No such restriction is present for the London component. Therefore this latter quantity is present between molecules of all substances, whether or not they have a permanent dipole. These are the same forces that we considered in Chap. 6 when we discussed the Girifalco–Good–Fowkes equation.

Figure 11.4 A linear arrangement of two dipoles used to define the potential energy in the Schrödinger equation for the London interaction energy.

The London equation for the attraction between a pair of induced dipoles is a quantum-mechanical result which represents one of the contributions to the "bond" between a pair of particles. Like other quantum-mechanical results, the interaction energy emerges as part of the solution to the Schrödinger equation. We shall dispense with a rigorous examination of the situation and consider only the physical model and the final results.

Figure 11.4 represents the situation we wish to consider. It represents two dipoles whose length l_i is negligible compared to the distance between their centers. The dipoles are formed by the symmetrical vibration of electrons in the two particles. According to Table 11.1, the potential energy of two dipoles in this arrangement is $\pm 2\mu^2 (4\pi\varepsilon_0)^{-1} x^{-3}$ or $\pm (el_1)(el_2)(4\pi\varepsilon_0)^{-1} x^{-3}$, since $\mu_i = el_i$. In addition, each of the vibrating dipoles may be regarded as a harmonic oscillator for which the potential energy is given by

$$\Phi = \tfrac{1}{2}k l_i^2 \tag{14}$$

in which

$$k = \frac{e^2}{\alpha_0} \tag{15}$$

Combining these various energy contributions gives the following expression to be used as the potential energy of this system:

$$\Phi_T = \tfrac{1}{2}k_1 l_1^2 \pm 2(el_1)(el_2)(4\pi\varepsilon_0)^{-1} x^{-3} + \tfrac{1}{2}k_2 l_2^2 \tag{16}$$

When this net potential energy function is substituted into the one-dimensional Schrödinger equation and the suitable mathematical operations are carried out, the allowed energies E are found to be

$$E = (n_i + \tfrac{1}{2})h\nu_i + (n_j + \tfrac{1}{2})h\nu_j \tag{17}$$

in which

$$v_i = v\left(1 - \frac{2\alpha_0}{4\pi\varepsilon_0 x^3}\right)^{1/2} \tag{18}$$

and

$$v_j = v\left(1 + \frac{2\alpha_0}{4\pi\varepsilon_0 x^3}\right)^{1/2} \tag{19}$$

and the n terms are integers. Note that the terms in Eq. (17) are formally identical to the energy of quantized harmonic oscillators of frequency v_i and v_j. In addition, we observe that v_i and v_j approach v as $x \to \infty$. Thus v is identified as the frequency of vibration for the system in the case where the electrons vibrate independently.

The lowest energy for the two coupled oscillators is the situation in which $n_i = n_j = 0$, in which case Eq. (17) becomes

$$E = \tfrac{1}{2}h(v_i + v_j) \tag{20}$$

On the other hand, the energy of two independent oscillators in their ground state is given by

$$E = 2(\tfrac{1}{2}hv) \tag{21}$$

The difference between Eqs. (20) and (21) gives the contribution of the interaction to the potential energy:

$$\Phi_L = \tfrac{1}{2}h[(v_i + v_j) - 2v] \tag{22}$$

This is one way of writing the London (subscript L) attraction between two induced dipoles.

If the expressions for v_i and v_j given by Eqs. (18) and (19) are substituted into Eq. (22), the following result is obtained:

$$\Phi_L = \tfrac{1}{2}hv\left[\left(1 - \frac{2\alpha_0}{4\pi\varepsilon_0 x^3}\right)^{1/2} + \left(1 + \frac{2\alpha_0}{4\pi\varepsilon_0 x^3}\right)^{1/2} - 2\right] \tag{23}$$

Expanding the square roots by the binomial expansion (see Appendix A) and retaining no terms higher than second order yields

$$\Phi_L = -\frac{hv\alpha_0^2}{2(4\pi\varepsilon_0)^2}x^{-6} \tag{24}$$

Several modifications of Eq. (24) are also important:

1. When the molecules are capable of vibration in all three dimensions, the numerical constant in Eq. (24) becomes $\frac{3}{4}$ rather than $\frac{1}{2}$:

$$\Phi_L = -\frac{3h\nu\alpha_0^2 x^{-6}}{4(4\pi\varepsilon_0)^2} \tag{25}$$

2. When unlike molecules are involved, their individual frequencies and polarizabilities are involved, and the expression equivalent to Eq. (24) is

$$\Phi_L = -\frac{3}{2}h\left(\frac{\nu_1\nu_2}{\nu_1 + \nu_2}\right)\frac{\alpha_{0,1}\alpha_{0,2}}{(4\pi\varepsilon_0)^2}x^{-6} \tag{26}$$

The latter is the result shown in Table 11.1. Note that this result becomes identical to the three-dimensional version of Eq. (24) when the atoms are identical.

3. The quantity $h\nu$ in Eq. (25) may be regarded as some energy which characterizes the system and is sometimes approximated by the ionization energy I:

$$\Phi_L \simeq -\frac{3}{2}\left(\frac{I_1 I_2}{I_1 + I_2}\right)\frac{\alpha_{0,1}\alpha_{0,2}}{(4\pi\varepsilon_0)^2}x^{-6} \tag{27}$$

4. The frequency of a harmonic oscillator, the model for the two dipoles, equals $(1/2\pi)\sqrt{k/m_e}$, where m_e is the mass of the electron. Substituting the value of k given by Eq. (15) yields

$$\Phi_L = -\frac{3}{8}\frac{he}{\pi}\frac{\alpha_0^{3/2}}{m_e^{1/2}(4\pi\varepsilon_0)^2}x^{-6} \tag{28}$$

for two identical molecules, since

$$\nu = \frac{1}{2\pi}\sqrt{\frac{e^2}{\alpha_0 m_e}} \tag{29}$$

Equations (25)–(28) are widely encountered expressions for the London attraction between two molecules.

In examining the Debye, Keesom, and London equations we see that (1) they share as a common feature an inverse sixth-power dependence on the separation and (2) the molecular parameters which describe the polarization of a molecule, polarizability and dipole moment, serve as proportionality factors in these expressions. For a full discussion of the experimental determination of α_0 and μ, a textbook of physical chemistry should be consulted. For our purposes, it is sufficient to note that the Clausius–Mosotti equation relates the molar polarization of a substance to its relative dielectric constant ε_r:

$$P = \frac{M}{\rho} \frac{\varepsilon_r - 1}{\varepsilon_r + 2} \tag{30}$$

where M and ρ are the molecular weight and density. Combining Eqs. (12) and (30) gives the general result

$$\frac{M}{\rho} \frac{\varepsilon_r - 1}{\varepsilon_r + 2} = \frac{N_A}{3\varepsilon_0} \left(\alpha_0 + \frac{1}{3} \frac{\mu^2}{k_B T} \right) \tag{31}$$

Thus studies of ε_r as a function of T may be analyzed to yield values of both α_0 and μ. For substances with no permanent dipole moment, $\mu = 0$ and $\varepsilon_r = n^2$, where n is the refractive index at long wavelengths. For such a system Eq. (31) becomes

$$\frac{M}{\rho} \frac{n^2 - 1}{n^2 + 2} = \frac{N_A \alpha_0}{3\varepsilon_0} \tag{32}$$

To use Eq. (32) it is necessary to extrapolate to infinite wavelength (or zero frequency) to obtain the unperturbed polarizability, since the electric field of the light also alters the molecule. Failure to carry out such an extrapolation introduces far less error, however, than is introduced by an approximation such as Eq. (27).

In general, we may think of any molecule as possessing a dipole moment and a polarizability. This means that each of the three types of interaction may operate between any pair of molecules. Of course, in nonpolar molecules where $\mu = 0$, two of the three sources of attraction make no contribution.

As we have already noted, all molecules display the dispersion component of attraction, since all are polarizable and that is the only requirement for the London interaction. Not only is the dispersion component the most ubiquitous of the attractions, but it is also the most important in almost all cases. Only in the case of highly polar molecules such as water is the dipole–dipole interaction greater than the dispersion component. Likewise, the mixed interaction described by the Debye equation is generally the smallest of the three.

For a pair of identical molecules, Eqs. (11), (13), and (25) may be combined to give the *net* van der Waals attraction (subscript A) Φ_A:

$$\Phi_A = - \frac{1}{(4\pi\varepsilon_0)^2} \left(2\alpha_{0,1}\mu_1^2 + \frac{2}{3} \frac{\mu_1^4}{k_B T} + \tfrac{3}{4}h\nu_1\alpha_{0,1}^2 \right) x^{-6} = -\beta_{11} x^{-6} \tag{33}$$

The interaction parameter β_{11} is defined

$$\beta_{11} = \frac{1}{(4\pi\varepsilon_0)^2} \left(2\alpha_{0,1}\mu_1^2 + \frac{2}{3}\frac{\mu_1^4}{k_B T} + \tfrac{3}{4}h\nu_1\alpha_{0,1}^2 \right) \tag{34}$$

where the subscript 11 has been added to β as a reminder that this result applies to a pair of identical molecules.

We have gone through a rather complicated list of equations without looking at the magnitude of the various energy contributions. This is the subject of the following example.

Example 11.1 The parameter β_{11} must have units energy length^6 in order to satisfy Eq. (33). Verify these units as well as the dimensional consistency of each of the three terms in Eq. (34). Taking $\mu = 1.0$ debye and $\alpha = 10^{-39}$ C^2 m^2 J^{-1}, calculate the amount of energy needed to separate a pair of molecules from 0.3 nm to ∞. Scaled up by Avogadro's number, how does this energy compare with typical enthalpies of vaporization?

Solution First, we examine the units of each of the terms in Eq. (34). Example 5.1 makes it easy to analyze the units of terms containing α. That example shows that $\alpha/4\pi\varepsilon_0$ has units length^3. Polarizabilities are often tabulated as $\alpha/4\pi\varepsilon_0$, thereby having volume units.

The dipole moment has units of charge length, or C m in SI. The square of these units divided by the units of $4\pi\varepsilon_0$ therefore has units C^2 m^2/C^2 J^{-1} m^{-1}, or J m^3.

The first term in Eq. (34) involves the product of α and μ^2, each expressed in multiples of $4\pi\varepsilon_0$ and therefore has units J m^6 as required.

Since $k_B T$ has units of energy, the second term in Eq. (34) is seen to have units (J m^3)^2/J = J m^6.

Since $h\nu$ has units of energy, the third term has units J (m^3)^2 = J m^6. The debye is widely used as a unit of dipole moment. It is equal to 10^{-18} esu cm. To convert this to SI we write

$$1 \text{ debye} = 10^{-18} \text{ esu cm} \times \frac{1 \text{ m}}{100 \text{ cm}} \times \frac{1.60 \times 10^{-19}\text{C}}{4.80 \times 10^{-10} \text{ esu}}$$

$$= 3.336 \times 10^{-30}\text{C m}$$

In Eq. (34) the Debye term is given by

$$\frac{2\alpha\mu^2}{(4\pi\varepsilon_0)^2} = \frac{2(10^{-39})(3.34 \times 10^{-30})^2}{(1.11 \times 10^{-10})^2} = 1.81 \times 10^{-78} \text{ J m}^6$$

and the Keesom term by

$$\frac{2}{3}\frac{\mu^4}{k_B T(4\pi\varepsilon_0)^2} = \frac{2(3.34 \times 10^{-30})^4}{3(1.38 \times 10^{-23})(293)} = 1.67 \times 10^{-78} \text{ J m}^6$$

To evaluate the London term we need the characteristic frequency ν. Using Eq. (29), we obtain for this frequency

$$\nu = \frac{1}{2\pi}\sqrt{\frac{(1.60 \times 10^{-19})^2}{(10^{-39})(9.11 \times 10^{-31})}} = 8.44 \times 10^{14}$$

in terms of which the London energy is given by

$$\frac{3h\nu\alpha^2}{4(4\pi\varepsilon_0)^2} = \frac{3(6.63 \times 10^{-34})(8.44 \times 10^{14})(10^{-39})^2}{4(1.11 \times 10^{-10})^2}$$

$$= 3.41 \times 10^{-77} \text{ J m}^6$$

The sum of these three contributions gives the β_{11} value for this system: $\beta_{11} = 3.76 \times 10^{-77}$ J m^6.

When the separation of the molecules is 0.3 nm, the energy is $3.76 \times 10^{-77}/(0.3 \times 10^{-9})^6 = 5.16 \times 10^{-20}$ J. When the distance of separation is infinity, the interaction energy is zero. Therefore scaling up 5.16×10^{-20} J by Avogadro's number gives an estimate of the energy required to separate a mole of these molecules from a separation typical of a liquid to one that represents a gas: $(5.16 \times 10^{-20})(6.02 \times 10^{23}) = 31,000$ J. This final result is of the correct order of magnitude of heats of vaporization.

•

Dividing Eq. (34) through by β_{11} gives the fractional contribution made to the total attraction by the Debye (D), Keesom (K), and London (L) components of potential energy:

$$f_D + f_K + f_L = 1 \tag{35}$$

Table 11.2 shows these fractions calculated for a variety of molecules. In virtually all cases except the highly polar water molecule the London or dispersion component is the largest of the contributions to attraction. In the case of water hydrogen bonding is also possible and contributes an additional strong interaction, so the role of dispersion is even less than shown in Table 11.2.

In this section we have examined the three major contributions to what is generally called the van der Waals attraction between molecules. All three show the same functional dependence on the molecular separation and are therefore conveniently considered together. So far we have only considered the interaction between pairs of molecules. In the

Table 11.2 Percentage of the Debye, Keesom, and London Contributions to the Van der Waals Attraction Between Various Molecules.[a]

Compound	μ (debye)	$\dfrac{\alpha}{4\pi\varepsilon_0} \times 10^{30}$ (m^3)	$\beta \times 10^{77}$ ($J\ m^6$)	Percentage contribution of		
				Keesom (permanent–permanent)	Debye (permanent–induced)	London (induced–induced)
CCl_4	0	10.7	4.41	0	0	100
Ethanol	1.73	5.49	3.40	42.6	9.7	47.6
Thiophene	0.51	9.76	3.90	0.3	1.3	98.5
t-Butanol	1.67	9.46	5.46	23.1	9.7	67.2
Ethyl ether	1.30	9.57	4.51	10.2	7.1	82.7
Benzene	0	10.5	4.29	0	0	100
Chlorobenzene	1.58	13	7.57	13.3	8.6	78.1
Fluorobenzene	1.35	10.3	5.09	10.6	7.5	81.9
Phenol	1.55	11.6	6.48	14.5	8.6	76.9
Aniline	1.56	12.4	7.06	13.6	8.5	77.9
Toluene	0.43	11.8	5.16	0.1	0.9	99.0
Anisole	1.25	13.7	7.22	5.5	6.0	88.5
Diphenylamine	1.08	22.6	14.25	1.5	3.7	94.7
Water	1.82	1.44	2.10	84.8	4.5	10.5

[a]Dipole moments and polarizabilities from A. L. McClellan, *Tables of Experimental Dipole Moments*, Freeman, San Francisco, 1963.

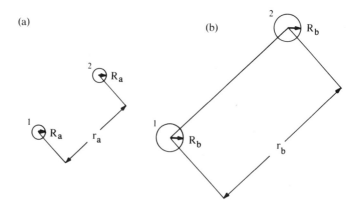

Figure 11.5 Two spheres of equal radius R separated by a distance r. All linear dimensions in (b) (i.e., R and r) are larger than those in (a) by a constant factor f.

next section we take a preliminary look at the way van der Waals attraction scales up for macroscopic (i.e., colloidal) bodies.

11.5   Interparticle Attraction and Rapid Flocculation

The interaction between individual molecules obviously plays an important role in determining, for example, the nonideality of gas. It is less clear how to apply this insight to dispersed particles in the colloidal size range. If atomic interactions are assumed to be additive, however, then the extension to macroscopic particles is not particularly difficult.

We begin by considering two spherical particles of the same composition and the same radius R. As Fig. 11.5 shows, we shall consider two situations. In case a we shall consider spheres of radius R_a, and in case b spheres of radius R_b. The two radii are related as follows:

$$R_b = fR_a \qquad (36)$$

where f is a factor greater than unity.

Assume that every atom in sphere 1a attracts every atom in sphere 2a with an energy given by Eq. (33). If $\rho N_A/M$ is the number of atoms per cubic centimeter in the material, then there are $\rho N_A/M \, dV_{1a}$ atoms in a volume element of sphere 1a and $\rho N_A/M \, dV_{2a}$ atoms in a volume element of sphere 2a. The number of pairwise interactions between the two volume elements is $\frac{1}{2}(\rho N_A/M)^2 \, dV_{1a} \, dV_{2a}$. The factor $\frac{1}{2}$ enters, since each pair is counted twice. This number times the interaction per pair gives the increment in potential energy for the two interacting volume elements in case a:

$$d\Phi_a = -\frac{1}{2}\left(\frac{\rho N_A}{M}\right)^2 \beta r_a^{-6}\, dV_{1a}\, dV_{2a} \tag{37}$$

By including additional geometrical considerations, the volume elements and their separation may be expressed in a common set of variables and Eq. (37) can be integrated over the volume of both spheres. Rather than complicate the issue by specific mathematical substitutions at this time, we indicate this procedure as follows:

$$\Phi_a = -\frac{1}{2}\left(\frac{\rho N_A}{M}\right)^2 \beta \iint_{1\text{ and }2} \frac{dV_{1a}\, dV_{2a}}{r_a^6} \tag{38}$$

Now let us turn our attention to Fig. 11.5b. In this case the spheres are larger by a factor f as just noted. In addition to this, the distance between their centers is also assumed to be larger than that in case a by the factor f. If we let the separation of centers be represented by r, then

$$r_b = f r_a \tag{39}$$

Following the same procedure as used for case a, we can evaluate the total interaction potential in case b. It is given by

$$\Phi_b = -\frac{1}{2}\left(\frac{\rho N_A}{M}\right)^2 \beta \iint_{1\text{ and }2} \frac{dV_{2a}\, dV_{2b}}{r_b^6} \tag{40}$$

Since the linear scale in case b is larger by the factor f than that in case a, Eq. (40) can also be written

$$\Phi_b = -\frac{1}{2}\left(\frac{\rho N_A}{M}\right)^2 \beta \iint_{1\text{ and }2} \frac{(f^3\, dV_{1a})(f^3\, dV_{2a})}{(f r_a)^6} \tag{41}$$

Note that the factor f cancels out of this expression entirely, so that

$$\Phi_a = \Phi_b \tag{42}$$

That is, the potential energy of attraction is identical in the two cases. This is an important result as far as the extension of molecular interactions to macroscopic spherical bodies is concerned. What it says is that two molecules, say, 0.3 nm in diameter and 1.0 nm apart, interact with exactly the same energy as two spheres of the same material which are 30.0 nm in diameter and 100 nm apart. Furthermore, an inspection of Eq. (41) reveals that this is a direct consequence of the inverse sixth-power dependence of the energy on the separation. Therefore the conclusion applies equally to all three contributions to the van der Waals attraction. Precisely the same forces that are responsible for the association of individual gas molecules to form a condensed phase operate—over a suitably enlarged range—between colloidal particles and are responsible for the flocculation of the latter.

The analogy between the condensation of a gas and the flocculation of a dispersion is important. It points out that under certain circumstances the aggregation of the separated units is inevitable. However, there are also conditions under which the dispersed state is stable. We know from kinetic molecular theory that thermal energy—measured by $k_B T$ per molecule, or RT per mole—supplies a reference level against which all energies are judged to be large or small. Thus a gas will condense if the energy of attraction between molecules is large compared to $k_B T$. Conversely, it remains dispersed if $k_B T$ is larger. The same is true for colloidal particles also, at least if they are not too large. For larger particles externally applied energy—such as mechanical stirring—joins thermal energy in establishing a reference level of energy. In general, however, the height of the barriers or the depth of the minima in the potential energy curves we discussed in Sect. 11.2 are considered large or small relative to $k_B T$. The argument leading to Eq. (42) shows that van der Waals attraction is as important for colloidal particles as it is for individual molecules.

In reaching this conclusion we have assumed that no time lag affects the field that establishes the attraction between the particles. We have also considered particles under vacuum so no intervening medium enters the picture. Each of these simplifying approximations has the effect of overestimating the van der Waals attraction between particles at large separations from one another and embedded in a medium. We will consider presently the effect of a time lapse between the interaction of a field with two different particles; the effect of the medium is taken up in Sect. 11.9.

The electric field which is responsible for the London attraction between molecules propagates itself with the speed of light between the particles. Thus, if a pair of molecules are widely separated, a time lag or a phase difference develops between vibrations at the two locations. The situation is analogous in many ways to the scattering of light by particles whose dimensions are large compared to the wavelength of the light (see

Sect. 5.9). In the present situation we find that the importance of this time lag or retardation increases as the separation becomes comparable to the wavelength of the propagating field.

Equation (29) provides us with an expression for the frequency of the interaction. Therefore its wavelength is given by

$$n\lambda = \frac{c}{v} = 2\pi c \sqrt{\frac{\alpha_0 m_e}{e^2}} \tag{43}$$

where n is the refractive index. For typical values of n and α_0, λ is about 200 nm. As was the case with light scattering, separations less than about $\frac{1}{20}$ of this distance may be considered "small compared to the wavelength." This means that at separations of about 10 nm or so the effects of retardation begin to enter the picture. Our reason for interest in this is the fact that the comparison of attraction and repulsion between colloidal particles is made at this distance in some cases.

We shall not present the detailed analysis of this complication. In essence, it involves the time-dependent Schrödinger equation rather than the time-independent equation which resulted in Eq. (22). H. B. G. Casimir and D. Polder have investigated this situation. They found that for values of $r \gg \lambda$, the potential energy of attraction according to the modified London treatment is given by

$$\Phi_A = -\frac{23}{8\pi^2(4\pi\varepsilon_0)^2} \frac{hc\alpha_0^2}{r^7} \tag{44}$$

This is the one entry in Table 11.1 that has not yet been discussed. Direct measurement of the force of attraction between macroscopic bodies reveals a crossover from an inverse sixth- to an inverse seventh-power law at separations in the range 10–100 nm.

Equation (41) is particularly convenient to show the effects of retardation on the attraction between spherical particles at large separations. Correcting the potential function for retardation according to Eq. (44), Eq. (41) becomes

$$\Phi_b \propto \iint_{1 \text{ and } 2} \frac{(f^3 \, dV_{1a})(f^3 \, dV_{2a})}{(fr_a)^7} = \frac{1}{f} \Phi_a \tag{45}$$

at large separations. This result shows that the scale factor f does not drop out of the expression in the case of the retarded van der Waals forces. Therefore the potential energy of the attraction decreases as the scale increases (i.e., as f increases).

In subsequent discussions we shall not consider the effect of retardation any further. Additional details are given by Israelachvili and Tabor [3] and Sonntag and Strenge [8].

The foregoing analysis establishes the physical basis for the attraction component of the potential energy curves that we discussed in Sect. 11.2. Since we have introduced no other mode of interaction, we have reached a situation represented by Fig. 11.1b in which attraction predominates and flocculation is expected; that is, colloidal particles meandering through the dispersion as a result of Brownian motion encounter one another from time to time. Because of interparticle attraction, these encounters result in the two particles being trapped within one another's influence and behaving henceforth as a single kinetic unit. The rate at which such a process occurs is identical to a diffusion-controlled, second-order chemical reaction. Such processes are a standard part of the discussion of chemical kinetics in physical chemistry courses. Therefore we only review the highlights of the topic; the reader who desires additional information will find this subject discussed in all physical chemistry texts.

The rate of disappearance of individual particles is the same as the rate at which two-particle encounters occur according to the model we have established. This, in turn, is proportional to the square of the particle concentration. Representing the number of particles per unit volume by N, we write

$$-\frac{dN}{dt} = kN^2 \tag{46}$$

where k is the rate constant for the flocculation step. If N_0 gives the particle concentration at $t = 0$, then Eq. (46) is readily integrated to

$$\frac{1}{N} - \frac{1}{N_0} = kt \tag{47}$$

The most reliable way to evaluate a rate constant for flocculation, therefore, is to measure N, the number of independent kinetic units per unit volume, as a function of time. Although this is an easy statement to make, it is not an easy thing to do experimentally. One technique for doing this is literally to count the particles microscopically. In addition to particle size limitations, this is an extraordinarily tedious procedure. Light scattering is particularly well suited to kinetic studies, since, in principle, experimental turbidities can be interpreted in terms of the number and size of the scattering centers. A variety of additional techniques for following the rate of particle disappearance have been developed for specific systems. We will not pursue these but merely note that experimental rate constants for flocculation can be determined.

In the absence of any resistance, flocculation occurs rapidly and our next interest is in establishing this diffusion-controlled rate. This is all we are prepared to deal with in terms of the development (until now) of the potential energy curves. In addition, knowing the rate predicted for rapid flocculation will prepare us to recognize the slower flocculation that occurs when a potential energy barrier exists for interparticle approach.

We begin by considering an array of uniform spherical particles whose motion is totally governed by Brownian movement. We assume the spheres to interact on contact, in which case they adhere, forming a doublet. Although this is a highly oversimplified picture, it provides a model from which more realistic models can be developed in subsequent stages of the presentation. Figure 11.6 shows a schematic illustration of the formation of a doublet. Since the primary particles adhere on contact, the rate at which these particles disappear equals the rate at which they diffuse across the dashed surface in the figure. The latter corresponds to a spherical surface of radius $2R$ inscribed around one of the spheres, which, for the present, is assumed to be stationary. After flocculation the number of independent kinetic units is locally decreased in the neighborhood of this flocculation site. Therefore we may imagine a concentration gradient around the fixed particle as responsible for the diffusion toward it.

In 1917 M. Smoluchowski applied the theory of diffusion to this situation to evaluate the rate of doublet formation. According to Fick's

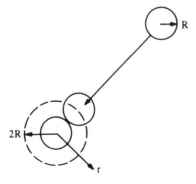

Figure 11.6   Flocculation of uniform spherical particles to form a doublet.

first law [Eq. (2.22)] the number of particles crossing a unit area toward the reference particle per unit of time, J, is given by

$$J = -D \frac{dN}{dr} \qquad (48)$$

where D is the diffusion coefficient of the spheres and N is their total number per unit volume. The total number crossing a spherical surface of radius r and area A around the reference particle per unit time, JA, is given by

$$\frac{\text{total number}}{\text{time}} = JA = -(4\pi r^2)D \frac{dN}{dr} \qquad (49)$$

Under steady-state conditions, the left-hand side of this equation is a constant, so the expression is readily integrated once the appropriate boundary conditions are considered. We assume that $N = N_0$, the initial concentration of the spheres, at $r = \infty$ and $N = 0$ at $r = 2R$, since the particles lose their kinetic identity on contact. With these conditions in mind, Eq. (49) is integrated to give

$$JA = -8\pi RDN_0 \qquad (50)$$

This equation gives the number of particles which move toward the fixed particle per unit time. If we remove the restriction that the "target" particle is stationary, the effect is to replace the diffusion coefficient of the single moving particle with the relative diffusion coefficient of the two, which is simply $2D$ when the particles are the same size. Thus the frequency with which uniform spheres collide by diffusion with a moving reference sphere (primed) is given by

$$(JA)' = -16\pi RDN_0 \qquad (51)$$

Instead of looking at a single sphere, we may regard any one of the N_0 spheres initially present as the "reference" particle. Therefore the total rate at which particles are lost is given by

$$\text{rate} = -(16\pi RDN_0)N_0 = -k_r N_0^2 \qquad (52)$$

where k_r is effectively a second-order rate constant for rapid (subscript r) flocculation. It is important to realize that Eq. (52) applies strictly only at the outset of flocculation when the only particles present are N_0 uniform spheres. In applying these ideas to actual kinetic experiments, therefore, we are limited to the initial rates of flocculation.

We saw in Chap. 2 [Eq. (2.32)] that for uniform spheres

$$D = \frac{k_B T}{f} = \frac{k_B T}{6\pi\eta R} \qquad (53)$$

Substituting this result into Eq. (52) yields

$$k_r = \frac{8 \, k_B T}{3\eta} \tag{54}$$

Note that the size of the reactant units drops out of the final expression for k_r; therefore the expression is equally valid for small molecules or colloidal particles so long as the various assumptions of the model apply. This constant describes the rate of diffusion-controlled reactions between molecules of the same size. In the following example we examine the numerical magnitude of the rate for the process we have been discussing.

Example 1 1.2   An aqueous dispersion initially contains 10^9 particles cm^{-3}. Assuming rapid flocculation, calculate the time required for the concentration of the dispersed units to drop to 90% of the initial value. The viscosity of water is 0.010 P at 20°C, which may be used for the temperature of the experiment.

Solution   First we evaluate k_r, using Eq. (54). It is convenient to use cgs units for this calculation; therefore we write $k_r = 8(1.38 \times 10^{-16})(293)/(3)(0.010) = 1.08 \times 10^{-11}$ cm^3 s^{-1}. Recall that the coefficient of viscosity has units mass length^{-1} time^{-1}, so the cgs unit, the poise, is the same as g cm^{-1} s^{-1}. As a second-order rate constant, k_r has units concentration^{-1} time^{-1}, so we recognize that the value calculated for k_r gives this quantity per particle, or $k_r = 1.08 \times 10^{-11}$ cm^3 particle^{-1} s^{-1}. Note that multiplication by Avogadro's number of particles per mole and dividing by 10^3 cm^3 per liter gives $k_r = 6.5 \times 10^9$ 1 mole^{-1} s^{-1} for the more familiar diffusion-controlled rate constant.

The 90% time is analogous to the half-life of the reaction. By considering a smaller extent of reaction, the assumptions of the model are more apt to remain valid. Substituting $N = 0.90 N_0$ into Eq. (47), we obtain $(0.90 N_0)^{-1} - (N_0)^{-1} = k_r t_{0.90}$ or $t_{0.90} = (1.00 - 0.90)/0.90 k_r N_0 = 0.10/(0.90)(10^9)(1.08 \times 10^{-11}) = 10.3$ s. Note the cancellation of concentration units in this last step.

•

Examples are readily found in which the observed rate constant for flocculation is several orders of magnitude smaller than the rate constant we have been discussing. These are cases of slow flocculation and imply a component of repulsion as well as attraction between the dispersed particles. We will take up the physical basis for this repulsion later, but it is convenient to continue with the analysis of the kinetics of slow flocculation at this point. This is the topic of the next section.

11.6   Slow Flocculation and Potential Energy Barriers

The flocculation process that we discussed in the last section is one in which no resistance opposes the flocculation. In Sect. 11.2 we indicated that repulsion components could also be present in the potential energy of interaction between particles. As illustrated schematically in Fig. 11.1c, it is possible for the resultant of van der Waals attraction and these yet unspecified modes of repulsion to generate a potential energy barrier to flocculation. Under certain conditions the height of this barrier relative to thermal energy may be such that flocculation occurs, but at a much slower rate than the unimpeded process we discussed in the last section. An experimental study of flocculation in this kind of system offers the possibility for measuring the height of this barrier and thus learning something about the net potential energy profile. These statements are made by analogy with the situation that prevails in ordinary reaction kinetics. In the latter activation energies are measured which are interpreted as giving the height of an energy barrier along the reaction coordinate. Let us now consider how to extend these ideas to flocculation.

To handle this situation we retrace our steps through the preceding derivation, this time including a term for the resistance to the flocculation process. Thus Eq. (49) becomes

$$JA = -(4\pi r^2)D\frac{dN}{dr} + \text{resistance term} \qquad (55)$$

for the case of a sphere diffusing toward a stationary particle. The problem now is to define the resistance term.

The following considerations will permit us to arrive at the proper form for the flux away from the stationary particle due to the potential energy barrier:

1. The derivative of the potential gives the force which opposes the approach of the particles: $d\Phi/dr$.
2. This force divided by the friction factor gives the velocity of "rebound" off the potential barrier: $(1/f)(d\Phi/dr)$.
3. This velocity times the concentration of the particles at r gives the flux of particles away from the stationary particle: $(N/f)(d\Phi/dr)$.
4. This flux times the area of the spherical shell at r gives the resistance term: $(4\pi r^2 N/f)(d\Phi/dr)$.
5. The sign of this term must be opposite to the sign of the first term. Since r is measured from the stationary particle, dN/dr is positive and $d\Phi/dr$ is negative beyond the maximum, which is the region of interest, since it is the domain of the second particle. Therefore the

required difference in sign is covered by the signs of the gradients of N and Φ.

Incorporating these five considerations into Eq. (55) yields

$$JA = -4\pi r^2 \left(D \frac{dN}{dr} + \frac{N}{f} \frac{d\Phi}{dr} \right) \tag{56}$$

Although it was derived for the case of repulsion, there is nothing to prevent this expression from being applied to situations in which the energy is attractive, in which case the sign of $d\Phi/dr$ would be reversed and the flux toward the stationary particle would be augmented. Since the particles are spheres, we again substitute Eq. (2.32) for f to obtain

$$JA = -4\pi r^2 \left(D \frac{dN}{dr} + \frac{ND}{k_B T} \frac{d\Phi}{dr} \right) \tag{57}$$

Removing the restriction that the reference particle is stationary again requires us to replace D by $2D$ to obtain

$$(JA)' = -8\pi r^2 D \left(\frac{dN}{dr} + \frac{N}{k_B T} \frac{d\Phi}{dr} \right) \tag{58}$$

or

$$\frac{dN}{dr} + \frac{N}{k_B T} \frac{d\Phi}{dr} = -\frac{(JA)'}{8\pi r^2 D} \tag{59}$$

This is the differential equation that must be solved to give N as a function of r for steady-state diffusion in an energy gradient.

The solution to this equation is easily verified to be

$$N \exp\left(\frac{\Phi}{k_B T} \right) = -\int \exp\left(\frac{\Phi}{k_B T} \right) \frac{(JA)'}{8\pi D} r^{-2} \, dr + \text{const.} \tag{60}$$

We eliminate the constant of integration as follows: At $r = \infty$, $N = N_0$ and $\Phi = 0$; therefore Eq. (60) becomes

$$N_0 = -\left[\int \exp\left(\frac{\Phi}{k_B T} \right) \frac{(JA)'}{8\pi D} r^{-2} \, dr \right]_{r=\infty} + \text{const.} \tag{61}$$

At $r = 2R$, $N = 0$; therefore

$$0 = -\left[\int \exp\left(\frac{\Phi}{k_{\rm B}T}\right) \frac{(JA)'}{8\pi D} r^{-2}\, dr\right]_{r=2R} + {\rm const.} \tag{62}$$

In writing this last result it is implicitly recognized that the primary minimum has a finite depth and does not drop to $\Phi = -\infty$ as implied earlier. Recall that at very small separations the interaction between molecules becomes one of repulsion; we have simply ignored this region until now. Subtracting Eq. (62) from Eq. (61) gives

$$N_0 = -\frac{(JA)'}{8\pi D} \int_{2R}^{\infty} \exp\left(\frac{\Phi}{k_{\rm B}T}\right) r^{-2}\, dr \tag{63}$$

We may verify that we have proceeded correctly by noting that Eq. (63) reduces to Eq. (51) in the event that $\Phi = 0$.

As before, the product $(JA)'$ times N_0 gives the initial rate at which particles are lost:

$$\text{rate} = (JA)'N_0 = \frac{-8\pi D N_0^2}{\int_{2R}^{\infty} \exp(\Phi/k_{\rm B}T) r^{-2}\, dr} = -k_{\rm s}N_0^2 \tag{64}$$

where $k_{\rm s}$ is the rate constant for slow flocculation. Comparing Eqs. (52) and (64) reveals the following relationship between the rate constants for rapid and slow flocculation:

$$k_{\rm s} = \frac{k_{\rm r}}{2R \int_{2R}^{\infty} \exp(\Phi/k_{\rm B}T) r^{-2}\, dr} = \frac{k_{\rm r}}{W} \tag{65}$$

Equation (65) shows that the effect of the energy barrier is to decrease the rate constant for flocculation by a factor called the stability ratio W:

$$W = 2R \int_{2R}^{\infty} \exp\left(\frac{\Phi}{k_{\rm B}T}\right) r^{-2}\, dr \tag{66}$$

Equation (66) shows that the stability ratio can be evaluated—say, by graphical integration—if the net interaction potential energy is known as a function of particle separation. The potential energy curves with which we began this chapter describe precisely this sort of information. As yet, we have considered no mechanism for repulsion, so the very existence of net curves with barriers is somewhat academic. This situation will be remedied in due time, however, so the type of calculation suggested by Eq. (66) should not be forgotten.

We have already seen that it is possible to measure experimental flocculation rate constants, and Eq. (65) shows how the stability ratio of an experimental system can be evaluated by comparing k_s with k_r. Unfortunately, the form of Eq. (66) is sufficiently complicated that it is not possible to extract the function $\Phi(r)$ from experimental W values. The following example considers an approximation wherein Eq. (66) is developed in terms of the potential energy at the maximum of the barrier.

Example 11.3 By replacing r by r_m, the separation at the maximum in the potential energy curve, and using a truncated Taylor series about the maximum to estimate $\Phi(r)$, show that $W \propto \exp(\Phi_m/k_BT)$, where Φ_m is the height of the potential energy barrier at the maximum. Comment briefly on these and any other assumptions or approximations involved.

Solution The function $(r^{-2})\exp(\Phi/k_BT)$ has its maximum at Φ_m and drops off rapidly for $\Phi < \Phi_m$. This justifies focusing attention on the maximum. Furthermore, the exponential term is more important than the r^{-2} factor; hence the latter is replaced by r_m^{-2}.

The Taylor series expansion of Φ can be written

$$\Phi \simeq \Phi_m + (r - r_m)\left(\frac{\partial\Phi}{\partial r}\right)_m + \frac{(r - r_m)^2}{2}\left(\frac{\partial^2\Phi}{\partial r^2}\right)_m + \cdots$$

according to Appendix A. The subscript m reminds us that the derivatives are evaluated at the maximum, and for this reason the term $(\partial\Phi/\partial r)_m$ equals zero. With these substitutions Eq. (66) may be written

$$W = \frac{2R}{r_m^2}e^{\Phi_m/k_BT}\int_{2R}^{\infty}\exp\left(\frac{(\partial^2\Phi/\partial r^2)(r - r_m)^2}{2k_BT}\right)dr$$

$$= \frac{2R}{r_m^2}e^{\Phi_m/k_BT}\int_{2R - r_m}^{\infty}\exp(-p^2\,\delta r^2)\delta r$$

In the second version of this, the variable has been changed to δr with the limits adjusted accordingly and $p^2 = -(\partial^2\Phi/\partial r^2)_m/2k_BT$. If the maximum is close to the surface of the particles, $2R - r_m$ can be set equal to zero, in which case the integral is one of those shown in Table 2.2. The fact that the function in question drops off with distance from the maximum further assures that no serious error is made in shifting the integration limit. The integral gives only half the area under the maximum, so an additional factor of 2 enters the final expression for W:

$$W = \frac{2\pi^{1/2}R}{r_m^2 p} \exp\left(\frac{\Phi_m}{k_B T}\right)$$

Note that $(\partial^2\Phi/\partial r^2)_m$ is negative, since the point of evaluation is a maximum; therefore there is no sign problem with p. If p can be regarded as a constant, then W plays a role in Eq. (65) that converts the latter to the form of the Arrhenius expression for ordinary chemical rate constants.

•

The development sketched in Example 11.3 is not valid under all circumstances, but when it applies, it allows Eq. (66) to be approximated by

$$k_s \propto k_r \exp\left(-\Phi_m/k_B T\right) \tag{67}$$

This is the Arrhenius form to which the example refers. In it the height of the maximum in a net potential energy curve plays the role of the activation energy. In the next chapter we shall see how this method has been used to evaluate W for systems in which the overlapping ion atmospheres of approaching colloidal particles provides the repulsion needed to give slow flocculation.

The development that led to Eq. (42) was conducted without any actual integrations being performed. In order to generate exact expressions for the van der Waals attraction between bodies of specific geometry, it is necessary to carry out these operations. This is taken up in the next section.

11.7   Attraction Between Macroscopic Bodies

The strategy for scaling up the van der Waals attraction to macroscopic bodies requires that all pairwise combinations of intermolecular attraction between the two bodies be summed. This has been done for several different geometries by H. C. Hamaker. We shall consider only one example of the calculations involved, namely, the case of blocks of material with planar surfaces. This example serves to illustrate the method and also provides a foundation for connecting van der Waals forces with surface tension, the subject of the next section.

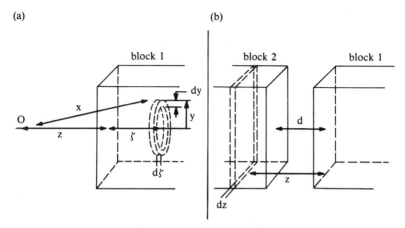

Figure 11.7 Schematic illustration showing the interaction (a) between a molecule and a block of material and (b) between two blocks of material. [Reprinted with permission from P. C. Hiemenz, *J. Chem. Educ.*, 49:164 (1972), copyright by the American Chemical Society.]

Figure 11.7a represents a molecule at O located a normal distance z from the surface of a bulk sample of the same material. The bulk portion is assumed to have a planar face but otherwise be of infinite extension. The molecule is located a distance x from all the molecules in the ring-shaped volume element shown in the figure. The volume of this ring is given by $dV = 2\pi y\, dy\, d\zeta$. Therefore the increment of interaction between the molecule and the block due to the molecules a distance x from the point O is given by

$$d\Phi = -\frac{\rho N_A}{M}\beta\,\frac{2\pi y\, dy\, d\zeta}{x^6} \tag{68}$$

We assume the ring is located a distance ζ inside the surface of the block; then

$$x^2 = (z + \zeta)^2 + y^2 \tag{69}$$

Combining Eqs. (68) and (69) gives

$$d\Phi = -\frac{\rho N_A}{M}\beta 2\pi\,\frac{y\, dy\, d\zeta}{[(z + \zeta)^2 + y^2]^3} \tag{70}$$

Equation (70) is now integrated over the volume of the block, that is, for $0 < y < \infty$ and $0 < \zeta < \infty$.

Integration over y yields

$$\int_0^\infty \frac{y \, dy}{[(z + \zeta)^2 + y^2]^3} = \frac{1}{2} \int_0^\infty \frac{du}{[(z + \zeta)^2 + u]^3}$$

$$= -\frac{1}{2} \left\{ \frac{1}{2[(z + \zeta)^2 + u]^2} \right\}_0^\infty = \frac{1}{4} \frac{1}{(z + \zeta)^4} \qquad (71)$$

and integration over ζ yields

$$\frac{1}{4} \int_0^\infty \frac{d\zeta}{(z + \zeta)^4} = -\frac{1}{4} \left[\frac{1}{3(z + \zeta)^3} \right]_0^\infty = \frac{1}{12z^3} \qquad (72)$$

Therefore the integration of Eq. (70) over the block gives

$$\Phi = -\frac{\rho N_A}{M} \frac{\beta \pi}{6z^3} \qquad (73)$$

Now suppose point O is located inside a second block of material, as shown in Fig. 11.7b. We recognize that all atoms in a slice of the second block a distance z from the first block will be attracted toward the latter with an energy given by Eq. (73). If we position a volume element of thickness dz at this location in the second block, we realize that it contains $\rho N_A / M$ molecules per unit area. This is $(\rho N_A / M) dz$ molecules; therefore the increment of attraction per unit area due to this slice is

$$d\Phi = -\left(\frac{\rho N_A}{M} \right)^2 \frac{\beta \pi}{6} z^{-3} \, dz \qquad (74)$$

Equation (74) may now be integrated over values of z between the distance of closest approach d and infinity. The result of this integration gives the potential energy of attraction per unit area between two blocks of infinite extension:

$$\Phi_A = -\left(\frac{\rho N_A}{M} \right)^2 \frac{\beta \pi}{12} d^{-2} \qquad (75)$$

where the subscript A has been appended to Φ to emphasize the fact that this energy is always attractive.

It is traditional to designate the cluster of contents $(\rho N_A/M)^2 \pi^2 \beta$ as the Hamaker constant A:

$$A = \left(\frac{\rho N_A \pi}{M} \right)^2 \beta \tag{76}$$

With this change of notation, Eq. (75) becomes

$$\Phi_A = - \frac{A}{12\pi} d^{-2} \tag{77}$$

The Hamaker constant has energy units, since β has units energy length[6] and the term in parentheses in Eq. (76) has units (volume^{-1})^2. The potential energy of attraction between blocks as calculated by Eq. (77) is expressed per unit area of the facing surfaces. Note also that this attraction grows weaker as the distance increases. This is different from the conclusion drawn for spheres in Eq. (42). The scaling up of van der Waals forces, therefore, depends on the geometry of the bodies involved.

Although we have considered numerical values of β before, we have not encountered the Hamaker constant previously and hence have no feel for its magnitude. A particularly convenient form for an order of magnitude estimation of A is obtained by making the following stipulations:

1. Assume that the dispersion component is the dominant contributor to the attraction; therefore $\beta = \frac{3}{4} h\nu (\alpha/4\pi\varepsilon_0)^2$.
2. Recognize that $\rho N_A/M$ is the reciprocal of the volume per molecule and that $\alpha/4\pi\varepsilon_0$ is typically about 10% the magnitude of atomic volume; therefore $A \simeq \frac{3}{4}\pi^2 h\nu(0.1)^2$.
3. The quantity $h\nu$ is the same order of magnitude as the ionization potential, typically about 10^{-18} J.

Combining these results leads us to estimate the Hamaker constant to lie in the range 10^{-20}-10^{-19} J. If we take the midpoint value of 5×10^{-20} J for A, Eq. (77) predicts that Φ_A equals 1.33 and 1.33×10^{-2} mJ m^{-2} when d equals 1.0 and 10 nm, respectively.

The interaction of two infinite blocks separated at the surface by a distance d is one of the easiest possible situations to consider and therefore was chosen as the example to develop in detail. Bodies of different geometries have also been analyzed in much the same way that we have done for the blocks. The results of several such derivations are shown in Table 11.3. The expressions for Φ_A become more complicated for more complex geometrical situations, but considerable simplification

Table 11.3 Potential Energy of Attraction Between Two Particles with the Indicated Geometries

Particles	Φ_A	Definitions/limitations
Two spheres	$-\dfrac{A}{6}\left[\dfrac{2R_1R_2}{s^2+2R_1s+2R_2s}+\dfrac{2R_1R_2}{s^2+2R_1s+2R_2s+4R_1R_2}\right.$ $\left.+\ln\left(\dfrac{s^2+2R_1s+2R_2s}{s^2+2R_1s+2R_2s+4R_1R_2}\right)\right]$	R_1, R_2 = radii; s = separation of surfaces along line of centers
Two spheres of equal radius	$-\dfrac{A}{6}\left[\dfrac{2R^2}{s^2+4Rs}+\dfrac{2R^2}{s^2+4Rs+4R^2}+\ln\left(\dfrac{s^2+4Rs}{s^2+4Rs+4R^2}\right)\right]$	$R_1=R_2=R$
Two spheres of equal radius	$-\dfrac{AR}{12s}$	$R \gg s$
Two spheres of unequal radius	$-\dfrac{AR_1R_2}{6s(R_1+R_2)}$	R_1 and $R_2 \gg s$
Two plates of equal thickness	$-\dfrac{A}{12\pi}\left(\dfrac{1}{d^2}+\dfrac{1}{(d+2\delta)^2}-\dfrac{2}{(d+\delta)^2}\right)$	δ = thickness of plates
Identical blocks	$-\dfrac{A}{12\pi d^2}$	$\delta \to \infty$

results for spheres in which the separation is much less than the radius. The feature described by Eq. (42) for interacting spheres is also evident in the results given in Table 11.3. It is most readily seen by examining the case of equal spheres separated by small distances. Note that both R and s can be multiplied by any common factor without changing the magnitude of the interaction energy. The expression for interacting plates of thickness δ in Table 11.3 is also seen to reduce to Eq. (77) as $\delta \to \infty$.

11.8 Van der Waals Attraction and Surface Tension

Adding together molecular interactions to account for macroscopic attractions is undoubtedly an oversimplification. There are several things that this procedure overlooks which limit its validity. For example, even for two bodies under vacuum, those molecules nearer the surface screen the interactions of molecules buried more deeply in the material. Because of the inverse power dependence of the attraction, those molecules nearest the faces of the blocks make the predominant contribution to the interaction. If these molecules have permanent dipoles, they may experience orientation effects under the influence of the surface that are not described by the Debye and Keesom models. From a practical point of view, dipole moments and polarizabilities may not always be available for all substances of interest. The possibility of surface heterogeneity confuses the choice of polarization parameters even further. Although we continue (for now) to assume a vacuum separates the bodies under consideration, the presence of an intervening medium can only aggravate all of the preceding points.

For these reasons a theory based entirely on measurable bulk properties rather than molecular parameters is a more powerful way of dealing with the interaction of macroscopic bodies. Dzyaloshinskii, Lifshiftz, and Pitaevskii (DLP) have developed such a macroscopic theory in a very general form from quantum field theory. Bulk dielectric properties of matter are the basis for this theory, so screening effects are built into the solution. Unfortunately, in its general form, the DLP theory is complicated and difficult to apply. An indication of this is found in the following expression for the Hamaker constant A_{213}—as it would be used in Eq. (77)—for the interaction of blocks of material 2 and material 3 when the two are separated by material 1:

$$A_{213} = \frac{3}{8\pi^2} h \int_0^\infty \frac{\varepsilon_2(i\xi) - \varepsilon_1(i\xi)}{\varepsilon_2(i\xi) + \varepsilon_1(i\xi)} \frac{\varepsilon_3(i\xi) - \varepsilon_1(i\xi)}{\varepsilon_3(i\xi) + \varepsilon_1(i\xi)} \, d\xi \qquad (78)$$

In this expression $\varepsilon(i\xi)$ is the dielectric "constant"—it is a function of frequency—along the imaginary frequency axis, $i\xi$; it is measurable as the dissipative part of the spectrum of dielectric constant for any material. The latter is an experimentally determined function of frequency for each of the three components, and the complicated expression in Eq. (78) is integrated over all frequencies.

Although very general, the difficulty in applying the DLP theory has limited its use and encouraged the development of various approximations and special cases. The only use we shall make of the DLP theory is to point out that the differences between the $\varepsilon(i\xi)$ values for the blocks and the medium appear in the numerator of Eq. (78). If these dissipative components of dielectric constant matched at all frequencies for substances 1, 2, and 3, then the Hamaker constant would equal zero according to Eq. (78). Although an exact match seems unlikely, this is the first indication we have had that an intervening medium might have a compensating effect on van der Waals attraction. This idea will be developed more fully in the next section.

Because of the complexity of the DLP theory, several semiclassical treatments have been developed to simplify and extend its findings. One such extension is applicable to blocks of nonpolar materials at small separations and is of interest to us because of the possibility it offers for relating surface tension to intermolecular forces.

For now we consider the interaction of two identical blocks of material separated by their own equilibrium vapor. From the point of view of modifying the force of attraction, the vapor is assumed to have a negligible effect, behaving essentially like a vacuum. The facing planes become equilibrium surfaces characterized by an interfacial free energy γ.

Our starting point in this discussion is the resemblance between Fig. 6.7a and Fig. 11.7b. The former illustrates the work of cohesion, and the latter the interaction between two blocks of material. Suppose we identify by d_0 the equilibrium spacing between molecules of a bulk sample of the material under consideration. Then the cohesion process represented by Fig. 6.7a can be viewed as one in which two blocks of material are separated from $d = d_0$ to $d = \infty$. In terms of Eq. (77),

$$\Delta\Phi = \Phi_f - \Phi_i = \frac{A}{12\pi}\frac{1}{d_0^2} \tag{79}$$

Equating this with the work of cohesion given by Eq. (6.56), we obtain

$$2\gamma = A/12\pi d_0^2 \tag{80}$$

or

$$A = 24\pi\gamma d_0^2 \tag{81}$$

A system for which this model is apt to work best is a nonpolar material such as an alkane for which $\gamma \simeq 25$ mJ m^{-2} as seen from Table 6.1. Using 0.2 nm for d_0, we calculate $A = 24\pi(0.025)(0.2 \times 10^{-9})^2 = 7.5 \times 10^{-20}$ J, which is very close to the value predicted for this quantity from molecular parameters in the last section.

The same logic that we used to obtain the Girifalco-Good-Fowkes equation in Sect. 6.12 suggests that the dispersion component of the surface tension γ^d may be better to use than γ itself when additional interactions besides London forces operate between the molecules. Also, it has been suggested that intermolecular spacing should be explicitly considered within the bulk phases, especially when the interaction at $d = d_0$ is evaluated. The Hamaker approach, after all, treats matter as continuous, and at small separations the graininess of matter can make a difference in the attraction. The latter has been incorporated into one model which results in the expression

$$A = \frac{4\pi}{1.2} \gamma^d d_0^2 \tag{82}$$

where d_0 is the intermolecular spacing in the bulk material. The contention is that this is more suitable than Eq. (81) for relating the Hamaker constant to surface tension for materials with interactions other than London forces. In addition to only a fraction of the full value of γ being used, the numerical coefficient in Eq. (82) is only about one-seventh that in Eq. (81). Both of these considerations reinforce the idea that Eq. (81) gives an upper limit for A.

Equations (81) and (82) provide alternatives to Eqs. (34) and (76) for the evaluation of the Hamaker constant. Although the last approach uses macroscopic properties and hence avoids some of the objections cited at the beginning of the section, the practical problem of computation is not solved by substituting one set of inaccessible parameters (γ^d and d_0) for another (α and μ).

The primary objective of the present discussion is to show the intrinsic connection between surface tension and the van der Waals energy of attraction between macroscopic bodies. The connection not only provides computational options but also—and more importantly— unites two apparently separate phenomena and strengthens our confidence in the correctness of our understanding.

Table 11.4 shows a few numerical examples of how well this attempt at unification succeeds. Equations (28), (32), and (76) have been used to calculate values of the Hamaker constant from refractive index data at

Table 11.4 Calculations Intended to Show that Eqs. (81) and (82) Predict Values of the Hamaker Constant Which are Compatible with Those Evaluated from Eqs. (28), (32), and (76)[a]

Compound	M (g mole^{-1})	ρ (g cm^{-3})	n	A (J)	γ^d (mJ m^{-2})	d_0 (nm)
Heptane	100.2	0.684	1.39	1.05×10^{-20}	20.3	0.22
Dodecane	170.3	0.749	1.42	9.49×10^{-21}	25.4	0.18
Eicosane	282.5	0.789	1.44	2.07×10^{-20}	29.0	0.26
SiO$_2$ (quartz)	60	2.65	1.54	4.14×10^{-20}	78.0	0.22
Polystyrene	(104)[b]	1.05	1.59	2.2×10^{-20}	41.0	0.23
Water	18	1.00	1.33	2.43×10^{-20}	21.3	0.33

[a]Equations (28), (32), and (76) are used to evaluate A, then A and γ^d values from Chap. 6 are used to evaluate δ_0.
[b]Monomer.

visible wavelengths. These values have then been used along with γ^d values from Chap. 6 to calculate d_0 values according to Eq. (82). The resulting values of d_0 are seen to be physically reasonable. That such plausible values for d_0 are obtained is especially noteworthy in view of the approximations made in the calculations.

The applicability of this procedure receives a far more stringent test in the case of water and SiO_2 than for the hydrocarbons. London forces are assumed to be the only contributors to γ (i.e., $\gamma^d = \gamma$) for hydrocarbons. This is definitely not the case for quartz or water, so the d_0 values obtained for these substances are quite satisfactory.

It is extremely difficult to measure the Hamaker constant directly, although this has been the object of considerable research effort. Direct evaluation, however, is complicated either by experimental difficulties or by uncertainties in the values of other variables which affect the observations.

The direct measurement of van der Waals forces has been undertaken by literally measuring the force between macroscopic bodies as a function of their separation. The distances, of course, must be very small, so optical interference methods may be used to evaluate the separation. The force has been measured from the displacement of a sensitive spring and also from capacitance-type measurements. The two principal sources of difficulty in these methods are external vibrations and surface roughness. Nevertheless, it has been possible to verify directly the functional dependence on radii for the attraction between dissimilar spheres (see Table 11.3), to determine the retardation of van der Waals forces (see Table 11.1), and to evaluate the Hamaker constant for several solids, including quartz. Values in the range of 6×10^{-20} to 7×10^{-20} J have been found for quartz by this method. This is remarkably close to the value listed in Table 11.4 for SiO_2.

A more common situation than two bodies interacting across a vacuum is the case where some medium intervenes between the interacting bodies. Before proceeding any further, then, let us examine the effect of this medium on particle interactions.

11.9 Effect of the Medium on the Van der Waals Attraction

Until now we have considered the interaction between isolated molecules or macroscopic bodies when the particles are separated by a vacuum. The former may be reasonably applied to molecules in the gas phase. However, for dispersions of one phase in another, the effect of the medium must be taken into account. The easiest way to do this is to

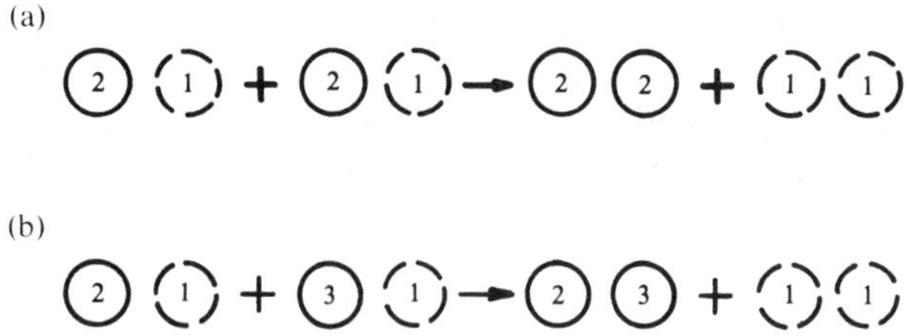

(a)

(b)

Figure 11.8 The flocculation process as a pseudochemical reaction. The solid lines indicate particles of the dispersed phase, and the dashed lines satellite particles of the solvent. In (a) the dispersed particles are chemically identical and in (b) the dispersed particles are different substances.

consider the pseudochemical reaction illustrated in Fig. 11.8a. The particles numbered 2 (solid lines) represent the dispersed phase, and those numbered 1 (dashed lines) are the solvent. Note that both of the dispersed particles are of the same material in this reaction. In the initial condition, each dispersed particle and its satellite solvent particle comprise an independent kinetic unit. Figure 11.8a represents the process in which the two dispersed particles come together to form a doublet and the two solvent particles form a kinetically independent doublet.

The change in the potential energy which accompanies this process is given by

$$\Delta\Phi = \Phi_{11} + \Phi_{22} - 2\Phi_{12} \tag{83}$$

where the subscripts apply to the two types of particles. Each of the terms for Φ on the right-hand side of Eq. (83) depends in the same way on the size and distance parameters and differs only in molecular parameters which are fully contained in the Hamaker constant. Therefore $\Delta\Phi$ follows the appropriate function for the interaction from Table 11.3 with the following value of the Hamaker constant:

$$A_{212} = A_{11} + A_{22} - 2A_{12} \tag{84}$$

The subscript 212 indicates two particles of type 2 separated by the medium of type 1.

An approximation that results in a useful simplification is

$$A_{12} \simeq \sqrt{A_{11}A_{22}} \tag{85}$$

Equation (85) says that the interaction between dissimilar bodies is given by the geometrical mean of the homogeneous interactions for the two species considered separately. This geometrical mixing rule is widely used in solution theory to calculate heterogeneous interactions. We invoked this type of averaging procedure in Sects. 3.7 and 6.12 when problems arose which required a 1–2 interaction to be expressed in terms of 1–1 and 2–2 interactions.

Combining Eqs. (84) and (85) leads to

$$A_{212} = (A_{11}^{1/2} - A_{22}^{1/2})^2 \tag{86}$$

This is the effective value of the Hamaker constant to be used in evaluating the attraction between (like) particles embedded in a medium. Equation (86) leads to three important generalizations about the value of A_{212}:

1. The effective Hamaker constant A_{212} is always positive, regardless of the relative magnitudes of A_{11} and A_{22}. Thus identical particles exert a net attraction on one another due to van der Waals forces in a medium as well as under vacuum.
2. Embedding particles in a medium generally diminishes the van der Waals attraction between them. Table 11.4 shows that the Hamaker constants for homogeneous interactions, A_{ii}, are generally of the same order of magnitude for different substances. Therefore the effective Hamaker constant—which depends on their difference according to Eq. (86)—will be smaller than A_{ii} for either of the homogeneous interactions. For example, if A_{11} and A_{22} equal 8.1×10^{-20} and 6.4×10^{-20} J, then A_{212} is 10^{-21} J, almost two orders of magnitude less than the individual A_{ii} values.
3. For $A_{11} = A_{22}$, $A_{212} = 0$, and $\Phi_A = 0$. From the viewpoint of van der Waals forces, this condition corresponds to no net interaction between particles. By using experimental criteria to identify this state of affairs, it is possible to vary the medium in a disperse system until this condition is met and then use the surface tension of the medium [via Eq. (81)] to evaluate A_{11} and, therefore A_{22}. Going further, Eq. (81) can be applied again to estimate γ_{22} for the dispersed particles. This strategy implies that suitable values for d_0 are available.

Since van der Waals forces are responsible for the flocculation of lyophobic colloids, the mitigating influence of the continuous phase on the attraction between dispersed particles imparts a measure of stability to the system. This result was anticipated in our remarks about the DLP theory in Sect. 11.8. An even more dramatic modification of van der Waals attraction results when three different substances are involved: dispersed particles of 2 and 3 separated by medium 1. Figure 11.8b shows the flocculation of the dispersed particles for this situation. By analogy with the flocculation of identical particles, we write the following:

1. For the change in potential energy

$$\Delta\Phi = \Phi_{11} + \Phi_{23} - \Phi_{12} - \Phi_{13} \qquad (87)$$

2. For the contribution of molecular properties

$$A_{312} = A_{11} + A_{23} - A_{12} - A_{13} \qquad (88)$$

3. With the A_{ij} values replaced by $(A_{ii}A_{jj})^{1/2}$

$$A_{312} = (A_{11}^{1/2}A_{11}^{1/2}) + (A_{22}^{1/2}A_{33}^{1/2}) - (A_{11}^{1/2}A_{22}^{1/2}) - (A_{11}^{1/2}A_{33}^{1/2}) \qquad (89)$$

which factors to

$$A_{312} = (A_{33}^{1/2} - A_{11}^{1/2})(A_{22}^{1/2} - A_{11}^{1/2}) \qquad (90)$$

What makes this result particularly interesting is that one of the factors in Eq. (90) can be positive and one negative, in which case the effective Hamaker constant itself becomes negative. A negative proportionality factor in an expression based on attraction makes the potential energy change positive. The pairwise attraction between bodies results in a net repulsion between dissimilar particles. Since the factors in Eq. (90) must have different signs, all that is required for this condition of a negative effective Hamaker constant is that the A value for the continuous phase be intermediate between the Hamaker constants of the two types of dispersed particles: $A_{22} > A_{11} > A_{33}$ or $A_{33} > A_{11} > A_{22}$.

For each of the A terms in Eq. (90) we may substitute the corresponding version of Eq. (81). Strictly speaking, each material is characterized by its own intermolecular spacing, and this as well as its γ value should be used in Eq. (81). In a number of systems that have been investigated, however, the observed range of d_0 values is quite narrow, suggesting that d_0 can be regarded as a constant, at least as a first approximation. For a variety of polymers a value of about 0.2 nm appears to be a reasonable estimate for d_0. With this (assumed constant) value factored out, Eq. (90) becomes

$$A_{312} = 24\pi d_0^2(\gamma_3^{1/2} - \gamma_1^{1/2})(\gamma_2^{1/2} - \gamma_1^{1/2}) \qquad (91)$$

With the assumed uniformity of d_0 values, Eq. (91) shows that the flocculation of dissimilar particles becomes energetically unfavorable when the surface tension of the medium is intermediate between the surface tensions of the two kinds of dispersed units.

An interesting variation on this idea is the study of particle engulfment by an advancing solidification front. In such an experiment solid particles are dispersed in an appropriate medium which is then allowed to solidify in a channel along which a suitable temperature gradient is maintained. The fate of the dispersed units is monitored microscopically as the solidification front advances. What is observed is that the dispersed particles are either engulfed by the solid or pushed along in the liquid by the advancing front. These observations can be interpreted in terms of the formalism we have developed by considering the solid and the dispersed units as interacting through a medium comprised of the melt. The fact that the solid and the melt are chemically identical is immaterial, since they are in different physical states.

If we designate the melt as component 1, the dispersed particles as 2, and the solid front as 3, then engulfment is equivalent to flocculation and occurs spontaneously when A_{312} is positive. In this case $\Delta\Phi$ as given by Eq. (87) provides ΔG for the engulfment process and, since the Φ's themselves are negative, A_{312} must be positive for spontaneous engulfment. Conversely, rejection by the front requires a negative value for A_{312}. Again under the assumption of the constancy of d_0, rejection is predicted so long as the surface tension of thet melt is intermediate between the values of the solid and the dispersed particles. The following example considers an illustration of this kind of data.

Example 11.4   Experiments were conducted at 80°C in which the solidification front of naphthalene was observed to either engulf or reject dispersed particles of several solids. Table 11.5 lists the observed engulfment (E) or rejection (R) behavior for various systems as well as the

Table 11.5   Results of Engulfment Experiments Involving Various Solids Dispersed in Naphthalene[a]

Dispersed particles	Observed behavior[b]	γ (mJ m^{-2})	$(\gamma_3^{1/2} - \gamma_1^{1/2})(\gamma_2^{1/2} - \gamma_1^{1/2}) \times 10^4$ (mJ m^{-2})
Acetal	R	41.9	−4.39
Nylon-6	R	41.7	−4.30
Nylon-6,6	R	40.8	−3.89
Nylon-12	R	38.4	−2.77
Nylon-6,10	R	36.0	−1.60
Nylon-6,12	R	32.0	+0.41
Polystyrene	E	27.6	+2.79
Teflon	E	15.5	+10.53
Siliconed glass	E	11.5	+13.75

Source: Data of A. W. Neumann, S. N. Omenyi, and C. J. van Oss, *Colloid Polym. Sci.,* *257*:413 (1979).
[a]Data discussed in Example 11.4
[b]E, engulfment; R, rejection.

surface tensions of the various substances. The surface tensions of solid and liquid naphthalene at 80°C are 26.4 and 32.8 mJ m^{-2}, respectively. Is the surface tension criterion cited above consistent with these observations? How might any inconsistencies be explained? Evaluate the product $(\gamma_3^{1/2} - \gamma_1^{1/2})(\gamma_2^{1/2} - \gamma_1^{1/2})$ for these systems.

Solution   The surface tension of liquid (1) naphthalene is greater than that of solid (3) naphthalene. Therefore A_{312} is expected to be negative for all systems having γ values greater than γ_1. This is the case for the first six compounds listed in Table 11.5. Therefore these substances are expected to display rejection by the solidification front. This is indeed observed for five of the six cases. The case of nylon-6,12 which deviates from the predicted behavior is best understood by examining the product $(\gamma_3^{1/2} - \gamma_1^{1/2})(\gamma_2^{1/2} - \gamma_1^{1/2})$. Values of this product for the various systems considered are listed in Table 11.5. The factor arising from the solid–liquid (3–1) naphthalene has the constant value -0.0186 for all cases, but differs when various solids are used as component 2. For nylon-6,12, the second factor becomes -0.0022, and the product of the two, 0.41×10^{-4} mJ m^{-2}, is the smallest of all such products listed in the table. As the surface tension difference decreases, the sensitivity of the behavior to variations in d_0 increases.

•

It is apparent that a single dispersed phase could be investigated with an assortment of matrix materials having known properties. The engulf-ment–rejection behavior of these systems may then be used to establish bracketing values of γ and A for the dispersed phase.

   Although we have not stopped dealing with van der Waals attraction, we have come a long way from our initial point of view. Without any help from a repulsive mode of interaction, we are able to account for situations in which the dispersed state of a system is energetically favored over the flocculated state. The dispersion medium clearly plays an important role in this, since the process involves breaking "bonds" between the dispersed particles and the medium and replacing them with new "bonds." Just as in a metathesis reaction between chemical compounds, it is the difference in the "bond strength" between the final and initial states that determines the net interaction. If the attraction between molecules can thus override itself, this may be helped along by other mechanisms which are based on actual repulsion between approaching surfaces. In the remainder of this chapter we will examine such a mechanism that arises from the mutual repulsion of polymer molecules adsorbed on the approaching surfaces. In the next chapter we consider the overlap of ion atmospheres as the basis for such a repulsion.

11.10 Steric Stabilization: General

Polymers have been used to stabilize dispersions of solids in liquids against flocculation since antiquity. Paints and inks used by ancient civilizations were prepared by dispersing suitable pigments in water and "protecting" the resulting system by additives such as gum arabic, egg albumin, or casein. Gelatin has also been used extensively as a stabilizing agent. In molecular terms, these substances are charged polymers, polyelectrolytes, and their stabilizing influence is traceable to both electrostatic and polymeric effects. Each of these contributions is complicated in its own right, so we divide the discussion into two parts. Charge effects are taken up in the next chapter in terms of low molecular weight electrolytes. In this section and the next, we consider the stabilizing effects of non-ionic polymers.

The first requirement for dispersion stabilization by polymers is that the polymer adsorb at the solid–solution interface, since it is the resulting "fringe" on the solid particles that produces the desired result. In general, the adsorbed polymer is considered to reside partially at surface sites and partially in loops or tails in the adjoining solution. The distribution of polymer segments between these two states depends on the relative strength of the interactions between the polymer and the solid compared to those between the polymer and the solvent. Synthetic polymers designed specifically as stabilizers are often block copolymers which contain two different kinds of repeat units, clustered in long sequences of one kind. In this type of polymer one sequence is designed to optimize the adsorption, while the other gives maximum extension from the surface. In polymers comprised of a single kind of repeat unit, these two considerations tend to work in opposition.

Attachment of a single segment of the polymer chain is sufficient to confine the molecule to the layer of solution adjacent to the adsorbing surface. The solid exerts very little influence on the polymer molecule as a whole in such a case. In fact, the overall spatial extension of the polymer chain is expected to be about the same as that of an isolated molecule in this situation. The polymer–solvent interaction plays a more important role than the polymer–surface interaction in determining the thickness of the adsorbed layer in this case. This is only one of the relative interaction combinations possible, but it is the one which we shall consider in the greatest detail. As the number of polymer segments actually attached to the surface increases, the influence of the surface causes the spatial extension of the adsorbed chains to decrease.

The picture that emerges from this visualizes the layer adjacent to the solid surface as a polymer solution characterized by some average volume fraction of polymer ϕ^* and having an average thickness δR. If the interaction with the surface is not too strong, δR may be on the order of $2R_g$, twice the radius of gyration of the polymer in the solution under consideration. The following example considers how the thickness of such a layer can be determined experimentally.

Example 11.5 An adsorbed layer of thickness δR on the surface of spherical particles increases the volume fraction occupied by the spheres and therefore makes the intrinsic viscosity of the dispersion greater than predicted by the Einstein theory. Derive an expression which allows the thickness of the adsorbed layer to be calculated from experimental values of intrinsic viscosity.

Solution   Equation (4.41) gives the Einstein relationship between $[\eta]$ and ϕ, the volume fraction occupied by the dispersed spheres. The volume fraction that should be used in this relationship is the value that describes the particles as they actually exist in the dispersion. In this case this includes the volume of the adsorbed layer. For spherical particles of radius R covered by a layer of thickness δR, the total volume of the particles is $\frac{4}{3}\pi R^3 + 4\pi R^2\, \delta R$. Factoring out the volume of the "dry" particle gives $V_{dry}(1 + 3\, \delta R/R)$, which shows by the second term how the volume is increased above the core volume by the adsorbed layer. Since it is the "dry" volume fraction that is used to describe the concentration of the dispersion and hence to evaluate $[\eta]$, the Einstein coefficient is increased above 2.5 by the factor $(1 + 3\, \delta R/R)$ by the adsorbed layer. The thickness of adsorbed layers can be extracted from experimental $[\eta]$ values by this formula.

•

There are several additional points to be noted about this adsorbed layer:

1. The polymer concentration ϕ^* in the adsorbed layer is different than that in the bulk solution. The two are related through the adsorption isotherm of the polymer.
2. The concentration ϕ^* is not expected to be uniform at all distances outward from the wall, although we shall use some average value of this quantity as if it were uniform.

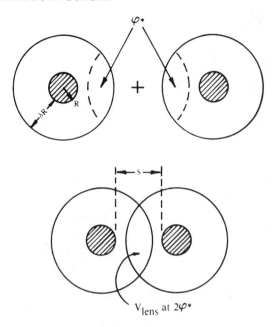

Figure 11.9 Overlap of adsorbed polymer layers upon close approach of dispersed solid particles.

3. As far as the flocculation behavior of the core particles carrying the adsorbed polymer is concerned, the surface layer rather than the bulk solution is the region of interest.

As the distance of separation between the core particles decreases in the flocculation step, the adsorbed layers begin to overlap as shown in Fig. 11.9. Ultimately, it is the crowding of the polymer chains within this overlap volume that produces any stabilizing effect observed. Consequently, this mechanism for protecting against flocculation is called steric stabilization. In the next section we consider a specific model whereby this can be accomplished. For now we take a more phenomenological point of view. It is an experimental fact that, at least under some circumstances, an adsorbed polymer layer stabilizes a dispersion against flocculation. This implies that the approaching particles in a flocculation step experience an increase in free energy ΔG_R (subscript R for repulsion) that prevents the completion of the step. Although the details of this repulsion are irrelevant for now, we associate ΔG_R with the overlap shown in Fig. 11.9; that is, ΔG_R is zero when the overlap is zero and increases as the volume of the lens-shaped overlap region increases. As a free energy change ΔG_R can be either positive or negative. In keeping with the usual sign conventions, a positive value for this quantity indicates repulsion (i.e., protection against flocculation), while a negative value contributes (along with van der Waals forces) to spontaneous flocculation.

As usual, ΔG_R can be broken into enthalpy and entropy contributions:

$$\Delta G_R = \Delta H_R - T \, \Delta S_R \tag{92}$$

where the individual terms describe changes in the respective property arising from the overlap. In principle, ΔH_R and ΔS_R can both be either positive or negative; therefore the possibility exists for ΔG_R to change sign with changes in temperature. In fact, it is observed experimentally that some sterically stabilized dispersions are caused to flocculate by increases in temperature, and others by decreasing T. The threshold temperature for the onset of flocculation in these systems is called the critical flocculation temperature (CFT). Thermodynamics offers a formalism for interpreting these observations. With this in mind, let us consider some different sign combinations for ΔH_R and ΔS_R:

1. Suppose ΔH_R and ΔS_R are both positive. In this case the enthalpy change arising from the close approach of particles with adsorbed layers opposes flocculation while ΔS_R favors it. Since ΔS_R is weighted by T in ΔG_R, it follows that increasing T causes the entropy effect to become more important.
2. The situation described in item 1 is one in which increasing temperature is expected to cause flocculation. Polyisobutylene adsorbed from 2-methyl butane and polyoxyethylene adsorbed from aqueous electrolyte are examples of systems which show a CFT with increasing temperature.
3. Since it is the positive ΔH_R that is responsible for the stability in this case, such a system is said to display enthalpic stabilization.
4. Suppose ΔH_R and ΔS_R are both negative. In this case ΔS_R opposes flocculation while ΔH_R favors it.
5. Since the resistance to flocculation decreases with decreasing temperature, flocculation is expected as T is lowered. Poly(12-hydroxystearic acid) adsorbed from n-heptane and polyoxyethylene adsorbed from methanol are examples of systems which display a CFT with decreasing temperature.
6. Since ΔS_R is the source of the stabilization in these cases, this mechanism is called entropic stabilization.

7. If ΔH_R is positive and ΔS_R is negative, ΔG_R is positive at all temperatures and the system is not subject to flocculation by changes in temperature.

Several things should be noted about the foregoing. First, both ΔH_R and ΔS_R are generally functions of temperature and may change signs themselves as T varies. Second, a given polymer may be governed by ΔH_R in one solvent and by ΔS_R in another. Finally, the addition of a second solvent to a dispersion can also induce flocculation in a polymer-stabilized system. We saw in Chap. 3 that changes of solvent goodness (as seen by the properties of polymer solutions) could be brought about by both temperature changes and addition of diluents.

Figure 11.10 illustrates the sort of data that can be used to determine the CFT for a dispersion. The absorbance of the system is measured as a function of temperature. A discontinuity in absorbance is observed to develop over a relatively narrow range of temperatures. The system shown in Fig. 11.10 is an aqueous latex dispersion, stabilized by polyoxyethylene. The threshold for flocculation is about 291 K. One characteristic of steric stabilization is that flocculated systems usually redisperse spontaneously if the goodness of the solvent is subsequently improved. In terms of the potential energy diagrams of Fig. 11.1, this shows that flocculation in these systems does not occur by particles dropping into a deep primary minimum but, rather, occurs in a shallow minimum at a larger distance from the surface.

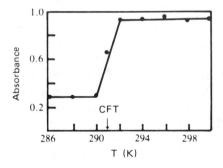

Figure 11.10 Critical flocculation temperature of aqueous poly(vinyl acetate) dispersion stabilized by poly(oxyethylene) indicated by a sharp change in absorbance with temperature. [Reprinted with permission from D. H. Napper, in *Colloid and Interface Science* (M. Kerker, R. L. Rowell, and A. C. Zettlemoyer, eds.), Academic, New York, 1977.]

Before we turn to a more detailed look at the origin of steric stabilization, it is informative to consider the assortment of observations that are encountered if the polymer content of the continuous phase is varied from almost pure solvent to pure liquid polymer. At very low polymer concentrations the solute can induce flocculation in a dispersion; at somewhat higher concentrations it can stabilize the dispersion against flocculation. At still higher concentrations there is a wide range of polymer concentrations in which the dispersion flocculates. Dispersions which are stable against flocculation have been prepared in polymer melts. This variety of behaviors with increasing polymer concentration suggests that the process under consideration is too complicated for explanation by a single model. In the next section we examine a mechanism for the stabilization that is observed at relatively low polymer concentrations.

It is not difficut to understand how polymers can be flocculating agents at very low concentrations. At these concentrations the solid surfaces have some vacant adsorption sites. Polymer-induced flocculation is interpreted to mean that polymer chains bridge core particles together by adsorbing on the surfaces of two different dispersed particles.

In the next section we assume the adsorbed polymer molecule is not very different from an isolated molecule of the same polymer in solution. This assumption limits the applicability of our discussion but permits us to invoke some of the concepts of Chapter 3 for the problem at hand.

11.11   Steric Stabilization: The Flory-Krigbaum Approach

The adsorbed layers on separate core particles first begin to overlap at the outermost extreme of the "fringe," where the surface exerts the least influence. As a first approximation, then, the initial encounter between two approaching core particles is comparable to the approach of two polymer coils in solution. In Sect. 3.6 we saw that the concept of excluded volume could be used to account for the initial nonideality in a polymer solution. Remember that the excluded volume is that region of space from which one molecule is denied access because of the presence of a second molecule. This also describes what happens as the adsorbed layers on two flocculating particles approach one another. To quantify this idea, we are interested in an expression for ΔG_R as a function of the distance of separation between core particles as shown in Fig. 11.9.

It is important to remember that the overlapping layers in which we are interested are solutions of pendant or looped polymer chains surrounded by solvent. There is room for interpenetration of the two domains, but the process is uphill thermodynamically. One way to illustrate this and to review some pertinent results from earlier chapters is the following argument against the overlap of coil domains:

1. If we treat the domain of the chain as a random coil, Eq. (2.73) or (2.77) shows that the radius of this domain is proportional to the square root of the degree of polymerization n. This means the volume of the domain varies with $n^{3/2}$.
2. Since n is the number of chain segments in this volume, the volume fraction ϕ^* of the domain which is actually occupied by polymer is proportional to $n/n^{3/2} = n^{-1/2}$.
3. The fraction of the domain which is *not* occupied by polymer is $1 - \phi^* = 1 - n^{-1/2}$. For large n this fraction is close to unity, which shows that there is plenty of space within the domain for additional polymer segments.
4. Those sites that are not occupied by chain segments are potential placement sites for the segments of a second polymer molecule.
5. Following the kind of statistical argument used in Sect. 3.6, we argue that there are $(1 - n^{-1/2})^n$ ways of placing a second molecule of n segments within the domain of the first.
6. This last expression is approximately equal to $\exp(-n^{1/2})$, which shows that the probability of a second molecule entering the domain of the first decreases rapidly with increasing n.

Flory and Krigbaum developed these ideas into a theory of solution nonideality which is useful in the present context. As the two coil domains overlap as shown in Fig. 11.9, the concentration of chain segments in the lens-shaped volume of overlap doubles compared to its value in the separate layers: $\phi^* \rightarrow 2\,\phi^*$. Solutions tend to dilute spontaneously, and not become more concentrated; therefore we expect ΔG in the lens to be positive for this process. The total free energy change associated with the overlap depends on both ΔG_{lens} and the volume of the lens; that is,

$$\Delta G_R = (\Delta G_{\text{lens}})(V_{\text{lens}}) = (\text{concentration effect}) \times (\text{geometrical effect}) \tag{93}$$

Next we consider these two contributions separately.

The Flory–Huggins theory of Sect. 3.7 provides the components from which we can assemble expressions for $\Delta G_{m,\phi^*}$ and $\Delta G_{m,2\phi^*}$, the weighted difference of which gives ΔG_{lens}. The strategy is to apply these expressions to the domain of the individual polymer coil, which we define to be V_d. If it were necessary to arrive at a numerical estimate for this volume, we could treat the domain as a sphere of radius R_g and write $V_d = \frac{4}{3}\pi R_g^3$. Within such a domain, the following applies:

1. The volume fraction of polymer is ϕ^* and the volume fraction solvent is $1 - \phi^*$.
2. The number of polymer molecules N_2 is unity and the number of solvent molecules is $(1 - \phi^*)V_d/(\bar{V}_1/N_A)$, where $\bar{V}_1$ is the partial molar volume of the solvent. Therefore $N_1 = (1 - \phi^*)V_d N_A/\bar{V}_1$.
3. The Flory–Huggins expression for the enthalpy of mixing is given by Eq. (3.71). In the present notation it becomes

$$\Delta H_m = \tfrac{1}{2}z N_1 \phi_2 \, \Delta w = \chi \, RTN_1 \phi^*$$

4. The Flory–Huggins expression for the entropy of mixing is given by Eq. (3.65). In the present notation it becomes

$$-T \, \Delta S_m = RT(N_1 \ln \phi_1 + N_2 \ln \phi_2)$$
$$= RT[N_1 \ln (1 - \phi^*) + N_2 \ln \phi^*]$$

Since $N_2 \ll N_1$, the second term can be neglected compared to the first to give

$$-T\Delta S_m \simeq RTN_1 \ln(1 - \phi^*)$$

5. Within the volume of the coil domain we can write

$$\Delta G_m = \Delta H_m - T \, \Delta S_m = RTN_1[\chi\phi^* + \ln(1 - \phi^*)]$$

Substituting for N_1 from item 2, we then have

$$\Delta G_m = (RT \, V_d N_A/\bar{V}_1)(1 - \phi^*)[\chi \, \phi^* + \ln(1 - \phi^*)]$$

which may be approximated as

$$\Delta G_m = RT \frac{V_d N_A}{\bar{V}_1} [(\chi - 1)\phi^* + (\tfrac{1}{2} - \chi)\phi^{*2}]$$

by expanding the logarithm (see Appendix A) and retaining no terms higher than second order in ϕ^*.

6. For the process $\phi^* \rightarrow 2 \phi^*$, $\Delta G_{lens} = \Delta G_{m,2\phi^*} - 2 \Delta G_{m,\phi^*}$. Therefore

$$\Delta G_{lens} = RT \frac{V_d N_A}{\bar{V}_1} \{[(\chi - 1)(2\phi^*) + (\tfrac{1}{2} - \chi)(2\phi^*)^2]$$

$$-2[(\chi - 1)\phi^* + (\tfrac{1}{2} - \chi)\phi^{*2}]\}$$

$$= \frac{2RTV_d N_A}{\bar{V}_1} (\tfrac{1}{2} - \chi)\phi^{*2}$$

7. We can write $\phi^* = \bar{V}_2/N_A V_d$, where $\bar{V}_2$ is the partial molar volume of the polymer. With this substitution

$$\Delta G_{\text{lens}} = 2k_B T \, \frac{\bar{V}_2^2}{\bar{V}_1 \, V_d} \, (\tfrac{1}{2} - \chi)$$

8. Finally, we give without proof the expression for the volume of the lens which experiences the free energy change given by item 7:

$$V_{\text{lens}} = \frac{2\pi}{3} \left(\delta R - \frac{s}{2} \right)^2 \left(3R + 2\,\delta R + \frac{s}{2} \right)$$

Substituting the results in items 7 and 8 into Eq. (93) gives

$$\Delta G_R = \frac{4\pi}{3} k_B T \, \frac{\bar{V}_2^2}{\bar{V}_1 \, V_d} \left(\frac{1}{2} - \chi \right) \left(\delta R - \frac{s}{2} \right)^2 \left(3R + 2\,\delta R + \frac{s}{2} \right) \quad (94)$$

which is the result toward which we have been working.

Two of the terms in Eq. (94) are easy to interpret in relation to dispersion stability. Although it is not the entire geometrical effect, the term containing $\delta R - s/2$—which is half the maximum width of the lens—shows that the repulsion increases with the square of this quantity. This shows how the repulsion develops as the overlap increases.

The term $\tfrac{1}{2}$-χ is the only factor in Eq. (94) that can have either positive or negative values, depending on whether χ is less than or greater than $\tfrac{1}{2}$. Again we review some ideas from Sect. 3.7 pertaining to the Flory-Huggins interaction parameter χ:

1. In units of $k_B T$, χ measures the energy change per pair for the process $(1,1) + (2,2) \rightarrow 2(1,2)$ in which the indices refer to different pairwise interactions.
2. $\chi = \tfrac{1}{2}$ corresponds to the critical point on a miscibility diagram for a polymer of infinite molecular weight.
3. $\chi < \tfrac{1}{2}$ corresponds to a good solvent and $\chi > \tfrac{1}{2}$ corresponds to a poor solvent.
4. According to Eq. (3.81),

$$\frac{1}{2} - \chi = \psi \left(1 - \frac{\Theta}{T} \right)$$

so $T = \Theta$ divides the temperature range into good ($T > \Theta$) and poor ($T < \Theta$) regions.

This review indicates that good solvent conditions (in terms of either χ or Θ) result in a positive value for ΔG_R. This is what would be expected from a model which assumes that the first encounter between particles with adsorbed layers is dominated by the polymers. Conversely, in a poor solvent ΔG_R is negative and amounts to a contribution to the attraction between the core particles as far as flocculation is concerned. Under these conditions the polymer itself is at the threshold of phase separation. Van der Waals attraction between the core particles further promotes aggregation, but it is possible that flocculation could be induced in a poor solvent even if the medium decreases the effective Hamaker constant to zero.

Replacing $\frac{1}{2} - \chi$ by $1 - \Theta/T$ offers an opportunity to test the ideas of the Flory–Krigbaum theory against experimental observations of the CFT. It is apparent from the material presented here that the CFT and the Θ temperature correspond to the same condition. Table 11.6 lists CFT values and independently determined Θ temperatures for several systems. The agreement between the two is quite satisfactory for these systems. Incidentally, electrolytes are added to the aqueous media in Table 11.6 to supress the ion atmosphere mechanism for stabilization. We shall see how this is accomplished in the next chapter.

There is another type of free energy change that can be considered within the overlap volume in addition to the concentration effect considered by the Flory–Krigbaum theory. This additional contribution to ΔG_{lens} is likely to be more important for $s < \delta R$ and should be considered when the outcome of the encounter is not determined by the initial approach of the colliding particles. This contribution arises from an elastic response by the adsorbed polymer, effectively pushing the approaching particles apart.

The elasticity of polymer coils is a well-known phenomenon and is involved in many important mechanical properties of bulk polymers. Stated briefly, it arises from a difference in conformational entropy between stretched and randomly jumbled chains. A statistical theory which counts the number of ways the two conformations can come about can be combined with the Boltzmann entropy equation [Eq. (3.45)] to give an expression for the entropy change associated with stretching a chain or its "snapping" back to equilibrium. At small deformations, individual chains obey Hooke's law with the force constant

$$k_H = 3k_B T/n l_0^2 \tag{95}$$

where n is the degree of polymerization and l_0 is the step length in the random walk analysis of Sect. 2.12. By Eq. (2.77), the denominator in this expression can be replaced with $6R_g^2$. In the context of steric stabilization,

Table 11.6 Critical Flocculation Temperatures for Various Polymer–Solvent Systems, with Θ Temperatures Included for Comparison.

Stabilizer	Molecular weight	Dispersion medium	CFT (K)	Θ (K)
Poly(ethylene oxide)	10,000	0.39 M MgSO$_4$	318	315
	96,000		316	315
	1,000,000		317	315
Poly(acrylic acid)	9,800	0.2 M HCl	287	287
	51,900		283	287
	89,700		281	287
Poly(vinyl alcohol)	26,000	2 M NaCl	302	300
	57,000		301	300
	270,000		312	300
Polyacrylamide	18,000	2.1 M (NH$_4$)$_2$SO$_4$	292	—
	60,000		295	—
	180,000		280	—
Polyisobutylene	23,000	2-Methylbutane	325	325
	150,000		325	325

Source: Reprinted with permission from D. H. Napper in *Colloid and Interface Science*, Vol. 1 (M. Kerker, R. L. Rowell, and A. C. Zettlemoyer, eds.), Academic, New York, 1977.

we can readily picture the close approach of two core particles as compressing a layer of polymeric "springs." According to Hooke's law, these springs will oppose compression with a restoring force that is proportional to the extent of compression with a force constant given by Eq. (95). This is the origin of the elastic mechanism for interparticle repulsion from adsorbed polymers.

One of the first theoretical attempts to understand steric stabilization of dispersions was based on an entropic mechanism which resembles the elastic contribution to ΔG_R. We consider this mechanism in the following example.

Example 11.6 Picture a flat surface to which rigid rods are attached by ball-and-socket type joints. The free ends of the rods can lie anywhere on the surface of a hemisphere. The approach of a second surface blocks access to some of the sites on the cap of the hemisphere. Outline the qualitative argument that converts this physical picture to a theory for stabilization. What are some of the shortcomings of the model?

Solution When the two solid surfaces are far apart ($d = \infty$), the free end of each adsorbed rod has access to Ω_∞ sites, where $\Omega_\infty = 2\pi L^2$ and L is the length of the rod. When the separation is such that the second surface cuts off access to some of these sites, the number of accessible sites becomes Ω_d. The subscript here indicates a separation less than some critical distance which is the threshold for interaction. The exact form of Ω_d and the critical separation at which it begins to apply depend on whether one or both surfaces carry the adsorbed rods. The fraction of the area of the hemisphere that remains accessible to the free ends of the rods could be calculated from geometrical considerations. Using these Ω values as substitutions in Eq. (3.46), we obtain $\Delta S_R = k \ln (\Omega_d/\Omega_\infty)$. Since $\Omega_d < \Omega_\infty$, ΔS_R is negative. This gives the effect per rod; if there are N such rods per unit area and if $\Delta H_R = 0$, then we obtain $\Delta G_R = N k_B T \ln (\Omega_d/ \Omega_\infty)$, which gives a positive repulsion as required.

This model has several deficiencies:

1. It is unrealistic for polymers, although it may be a reasonable approximation for linear amphipathic molecules.
2. It assumes very low surface coverage so that the individual rods do not interfere laterally.
3. It is a purely entropic mechanism and makes no provision for enthalpy effects.

•

References

1. F. M. Fowkes, Calculation of Work of Adhesion by Pair Potential Summation, in *Hydrophobic Surfaces* (F. M. Fowkes, ed.), Academic, New York, 1969.
2. E. D. Goddard and B. Vincent, *Polymer Adsorption and Dispersion Stability*, American Chemical Society, Washington, D.C., 1984.
3. J. N. Israelachvili and D. Tabor, Van der Waals Forces: Theory and Experiment, in *Progress in Surface and Membrane Science*, Vol. 7 (J. F. Danielli, M. D. Rosenberg, and D. A. Cadenhead, eds.), Academic, New York, 1973.
4. D. H. Kaelble, *Physical Chemistry of Adhesion*, Wiley, New York, 1971.
5. E. A. Moelwyn-Hughes, *Physical Chemistry*, 2nd ed., MacMillan, New York, 1964.
6. D. H. Napper, Steric Stabilization, in *Colloid and Interface Science*, Vol. 1 (M. Kerker, R. L. Rowell, and A. C. Zettlemoyer, eds.), Academic, New York, 1977.
7. J. Th. G. Overbeek, in *Colloid Science*, Vol. 1 (H. Kruyt, ed.), Elsevier, Amsterdam, 1952.
8. H. Sonntag and K. Strenge, *Coagulation and Stability of Disperse Systems*, Halsted, New York, 1964.
9. E. J. W. Verwey and J. Th. G. Overbeek, *Theory of the Stability of Lyophobic Colloids*, Elsevier, Amsterdam, 1948.

Problems

1. The parameter β_{12} for heterogeneous (12) interactions plays a similar role as β_{11} [Eq. (34)] does for homogeneous (11) interactions. Use entries from Table 11.1 to write an expression for β_{12}. Debye interaction makes a negligible contribution to β_{12} and $v_1 v_2 / (v_1 + v_2) \simeq \frac{1}{2}(v_1 v_2)^{1/2}$, show that

    $$\beta_{12} \simeq (f_{11L}\beta_{11})^{1/2}(f_{22L}\beta_{22})^{1/2} + (f_{11K}\beta_{11})^{1/2}(f_{22K}\beta_{22})^{1/2}$$

    where the f terms are defined by Eq. (35). If $f_{11L} = f_{22L}$ and $f_{11K} = f_{22K}$, show that this last result becomes

    $$\beta_{12} \simeq (\beta_{11}\beta_{22})^{1/2}$$

    Comment on the relevancy of this result to Eq. (85). Criticize or defend the following proposition: The geometrical mixing rule does not require the absence of permanent dipoles, only that 11 and 22 interactions both consist of the same fraction of London and permanent dipole contributions. Specific interactions, such as hydrogen bonding, must also be absent in the 11, 22, and 12 systems.

2. Pressure is a manifestation of the kinetic energy of gas molecules. According to the van der Waals equation of state [see Eq. (7.15)], the pressure of 1 mole of gas must be increased by an amount a/V^2 due to intermolecular attractions which decrease the pressure from what it would be if ideal. Use the term a/V^2 as a general expression for the attraction between a pair of molecules and, based on this, reconstruct the argument leading to Eq. (44).

3. H. Müller studied by dark-field microscopy the flocculation of colloidal gold upon the addition of NaCl to the aqueous sol. For a sample in which the gold particles have a 36.9-Å radius, the following particle counts were observed at different times after the colloid was made about 0.2 M with NaCl*:

$t(s)$	120	195	270	390	450	570
$N \times 10^{-8}$ (cm^{-3})	11.2	7.3	5.4	4.5	3.7	2.7

    Determine the second-order rate constant k_{exp} which describes this flocculation process. How does k_{exp} compare with k_r as given by Eq. (54)?

4. Polystyrene latex particles were flocculated by the addition of $Ba(NO_3)_2$. The number of dispersed particles deposited onto a planar polystyrene surface was determined 15 min after the addition of salt by optical microscopy. The light microscope does not permit the aggregation of the deposited particles to be determined; subsequent examination by the electron microscope gives

---

*H. Müller, *Kolloid Z.*, *38*:1 (1926).

this information. G. E. Clint et al.* obtained the following results:

Ba(NO$_3$)$_2$ concentration $\times 10^3$ (mole liter^{-1})	Total deposition cm$^{-2} \times 10^{-5}$ (after 15 min)	Percent deposit		
		Single	Double	Triple
9.1	8.04	94.7	3.3	0.3
15.4	14.25	95.0	4.3	0.4
22.7	14.43	82.9	11.3	3.5
57.0	11.25	75.0	15.8	5.6

Discuss these data in terms of the following points:

(a) For all salt concentrations the order of particle abundance in the deposit is single > double > triple.

(b) The decrease in total deposition with increasing concentration is not offset by the higher aggregation state of the deposit, but arises from the slower diffusion of more highly aggregated kinetic units.

5. If flocculation involves two noninteracting spheres of different radii R_i and R_j, Eq. (52) predicts

$$k_r = \frac{2}{3}\frac{kT}{\eta}(R_i + R_j)\left(\frac{1}{R_i} + \frac{1}{R_j}\right)$$

Retrace the development of Eq. (52) to verify this result. Show that this expression is identical to

$$k_r = \frac{2}{3}\frac{kT}{\eta}\left[4 + \left(\sqrt{\frac{R_i}{R_j}} - \sqrt{\frac{R_j}{R_i}}\right)^2\right]$$

Estimate the ratio R_i/R_j needed to account for a k_r value of 2.9×10^{-11} cm^3 s^{-1} as observed for arachidic acid sols.† Does this expression reduce to the proper limit when $R_i = R_j$?

6. Verify (a) that Eq. (60) is a solution to Eq. (59), (b) that Eq. (63) reduces to Eq. (51) if $\Phi = 0$, and (c) that Eq. (64) leads to a rate constant larger than k_r if $\Phi = 0$ for $r > \Delta$ and $\Phi = -\infty$ for $r \leqslant \Delta$, where $\Delta > 2R$. Show that the experimental k_r value cited in Problem 5 is consistent with a value of Δ equaling $5.4R$ for an aqueous colloid at 20°C ($\eta = 0.01$ P). Discuss the relevancy of this last result to the rapid flocculation of particles between which van der Waals attraction exists.

7. A gas adsorption isotherm may be derived by comparing the adsorbed layer around a solid particle to a planetary atmosphere, with an equilibrium pressure p_0 at the surface. The change in free energy for a molecule going from the bulk pressure p to the surface is $kT \ln (p_0/p)$. Equating this with Eq.

---

*G. E. Clint, J. H. Clint, J. M. Corkill, and T. Walker, *J. Colloid Interface Sci.,* **44**:121 (1973).

†R. H. Ottewill and D. J. Wilkins, *Trans. Faraday Soc.,* **58**:608 (1962).

(73), the potential energy of attraction which is responsible for the adsorption, gives

$$kT \ln\left(\frac{p_0}{p}\right) = \frac{\rho N_A}{M} \frac{\beta\pi}{6} z^{-3}$$

Since $z \propto V$, this may be written $\ln(p_0/p) = \text{const.} \, V^{-3}$, where V is the volume of gas adsorbed. A more general form of this isotherm, called the Frenkel–Halsey–Hill (FHH) isotherm, treats the power dependence of V as an unknown n and writes

$$\left(\frac{V}{V_m}\right)^n = \frac{K}{\ln(p_0/p)}$$

Prepare a plot of the FHH isotherm using $n = 3$ and $K = 0.1$ and comment on the resemblance of this isotherm to actual gas adsorption isotherms as shown in Chap. 9.

8. The derivation of Eq. (82) follows the same argument that leads to Eq. (75), except that the blocks are assumed to be composed of stacks of matter of density ρ' in slices having a thickness δ and separated by a distance d. A molecule at O interacts with the ith slice in such a stack with an energy that is the analog of Eq. (70):

$$-d\Phi_i = \frac{\rho' N_A \beta \delta}{M} \frac{2\pi y \, dy}{\{[z + i(d + \delta)]^2 + y\}^3}$$

Evaluate the attraction between the molecule and the ith layer by integrating this expression over all values of y between 0 and ∞. The assumption that $\rho'\delta = \rho(d + \delta)$ ensures that the blocks have the correct macroscopic density. If the separation between layers is the same as the separation of the surfaces of the blocks (i.e., $z = d$), then the equivalent of Eq. (73) results from summing the Φ_i values for all i's between 0 and ∞ and taking the limit of $\delta \to 0$. Derive the analog of Eq. (73) for matter with this hypothetical structure. Note that $(1 + i)^{-4} = 1.082$. Equation (82) is obtained by assuming a similar structure for the second block and continuing along these lines.

9. Neumann et al.* have tabulated average values of experimentally determined Hamaker constants and then used surface tension data for the same systems to calculate d_0 values for the following materials:

Substance	$A_{22} \times 10^{20}$ (J)	γ (mJ m^{-2})
Polystyrene	8.47	33.0
Teflon	5.23	20.0
Nylon-6,6	12.05	46.0
Poly(methyl methacrylate)	8.83	39.0

---

*A. W. Neumann, S. N. Omenyi, and C. J. Van Oss, *Colloid Polym. Sci.*, *257*: 413 (1979).

Substance	$A_{22} \times 10^{20}$ (J)	γ (mJ m^{-2})
n-Decane	5.13	23.9
Polyethylene	8.43	31.0
Poly(vinyl alcohol)	8.84	41.0
Poly(hexafluoropropylene)	5.2	17.0

Verify that the d_0 values thus calculated show a relatively narrow distribution around a mean value close to 0.2 nm. Criticize or defend the following proposition: As a mean center-to-center intermolecular spacing, this value is on the low side; as a back-calculated parameter, however, it probably compensates for deviations from the assumed geometry, breakdown of Eq. (33) at short distances, or other shortcomings of the molecular additivity principle.

10. An extreme in sediment volume was used[*] as a criterion for the effective cancellation of interparticle attraction by the continuous phase. Nylon-6,6 dispersions consisting of 1.0 g of solid in 10 ml of n-propanol–thiodiethanol mixtures of various compositions were allowed to settle to sedimentation equilibrium. Listed here are the equilibrium sediment volumes, the volume/ volume compositions, and the surface tensions of the media:

n-Propanol–thiodiethanol
(vol/vol)

	20/20	60/40	45/55	35/65	25/75	20/80	15/85	10/90	7.5/92.5	5/95
V_{sed} (cm^3)	2.15	2.40	2.40	2.50	2.60	2.60	2.70	2.48	2.40	2.35
γ (mJ m^{-2})	26.2	28.7	30.9	33.6	36.2	38.2	39.6	40.7	44.0	47.1

Plot the sediment volume versus the surface tension of the continuous phase for these dispersions. What is the apparent surface tension of the nylon-6,6? Briefly describe some precautions that must be observed in interpreting results such as these in terms of γ values for the dispersed solid.

11. African green monkey kidney cells (component 2) were cultured on suspensions of collagen-coated dextran particles (component 3). Harvesting such cells is traditionally accomplished by scraping, ultrasonication, or the use of chelating agents to disrupt cation bridging between surfaces 2 and 3. Van Oss and collaborators[†] reasoned that the cells might be eluted by lowering the surface tension of the suspending medium (component 1). What is the basis for this expectation? Dimethyl sulfoxide (DMSO) was incrementally added to cell-carrier particle dispersions in buffered aqueous solution. The surface tension of the eluting liquids and the percent yield of the harvested cells are tabulated:

---

[*]A. W. Neumann, J. Visser, R. P. Smith, S. N. Omenyi, D. W. Francis, J. K. Spelt, E. B. Vargha-Butler, W. Zingg, C. J. Van Oss, and D. R. Absolom, *Powder Technol.*, 37:229 (1984).

[†]C. J. Van Oss, C. K. Charney, D. R. Absolom, and T. J. Flanagan, *BioTechniques*, Nov./Dec., 194 (1983).

Vol % DMSO	0.0	7.5	10.0	12.5	15.0	17.0	18.0	19.0
γ (mJ m^{-2})	73.0	67.0	65.0	63.6	62.1	61.2	60.7	60.3
Percent yield	8.4	15.1	16.5	35.6	42.3	59.3	66.7	80.1

For the solids γ_2 and γ_3 have been estimated to be 68.9 and 32 mJ m^{-2}, respectively. Is the observed behavior qualitatively consistent with expectations? Suggest some factors that might be responsible for any quantitative discrepancy.

12. Ahmed et al.* measured η_{red}/ϕ for polystyrene latexes with adsorbed layers of commercial poly(vinyl alcohol) (PVA) samples of different molecular weights. The latex particles were 190 nm in diameter and the limiting values of η_{red}/ϕ as $\phi \rightarrow 0$ had the following values for PVA samples of the indicated molecular weight:

M_{PVA} (g mole^{-1}):	None	26,400	23,000	80,000	79,100
R_g (nm)	—	6.5	7.4	13.2	13.0
[η]	3.0	5.0	5.5	8.2	9.1

Use the experimental value of [η] for the bare particles and the relationship given in Example 11.5 to estimates δR for these particles. How do the layer thicknesses compare with $2R_g$, where the given radii of gyration were determined for the polymers in bulk solution?

13. The parameter χ is proportional to the energy of interaction per 1–2 pair. To allow for solvent molecules of various sizes, we can write $\chi \propto n_1^* \Delta w$, where n_1^* is the number of segments in a solvent molecule, or $\bar{V}_1/V_1^*$, where V_1^* is the molar volume of a segment. The ratio $(\frac{1}{2} - \chi)/\bar{V}_1$ which appears in the expression for ΔG_R in Eq. (94) can be written $(1/2\bar{V}_1) - (\text{const.} \ \Delta w/V_1^*)$. Criticize or defend the following proposition: For a specific polymer in a homologous series of solvent molecules of different sizes, the second term in this expression should remain constant while the first decreases with increasing $\bar{V}_1$. When the "solvent" is the melt of the polymer under consideration, the first term is negligible because $\bar{V}_1$ becomes the partial molar volume of the polymer and is very large. Therefore ΔG_R becomes negative for such a system and flocculation is predicted. Since this contradicts experimental evidence, the Flory–Krigbaum model is seen to break down at this limit.

14. Mackor† used the model outlined in Example 11.6 to derive the expression $\Delta G_R = N k_B T(1 - d/L)$ for the repulsion per unit area of particles carrying N rods of length L when the surfaces are separated by a distance d. Assuming this repulsion equals the van der Waals attraction when the particle

---

*M. S. Ahmed, M. S. El-Aasser, and J. W. Vanderhoff, in Ref. 2.

†E. L. Mackor, *J. Colloid Sci.*, 6:492 (1951).

separation is 1.5 nm, calculate the effective Hamaker constant in this system if $L = 2.5$ nm. Select a reasonable value for N in this calculation and justify your choice.

15. In view of the model used in the last problem, criticize or defend the following proposition: If one surface carries adsorbed rods and the other is bare, the system could be stabilized against flocculation by dispersing the particles in a medium of intermediate γ. Such a system would remain dispersed indefinitely, since both steric considerations and a negative Hamaker constant oppose flocculation.

12
THE ELECTRICAL DOUBLE LAYER

Suppose that I had the power of passing through ... things, so that I could penetrate my subjects, one after another, even to the number of a billion, verifying the size and distance of each by the sense of feeling.

[From Abbott's *Flatland*]

12.1 Introduction

When ions are present in a system that contains an interface, there will be a variation in the ion density near that interface which is described by a profile like that shown in Fig. 7.11. The boundary we identify as *the* surface defines the surface excess charge. Suppose that it were possible to separate the two bulk phases at this boundary in the manner shown in Fig. 6.7. Then each of the separated phases would carry an equal and opposite charge. These two charged portions of the interfacial region are called the electrical double layer.

The first topic we take up in this chapter is the origin of the charge at certain surfaces through ion adsorption. A straightforward application of thermodynamics enables us to quantify this effect in terms of an electrical potential associated with the adsorbed ions.

After these phenomenological considerations, we shall turn our attention to various models for the distribution of charge near the surface. Although we shall examine several different models under several limiting conditions, most of the theoretical developments of this chapter will involve the following assumptions: (1) planar surfaces, (2) isolated surfaces, and (3) constant potential surfaces, examined specifically for (4) the variation with distance from the surface of the potential and (5) the effect of added electrolyte on the potential.

To apply these ideas to flocculation phenomena, we must consider what happens to these distributions of potential when two similar surfaces approach one another. By the time we reach this point we shall also want to compare the electrostatic effects of particle approach with the van der Waals effects discussed in the last chapter. This is done in terms of potential energy curves as discussed in Sect. 11.2. As we move through the chapter, our interest shifts from potential (volts) to potential energy (joules). It is important to keep track of the difference between the two as the development progresses.

12.2 Reversible Electrodes: The Silver Iodide Electrode

To arrive at an understanding of the distribution of charge and potential near an interface, it is helpful to consider an electrode. A reversible electrode is one in which each of the phases contains a common ion which is free to cross the interface. The system Ag–AgI–aqueous solution is an example of a reversible electrode. A polarizable electrode, on the other hand, is impermeable to charge carriers, although charge may be brought to the surface by the application of an external potential. The system metallic Hg–aqueous solution is an example of a polarizable electrode; we discussed the relationship between the applied potential, the interfacial tension, and the adsorption of ions in Sect. 7.13.

It is also convenient to divide ions into two categories: potential-determining and indifferent ions. The terminology here is self-explanatory. For example, we can say that Ag^+ is potential determining for the $Ag–Ag^+$ electrode and that $NaNO_3$ is an indifferent electrolyte as far as this potential is concerned. This obviously neglects any effects of $NaNO_3$ on the activity of the Ag^+. Such an approximation increases in accuracy as the concentration of electrolyte decreases. We shall consistently neglect activity corrections in this chapter.

The solubility product constant for AgI is about 7.5×10^{-17} at 25°C. This means that the equilibrium concentration of Ag^+ and I^- in a saturated solution of AgI in pure water equals about 8.7×10^{-9} mole liter^{-1}. Electrokinetic experiments (Chap. 13) on AgI particles under these conditions reveal that the particles carry a negative charge in this case. Common ion sources such as $AgNO_3$ or KI may be added to the solution to vary the proportions of the Ag^+ and I^- ions in solution, subject to the condition that the ion product equals K_{sp}. When this is done, it is found that the AgI particles reverse charge at a Ag^+ concentration of about 3.0×10^{-6} mole liter^{-1}. When the concentration of Ag^+ is greater than this, the particles are positively charged. For Ag^+ concentrations less than 3.0×10^{-6} M, they are negatively charged.

One way of understanding these results is to consider the Ag^+ and I^- ions competing for adsorption sites on the surface. The tendency of both kinds of ions to adsorb at the AgI interface is not hard to understand. After all, the solid crystals would continue to grow if more ions were present. At the point of zero charge the two kinds of ions are adsorbed equally (in stoichiometric proportion). Negatively charged particles imply the adsorption of excess I^- ions, whereas positively charged particles imply excess Ag^+ adsorption. Since the zero point of charge and the saturation concentration in pure water do not coincide, we infer that the I^- ions have a greater affinity for the surface.

The Nernst equation provides us with a relationship which permits an electrical potential difference to be associated with a concentration difference. We shall adopt the convention that the potential at the AgI–solution interface is zero at the zero point of charge (zp), a point at which the ion molarities will be symbolized c_{zp}. Our interest is to express the potential at the interface ψ_0 in terms of the concentration of ions in solution for conditions other than the zero point of charge. The Nernst equation gives

$$\psi_0 = \frac{kT}{e} \ln\left(\frac{c}{c_{zp}}\right) = \frac{2.303RT}{F} \log\left(\frac{c}{c_{zp}}\right) \tag{1}$$

where F is the Faraday constant. We are accustomed to using the second form of Eq. (1) in physical and analytical chemistry. The quantity $2.303RT/F$ has the familiar numerical value 0.05917 V at 25°C. Multiplying this by 10^3 and dividing by 2.303 gives 25.7 mV as the value of kT/e. We shall verify this result shortly, but this is a convenient way to relate a familiar numerical constant to the units which are most often used in surface and colloid chemistry.

Suppose we apply Eq. (1) to AgI in "pure water" (i.e., no common ion source present). We use $c_{Ag} = 8.7 \times 10^{-9}$ and $c_{Ag,zp} = 3.0 \times 10^{-6}$ to calculate

$$\psi_0 = 25.7 \ln \frac{8.7 \times 10^{-9}}{3.0 \times 10^{-6}} = -150 \text{ mV} \tag{2}$$

Identical results would be obtained if the calculation had been based on I^- concentrations rather than Ag^+. As noted, the surface is negatively charged at this concentration, since I^- is preferentially adsorbed.

In principle, part of the potential of any cell may be attributed to each interface; that is, if ψ_i is the potential drop associated with the ith interface, we can write for the total potential difference ψ_T

$$\psi_T = \psi_1 + \psi_2 + \cdots + \psi_i \tag{3}$$

Any electrochemical cell containing a Ag–AgI electrode automatically includes the AgI–solution interface and the potential associated therewith. Generally speaking, we are not able to assign absolute numerical values to the various contributions in Eq. (3). We can, however, design cells such that only one of the interfaces is sensitive to a particular ion. Clearly, Ag^+ and I^- are the potential-determining ions at the AgI-solution interface. If none of the other interfaces in the cell are appreciably affected by changes in the concentrations of these ions, then variations in the experimental cell potential ψ_T measure changes in ψ_0. This may be expressed

$$d\psi_T = d\psi_0 = \frac{kT}{e} \frac{dc}{c} \qquad (4)$$

where c is the concentration of the potential-determining ion. This result is important because it shows how *changes* in ψ_0 can be measured even if ψ_0 itself is unknown. That is, to integrate Eq. (4) back to an absolute value of ψ_0, an integration constant must be evaluated. According to our convention, this involves knowing the zero point of charge ($\psi_0 = 0$ at $c = c_{zp}$).

Although we have approached the potential ψ_0 from the perspective of electrodes, the discussion makes it clear that the potential given by Eq. (1) applies to any AgI–aqueous solution interface, and not just to electrode surfaces. We may not always know the concentrations required to use Eq. (1) numerically, but so long as the bulk concentration of potential-determining ions differs from c_{zp}, a potential difference exists at the surface. The ions of water itself are potential determining for many surfaces. These as well as added solutes or ions in equilibrium with the solid mean that surface potentials at (especially, but not exclusively) aqueous interfaces are the norm rather than something exceptional.

The potential at an interface can be related through the abundant relationships of thermodynamics to the concentration and adsorbability of ions, but thermodynamics provides no information as to how the potential varies as we move through a small distance perpendicular to the surface. This observation reminds us of Fig. 7.11, in which the profile of the variation in some general property in the immediate vicinity of a surface is shown. Figure 12.1a is essentially the same drawing with the property under discussion specified to be the potential.

The question we shall consider in the next few sections is this: How does the potential vary with distance across an interface? This question cannot be answered by thermodynamics alone, but it can be examined in terms of various models. We shall consider a succession of models for a planar surface between two phases. The models will become progres-

(a)   (b)

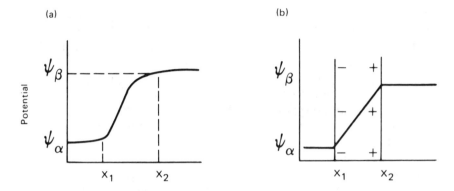

Figure 12.1   The variation of electrochemical potential in the vicinity of the interface between two phases, α and β: (a) according to a schematic profile and (b) according to the parallel plate capacitor model.

sively more complex—and therefore realistic—as we proceed. As far as this presentation is concerned, models will be proposed and modified in an intuitive way, rather than by critique of each in terms of experimental results.

12.3   The Capacitor Model of the Double Layer

Figure 12.1a shows schematically the situation in which the potential equals ψ_α at a position x_1, a small distance into the α phase, and equals ψ_β at x_2, a small distance into the β phase. One of our major goals in this chapter is to study the details of the potential variation between x_1 and x_2.

A vastly oversimplified model of how this potential variation might occur on a molecular scale (remember that the distance between x_1 and x_2 is on the order of molecular dimensions) is shown in Fig. 12.1b. In this representation two smeared-out planes of charge are situated at x_1 and x_2. Note that the model shown in Fig. 12.1b resembles a parallel plate capacitor in which two charged conducting surfaces separated by a dielectric occur with a potential difference $\Delta\psi$ between them. Although this is certainly not a realistic picture of the actual distribution of charge at a solution interface, Fig. 12.1b allows us to visualize a double layer of charge at an interface. Elsewhere in this book we have referred to the counter-ion atmosphere that adjoins a charged surface. Figure 12.1b represents this situation, and one of the layers may be regarded as a crude model of this ion atmosphere. The term *electrical double layer* is generally used to describe this physical situation and we shall use this terminology from now on in preference to *ion atmosphere*.

Our interest in this chapter and in Chap. 13 is centered primarily on that part of the double layer which extends into the aqueous solution, which is the continuous phase in many important systems. There may be some interfaces between water and a second phase in which the charge on the nonaqueous phase is essentially concentrated on the surface plane. The rigid alignment of a second layer of counter-ions in the aqueous solution is implausible, however, because of thermal agitation, which tends to diffuse the ions throughout the solution. For now, the parallel plate capacitor model will get us started by reviewing some basic relationships and units from elementary physics. The diffuse model of the double layer will be taken up in later sections.

Coulomb's law is the basic point of departure. It may be written

$$F = \frac{1}{4\pi\varepsilon_0} \frac{qq'}{\varepsilon_r r^2} \tag{5}$$

to describe the force operating between two charges q and q' separated by a distance r. The factor ε_r is the dielectric constant of the medium and the proportionality factor $1/4\pi\varepsilon_0$ implies that SI units are being used. Remember that ε_0 has the value 8.85×10^{-12} C^2 J^{-1} m^{-1}, $4\pi\varepsilon_0 = 1.11 \times 10^{-10}$ C^2 J^{-1} m^{-1}, and $1/4\pi\varepsilon_0 = 8.99 \times 10^9$ J m C^{-2}. This factor does not appear when cgs units are used. Appendix B contains some additional remarks about these two systems of units which can be especially troublesome in electrical calculations.

Next let us consider the definition of the strength of an electric field $\bar{E}$. The field $\bar{E}$ describes the force per unit charge in an electrically influenced environment,

$$\bar{E} = \frac{F}{q} \tag{6}$$

Now suppose we bring two identical $+q$ charges toward one another to a distance of separation r. Combining Eqs. (5) and (6) enables us to calculate the field at that separation:

$$\bar{E} = \frac{1}{4\pi\varepsilon_0} \frac{q}{\varepsilon_r r^2} \tag{7}$$

This is precisely the same as the force that a unit positive charge would experience at the same location. Since force is the negative gradient of the potential, Eq. (7) also supplies a second definition of field:

$$\bar{E} = -\frac{d\psi}{dx} \tag{8}$$

where ψ is the potential and x is the separation of the plates.

A fiction which helps us understand electric fields is the notion of lines of force. Suppose we imagine one line of force as emanating from each unit of positive charge. If the charge has a magnitude of $+q$, then there would be q lines of force produced by this particular charge.

A radial distance r from this central charge, the lines of force cut across a spherical surface of area $4\pi r^2$. If we divide the number of lines of force by the cross-sectional area, we obtain

$$\frac{\text{number of lines}}{\text{area}} = \frac{q}{4\pi r^2} = \varepsilon_0 \bar{E} \tag{9}$$

Equation (9) shows that the field $\bar{E}$ and the number of lines per area are directly proportional, with ε_0 the factor of proportionality. In the presence of a dielectric, ε_r is inserted into Eq. (9) to bring the latter into conformity with Eq. (7). Now suppose we apply this idea to a parallel plate capacitor.

In the case of a capacitor—taken here as a prototype of a double layer—the charges are not isolated. Instead, the lines of force emanating from one charged surface terminate at an opposite charge on the other plate of the capacitor. Figure 12.2a represents such a situation when the plates are separated by a vacuum. Suppose a plate of area A carries q charges; then we define the charge density σ^* as

$$\sigma^* = \frac{q}{A} \tag{10}$$

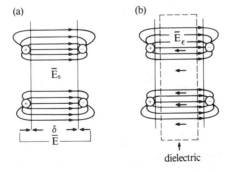

Figure 12.2 The electric field in a parallel plate capacitor: (a) The dielectric is a vacuum and (b) a material of dielectric constant ε_r is present.

Since a line of force is associated with each unit of charge, there are q/A lines of force crossing the evacuated gap between the two plates of the capacitor. As we have already seen, the number of lines of force measures the field; therefore we write for a capacitor which contains a vacuum

$$\bar{E}_0 = \frac{q}{\varepsilon_0 A} = \frac{\sigma^*}{\varepsilon_0} \tag{11}$$

If a substance with a relative dielectric constant ε_r is placed between the plates, the field will be less by this factor. This is because of the partially compensating field which is induced within the dielectric by dipole orientation, as suggested by Fig. 12.2b. Therefore, in the presence of the dielectric, the field is given by

$$\bar{E}_\varepsilon = \frac{q}{A} = \frac{\sigma^*}{\varepsilon_r \varepsilon_0} \tag{12}$$

Next we ignore the directional (sign) aspect of the field and equate Eqs. (8) and (12) to obtain

$$\frac{d\psi}{dx} = \frac{\Delta\psi}{\delta} = \frac{\sigma^*}{\varepsilon_r \varepsilon_0} \tag{13}$$

where $\Delta\psi$ is the potential drop between plates separated by a distance δ. This equation relates the charge density, voltage difference, and distance of separation of the capacitor. Since this is the model we are using for the double layer, it is of interest to check whether Eq. (13) agrees—at least qualitatively—with what we know about the double layer.

We saw in Chap. 7 that charged monolayers are likely to obey the two-dimensional ideal gas law, and we also saw that areas per molecule of 10 nm^2 or so were also required for this ideal law to apply. Hence we may estimate σ^* for a monovalent ion to be

$$\sigma^* = \frac{\text{ion}}{10 \text{ nm}^2} \times \frac{10^{19} \text{ nm}^2}{1 \text{ m}^2} \times \frac{1.60 \times 10^{-19} \text{ C}}{\text{ion}} = 1.6 \times 10^{-2} \text{ C m}^{-2} \tag{14}$$

Taking the dielectric constant of water to be about 80, its bulk value, Eq. (12) permits the field strength to be estimated:

$$\bar{E} = \frac{1.6 \times 10^{-2} \text{ C m}^{-2}}{(80)(8.85 \times 10^{-12} \text{ C}^2 \text{ J}^{-1} \text{ m}^{-1})} = 2.26 \times 10^7 \text{ V m}^{-1} \tag{15}$$

Even allowing for an order of magnitude error in this estimate, we see that there is an exceptionally strong field in the vicinity of a charged interface. In Sect. 13.8 we shall examine this in greater detail and consider whether we are justified in using bulk values for such parameters as ε and η within the double layer.

If we estimate the potential drop between the two phases, we may determine the distance over which the potential drop occurs from the value of $\bar{E}$ given by Eq. (15). If we take the potential difference to be 0.10 V—an arbitrary but reasonable value—Eq. (13) shows that the plate separation of an equivalent capacitor is

$$\delta = \frac{\Delta \psi}{\bar{E}} = \frac{0.10}{2.26 \times 10^7} = 4.4 \times 10^{-9} \text{ m} = 4.4 \text{ nm} \tag{16}$$

Considering the simplicity of the model and the arbitrariness of the numerical estimates made in this calculation, 4.4 nm seems like a reasonable estimate of the distance over which surface charge neutralization is accomplished.

Throughout this section we have examined the distribution of charge at an interface as if the charge were constrained to two planes. When one of the phases is an aqueous electrolyte solution, the inadequacy of this model is apparent. An immediate improvement of the model is anticipated if we allow for a diffuse double layer, that is, a situation in which the charge density varies with distance from the interface, as shown in Fig. 12.3a. Alternatively, we might combine features from both the parallel plate distribution and the diffuse distribution to give a still more elaborate picture of the double layer, as shown in Fig. 12.3b. We shall consider this latter situation in Sect. 12.10. According to this picture, each part of the double layer is analyzed independently, and the effects

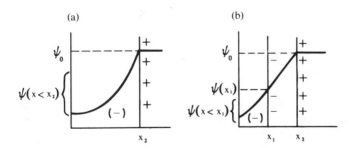

Figure 12.3 Two models for the double layer: (a) a diffuse double layer and (b) charge neutralization due partly to a parallel plate charge distribution and partly to a diffuse layer.

combined according to the rule for adding capacitors in series:

$$\frac{1}{C_T} = \frac{1}{C_1} + \frac{1}{C_2} + \cdots + \frac{1}{C_i} \tag{17}$$

where C_i is the capacitance of the ith element in series and C_T is the total capacitance.

12.4 The Diffuse Double Layer: The Debye-Hückel Approximation

In the preceding section we discussed the problem of the variation of potential with distance from an interface from the highly artificial perspective of a parallel plate capacitor. The variation of potential with distance from a charged surface of arbitrary shape is a classical electrostatic problem. The general problem is described by the Poisson equation,

$$\frac{\partial^2 \psi}{\partial x^2} + \frac{\partial^2 \psi}{\partial y^2} + \frac{\partial^2 \psi}{\partial z^2} = -\frac{\rho^*}{\varepsilon} \tag{18}$$

or in terms of the Laplacian operator ∇^2,

$$\nabla^2 \psi = -\frac{\rho^*}{\varepsilon} \tag{19}$$

where we define ε as the product $\varepsilon_r \varepsilon_0$ and thereby provide for the effect of the medium. In these expressions, ρ^* is the charge density (i.e., C m^{-3}) in the system, a quantity which itself is a function of x, y, and z. The solution to this differential equation, therefore, is an expression for the potential ψ which satisfies Eq. (18) and also the boundary conditions of the specific problem. We shall adopt the convention of measuring all distances outward from the interface where the potential has the value ψ_0. As the distance from an isolated surface increases to infinity, the value of ψ approaches zero. The stipulation of an "isolated" surface means that we are concerned with only one interface at this time. In Sect. 12.7 we shall consider the case in which electrical double layers overlap.

The Poisson equation is a fundamental relationship of classical electrostatics and really need not be proved here. However, since we are using it as a starting point, it seems desirable to explore the meaning of this important equation to some extent.

Equation (7) describes the field a distance r from a charge $+q$. A basic law of electrostatics is that this field describes any distribution of charge which results in q units of positive charge being enclosed by a sphere of radius r. It is not critical that a single $+q$ charge be situated at the center for this expression to apply. Suppose, therefore, we consider a portion of solution in which the charge is distributed with a uniform density ρ^*. In this case

$$q = \tfrac{4}{3}\pi r^3 \rho^* \tag{20}$$

and

$$\bar{E} = \frac{r\rho^*}{3\varepsilon} \tag{21}$$

Next we multiply both sides of Eq. (21) by r^2 and then differentiate with respect to r:

$$\frac{d}{dr}(r^2\bar{E}) = \frac{r^2\rho^*}{\varepsilon} \tag{22}$$

In the present notation, Eq. (8) becomes

$$\bar{E} = -\frac{d\psi}{dr} \tag{23}$$

where the minus sign is included since ψ decreases as r increases. Substituting this result into Eq. (22) gives

$$\frac{1}{r^2}\frac{\partial}{\partial r}\left(r^2\frac{\partial\psi}{\partial r}\right) = -\frac{\rho^*}{\varepsilon} \tag{24}$$

Remember that the operator $\nabla^2\psi$ in Eq. (19) transforms into the following form in spherical coordinates:

$$\nabla^2\psi = \frac{1}{r^2}\frac{\partial}{\partial r}\left(r^2\frac{\partial\psi}{\partial r}\right) + \frac{1}{r^2\sin\theta}\frac{\partial}{\partial\theta}\left(\sin\theta\frac{\partial\psi}{\partial\theta}\right)$$

$$+ \frac{1}{r^2\sin^2\theta}\left(\frac{\partial^2\psi}{\partial\phi^2}\right) \tag{25}$$

Thus the left-hand side of Eq. (24) is seen to be identical to $\nabla^2\psi$ for the case (spherical symmetry) in which ψ is independent of θ and ϕ. Although this presentation does not constitute the most general proof of the Poisson equation, it does give it some plausibility.

There are many situations in which the spherically symmetrical case is specifically invoked, as in the Debye–Hückel theory of electrolyte nonideality, for example. We shall consider situations for which this is the case in Chapter 13. For now, however, we consider the potential distribution adjacent to a planar wall which carries a positive charge. We shall define the direction perpendicular to the wall as the x direction and shall consider the wall as extending to infinity in the positive and negative y and z directions. In this case the operation $\nabla^2 \psi$ becomes $d^2\psi/dx^2$, and Eq. (18) is written

$$\frac{d^2\psi}{dx^2} = -\frac{\rho^*}{\varepsilon} \tag{26}$$

The next problem is to express the charge density as a function of the potential so the differential equation (26) can be solved for ψ. The procedure is to describe the ion concentrations in terms of the potential by means of a Boltzmann factor in which the work required to bring an ion to a position where the potential is ψ is given by $z_i e \psi$. The probability of finding an ion at this position is given by the Boltzmann factor, with this work appearing as the exponential energy:

$$\frac{n_i}{n_{i0}} = \exp\left(\frac{-z_i e \psi}{kT}\right) \tag{27}$$

In this expression n_i is the number of ions of type i per unit volume near the surface, and n_{i0} is the concentration far from the surface, that is, the bulk concentration. The valence number z_i is either a positive or negative integer.

The charge density is related to the ion concentrations as follows:

$$\rho^* = \sum_i z_i e n_i = \sum_i z_i e n_{i0} \exp\left(\frac{-z_i e \psi}{kT}\right) \tag{28}$$

Combining Eqs. (26) and (28) gives a result known as the Poisson–Boltzmann equation:

$$\frac{d^2\psi}{dx^2} = -\frac{e}{\varepsilon}\sum_i z_i n_{i0} \exp\left(\frac{-z_i e \psi}{kT}\right) \tag{29}$$

This same relationship is the starting point of the Debye–Hückel theory of electrolyte nonideality, except that the latter uses the value of $\nabla^2 \psi$ required for spherical symmetry. It is interesting to note that G. Gouy (in 1910) and D. L. Chapman (in 1913) applied this relationship to the diffuse double layer a decade before the Debye–Hückel theory appeared.

The derivation of the Poisson equation implies that the potentials associated with various charges combine in an additive manner. The

Boltzmann equation, on the other hand, involves an exponential relationship between the charges and the potential. In this way a fundamental inconsistency is introduced when Eqs. (26) and (28) are combined. Equation (29) does not have an explicit general solution anyhow and must be solved for certain limiting cases. These involve approximations which—at the same time—overcome the objection just stated.

We introduce the first of these approximations by considering only those situations in which $z_i e\psi < kT$. In this case the exponentials in Eq. (28) may be expanded (see Appendix A) as a power series. If only first-order terms in $z_i e\psi / kT$ are retained, Eq. (28) becomes

$$\rho^* = \sum_i z_i e n_{i0} \left(1 - \frac{z_i e\psi}{kT} \right) \tag{30}$$

Because of electroneutrality, two of the terms in Eq. (30) cancel:

$$\sum_{+\text{and}-} z_i e n_{i0} = 0 \tag{31}$$

so that Eq. (30) becomes

$$\rho^* = \sum_i \frac{z_i^2 n_{i0} e^2 \psi}{kT} \tag{32}$$

In this approximation, the ion potentials are additive, so Eq. (32) may be consistently substituted into Eq. (26) to give

$$\frac{d^2\psi}{dx^2} = \frac{e^2\psi}{\varepsilon kT} \sum_i z_i^2 n_{i0} \tag{33}$$

The assumption of low potentials made in reaching this result is also made in the Debye–Hückel theory and prompts us to call this model the Debye–Hückel approximation. Equation (33) has an explicit solution. Since potential is the quantity of special interest in Eq. (33), let us evaluate the potential at 25°C for a monovalent ion which satisfies the condition $ze\psi = kT$:

$$\psi = \frac{kT}{e} = \frac{(1.38 \times 10^{-23})(298)}{1.60 \times 10^{-19}} = 0.0257 \text{ V} = 25.7 \text{ mV} \tag{34}$$

Thus at 25°C potentials may be regarded as low or high, depending on whether they are less or more than about 25 mV. The factor $e\psi/kT$ appears often in double layer calculations, so this conversion factor is worth remembering. The relationship of kT/e to RT/F was noted in Sect. 12.2.

It is convenient to identify the cluster of constants in Eq. (33) by the symbol κ^2, which is defined as follows:

$$\kappa^2 = \frac{e^2 \Sigma_i z_i^2 n_{i0}}{\varepsilon k T} \tag{35}$$

With this change in notation, Eq. (33) becomes simply

$$\frac{d^2\psi}{dx^2} = \kappa^2\psi \tag{36}$$

Equation (36) has the solution

$$\psi = \psi_0 \exp(-\kappa x) \tag{37}$$

Note that Eq. (37) satisfies the required boundary conditions inasmuch as $\psi \to \psi_0$ as $x \to 0$ and $\psi \to 0$ as $x \to \infty$. In the following section we shall examine the implications of Eq. (37) in detail.

12.5 The Debye–Hückel Approximation: Results

The Debye–Hückel approximation is strictly applicable only in the case of low potentials. Nevertheless, there are several reasons why the significance of Eq. (37) should be fully appreciated:

1. It is simpler to understand than any of the modifications we shall consider subsequently.
2. It is a limiting result to which all equations that are more general must reduce in the limit of low potentials.
3. The effects of electrolyte concentration and valence in this approximation are qualitatively consistent with the results of more elaborate calculations.

One of the most important quantities to emerge from the Debye–Hückel approximation is the parameter κ. This quantity appears throughout double layer discussions and not merely at this level of approximation. Since the exponent κx in Eq. (37) is dimensionless, κ must have units of reciprocal length. This means that κ^{-1} has units of length. This latter quantity is often (imprecisely) called the "thickness" of the double layer. All distances within the double layer are judged large or small relative to this length. Note that the exponent κx may be written x/κ^{-1}, a form which emphasizes the notion that distances are measured relative to κ^{-1} in the double layer.

Since κ and κ^{-1} are such important quantities, we shall examine them in greater detail, first verifying their dimensions and then considering

their numerical magnitude. Especially important is the dependence of κ and κ^{-1} on the concentration and valence of the electrolyte in solution.
It is an easy matter to verify that κ^2 as defined by Eq. (35) does indeed have units of length^{-2}, or m^{-2} in SI. This is seen by writing the SI units for the various factors appearing in Eq. (35) as follows:

$$\kappa^2 = \frac{(C)^2(m^{-3})}{(C^2\ J^{-1}\ m^{-1})(J\ K^{-1})(K)} = m^{-2}$$

The parameter κ is concentration dependent; accordingly, we must express it in practical concentration units. If n_i is expressed as the number of ions per cubic meter, then n_i is related to the molar concentration M_i of the ions by

$$n_i = 1000 M_i N_A \tag{38}$$

since 1000 dm^3 = 1 m^3. Therefore Eq. (35) yields

$$\kappa = \left(\frac{1000 e^2 N_A}{\varepsilon k T} \sum_i z_i^2 M_i \right)^{1/2} \tag{39}$$

The summation in this expression is twice the ionic strength of the solution. We examine the numerical substitutions into Eq. (39) in the following example.

Example 12.1 Evaluate the numerical factor in Eq. (39) for aqueous solutions at 25°C. At this temperature $\varepsilon_r = 78.54$ for water. Calculate κ and κ^{-1} for 0.01 M solutions of 1:1, 2:1, and 3:1 electrolytes. Suggest how these values can be adapted to other temperatures (or media) without complete recalculation.

Solution Recalling that $\varepsilon = \varepsilon_r \varepsilon_0$, we write

$$\kappa = \left(\frac{(1000)(1.60 \times 10^{-19})^2(6.02 \times 10^{23})(2I)}{(78.54)(8.85 \times 10^{-12})(1.38 \times 10^{-23})(298)} \right)$$

$$= 2.32 \times 10^9 (2I)^{1/2}\ m^{-1}$$

and

$$\kappa^{-1} = 4.31 \times 10^{-10}(2I)^{-1/2}\ m$$

where I is the ionic strength and we have written $2I$ in place of the summation. For a symmetrical (z:z) electrolyte, the ionic strength equals $z^2 M$ (or $I^{1/2} = |z| M^{1/2}$). Therefore in a 0.01 M solution of 1:1 electrolyte

$\kappa = 3.29 \times 10^8 \text{ m}^{-1}$ and $\kappa^{-1} = 3.04 \times 10^{-9}$ m = 3.04 nm

For asymmetrical electrolytes the ionic strength is the same for, say, 1:2 as for 2:1 solutes; namely, $3 \times M$ as verified by substitution into the summation (and remembering the stoichiometry of the dissociation!). therefore in a 0.01 M solution of 2:1 electrolyte

$\kappa = 5.68 \times 10^8 \text{ m}^{-1}$ and $\kappa^{-1} = 1.76 \times 10^{-9}$ m = 1.76 nm

and in a 0.01 M solution of 3:1 electrolyte ($I = 6 \times M$)

$\kappa = 8.04 \times 10^8 \text{ m}^{-1}$ and $\kappa^{-1} = 1.24 \times 10^{-9}$ m = 1.24 nm

It is easy enough to evaluate κ and κ^{-1} for different concentrations and valence types; it is more of a nuisance to recalculate these quantities at different temperatures and/or in different media. An easy way to do this using the expressions given is to factor out $\varepsilon_{r,298}$ and T_{298} and replace them with quantities which are pertinent to the problem at hand. For example, at 90°C, $\varepsilon_r = 57.98$ for water; therefore the value of κ at this temperature is given by

$$\kappa = 2.32 \times 10^9 \left(\frac{(78.54)(298)}{(57.98)(363)} \right)^{1/2} (2I)^{1/2} = 2.32 \times 10^9 (1.05)(2I)^{1/2}$$

$$= 2.45 \times 10^9 (2I)^{1/2} \text{ m}^{-1}$$

•

Both κ and κ^{-1} are used extensively in this chapter and the next. Table 12.1 lists numerical values for these quantities and the pertinent equations for their calculation for aqueous solutions at 25°C. This table may be consulted as a source for κ and κ^{-1} values when these are required for exercises in these chapters.

Several things should be noted about Table 12.1:

1. The tabulated values of κ^{-1} multiplied by 10^9 give the double layer "thicknesses" in nanometers. For example, in a 0.01 M solution of a 1:1 electrolyte, κ^{-1} equals 3.04 nm.
2. This "thickness" is the same magnitude as the prediction based on the capacitor model [Eq. (16)]. The diffuse model is clearly superior, however, since it shows how the double layer "thickness" depends on the concentration and valence of the ions in the solution.
3. The "thickness" of the double layer varies inversely with z and inversely with $M^{1/2}$ for a symmetrical $z{:}z$ electrolyte. Therefore κ^{-1}

Table 12.1 Values of κ and κ^{-1} for Several Different Electrolyte Concentrations and Valences with Numerical Formulas for These Quantities also Given for Aqueous Solutions at 25°C

	Symmetrical electrolyte			Asymmetrical electrolyte		
	$z_+{:}z_-$	$\kappa\ (m^{-1}) =$ $3.29 \times 10^9 |z| M^{1/2}$	$\kappa^{-1}\ (m) =$ $3.04 \times 10^{-8} |z|^{-1} M^{-1/2}$	$z_+{:}z_-$	$\kappa\ (m^{-1}) =$ $2.32 \times 10^9 (\Sigma z_i^2 z_j^2 M_i)^{1/2}$	$\kappa^{-1}\ (m) =$ $4.30 \times 10^{-10} (\Sigma z_i^2 z_j^2 M_i)^{-1/2}$
Molarity						
0.001	1:1	1.04×10^8	9.61×10^{-9}	1:2, 2:1	1.80×10^8	5.56×10^{-9}
	2:2	2.08×10^8	4.81×10^{-9}	3:1, 1:3	2.54×10^8	3.93×10^{-9}
	3:3	3.12×10^8	3.20×10^{-9}	2:3, 3:2	4.02×10^8	2.49×10^{-9}
0.01	1:1	3.29×10^8	3.04×10^{-9}	1:2, 2:1	5.68×10^8	1.76×10^{-9}
	2:2	6.58×10^8	1.52×10^{-9}	1:3, 3:1	8.04×10^8	1.24×10^{-9}
	3:3	9.87×10^8	1.01×10^{-9}	2:3, 3:2	1.27×10^9	7.87×10^{-10}
0.1	1:1	1.04×10^9	9.61×10^{-10}	1:2, 2:1	1.80×10^9	5.56×10^{-10}
	2:2	2.08×10^9	4.81×10^{-10}	1:3, 3:1	2.54×10^9	3.93×10^{-10}
	3:3	3.12×10^9	3.20×10^{-10}	2:3, 3:2	4.02×10^9	2.49×10^{-10}

equals 1.0 nm for a 0.01 M solution of a 3:3 electrolyte and is about 10 nm for a 0.001 M solution of a 1:1 electrolyte.

Figure 12.4 shows how the potential drops with distance from the surface according to Eq. (37). The curves in this figure are drawn for two different variations in κ: in (a) a 1:1 electrolyte at 0.1, 0.01, and 0.001 M concentrations, and in (b) a 0.001 M solution of 1:1, 2:2, and 3:3 electrolytes. Again, it is important to recognize that the curves drop off more rapidly for either higher concentrations or higher valences of the electrolyte.

The curves in Fig. 12.4 are marked at the x value that corresponds to κ^{-1}. Note that the potential has dropped to the value ψ_0/e at this point. Calling κ^{-1} the double layer "thickness" is clearly a misnomer. We shall see presently, however, that there is some logic underlying this terminology.

Although the potential is fundamentally a more important quantity than charge density, examining the latter will enable us to compare the capacitor and diffuse models for the double layer.

The condition of electroneutrality at a charged interface requires that the density of charge at the two faces be equal. Note that this does not require the charges to be physically situated at the interface. When one of the phases contains a diffuse layer, the total charge contained in a volume element of solution of unit cross section and extending from the wall to infinity must contain the same amount of charge—although of opposite sign—as a unit area of wall contains. Stated in formula, this becomes

$$\sigma^* = -\int_0^\infty \rho^* \, dx \tag{40}$$

We shall now examine the implications of Eq. (40) for the situation in which one of the adjoining phases contains the diffuse half of a double layer. Combining Eqs. (26) and (40) gives

$$\sigma^* = \varepsilon \int_0^\infty \frac{d^2\psi}{dx^2} \, dx \tag{41}$$

a result which is easily integrated to yield

$$\sigma^* = \varepsilon \left. \frac{d\psi}{dx} \right|_0^\infty \tag{42}$$

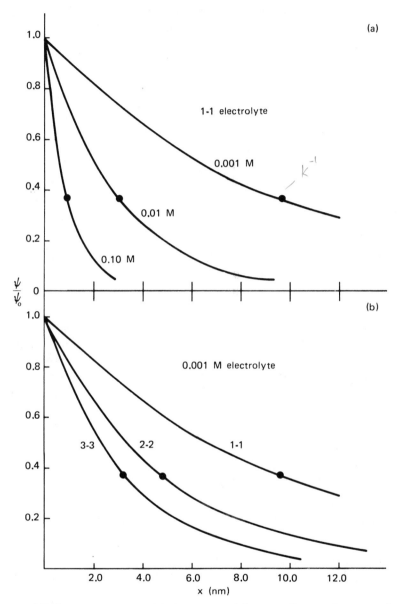

Figure 12.4 Fraction of double layer potential versus distance from a surface according to the Debye–Hückel approximation, Eq. (37): (a) curves drawn for 1:1 electrolyte at three concentrations and (b) curves drawn for 0.001 M symmetrical electrolytes of three different valence types.

The quantity $d\psi/dx$ is zero at infinity, and we shall define its value at the wall by $(d\psi/dx)_0$; therefore Eq. (42) becomes

$$\sigma^* = -\varepsilon \left(\frac{d\psi}{dx} \right)_0 \qquad (43)$$

Next, we turn to Eq. (37)—the Debye-Hückel approximation for ψ—to evaluate $(d\psi/dx)_0$. Differentiation leads to the value

$$\left(\frac{d\psi}{dx} \right)_0 = \lim_{x \to 0} - \kappa\psi_0 \exp(-\kappa x) = -\kappa\psi_0 \qquad (44)$$

Substituting Eq. (44) into Eq. (43) gives

$$\sigma^* = \varepsilon\kappa\psi_0 \qquad (45)$$

Rewriting Eq. (45) in terms of κ^{-1}, the double layer "thickness," yields

$$\sigma^* = \varepsilon \frac{\psi_0}{\kappa^{-1}} \qquad (46)$$

Equation (46) is identical to Eq. (13) for a parallel plate capacitor, with ψ_0 replacing $\Delta\psi$ and κ^{-1} replacing δ. This result shows that a diffuse double layer at low potentials behaves like a parallel plate capacitor in which the separation between the plates is given by κ^{-1}. This explains why κ^{-1} is called the double layer thickness. It is important to remember, however, that the *actual* distribution of counterions in the vicinity of a charged wall is diffuse and approaches the unperturbed bulk value only at large distances from the surface.

Even allowing for the fact that the Debye-Hückel approximation applies only for low potentials, this analysis reveals some features of the electrical double layer that are general and of great importance as far as stability with respect to flocculation of dispersions and electrokinetic phenomena are concerned. In summary three specific items might be noted:

1. The distance away from the wall that an electrostatic potential persists may be comparable to the dimensions of colloidal particles themselves.
2. The distance over which significant potentials exist decreases with increasing electrolyte concentration.
3. The range of electrostatic potentials decreases as the valence of the ions in solution increases.

The Debye–Hückel approximation to the diffuse double layer problem produces a number of relatively simple equations which introduce a variety of double layer topics as well as a number of qualitative generalizations. In order to extend the range of the quantitative relationships, however, it is necessary to return to the Poisson–Boltzmann equation and the unrestricted Gouy–Chapman theory.

12.6 The Electric Double Layer: Gouy–Chapman Theory

The theoretical inconsistencies inherent in the Poisson–Boltzmann equation were shown in Sect. 12.4 to vanish in the limit of very small potentials. It may also be shown that errors arising from this inconsistency will not be too serious under the conditions that prevail in many colloidal dispersions, even though the potential itself may no longer be small. Accordingly, we return to the Poisson–Boltzmann equation as it applies to a planar interface, Eq. (29), to develop the Gouy–Chapman result without the limitations of the Debye–Hückel approximation.

If both sides of Eq. (29) are multiplied by 2 $d\psi/dx$, we obtain

$$2 \frac{d\psi}{dx} \frac{d^2\psi}{dx^2} = -\frac{2e}{\varepsilon} \sum_i z_i n_{i0} \exp\left(\frac{-z_i e \psi}{kT}\right) \frac{d\psi}{dx} \tag{47}$$

The left-hand side of this equation is the derivation of $(d\psi/dx)^2$; therefore

$$\left(\frac{d\psi}{dx}\right)^2 = \frac{2kT}{\varepsilon} \sum_i z_i n_{i0} \exp\left(\frac{-z_i e \psi}{kT}\right) + \text{const.} \tag{48}$$

The integration constant in this expression is easily evaluated if we define the potential in the solution at $x = \infty$ to be zero. At the same limit $d\psi/dx$ also equals zero. In view of these conventions, Eq. (48) becomes

$$\left(\frac{d\psi}{dx}\right)^2 = \frac{2kT}{\varepsilon} \sum_i n_{i0} \left[\exp\left(\frac{-z_i e \psi}{kT}\right) - 1\right] \tag{49}$$

This result may be integrated further if we restrict the electrolyte in solution to the symmetrical $z{:}z$ type. In that case, Eq. (49) can be written

$$\left(\frac{d\psi}{dx}\right)^2 = \frac{2kT n_0}{\varepsilon} \left[\exp\left(\frac{-ze\psi}{kT}\right) + \exp\left(\frac{ze\psi}{kT}\right) - 2\right] \tag{50}$$

in which z is the absolute value of the valence number, the sign having been incorporated into the algebraic form. The bracketed term is readily seen to equal $[\exp(-ze\psi/2kT) - \exp(ze\psi/2kT)]^2$; therefore Eq. (50) may be written

$$\left(\frac{d\psi}{dx}\right)^2 = \frac{2kTn_0}{\varepsilon}\left[\exp\left(\frac{-ze\psi}{2kT}\right) - \exp\left(\frac{ze\psi}{2kT}\right)\right]^2 \tag{51}$$

Identifying $ze\psi/kT$ as y permits the simplification of notation to

$$\frac{dy}{dx} = \left(\frac{2e^2z^2n_0}{\varepsilon kT}\right)^{1/2}(e^{-y/2} - e^{y/2}) = \kappa(e^{-y/2} - e^{y/2}) \tag{52}$$

This last result may be written in an integrable form by defining u as $e^{y/2}$, in which case $dy = 2e^{-y/2}\,du$, and the following relationships hold:

$$\frac{dy}{e^{-y/2} - e^{y/2}} = \frac{2\,du}{e^{y/2}(e^{-y/2} - e^{y/2})} = \frac{2\,du}{1 - e^y} = \frac{2\,du}{1 - u^2} = \frac{du}{1 + u} + \frac{du}{1 - u} \tag{53}$$

Combining Eqs. (52) and (53) gives

$$\frac{du}{1 + u} + \frac{du}{1 - u} = \kappa\,dx \tag{54}$$

which is easily integrated to yield

$$\ln\left(\frac{1 + u}{1 - u}\right) = \kappa x + \text{const.} \tag{55}$$

The integration constant is evaluated from the fact that $\psi = \psi_0$, $y = y_0$, and $u = u_0$ at $x = 0$; therefore

$$\ln\left[\left(\frac{1 + u}{1 - u}\right)\left(\frac{1 - u_0}{1 + u_0}\right)\right] = \kappa x \tag{56}$$

In terms of the physical variables, Eq. (56) may be written

$$\exp\left(\frac{[\exp(ze\psi/2kT) + 1][\exp(ze\psi_0/2kT) - 1]}{[\exp(ze\psi/2kT) - 1][\exp(ze\psi_0/2kT) + 1]}\right) = \kappa x \tag{57}$$

Equation (57) describes the variation in potential with distance from the surface for a diffuse double layer without the simplifying assumption of low potentials. It is obviously far less easy to gain a "feeling" for this relationship than for the low-potential case. Anticipation of this fact is why so much attention was devoted to the Debye–Hückel approximation in the first place.

Note that Eq. (57) may be written

$$\frac{\exp(ze\psi/2kT) - 1}{\exp(ze\psi/2kT) + 1} = \frac{\exp(ze\psi_0/2kT) - 1}{\exp(ze\psi_0/2kT) + 1} \exp(-\kappa x) \tag{58}$$

Equation (58) is the Gouy–Chapman expression for the variation of potential within the double layer. For simplicity, Eq. (58) may be written

$$\Upsilon = \Upsilon_0 \exp(-\kappa x) \tag{59}$$

where Υ is defined by the relationship

$$\Upsilon = \frac{\exp(ze\psi/2kT) - 1}{\exp(ze\psi/2kT) + 1} \tag{60}$$

Equation (54) shows that it is the complex ratio Υ that varies exponentially with x in the Gouy–Chapman theory rather than ψ, as is the case in the Debye–Hückel approximation. Some values of Υ calculated for a variety of ψ_0 values are listed in Table 12.2.

As a check on the consistency of our mathematics, it is profitable to verify that Eq. (58) reduces to Eq. (37) in the limit of low potentials.

Table 12.2   Variation of the Parameter Υ_0 with ψ_0 at 25°C [Υ is Defined by Eq. (60)]

ψ_0 (mV)	Υ_0
260	0.9874
240	0.9814
220	0.9727
200	0.9600
180	0.9415
160	0.9149
140	0.8765
120	0.8230
100	0.7500
80	0.6528
60	0.5249
40	0.3711
20	0.1968

Expanding the exponentials in Υ and truncating the series so that only one term survives in both the numerator and denominator results in the Debye-Hückel expression, Eq. (37).

Another situation of interest in which Eq. (58) simplifies considerably is the case of large values of x at which ψ has fallen to a small value regardless of its initial value. Under these conditions the exponentials of the left-hand side are expanded to give

$$\frac{ze\psi}{4kT} = \Upsilon_0 \exp(-\kappa x) \tag{61}$$

or

$$\psi = \frac{4kT\Upsilon_0}{ze} \exp(-\kappa x) \tag{62}$$

For very large values of ψ_0, $\Upsilon_0 \rightarrow 1$. In this case, Eq. (62) becomes

$$\psi = \frac{4kT}{ze} \exp(-\kappa x) \tag{63}$$

which shows that the potential in the outer (i.e., well removed from the wall) portion of the diffuse double layer is independent of the potential at the wall for larger potentials. In colloidal dispersions $ze\psi_0/kT$ is generally greater than unity, but not too much greater. This means that approximations (37) and (62) will generally bracket the true potential versus distance curve given by Eq. (58). The situation is shown graphically in Fig. 12.5. In drawing these curves, we chose a value of ψ_0 equal to 77.1 mV and arbitrarily selected the electrolyte to be a 0.01 M solution of a 1:1 electrolyte for which κ^{-1} is 3.04 nm. Equation (63) is a poor approximation in this case, because ψ_0 is not large enough. Values of the abscissa are readily converted into dimensionless variables which apply to any solution by dividing the x coordinate in nanometers by 3.04 nm.

We conclude this section by considering the expression for charge density, Eq. (40), as it applies in the Gouy-Chapman model.

As we saw in the preceding section, the charge density expression integrates to Eq. (43), with no assumptions as to the nature of the potential function. Accordingly, we may combine Eqs. (43) and (51) to obtain

$$\sigma^* = \varepsilon \left(\frac{2kTn_0}{\varepsilon}\right)^{1/2} \left[\exp\left(\frac{ze\psi_0}{2kT}\right) - \exp\left(\frac{-ze\psi_0}{2kT}\right) \right] \tag{64}$$

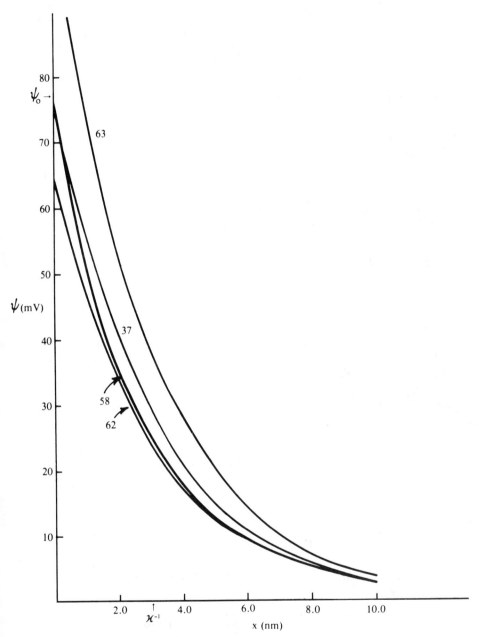

Figure 12.5 Variation of the double layer potential versus distance from the surface according to four expressions from this chapter for $\psi_0 = 77.1$ mV and $\kappa = 3.29 \times 10^8$ m^{-1} (or 0.01 M solution of 1:1 electrolyte). Curves are drawn according to Eqs. (37), (58), (62), and (63).

Equation (64) describes the variation of charge density with potential at the surface with no limitations as to the value of the potential. The following example considers an application of this relationship.

Example 12.2   Show that for a 1:1 electrolyte in water at 25°C, Eq. (64) can be rearranged to give

$$\psi_0 = 51.4 \, \sinh^{-1}\left(\frac{137}{\sigma^0 \sqrt{c}}\right)$$

in which ψ_0 is expressed in millivolts, c is in moles per liter, and σ^0 is the area (in Å^2) per charge at the surface. Davies* measured the potential across an air-aqueous NaCl interface which carried a monolayer of $C_{18}H_{37}N(CH_3)_3^+$. When the quaternary octadecylamine was at a pressure corresponding to $\sigma^0 = 85 \, \text{Å}^2$, the following potentials were measured at different concentrations of NaCl:

E (mV)	240	280	325	340	380
c_{NaCl} (M)	2.0	0.5	0.1	0.033	0.01

About 200 mV of these potential differences arises from dipole effects at the interface and should be subtracted from each value to give the double layer contribution to the measured potentials. Compare these corrected values with the values of ψ_0 calculated by the equation given.

Solution   First recognize that $2 \sinh x = e^x - e^{-x}$; therefore Eq. (64) can be written

$$\sigma^* = (8\varepsilon kTn_0)^{1/2}\sinh\left(\frac{ze\psi_0}{2kT}\right) \quad \text{or} \quad \sinh^{-1}[\sigma^*(8\varepsilon kTn_0)^{-1/2}] = \frac{ze\psi_0}{2kT}$$

For a 1:1 electrolyte this yields

$$\psi_0 = \left(\frac{2kT}{e}\right)\sinh^{-1}[\sigma^*(8\varepsilon kTn_0)^{-1/2}]$$

To assure proper units, we assemble the following substitutions for this expression:

$$\frac{2kT}{e} = \frac{2(1.38 \times 10^{-23})(298)}{1.60 \times 10^{-19}} = 0.0514 \, \text{V} = 51.4 \, \text{mV}$$

---

*J. T. Davies, *Proc. R. Soc.*, *208A*:224 (1951).

which gives the desired coefficient. Next we use Eq. (38) to obtain n_0: $n_0 = (10^3)(6.02 \times 20^{23})c = 6.02 \times 10^{26}c$. With this the factor $8\varepsilon kTn_0$ becomes

$$8(78.54)(8.85 \times 10^{-12})(1.38 \times 10^{-23})(298)(6.02 \times 10^{26}c) = 1.38 \times 10^{-2}c.$$

Finally, σ^* is related to σ^0 as follows: $\sigma^* = 1.60 \times 10^{-19}/\sigma^0(10^{-10})^2 = 16/\sigma^0$.

From these components the argument of the sinh can be evaluated to be $137/\sigma^0 \sqrt{c}$, which is the desired result.

By substituting the value of σ^0 into this expression and using the various NaCl concentrations given, ψ_0 values are readily calculated; these are to be compared with $E_{obs} - 200$. The following values are obtained:

ψ_0 (mV)	50.4	80.2	119.8	148.0	178.9
$E_{obs} - 200$ (mV)	40	80	125	140	180

The agreement between theory and experiment is seen to be quite satisfactory.

•

12.7 Overlapping Double Layers and Interparticle Repulsion

From the viewpoint of the stability of lyophobic colloids, this and the following section are of central importance. In this section we shall examine the force per unit area—that is, the pressure—that operates on two charged surfaces as a result of the overlapping of their double layers. As we saw in Chap. 11, it is more convenient to compare attraction and repulsion between particles in terms of potential energy rather than force. In Sect. 12.8 we shall see how to express double layer repulsion in terms of potential energy. As was the case with a single double layer, it is easier to treat overlapping double layers if the potential is low. Accordingly, the detailed derivation we consider will assume this condition. We shall generalize to the case of higher potentials later, but without presenting all the mathematical details of that situation.

Figure 12.6 is a schematic representation of the situation with which we are concerned. It shows two plates of unspecified thickness; the planar faces of these plates are separated by a distance d. The plates are immersed in an infinite reservoir of electrolyte, the bulk concentration of which is n_0. The potential at the surface of the plates is defined to be ψ_0. It will be convenient to distinguish between the inner and outer regions of

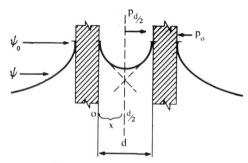

Figure 12.6 Schematic representation of the overlap of two double layers when a pair of plates are brought to a surface separation d.

the solution. By the former we refer to the region between the plates, and by the latter the region influenced by only one of the faces.

When the distance between the plates is large, the potential on both the inner and outer faces will drop with distance from the surface according to Eq. (58) or one of its approximations. The profile of the potential drop in this case is shown by the solid curve in the outer region of the figure and by the dashed line in the inner region. As the distance between the plate decreases, the potential from each of the inner faces begins to overlap in the inner region. Therefore the net potential in the inner region varies as the solid line in the figure for this region. The potential in the outer region is unaffected by the separation of the plates. Since we already know how to handle the potential in the outer region, our interest now focuses on the potential in the inner region where double layer overlap occurs.

At equilibrium all forces on any volume element of a solution must balance. Suppose we apply this equilibrium criterion to a volume element of solution which lies in the plane parallel to the face of the plates in Fig. 12.6 and lies a distance x from the face of one of these plates. Two kinds of force must operate on this volume element: pressure and electrostatic forces. According to Eqs. (4.25) and (4.27), the x component of the pressure force acting on the volume element (i.e., per unit volume) is given by

$$F_x = \frac{\partial p}{\partial x} \tag{65}$$

The electrical force per unit volume is given by the product of the charge density times the field strength according to Eq. (6):

$$F_{el} = \rho^* \frac{d\psi}{dx} \tag{66}$$

Combining these results with the criterion for equilibrium leads to the expression

$$\frac{dp}{dx} + \rho^* \frac{d\psi}{dx} = 0 \tag{67}$$

When Eq. (26) is substituted for ρ^*, this becomes

$$\frac{dp}{dx} - \varepsilon \frac{d^2\psi}{dx^2} \frac{d\psi}{dx} = 0 \tag{68}$$

Since

$$\frac{d^2\psi}{dx^2} \frac{d\psi}{dx} = \frac{1}{2} \frac{d}{dx} \left(\frac{d\psi}{dx} \right)^2$$

Eq. (68) can be written

$$\frac{d}{dx} \left[p - \frac{\varepsilon}{2} \left(\frac{d\psi}{dx} \right)^2 \right] = 0 \tag{69}$$

This result shows that the condition for equilibrium is equivalent to requiring that

$$p - \frac{\varepsilon}{2} \left(\frac{d\psi}{dx} \right)^2 = \text{const.} \tag{70}$$

at all locations in the solution. Equation (70) shows that two influences—the pressure and the electric field—operate within the solution, but that the difference between them is a constant that still remains to be evaluated. We are specifically interested in evaluating this constant in the inner region.

At this point, the symmetry of the situation shown in Fig. 12.6 becomes helpful. We realize that ψ goes through a minimum at the midpoint position; that is, $d\psi/dx = 0$ at $x = d/2$. Thus the constant in Eq. (70) equals the pressure at the midpoint $p_{d/2}$. The difference between the pressure and the field effect is equal to this quantity at all locations between the plates. Because of this constancy, the entire inner region is characterized by the parameters that apply at the midpoint. Therefore $\psi_{d/2}$ is the potential that governs the repulsion between the surfaces. Next, we write Eq. (67) as

$$dp = -\rho^* \, d\psi \tag{71}$$

Now Eq. (28) may be substituted for ρ^* for a $z{:}z$ electrolyte, giving

$$dp = -zen_0 \left[\exp\left(-\frac{ze\psi}{kT} \right) - \exp\left(\frac{ze\psi}{kT} \right) \right] d\psi \tag{72}$$

Since $e^x - e^{-x} = 2 \sinh x$, Eq. (72) may also be written

$$dp = 2zen_0 \sinh\left(\frac{ze\psi}{kT} \right) d\psi \tag{73}$$

Equation (73) is easily integrated between the following limits: $p = p_0$ (the outer reference pressure) at $\psi = 0$ and $p = p_{d/2}$ at $\psi = \psi_{d/2}$. Integration of Eq. (73) gives

$$p_{d/2} - p_0 = 2kTn_0 \left[\cosh\left(\frac{ze\psi_{d/2}}{kT} \right) - 1 \right] = F_R \tag{74}$$

Equation (74) gives the excess pressure at $x = d/2$ and therefore the force per unit area with which the plates are pushed apart, F_R.

This analysis is one of several treatments of double layer repulsion presented in Verwey and Overbeek's classic book [8]. The reader will find the topics of this chapter developed in great detail in that source.

Although Eq. (74) is correct, it is not particularly helpful. The problem is that F_R is expressed in terms of the potential at the midpoint, which is itself an unknown quantity. For the special case in which $d/2$ is relatively large, the approximation given by Eq. (62) may be applied to the potential from each of the two approaching surfaces. The potential at the midpoint then becomes

$$\psi_{d/2} = \psi_1 + \psi_2 \simeq 2 \left(\frac{4kT\Upsilon_0}{ze} \right) \exp(-\kappa d/2) \tag{75}$$

according to this approximation. Since this result applies well away from the surface, the potential is low when Eq. (75) holds. Therefore $\cosh(ze\psi_{d/2}/kT)$ may be expanded as a power series (see Appendix A), with only the leading terms retained. This leads to the result

$$F_R \simeq n_0 kT \left(\frac{ze\psi_{d/2}}{kT} \right)^2 = n_0 kT [8\Upsilon_0 \exp(-\kappa d/2)]^2 \tag{76}$$

or

$$F_R \simeq 64 n_0 k T \Upsilon_0^2 \exp\left(\frac{-d}{\kappa^{-1}}\right) \qquad (77)$$

An issue of considerable practical importance is how the force of repulsion described by Eq. (74) and its approximation Eq. (71) varies with the electrolyte content of a solution. Since κ varies with $n_0^{1/2}$, Eq. (71) is of the form

$$F_R = (\text{const. } 1) n_0 \exp[-(\text{const. } 2) n_0^{1/2}] \qquad (78)$$

The exponential factor is clearly the more sensitive involvement of n_0 in Eq. (78). Therefore this expression shows that the force of repulsion decreases with increasing electrolyte concentration between two surfaces compared at the same separation, at least at relatively large separation. The addition of indifferent electrolyte to an aqueous dispersion of a lyophobic colloid may induce the flocculation of that colloid. Equation (78) is therefore an important step toward understanding this behavior. One interesting system to which these ideas have been applied is the case of liquid films. An aqueous soap bubble, for example, reaches an equilibrium thickness at which the various forces acting on it balance out. Such forces are readily enumerated:

1. Gas pressure tends to squeeze the two faces of the film together.
2. Van der Waals attraction between the two gas masses is transmitted across the liquid surface.
3. Amphipathic surfactant ions adsorb at the liquid surface—tails out—so counter-ions form diffuse double layers which can overlap in a manner resembling Fig. 12.6.
4. The equilibrium thickness of most bubbles is so much less than the radius of curvature of the bubble that the air masses can be regarded as blocks with planar faces like those in our models. In research studies on such systems, the bubbles are allowed to equilibrate on frames which make their compliance with the model even better.

In the following example we discuss the results of such a study with soap bubbles.

Example 12.3 Lyklema and Mysels* measured the equilibrium thickness of a soap bubble to be 73.1 when the bubble was stabilized by 8.7×10^{-4} M sodium dodecyl sulfate and the hydrostatic pressure on the surface of the film was 66 N m^{-2}. In this experiment the thickness satisfies

---

*J. Lyklema and K. J. Mysels, *J. Am. Chem. Soc.*, 87:2539 (1965).

the condition of equilibrium between the hydrostatic pressure and the force of repulsion between double layers on the adjacent faces of the film, as given by Eq. (77). Assuming that the adsorption of dodecyl sulfate ions at the air–solution interface gives a very high value of ψ_0, calculate the equilibrium bubble thickness predicted by this model. Criticize or defend the following propositions: If an appreciable concentration of indifferent electrolyte is added to the soap solution, the calculation just given is no longer feasible because (1) Eq. (77) ceases to be valid for F_R and (2) van der Waals attraction between the two air masses also promotes film thinning.

Solution   The strategy here is to neglect van der Waals forces and to solve Eq. (77) for d when $F_R = 66$ N m^{-2}. If ψ_0 is large enough, $\Upsilon_0 = 1$; furthermore, by Eq. (38), $n_0 = 1000\ MN_A = 5.24 \times 10^{23}$ liter^{-1}. Assuming the bulk concentration is undiminished by adsorption, we obtain $\kappa = 3.29 \times 10^9\ M^{1/2} = 9.70 \times 10^7$ m^{-1}. Therefore $66 = 64(5.24 \times 10^{23})$ $(1.38 \times 10^{-23})(298)$ exp$[-(9.70 \times 10^7)d]$, from which we find $d = 7.88 \times 10^{-8}$ m $= 78.8$ nm. Considering that this calculation is insensitive to the actual potential at the surface, the agreement between this quantity and the experimental film thickness is acceptable.

The quantity F_R continues to be valid in the presence of indifferent electrolyte as long as the latter is a $z{:}z$ type of compound. As a matter of fact, added electrolyte helps meet the requirement that $\psi_{d/2}$ be low, which actually improves the applicability of Eq. (77).

Adding electrolyte will cause the film to thin by shortening the range of the repulsive force. At smaller distances of separation the van der Waals attraction is definitely increased and should be added (as force area^{-2}) to the pressure before attempting this kind of calculation. Even without added electrolyte, the air masses attract each other; neglecting this is another possible source of discrepancy between theory and experiment in this example.

•

In order to evaluate fully the effect of electric charge on the stability of a colloidal dispersion, it is necessary to compare the magnitude of the repulsion between particles with the magnitude of the attraction between them. The attraction is most readily described in terms of potential energy; therefore the repulsion should be expressed in this form as well. For the approximation we have just discussed, this is not particularly difficult to evaluate. Since potential energy is given by the force times the distance through which it operates, we may write

$$d\Phi_R = -F_R d(d) \tag{79}$$

In this expression, $d\Phi_R$ is the increment in potential energy arising from a change in the separation. The minus sign arises from the fact that the potential energy decreases with increasing separation.

Substituting the approximation given by Eq. (77) for F_R gives

$$d\Phi_R = -64n_0 kT \Upsilon_0^2 \exp\left(\frac{-d}{\kappa^{-1}}\right) d(d) \tag{80}$$

a result that is easily integrated by recalling that $\Phi_R = 0$, when $d = \infty$. Integration yields

$$\Phi_R = \frac{64n_0 kT \Upsilon_0^2}{\kappa} \exp\left(\frac{-d}{\kappa^{-1}}\right) \tag{81}$$

This particular form is limited in applicability to situations in which the separation of the surfaces is large compared to κ^{-1} and ψ_0 is large so that $\Upsilon_0 \simeq 1$. As we did with the force of repulsion, we may write Φ_R as

$$\Phi_R = (\text{const. } 1)n_0^{1/2} \exp[-(\text{const. } 2)n_0^{1/2}] \tag{82}$$

since κ varies with $n_0^{1/2}$. Again we see that at large separations the potential energy of repulsion decreases with increasing electrolyte concentration. The continued emphasis on large separations is formally imposed by the use of approximation (62) in the development of this result. There are practical reasons for interest in this limit also. As colloidal particles approach one another, it is the outermost portions of their double layers that first interact. The outcome of such an encounter, then, is influenced by the interaction between the particles at large separations.

The derivation of Eq. (81) is possible only because relatively simple approximations are available which permit $\psi_{d/2}$ to be solved explicitly and generate an integrable expression for Φ_R. This is not generally the case, however, so the approach used to reach Eq. (81) is not applicable to most situations. E. J. W. Verwey and J. Th. G. Overbeek have found another method for evaluating Φ_R, but the mathematics are tedious by this approach, involving numerical integrations. The method consists of calculating the free energy difference between particles separated by a distance d and infinitely separated. The interested reader will find details of this method discussed by Verwey and Overbeek [8]. As far as we are concerned, it is sufficient to note the following conclusions from the general theory:

1. A potential energy of repulsion may extend appreciable distances from surfaces, but its range is compressed by increasing the electrolyte content of the system.

2. The conditions under which approaching particles first influence one another are at large distances of separation for which the approximate relationship given by Eq. (81) holds.

3. The sensitivity of aqueous lyophobic colloids to electrolyte content is due to the dependence of interparticle repulsion on this concentration.

What makes these generalizations significant is their relation to the discussion of the potential energy of attraction as developed in the last chapter. Item 1 means that, at least under some conditions, double layer repulsion competes with van der Waals attraction in both magnitude and range to govern particle interactions. Item 2 reminds us of the attitude we took in discussing steric stabilization in Sect. 11.11, namely, that the interactions which the particles experience at their first encounter are the easiest to handle and may determine their subsequent behavior. Item 3 indicates that the location and shape of the repulsion curves in Fig. 11.1 are governed by the electrolyte concentration in the system. There are two aspects to this last point: The concentration of the potential-determining ions determines the potential at the wall via Eq. (1) and the indifferent electrolyte content determines κ^{-1}, which, in turn, measures the range over which Φ_R decays. As we discussed in Sect. 11.2, the net potential energy curve—the resultant of attractive and repulsive components—is a convenient way to analyze flocculation phenomena. In the following sections we shall combine various results from this chapter and the last to arrive at a quantitative theory for electrostatic stabilization against flocculation. The resulting theory is generally called the DLVO theory, after B. Derjaguin, L. D. Landau, E. J. W. Verwey, and J. Th. G. Overbeek, who initially brought the diverse elements of this theory together.

12.8 Potential Energy Curves and the DLVO Theory

The exact shape of potential energy curves depends on the physical factors responsible for the interaction and also on the assumed geometry of the particles. We shall be mainly concerned with the case of interacting blocks, since the expressions describing the various interactions are simpler in this case than for the more realistic case of interacting spheres. The more complicated spherical geometry contains no fundamental insights beyond those already obtained from considering the blocks.

A quantitative expression for the net interaction of two blocks of material separated by a distance d at their surface is obtained by combining Eqs. (11.77) and (81) to give

$$\Phi = \frac{64 n_0 k T \Upsilon_0^2}{\kappa} \exp(-\kappa d) - \frac{A}{12\pi} d^{-2} \tag{83}$$

In the next few paragraphs the effect on the net potential energy curves of the Hamaker constant, the surface potential, and the electrolyte content—considered separately and in this order—will be examined.

1. The effect of A. It is understood that A in Eq. (83) is the effective Hamaker constant A_{212} for the system. Of the variable parameters in this equation, it is the one over which we have least control; its value is determined by the chemical nature of the dispersed and continuous phases. The presence of small amounts of solute in the continuous phase leads to a negligible alteration of the value of A for the solvent.

The effect of variations in the value of A_{212} on the net potential energy is shown in Fig. 12.7. Each of the curves in the figure is drawn for a different value of A, but at identical values of κ (10^9 m^{-1} or 0.093 M for 1:1 electrolyte and ψ_0 (103 mV). As might be expected, the height of the potential energy barrier decrease and the depth of the secondary minimum increases with increasing values of A. If the cross-sectional area of interaction is 4.0 nm^2, each unit on the ordinate scale corresponds to kT at 25°C. This is the unit of thermal energy against which all interactions are judged to be large or small. Thus for the curves shown in Fig. 12.7, the depth of the secondary minimum is slight, and only for the smallest A value is the barrier height significant for particles with this interaction cross section.

2. The effect of ψ_0. The potential at the inner limit of the diffuse part of the double layer enters Eq. (83) through Υ_0, defined by Eq. (60). For large values of ψ_0, $\Upsilon_0 \simeq 1$, so sensitivity to the value of ψ_0 decreases as ψ_0 increases. Figure 12.8 shows the effect of variations in the value of ψ_0 on the total interaction potential energy with κ (10^9 m^{-1} or 0.093 M for a 1:1 electrolyte) and A (2×10^{-19} J) constant. The height of the potential energy barrier is seen to increase with increasing values of ψ_0, as would be expected in view of the increase of repulsion with this quantity. For some systems ψ_0 is adjustable by varying the concentration of potential-determining ions, as described in Sect. 12.2 for the AgI surface. We shall see in Sect. 12.10:10 that this quantity can be complicated by other adsorption phenomena; this is why we described it above as the potential at the "inner limit of the diffuse part of the double layer" rather than simply "at the wall." As we shall see in the following chapter, electrokinetic experiments measure a potential within the double layer—the ζ potential—but it is not entirely clear at which location within the double layer this potential applies. However, the experimental ζ potential does establish a lower limit for ψ_0.

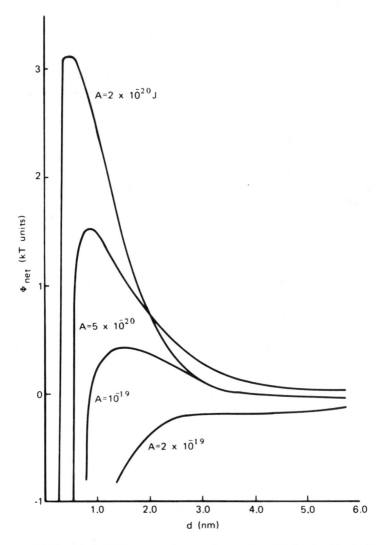

Figure 12.7 Plot of Φ_{net} versus d according to Eq. (83) for flat blocks. Curves are drawn for different values of A_{212} with constant values of κ (10^9 m^{-1}) and ψ_0 (103 mV). Units of ordinate: multiples of kT at 25°C for an interaction area of 4.0 nm^2.

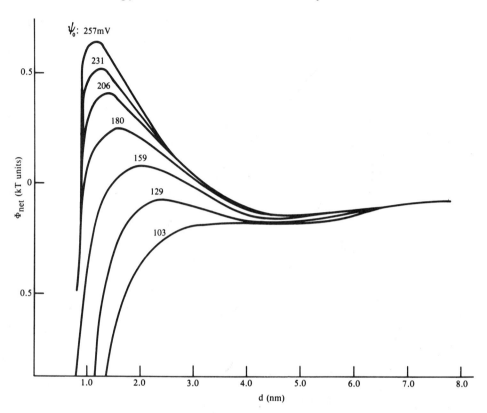

Figure 12.8   Plot of Φ_{net} versus d according to Eq. (83) for flat blocks. Curves are drawn for different values of ψ_0 with constant values of κ (10^9 m^{-1}) and A (2×10^{-19} J). Units of ordinate: multiples of kT at 25°C for an interaction area of 4.0 nm^2.

3. The effect of electrolyte concentration. Of the various quantities which affect the shape of the net interaction potential curve, none is as accessible to empirical adjustment as κ. This quantity depends on both the concentration and valence of the indifferent electrolyte as shown by Eq. (39). For the present we examine only the consequences of concentration changes on the total potential energy curve. We shall consider the valence of electrolytes in the following section. To consider the effect of electrolyte concentration on the potential energy of interaction, it is best to use the more elaborate expressions for interacting spheres. Figure 12.9 is a plot of Φ_{net} for this situation as a function of separation of surfaces with κ as the parameter which varies from one curve to another.

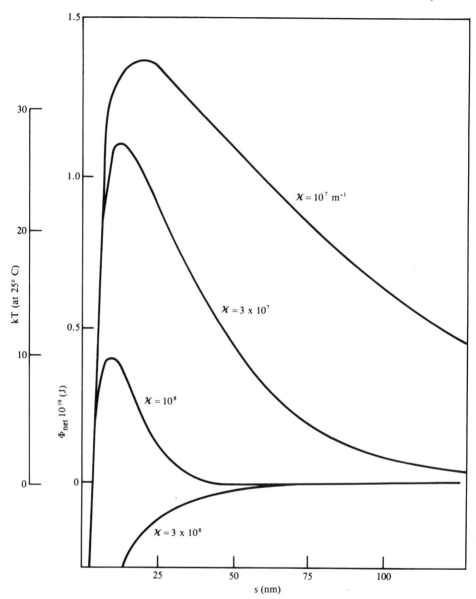

Figure 12.9 Plot of Φ_{net} versus s, the separation of surfaces, for two spheres of equal radius (100 nm). Curves are drawn for different values of κ with constant values of A (10^{-19} J) and ψ_0 (25.7 mV). (From Ref. 8, used with permission.)

Figure 12.9 has been drawn for spheres with $R = 100$ nm, $A = 10^{-19}$ J, and $\psi_0 = 25.7$ mV. The ordinate in this figure has been labeled both in joules and in multiples of kT at $25°C$. For the system described by these curves a significant energy barrier is present at all concentrations of a 1:1 electrolyte less than about 10^{-3} M. For concentrations between 10^{-3} and 10^{-2} M, however, the barrier vanishes. This particular colloid is thus expected to undergo a transition from a stable dispersion to a flocculated one with additions of an indifferent 1:1 electrolyte to a concentration in this range. An indication of how the repulsion is calculated between spherical particles is taken up in the following example.

Example 12.4 Spherical particles can be approximated by a stack of circular rings with planar faces as shown in Fig. 12.10. Use Eq. (81) to describe the repulsion between rings separated by a distance z and derive an expression for the repulsion between the two spheres. Assume that the strongest interaction occurs along the line of centers and make any approximations consistent with this to obtain the final result.

Solution Indexing the various tiers of rings by i, we note that the increment of the interaction due to the ith ring is $d\Phi_R = \Phi_i dA_i$, where the area of the ring is $2\pi\, h_i dh$. We eliminate dh as follows: The separation of the ith ring z_i is related to the distance of closest approach by the formula $\tfrac{1}{2}(z_i - s) = R - (R^2 - h^2)^{1/2}$, in which the factor $\tfrac{1}{2}$ arises because part of the effect occurs at each surface. Since R is a constant, this last result may be differentiated and rearranged to give $R(1 - h^2/R^2)^{1/2}dz = 2h\, dh$. This may be combined with the expression for $d\Phi_R$ to yield $d\Phi_R = \pi R(1 - h^2)^{1/2}\Phi(z)\, dz$, where the functional notation has been added to Φ as a reminder that it is different for each ring. Substituting Eq. (81) for $\Phi(z)$, we obtain

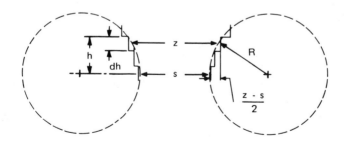

Figure 12.10 Schematic illustration which shows how the repulsion between spheres may be calculated from the interaction between flat plates.

$$d\Phi_R = \pi R \left(1 - \frac{h^2}{R^2} \right)^{1/2} \frac{64 n_0 k T \Upsilon_0^2}{\kappa} \exp\left(-\frac{z}{\kappa^{-1}} \right) dz$$

If we further assume that $h^2/R^2 \ll 1$, this simplifes to

$$d\Phi_R = \frac{64 \pi R n_0 k T \Upsilon_0^2}{\kappa} \exp(-\kappa z) dz$$

The total potential energy of repulsion is given by integrating over all values of z. This is most readily done by assuming that z varies between s and infinity. This upper limit is justified because the potential function drops off exponentially with distance. Therefore large separations make a negligible contribution to the total. Integrating between these limits, we obtain

$$\Phi_R = \frac{64 \pi R n_0 k T \Upsilon_0^2}{\kappa^2} \exp(-\kappa s)$$

Under various circumstances, it may be justified to replace either the exponential or Υ_0 by their series approximations, thereby simplifying this result still further.

•

It has long been known that the addition of an indifferent electrolyte can cause a lyophobic colloid to undergo flocculation. The DLVO theory provides a quantitative explanation for this fact. Furthermore, it is known that for a particular salt a fairly sharply defined concentration is needed to induce flocculation. This concentration may be called the critical flocculation concentration. The DLVO theory in general and Figs. 12.7–12.9 in particular can be summarized by the following statements:

1. The higher the potential at the surface of a particle—and therefore throughout the double layer—the larger the repulsion between the particles will be.
2. The lower the concentration of indifferent electrolyte, the longer is the distance from the surface before the repulsion drops significantly.
3. The larger the Hamaker constant, the larger is the attraction between macroscopic bodies.
4. Remember, the point of these figures is to see the effect on the potential energy curves of systematically varying one parameter at a time. It is the trends of behavior rather than the parameters themselves that are of greatest interest. In the following section we shall discuss the critical flocculation concentration as a simple

quantitative test of the theory. In subsequent sections we shall see how studies of the rate of flocculation provide still more stringent tests of the theory and means for evaluating parameters of interest.

12.9 The Critical Flocculation Concentration and the Schulze-Hardy Rule

One of the easiest tests that can be performed on an aqueous colloid is to determine the critical concentration of electrolyte required to flocculate the colloid. We shall use the notation CFC (for critical flocculation concentration) to indicate this quantity. This experiment is conducted by introducing the dispersion into a series of test tubes and adding to each various proportions of water and electrolyte solution. In this way the total dilution of the dispersed particles is held constant while different amounts of salt are added to each. After mixing and waiting an arbitrary but consistent length of time, we visually inspect the tubes for evidence of the effect of the added salt. There will generally be clear evidence of flocculation (e.g., the settling out of the dispersed phase) in some of the tubes, whereas other tubes appear unchanged. Thus the highest concentration of salt which leaves the colloid unchanged and the lowest concentration which causes flocculation bracket the CFC. A second series of experiments may be conducted within this range to narrow the range of the CFC still further.

The actual concentration of electrolyte at the CFC depends on the following:

1. The time allowed to elapse before the evaluation is made
2. The uniformity or, more likely, the polydispersity of the sample
3. The potential at the surface
4. The value of A
5. The valence of the ions

In a series of tests on any particular system, items 1–4 remain constant, so the CFC is a quantitative measure of the effect of the valence of the added ions. Table 12.3 summarizes some experimental results of this sort.

The results in Table 12.3 have been collected for colloids bearing both positive and negative surface charges. One of the earliest (1900) generalizations about the effect of added electrolyte is a result known as the Schulze-Hardy rule. This rule states that it is the valence of the ion of opposite charge to the colloid that has the principal effect on the stability of the colloid. The CFC value for a particular electrolyte is essentially determined by the valence of the counter-ion regardless of the nature of the ion with the same charge as the surface. The numbers listed in

Table 12.3 CFC Values (in Moles Liter^{-1}) for Mono-, Di-, Tri-, and Tetravalent Ions Acting on Both Positive and Negative Colloids (numbers in parentheses) and CFC Values Relative to the Value for Monovalent Electrolytes on the Same System (numbers outside parentheses)^{a}

Valence of counter-ion	Negative colloids			Positive colloids		Theory
	As$_2$S$_3$	Au	AgI	Fe$_2$O$_3$	Al$_2$O$_3$	
1	(5.5×10^{-2}) 1	(2.4×10^{-2}) 1	(1.42×10^{-1}) 1	(1.18×10^{-2}) 1	(5.2×10^{-2}) 1	1
2	(6.9×10^{-4}) 1.3×10^{-2}	(3.8×10^{-4}) 1.6×10^{-2}	(2.43×10^{-3}) 1.7×10^{-2}	(2.1×10^{-4}) 1.8×10^{-2}	(6.3×10^{-4}) 1.2×10^{-2}	1.56×10^{-2}
3	(9.1×10^{-5}) 1.7×10^{-3}	(6.0×10^{-6}) 0.3×10^{-3}	(6.8×10^{-5}) 0.5×10^{-3}	—	(8×10^{-5}) 1.5×10^{-3}	1.37×10^{-3}
4	(9.0×10^{-5}) 17×10^{-4}	(9.0×10^{-7}) 0.4×10^{-4}	(1.3×10^{-5}) 1×10^{-4}	—	(5.3×10^{-5}) 10×10^{-4}	2.44×10^{-4}
Potential-determining ion	S^{2-}	Cl^{-}	I^{-}	H^{+}	H^{+}	

Source: Data from Ref. 6.
^{a}Theoretical values given by Eq. (92).

parentheses in Table 12.3 are the CFC values in moles per liter for counter-ions of the indicated valence. That is, about 7×10^{-4} M of divalent cation is needed to flocculate the negative As_2S_3 sols, whereas about 6×10^{-4} M of divalent anion is required to flocculate positive Al_2O_3 sols.

The actual values of these concentrations depend on a whole array of unknown parameters, but their relative values depend only on the valence of the counter-ions. The entries outside of parentheses in Table 12.3 are the values of the CFC relative to the value for the monovalent electrolyte in the same set of experiments. These are seen to be remarkably consistent for the divalent ions and acceptably close together for trivalent and tetravalent counterions.

Now let us see how this result is to be understood in terms of the DLVO theory. At first glance, it seems remarkable that any consistency at all can be found in tests as arbitrary as the CFC determination. It is not difficult, however, to show that these results are quite close to the values predicted in terms of the DLVO model for interacting blocks with flat faces. From an inspection of Fig. 12.9, we concluded that the system at $\kappa = 10^8$ m^{-1} would be stable with respect to flocculation, whereas the one at $\kappa = 3 \times 10^8$ m^{-1} would flocculate. Furthermore, we examined the energy barrier to draw these conclusions. Next we must ask how the qualitative criteria we used in discussing the curves can be translated into an analytical expression.

One way of doing this is to assume that the demarcation between stable and unstable colloids occurs at the value of κ for which the height of the "barrier" is zero. Physically, this is a somewhat arbitrary choice: Thermal energy is sufficient to allow particles to overcome a barrier of low but nonzero height. Mathematically, however, the assumption that the maximum in the potential energy curve occurs at zero permits us to write

$$\Phi_{net} = 0 \tag{84}$$

and

$$\frac{d\Phi_{net}}{d(d)} = 0 \tag{85}$$

as the conditions for stability.

Applying Eqs. (84) and (85) to Eq. (83) gives

$$\frac{64 n_0 k T \Upsilon_0^2}{\kappa} \exp(-\kappa d_m) = \frac{A}{12\pi} d_m^{-2} \tag{86}$$

and

$$64n_0 kT\Upsilon_0^2 \exp(-\kappa d_m) = \frac{A}{6\pi} d_m^{-3} \tag{87}$$

where the subscript m reminds us that this describes the maximum. From these equations it is readily apparent that

$$\kappa d_m = 2 \tag{88}$$

is the criterion for stability according to this model. This may also be written

$$d_m = 2\kappa_m^{-1} \tag{89}$$

in terms of the "thickness" of the double layer. Again we see an important distance measured in terms of κ^{-1}.

It is the dependence of the CFC values on the valence of the electrolyte that we seek to obtain rather than the absolute value of the CFC. Therefore it is sufficient to proceed from this point by merely retaining those factors which involve either the concentration ($n_0 \propto M$) or the valence (z). Substituting Eq. (88) back into Eq. (87) yields

$$n_0 \propto \kappa^3 \tag{90}$$

The dependence of Υ_0 on z has been neglected in writing this result, a procedure that is entirely justified for the level of approximation involved here (recall $\Upsilon_0 \simeq 1$). From the definition of κ [Eq. (39)], we obtain

$$n_0 \propto z^3 n_0^{3/2} \tag{91}$$

or

$$M \propto z^{-6} \tag{92}$$

which is the desired result. According to Eq. (92), the CFC value varies inversely with the sixth power of the valence of the ions in solution. The column of numbers in Table 12.3 labeled "theory" follows the progression z^{-6}: 1, 2^{-6}, 3^{-6}, 4^{-6}. The actual CFC values are seen to be in quite reasonable accord with these predictions.

Note that if ψ_0 is not assumed to be large, Υ_0 depends on both z and ψ_0. In this case the CFC is found to show a less sensitive dependence on the counter-ion valence than predicted by Eq. (92). The following example examines this point.

Example 12.5 Use the accompanying data* to criticize or defend the

---

*Data cited by E. Matijević, *J. Colloid Interface Sci.*, *43*:217 (1973).

following proposition: Since the CFC for positively charged AgBr is less (regardless of counter-ion valence) than that for poly(vinyl chloride) latex (PVC), ψ_0 must be less for AgBr.

	CFC (mole liter^{-1}) of ions opposite in charge to ψ_0	
Colloid	$z = 1$	$z = 2$
PVC latex	2.3×10^{-1}	1.2×10^{-2}
AgBr	1.6×10^{-2}	2.3×10^{-4}

Solution Because of the arbitrary time of observation in CFC experiments, the absolute values of the CFCs have no significance. In this regard the proposition is wrong. It may be possible to rank the two colloids with respect to surface potential, however, by examining the order of the dependence of CFC on ion charge for each of the colloids:

For AgBr: $1.6 \times 10^{-2}/2.3 \times 10^{-4} = 69.6 = (\frac{1}{2})^{-n}$ so $n = 6.12$

For PVC: $2.3 \times 10^{-1}/1.2 \times 10^{-2} = 19.2 = (\frac{1}{2})^{-n}$ so $n = 4.26$

The fact that AgBr agrees with the prediction of Eq. (92) which applies at high potentials suggests that ψ_0 is greater for this colloid than for PVC.

•

Although this test of the DLVO theory itself introduces some additional approximations, it is a workable unification of experimental and theoretical points of view. The threshold of stability in terms of the concentration and valence of indifferent electrolyte is easily measured. Theoretical models describe the interaction between a pair of particles in terms of potential energy diagrams. The reconciliation of these two approaches constitutes an important step toward obtaining still more quantitative information from the study of flocculation. This is taken up in the following sections concerned with the kinetics of flocculation. First, however, a few remaining comments about the critical flocculation concentrations must be made.

In Sects. 12.4 and 12.6 we implicitly anticipated that the ion opposite in charge from the wall plays the predominant role in the double layer, the central observation of the Schulze–Hardy rule. This enters the mathematical formalism of the Gouy–Chapman theory in Eq. (47) in which a Boltzmann factor is used to describe the relative concentration of the ions in the double layer compared to the bulk solution. For those ions which have the same charge as the surface (positive), the exponent in the Boltzmann factor is negative. This reflects the coulombic repulsion of these ions from the wall. Ions with the same charge as the surface are thus present at lower concentration in the double layer than in the bulk solution. The signs are reversed for oppositely charged ions; therefore the concentration of the latter is increased in the double layer. It may be shown—at least for high ψ_0 values—that the result of these considerations is essentially equivalent to emptying a region $2\kappa^{-1}$ thick of ions having the same charge as the wall. Thus, in terms of the model for flocculation just presented, it is essentially only the counter-ions that contribute to the diffuse double layer at the critical separation for flocculation.

The CFC values reported in Table 12.3 in many cases are average values for several compounds of similar valence. The use of averages to compare CFC values is justified, since the valence primarily determines the CFC for an electrolyte.

However, a closer inspection of the data reveals that there are second-order differences between different ions. For example, 0.058 and 0.051 M are the Li^+ and Na^+ concentrations required to flocculate As_2S_3 sols, and 0.165 and 0.140 M are the concentrations required to flocculate AgI sols. Although both sets of values are acceptably close to the mean (which includes a number of other compounds), it is also clear that Li^+ is consistently slightly less effective than Na^+ in inducing flocculation. A more complete sequence of these variations in effectiveness is as follows:

1. For monovalent cations

$$Cs^+ > Rb^+ > NH_4^+ > K^+ > Na^+ > Li^+$$

2. For monovalent anions

$$F^- > Cl^- > Br^- > NO_3^- > I^- > SCN^-$$

It is apparent from the correlation between the rankings of monatomic ions and their placement in the periodic table that some systematic effect is responsible for this ordering. This is the topic of the next section.

12.10 "Not-Quite-Indifferent" Electrolytes: Stern Adsorption

At the beginning of this chapter we divided electrolytes into categories of potential determining and indifferent. We saw in Sect. 12.2 that the potential-determining ions are adsorbed at surfaces and determine the value of ψ_0 according to Eq. (1). Throughout our discussion of the diffuse double layer, we have treated indifferent electrolytes as point charges with no chemical uniqueness except for their valence number. While this is a useful simplifying approximation, it underestimates the complexity of the "real world." The assumption that ions have no volume is acceptable for the bulk region of dilute solutions, but real ions cannot be drawn toward charged surfaces without crowding becoming a problem. At the inner edge of the diffuse part of the double layer some sort of saturation limit must be approached.

One way of handling this—due to O. Stern—is to divide the aqueous part of the double layer by a hypothetical boundary known as the Stern surface. The Stern surface is situated a distance δ from the actual surface. Figure 12.11 schematically illustrates the way this surface intersects the double layer potential and how it divides the charge density of the double layer.

The Stern surface is drawn through the ions which are assumed to be adsorbed on the charged wall. There are several consequences of assuming an adsorbed layer of ions at the surface:

1. An adsorption isotherm may be written for these ions which allows for surface saturation and thus introduces the idea of finite ionic size.

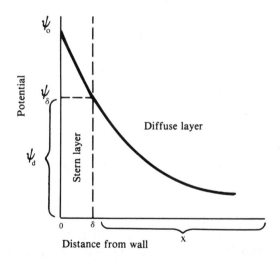

Figure 12.11 Schematic illustration of the variation of potential with distance from a charged wall in the presence of a Stern layer. Significance of the subscripts: 0 at wall, δ at Stern surface, d in diffuse layer.

The Langmuir isotherm [Eq. (7.67)] is one expression that can be used for this purpose:

$$\theta = \frac{Kn_0}{1 + Kn_0} \tag{93}$$

In this expression θ is the fraction of surface adsorption sites occupied, n_0 is the concentration of the adsorbed ions in the solution, and K is a constant.

2. The constant in Eq. (93) is easily shown to be proportional to a Boltzmann factor in which the exponential energy consists of two contributions: $ze\psi_\delta$, the electrical energy associated with the ion in the Stern layer, and ϕ, the specific chemical energy associated with the adsorption:

$$K \simeq \exp\left(\frac{ze\psi_\delta + \phi}{kT}\right) \tag{94}$$

3. The Stern layer resembles the parallel plate capacitor model for the double layer. Therefore Eq. (13) may be applied to this region:

$$\frac{\psi_0 - \psi_\delta}{\delta} = \frac{\sigma_{\delta 0}^*}{\varepsilon_\delta} \tag{95}$$

and ε_δ is the product of ε_0 and the relative dielectric constant that applies within the Stern layer.

4. The fraction of surface sites occupied equals the ratio $\sigma_\delta^*/\sigma_{\delta 0}^*$, where $\sigma_{\delta 0}^*$ is the charge density at surface saturation. Therefore, Eqs. (93) and (95) may be combined to give

$$\frac{\psi_0 - \psi_\delta}{\delta} = \frac{1}{\varepsilon_\delta} \frac{\sigma_{\delta 0}^* K n_0}{1 + K n_0} \tag{96}$$

This equation shows that the potential drop in the Stern layer increases with the concentration of the adsorbed ion and ultimately approaches a constant value when the surface is saturated.

Outside the Stern surface the double layer continues to be described by Eq. (58) or one of its approximations. The only modifications of the analysis of the diffuse double layer required by the introduction of the Stern surface are that x be measured from δ rather than from the wall and that ψ_δ be used instead of ψ_0 as the potential at the inner boundary of the diffuse layer.

The Stern theory is difficult to apply quantitatively because several of the parameters it introduces into the picture of the double layer cannot be evaluated experimentally. For example, the dielectric constant of the water is probably considerably less in the Stern layer than it would be in bulk, because the electric field is exceptionally high in this region. This effect is called dielectric saturation and has been measured for macroscopic systems, but it is difficult to know what value of ε_δ applies in the Stern layer. The constant K is also difficult to estimate quantitatively, principally because of the specific chemical interaction energy ϕ. Some calculations have been carried out, however, in which the various parameters in Eq. (96) were systematically varied to examine the effect of these variations on the double layer. The following generalizations are based on these calculations.:

1. As ϕ increases, K increases and the equilibrium amount adsorbed for any n_0 value short of saturation also increases.
2. As the electrolyte concentration increases, increasing amounts of the potential drop occur in the Stern layer. This is true even if $\phi = 0$, which shows that specific chemical effects are not necessary for this result.
3. Values of ψ_δ which are much less than ψ_0 are possible in dilute solutions only if ϕ is relatively large.
4. The quantity ψ_δ varies only slightly with ϕ—although it is highly sensitive to n_0—until ϕ is relatively large.

Values of the parameter ϕ may be experimentally evaluated for the mercury–water surface from electrocapillary studies. The displacement of the coordinates of the electrocapillary maxima in Fig. 7.19 reflects differences in the intrinsic adsorbability of various ions. Electrocapillary studies reveal that the strength of specific adsorption at the mercury–water interface for some monovalent anions follows the order

$$I^- > SCN^- > Br^- > Cl^- > OH^- > F^-$$

whereas for some monovalent cations the order is

$$N(C_2H_5)_4^+ > N(CH_3)_4^+ > Tl^+ > Cs^+ > Na^+$$

In general, the specific adsorption of an ion is enhanced by larger size—and therefore larger polarizability—and lower hydration, which itself is a function of ion size. For example, among the ions just listed, the large I^- ion is the most strongly adsorbed, and the small but highly hydrated Na^+ ion is adsorbed least.

By allowing for surface saturation, the Stern theory overcomes the objection to the Gouy–Chapman theory of excessive surface concentrations. In so doing, however, it trades off one set of difficulties for another. In the Gouy-Chapman theory the functional dependence of ψ

on x involves only the parameters κ and ψ_0. The former is known and the latter may be evaluated—at least for some surfaces—by Eq. (1) when the point of zero charge is known. The Stern modification of the double layer picture introduces parameters which are not only difficult to estimate— such as δ and K (or ϕ)—but also specific characteristics of different ions. The generality of the Gouy–Chapman model is thus lost when the specific adsorption effects of the Stern theory are considered.

It is the outer portion of the double layer that interests us most as far as colloidal stability is concerned. The existence of a Stern layer does not invalidate the expressions for the diffuse part of the double layer. As a matter of fact, by lowering the potential at the inner boundary of the diffuse double layer, we enhance the validity of low-potential approx-imations. The only problem is that specific adsorption effects make it difficult to decide what value to use for ψ_δ.

In subsequent discussions it will be the potential in the diffuse double layer that concerns us. It can be described relative to its value at the inner limit of the diffuse double layer which may be either the actual surface or the Stern surface. We shall continue to use the symbol ψ_0 for the potential at this inner limit. It should be remembered, however, that specific adsorption may make this quantity lower than the concentration of potential-determining ions in the solution would indicate. We shall see in Chap. 13 how the potential at some (unknown) location close to this inner limit can be measured. It is called the zeta potential.

12.11 DLVO Theory and Flocculation Kinetics

Using the approach developed in Example 11.3 and potential energy expressions for spherical particles, it has been possible to predict how the stability ratio W varies with electrolyte concentration according to the DLVO theory. Since W can be measured by experimental studies of the rate of flocculation, this approach allows an even more stringent test of the DLVO theory than CFC values permit. We shall not bother with algebraic details, but instead go directly to the final result:

$$\log W = K_1 \log c + K_2 \tag{97}$$

where K_1 and K_2 are constants and c is the concentration of the ions in moles per liter. For water at 25°C the value of K_1 has been calculated to be $-2.15 \times 10^9 \, \Upsilon_0^2 R/z^2$, where Υ_0 is given by Eq. (60) and z is the valence of the counter-ions.

Figure 12.12 is a plot of $\log W$ versus $\log c$ for AgI sols of several different particle sizes. The experimental W values in this figure were determined from absorbance measurements. According to the preceding

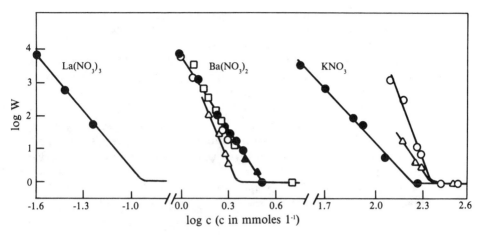

Figure 12.12 log W versus log c for AgI sols of five different particle sizes flocculated with the electrolytes shown. The mean particle radii in the different sols are ●, 52.0 nm; ○, 22.5 nm; □, 53.5 nm; ▲, 65.0 nm and △, 158.0 nm. [From H. Reerink and J. Th. G. Overbeek, *Discuss. Faraday Soc., 18*:74 (1954).]

section, data of this sort not only test the DLVO theory but also permit the evaluation of several important colloidal parameters. From the data in Fig. 12.12 the following conclusions can be drawn:

1. A plot of log W versus log c is linear as required by Eq. (97).
2. The concentrations at which $W = 1$ (where the breaks in the curves appear) measure the CFC values for the electrolyte involved. The CFC values for mono-, di-, and trivalent ions are about 0.199, 2.82×10^{-3}, and 1.3×10^{-4} mole liter^{-1}, respectively. These are in the ratio $1:1.42 \times 10^{-2}:0.7 \times 10^{-3}$. These figures compare very favorably with the other experimental data for AgI and the theoretical values presented in Table 12.3.
3. Slow flocculation is observed for log $W < 4$, or $W < 10^4$. For a typical potential energy curve, this corresponds to a value of Φ_m of about 15 kT. From this we may conclude that the height of an energy barrier must be *at least* 15 kT if the colloid is to have any appreciable stability. Likewise, we may assume that unless the secondary minimum is approximately this deep, particles will be able to "escape" from it. In view of the general shape of the potential energy curves, the retardation effect, and this assessment of what constitutes a "high" barrier or a "deep" well, it seems likely that rigid aggregates are not formed in the secondary minimum.

4. Equation (97) can be used to analyze the slopes of the curves in Fig. 12.12, since the mean size of the AgI particles is known. In this way, Reerink and Overbeek found ψ_0 values in the range 12–53 mV and A values in the range 0.2×10^{-20} to 10×10^{-20} J. Both of these are of the proper order of magnitude—no minor accomplishment in itself in light of the diverse assumptions required to get to this point.

5. The values of A and ψ_0 obtained from this analysis are slightly less satisfying in detail: The values of A show a lot of scatter and ψ_0 appears to be too low. Recall that the variation of the CFC with z^{-6} implies large values of ψ_0; for lower ψ_0 values a different dependence on z is expected.

6. Least satisfactory of all is the correlation with particle size. The results shown in Fig. 12.12 were determined for AgI sols covering a 10-fold range of particle sizes.

It is evident from Fig. 12.12 that the slopes do not vary over a similar range, as required by Eq. (97). As a matter of fact, there are examples for which the steepest slope is associated with the coarsest particles (as required by theory) and others where it occurs with the smallest particles. The quantitative predictions fail on this particular point, but, as we shall see presently, there are some discrepancies between the theoretical model and the actual experimental system that may account for this apparent insensitivity to particle size. The following example considers another application of Eq. (97) to an experimental system.

Example 12.6 Colloidal gold stabilized by citrate ions and having a mean particle radius of 103 Å was flocculated by the addition of $NaClO_4$. The kinetics of flocculation were studied colorimetrically and the stability ratio W for different $NaClO_4$ concentrations was determined*:

$c \times 10^3$ (mole liter^{-1})	2	3	5	8	10.5
W	48	31	17	8.9	0.84

when these data are plotted in the manner suggested by Eq. (97), a straight line is obtained up to about 10^{-2} M, at which point a precipitous deviation from linearity sets in. The slope of the linear portion is about -1.20; estimate Υ_0 from this slope. Verify that this value of Υ_0 corresponds to a value of ψ_0 equal to about 25 mV. Estimate what the CFC value would be for this system if W continued to vary according to the same function of c both above and below 10^{-2} M. Suggest an explanation for the abrupt decrease in W near 10^{-2} M.

---

*B. V. Enüstün and J. Turkevich, *J. Am. Chem. Soc.,* **85**: 3317 (1963).

Solution   The slope of the linear portion equals K_1, whose theoretical value is given above. Since R is known for these particles, Υ_0 may be calculated as follows:

$$\Upsilon_0^2 = \frac{1.20}{(2.15 \times 10^9)(103 \times 10^{-10})} = 0.054 \quad \text{or} \quad \Upsilon_0 = 0.23$$

Table 12.2 shows that this value of Υ_0 corresponds to a value of ψ_0 between 20 and 40 mV. For $\psi_0 = 25$ mV, Eq. (60) shows that $\Upsilon_0 = \exp\{[25/2(25.7)] - 1\}/\{\exp[25/2(25.7)] + 1\} = 0.24$, which is very close to the experimental value. If the linear portion of the plot is extrapolated to log $W = 0$, the value for log CFC is found to be -1.25, from which CFC $= 0.055$ M. Apparently significant Na^+ adsorption begins to occur at about 10^{-2} M, and W begins to decrease rapidly above this concentration.

•

Before concluding this section, it seems desirable to comment a bit more fully on some possible sources of the discrepancy between the predictions of Eq. (97) and the data shown in Fig. 12.12. It is convenient to divide these remarks into those which involve the interaction potential explicitly and those which pertain to the kinetic part of the discussion.

As far as the interaction potential is concerned, we can state the following:

1. It may not be adequate to describe the interaction between AgI particles—especially at relatively close range—in terms of the radii of the dispersed units. In fact, the radii of surface protuberances rather than the dimensions of the particle as a whole may affect the short-range interaction.

2. Throughout this discussion only nonspecific effects have been considered; that is, we have totally neglected to consider ion adsorption and the contribution of the Stern layer to the overall picture. This can be a serious source of complication, at least in some systems. This is evident from the fact that some dispersions show a reversal of charge (from negative to positive) with the addition of La^{3+} and Th^{4+}, indicating the adsorption of these ions.

Several aspects of the kinetic part of this discussion also warrant additional comment:

3. For highly asymmetrical particles, the probability of collision is greater than that predicted by Eq. (11.51) or (11.58). This may be understood by thinking that the diffusion coefficient is most influenced by the smaller dimensions of the particles (therefore increased) and the "target radius" is most influenced by the longer

dimension (also increased, relative to the case of symmetrical particles).

4. The frequency of collisions is also expected to be greater in a polydisperse system than in a monodisperse system by the same logic as presented in item 3.

5. The presence of velocity gradients in the system may also increase the rate of flocculation above the value given by Eq. (11.51) or (11.58).

The ratio of the probability of a collision induced by a velocity gradient (mechanical agitation) to the collision probability under the influence of Brownian motion (thermal) has been shown to be

$$\text{ratio} = \frac{\eta R^3 (dv/dx)}{2kT} \tag{98}$$

Since this increases with the cube of particle size, it may be the dominant mechanism for the flocculation of larger particles. Note that all the kinetic complications—items 3 through 5—tend to cancel out of the evaluation of the stability ratio, since the rates of both rapid and slow flocculation are determined on the same colloid. In spite of these complications, the value of the kinetic approach over, say, using Eq. (1) or electrophoresis results (see Chap. 13) to determine ψ_0 is that the potential is evaluated in the actual flocculating system and hence includes any contribution from Stern adsorption. Once a suitable value for ψ_0 is known, potential energy curves for that value of ψ_0 and a measured concentration of electrolyte (i.e., κ) can be constructed for different values of A_{212}. That plot of Φ_{net} versus d which is most consistent with W (e.g., by the method of Example 11.3) may be used to identify A_{212} for the system under consideration.

References

1. N. K. Adam, *The Physics and Chemistry of Surfaces*, Dover, New York, 1968.
2. C. A. Barlow, Jr., The Electrical Double Layer, in *Physical Chemistry*, Vol. IXA (H. Eyring, D. Henderson, and W. Jost, eds.), Academic, New York, 1970.
3. P. Delahay, *Double Layer and Electrode Kinetics*, Wiley, New York, 1965.
4. D. A. Haydon, Electrical Double Layers and Electrokinetics, in *Recent Progress in Surface Science*, Vol. 1 (J. F. Danielli, K. A. G. Parkhurst, and A. C. Riddiford, eds.), Academic, New York, 1964.
5. A. L. Loeb, J. Th. G. Overbeek, and P. H. Wiersema, *The Electrical Double Layer around a Spherical Colloid Particle*, MIT Press, Cambridge, Mass., 1960.
6. J. Th. G. Overbeek, in *Colloid Science*, Vol. 1 (H. R. Kruyt, ed.), Elsevier, Amsterdam, 1952.
7. H. van Olphen, *An Introduction to Clay Colloid Chemistry*, Wiley, New York, 1963.

8. E. J. W. Verwey and J. Th. G. Overbeek, *Theory of the Stability of Lyophobic Colloids*, Elsevier, Amsterdam, 1948.

Problems

1. It has been observed that AgI sols at pH 3.5 are flocculated by 4.5×10^{-5} M Al $(NO_3)_3$ in the absence of a second salt and by 1.7×10^{-4} M Al$(NO_3)_3$ in the presence of 0.009 M K_2SO_4. Calculate the value of κ for each of these solutions. Is the threshold of instability consistent with your expectations in terms of the values of κ for these two systems? The authors of this research suggest that $K = 370$ for the equilibrium $Al^{3+} + SO_4^{2-} \rightleftharpoons AlSO_4^+$. Calculate the concentration of Al^{3+} in the system containing K_2SO_4. Does the behavior of the AgI appear to be correlated with the concentration of "free" Al^{3+}? Is the specific adsorption of Al^{3+} expected on AgI in the presence of 4×10^{-4} M excess KI? Discuss in terms of ψ_0 and ϕ.

2. The viscosity of negatively charged colloidal agar (0.14% at 50°C) was studied with a variety of different electrolytes added.† The ratio of the specific viscosity ($\eta_{sp} = \eta/\eta_0 - 1$) in the presence of salt to η_{sp} without salt is given below for some of these salts at several concentrations:

$$(\eta_{sp})_{salt}/(\eta_{sp})_{H_2O}$$

c (mEq liter^{-1})	KCl	K_2SO_4	$K_4Fe(CN)_6$	$BaCl_2$	$SrCl_2$	$MgSO_4$	$La(NO_3)_3$	$Pt(en)_3(NO_3)_4^a$
0.25	—	—	—	—	—	—	—	79.2
0.50	90.8	90.7	90.5	—	—	—	76.6	69.7
1.00	—	—	—	78.2	77.9	78.6	—	64
2.00	81.2	81.4	81.1	—	—	—	68.1	60.6
4.00	77.3	77.5	76.7	70.6	70.9	71.6	67.2	60

^{a}en = ethylenediamine.

Discuss this electroviscous effect in terms of the concepts of this chapter and Chapter 4.

3. The deficiency of positive ions ($\Gamma_+ < 0$) adjacent to a positively charged planar surface may be evaluated as follows ($y = ze\psi/kT$):

$$\Gamma_+ \overset{(1)}{=} \int_0^\infty (c_0 - c_x)dx \overset{(2)}{=} c_0 \int_0^\infty (1 - e^{-y})dx \overset{(3)}{=} c_0 \int_0^\infty (1 - e^{-y})\frac{dx}{dy}dy$$

$$\overset{(4)}{=} c_0 \int_{\Psi_0}^0 \frac{(1 - e^{-y})dy}{dy/dx} \overset{(5)}{=} \frac{c_0}{\kappa} \int_{\Psi_0}^0 \frac{1 - e^{-y}}{e^{-y/2} - e^{y/2}}dy$$

---

*L. J. Stryker and E. Matijević, *J. Phys. Chem.*, **73**:1484 (1969).
† H. R. Kruyt and H. G. deJong, *Kolloid Z.*, **100**:250 (1922).

$$\overset{(6)}{=} - \frac{c_0}{\kappa} \int_{\Psi_0}^{0} e^{-y/2} \, dy \overset{(7)}{=} \frac{2c_0}{\kappa} \left[1 - \exp\left(\frac{ze\psi_0}{2kT} \right) \right]$$

Present the physical and/or mathematical justification for equalities (1)–(7) in this sequence. Show that the final result is equivalent, at high surface potentials, to emptying a region of thickness $2\kappa^{-1}$ of ions possessing the same charge as the wall.

4. If a soap film is sufficiently thin, its equilibrium thickness is the resultant of the double layer repulsion, given by Eq. (77), and van der Waals attraction, given by

$$F_A = \frac{\partial \Phi_A}{\partial d} = \frac{A}{6\pi} d^{-3}$$

according to Eq. (11.77). These conditions are satisfied by certain films studied by Lyklema and Mysels,* who obtained the following results with 1:1 electrolyte:

Concentration (mole liter^{-1})	0.103	0.066	0.0197
Thickness of aqueous layer (Å)	91	94	153
Thickness of entire film (Å)	123	126	185

Use the thickness of the aqueous layer in Eq. (77) to calculate F_R per unit area (assume $\Upsilon_0 = 1$). By equating this quantity to F_A per unit area (just given) and using the total film thickness, estimate A for each of these data. Explain why two different values are used for d in this calculation.

5. Using the average value of A you determined in Problem 4, criticize or defend the proposition that van der Waals forces are negligible compared to hydrostatic forces when the latter equal 660 dyne cm^{-2} in a film for which the total thickness is 763 Å. Note that this is the assumption made in Example 12.3. Qualitatively re-examine the latter question in light of the results of this problem.

6. Criticize or defend the following propositions: The DLVO theory should apply to particles dispersed in nonaqueous media once ε and A for the solvent have been included in the relevant expressions. Since ion concentrations are low in media with low dielectric constant, κ^{-1} will be very large for such systems. For a concentrated colloid the mean interparticle spacing may be less than κ^{-1}. In such a case it is more plausible to picture a particle approaching a "target" as traveling along a potential energy plateau rather than facing a potential energy barrier.

    Comment on the relevancy of these propositions to the observation † that a 15% (by volume) water-in-benzene emulsion stabilized by the calcium salt

---

*J. Lyklema and K. J. Mysels, *J. Am. Chem. Soc.,* *87*:2539 (1965).
† W. Albers and J. Th. G. Overbeek, *J. Colloid Sci.,* *14*:501 (1959).

of didodecylsalicylic acid has a ψ_0 value of ~130 mV, yet breaks immediately after preparation.

7. Verify that combining Eqs. (87) and (88) with the definition of κ [Eq. (35)] leads to the following expression (purely numerical factors may be omitted):

$$CFC \propto n \propto \frac{\varepsilon^3 (kT)^5 \Upsilon_0^4}{e^6 A^2 z^6}$$

By the series expansion of Υ_0 verify that $\Upsilon_0^4 \propto (ze\psi_0/kT)^4$ if ψ_0 is low. Use these two results to predict the dependence of the CFC on the ionic valence if ψ_0 is small. Compare this result with the same quantity in the limit of large ψ_0.

8. The interfacial tension at the electrocapillary maximum for several electrolytes in dimethylformamide (DMF) solutions has been measured as a function of the electrolyte concentration*:

	γ_{max}(dyne cm^{-1})		
log c	KI	LiCl	KSCN
0	354	366	370
−0.3	356	367	371
−1.0	361	369	373
−1.3	364	370	374

Use these results to estimate the relative adsorbabilities of the I$^-$, Cl$^-$, and SCN$^-$ ions from DMF. How does the sequence of anion adsorbabilities compare with that from aqueous solution as given in Sect. 12.10? More comprehensive electrocapillary data suggest that SCN$^-$ is more solvated in DMF than in water. Is this consistent with the adsorbability series just compared?

9. A negatively charged AgI dispersion was caused to flocculate by the addition of various electrolytes. The concentrations of several divalent metal nitrates needed to produce flocculation are as follows†:

Salt	$Mg(NO_3)_2$	$Ca(NO_3)_2$	$Sr(NO_3)_2$
$c \times 10^3$ (mole liter^{-1})	2.60	2.40	2.38
Salt	$Ba(NO_3)_2$	$Zn(NO_3)_2$	$Pb(NO_3)_2$
$c \times 10^3$ (mole liter^{-1})	2.26	2.50	2.43

---

*V. D. Bezuglyi and L. A. Korshikov, *Electrokhimiya*, 3:390 (1967).
†H. R. Kruyt and M. A. Klompe, *Kolloid Beihefte*, 54:484 (1942).

That these different compounds produce the same effect at so nearly the same concentration argues that the principal cause of the effect is electrostatic. Use the average of these concentrations to calculate (a) the value of κ at which this system flocculates, (b) the force of repulsion [Eq. (77)], and (c) the potential energy of repulsion [Eq. (81)] when two planar surfaces are separated by a distance of 10 nm. For the purpose of calculation in parts (b) and (c), ψ_0 may be taken as 100 mV. Comment on the applicability of these equations to the physical system under consideration.

10. The slight differences in the concentrations of divalent cations required to flocculate the AgI sol of the preceding problem may be attributed to differences in the adsorbability of these cations at the AgI-solution interface. Use the data of Problem 9 to rank the cations with respect to their tendency to adsorb. Is there a correlation between adsorbability and ion size and/or hydration? List any references consulted for data concerning the latter two quantities.

11. Once the significance of the midpoint between two parallel plates for the force between those plates is established, there are several ways of arriving at Eq. (74). One argument is that the plates shown in Fig. 12.6 function as a semipermeable membrane, sustaining a concentration difference between $x = d$ and the outer region of the solution. Use Eqs. (3.25) and (27) to show that the osmotic pressure across this "semipermeable membrane" is given by Eq. (74).

12. Arachidic acid sols were studied with different concentrations of La^{3+} added. The stability ratio W and the direction of particle migration in an electric field (i.e., particle charge) were observed* and the following results obtained:

c (mole La^{3+} liter^{-1})	10^{-5}	3×10^{-5}	10^{-4}	3×10^{-3}	10^{-3}
W	7.9	4.5	~1	1.6	15.8
Particle charge	−	−	~0	+	+

Taking 10^{-4} M as the CFC value for La^{3+}, CFC values of 7.29×10^{-2} M and 1.1×10^{-3} M would be predicted for monovalent and divalent cations, respectively, according to Eq. (92). In view of the observed behavior of La^{3+}, would you expect these calculated CFC values to be correct or too low or too high? Explain briefly.

13. A. Kitahara and H. Ushiyama† flocculated a polystyrene latex of radius 665 Å with KCl. The stability ratio W was found to vary with the KCl concentration as follows:

$\log c$ (c in mole liter^{-1})	0	−0.13	−0.33	−0.44	−0.60
$\log W$	0	0	0.30, 0.46	0.73	1.20

---

*R. H. Ottewill and D. J. Wilkins, *Trans. Faraday Soc.*, 58:608 (1962).
†A. Kitahara and H. Ushiyama, *J. Colloid Interface Sci.*, 43:73 (1973).

From a plot of log W versus log c determine the CFC value and Υ_0 [by means of Eq. (97)]. Use the approximation for Υ_0 given in Problem 7 to estimate ψ_0 for this colloid. Use the values of the CFC and Υ_0 determined in Eqs. (87) and (88) to estimate the effective Hamaker constant A_{212} for polystyrene dispersed in water. Describe how A might be estimated using a more realistic model than that used in the derivation of Eqs. (87) and (88).

13
ELECTROPHORESIS AND OTHER ELECTROKINETIC PHENOMENA

There is a constant attraction to the South ... yet the hampering effect of the southward attraction is quite sufficient to serve as a compass in most parts of our earth.

[From Abbott's *Flatland*]

13.1 Introduction

The word *electrokinetic* implies the combined effects of motion and electrical phenomena. Specifically, our interest in this chapter centers on those processes in which a relative velocity exists between two parts of the electrical double layer. This may arise from the migration of a particle relative to the continuous phase which surrounds it. In this case the resulting electrokinetic effect is called electrophoresis. Alternatively, it could be the solution phase which moves relative to stationary walls, in which case either electro-osmosis or streaming potential is the phenomenon observed.

These three electrokinetic processes are our concern in this chapter, with the emphasis on electrophoresis. In each case the electrokinetic measurements can be interpreted to yield a quantity known as the zeta (ζ) potential. It is important to note that this is an *experimentally* determined potential measured in the double layer. Therefore it is the empirical equivalent to the double layer potentials discussed in the last chapter. We saw in Chap. 11 how the stability of a hydrophobic colloid depends on the relative magnitude of the potential energies of attraction and repulsion between a pair of particles approaching a collision with each other. Therefore the electrokinetic or ζ potential has a direct bearing on the material of the preceding two chapters as far as the theory and practice of colloid stability are concerned.

737

Although the ζ potential is undoubtedly an important quantity in colloid chemistry, it is not totally free of ambiguity. The problem is this: It is not clear at what location within the double layer the potential is measured. The derivations of this chapter will show that the ζ potential is the double layer potential close to the surface, but the precise quantitative meaning of *close* cannot be defined.

13.2 Comparison of Small Ions and Macroions

The fact that positive ions migrate toward the cathode and negative ions migrate toward the anode is so well known as to be virtually self-evident. It seems equally evident, therefore, that positively and negatively charged colloidal particles should display similar migrations. Indeed, this is the case. Because we are relatively familiar with the conductivity of simple electrolytes, we shall start our discussion of electrokinetic phenomena with a comparison of the migrations of the particles in these two different size domains.

An isolated ion in an electric field experiences a force directed toward the oppositely charged electrode. This force is given by the product of the charge of the ion q times the electric field $\bar{E}$:

$$F_{el} = q\bar{E} \tag{1}$$

In SI units $\bar{E}$ is expressed in volt meter^{-1} and q in coulomb, so F_{el} is correctly given in Newtons, since C V = J = N m. The cgs unit system which is widely encountered in older references requires dividing the right-hand side of Eq. (1) by a factor of about 300, since 299.8 V = 1.0 statvolt. Use of Eq. (1) is limited to situations in which the electric field at the ion is due to the applied potential gradient only, undisturbed by the effects of other ions in the solution (i.e., infinite dilution).

An ion in an electric field thus experiences an acceleration toward the oppositely charged electrode. However, its velocity does not increase without limit. An opposing force due to the viscous resistance of the medium increases as the particle velocity increases:

$$F_{vis} = fv \tag{2}$$

where f is the friction factor [see Eq. (2.2)]. A stationary state velocity is established quite rapidly in which these two forces are equal:

$$v = \frac{q\bar{E}}{f} \tag{3}$$

The situation is thus very much like the sedimentation velocity discussed in Chap. 2 in which the gravitational forces on a particle are opposed by viscous resistance.

As a further development, we may tentatively substitute the value for f given by Stokes' law [Eq. (2.7)] to obtain

$$v = \frac{q\bar{E}}{6\pi\eta R} \tag{4}$$

where R is the radius of the particle, assumed to be a sphere by this substitution. The charge of a simple ion can be written as the product of its valence z times the electron charge e:

$$q = ze \tag{5}$$

Substitution of this result into Eq. (4) yields

$$v = \frac{ze\bar{E}}{6\pi\eta R} \tag{6}$$

The velocity per unit field is defined to be the mobility u of the ion:

$$u = \frac{v}{\bar{E}} \tag{7}$$

For simple ions mobilities are typically on the order of 10^{-8} m s^{-1}/V m^{-1} (m^2 V^{-1} s^{-1}). It is shown in physical chemistry that the mobility of an ion is directly proportional to its equivalent conductance λ_{i0}:

$$u_i = \frac{\lambda_{i0}}{F} \tag{8}$$

where F is the Faraday constant. We have stipulated the conductance at infinite dilution (subscript 0) as a reminder that these relationships all refer to isolated ions. When ion mobilities are analyzed by Eq. (6), quite reasonable values for the radii of the hydrated ions are obtained.

The success and relative simplicity of conductivity as a method of study for small ions prompt us to extend these ideas to particles in the colloidal size range. For certain colloids the experimental aspects of this are simpler than for small ions, because of the possibility of measuring the velocity of high contrast particles by direct microscopic observation. If the velocity and the field responsible for the migration are known, the mobility of the colloid may be evaluated directly from Eq. (7). When the term is applied to colloidal particles, the mobility is known specifically as the electrophoretic mobility. In this case the overall phenomenon is known as electrophoresis and the specific experimental technique of direct microscopic observation of the electrophoretic mobility is called microelectrophoresis. This and other electrophoretic techniques are described in more detail in Sect. 13.9.

Although the electrophoretic mobility is—at least in some cases—a readily measured quantity, its interpretation is considerably more difficult for colloidal particles than for small ions. First, we realize that the charge carried by a colloidal particle is not a constant known quantity as is the case for simple ions. This prevents us from using Eq. (6) to evaluate R, but suggests instead a method whereby the charge might be determined. Suppose, for example, we substitute Eq. (2.32) for f rather than use Stokes' law for this quantity. Then the electrophoretic mobility is given by

$$u = \frac{ze}{kT/D} = \frac{zeD}{kT} \tag{9}$$

It appears that the combination of electrophoresis and diffusion experiments would allow for the evaluation of the charge carried by the macroion. Again, the situation is reminiscent of the procedures described in Chap. 2 in which sedimentation and diffusion experiments were combined. However, this is only the beginning of the difficulty. The validity of Eq. (9) is limited to the situation in which a charged particle is considered in isolation from other ions. A charged colloid will be surrounded by an electrical double layer as we saw in Chap. 12. Thus the field at the particle is modified by the potential of the double layer; that is, the migrating unit is the charged colloidal particle *along with* its electrical double layer just as the same composite is the kinetic unit in flocculation.

Therefore this strategy for determining the charge of a colloid from electrophoresis measurements is invalid except for the rather special case of determining the conditions of zero charge for the colloid. We shall return to a discussion of this point in Sect. 13.10.

In Chapter 12 we discussed the structure of the double layer in terms of the potential of the surface. This background plus the realization that the ion atmosphere also contributes to the electrophoretic mobility of a colloid suggests that potential rather than charge is the more useful parameter to pursue. This is the topic of the following section. In discussing the migration of charged colloidal particles through a solution containing small ions, it is convenient to begin by distinguishing between two extremes of particle size. We saw in Chap. 12 that the parameter κ^{-1} (see Table 12.1) is a convenient way to characterize the "thickness" of the ion atmosphere near a surface. Distances are regarded as large or small

(a)   (b)

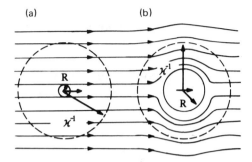

Figure 13.1 Streamlines (which also represent the electric field) around spherical particles of radius R. The dashed lines are displaced from the surface of the spheres by the double layer thickness, κ^{-1}. In (a) κR is small; in (b) κR is large.

relative to this quantity. For simplicity we shall restrict our consideration to spherical, nonconducting particles and shall begin by examining the two extremes of very small and very large particles. These designations acquire specific meaning when compared to κ^{-1}, taken as a standard length. Thus the two cases we consider first are those in which R/κ^{-1} (or simply κR) is small and large.

Figure 13.1 shows schematically the shape of the flow streamlines around the particle in the two cases. The dashed line in the figure is displaced from the surface of the spherical particles by an amount κ^{-1}. In Fig. 13.1a, R is small (compared to κ^{-1}) and the streamlines undergo negligible displacement. In Fig. 13.1b, on the other hand, the streamlines follow the contours of the particle nearly tangentially. Since matter is conserved, a flow streamline which carries matter into a volume element must also carry matter out of that volume element. Charge is also conserved, so the same number of lines of force must enter and leave a volume element. Accordingly, the streamlines shown schematically in Fig. 13.1 may also be regarded as describing the electric field in the neighborhood of small and large particles.

13.3 The Zeta Potential for Small Values of κR

We know from Chapter 12 that the potential drops off gradually with distance from a charged surface, its range decreasing with increasing electrolyte content. Most of the expressions developed in Chap. 12, however, describe the potential situation adjacent to a planar wall. In the present context we need to know how the potential varies with distance from the surface of a sphere. The Poisson equation (see Sect. 12.4) gives the fundamental differential equation for potential as a function of charge density. The Debye–Hückel approximation may be used to express the charge density as a function of potential as in Eq. (12.28) if the potential is low. Combining Eqs. (12.24) and (12.32) gives

$$\frac{1}{r^2}\frac{d}{dr}\left(r^2\frac{d\psi}{dr}\right) = \frac{e^2}{\varepsilon kT}\left(\sum_i z_i^2 n_i\right)\psi = \kappa^2\psi \qquad (10)$$

Remember that ε equals the product of ε_0 times ε_r with $\varepsilon_0 = 8.85 \times 10^{-12}$ $C^2\, J^{-1}\, m^{-1}$; for water at 25°C, $\varepsilon_r = 78.54$. Because of its importance to the present material, we repeat the defining expression for κ as given originally by Eq. (12.39):

$$\kappa = \left(\frac{1000e^2 N_A}{\varepsilon kT}\sum_i z_i^2 M_i\right)^{1/2} \qquad (11)$$

or, in terms of ionic strength $I = \frac{1}{2}\sum_i z_i^2 M_i$,

$$\kappa = \left(\frac{2000e^2 N_A I}{\varepsilon kT}\right)^{1/2} \qquad (12)$$

Remember that Table 12.2 contains some useful numerical values of κ at different concentrations of various electrolytes. Equation (10) is the basic relationship of the Debye–Hückel theory and may be integrated as follows. The variable x is introduced with the following definition:

$$x = r\psi \qquad (13)$$

Thus Eq. (10) may be written

$$\frac{d}{dr}\left(r^2\frac{d\psi}{dr}\right) = \kappa^2 rx \qquad (14)$$

Now let us consider the incorporation of Eq. (13) into the left-hand side of Eq. (14):

$$\frac{d\psi}{dr} = \frac{d(x/r)}{dr} = \frac{1}{r}\frac{dx}{dr} - \frac{x}{r^2} \qquad (15)$$

and

$$\frac{d}{dr}\left(r^2\frac{d\psi}{dr}\right) = \frac{d}{dr}\left(r\frac{dx}{dr} - x\right) = r\frac{d^2x}{dr^2} \tag{16}$$

Combining Eqs. (14) and (16) gives

$$\frac{d^2x}{dr^2} = \kappa^2 x \tag{17}$$

for which a general solution is

$$x = A\,\exp(-\kappa r) + B\,\exp(\kappa r) \tag{18}$$

as may be readily verified by differentiation. Replacing x in this equation by its definition in Eq. (13) gives

$$\psi = \frac{A\,\exp(-\kappa r)}{r} + \frac{B\,\exp(\kappa r)}{r} \tag{19}$$

Since $\psi \to 0$ as $r \to \infty$, it is apparent that $B = 0$.

To evaluate A we proceed as follows. In the limit of infinite dilution—that is, as $\kappa \to 0$—the potential around the charged particle is given by the expression for the potential of an isolated charge. Elementary physics gives this as

$$\psi = \frac{1}{4\pi\varepsilon}\frac{q}{r} \tag{20}$$

a distance r from a charge q. As $\kappa \to 0$, Eqs. (19) and (20) must converge; therefore A must equal $q/4\pi\varepsilon$. The general expression for potential around a spherical particle at low potential may be written

$$\psi = \frac{q}{4\pi\varepsilon r}\,\exp(-\kappa r) \tag{21}$$

As a reminder that the level of approximation in Eq. (21) is the same as that of the Debye–Hückel limiting law, the following example continues from this last result to the Debye–Hückel expression for the mean ionic activity coefficient of an electrolyte solution.

Example 13.1   The Debye–Hückel limiting law attributes all of the nonideality of an electrolyte solution to electrostatic effects associated with the diffuse double layer. As a way to isolate this effect, consider the hypothetical process of discharging an ion in a solution of concentration c_1, moving it to a solution of concentration c_2, and then recharging it in the new solution. The individual steps and general expressions for the associated free energy changes are listed below, with the subscripts 1 and 2 indicating the two concentration conditions:

$$q = ze \text{ at } c_1 \rightarrow q = 0 \text{ at } c_1 \qquad \Delta G = N_A \int_{ze}^{0} \psi_1 \, dq$$

$$q = 0 \text{ at } c_1 \rightarrow q = 0 \text{ at } c_2 \qquad \Delta G = RT \ln(c_2/c_1)$$

$$q = 0 \text{ at } c_2 \rightarrow q = ze \text{ at } c_2 \qquad \Delta G = N_A \int_{0}^{ze} \psi_2 \, dq$$

$$\text{net: } q = ze \text{ at } c_1 \rightarrow q = ze \text{ at } c_2 \qquad \Delta G = RT \ln(a_2/a_1)$$

Derive the Debye–Hückel expression for the activity coefficient γ, assuming Eq. (21) describes ψ and that solution 2 is dilute and solution 1 very dilute.

Solution The activities in the expression for ΔG_{net} can be replaced by the product of concentrations and activity coefficients:

$$\Delta G_{net} = RT \ln\left(\frac{\gamma_2 c_2}{\gamma_1 c_1}\right) = RT \ln\left(\frac{\gamma_2}{\gamma_1}\right) + RT \ln\left(\frac{c_2}{c_1}\right) = N_A \int_{ze}^{0} \psi_1 \, dq$$

$$+ RT \ln\left(\frac{c_2}{c_1}\right) + N_A \int_{0}^{ze} \psi_2 \, dq$$

Since solution 1 is very dilute, γ_1 can be set equal to unity, in which case

$$RT \ln \gamma_2 = N_A \int_{ze}^{0} \psi_1 \, dq + N_A \int_{0}^{ze} \psi_2 \, dq$$

The activity coefficient for (dilute) solution 2 is therefore obtained by evaluating the integrals based on Eq. (21). Using a series approximation (see Appendix A) for the exponential in Eq. (21) and retaining only the leading term for solution 1, where κ_1 (i.e., c_1) is very small, and the first two terms for solution 2, where κ_2 is small but larger than for 1, we obtain $\psi_1 \simeq q/4\pi\varepsilon r$ and $\psi_2 \simeq (q/4\pi\varepsilon r)(1 - \kappa_2 r)$. With these substitutions the integrals can be evaluated as follows:

$$RT \ln \gamma_2 = \frac{N_A}{4\pi\varepsilon r} \left(\int_{ze}^{0} q \, dq + \int_{0}^{ze} q \, dq - \kappa_2 r \int_{0}^{ze} q \, dq \right) = \frac{N_A \kappa_2 (ze)^2}{8\pi\varepsilon}$$

Substituting Eq. (12) for κ yields

$$\ln \gamma_2 = -\frac{N_A z^2 e^2}{8\pi\varepsilon RT}\left(\frac{2000 e^2 N_A}{\varepsilon kT}\right)^{1/2} I^{1/2}$$

For aqueous solutions at 25°C this becomes $\log_{10} \gamma_2 = 0.0509 z^2 I^{1/2}$, or, with a bit more argumentation, $\log_{10} \gamma_\pm = -0.0509 z_+ z_- I^{1/2}$, which is the result sought.

•

Next, let us consider the application of Eq. (21) to a particle migrating in an electric field. We recall from Chap. 4 that the layer of liquid immediately adjacent to a particle moves with the same velocity as the surface; that is, whatever the relative velocity between the particle and the fluid may be some distance from the surface, it is zero at the surface. What is not clear is the actual distance from the surface at which the relative motion sets in between the immobilized layer and the mobile fluid. This boundary is known as the surface of shear. Although the precise location of the surface of shear is not known, it is presumably within a couple of molecular diameters of the actual particle surface for smooth particles. Ideas about adsorption from solution (e.g., Sect. 7.8) in general and about the Stern layer (Sect. 12.10) in particular give a molecular interpretation to the stationary layer and lend plausibility to the statement about its thickness. What is most important here is the realization that the surface of shear occurs well within the double layer, probably at a location roughly equivalent to the Stern surface. Rather than identify the Stern surface and the surface of shear, we define the potential at the surface of shear to be the zeta potential ζ. It is probably fairly close to the Stern potential ψ_δ in magnitude, and definitely less than the potential at the surface, ψ_0. The relative values of these different potentials are shown in Fig. 13.2.

Distances within the double layer are considered large or small, depending on their magnitude relative to κ^{-1}. Thus in dilute solutions where κ^{-1} is large, the surface of shear—which is close to the particle surface even in absolute units—may be safely regarded as coinciding with the surface in units which are relative to the double layer thickness. Therefore, in the case where κ^{-1} is large (or κ small), Eq. (21) becomes

$$\zeta = \frac{q}{4\pi\varepsilon R} \exp(-\kappa R) \tag{22}$$

where R is the actual radius of the particle.

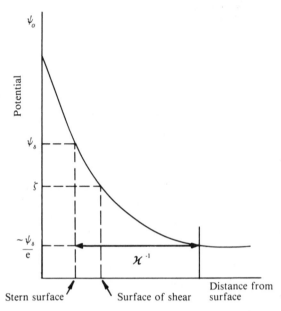

Figure 13.2   The relative magnitudes of various double layer potentials of interest.

Since this result applies only when κ is small, the exponential may be expanded (see Appendix A) to give

$$\zeta \simeq \frac{q}{4\pi\varepsilon R}\frac{1}{\exp(\kappa R)} \simeq \frac{q}{4\pi\varepsilon R}\left(\frac{1}{1 + \kappa R}\right) \tag{23}$$

This result may also be written

$$\zeta = \frac{q}{4\pi\varepsilon R}\frac{\kappa^{-1}}{R + \kappa^{-1}} \tag{24}$$

which is the same as

$$\zeta = \frac{q}{4\pi\varepsilon R} - \frac{q}{4\pi\varepsilon(R + \kappa^{-1})} \tag{25}$$

This last result is interesting because it may be interpreted as the sum of two superimposed potentials: one arising from a charge q on a surface of radius R and a second arising from a charge $-q$ on a sphere of radius $R + \kappa^{-1}$. This is the net potential between two concentric spheres carrying equal but opposite charges and differing in radius by an amount κ^{-1}.

Such a situation corresponds to a concentric sphere capacitor. As in Chapter 12, we again see the double layer behaving as if it were a capacitor with a characteristic spacing κ^{-1}.

Having explored the capacitor analogy, we no longer need to retain the second term in the series expansion of the exponential in Eq. (23). For our present purposes it is sufficient to note that for small values of κR, Eq. (22) becomes

$$\zeta = \frac{q}{4\pi\varepsilon R} \tag{26}$$

Solving this result for q and substituting into Eq. (6), we obtain

$$u = \frac{2\varepsilon\zeta}{3\eta} \tag{27}$$

The possible usefulness of this relationship—which is known as the Hückel equation—should not be overlooked. Throughout Chapter 12 we were concerned with the potential surrounding a charged particle. Equation (12.1) provides a way of evaluating the potential at the surface, ψ_0, in terms of the concentration of potential-determining ions. Owing to ion adsorption in the Stern layer, this may not be the appropriate value to use for the potential at the inner limit of the diffuse double layer. Although ζ is not necessarily identical to ψ_δ, it is nevertheless a quantity of considerable interest.

More elaborate theory shows that Eq. (27) is valid for spheres when κR is less than about 0.1. This imposes a rather severe restriction on the applicability of this result in aqueous systems since for $R = 10^{-8}$ m the corresponding concentration is about 10^{-5} M for a 1:1 electrolyte. In nonaqueous media, however, ion concentrations may be very low and this result assumes increasing importance.

Example 13.2 In many references the Hückel equation is written $u = \varepsilon_r\zeta/6\pi\eta$. How do you account for the difference between this expression and Eq. (27)? What is the ζ potential of a particle which displays a mobility of 10^{-4} cm^2 V^{-1} s^{-1} in water at 20°C for which $\eta = 0.010$ P and $\varepsilon_r = 80.4$, assuming the Hückel potential applies?

Solution The discrepancy between the equation given here and Eq. (27) arises from the fact that the equation above is written for cgs units whereas Eq. (27) applies to SI. Remember that $\varepsilon = \varepsilon_r\varepsilon_0$ in Eq. (27) and that the vacuum permittivity ε_0 usually appears with the factor 4π. When we combine Eqs. (6) and (26), the ratio $4\pi/6\pi$ reduces to $\frac{2}{3}$. Many electrokinetic formulas differ by a factor 4π, depending on the system of units being used, and the reader is cautioned to be aware of this difference. The presence or absence of the vacuum permittivity in the equation is the key to the system being used, although this factor is sometimes hidden, as in the case of Eq. (27).

Either system of units can be used to calculate ζ from the mobility given; in either case some unit conversion must be done on the mobility, since the units given are a hybrid of SI and cgs units.

In SI the mobility is given by

$$10^{-4} \ \mathrm{cm^2 \ V^{-1} \ s^{-1}} \times \left(\frac{1 \ \mathrm{m}}{100 \ \mathrm{cm}} \right)^2 = 10^{-8} \ \mathrm{m^2 \ V^{-1} \ s^{-1}}$$

therefore

$$\zeta = \frac{(10^{-8})(3)(0.010 \ \mathrm{P} \times 1 \ \mathrm{kg \ m^{-1} \ s^{-1}/10 \ P})}{2(80.4)(8.85 \times 10^{-12})} = 2.11 \times 10^{-2} \ \mathrm{V} = 21.1 \ \mathrm{mV}$$

In cgs the mobility is given by

$$10^{-4} \ \frac{\mathrm{cm/s}}{\mathrm{V/cm}} \times \frac{300 \ \mathrm{V}}{\mathrm{statV}} = 3 \times 10^{-2} \ \mathrm{cm^2 \ statV^{-1} \ s^{-1}}$$

therefore

$$\zeta = \frac{6\pi(3 \times 10^{-2})(0.010)}{80.4} = 7.03 \times 10^{-5} \ \mathrm{statV}$$

or

$$\zeta = 7.03 \times 10^{-5} \ \mathrm{statV} \times \frac{300 \ \mathrm{V}}{\mathrm{statV}} = 2.11 \times 10^{-2} \ \mathrm{V}$$

Note that the factor 300 V/statvolt enters the calculation twice if cgs units are used.

•

The next question to be considered is the relationship between u and ζ for the case where κR is not small.

13.4 The Zeta Potential for Large Values of κR

In this section we consider the situation in which the thickness of the double layer is negligible compared to the radius of curvature of the surface. The derivation is not limited to any particle geometry, as long as the radius of curvature R is large compared to κ^{-1}. This situation may be brought about by making κ^{-1} small (i.e., κ large), which is equivalent to dealing with relatively high concentrations of electrolyte or with flat or slightly curved surfaces. For our purposes it is convenient to consider a planar surface, but the results will apply equally to any case for which the product κR is large.

Suppose we consider a volume element of area A and thickness dx situated a distance x from a planar surface as shown in Fig. 13.3. The viscous force on the face nearest the surface is given by

$$F_x = \eta A \left(\frac{dv}{dx} \right)_x \qquad (28)$$

and the force exerted on the face farther from the surface is given by

$$F_{x+dx} = \eta A \left(\frac{dv}{dx} \right)_{x+dx} \qquad (29)$$

In these equations v is the relative velocity between the particle and the surrounding medium. The difference between Eqs. (28) and (29) therefore equals the net viscous force on the volume element:

$$F_{\text{vis}} = \eta A \left[\left(\frac{dv}{dx} \right)_{x+dx} - \left(\frac{dv}{dx} \right)_x \right] \qquad (30)$$

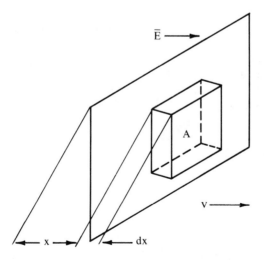

Figure 13.3 Location of a volume element of solution adjacent to a planar wall.

Equation (4.12) can be used to relate $(dv/dx)_x$ to $(dv/dx)_{x+dx}$, so Eq. (30) becomes

$$F_{vis} = \eta A \frac{d^2v}{dx^2}\, dx \tag{31}$$

Under stationary state conditions an equal and opposite force is exerted on the volume element by the electric field acting on the ions contained in the volume element. The force on the ions is given by the product of the field strength times the total charge. The latter equals the charge density ρ^* times the volume of the element; therefore

$$F_{el} = \bar{E}\rho^* A\, dx \tag{32}$$

Poisson's equation [Eq. (12.26)] may now be used as a substitution for ρ^* to yield

$$\rho^* = -\varepsilon\, \nabla^2\psi = -\varepsilon\, \frac{d^2\psi}{dx^2} \tag{33}$$

where the second result applies specifically to the region adjacent to the planar surface and where ψ is the potential a distance x from the surface.

Setting Eqs. (31) and (32) equal to each other, substituting Eq. (33), and simplifying leads to the equation

$$\eta\, \frac{d^2v}{dx^2} = -\varepsilon\bar{E}\, \frac{d^2\psi}{dx^2} \tag{34}$$

With certain assumptions this result may be integrated twice to give the relation between v and ψ.

The integration of Eq. (34) is carried out by assuming that both η and ε are constants in the vicinity of the surface. We shall return to a discussion of this assumption in Sect. 13.8. Making this assumption, we can write Eq. (34) as

$$\frac{d}{dx}\left(\eta\, \frac{dv}{dx}\right) = -\bar{E}\, \frac{d}{dx}\left(\varepsilon\, \frac{d\psi}{dx}\right) \tag{35}$$

In this form the first integration is readily found to give

$$\eta\, \frac{dv}{dx} = -\varepsilon\bar{E}\, \frac{d\psi}{dx} + C_1 \tag{36}$$

The constant of integration C_1 is evaluated by noting that both dv/dx and $d\psi/dx$ must equal zero at large distances from the surface; therefore $C_1 = 0$.

The resulting expression is easily integrated again with the following limits: (1) at the surface of shear $\psi = \zeta$ and $v = 0$; (2) at the outside edge of the double layer $\psi = 0$ and v equals the observed velocity of particle migration. Therefore

$$\eta \int_{v}^{0} dv = -\varepsilon \bar{E} \int_{0}^{\zeta} d\psi \tag{37}$$

or

$$\eta v = \varepsilon \bar{E} \zeta \tag{38}$$

In terms of electrophoretic mobility, Eq. (38) can be written

$$u = \frac{v}{\bar{E}} = \frac{\varepsilon \zeta}{\eta} \tag{39}$$

Equation (39) is known as the Helmholtz–Smoluchowski equation. No assumptions are made in its derivation as to the actual structure of the double layer, only that the Poisson equation applies and that bulk values of η and ε apply within the double layer. It has been shown that this result is valid for values of κR larger than about 100.

We have now reached the position of having two expressions—Eq. (27) and (39)—to describe the relationship between the mobility of a particle (an experimental quantity) and the zeta potential (a quantity of considerable theoretical interest). The situation may be summarized by noting that both the Hückel and the Helmholtz–Smoluchowski equations may be written

$$u = C \frac{\varepsilon \zeta}{\eta} \tag{40}$$

where C is a constant whose numerical value depends on the magnitude of κR. In the limit of both large and small values of κR, the value of C becomes independent of κR:

1. For κR > 0.1

$$C = \tfrac{2}{3} \tag{41}$$

2. For κR > 100

$$C = 1 \tag{42}$$

In view of the widely different pictures of the electric field surrounding the particles in the two extremes—as shown schematically in Fig. 13.1—it is not surprising that different results are obtained in the two limits.

A major remaining problem is that many systems of interest in colloid chemistry do not correspond to either of these two limiting cases. The

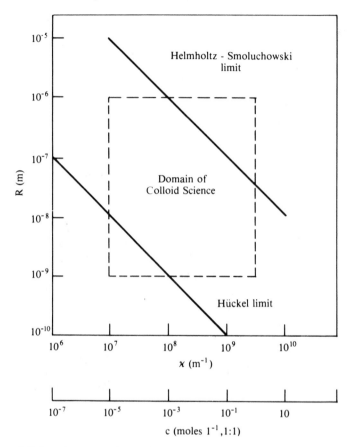

Figure 13.4 The domain within which most investigations of aqueous colloidal systems lie in terms of particle radii and 1:1 electrolyte concentration. The diagonal lines indicate the limits of the Hückel and the Helmholtz–Smoluchowski equations. (From Ref. 5, used with permission.)

situation is summarized in Fig. 13.4, which maps the particle radii and 1:1 electrolyte concentrations which correspond to various κR values. Clearly, there is a significant domain of particle size and/or electrolyte concentration for which neither the Hückel nor the Helmholtz–Smoluchowski equations can be used to evaluate ζ from experimental mobility values. The relationship between ζ and u for intermediate values of κR is the topic of the following section.

13.5 The Zeta Potential: General Theory for Spherical Particles

It is apparent from the preceding sections that the understanding of electrophoretic mobility involves both the phenomena of fluid flow as discussed in Chap. 4 and the double layer potential as discussed in Chap. 12. In both places we see that theoretical results are dependent on the geometry chosen to describe the boundary conditions of the system under consideration. This continues to be true in discussing electrophoresis, where these two topics are combined. As was the case in Chaps. 4 and 12, solutions to the various differential equations which arise are possible only for rather simple geometries, of which the sphere is preeminent.

The generalized electrophoresis problem has been solved for spherical and rod-shaped particles and, more approximately, for random coils. In this section we shall restrict our attention to spheres, although in the limit of large values of κR the Helmholtz–Smoluchowski equation is obtained, a result which is independent of particle shape. In the general theory the conductivity of the particle is one of the parameters that must be considered. We shall discuss only the case of nonconducting spheres. It has been shown experimentally that mercury droplets for which κR is large follow Eq. (39), even though—as conductors—the full theory predicts they should show zero mobility. The explanation of this anomaly is that the surface of the metallic drops becomes sufficiently polarized to block the passage of current through the particle. Thus even a metallic particle may behave as an insulator, thereby justifying our choice of the nonconducting particle as the model for consideration.

In addition, we shall consider only the case in which the colloid is present in small concentration so that colloid–colloid interactions can be ignored. We shall assume that the diffuse part of the double layer is adequately described by the Gouy–Chapman theory. Since the surface of shear more or less coincides with the Stern surface, it is the diffuse part of the double layer and not the Stern layer (where specific adsorption occurs) in which we are interested. Specific adsorption in the Stern layer may have a large effect on the zeta potential itself, but should be unimportant when it comes to establishing the connection between u and ζ. The Gouy–Chapman theory ignores the actual discreteness of electrical charges and is also subject to the objections against the Poisson–Boltzmann equation (see Sect. 12.4). An extensive body of research has been devoted either to circumventing these limitations or to estimating the approximation introduced by their use. Overbeek and Wiersma [8] have rightly noted that it is rather futile to introduce one or two corrections to the theory while neglecting other approximations that are probably of the same magnitude. A safer procedure, they note, is to use the simpler theory, keeping in mind the semiquantitative nature of the result.

By assuming that the external field—deformed by the presence of the colloidal particle—and the field of the double layer are additive, D. C. Henry derived the following expression for mobility:

$$u = \frac{\varepsilon}{\eta} \left(\zeta + 5R^5 \int_\infty^R \frac{\psi}{r^6} \, dr - 2R^3 \int_\infty^R \frac{\psi}{r^4} \, dr \right) \tag{43}$$

where r is the radial distance from the center of the particle. To go beyond Eq. (43), it is necessary to know ψ as a function of r. The resulting expressions are mathematically intractable unless a relatively simple expression is used for ψ. We may use the Debye–Hückel approximation given by Eq. (19) for this, but the constant in that equation is best evaluated somewhat differently before proceeding.

We return to the solution of the Poisson–Boltzmann equation for a spherical particle, Eq. (19), with $B = 0$:

$$\psi = \frac{A \, \exp(-\kappa r)}{r} \tag{44}$$

In the present development we evaluate A by recalling that $\psi = \zeta$ when $r = R$. Therefore

$$A = R\zeta \, \exp(\kappa R) \tag{45}$$

and Eq. (44) becomes

$$\psi = \frac{R\zeta}{r} \exp[-\kappa(r - R)] \tag{46}$$

Combining Eqs. (43) and (46) and integrating leads to the result

$$u = \frac{2\varepsilon\zeta}{3\eta} \left(1 + \tfrac{1}{16}(\kappa R)^2 - \tfrac{5}{48}(\kappa R)^3 - \tfrac{1}{96}(\kappa R)^5 \right.$$

$$\left. - [\tfrac{1}{8}(\kappa R)^4 - \tfrac{1}{96}(\kappa R)^6] \, \exp(\kappa R) \int_\infty^{\kappa R} \frac{e^{-t} \, dt}{t} \right) \tag{47}$$

Equation (47) is called Henry's equation. Two specific assumptions underlying its derivation should be pointed out: (1) The ion atmosphere is undistorted by the external field and (2) the potential is low enough to justify writing $e\psi/kT < 1$, which is equivalent to requiring that $\psi < 25$ mV (Sect. 12.2). It should also be noted that in the limit of $\kappa R \to 0$, Eq. (47) reduces to the Hückel equation, and in the limit of $\kappa R \to \infty$, it reduces to the Helmholtz–Smoluchowski equation. Thus the general theory confirms the idea introduced in connection with the discussion of Fig. 13.1, that the amount of distortion of the field surrounding the particles will be totally different in the case of large and small particles. The two values of C in Eq. (40) are a direct consequence of this difference. Figure 13.5a shows how the constant C varies with κR (shown on a logarithmic scale) according to Henry's equation.

We noted earlier that many systems of interest in colloid chemistry involve intermediate values of κR, so Henry's equation fills an important gap. At the same time it explicitly introduces additional restrictions: low potentials and undistorted double layers. A topic of considerable importance is the *actual* distortion of the double layer which accompanies particle migration. The consequences of this distortion—known as the relaxation effect—are known to be important in the conductivity of simple electrolytes. A remaining development, therefore, is to consider the relaxation effect in colloidal systems.

Because the charged particle and its ion atmosphere move in opposite directions, the center of positive charge and the center of negative charge do not coincide. If the external field is removed, this asymmetry disappears over a period of time known as the relaxation time. Therefore, in addition to the fact that the colloid and its atmosphere move countercurrent with respect to one another (which is called the retardation effect), there is a second inhibiting effect on the migration which arises from the tug exerted on the particle by its *distorted* atmosphere. Retardation and relaxation both originate with the double layer, then, but describe two different consequences of the ion atmosphere. The theories we have discussed until now have all correctly incorporated retardation, but relaxation effects have not been included in any of the models considered so far.

A number of workers have tackled the problem of relaxation. The use of computers has greatly assisted this area of research because of the complexity of the mathematics involved. Loeb et al. listed in Chap. 12, Ref. 5, report the results of some numerical solutions to the mobility problem with relaxation specifically considered. Figure 13.5b summarizes some results from these studies for the case of a 1:1 electrolyte. The various curves correspond to values of zeta equaling 25.7, 51.4, 77.1, and 102.8 mV at 25°C. It will be noted that the restriction to low potentials no longer applies in the theory from which these curves were evaluated. It

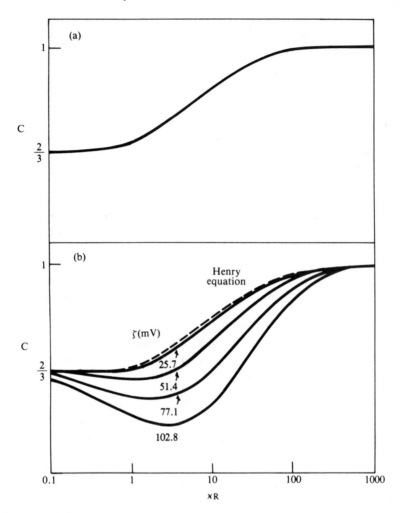

Figure 13.5 Variation of the constant C [Eq. (40)] with R (log scale): (a) at low potentials according to Henry's equation and (b) for various potentials. [Data from P. H. Wiersma, A. L. Loeb, and J. Th. G. Overbeek, *J. Colloid Interface Sci.,* *22*:78 (1966), redrawn from Ref. 9 with permission.]

is evident from the figure that the relaxation effect is negligible when $\zeta < 25$ mV, regardless of the value of κR, and in the limit of both large and small values of κR, regardless of the value of ζ. That is, intermediate values of κR and large potentials correspond to the condition of maximum resistance to flow arising from relaxation.

A family of curves qualitatively similar in appearance to those shown in Fig. 13.5b results when C is plotted versus κR at constant ζ with the valence of the electrolyte taken as the variable parameter. In that case the relaxation effect is found to increase with the valence of the counter-ions. As the valence of those small ions which have the same charge as the macroion increases, the relaxation effect leads to a higher mobility (at constant ζ) than would be predicted from Henry's equation.

In this section we have considered the relationship between u and ζ under conditions of intermediate κR values, a wide range of ζ values, and a number of ionic valence possibilities. The relationship is seen to be quite complex, except in the Hückel and Helmholtz–Smoluchowski limits. When the particle size–electrolyte concentration conditions are such that one of these limits clearly applies, ζ can be evaluated unambiguously from experimental mobilities. The Helmholtz–Smoluchowski limit is independent of particle shape. The Hückel equation is equally free from ambiguity, although it does require spherical particles and—as already noted—the circumstances under which it holds are not especially useful for aqueous colloids. If a particle is of intermediate size with definite, known values of κ and R and with ζ known to be small, Henry's equation (or Fig. 13.5a) could be used to evaluate ζ from mobility measurements. As the complexity (i.e., higher potentials, mixed electrolyte valences) of the system increases, however, the feasibility of evaluating ζ from experimental mobilities becomes increasingly tenuous. In these circumstances precise experimental results are best reported as mobilities with the corresponding value of ζ only an approximation.

13.6 Electro-Osmosis

In all the sections of this chapter until now we have focused attention on electrophoresis. We have seen that the potential at the surface of shear can be measured from electrophoretic mobility measurements, provided the system complies with the assumptions of a manageable model. One feature that has been conspicuously lacking from our discussions is any comparison between electrophoretically determined values of ζ and potential values determined by another method. The reasons for this are twofold:

1. Other techniques for measuring ζ are contingent on the same set of assumptions associated with electrophoresis and therefore do not constitute an independent determination.
2. Uncertainty as to the location within the double layer at which the shear surface is located makes it difficult to relate ζ to other double layer potentials, such as ψ_0 as determined from knowledge of the concentration of potential-determining ions [see Eq. (12.1)].

In this section we shall describe electro-osmosis, and in the following section the streaming potential. These two electrokinetic techniques also permit the evaluation of ζ, but are subject to objection 1. In Sect. 13.8 we shall examine in greater detail the location of the surface of shear which is the essence of objection 2.

Earlier, we defined electrokinetic phenomena as arising from the relative motion of a charged surface and its associated double layer. In electrophoresis it is the dispersed phase that moves, with the continuous phase remaining (more or less) stationary. It is apparent that the required relative motion between a surface and its double layer could also be brought about by causing the electrolyte solution to flow past a stationary charged wall. The complements of electrophoresis are electroosmosis and streaming potential. The latter two measurements differ from each other as follows: (1) in electro-osmosis it is an applied potential which induces the flow of solution; (2) in streaming potential the solution is made to flow by applying a pressure and a potential is induced as a result. Cause and effect are thus interchanged in electro-osmosis and streaming potential.

The electro-osmosis apparatus shown in Fig. 13.6a consists of two capillaries in parallel attached at either end to reservoirs of electrolyte solution. One of the capillaries—the working capillary—is arranged with reversible electrodes at either end, while the measuring capillary contains an air bubble to indicate fluid displacement. It is the glass–solution interface in the working capillary at which the electro-osmosis phenomenon originates. Substances other than glass may also be investigated by this method, a particularly useful variation being the replacement of the capillary by a plug of powdered material which cannot be fabricated into a cylindrical tube. For the purpose of discussion, we shall continue to refer to the capillary. The conditions under which the same analysis applies to a plug will be clear from the following discussion.

When an electric field is applied across the working capillary, the double layer ions begin to migrate and soon reach the stationary state

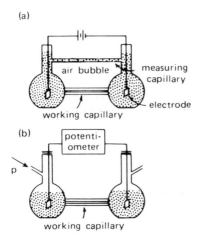

Figure 13.6 Schematic illustrations of the apparatus used to measure (a) electro-osmosis and (b) streaming potential.

velocity. In the stationary state electrical and viscous forces balance one another. The forces exerted on the ions by the medium are equal and opposite to the forces exerted on the medium by the ions; consequently, the liquid also attains a stationary state velocity. The tangential displacement of the fluid relative to the wall defines a surface of shear at which the potential equals ζ.

Although the Helmholtz–Smoluchowski equation was derived in reference to electrophoretic mobility, it clearly applies to electro-osmosis as well, since the displacement of one part of the double layer relative to another part is common to both. Figure 13.3, for example, may be taken as an illustration of either electrophoresis or electro-osmosis. The condition of the Helmholtz–Smoluchowski equation—that R is large compared to κ^{-1}—is clearly applicable to capillaries of macroscopic dimensions. We noted earlier that the Helmholtz–Smoluchowski equation applies to the electrophoresis of nonspherical particles as long as κR is large; the same logic permits Eq. (39) to be applied to cylindrical capillaries, as well as pores of irregular shape. It is this latter application that allows the replacement of a well-defined capillary by a porous plug of material in an apparatus such as that shown in Fig. 13.6a.

Equation (39) may therefore be used to describe the relationship between the potential at the capillary wall and at the velocity of electro-osmotic flow. The volume of liquid displaced per unit time V is given by multiplying both sides of Eq. (39) by the cross-sectional area of the capillary:

$$V = \upsilon A = \frac{\varepsilon \zeta \bar{E} A}{\eta} \tag{48}$$

Now suppose we apply Ohm's law to the capillary. The electric field is related to the current I and the conductivity k of the electrolyte solution as follows:

$$\bar{E} = \frac{I}{Ak} \tag{49}$$

This result may be substituted into Eq. (47) to yield

$$V = \frac{\varepsilon \zeta I}{\eta k} \tag{50}$$

This equation permits ζ to be evaluated from measurements of the rate of volume flow through the capillary; the latter are made by observing the rate of displacement of the air bubble in the measuring capillary of Fig. 13.6a.

The preceding equations are the first we have encountered in which conductivity plays a role. What is troublesome about this quantity is the fact that it is a property of bulk solutions and we are considering here an effect which arises precisely as a result of the uneven distribution of ions near a charged wall. It is essential, therefore, to examine the current carried by the ions in the double layer. Toward this end, current may be written as the sum of two contributions:

$$I = I_b + I_s \tag{51}$$

where the subscripts refer to bulk and surface contributions. Equation (49) may be used as a substitution for I_b, with πR^2 as the area of a cylindrical capillary of radius R. An analogous expression may be written for the current carried by the surface layer. In this case the bulk conductivity is replaced by surface conductivity and the cross-sectional area is replaced by the perimeter of the capillary. With these substitutions Eq. (51) becomes

$$I = \bar{E}(\pi R^2 k_b + 2\pi R k_s) = \bar{E} A \left(k_b + \frac{2k_s}{R} \right) \tag{52}$$

According to this relationship, the product $\bar{E}A$ in Eq. (48) should be replaced by

$$\bar{E}A = \frac{I}{k_b + 2k_s/R} \qquad (53)$$

to give

$$V = \frac{\varepsilon\zeta I}{\eta(k_b + 2k_s/R)} \qquad (54)$$

It will be noted that the importance of the correction for surface conductivity increases as R decreases and vanishes as $R \to \infty$. Equation (54) also suggests that the numerical evaluation of k_s may be accomplished by studying electro-osmosis in a set of capillaries identical in all respects except for variability in R. Finally, the expansion of Eq. (50) to Eq. (54) in correcting for surface conductivity explicitly assumes a cylindrical capillary. Experiments made with porous plugs cannot be corrected for surface conductivity by Eq. (54), but the qualitative conclusion that the effect of surface conductivity increases as the pore radius decreases is valid in this case also.

It has already been noted that there is a close similarity between electro-osmosis and streaming potential. Therefore we shall consider this additional electrokinetic phenomenon next.

13.7 Streaming Potential

Figure 13.6b is a sketch of an apparatus that may be used to measure streaming potential. As was the case with electro-osmosis, the capillary can be replaced by a plug of powdered material between perforated electrodes. An applied pressure difference p across the capillary causes the solution to flow through the capillary, thereby tangentially displacing the part of the double layer in the mobile phase from the stationary part.

The relationships developed in Chap. 4 for fluid flow through a capillary can be applied to this situation as follows:

1. The velocity of a cylindrical shell of radius r in a capillary of radius R and length l is given by Eq. (4.18):

$$v = \frac{p}{4\eta l}(R^2 - r^2) \qquad (55)$$

2. The rate of volume flow from this cylindrical volume element is given by Eq. (4.19):

$$\frac{dV}{dt} = \frac{p}{4\eta l}(R^2 - r^2)2\pi r \, dr \qquad (56)$$

3. The current associated with this volume element is

$$dI = \rho^* \frac{dV}{dt} = \frac{\rho^* p}{4\eta l} (R^2 - r^2) 2\pi r \, dr \tag{57}$$

where ρ^* is the charge density.

4. Next a change of variable is helpful. We replace r by a distance measured from the surface of shear, x, where

$$x = R - r \tag{58}$$

In terms of this substitution, Eq. (57) becomes

$$dI = - \frac{\rho^* p}{4\eta l} (2Rx - x^2) 2\pi (R - x) dx \tag{59}$$

Our specific interest is in the region near the walls of the capillary where $x \ll R$. In this region Eq. (59) may be approximated as

$$dI \simeq - \frac{\rho^* p}{\eta l} R^2 x \, dx \tag{60}$$

5. Substituting Eq. (12.26) for ρ^* yields

$$dI = \frac{\pi \varepsilon p R^2}{\eta l} \frac{d^2 \psi}{dx^2} x \, dx \tag{61}$$

6. The total current carried by the capillary is obtained by integrating x over the radius of the cylinder. Integration by parts yields

$$I = \frac{\pi \varepsilon p R^2}{\eta l} \left(x \frac{d\psi}{dx} \Big|_0^R - \int_0^R \frac{d\psi}{dx} dx \right) = \frac{\pi \varepsilon p R^2 \zeta}{\eta l} \tag{62}$$

since $\psi = \zeta$ at $x = 0$ and $\psi = d\psi/dx = 0$ at $x = R$.

The quantity calculated by Eq. (62) is known as the streaming current. It is specifically due to the net displacement of the mobile part of the double layer relative to the stationary part of the double layer. The field associated with this current is given by combining Eqs. (52) and (62):

$$\bar{E} = \frac{\varepsilon \zeta}{\eta} \left(\frac{1}{k_b + 2k_s/R} \right) \frac{p}{l} \tag{63}$$

If both sides of Eq. (63) are multiplied by the length of the capillary l, the potential difference between the measuring electrodes—the streaming potential E (not to be confused with the field $\bar{E}$)—is obtained:

$$E = \frac{\varepsilon \zeta p}{\eta(k_b - 2k_s/R)} \tag{64}$$

The conditions under which Eq. (64) for streaming potential and Eq. (54) for electro-osmosis were derived are comparable inasmuch as each applies to the case of large κR. Comparison of Eqs. (54) and (64) in the limit of large R shows that

$$\frac{E}{p} = \frac{V}{I} = \frac{\varepsilon \zeta}{\eta k} \tag{65}$$

The coupling of two different electrokinetic ratios (E/p and V/I) through Eq. (65) is an illustration of a very general law of reciprocity due to L. Onsager (Nobel Prize, 1968). The general theory of the Onsager relations, of which Eq. (65) is an example, is an important topic in nonequilibrium thermodynamics.

If the relationships shown in Eq. (65) are to be used in computations, it is essential that proper units be used. The following example considers some numerical substitutions into Eq. (65).

Example 13.3 Show that E/p, V/I, and $\varepsilon_0 \zeta/\eta k$ all have units m^3 C^{-1}. For water at 25°C, $\eta = 8.937 \times 10^{-4}$ kg m^{-1} s^{-1} and $\varepsilon_r = 78.54$. Evaluate the proportionality factor between V/I and ζ/k if V is expressed in cm^3 s^{-1} and I is expressed in milliamperes with all other quantities in SI units. Evaluate the proportionality factor between E/p and ζ/k if p is expressed in Torr with all other quantities in SI units.

Solution Examine the SI units of each term in Eq. (65). The units of V/I are $(m^3 \ s^{-1})/(C \ s^{-1}) = m^3 \ C^{-1}$. The units of E/p are $V/(N \ m^{-2})$; when multiplied by the conversion factor $(J \ C^{-1})/V$, this becomes $m^3 \ C^{-1}$. The units of $\varepsilon_0 \zeta/\eta k$ are $(C^2 \ J^{-1} \ m^{-1})(V)/(kg \ m^{-1} \ s^{-1})(ohm^{-1} \ m^{-1})$; when multiplied by the conversion factor $V (C \ s^{-1})^{-1}/ohm$, this becomes $J \ C^{-1} \ s/kg \ m^{-1} = m^3 \ C^{-1}$.

Substituting the SI values for all quantities, we obtain the following for $\varepsilon \zeta/\eta k$:

$$\frac{(78.54)(8.85 \times 10^{-12})\zeta}{(8.937 \times 10^{-4})k} = 7.777 \times 10^{-7} \frac{\zeta}{k} \ m^3 \ C^{-1}$$

from which

$$\frac{V}{I} = 7.777 \times 10^{-7} \frac{\zeta}{k} \ m^3 \ C^{-1} \times \left(\frac{10^2 \ cm}{m}\right)^3 \times \frac{1 \ C \ s^{-1}}{10^3 \ mA}$$

$$= 7.777 \times 10^{-4} \frac{\zeta}{k} \frac{\text{cm}^3\,\text{s}^{-1}}{\text{mA}}$$

and

$$\frac{E}{p} = 7.777 \times 10^{-7} \frac{\zeta}{k} \text{ m}^3\,\text{C}^{-1} \times \frac{10^3\,\text{liter}}{\text{m}^3} \times \frac{1\,\text{C V}}{1\,\text{J}} \times \frac{8.314\,\text{J}}{0.08205\,\text{liter atm}}$$

$$\times \frac{1\,\text{atm}}{760\,\text{Torr}} = 1.037 \times 10^{-4} \frac{\zeta}{k} \text{ V Torr}^{-1}$$

•

For 10^{-3} M NaCl, $k = 1.26 \times 10^{-2}$ ohm^{-1} m^{-1} and a surface with a ζ potential of 50 mV will displace about 11 cm^3 h^{-1} if a current of 1.0 mA flows through an electro-osmosis apparatus. With the same electrolyte and the same value of ζ, an applied pressure of 760 mm Hg would produce a streaming potential of 313 mV.

In hydrocarbons the specific conductivity may be lower than that of aqueous solutions by many orders of magnitude, so the streaming potentials generated by the high-pressure pumping of these materials may be quite spectacular. The danger of sparking at such voltages plus the flammability of these substances makes the petroleum industry an area in which streaming potential finds important applications. For example, gasoline (for which the specific conductivity would be as low as 10^{-12} ohm^{-1} m^{-1} if untreated) pumping equipment must be grounded. In addition, a variety of organic–soluble electrolytes have been developed as antistatic additives for petroleum. Examples of two such compounds are tetraisoamyl ammonium picrate and calcium diisopropyl salicylate. Crude petroleum is less troublesome in this regard than refined products, since the crude contains oxidation products, asphaltenes, and so on, which impart a natural conductivity to this material.

The objective of comparing values of ζ determined from electrophoresis with those determined by other electrokinetic methods was stated at the beginning of Sect. 13.6. Enough experiments have been conducted in which at least two of the electrokinetic methods we have discussed are compared to leave no doubt as to the self-consistency of ζ as determined by these different methods. There is no guarantee, however, that self-consistent ζ potentials are correct. Consistency means only that ζ has been extracted from experimental quantities by a self-consistent set of approximations. It should be emphasized, however, that the existence of a potential at the surface of shear—which is the common component in all the electrokinetic analyses we have discussed—is more than amply confirmed by these observations.

Two conditions must be met to justify comparisons between ζ values determined by different electrokinetic measurements: (1) The effects of relaxation and surface conductivity must be either negligible or taken into account and (2) the surface of shear must divide comparable double layers in all cases being compared. This second limitation is really no problem when electro-osmosis and streaming potential are compared, since, in principle, the same capillary can be used for both experiments. However, obtaining a capillary and a migrating particle with identical surfaces may not be as readily accomplished. One means by which particles and capillaries may be compared is to coat both with a layer of adsorbed protein. It is an experimental fact that this procedure levels off differences between substrates: The surface characteristics of each are totally determined by the adsorbed protein. This technique also permits the use of microelectrophoresis for proteins, since adsorbed and dissolved proteins have been shown to have nearly identical mobilities.

13.8   The Surface of Shear

The surface of shear is the location within the electrical double layer at which the various electrokinetic phenomena measure the potential. We saw in Chap. 12 how the double layer extends outward from a charged wall. Its value at any particular distance from the wall can, in principle, be expressed in terms of the potential at the wall and the electrolyte content of the solution. In terms of electrokinetic phenomena, the question is: How far from the interface is the surface of shear situated?

The very existence of a surface of shear implies some interesting behavior within the fluid phase of the system under consideration. In our discussion of all electrokinetic phenomena until now, we have assumed that the viscosity of the medium has its bulk value right up to the surface of shear. In addition, it has been implicitly assumed that the viscosity abruptly becomes infinite at the surface of shear.

At this point, it is convenient to recall Fig. 7.11 and the discussion thereof. In that context we observed that there is generally a variation of properties in the vicinity of an interface from the values which characterize one of the adjoining phases to those which characterize the other. This variation occurs over a distance τ measured perpendicular to the interface. In the present discussion viscosity is the property of interest and the surface of shear—rather than the interface per se—is the boundary of interest. The model we have considered until now has implied an infinite jump in viscosity, occurring so sharply that τ is essentially zero. From a molecular point of view such an abrupt transition is highly unrealistic. A gradual variation in η over a distance comparable to molecular dimensions is a far more realistic model. With these ideas in mind, it is evident that we would do better to think of a *zone* of shear rather than a surface of shear. Although we shall continue to speak of the shear "surface," the term is not used in the mathematical sense of possessing zero thickness but, rather, in the broader sense of Chap. 7.

How must the expressions derived in the earlier sections of this chapter be modified to take into account the finite distance over which η increases? The answer is that η—the viscosity within the double layer—must be written as a function of location. Our objective in discussing this variation is not to examine in detail the efforts that have been directed along these lines. Instead, it is to arrive at a better understanding of the relationship between ζ and the potential at the inner limit of the diffuse double layer and a better appreciation of the physical significance of the surface of shear.

Measurements of the viscosity of organic liquids in the presence of an electric field reveal that there is an increase in viscosity in high electric fields which is described by the expression

$$\frac{\eta_{\bar{E}} - \eta_0}{\eta_0} = f\bar{E}^2 \tag{66}$$

where the subscripts indicate the presence ($\bar{E}$) or absence (0) of a field. The factor f is called the viscoelectric constant and has a value of about 2×10^{-16} V^{-2} m^2 for several organic liquids. Thus a 10% increase in viscosity may be anticipated for a field strength of about 2×10^7 V m^{-1}.

An expression such as Eq. (12.51) may be used to estimate $\bar{E}(= d\psi/dx)$ in the double layer. Table 13.1 shows values of $\bar{E}$ evaluated by means of this equation for a variety of ψ_0 values and 1:1 electrolyte concentrations. It will be noted that for high values of ψ_0 and high ionic strengths the field in the double layer may be large enough to produce a very significant viscoelectric effect.

Now suppose we re-examine the derivation of the Helmholtz-Smoluchowski equation as given in Sect. 13.4. Returning to Eq. (37), we note that the relationship between u and ζ is given by

$$u = \varepsilon \int_0^\zeta \frac{d\psi}{\eta} \tag{67}$$

where η has been left inside the integral this time, since its value is assumed to vary with ψ. We continue to assume that ε is a constant, since the effect of the field is known to be less for this quantity than for η.

Table 13.1   Values of the Electric Field (in V m^{-1}) Calculated in the Double Layer by Eq. (12.51) for Various ψ_0 Values and Concentrations of 1:1 Electrolyte

ψ_0 (mV)	c (mole liter^{-1})		
	10^{-3}	10^{-2}	10^{-1}
50	6.36×10^6	2.01×10^7	6.36×10^7
100	1.98×10^7	6.24×10^7	1.98×10^8
150	5.49×10^7	1.74×10^8	5.49×10^8
200	1.51×10^8	4.77×10^8	1.51×10^9

Now we substitute $\eta_{\bar{E}}$ from Eq. (66) for the viscosity in the double layer in Eq. (67) to obtain

$$u = \frac{\varepsilon}{\eta_0} \int_0^\zeta \frac{d\psi}{1 + f(d\psi/dx)^2} \tag{68}$$

Finally, Eqs. (12.38) and (12.51) may be used to evaluate $d\psi/dx$ in the double layer:

$$u = \frac{\varepsilon}{\eta_0} \int_0^\zeta \frac{d\psi}{1 + [f(8000cRT/\varepsilon)] \sinh^2(ze\psi/2kT)} \tag{69}$$

where c is in mole liter^{-1} and 2 sinh x has been substituted for $e^x - e^{-x}$. If we define $A = 8000fcRT/\varepsilon$ and $B = ze/2kT$, this result is more concisely written as

$$u = \frac{\varepsilon}{\eta_0} \int_0^\zeta \frac{d\psi}{1 + A \sinh^2(B\psi)} \tag{70}$$

If c (and therefore A) and ψ are small, Eq. (70) becomes approximately

$$u \simeq \frac{\varepsilon}{\eta_0} \int_0^\zeta [1 - A \sinh^2(B\psi)] d\psi \tag{71}$$

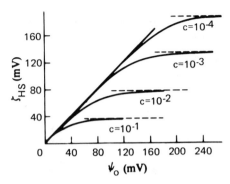

Figure 13.7   Plot of $\eta u/\varepsilon$ versus ψ_0 or, in words, the zeta potential according to the Helmholtz–Smoluchowski equation, Eq. (39), versus the potential at the inner limit of the diffuse part of the double layer. Curves are drawn for various concentrations of 1:1 electrolyte with $f = 10^{-15}$ V^{-2} m^2. [From J. Lyklema and J. Th. G. Overbeek, *J. Colloid Sci.*, *16*:501 (1961), used with permission.]

Under the same conditions the hyperbolic sine function may be expanded (see Appendix A), with only the leading term retained, to obtain

$$u \simeq \frac{\varepsilon}{\eta_0} \int_0^{\zeta} [1 - A(B\psi)^2]\,d\psi \tag{72}$$

This equation is readily integrated to yield

$$u \simeq \frac{\varepsilon}{\eta_0}\left(\zeta - \frac{AB^2}{3}\zeta^3\right) = -\frac{\varepsilon\zeta}{\eta_0}\left(1 + \frac{AB^2}{3}\zeta^2\right) \tag{73}$$

Under conditions in which the second term is negligibly small, Eq. (73) becomes identical to Eq. (39), the Helmholtz–Smoluchowski result. On the other hand, when the concentration and ζ increase, the value of ζ which would be associated with an observed mobility is larger than the Helmholtz–Smoluchowski equation would indicate.

Equation (69) may also be integrated analytically. Although we shall not consider the actual solutions, which are rather complex, Fig. 13.7 shows graphically the results of these integrations drawn for water at 25°C, assuming $f = 10^{-15}$ V^{-2} m^2. The abscissa shows values of ψ_0, the potential at the inner limit of the diffuse double layer, with $\eta u/\varepsilon$ plotted on the ordinate. It must be remembered that this latter quantity equals ζ

according to Eq. (39)—which we shall designate ζ_{HS}—when the viscosity is assumed to be the bulk value throughout the double layer. The figure shows that $\zeta_{HS} = \psi_0$ at low values of the potential. As the potential increases, however, ζ_{HS} begins lagging behind ψ_0, the effect indicated by Eq. (73) in a limiting approximation. At still higher potentials ζ_{HS} eventually reaches a constant value which is independent of the actual value of ψ_0. Note, further, that this leveling off occurs at progressively lower potentials as the concentration of electrolyte increases. Increasing both the potential and the electrolyte concentration tends to increase the field in the double layer (see Table 13.1), which, in turn, increases the viscosity of solvent in the double layer. As the effective viscosity of the medium increases, the surface of shear occurs progressively further from the surface. This accounts for the fact that ζ_{HS} falls behind ψ_0 as the latter increases. These conclusions are consistent with the experimental observation that ζ_{HS} for AgI becomes independent of the concentration of the potential-determining Ag^+ and I^- ions once the concentrations of these ions are well removed from the conditions at which the particles are uncharged.

The results shown in Fig 13.7 illustrate quite clearly the relationship between ζ and ψ_0 and in this way reveal the dependence of the location of the surface of shear on the structure of the double layer. It might appear that one could consult curves such as those shown in Fig. 13.7 to read from the appropriate plot that value of ψ_0 which corresponds to a particular ζ, at least for values of ζ which are less than the limiting value. Although semiquantitative interpretations based on this figure may be trusted, some caution must be expressed about the numerous assumptions and approximations inherent in Fig. 13.7. In summary, the following may be cited as examples of such constraints:

1. The possible immobilization of solvent near the surface due to either chemical or mechanical (as opposed to viscoelectric) interaction with the solid phase has not been considered.
2. Use of the Gouy–Chapman theory [Eq. (69)] overlooks any specific effects arising from differences between ions, especially with regard to hydration.
3. The validity of Eq. (66) in electrolyte solutions, especially the dependence of f on concentration, has not been investigated as fully as might be desired.

13.9 Experimental Aspects of Electrophoresis

Of the electrokinetic phenomena we have considered, electrophoresis is by far the most important. Until now our discussion of experimental techniques of electrophoresis has been limited to a brief description of microelectrophoresis. The latter is easily visualized and has provided sufficient background for our considerations up to this point. Microelectrophoresis itself is subject to some complications that can be discussed now that we have some background in the general area of electrical transport phenomena. In addition, the methods of moving boundary electrophoresis and zone electrophoresis are sufficiently important to warrant at least brief summaries.

Microelectrophoresis depends on the visibility of the migrating particles under the microscope. As such, it is inapplicable to molecular colloids such as proteins. By adsorbing the protein molecules on suitable carrier particles, however, the range of utility for microelectrophoresis can be extended. The use of dark-field illumination (see Sect. 1.5) can sometimes be used to advantage to extend microelectrophoresis observations to small, high-contrast particles.

The migrating particles are observed in a cell which may be either cylindrical or rectangular. The walls must be optically uniform for observation and fewer optical corrections, and thermostating difficulties are encountered if the walls are thin. The working part of the apparatus is thus fragile and auxiliary connecting rods are generally incorporated into the design to increase the mechanical strength of the cell. Figure 13.8 is a sketch of an electrophoresis apparatus with a rectangular working compartment.

The electric field in the cell is best established by means of reversible electrodes such as Ag–AgCl or Cu–$CuSO_4$. Care must be taken to prevent the electrolyte of the electrode from contaminating the dispersion. Platinized electrodes behave reversibly with low currents, but gas evolution causes troubles at higher currents.

The field strength is best obtained by including an accurate ammeter in the circuit to determine the current. Independent conductivity measurements in the cell with standard solutions permit the determination of the field through Eq. (49).

The rate of particle migration is determined by measuring with a stopwatch the time required for a particle to travel between the marks of a calibrated graticule in the microscope eyepiece. If the objective of the microscope is immersed during the electrophoresis measurement, the calibration of the graticule should be made with the same immersion liquid.

Electrophoretic migrations are always superimposed on other displacements, which must either be eliminated or corrected to give accurate values for mobility. Examples of these other kinds of movement are Brownian motion, sedimentation, convection, and electro-osmotic flow.

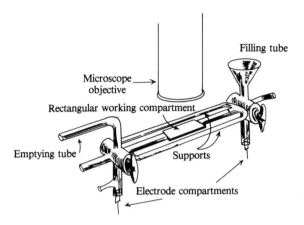

Filling tube

Microscope objective

Rectangular working compartment

Emptying tube

Supports

Electrode compartments

Figure 13.8 Schematic illustration of a microelectrophoresis cell with a rectangular working compartment.

Brownian motion, being random, is eliminated by averaging a series of individual observations. Sedimentation and convection, on the other hand, are systematic effects. Corrections for the former may be made by observing a particle with and without the electric field, and the latter may be minimized by effective thermostating and working at low current densities.

Even in the absence of a colloid, an electrolyte solution will display electro-osmotic flow through a chamber of small dimensions. Therefore the observed particle velocity is the sum of two superimposed effects, namely, the true electrophoretic velocity relative to the stationary liquid and the velocity of the liquid relative to the stationary chamber. Figure 13.9a shows the results of this superpositioning for particles tracked at different depths in the cell. The particles used in this study are cells of the bacterium *Klebsiella aerogenes* in phosphate buffer. Rather than calculated velocities or mobilities, Fig. 13.9a shows the reciprocal of the time required for the particles to travel a fixed distance. Since the electrophoretic velocity is a single-valued property of the (uniform) particles under consideration, Fig. 13.9a raises the question of which, if any, of these "velocities" represents the true mobility of these particles. The following example considers a possible interpretation of these results.

Example 13.4 It is proposed to evaluate the electrophoretic mobility of the bacteria cells shown in Fig. 13.9a by multiplying the appropriate value of time^{-1} by the distance of particle displacement and then dividing

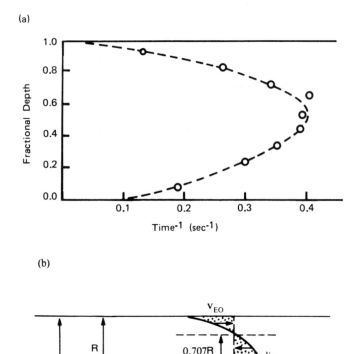

Figure 13.9 (a) Velocity (as time^{-1}) of *Klebsiella aerogenes* particles as a function of their location in a rectangular electrophoresis cell [reprinted with permission from A. M. James, in *Surface And Colloid Science*, Vol. 11 (R. J. Good and R. R. Stromberg, eds.), Plenum, New York, 1979]. (b) Location of the surface of zero liquid velocity in a cylindrical capillary.

by $\bar{E}$. Criticize or defend the following proposition: It is appropriate to use the maximum apparent velocity, since this is measured at the center of the cell and is therefore subject to the least interference by wall effects.

Solution The observed effect is the sum of two contributions, one of which is the electro-osmotic flow of the medium through the cell. The latter has its maximum value at the center, since the layer of fluid adjacent to the walls is stationary. The particles tracked at the center of the cell therefore possess the maximum increment in velocity due to electro-osmotic flow. Since the cell is a closed compartment, the liquid displaced by electro-osmosis along the walls must circulate by a backflow down the center of the tube. Since the total liquid flow in a closed cell must be zero, the appropriate value from Fig. 13.9a to use for the velocity is the average of observations made at all depths.

•

Since the liquid circulates, there must be certain locations in the cell at which the forward and backward flows of the liquid are equal. An alternative to the averaging procedure suggested in the example is to do the particle tracking at a location where the medium experiences no net flow.

The analysis of this effect in a closed cyclindrical cell is obtained by subtracting from the electro-osmotic velocity, v_{EO}, the velocity of flow through a capillary given by Poiseuille's equation [Eq. (4.18), v_P]:

$$v_L = v_{EO} - v_P = v_{EO} - C(r^2 - R^2) \tag{74}$$

where v_L is the velocity of the liquid and C is a constant. The requirement of no net displacement of liquid is incorporated by integrating Eq. (74) over the cross section of the cylinder and setting the result equal to zero:

$$\int_0^R v_L(2\pi r)dr = 0 \tag{75}$$

In this expression R is the radius of the capillary and r is the radial distance from the capillary axis as shown in Fig. 13.9b. Substitution of Eq. (74) into Eq. (75) and integration gives

$$C = -\frac{2v_{EO}}{R^2} \tag{76}$$

This result may be substituted back into Eq. (75) to evaluate that location in the cylinder where the net liquid displacement is zero:

$$v_L = 0 = v_{EO}\left(1 + \frac{2}{R^2}(r^2 - R^2)\right) = v_{EO}\left(\frac{2r^2}{R^2} - 1\right) \tag{77}$$

This result shows that electro-osmotic flow and backflow in the capillary cancel when the factor $2r^2/R^2 - 1$ equals zero. This condition corresponds to $r/R = 0.707$. Thus at 70.7% of the radial distance from the center of the

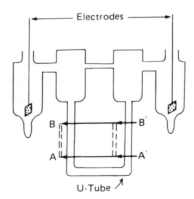

Figure 13.10 Schematic illustration of a Tiselius-type moving boundary electrophoresis apparatus.

capillary lies a circular surface of zero liquid flow. Any particle tracked at this position in the capillary will display its mobility uncomplicated by the effects of electro-osmosis. This location may also be described as lying 14.6% of the cell diameter inside the surface of the capillary. Experimentally, then, one establishes the inside diameter of the capillary and focuses the microscope 14.6% of this distance inside the walls of the capillary. Corrections for the effect of the refractive index must also be included. Additional details on this correction can be found in the book by Shaw [9].

The location of the surface of zero liquid flow in cells of rectangular cross section has also been worked out. For a cell in which the direction of migration is very long compared to the width of the cell, the surface where $v = 0$ lies 21.1% of the cell depth above the bottom and below the top of the working compartment.

In addition to microelectrophoresis, another important method for the determination of mobility is the moving boundary method. In essence, this is no different from the moving boundary method as applied to simple ions. The apparatus most commonly used is that of A. Tiselius (Nobel Prize, 1948), which is illustrated schematically in Fig. 13.10. The Tiselius cell consists of a U-tube of rectangular cross section which is segmented in such a way that the sections between the lines AA' and BB' in the figure can be laterally displaced with respect to the rest of the apparatus. The offset segments of the U-tube are filled with the colloidal dispersion and, after thermal equilibration with the buffer solution contained in other parts of the apparatus, the various sections are aligned so that sharp boundaries are obtained. The location of the boundaries is usually observed by schlieren optics which identify refractive index gradients (see Sect. 2.5). As the macroions migrate in the electric field, the schlieren peak becomes displaced and the mobility of a colloidal component may be determined by measuring the rate of boundary movement per unit electric field. Relatively longer times are required for accurate mobility experiments than for microelectrophoresis, since the particles must migrate over macroscopic distances rather than microscopic ones. To avoid contamination of the electrolyte in the U-tube with electrode products, the electrodes are generally located near the bottom of large reservoirs as shown in Fig. 13.10. Relatively concentrated salt solution is used to cover the electrodes, with the buffer solution layered on top.

Under optimum conditions the dimensions of the cross section of the cell are such that the effects of electro-osmosis are minimal. The rectangular profile of the cross section allows for both good thermal equilibration (because one dimension is short) and good optical precision (because the other dimension is longer).

Moving boundary electrophoresis is most widely applied to protein mixtures. In such a case each molecular species travels with a characteristic velocity. After sufficient time the various components in a mixture become effectively separated, and the percentage of each may be determined by measuring the areas under the schlieren peaks. Figure 13.11a shows a typical electrophoresis pattern for human blood serum. In this figure the protein albumin (A), α_1-, α_2-, $\beta-$, and γ-globulin, and fibrinogen (ϕ) are fairly clearly resolved. The remaining peak in the figure is the boundary between the original buffer and the colloid. This "false boundary" moves little in an electrophoresis experiment and is obviously not considered in determining the percentages of different proteins in a mixture.

When separation rather than determination of mobility is the primary objective of an electrophoresis experiment, a technique called zone electrophoresis is quite widely employed. In zone electrophoresis a supporting medium such as moist filter paper or a gel such as polyacrylamide is the location of the particle migration. The method thus resembles solid–liquid chromatography, and many of the substrates and analytical methods of the latter are used in this electrophoretic procedure as well. As with chromatography, a spot or band of a mixture is applied to one end of the support medium. As the electrophoresis proceeds, spots or bands of the individual components appear at different locations along the axis of the voltage gradient. Sometimes the resolution is improved by following the electrophoresis by a chromatographic separation at right angles to the direction of the initial separation.

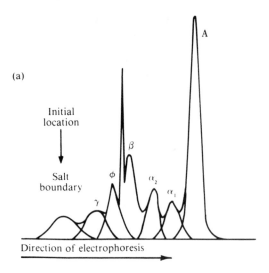

Figure 13.11 Electrophoresis patterns for human serum. (a) Schematic of schlieren profiles. (b) Semilog plot of protein molecular weight versus electrophoretic mobility for particles electrophoresed on crosslinked polyacrylamide. [Reprinted with permission from K. Weber and M. Osborn, *J. Biol. Chem.*, *244*:4404 (1969).]

   Zone electrophoresis is influenced by adsorption and capillarity, as well as by electro-osmosis. Therefore evaluation of mobility (and ζ) from this type of measurement is considerably more complex than from either microelectrophoresis or moving boundary electrophoresis. Nevertheless, zone electrophoresis is an important technique which is widely used in biochemistry and clinical chemistry. One particularly important area of application is the field of immunoelectrophoresis, which is described briefly in Sect. 13.11.

   Figure 13.11b illustrates the application of gel electrophoresis to protein characterization. In this work a crosslinked polyacrylamide gel is the site of the electrophoretic migration of proteins which have been treated with sodium dodecyl sulfate. The surfactant dissociates the protein molecules into their constituent polypeptide chains. The results shown in Fig. 13.11b were determined with well-characterized polypeptide standards and serve as a calibration curve in terms of which the mobility of an unknown may be interpreted to yield the molecular weight of the protein. As with any experiment that relies on prior calibration, the successful application of this method requires that the unknown and the

(b)

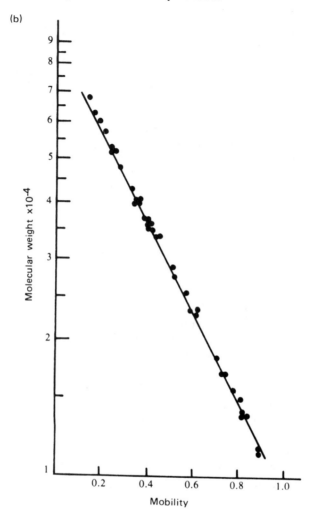

standard be treated in the same way. This includes such considerations as the degree of crosslinking in the gel, the pH of the medium, and the sodium dodecyl sulfate concentration. The last two factors affect the charge of the protein molecules by dissociation and adsorption, respectively. The following example considers a similar application of electrophoresis.

Example 13.5 Synthetic DNA standards and RNA molecules were electrophoresed in 7 M urea solution on crosslinked polyacrylamide gels.* A semilog plot of the number of nucleotides versus the mobility relative to xylene cyanol FF dye is linear and includes the points (N = 100, u_{rel} = 0.33) and (N = 50, u_{rel} = 0.55). Estimate the number of nucleotides in the glycine tRNA molecule of *Staphylococcus epidermidis* if the latter shows a relative mobility of 0.16.

Solution The linear semilog plot means that these data follow the equation ln $N = b + mu_{rel}$. Since we know two points from this plot, we can evaluate the constants m and b by simultaneous equations. This procedure yields b = 5.63 and m = −3.09. Combining these constants with the observed mobility allows the number of nucleotides in the unknown to be calculated by the formula ln N = 5.63 − 3.09(0.16) = 170.

•

13.10 The Charge of Protein Molecules

In the quantitative sections of this chapter the primary emphasis has been on establishing the relationship between the electrophoretic properties of the system and the zeta potential. We saw in Chap. 12 that potential is a particularly useful quantity for the characterization of lyophobic colloids. In this context, then, the ζ potential is a valuable property to measure for a lyophobic colloid. For lyophilic colloids such as proteins, on the other hand, the charge of the particle is a more useful way to describe the molecule. In this section we shall consider briefly what information may be obtained about the charge of a particle from electrophoresis measurements.

We have lamented the fact that electrokinetic potentials cannot be evaluated independently to check the correctness of various theories. However, the charge of a protein can be evaluated from its titration curve. Therefore, if we can find a way of evaluating particle charge from electrokinetic data, the long-sought independent verification will be established. The net charge of a particle q is equal and opposite to the total charge in the double layer. The increment of charge in a spherical shell of radius r and thickness dr in the double layer is given by the area of the shell times its thickness times the charge density:

$$dq = 4\pi r^2 \rho^* \, dr \tag{78}$$

Integrating this expression over the entire double layer gives

---

*T. Maniatis, A. Jeffrev, and H. van de Sande, *Biochemistry*, *14*:3787 (1975).

$$q = - \int_{R}^{\infty} 4\pi r^2 \rho^* \, dr = \int_{R}^{\infty} \varepsilon \, \nabla^2 \psi 4\pi r^2 \, dr = 4\pi\varepsilon \int_{R}^{\infty} \frac{d}{dr}\left(r^2 \frac{d\psi}{dr} \right) dr$$

$$(79)$$

where the Poisson equation (12.24) has been substituted for ρ^*. Integration yields

$$q = 4\pi\varepsilon \left(r^2 \frac{d\psi}{dr} \right) \Bigg|_{R}^{\infty} = -4\pi\varepsilon R^2 \frac{d\psi}{dr} \Bigg|_{R}$$

$$(80)$$

where the derivative is evaluated at $r = R$.

Now Eq. (46) is used to evaluate the derivative in Eq. (80):

$$\frac{d\psi}{dr}\Bigg|_{R} = -\frac{\zeta}{R}(1 + \kappa R)$$

$$(81)$$

Substituting this result into Eq. (80) gives

$$q = \varepsilon\zeta R(1 + \kappa R)$$

$$(82)$$

for the charge enclosed by the surface of shear.

This discussion shows that the evaluation of charge from electrokinetic measurements involves all the complications inherent in the evaluation of ζ plus the additional restrictions of low potentials and spherical particles. Additional relationships have been developed which permit these restrictions to be relaxed, but we shall not discuss these here.

We conclude this section by comparing briefly the charge on protein molecules as determined by electrophoresis measurements through Eq. (82) and as determined by titration. Protein molecules carry acid and base functions in side groups along the macromolecule. In a strongly acidic solution amine groups will be protonated and the protein will carry a positive charge. Addition of a known number of equivalents of strong base to a measured volume of protein solution results in a change of pH and a change in the state of charge of the protein. From the volume of the solution, the change of pH, and a knowledge of activity coefficients, the number of added equivalents of base which react with the protein may be determined. It should be noted that the added base may remove protons from either neutral groups or cationic groups. Thus in acid solution a protein may have a charge corresponding to the binding of z H^+ ions: $+z$. After z OH^- ions have reacted with it, the molecule will have a *net* charge of zero. If the reactions consist exclusively in the removal of bound H^+ ions, the net charge (zero) would correspond to the *actual* charge of the particle. If all the reacting OH^- ions remove H^+ ions from neutral groups, on the other hand, the molecule would be twice as highly charged as the initial species: an equal number of positive and negative charges. In reality, both processes occur together, so it cannot be inferred that the point of equivalency—called the isoionic point—corresponds to an uncharged state. All that can be said is that the *net* charge is zero at the isoionic point. Addition of more base beyond this point will increase (still by both processes) the negative charge of the molecule even further. The isoionic point corresponds to a point at which the polyelectrolyte changes sign. This discussion shows that the net charge relative to the initial condition of the colloid is readily determined from titration curves.

The electrophoretic mobility of a protein solution may also be measured as a function of pH. By this technique also it is observed that the colloid passes through a point of zero net charge where its mobility is zero. The point at which charge reversal is observed electrophoretically is called the isoelectric point.

Figure 13.12 shows the relationship between the charge of egg albumin as determined by titration and by electrophoresis. The points were determined electrophoretically, and the solid line was determined by titration. The titration curve has been shifted so that the isoionic point and the isoelectric point match. It will be observed that the two independent charge determinations lead to slightly different values. The charges determined electrophoretically are 60% of those determined analytically. If the titration results are multiplied by 0.60, the dashed line in Fig. 13.12 is obtained. This shows clearly that the two determinations are identical in pH dependence, but raises the question as to the origin of the constant percentage difference.

There are several minor corrections which tend to reduce the discrepancy between the two curves, for example, corrections for relaxation and finite ion size. It should also be remembered that electrophoresis measures the net charge inside the surface of shear. To the extent that this diverges from the "surface" of the molecule, the two techniques may very properly "see" different charges for the colloid. Additional studies in this area, therefore, might help to clarify the relationship between the actual surface and the surface of shear.

We noted earlier that proteins display essentially the same mobility both as free molecules and when adsorbed on carrier particles. Adsorption clearly increases the radius of the kinetic unit appreciably, so this effect on mobility is unexpected. One way to rationalize this result is to assume that the protein adsorbs on the surface with very little alteration

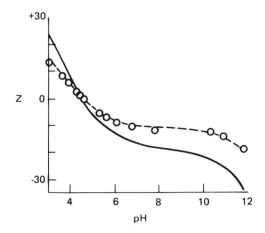

Figure 13.12 Net charge of egg albumin versus pH. The points were determined by electrophoresis, and the solid line by titration; the broken line represents 60% of charge from titration. [Data from L. G. Longsworth, *Ann. N.Y. Acad. Sci., 41*:267 (1941), redrawn from Ref. 5 with permission.]

of the shape it has in free solution. Next assume that it is the radius of these molecular protuberances rather than the overall radius of curvature of the carrier that governs the mobility.

13.11 Applications of Electrokinetic Phenomena

Throughout most of this chapter the emphasis has been on the evaluation of zeta potentials from electrokinetic measurements. This emphasis is entirely fitting in view of the important role played by the potential in the DLVO theory of colloidal stability. From a theoretical point of view, a fairly complete picture of colloidal stability can be built up from a knowledge of potential, electrolyte content, Hamaker constants, and particle geometry. From this perspective the fundamental importance of the ζ potential is evident.

In addition to fundamental principles, however, there are many practical situations in which flocculation is a process of considerable importance. Often all that is desired in these cases is either to maximize or minimize flocculation in some experimental system. Systems of practical interest are frequently so complex that theoretical models apply to them only qualitatively at best. In this context the concept of zeta potential emerges as a valuable practical parameter. If two systems of different ζ are compared—all other factors being equal—the one that has the higher ζ potential is expected to be more stable with respect to flocculation, and the one with the lower potential less stable. The second case is particularly important. At the isoelectric point electrophoretic mobility is zero, ζ is zero, and the potential energy of repulsion between particles is minimum. Thus electrophoresis measurements can be used as an indicator for optimum conditions for flocculation. In this type of application the technique is used as a null detector; hence it is independent of any model or equation for interpretation.

One important—if unattractive—example of this application is in sewage treatment. Industrial waste water and domestic sewage contain an enormous assortment of hydrophilic and hydrophobic debris of technological and biological origin. The concentration of surface-active materials in sewage from household detergents alone is about 10 ppm. In addition, sewage abounds in amphipathic materials of natural and biological origin. These substances tend to adsorb on and impart a charge to the suspended solid and liquid particles in the polluted water. Negative zeta potentials in the range of 10–40 mV are fairly typical for the suspended particles in sewage.

A typical purification scheme consists of adding $NaHCO_3$ and $Al_2(SO_4)_3$ (alum) to water with agitation. The aluminum ion undergoes hydrolysis and precipitates as a gelatinous, polymeric hydrated oxide. Suspended material is enmeshed in this amorphous precipitate which produces flocs by bridging the particles together. The polymeric nature of the "$Al(OH)_3$" precipitate permits us to compare it with protein in its ability to coat particles and impart to the carrier particles its own characteristic potential. Like proteins, $Al(OH)_3$ is also capable of reacting with both H^+ and OH^- so that these ions determine the charge of the suspended units, whether these are flocs formed by the $Al(OH)_3$ network or individual particles with an adsorbed layer of $Al(OH)_3$. In either case the charge is pH sensitive, the isoelectric point occurring near pH 6. It is under these pH conditions, then, that the flocculating effectiveness of the precipitate is optimum. In fact, the pH is often adjusted so that the hydrous aluminum oxide surface has a slightly positive value of ζ (about 5 mV). This promotes further interaction with slightly anionic polymeric materials which are also added to further build up and strengthen flocs. Once adequate flocculation has been accomplished, the dispersed particles are removed by sedimentation or filtration.

Numerous other applications could be listed in which electrokinetic characterization provides a convenient experimental way of judging the relative stability of a system to flocculation. Paints, printing inks, drilling muds, and soils are examples of additional systems whose properties are extensively studied and controlled by means of the ζ potential.

In addition to these applications in which ζ is used to monitor for optimum flocculation conditions, there are applications which explicitly depend on mobility or differences in mobility for their usefulness. We have already noted that zone electrophoresis is similar in many ways to chromatography. One important application of the ability of electrophoresis to segregate materials by mobility is immunoelectrophoresis. This technique uses known immunochemical reactions between antigen and antibody for the identification of proteins separated electrophoretically. Experimentally, an antigen mixture is subjected to electrophoresis on a suitable medium (usually agar gel). Next the antibody mixture is introduced into a slit cut in the gel parallel to the axis of the separation. The antigen and antibody components then diffuse toward one another, producing an arc-shaped precipitate where the two fronts meet.

Tests of this sort are particularly useful for comparing either two antigen preparations (against a single antibody) or two antibody preparations (against a single antigen). In such a comparison one of the samples serves as a control, and differences between the two are revealed by an unpaired arc of precipitate at a particular loction along the path of separation.

Alternatively, electrophoretic separation in one direction may be followed by a second electrophoresis in a perpendicular direction, the latter into a gel containing antibodies. This technique is called crossed immunoelectrophoresis and combines high resolution with the possibility of quantification by measuring the area of the precipitate formed. Figure 13.13 is a photograph of the peaks of antigen–antibody precipitates formed by crossed immunoelectrophoresis of human serum interacting with rabbit anti-human serum.

Electrodeposition is another direct application of electrophoretic mobility. In this process, as in electroplating with metals, the substance to be coated is made into an electrode of opposite charge from the particles to be deposited. At one time, natural rubber latex was extensively fabricated in this way. Paint coatings which are quite dense and coherent with little tendency to sag or run can be prepared by electrodeposition. If the deposited layer has insulating properties, this technique is also self-regulating, producing a uniform thin covering of very good quality.

Although electrophoresis is the most important of the electrokinetic methods, it is not the only one with practical applications. We already noted in Sect. 13.7 that streaming potentials could be quite hazardous in

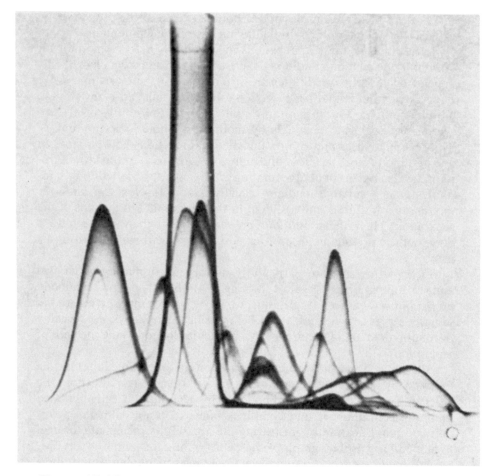

Figure 13.13   Crossed immunoelectrophoresis of human serum with rabbit anti-human serum. [From B. Weeke, in *A Manual of Quantitative Immunoelectrophoresis: Methods and Applications* (N. H. Axelsen, J. Krøll, and B. Weeke, eds.) Universitetsforlaget, Oslo, Norway, 1973.]

low-conductivity, highly flammable substances such as purified hydrocarbons. In this case our knowledge of the effect enables us to minimize it. Dewatering fine suspensions which are not amenable to filtration is an application of electro-osmosis. Peat, clay, and other minerals have been dewatered this way, and water may be removed from moist soil prior to excavation by electro-osmosis. Electrodes are driven into the ground, the cathode in the form of a perforated pipe. The surface of the soil particles carries a negative charge; therefore the diffuse part of the double layer is positive and the solution moves toward the cathode. The water which collects in the cathode is subsequently removed by pumping.

13.12   Wrap-Up: Chapters 11-13

The common theme connecting Chaps. 11-13 is the stability of some dispersed lyophobic systems with respect to flocculation. To understand this stability, it is important to appreciate the electrical potential near a charged wall. This is the central consideration of Chap. 12. In Chap. 13 we examined what happens if one part of the electrical double layer moves with respect to the other. At first glance, this seems to have nothing to do with flocculation, although it clearly depends on the structure of the double layer. However, further analysis reveals that the potential in the double layer at some (unknown) location close to the surface is measured by electrokinetics. This measured potential is part of the total potential which is responsible for stability.

The discussion of stability would be purely academic if double layer repulsion were the only interaction between particles. This is not the case, and Chap. 11 is concerned with the attraction which operates among all particles. The origin of these ubiquitous attractions can be traced all the way back to the molecular level, where they are seen to be the same forces responsible for gas nonideality, liquid miscibility, and so on.

Van der Waals attractions are electrostatic in origin, just as double layer interactions are. In this sense, Chapters 11-13 have all been concerned with the electrical interactions among colloidal particles. In one way or another, the same statement can be made for all topics in chemistry, however; so there is little advantage in promoting this viewpoint to unify the material of these chapters.

The primary variables which determine the stability of an aqueous dispersion have been shown to be the potential at the surface or at the inner limit of the diffuse part of the double layer, the valence and concentration of the electrolyte in the continuous phase, and the Hamaker constant of water and the dispersed phase. These same principles undoubtedly apply in nonaqueous electrolyte solutions as well, but we have emphasized aqueous hydrophobic colloids in our discussion because they are relatively well understood and contain many systems of practical importance.

References

1. H. A. Abramson, L. S. Moyer, and M. H. Gorin, *Electrophoresis of Proteins*, Hafner, New York, 1964.
2. C. C. Brinton, Jr., and M. A. Lauffer, The Electrophoresis of Viruses, Bacteria and Cells, and the Microscope Method of Electrophoresis, in *Electrophoresis*, Vol. 1 (M. Bier, ed.), Academic, New York, 1959.
3. D. A. Haydon, The Electrical Double Layer and Electrokinetic Phenomena, in *Recent Progress in Surface Science*, Vol. 1 (J. F. Danielli, K. G. A. Pankhurst, and A. C. Riddiford, eds.), Academic, New York, 1964.
4. L. G. Longsworth, Moving Boundary Electrophoresis—Practice, in *Electrophoresis*, Vol. 1 (M. Bier, ed.), Academic, New York, 1959.
5. J. Th. G. Overbeek, Quantitative Interpretation of the Electrophoretic Velocity of Colloids, in *Advances in Colloid Science*, Vol. 3 (H. Mark and E. J. W. Verwey, eds.), Wiley, New York, 1950.
6. J. Th. G. Overbeek, Electrokinetic phenomena, in *Colloid Science*, Vol. 1 (H. R. Kruyt, ed.), Elsevier, Amsterdam, 1952.
7. J. Th. G. Overbeek and J. Lyklema, Electric Potentials in Colloidal Systems, in *Electrophoresis*, Vol. 1 (M. Bier, ed.), Academic, New York, 1959.
8. J. Th. G. Overbeek and P. H. Wiersma, The Interpretation of Electrophoretic Mobilities, in *Electrophoresis*, Vol. 2 (M. Bier, ed.), Academic, New York, 1967.
9. D. J. Shaw, *Electrophoresis*, Academic, New York, 1969.
10. R. D. Vold and M. J. Vold, *Colloid And Interface Chemistry*, Addison-Wesley, Reading, Mass., 1984.

Problems

1. Particles of Fe_2O_3 with an average diameter of 1 μm were dispersed in xylene containing 5×10^{-3} mole liter^{-1} of copper(I) oleate. These showed an electrophoretic mobility of 0.110 μs^{-1} V^{-1} cm^{-1}. The conductivity of the solution was 4.7×10^{-10} ohm^{-1} cm^{-1}, indicating an ion concentration about 10^{-11} M.* Calculate κ^{-1} for this concentration. Which limiting form of Eq. (40) is most applicable in this system? Would the same conclusion be true for a 5×10^{-3} M aqueous solution of a 1:1 electrolyte? What is ζ for these particles? For xylene, $\varepsilon_r = 2.3$ and $\eta = 0.0065$ P.

2. Criticize or defend the following proposition: Zeta potentials for three different polystyrene latex preparations were calculated by the Helmholtz-Smoluchowski equation from electrophoresis measurements made in different concentrations of KCl.†

---

*H. Koelmans and J. Th. G. Overbeek, *Discuss. Faraday Soc.*, *18*:52 (1954).
†A. Kitahara and H. Ushiyama, *J. Colloid Interface Sci.*, *43*:73 (1973).

Latex designation	$R \times 10^8$ (cm)	ζ (mV)		
		10^{-1} M KCl	10^{-2} M KCl	10^{-3} M KCl
L	475	21	29	40
M	610	29	39	53
N	665	34	47	64

These zeta potentials are inaccurate because the range of κR values exceeds the range of validity for the Helmholtz–Smoluchowski equation. The nature of the error is such as to make the estimated values of ζ too low.

3. The electrophoretic mobility of sodium dodecyl sulfate micelles was determined by the moving boundary method after the micelles were made visible by solubilizing dye in them. This quantity was measured at the CMC in the presence of various concentrations of NaCl. The radius of the micelles was determined by light scattering.*

Moles of NaCl liter^{-1}	$u \times 10^4$ (cm^2 s^{-1} V^{-1})	κR
0	4.55	0.61
0.05	3.63	1.69

Estimate from Fig. 13.5a the appropriate value of C to be used in Eq. (40) according to Henry's equation. Calculate ζ using these estimated C values. Figure 13.5b shows that Henry's equation overestimates C (and therefore underestimates ζ). Estimate C from Fig. 13.5b using the curve for the ζ value which is nearest—on the high side—to the values obtained by using Henry's equation. Reevaluate ζ on the basis of these "constants." For this system $\varepsilon_r = 78.5$ and $\eta = 0.0089$ P.

4. The accompanying mobility data for colloidal SiO_2 at a constant ionic strength of 10^{-3} M reveal the superpositioning of specific chemical effects on general electrostatic phenomena. Adjustment of pH was made by addition of HNO_3 or KOH, maintaining the ionic strength. The following results were obtained†:

pH of solution:	2.0	3.0	4.0	5.0	6.0	7.0	8.0	9.0	10.0
SiO_2	0	−1.4	−1.7	−2.0	−2.3	−2.5	−2.6	−2.8	−3.0
$SiO_2 + 10^{-4}$ M La(NO$_3$)$_3$	0	−1.1	−1.2	−1.2	−1.1	−0.1	+2.2	+0.5	−1.2

$u \times 10^4$ (cm^{2}s^{-1}V^{-1})

Criticize or defend the following propositions: H^+ and OH^- are potential determining for SiO_2—in the absence of a hydrolyzable cation—with an

---

*D. Stigter and K. J. Mysels, *J. Phys. Chem.*, 59:45 (1955).
†R. O. James and T. W. Healy, *J. Colloid Interface Sci.*, 40:42 (1972).

isoelectric point of 2.0. For solid $La(OH)_3$ the zero point of charge is known (by independent studies) to be 10.4. The solid surface apparently becomes coated by $La(OH)_3$ at higher pH levels and goes through a transition from one character to another at intermediate pH values.

5. In their study of the effects of hydrolyzable cations on electrokinetic phenomena (see Problem 4), James and Healy compared the electrophoretic behavior of colloidal silica with the streaming potential through a silica capillary. In both sets of experiments the solution was 10^{-3} M KNO_3 and 10^{-4} M $Co(NO_3)_2$. The following results were obtained:

pH	6.0	7.0	7.5	8.0	9.0	10.0
ζ (mV) from streaming potential	−65	−55	−30	+10	+25	+20
$u \times 10^4$ (cm^{2}s^{-1}V^{-1})	−2.5	−2.5	−2.2	−1.8	+0.5	−0.3

The silica surface area-to-solution volume ratio was 2×10^{-3} m^{2}liter^{-1} for the streaming potential experiment and 1.0 m^{2}liter^{-1} for the electrophoresis experiment. Calculate ζ_{HS} at each pH from the electrophoresis data ($\eta = 0.00894$ P, $\varepsilon_r = 78.5$). Propose an explanation for the charge reversal behavior of the silica. Discuss the origin of the difference between ζ_{HS} and $\zeta_{St\ Pot}$ in terms of this model.

6. P. Somasundaran and R. D. Kulkarni* measured the streaming potential of 10^{-3} N KNO_3 against quartz at 25°C, obtaining the following results:

Φ (mV)	−9.0	−18.0	−26.0	−35.0
p (mm Hg)	50	100	150	200

Use these data to evaluate ζ/k. What would be the value of the ratio V/I for this system? What would be the rate of volume displacement if a current of 1.0 mA flowed through the apparatus? Evaluate ζ for the quartz–solution interface, assuming $\Lambda \simeq 145$ cm^2 Eq^{-1} ohm^{-1} for 10^{-3} N KNO_3.

7. It has been estimated† that a specific conductivity of 10^3 pmho m^{-1} would provide an ample margin of safety against electrokinetic explosions for the handling of refined petroleum products. These authors also measured the concentrations of various additives needed to reach this level of conductivity:

Solvent	Additive	Concentration (kmol m^{-3})
Benzene	Tetraisoamyl ammonium picrate	1×10^{-4}
Benzene	Calcium diisopropylsalicylate	5×10^{-3}
Gasoline	Ca salt of di-(2-ethylhexyl)sulfosuccinic acid	1×10^{-3}
Gasoline	Cr salt of mono- and dialkyl (C_{14}–C_{18}) salicylic acid	2.5×10^{-6}

---

*P. Somasundaran and R. D. Kulkarni, *J. Colloid Interface Sci.*, **45**:591 (1973).

†A. Klinkenberg and B. V. Poulston, *J. Inst. Pet.*, **44**:379 (1958).

Calculate the apparent value of the equivalent conductance Λ for each of these electrolytes in the conventional units $cm^2 \, Eq^{-1} \, ohm^{-1}$. How do the Λ values of these compounds compare with Λ_0 for simple electrolytes in aqueous solutions?

8. The pH variation of the electrophoretic mobility of solid $Th(OH)_4$ in 10^{-2} M HNO_3-KOH electrolyte is as follows*:

pH	7.6	8.0	9.0	9.6	10.0	10.6	11.3
$u \times 10^4$ $(cm^2 s^{-1} V^{-1})$	+2.4	+2.2	+1.3	+1.0	−0.1	−1.1	−1.5	−1.8

Use these data to evaluate the isoelectric point for $Th(OH)_4$. Since H^+ and OH^- appear to be potential determining, ψ may be estimated at various pH levels according to Eq. (12.1) if we identify the isoelectric point with the true point of zero charge. Compare these values with values of ζ calculated by means of the Helmholtz–Smoluchowski equation ($\eta = 0.0089$ P, $\varepsilon_r = 78.5$). Are the results qualitatively (quantitatively?) consistent with Fig. 13.7?

9. The aggregation number n and radius of sodium dodecyl sulfate micelles (by light scattering) and the zeta potential (from electrophoresis, by an accurate formula) were determined in the presence of various concentrations of NaCl†:

Moles of NaCl liter^{-1}	ζ (mV)	$R \times 10^8$ (cm)	κR	n
0.01	92.3	22.1	0.86	89
0.03	80.9	23.0	1.32	100
0.10	68.3	24.0	2.40	112

Use Eq. (82) to estimate the charge of the micelles. What approximation(s) in the derivation of Eq. (82) prevents this expression from applying exactly to this system? On the basis of the charges evaluated by Eq. (82), calculate the ratio of charge to aggregation number, the effective degree of dissociation, of these micelles. How do these results compare with the numbers given in Table 8.1?

10. An electrophoretic technique which is especially interesting for the study of proteins is called "isoelectric focusing." In this method electrophoresis is carried out across a medium which supports a pH gradient. The pH gradient and cell polarity are such that the cathode end of the column is relatively basic. Thus a positively charged protein gradually loses its charge as it migrates, finally coming to rest at a pH corresponding to its isoelectric point. A. Carlstrom and D. Vesterberg† used this method to study the hetero-geneity of peroxidase from cow's milk. After focusing was achieved, the column was drained and the pH and absorbance (at 280 nm) of successive fractions of eluent were measured:

---

* R. O. James and T. W. Healy, *J. Colloid Interface Sci.*, 40:42 (1972).
† D. Stigter and K. J. Mysels, *J. Phys. Chem.*, 59:45 (1955).
‡ A. Carlstrom and D. Vesterberg, *Acta Chem. Scand.*, 21:271 (1967).

Fraction number	Absorbance	pH	Fraction number	Absorbance	pH
12	0.9	9.83	28	1.4	9.49
14	3.0	9.80	30	0.9	9.45
16	2.1	9.75	32	0.6	9.38
18	1.2	9.70	34	0.8	9.31
20	2.7	9.69	36	0.5	9.30
21	2.2	9.685	38	0.3	9.28
22	2.8	9.68	40	0.4	9.23
24	1.6	9.60	41	0.5	9.16
26	1.2	9.55	42	0.4	9.10

How many components does this sample apparently contain? What are the values of the isoelectric points for each?

APPENDIX A

Examples of Expansions Encountered in this Book

(i) $1/(1-x) = 1 + x + x^2 + x^3 + \cdots$

(ii) $\ln(1+x) = x - \tfrac{1}{2}x^2 + \tfrac{1}{3}x^3 - \cdots$ $(-1 < x < +1)$

(iii)   Binomial: $(1 \pm x)^n = 1 \pm nx + [n(n-1)/2!]x^2$
$$\pm [n(n-1)(n-2)/3!]x^3 + \cdots \qquad (x^2 < 1)$$

(iv)   Taylor: $f(x) = f(x_0) + (x-x_0)f'(x_0) + (x-x_0)^2/2! f''(x_0) + \cdots$

(v) $\sin x = x - x^3/3! + x^5/5! - \cdots$

(vi) $\sinh x = x + x^3/3! + x^5/5! + \cdots$

(vii) $\cosh x = 1 + x^2/2! + x^4/4! + \cdots$

(viii) $e^x = 1 + x + x^2/2! + x^3/3! + \cdots$

APPENDIX B

Units: CGS–SI Interconversions

From time to time, probably all science students find themselves entangled in a problem of units. For those who have advanced through physical chemistry to the level of this book, these problems have obviously not been insurmountable. It is likely, however, that—along with feelings of frustration—these students have been left with the wish that everyone used the same units, specifically those units with which they are most comfortable. In response to the recognized need for uniformity, IUPAC recommends the use of *Système International* (SI) units, which are essentially standardized mks units.

The SI system is based on mutually consistent units assigned to the nine physical quantities listed in Table B.1. In addition to the SI units for these nine quantities, the table also lists cgs or other commonly encountered units, as well as the conversion factors between the two. In this table the headings at the top of the table indicate how the conversion factors are to be used in going from SI to cgs/common units, whereas the bottom headings indicate the use of these factors for calculations in the reverse direction.

From these nine basic quantities numerous other SI units may be derived. Table B.2 lists a number of these derived units, particularly those which are relevant to colloid and surface chemistry. The table is arranged alphabetically according to the name of the physical quantity involved. Note that instructions for the use of the conversion factors—depending on the direction of the conversion—are given in the top and bottom headings of the columns. Table B.2 is by no means an exhaustive list of the various derived SI units; Hopkins [1] reports on many additional conversions, as do most handbooks and numerous other references.

Table B.1 Basic SI Units and Their Relation to Cgs or Other Common Units[a]

Physical quantity	SI unit		× Conversion factor →	cgs/common unit	
	Name	Symbol		Name	Symbol
Length	meter	m	10^2	centimeter	cm
Mass	kilogram	kg	10^3	gram	g
Time	second	s	1	second	s
Electric current	ampere	A	2.998×10^9	statampere	statamp
Thermodynamic temperature	kelvin	K	1	kelvin	K
Luminous intensity	candela	cd	π	Lambert	(cm^2)
Amount of substance	mole	mole	1	mole	mole
Plane angle	radian	rad	$180/\pi$	degree (angle)	°
Solid angle	steradian	sr	1	steradian	sr
	SI unit		← Conversion factor ÷	cgs/common unit	

[a]Note that different column headings are given at the top and bottom of the table to facilitate conversions from SI to cgs and from cgs to SI, respectively.

Table B.2 Derived SI Units and Their Relation to Cgs or Other Common Units[a]

Physical quantity	SI unit	×	Conversion factor	→cgs/common unit
Acceleration	$m\ s^{-2}$		10^2	$cm\ s^{-2}$
Acceleration, angular	$rad\ s^{-1}$		1	$rad\ s^{-1}$
Area	m^2		10^4	cm^2
			10^{20}	$Å^2$
Capacitance (farad)	$F = m^{-2}\ kg^{-1}\ s^4\ A^2 = C\ V^{-1}$		8.99×10^{11}	statfarad
Charge (Coulomb)	$C = A\ s = J\ V^{-1}$		3.00×10^9	statcoulomb (esu)
Charge density, surface	$C\ m^{-2}$		3.00×10^5	$statcoul\ cm^{-2}$
Charge density, volume	$C\ m^{-3}$		3.00×10^3	$statcoul\ cm^{-3}$
Conductance (siemens)	$S = m^{-2}\ kg^{-1}\ s^3\ A^2 = ohm^{-1}$		8.99×10^{11}	statmho ($statohm^{-1}$)
Conductivity	$ohm^{-1}\ m^{-1}$		8.99×10^9	$statmho\ cm^{-1}$
			10^{-2}	$mho\ cm^{-1}$ ($ohm^{-1}\ cm^{-1}$)
Density	$kg\ m^{-3}$		10^{-3}	$g\ cm^{-3}$
Diffusion coefficient	$m^2\ s^{-1}$		10^4	$cm^2\ s^{-1}$
Dipole moment	$C\ m$		3.00×10^{11}	$statcoul\ cm$
Electric field	$V\ m^{-1}$		3.34×10^{-5}	$statvolt\ cm^{-1}$
			10^{-2}	$V\ cm^{-1}$
Electric potential (volt)	$V = m^2\ kg\ s^{-3}\ A^{-1} = J\ C^{-1}$		3.34×10^{-3}	statvolt
Energy (joule)	$J = m^2\ kg\ s^{-2} = N\ m$		10^7	erg
			0.2390	calorie
Entropy	$J\ K^{-1}$		0.2390	$cal\ K^{-1}$
Force (newton)	$N = m\ kg\ s^{-2}$		10^5	dyne
Frequency (hertz)	$Hz = s^{-1}$		1	s^{-1}
Friction factor	$kg\ s^{-1}$		10^3	$g\ s^{-1}$
Heat capacity	$J\ K^{-1}$		0.2390	$cal\ K^{-1}$
Molarity	$mole\ dm^{-3}$		1	$mole\ liter^{-1}$
Moment, dipole	$C\ m$		3.00×10^{11}	$statcoul\ cm$

SI unit	$\downarrow$	Conversion factor	$\div$	cgs/common unit
Moment, force	$N\,m$	10^7		$dyne\ cm$
Moment, inertia	$kg\,m^2$	10^7		$g\ cm^2$
Momentum	$N\,s$	10^5		$dyne\ s$
Momentum, angular	$J\,s$	10^7		$erg\ s$
Period	s	1		s
Permitivity	$F\,m^{-1}$	8.99×10^9		$statfarad\ cm^{-1}$
Polarization, electric	$C\,m^{-2}$	3.00×10^5		$statcoul\ cm^{-2}$
Potential (volt)	V	3.34×10^{-3}		$statvolt$
Power (watt)	$W = m^2\,kg\,s^{-3} = J\,s^{-1}$	10^7		$erg\ s^{-1}$
Pressure (pascal)	$Pa = m^{-1}\,kg\,s^{-2} = N\,m^{-2}$	10		$dyne\ cm^{-2}$
		9.87×10^{-6}		atm
Radius of gyration	m	10^2		cm
Resistance (ohm)	$ohm = m^2\,kg\,s^{-3}\,A^{-2} = VA^{-1}$	1.11×10^{-12}		$statohm$
Specific heat capacity	$J\,kg^{-1}\,K^{-1}$	2.39×10^{-4}		$cal\ g^{-1}\,K^{-1}$
Stress	$N\,m^{-2}$	10		$dyne\ cm^{-2}$
Surface energy	$J\,m^{-2}$	10^3		$erg\ cm^{-2}$
Surface tension	$N\,m^{-1}$	10^3		$dyne\ cm^{-1}$
Torque	$N\,m$	10^7		$dyne\ cm$
Velocity	$m\,s^{-1}$	10^2		$cm\ s^{-1}$
Velocity, angular	$rad\,s^{-1}$	1		$rad\ s^{-1}$
Viscosity	$N\,s\,m^{-2}$	10		$dyne\ s\ cm^{-2}\ (P)$
Volume	m^3	10^6		cm^3
		10^3		$dm^3\ (liter)$
Wave number	m^{-1}	10^{-2}		cm^{-1}
Weight	N	10^5		$dyne$

[a] Note that different column headings are given at the top and bottom of the table to facilitate conversions from SI to cgs and from cgs to SI, respectively.

Table B.3 Multiples of Units, Their
Names, and Symbols

Multiple	Prefix	Symbol
10^{12}	tera	T
10^9	giga	G
10^6	mega	M
10^3	kilo	k
10^2	hecto	h
10	deca	da
10^{-1}	deci	d
10^{-2}	centi	c
10^{-3}	milli	m
10^{-6}	micro	μ
10^{-9}	nano	n
10^{-12}	pico	p
10^{-15}	femto	f
10^{-18}	atto	a

One reason for the great diversity of units in existence is the fact that quantities of such diverse magnitudes are measured. A general rule is that the unit should be appropriate in magnitude to the quantity being measured. To obtain a dimension of convenient size in SI units, the SI unit is multiplied by a power of 10 and the prefixes listed in Table B.3 are affixed to the unit.

References

1. R. A. Hopkins, *The International (SI) Metric System and How It Works*, Polymetric Services, Reseda, California, 1973.
2. M. L. McGlashan, *Pure Appl. Chem., 21*:577 (1970).
3. C. H. Page and P. Vigoureux (eds.), *The International System of Units (SI)*, National Bureau of Standards, Special Publication 330, Washington, D.C., 1974.
4. M. A. Paul, *J. Chem. Educ., 48*:569 (1971).

INDEX

The notation F or T which accompanies certain entries refers to figures or tables, respectively, appearing on the indicated pages.

S